At a Glance

In Appreciation of

Hilton S. Read, M.D., and Kathryn S. Read

whose Ventnor Foundation program of internships (1951–1974)
allowed young medical doctors from postwar Germany to experience
a year of American medicine and the American way of life,
under the motto:

"Wage Peace through Brotherhood"

Preface

As for the previous three editions, the object of this book is to provide an overview of the field of genetics based on 189 color plates, each with an explanatory text. These plates provide a visual display of selected concepts and related facts, with a focus on genomics. Additional information is presented in the introduction, with a chronology listing important discoveries in the history of genetics; in an appendix with supplementary data; and in an extensive glossary explaining genetic terms. References and websites are provided with each plate for further studies in depth.

This book is written for two kinds of readers: for students of biology or medicine, as an introductory overview; and for their mentors, as a teaching aid. Other interested individuals will also be able to obtain selected information about current developments and achievements in this rapidly evolving field. The concommittant use of one of the standard textbooks of human or general genetics is highly recommended.

The term *Atlas* for a book was introduced in 1594 by Gerard de Kremer (1512–1594), the Flemish mathematician and cartographer also known as Mercator. His book, with a collection of 107 double page geographic maps with the title *Atlas sive Cosmographicae Meditationes de Fabrica Mundi et Fabrica Figura,* was published in 1595, a year after his death. Although Africa, Asia, the "New World", and the northern polar region are represented with only one map each, this was the first world atlas. Mercator explains in his introduction that he derived the term from the mythic king Atlas of Mauretania because of his outstanding astronomical knowledge. Earlier it was assumed that the term atlas referred to the titan Atlas of Greek mythology. In 1552 Mercator moved to the new University in Duisburg, near Essen, Germany. When Mercator's atlas appeared, many geographic regions were not yet known and had remained unmapped in his collection. Establishing genetic maps is an activity not unlike mapping new, unknown territories 500 years ago. Genetic maps are a leitmotif in genetics and a recurrent theme in this book.

This fourth edition has been extensively rewritten and updated. Every sentence and illustration was reviewed, and changed to improve clarity when necessary. Twelve entirely new plates have been added. About as many have been deleted because they have become outdated. Thus, the fourth edition is slightly smaller than the third edition from 2007. The general structure of the previous editions, which have appeared in twelve languages, has been maintained: Part I, *Fundamentals*; Part II, *Genomics*; Part III, *Genetics and Medicine.*

Each color plate is accompanied by an explanatory text on the opposite page. Each plate and its text together correspond to a small chapter, concentrating on the most important threads of information. The topic for each plate has been selected to illustrate a genetic or genomic set of concepts or facts. The emphasis is on the intersection of theoretical foundations and practical applications, in particular as relevant to human and medical genetics. Genetic diseases are selected as examples representing a genetic principle, but without the many details required in practice.

The reader should be aware that each plate and its text represent an abstract rather than a treatise, with many related details necessarily omitted. Therefore, this book is intended as a supplement to classic textbooks rather than a substitute.

New topics, represented by new plates, include overviews of human evolution, genomic disorders and genome-wide association studies, cancer genomes, genetic disorders of fibroblast growth factors receptors, ciliopathies, neurocristopathies, and other emerging topics.

A single-author book cannot provide all the details on which specialized scientific knowledge is based. However, it can present an individual perspective that is suitable as an introduction. This hopefully will stimulate interest.

Throughout the book I have emphasized the role of the evolution of genes, genomes, and organisms in understanding genetics. As noted by the great geneticist Theodosius Dobzhansky, "Nothing in biology makes sense except in the light of evolution." Indeed, genetics and the science of evolution are closely related. One could add "Nothing in evolution makes sense except in the light of genetics."

For the many young readers naturally interested in the future I have included a historical perspective. Whenever possible and appropriate, I have referred to the first description of a discovery. This is a reminder that the platform of knowledge today rests on previous advances.

All color plates were prepared by Jürgen Wirth, 1978–2005 Professor of Visual Communication at the Universities of Applied Sciences, Darmstadt and Schwaebisch Gmuend, Germany. He created the plates from the author's drafts, computer drawings or photographs. I am deeply indebted to Professor Jürgen Wirth for his most skillful work, which is a fundamental part of this book. I thank my wife, Mary Fetter Passarge, M.D., for helpful suggestions. At Thieme Publishers, Stuttgart, I was guided and supported by Angelika Findgott and Annie Hollins. I also wish to thank Sophia Hengst, Production Editor at Thieme, a house known for its superior quality, for the pleasant and effective collaboration during the production process.

Eberhard Passarge

Acknowledgments

In preparing this fourth edition, several colleagues from different countries again kindly provided me with photographs for illustrations and other information for new plates. In this regard I am grateful to Mohammad Reza Ahmadian (Institute of Biochemistry and Molecular Biology II, University Hospital Düsseldorf, Germany); Beate Albrecht, Tea Berulava, and Bernhard Horsthemke (all Institute for Human Genetics, Essen, Germany); Thomas Langmann (Pro Retina Professorship, Institute of Human Genetics, University of Regensburg, Germany), Maximilian Muenke (Medical Genetics Branch, National Human Genome Research Institute, National Institutes of Health, Bethesda, Maryland, USA), Heike Olbrich and Heymut Omran (Department of Pediatrics, University of Münster, Germany), and Martin Zenker (Institute of Human Genetics, University of Magdeburg, Germany). Their contributions are also acknowledged in the text accompanying the relevant plates.

In addition, the following colleagues have kindly contributed valuable advice and additional information:

Karin Buiting, Stephanie Gkalympoudis, Deniz Kanber, Dietmar Lohmann, Hermann-Josef Lüdecke, and Nicholas Wagner (all Essen, Germany), Nicholas Katsanis (Duke University, Durham, North Carolina, USA), James R. Lupski (Houston, USA), Maximilian Muenke (Medical Genetics Branch, National Human Genome Research Institute, National Institutes of Health, Bethesda, Maryland, USA), Arne Pfeuffer (Helmholtz-Zentrum, Munich, Germany), and Friedrich Stock (Leipzig, Germany).

I also would like to thank the following colleagues for photographs or drafts for illustrations received previously for earlier editions, which are also used in this fourth edition: Alireza Baradaran (Mashhad, Iran, now Vancouver, Canada), Dirk Bootsma (Rotterdam, Netherlands), Laura Carrel (Hershey, Pennsylvania, USA), Aravinda Chakravarti and Richard I. Kelley (Baltimore, USA), Thomas Cremer (Munich, Germany), Robin Edison (NIH, Bethesda, Maryland, USA), Evan E. Eichler (Seattle, USA), Wolfgang Engel (Göttingen, Germany), Reiner Johannisson (Lübeck, Germany), Nikolaus Konjetzko, Alma Küchler, Dietmar Lohmann, Axel Schneider, and Dagmar Wieczorek (all Essen, Germany), Nicole McNeil and Thomas Ried (NIH, Bethesda, Maryland, USA), Clemens Müller-Reible (Würzburg, Germany), Stefan Mundlos (Berlin, Germany), Helga Rehder (Marburg, Germany/Vienna, Austria), David L. Rimoin (Los Angeles, USA), Evelin Schröck (Dresden, Germany), Peter Steinbach (Ulm, Germany), Hans Hilger Ropers (Berlin, Germany), Sabine Uhrig and Michael Speicher (Graz, Austria), Michael Weis (Cleveland, Ohio, USA, and Eberhard Zrenner (Tübingen, Germany).

If errors of commission or omission are found I would appreciate receiving readers' comments and corrections.

About the Author

Eberhard Passarge, M.D., is a German human geneticist at the University Institutes of Human Genetics at Essen and Leipzig, Germany. He graduated with an M.D. degree from the University of Freiburg, Germany, in 1960, and received general medical training at the General Hospital Hamburg-Harburg, Germany, 1961–1962, and at The Worcester Memorial Hospital, Worcester, Massachusetts, USA (1962–1963), with a stipend from the Ventnor Foundation. His postgraduate education was in pediatrics and human genetics at Children's Hospital Medical Center Cincinnati, Ohio, USA, with Josef Warkany 1963–1966 and in human genetics at Cornell Medical Center New York, USA, with James German 1966–1968. After completing his training he established a new Division of Cytogenetics and Clinical Genetics at the Department of Human Genetics, University of Hamburg, Germany, in 1968 and directed it until 1976, when he became Professor of Human Genetics and founding Chairman of the Institute of Human Genetics, University of Essen, Germany. This he directed until his retirement in 2001. He has remained active in the field of human genetics. In February of 2010 he assumed duties as Interim Chairman at the Institute of Human Genetics at the University of Leipzig, Germany. Among his main scientific interests are the investigation of hereditary and congenital diseases, and the application of this knowledge in genetic diagnosis and counseling. He is author or co-author of more than 240 scientific articles in international, peer-reviewed journals, the author of chapters in several international textbooks and author of three books on human and medical genetics. His experience in teaching human genetics is reflected in this *Color Atlas of Genetics.* He has served on the editorial board of several international human genetic journals. He has been Secretary-General of the European Society of Human Genetics (1987–1990) and President of the German Society of Human Genetics (1990–1996), of which he became an honorary member in March 2011. He is a member of the American Society of Human Genetics and several other scientific societies, a founding member of the European Society of Human Genetics, and a corresponding member of the American College of Medical Genetics.

Table of Contents

Introduction

Genetics and Genomics

When, in 1906, the British biologist William Bateson (1861–1926) proposed the term *genetics* for a new biological field devoted to the scientific investigation of heredity and variation, no one could foresee that a century later the underlying biological structures and functions in genetics would be understood at the molecular level. Bateson had clearly recognized the importance of the Mendelian rules, which had been rediscovered in 1900 by Correns, Tschermak, and De Vries. "Among the biological sciences the study of genetics occupies a central position," is the opening sentence in his *Mendel's Principles of Heredity*, published in 1909. A long series of ensuing discoveries and technical advances culminated with the completion of the Human Genome Project (HGP) in 2003 and the publication of a reference human genome sequence (IHGSC, 2001, 2004). This represents an unprecedented milestone in biology and medicine.

William Bateson
(1861–1926)

The term *genomics*, introduced in 1987 by V. A. McKusick and F. H. Ruddle to define the new field of studying entire genomes rather than selected genes, reflects the dramatic progress during the past two decades. Genomics refers to the scientific study of the structure and function of genomes of different species of organisms. The genome of an animal, plant, or microorganism contains all biological information required for life and reproduction. It comprises: the entire nucleotide sequence; all genes, their structure and function, and their chromosomal localization; chromosome-associated proteins; and the architecture of the nucleus. Genomics integrates genetics, molecular biology, and cell biology. The scientific goals of genomics are manifold and all aimed at the entire genome of an organism: sequencing of the nucleotide bases of an organism, in particular all genes and gene-related sequences; analysis of all molecules involved in transcription and translation, and their regulation (the *transcriptome*); analysis of all proteins that a cell or an organism is able to produce (the *proteome*); identification of all genes and functional analysis (*functional genomics*); establishing genomic maps with regard to the evolution of genomes (*comparative genomics*); and assembly, storage, and management of data (*bioinformatics*).

Genetics and genomics are relevant to virtually all fields of medicine and biological disciplines, anthropology, evolution, biochemistry, physiology, psychology, ecology, and other fields of science. As both are a theoretical and an experimental science, genetics and genomics have broad practical applications in the understanding and control of genetic diseases, and understanding the biology of the living world in general. Truly a new era with vast consequences for human biology and medicine has begun.

Genetic Basis of Life

Each of the approximately 80 trillion (10^{12}) cells of an adult human contains a program with life-sustaining information in its nucleus (except for red blood cells, which do not have a nucleus). This information is hereditary, transmitted from each cell to its descendent cells, and from one generation to the next. About 200 different types of cells carry out the complex molecular transactions required for life.

Genetic information allows cells to convert atmospheric oxygen and ingested food into energy production, it regulates the synthesis and transport of biologically important molecules, it protects against unwarranted invaders, such as bacteria, fungi, and viruses, by means of an elaborate immune defense system, and it maintains the shape and mobility of bones, muscles, and skin. Genetically determined functions of the sensory organs enable us to see, to hear, to taste, to feel heat, cold, and pain, to communicate by speech, to support brain function with the ability to learn from experience, and to integrate the environmental input into cognate behavior and social interaction. Likewise, reproduction and detoxification of exogenous molecules are under genetic control. Yet, the human brain is endowed with the ability to make free decisions in daily life and to develop plans for the future.

The living world consists of two basic types of cells, the smallest membrane-bound units capable of independent reproduction: *prokaryotic* cells without a nucleus (represented by bacteria), and *eukaryotic* cells with a nucleus and complex internal structures (these cells make up higher organisms). Genetic information is transferred from one cell to both daughter cells at each cell division and from one generation to the next through specialized cells, the *germ cells*, *oocytes*, and *spermatozoa*.

The integrity of the genetic program must be maintained without compromise, yet it must be adaptable to respond to long-term changes in the environment. Consequently, errors in maintaining and transmitting genetic information occur in all living systems despite the existence of complex systems for damage recognition and repair.

Biological processes are mediated by biochemical reactions performed by biomolecules, called *proteins*. Each protein is made up of dozens to several hundreds of amino acids arranged in a linear sequence that is specific for its function. This primary sequence of amino acids is a *polypeptide*. Subsequently, it assumes a specific three-dimensional structure, often in combination with other polypeptides. Only this latter feature allows biological function. Genetic information is the blueprint for producing the proteins in a given cell. Most cells do not produce all possible proteins, but a selection, depending on the type of cell. The instructions are encoded in discrete units, the *genes*. Genes contain encoded information for the intracellular synthesis of either proteins or different types of small molecules made up of *ribonucleic acid* (RNA).

Genetic information is stored in a linear fashion like a text of individual letters and words in a defined sequence that alone makes biological sense. This text consists of the nucleotide bases of a large molecule, *deoxyribonucleic acid* (DNA). DNA is a read-only memory device of a genetic information system, called the *genetic code*. In contrast to the binary system of strings of ones and zeros used in computers ("bits," which are then combined into "bytes," which are eight binary digits long), the genetic code in the living world uses a quaternary system of four nucleotide bases with chemical names having the initial letters A, C, G, and T (see Part I, Fundamentals). The quaternary code used in living cells uses three building blocks, called a *triplet codon*. This genetic code is universal and is used by all living cells, including plants and viruses. A gene is a unit of genetic information. It is equivalent to a single sentence in a text. Thus, genetic information is highly analogous to a linear text and is amenable to being stored in computers.

Genes

Depending on the organizational complexity of an organism, its size and number of genes vary considerably. Their number ranges from about 500 to 5000 in most prokaryotes, and about 6000 to 40 000 in most eukaryotes. The minimal number of genes required to sustain independent cellular life is surprisingly small: about 250 for a prokaryote. Since many proteins are involved in related functions of the same pathway, they and their corresponding genes can be grouped into families of related function. It is estimated that the human genes form about 1000 gene families.

Genes are located on chromosomes. Chromosomes are individual, complex structures that are located in the cell nucleus, consisting of DNA and special proteins. Chromosomes in eukaryotes come in pairs of homologous chromosomes, one derived from the mother, and one from the father. Man has 23 pairs, consisting of chromosomes 1–22 and an X and a Y chromosome in males or two X chromosomes in females. The number and size of chromosomes vary in different organisms, but the total amount of DNA and the total number of genes are the same for a particular species. Genes are arranged in linear order along each chromosome. Each gene has a defined position, called a *gene locus*. In higher organisms, genes are structured into contiguous sections of coding and noncoding sequences, called *exons* (coding) and *introns* (noncoding), respectively. Genes in multicellular organisms vary with respect to size (ranging from a few thousand to over a million nucleotide base pairs), the number and size of exons, and regulatory DNA sequences. The latter determine the state of activity of a gene, called *gene expression*. Most genes in differentiated, specialized cells are permanently turned off. Remarkably, more than 90% of the 3 billion (3×10^9) base pairs of DNA in higher organisms do not carry known coding information (see Part II, Genomics).

The linear text of information contained in the coding sequences of DNA in a gene cannot be read directly. Rather, its total sequence is first transcribed into a structurally related molecule with a corresponding sequence of codons. This molecule is called RNA. RNA is processed by removing the noncoding sections (*introns*). Subsequently, the coding sections (*exons*) are spliced together into the final template, called *messenger RNA* (mRNA). This serves as a template to arrange the amino acids in the sequence specified by the genetic code. This process is called *translation*.

Each of the 21 amino acids used by living organisms is recognized by a specific sequence of three RNA molecules and lined up in the same sequence as determined in the DNA.

Genes and Evolution

Genes with comparable functions in different organisms share structural features. Occasionally they are nearly identical. This is the result of evolution. Living organisms are related to each other by their origin from a common ancestor. Cellular life was established about 3.5 billion years ago. Genes required for fundamental functions are similar or almost identical across a wide variety of organisms, e.g., in bacteria, yeast, insects, worms, vertebrates, mammals, and even plants.

Genes evolve within the context of the genome of which they are a part. Evolution does not proceed by accumulation of mutations. Most mutations are detrimental to function and usually do not improve an organism's chance of surviving. Rather, during the course of evolution existing genes are duplicated or parts of genes reshuffled and brought together in a new combination. The duplication event can involve an entire genome, a whole chromosome or a part of it, or a single gene or group of genes. All these events have been documented in the evolution of vertebrates. The human genome contains multiple sites that were duplicated during evolution (see Part II, Genomics).

Humans, *Homo sapiens*, are the only living species within the family of Hominidae. All data available are consistent with the assumption that ancestors of today's humans originated in Africa about 100000–300000 years ago, spread out over the earth, and populated all continents. As a result of regional adaptation to climatic and other conditions, and favored by geographic isolation, different ethnic groups evolved. Human populations living in different geographic regions differ in the color of the skin, eyes, and hair. This is often mistakenly referred to as human races. However, genetic data do not support the existence of human races. Genetic differences exist mainly between individuals regardless of their ethnic origin. In a study of DNA variation from 12 populations living on five continents of the world, 93%–95% of differences were between individuals; only 3%–5% of differences were between the populations (Rosenberg et al., 2002). Observable differences are literally superficial and do not form a genetic basis for distinguishing races. Genetically, *Homo sapiens* represents one single species of recent evolutionary origin.

Changes in DNA and Genes: Mutations and Polymorphism

In 1901, H. De Vries recognized that genes can change the contents of their information. For this new observation, he introduced the term *mutation*. The systematic analysis of mutations contributed greatly to the development of the science of genetics. In 1927, H. J. Muller determined the spontaneous mutation rate in *Drosophila* and demonstrated that mutations can be induced by roentgen rays. C. Auerbach and J. M. Robson in 1941 and, independently,

F. Oehlkers in 1943 observed that certain chemical substances also could induce mutations. However, it remained unclear what a mutation actually was, since the physical basis for the transfer of genetic information was not known. Genes of fundamental importance do not tolerate changes (mutations) that compromise function. As a result, deleterious mutations do not accumulate in any substantial number. All living organisms have elaborate cellular systems that can recognize and eliminate faults in the integrity of DNA and genes (*DNA repair*). Mechanisms exist to sacrifice a cell by programmed cell death (*apoptosis*) if the defect cannot be successfully repaired.

The sequence of DNA is not identical between unrelated individuals but differs (DNA polymorphism). Individual differences occur about once in 500 base pairs of human DNA between individuals. Usually this involves just one nucleotide (*single nucleotide polymorphism*, SNP). Other forms of DNA polymorphism involve small or large blocks of repeated nucleotide sequences (copy number variation, CNV). Such individual genetic differences are the basis of genetic individuality.

Archibald Garrod
(1857–1936)

Genetic Individuality

In 1902, Archibald Garrod (1857–1936), later Regius Professor of Medicine at Oxford University, demonstrated that four congenital metabolic diseases (albinism, alkaptonuria, cystinuria, and pentosuria) are transmitted by autosomal recessive inheritance. He called these *inborn errors of metabolism*. Garrod was also the first to recognize that subtle biochemical differences among individuals result from individual genetic differences. In 1931, he published a prescient monograph entitled *The Inborn Factors in Disease* (Garrod, 1931). He suggested that small genetic differences might contribute to the causes of diseases. Garrod, together with W. Bateson, introduced genetic concepts into medicine in the early years of genetics between 1902 and 1909. In late 1901, Garrod and Bateson began an extensive correspondence about the genetics of alkaptonuria and the significance of consanguinity, which Garrod had observed among the parents of affected individuals. In a letter to Bateson on 11 January 1902, Garrod wrote, "I have for some time been collecting information as to specific and individual differences of metabolism, which seems to me a little explored but promising field in relation to natural selection, and I believe that no two individuals are exactly alike chemically any more than structurally" (Bearn, 1993). However, Garrod's concept of the genetic individuality of man was not recognized at the time. One reason may have been that the structure and function of genes were totally unknown, despite early fundamental discoveries. Today we recognize that individual susceptibility to disease is an important factor in its causes (see Childs, 1999). Consequently, the causes of diseases are not viewed as random processes, but rather as the consequences of individual attributes of a person's genome and its encounter with the environment, as first proposed by Garrod.

Early Genetics between 1900 and 1910

The Mendelian rules are named after the Augustinian monk Gregor Mendel (1822–1884), who conducted crossbreeding experiments on garden peas in his monastery garden in Brünn (Brno, Czech Republic) in 1865. Mendel recognized that heredity is based on individual factors that are independent of each other. These factors are transmitted from one plant generation to the next in a predictable pattern, each factor being responsible for an observable trait. The trait one can observe is the *phenotype*. The underlying genetic information is the *genotype*.

Gregor Mendel
(1822–1884)

However, the fundamental importance of Mendel's conclusions was not recognized until 1900. The term *gene* for this type of a heritable factor was introduced in 1909 by the Danish biologist Wilhelm Johannsen (1857–1927). Beginning in 1901, Mendelian inheritance was systematically analyzed in animals, plants, and also in man. Some human diseases were recognized as having a hereditary cause. A form of brachydactyly (type A1, McKusick number OMIM 112500) observed in a large Pennsylvania sibship by W. C. Farabee (PhD thesis, Harvard University, 1903) was the first condition in man to be described as being transmitted by autosomal dominant inheritance (Haws & McKusick, 1963).

Chromosomes were observed in dividing cells (in mitosis by Flemming in 1879; in meiosis by Strasburger in 1888). Waldeyer coined the term chromosome in 1888. Before 1902, the existence of a functional relationship between genes and chromosomes was not suspected. Early genetics was not based on chemistry or cytology. An exception was the prescient work of Theodor Boveri (1862–1915), who recognized the genetic individuality of chromosomes in 1902. He wrote that not a particular number but a certain combination of chromosomes is necessary for normal development. This clearly indicated that the individual chromosomes possess different qualities.

Genetics became an independent scientific field in 1910 when Thomas H. Morgan introduced the fruit fly (*Drosophila melanogaster*) for systematic genetic studies at Columbia University in New York. Subsequent systematic genetic studies on *Drosophila* showed that genes are arranged on chromosomes in sequential order. Morgan summarized this in 1915 as the *chromosome theory of inheritance* (Morgan, 1926).

The English mathematician G. H. Hardy and the German physician W. Weinberg independently recognized in 1908 that Mendelian inheritance accounts for certain regularities in the genetic structure of populations. Their work contributed to the successful introduction of genetic concepts into plant and animal breeding. Although genetics was well established as a biological field by the end of the second decade of the last century, knowledge of the physical and chemical nature of genes was

Thomas H. Morgan
(1866–1945)

sorely lacking. Structure and function remained unknown.

A Misconception in Genetics: Eugenics

Eugenics, a term coined by Francis Galton in 1882, is the study of improvement of humans by genetic means. Such proposals date back to ancient times. Between about 1900 and 1935, many countries adopted policies and laws that were assumed to lead to the erroneous goals of eugenics. It was believed that the "white race" was superior to others, but proponents did not realize that genetically defined human races do not exist. Eugenics assumed that sterilizing individuals with diseases thought to be hereditary would improve human society. By 1935, sterilization laws had been passed in Denmark, Norway, Sweden, Germany, and Switzerland, as well as in 27 states of the United States. Individuals with mental impairment of variable degree or epilepsy, and criminals and homosexuals were prime targets. Although in most cases the stated purpose was eugenic, sterilizations were performed for social rather than genetic reasons.

The complete lack of knowledge of the structure and function of genes probably contributed to the eugenic misconceptions, which assumed that "bad genes" could be eliminated from human populations. However, the disorders targeted are either not hereditary or have a complex genetic background. Sterilization simply will not reduce the frequency of genes contributing to mental retardation and other disorders. In Nazi Germany, eugenics was used as a pretext for widespread discrimination and the murder of millions of innocent human beings claimed to be "worthless" (Müller-Hill, 1988; Vogel & Motulsky, 1997; Strong, 2003). All such reasons based on genetics are totally invalid. Modern genetics has shown that the ill-conceived eugenic approach to attempt to eliminate human genetic disease is impossible. Thus, incomplete genetic knowledge was applied to human individuals at a time when nothing was known about the structure of genes. Indeed, up to 1949 no fundamental advances in genetics had been obtained by studies in humans. Quite the opposite holds true today. It is evident that genetically determined diseases cannot be eradicated. Society has to adjust to their occurrence. No one is free from a genetic burden. Every individual carries about five or six potentially harmful changes in the genome that might manifest as a genetic disease in a child.

The Rise of Modern Genetics between 1940 and 1953

With the demonstration in the fungus *Neurospora crassa* that one gene is responsible for the formation of one enzyme ("one gene, one enzyme," Beadle and Tatum in 1941), the close relationship of genetics and biochemistry became apparent. This is in agreement with Garrod's concept of inborn errors of metabolism. Systematic studies in microorganisms led to other important advances in the 1940s. Bacterial genetics began in 1943 when Salvador E. Luria and Max Delbrück discovered mutations in bacteria. Other important advances were genetic recombination demonstrated in bacteria by Lederberg and Tatum in 1946, and in viruses by Delbrück and Bailey in 1947; as well as spontaneous mutations observed in bacterial viruses, the bacteriophages, by Hershey in 1947. The study of genetic phenomena in microorganisms turned out to be as significant for the further development of genetics as the analysis of *Drosophila* had been 35 years earlier (Cairns et al., 1978). A very influential, small

Max Delbrück and Salvador E. Luria at Cold Spring Habor (Photograph by Karl Maramorosch, from Judson, 1996)

book entitled *What is Life?* by the physicist E. Schrödinger (1944) postulated a molecular basis for genes. From then on, the elucidation of the molecular biology of the gene became a central theme in genetics.

Genetics and DNA

A major advance was the discovery by Avery, MacLeod, and McCarty, at the Rockefeller Institute in New York in 1944, that a chemically relatively simple, long-chained nucleic acid (DNA), which had been discovered in 1869 by Friedrich Miescher, is able to carry genetic information in bacteria (for historical reviews see Dubos, 1976; McCarty, 1985). Many years earlier in 1928, F. Griffith had observed that permanent (genetic) changes could be induced in pneumococcal bacteria by a cell-free extract derived from other strains of pneumococci (the *transforming principle*). Avery and his coworkers showed that DNA was this transforming principle. In 1952, Hershey and Chase proved that DNA alone carries genetic information and they excluded other molecules. With this discovery, the question of the structure of DNA took center stage in biology.

Oswald T. Avery in 1937 (wikipedia.org) (1877–1955)

This question was resolved most elegantly by James D. Watson, a 24-year-old American on a scholarship in Europe, and Francis H. Crick, a 36-year-old English physicist, at the Cavendish Laboratory of the University of Cambridge. On 25 April 1953, in a short article of one page in the journal *Nature*, they proposed the structure of DNA as a double helix (Watson & Crick, 1953). Although it was not immediately recognized as such, this discovery is the cornerstone of modern genetics in the 20th century. An earlier basis for recognizing the importance of DNA was the discovery by E. Chargaff in 1950 that of the four nucleotide bases guanine was present in the same quantity as cytosine, and adenine in the same quantity as thymine. However, this was not taken to be the result of pairing (Wilkins, 2003).

The structure of DNA as a double helix with the nucleotide bases inside explains two fundamental genetic mechanisms: storage of genetic information in a linear, readable pattern and replication of genetic information to ensure its accurate transmission from generation to generation. The DNA double helix consists of two complementary chains of alternating sugar (deoxyribose) and monophosphate molecules, oriented in opposite directions. Inside the helical molecule are paired nucleotide bases. Each pair consists of a pyrimidine and a purine, either cytosine (C) and guanine (G) or thymine (T) and adenine (A). The crucial feature is that the base pairs are inside the molecule, not outside. That the authors fully recognized the significance for genetics of the novel structure is apparent from the closing statement of their article, in which they state, "It has not escaped our notice that the specific pairing we have postulated immediately suggests a possible copying mechanism for the genetic material." Vivid, albeit different, accounts of their discovery have been given by the authors (Watson, 1968; Crick, 1988) and by Wilkins (2003).

The elucidation of the structure of DNA is regarded as the beginning of a new era of molecular biology and genetics. The description of DNA as a double-helix structure led directly to an understanding of the possible structure of genetic information. When F. Sanger determined the sequence of amino acids of insulin in 1955, he provided the first proof of the primary structure of a protein. This supported the

J. D. Watson and F. H. C. Crick

D.N.A.

BASE — SUGAR
PHOSPHATE
BASE — SUGAR
PHOSPHATE
BASE — SUGAR
PHOSPHATE
BASE — SUGAR
PHOSPHATE
BASE — SUGAR
PHOSPHATE

Fig. 1. Chemical formula of a single chain of deoxyribonucleic acid

Fig. 2. This figure is purely diagrammatic. The two ribbons symbolize the two phosphate-sugar chains, and the horizontal rods the pairs of bases holding the chains together. The vertical line marks the fibre axis

DNA structure 1953

Watson and Crick in 1953
(Photograph by Anthony Barrington Brown, Nature 421: 417, 2003)

Rosalind Franklin (Maddox, 2002)
(1920–1958)

Maurice Wilkins (Wilkins, 2003)
(1916–2004)

notion that the sequence of amino acids in proteins could correspond to the sequential character of DNA. The genetic code for the synthesis of proteins from DNA and mRNA was determined in the years 1963–1966 by Nirenberg, Mathaei, Ochoa, Benzer, Khorana, and others. Detailed accounts of these developments have been presented by several authors (Watson, 1968, 2000; Chargaff, 1978; Stent, 1981; Watson & Tooze, 1981; Crick, 1988; Judson, 1996; Wilkins, 2003).

With the structure of DNA known, the nature of the gene could be redefined in molecular terms. In 1955, Seymour Benzer provided the first genetic fine structure. He established a map of contiguous deletions of a region (*r*II) of the bacteriophage T4. He found that mutations fell into two functional groups: A and B. Mutants belonging to different groups could complement each other (eliminate the effects of the deletion); those belonging to the same group could not. This work showed that the linear array of genes on chromosomes also applied to the molecule of DNA. This defined the gene in terms of function and added an accurate molecular size estimate for the components of a gene.

New Methods in the Development of Genetics after 1953

From the beginning, genetics has been a field developing new concepts that are based on the development of new experimental methods. In the 1950s and 1960s, the groundwork was laid for *biochemical genetics* and *immunogenetics*. Relatively simple but reliable procedures for separating complex molecules by different forms of electrophoresis, methods of synthesizing DNA in vitro (Kornberg in 1956), and other approaches were applied to genetics. The introduction of cell culture methods was of particular importance for the genetic analysis of humans. G. Pontecorvo introduced the genetic analysis of cultured eukaryotic cells (*somatic cell genetics*) in 1958. The study of mammalian genetics, with increasing significance for studying human genes, was facilitated by methods of fusing cells in culture (*cell hybridization*; T. Puck, G. Barski, and B. Ephrussi in 1961) and the development of a cell culture medium for selecting certain mutants in cultured cells (*HAT medium*; Littlefield in 1964). The genetic approach that had been so successful in bacteria and viruses could now be applied in higher organisms, thus avoiding the obstacles of a long generation time and breeding experiments. A hereditary metabolic defect in man (galactosemia) was demonstrated for the first time in cultured human cells in 1961 (R. S. Krooth). The correct number of chromosomes in man was determined in 1956 (Tjio and Levan; Ford and Hamerton). Lymphocyte cultures were introduced for chromosomal analysis (Hungerford and coworkers in 1960). The replication pattern of human chromosomes was described (German in 1962). These and other developments paved the way for a new field, *human genetics*. Since the late 1970s, this field has taken root in all areas of genetic studies, in particular molecular genetics.

Molecular Genetics

The discovery of reverse transcriptase, independently by H. Temin and D. Baltimore in 1970, upset a central dogma in genetics that the flow of genetic information is in one direction only, from DNA to RNA and from RNA to a protein as the gene product. *Reverse transcriptase* is an enzyme complex in RNA viruses (*retroviruses*) that transcribes RNA into DNA. This is not only an important biological finding, but this enzyme can be used to obtain *complementary DNA* (cDNA) that corresponds to the coding regions of an active gene. This allows one to analyze a gene directly without knowledge of its gene product. Enzymes that cleave DNA at specific sites, and are known as *restriction endonucleases* or, simply, *restriction enzymes*, were discovered in bacteria by W. Arber in 1969, and by D. Nathans and H. O. Smith in 1971. They can be used to cleave DNA into fragments of reproducible and defined sizes, and this allowed the molecular analysis of a selected region of the DNA molecule to be studied. DNA fragments of different origin can be joined and their properties analyzed. Methods of probing for genes, producing multiple copies of DNA fragments (polymerase chain reaction, PCR), and determining the sequence of the nucleotide bases of DNA were developed between 1977 and 1985 (see Part I, Fundamentals). All these methods are collectively referred to as *recombinant DNA technology*.

In 1977, recombinant DNA analysis led to a completely new and unexpected finding about the structure of genes in higher organisms. Genes are not continuous segments of coding DNA, but are interrupted by noncoding segments. The size and pattern of coding DNA segments, called *exons*, and of the noncoding segments, called *introns* (two new terms introduced by W. Gilbert in 1978) are characteristic for each gene. This is known as the *exon/intron structure* of eukaryotic genes. Modern molecular genetics allows the determination of the chromosomal location of a gene and the analysis of its structure without prior knowledge of the gene product. The extensive homologies of genes that regulate embryological development in different organisms and the similarities of genome structures have removed the boundaries in genetic analysis that formerly existed between different organisms (e.g., *Drosophila* genetics, mammalian genetics, yeast genetics, bacterial genetics). Genetics has become a broad, unifying discipline in biology, medicine, and evolutionary research.

Transposable DNA

Certain DNA sequences can suddenly appear at a new location by one of several mechanisms, collectively called *transposition*. Such transposable elements or transposons are discrete sequences in the genome that are mobile owing to their ability to move to other locations within the genome (Lewin, 2008). This entirely new phenomenon was first described between 1950 and 1953 by Barbara McClintock, at Cold Spring Harbor Laboratory, New York. The author described genetic changes in Indian corn plants (maize) and their effect on the phenotype induced by a mutation in a gene that is not located at the site of the mutation. Surprisingly, such a gene can exert a type of remote control. In subsequent work, McClintock described the special properties of this group of genes, which she called *controlling genetic elements*. Different controlling elements could be distinguished according to their effects on other genes and the mutations caused. Originally her work was received with skepticism (Fox Keller, 1983; Fedoroff & Botstein, 1992). In 1983 she received the Nobel Prize (McClintock, 1984). Today we know that different types of transposition form families of transposons with different mechanisms. Transposition lends the genome flexibility during the course of evolution. Some transposable elements influence gene expression (suppressor-mutator elements, for details see Lewin, 2008).

Human Genetics

Human genetics deals with all human genes, normal and abnormal. However, it is not limited to humans, but applies knowledge and uses methods relating to many other organisms. These are mainly other mammals, vertebrates, yeast, fruit fly, and microorganisms. Arguably, human genetics was inaugurated when The American Society of Human Genetics and the first journal of human genetics, *The American Journal of Human Genetics*, were established in 1949. In addition, the first textbook of human genetics appeared in 1949, Curt Stern's *Principles of Human Genetics* (Stern, 1973).

The medical applications of human genetics contribute to the understanding of the underlying causes of diseases. This leads to improved precision in diagnosis. The concept of disease in human genetics differs from that in medicine. In medicine, diseases are mainly classified according to their phenotype, i.e., the manifestation according to organ systems, age, and gender. In human genetics, diseases are classified according to the genotype, i.e., gene loci, genes, types of mutations (*molecular pathology*). Some genetic diseases have a similar phenotype, although they result from rearrangements in different genes. On the other hand, different rearrangements in one and the same gene or different genes may result in the same phenotype. This is referred to as *etiological* (genetic) *heterogeneity*, an important principle to be observed when diagnosing human genetic disorders.

Two important discoveries in 1949 relate to a human disease that still poses a public health problem in tropical parts of the world. J. V. Neel showed that sickle cell anemia is inherited as an autosomal recessive trait. Pauling, Itano, Singer, and Wells demonstrated that a defined alteration in normal hemoglobin was the cause. This is the first example of a human molecular disease. An enzyme defect, glucose-6-phosphatase deficiency, in glycogen storage disease type I (von Gierke disease) was the first biochemical basis of a human disease demonstrated in liver tissue (Cori and Cori in 1952).

Important advances in the analysis of human metaphase chromosomes (*cytogenetics*) established chromosomal aberrations as important causes of human disorders. In 1959, the first chromosomal aberrations were discovered in three clinically well-known human disorders: trisomy 21 in Down syndrome (J. Lejeune, M. Gautier, R. Turpin); monosomy X (45,X) in females with Turner syndrome (Ford and coworkers); and an extra X chromosome (47,XXY) in males with Klinefelter syndrome (Jacobs and Strong). This led to the recognition of the central role of the Y chromosome in establishing gender in mammals, because it became apparent that individuals without a Y chromosome are female and individuals with a Y chromosome are male, irrespective of the number of X chromosomes present. Subsequently, other numerical chromosome aberrations were shown to cause recognizable diseases in man: trisomy 13 and trisomy 18, by Patau and coworkers and Edwards and coworkers in 1960, respectively. The loss of a specific region (a deletion) of a chromosome was shown to be associated with a recognizable pattern of severe developmental defects (Lejeune and coworkers in 1963, Hirschhorn in 1964, and Wolf in 1964). The Philadelphia chromosome, a characteristic structural alteration of a chromosome in bone marrow cells of patients with chronic myelogenous leukemia, which was discovered by Nowell and Hungerford in 1962, showed a connection to the origins of cancer.

Numerous subspecialties of human genetics have arisen, such as *biochemical genetics, immunogenetics, somatic cell genetics, cytogenetics, clinical genetics, population genetics, teratology*, mutational studies, and others. The development of human genetics has been well summarized by McKusick (1992, 2007) and Vogel & Motulsky (1997). The foundation of numerous new scientific journals dealing with human genetics since 1965 reflects the advances in this field: *The American Journal of Medical Genetics, European Journal of Human Genetics, Human Genetics* (prior to 1976 *Humangenetik*), *Clinical Genetics, Human Molecular Genetics, Journal of Medical Genetics, Genetics in Medicine, Annales de Génétique* (now *European Journal of Medical Genetics*), *Cytogenetics and Cell Genetics* (now *Chromosome Research*), *Genomics, Genomic Research, Prenatal Diagnosis, Clinical Dysmorphology, Molecular Syndromology* (founded in 2011), *Community Genetics, Genetic Counseling*, and others.

The enormous progress in the medical aspects of human genetics (*medical genetics*) is documented in *Mendelian Inheritance in Man* (MIM), a catalog of human genes and genetic disorders (McKusick, 1998). It was first established in 1966 by Victor A. McKusick (1921–2008) at Johns Hopkins University in Baltimore and went through 12 printed editions (1968–1998, see p. 398). It is available as a free online publication: *Online Mendelian Inheritance in Man* (OMIM). As of August 2011 it contains over 20000 entries with about 13600 gene descriptions, almost 3000 phenotypic descriptions with a known molecular basis, and more than 1600 phenotypic descriptions without a known molecular basis. More than 2600

Victor A. McKusick (www.hopkinsmedicine.org) (1921–2010)

human genes with mutations are known to cause a disease, and for more than 4400 disorders the molecular basis is known. The McKusick catalog OMIM also lists the genes and related disorders for each human chromosome. Throughout this book, for all human disorders mentioned, the OMIM number is given. This is a six-digit number. The first digit indicates the mode of inheritance or status of molecular knowledge (1, autosomal dominant; 2, autosomal recessive; 3, X-chromosomal; 4, Y-chromosomal; 5, mitochondrial; 6, additional molecular information (OMIM, Online Mendelian Inheritance in Man, see p. 398).

Epigenetics

This term defines a field of study that has attracted considerable interest in recent years. It refers to heritable changes in gene expression without concomitant changes in the DNA sequence. In 1942, C. H. Waddington derived the term from the words genetics and epigenesis. Epigenetic changes involve different molecular mechanisms. Addition of methyl groups to

DNA (DNA methylation) is an important form of an epigenetic change. It results from the addition of methyl groups to cytosine. Methylated DNA is associated with a genetically inactive state, whereas unmethylated DNA is found in genetically active regions. About 40 regions in the genome of man and mouse show a striking pattern of DNA methylation. Here, only one allele of a given gene or region is unmethylated and active, whereas the other allele is methylated and inactive. The methylation pattern is determined by the parental origin of the allele. Thus, either the allele of paternal origin or the allele of maternal origin is methylated. The pattern, called *genomic imprinting*, is transmitted to daughter cells and maintained. DNA methylation is an important control mechanism in gene expression. Errors in establishing or maintaining the correct methylation pattern are a cause of several human disorders (imprinting disorders; pp. 194 and 368).

Important epigenetic changes take place during embryologic development. The DNA-associated proteins, histone proteins, in the chromatin (the packaged DNA in the cell nucleus) are subject to many different epigenetic modifications. Special enzymes add or remove methyl groups, acetyl groups, or phosphate groups at specific sites. This alters the functional state in chromatin (p. 192).

The Human Genome and other International Projects

A new dimension was introduced into biomedical research in 1990 by the Human Genome Project (HGP) and related programs in many other organisms (see Part II, Genomics; Lander & Weinberg, 2000; Green & Guyer, 2011). It ended in 2003 with the publication of the sequence of DNA in a reference sequence (IHGSC, 2004). The HGP was an international organization representing several countries under the leadership of biomedical centers in the United States and the United Kingdom. The main goal of the HGP was to determine the entire sequence of the 3 billion nucleotide pairs in the DNA of the human genome and to find all the genes within it. At the time this was a daunting task, and it was comparable to deciphering each individual 1-mm-wide letter along a text strip 3000 km long. A first draft of a sequenced human genome covering about 90% of the genome was announced in June 2000 (IHGSC, 2001; Venter et al., 2001). The complete DNA sequence of man was published in 2004 (IHGSC, 2004). As of May 2006, all human chromosomes have been sequenced (see Selected Websites for Access to Genetic and Genomic Information: Nature Web Focus: Human Genome Collection; and OMIM, p. 17). Other international genomic projects also exist.

The International HapMap Project is an international multi-country project that officially began in 2002. It is aimed at identifying individual genetic variants in the DNA sequence. These may have important influences on the causes of diseases or responses to therapeutic drugs (see Selected Websites for Access to Genetic and Genomic Information: International HapMap Project).

ENCODE (Encyclopedia of DNA Elements) is an international consortium aimed at studying all functional elements in the human genome (see Selected Websites for Access to Genetic and Genomic Information: Encyclopedia of DNA Elements). It was initiated in 2003.

Ethical and Societal Issues, Education

The HGP also devoted attention and resources to ethical, legal, and social issues (the Ethical, Legal and Social Implications [ELSI] Research program). This was an important part of the HGP, in view of the far-reaching consequences of the current and expected knowledge about human genes and the genome. Depending on the family history and the type of disease, it is now often possible to obtain diagnostic information about a disease years or even decades prior to its manifestation. This widens the time frame of a diagnosis. Furthermore, not only the affected individual, the patient, but also other, unaffected family members may seek information about their own risk for a disease or the risk for a disease in their offspring. The possibility of presymptomatic or predictive genetic testing raises new questions about the use of genetic data. The decision to perform a genetic test has to take into account a person's view on an individual basis, and be obtained after proper counseling about the purpose, validity, and reliability, and the possible consequences of the test result. In some countries legal regulation has been introduced to assure that genetic information generated is used in the best interests of the individual involved, informed consent is obtained, and confidentiality of data is assured. Although genetic principles are quite straightforward, genetics is opposed by some and misunderstood by many. Scientists should seize any opportunity to inform the public about the goals of genetics and genomics and the principal methods employed. Genetics should be highly visible in elementary and high schools. Human genetics should be emphasized in teaching in medical schools.

Genetic and Genomic Causes of Human Disorders

A disease is genetically determined if it is mainly or exclusively caused by disorders in the genetic program of cells and tissues. Genetic disorders can be assigned to four broad categories: monogenic, chromosomal, complex (multigenic, with interaction with environmental influences), and somatic mutations (different forms of cancer). During the past two decades a newly recognized category is represented by genomic disorders (p. 242). Here different mechanisms resulting from certain structural features of the human genome are the causes. The total estimated frequency of genetically determined diseases of different categories in the general population is about 3%–5% (see **table**). Most disease processes result from environmental influences interacting with the individual genetic makeup of the affected individual. These are referred to as complex (multigenic or multifactorial) diseases. They include many relatively common chronic diseases, e.g., high blood pressure, hyperlipidemia, diabetes mellitus, gout, psychiatric disorders, and certain congenital malformations. Their cause is not necessarily one or more mutations, but rather specific variants present in the genome of an affected individual. Another common category is cancer, a large, heterogeneous group of nonhereditary genetic disorders resulting from mutations in somatic cells or hereditary changes in germ cells. It is now possible to characterize the genome of cancer cells (cancer genomes). Chromosomal aberrations are also an important category.

The completion of the HGP and the introduction of new tools for genomic research, in particular the relatively inexpensive high-capacity methods of sequencing DNA (massive parallel sequencing, "second generation," p. 68), since 2005 have ushered in a new era of genomic medicine (reviewed by Green & Guyer, 2011; Lupski et al., 2011). In addition, the whole genome can now be subjected to a search for contributing genetic factors in the causes of a given disorder (genome-wide association studies, p. 232). Genetic and nongenetic bases of disease can be distinguished and individual risk factors determined. Catalogs of germline variants conferring a risk for a disease can be established. Individual adverse responses to therapeutic agents (pharmacogenetics) can now be defined at the level of the whole genome (pharmacogenomics). The new, often unsuspected, genomic information has to be channeled into individual counseling and decision making. The scope of information along with associated uncertainties has to consider psychological and ethical issues.

Categories and frequency of genetically determined diseases

Category of disease	Frequency per 1000 individuals[a]
Monogenic diseases total	5–17
Autosomal recessive	2–7
Autosomal dominant	2–8
X-chromosomal	1–2
Chromosome aberrations (light microscopy)	5–7
Complex disorders (multigenic)	70–90
Genomic disorders	5–10
Somatic mutations (cancer)	200–250
Mitochondrial disorders	2–5

[a] Approximate estimates based on various sources.

Literature Cited in the Text

Bateson W. Mendel's Principles of Heredity. Cambridge: University of Cambridge Press; 1913

Bearn AG. Archibald Garrod and the Individuality of Man. Oxford: Oxford University Press, Oxford; 1993

Cairns J, Stent GS, Watson JD, eds. Phage and the Origins of Molecular Biology. New York: Cold Spring Harbor Laboratory Press; 1978

Chargaff E. Heraclitean Fire: Sketches from a Life before Nature. New York: Rockefeller University Press; 1978

Childs B. Genetic Medicine. A Logic of Disease. Baltimore: Johns Hopkins University Press; 1999

Crick F. What Mad Pursuit: A Personal View of Scientific Discovery. New York: Basic Books; 1988

Dubos RJ. The Professor, the Institute, and DNA: Oswald T. Avery, his Life and Scientific Achievements. New York: Rockefeller University Press; 1976

Fedoroff N, Botstein D, eds. The Dynamic Genome: Barbara McClintock's Ideas in the Century of Genetics. New York: Cold Spring Harbor Laboratory Press; 1992

Fox Keller EA. A Feeling for the Organism: the Life and Work of Barbara McClintock. New York: W. H. Freeman; 1983

Garrod AE. The Inborn Factors in Disease: an Essay. Oxford: Clarendon Press; 1931

Green ED, Guyer MS; National Human Genome Research Institute. Charting a course for genomic medicine from base pairs to bedside. Nature 2011;470:204–213

Haws DV, McKusick VA. Farabee's brachydactylous kindred revisited. Bull Johns Hopkins Hosp 1963;113:20–30

IHGSC (International Human Genome Sequencing Consortium). Initial sequencing and analysis of the human genome. Nature 2001;409:860–921

IHGSC (International Human Genome Sequencing Consortium. Finishing the euchromatic sequence of the human genome. Nature 2004;431:931–945. Available at: http://www.nature.com/nature/supplements/collections/humangenome/. Accessed January 24, 2012

Judson HF. The Eighth Day of Creation. Makers of the Revolution in Biology, expanded edition. New York: Cold Spring Harbor Laboratory Press; 1996

Lander ES, Weinberg RA. Genomics: journey to the center of biology. Science 2000;287:1777–1782

Lewin B. Genes IX. Sudbury: Jones & Bartlett; 2008

Lupski JR, Belmont JW, Boerwinkle E, Gibbs RA. Clan genomics and the complex architecture of human disease. Cell 2011;147:32–43

McCarty M. The Transforming Principle. New York: W. W. Norton; 1985

McClintock B. The significance of responses of the genome to challenge. Science 1984;226:792–801

McKusick VA. Human genetics: the last 35 years, the present, and the future. Am J Hum Genet 1992;50:663–670

McKusick VA. History of medical genetics. In: Rimoin DL, Connor JM, Pyeritz JE, eds. 5th ed. Philadelphia: Churchill Livingstone-Elsevier; 2007:3–32

McKusick VA. Mendelian Inheritance in Man: A Catalog of Human Genes and Genetic Disorders, 12th ed. Baltimore: Johns Hopkins University Press; 1998. Available at: http://www.ncbi.nlm.nih.gov/omim. Accessed January 24, 2012

Morgan TH: The Theory of the Gene. Enlarged and Revised Edition. New Haven: Yale Univ. Press;1926

Müller-Hill B. Murderous Science. Oxford: Oxford University Press; 1988

OMIM. Online Inheritance in Man. A Catalog of Human Genes and Genetic Disorders. Available at: http://www.ncbi.nlm.nih.gov/omim

Rosenberg NA, Pritchard JK, Weber JL, et al. Genetic structure of human populations. Science 2002; 298:2381–2385

Schrödinger E. What Is Life? The Physical Aspect of the Living Cell. New York: Penguin Books; 1944

Stent GS, ed. James D. Watson. The Double Helix: A Personal Account of the Discovery of the Structure of DNA. London: Weidenfeld & Nicolson; 1981

Stern C. Principles of Human Genetics. 3rd ed. San Francisco: W. H. Freeman; 1973

Strong C. Eugenics. In: Cooper DV, ed. Encyclopedia of the Human Genome. Vol. 2. London: Nature Publishing Group; 2003:335–340

Venter JC, Adams MD, Myers EW, et al. The sequence of the human genome. Science 2001;291:1304–1351

Vogel F, Motulsky AG. Human Genetics: Problems and Approaches. 3rd ed. Heidelberg: Springer-Verlag; 1997

Watson JD. The Double Helix. A Personal Account of the Discovery of the Structure of DNA. New York: Atheneum; 1968

Watson JD. A Passion for DNA. Genes, Genomes, and Society. New York: Cold Spring Harbor Laboratory Press; 2000

Watson JD, Crick FH. Molecular structure of nucleic acids; a structure for deoxyribonucleic acid. Nature 1953;171:737–738

Watson JD, Tooze J. The DNA Story: a documentary history of gene cloning. San Francisco: W. H. Freeman; 1981

Wilkins M. The Third Man of the Double Helix. Oxford: Oxford University Press; 2003

Selected Introductory Reading

Alberts B, Johnson A, Lewis J, Raff M, Roberts K, Walter P. Molecular Biology of the Cell. 5th ed. New York: Garland Publishing Co; 2008

Brown TA. Genomes. 3rd ed. New York: Garland Science; 2007

Dobzhansky T. Genetics of the Evolutionary Process. New York: Columbia University Press; 1970

Dunn LC. A Short History of Genetics. New York: McGraw-Hill; 1965

Epstein CJ, Erickson RP, Wynshaw-Boris A, eds. Inborn Errors of Development. The Molecular Basis of Clinical Disorders of Morphogenesis. 2nd ed. Oxford: Oxford University Press; 2008

Franklin RE, Gosling RG. Molecular configuration in sodium thymonucleate. Nature 1953;171:740–741

Gilbert SF. Developmental Biology. 10th ed. Sunderland: Sinauer; 2010

Harper PS. Practical Genetic Counselling. 7th ed. London: Edward Arnold; 2010

Griffith AJF, Wessler S, Lewontin R. Carroll: Introduction to Genetic Analysis. 9th ed. San Francisco: W. H. Freeman; 2007

Hirsch-Kauffmann M, Schweiger M, Schweiger M-R. Biologie und molekulare Medizin. 7th ed. Stuttgart: Thieme; 2009

Jameson JL, Kopp P. Principles of human genetics. In: Harrison's Principles of Internal Medicine. 18th ed. New York: McGraw-Hill Medical; 2012:486–509

Jobling MA, Hurles M, Tyler-Smith C. Human Evolutionary Genetics. Origins, Peoples, and Disease. New York: Garland Science; 2004

Jorde LB, Carey JC, Bamshad MJ. Medical Genetics. 4th ed. Philadelphia: Mosby Elsevier; 2010

King R, Rotter J, Motulsky AG, eds. The Genetic Basis of Common Disorders. 2nd ed. Oxford: Oxford University Press; 2002

King RC, Stansfield WD. A Dictionary of Genetics. 7th ed. Oxford: Oxford University Press; 2006

Klein J, Takahata N. Where do we come from? The Molecular Evidence for Human Descent. Heidelberg: Springer-Verlag; 2002

Lewin B. Genes IX. Sudbury: Jones & Bartlett; 2008

Lodish H, Berk A, Matsudaira P, et al. Molecular Cell Biology (with an animated CD-ROM). 7th ed. New York: W. H. Freeman; 2007

Lupski JR, Stankiewicz P, eds. Genomic Disorders. The Genomic Basis of Disease. Totowa: Humana Press; 2006

Maddox B. Rosalind Franklin. Dark Lady of DNA. London: HarperCollins; 2002

McClintock B. The origin and behavior of mutable loci in maize. Proc Natl Acad Sci U S A 1950;36:344–355

McClintock B. Chromosome organization and genic expression. Cold Spring Harb Symp Quant Biol 1951;16:13–47

McClintock B. Induction of instability at selected loci in maize. Genetics 1953;38:579–599

Morgan TH. The Theory of the Gene. New Haven: Yale University Press; 1926

Nussbaum RL, McInnes RR, Willard HF. Thompson & Thompson Genetics in Medicine. 7th ed. Philadelphia: W. B. Saunders; 2007

Passarge E, Kohlhase J. Genetik. In: Siegenthaler W, Blum HE, eds. Klinische Pathophysiologie. 9th ed. Stuttgart: Thieme; 2006:4–66

Rimoin DL, Connor JM, Pyeritz RE, Korf BR, eds. Emery and Rimoin's Principles and Practice of Medical Genetics. 5th ed. Edinburgh: Churchill-Livingstone; 2007

Scriver CR, Beaudet AL, Sly W, Valle D, eds. The Metabolic and Molecular Bases of Inherited Disease. 8th ed. McGraw-Hill; New York: 2001. Available at: http://www.ommbid.org. Accessed January 24, 2012

Stankiewicz P, Lupski JR. Structural variation in the human genome and its role in disease. Annu Rev Med 2010;61:437–455

Speicher MR, Antnonarakis SE, Motulsky AG, eds. Vogel and Motulsky's Human Genetics. Problems and Approaches. 4th ed. Heidelberg: Springer-Verlag; 2010

Stebbins GL. Darwin to DNA. Molecules to Humanity. San Francisco: W. H. Freeman; 1982

Stent G, Calendar R. Molecular Genetics. An Introductory Narrative. 2nd ed. San Francisco: W. H. Freeman; 1978

Strachan T, Read AP. Human Molecular Genetics. 4th ed. London: Garland Science; 2010

Sturtevant AH. A History of Genetics. New York: Harper & Row; 1965

Turnpenny PD, Ellard S. Emery's Elements of Medical Genetics. 14th ed. Edinburgh, Philadelphia: Elsevier-Churchill Livingstone; 2011

Watson JD, Baker TA, Bell SP, Gann A, Levine M, Losick R. Molecular Biology of the Gene. 6th ed. New York: Pearson/Benjamin Cummings and Cold Spring Harbor Laboratory Press; 2008

Weinberg RA. The Biology of Cancer. New York: Garland Science; 2006

Weatherall DJ. The New Genetics and Clinical Practice. 3rd ed. Oxford: Oxford University Press; 1991

Whitehouse HLK. Towards an Understanding of the Mechanism of Heredity. 3rd ed. London: Edward Arnold; 1973

Selected Websites for Access to Genetic and Genomic Information

Cancer Genomes. Available at: http://www.icgc.org and http://www.sanger.ac.uk/perl/genetics/CGP/cosmic. Accessed January 24, 2012

Deciphering human disease. Database of causes of illness. Available at: http://www.sanger.ac.uk/. Accessed January 24, 2012

Encyclopedia of DNA Elements. ENCODE. Available at: http://www.genome.gov/encode. Accessed January 24, 2012

Information, National Library of Medicine, Bethesda, Maryland. Available at: http://www.ncbi.nlm.nih.gov/sites/entrez?db=omim. Accessed January 24, 2012

GeneTests, a clinical information resource relating genetic testing to the diagnosis, management, and genetic counseling of individuals and families with specific inherited disorders. Available at: http://www.ncbi.nlm.nih.gov/sites/GeneTests/. Accessed January 24, 2012

Genome Bioinformatics UCSC Genome Browser. Available at: http://genome.ucsc.edu/. Accessed January 24, 2012

Genome-wide Association Studies. Available at: http://www.genome.gov/GWAStudies/. Accessed January 24, 2012

Information on Individual Human Chromosomes and Disease Loci. Human Chromosome Launchpad. Available at: https://public.ornl.gov/hgmis/launchpad/default.cfm. Accessed January 24, 2012

International HapMap Project. Available at: www.hapmap.ncbi.nlm.nih.gov. Accessed January 24, 2012

National Human Genome Research Institute. Available at: http://www.genome.gov/Planning/. Accessed January 24, 2012

National Center for Biotechnology Information Genes and Disease Map. Available at: http://www.ncbi.nlm.nih.gov/disease/. Accessed January 24, 2012

Medline. Available at: http://www.ncbi.nlm.nim.nih.gov/PubMed/. Accessed January 24, 2012

MITOMAP. A human mitochondrial genome database. Available at: http://www.gen.emory.edu/mitomap.html. Accessed January 24, 2012

National Center for Health Statistics at Centers for Disease Control and Prevention. Available at: http://www.cdc.gov/nchs/. Accessed January 24, 2012

Nature Web Focus: Human Genome Collection. Available at: http://www.nature.com/nature/supplements/collections/humangenome/. Accessed January 24, 2012

OMIM. Online Mendelian Inheritance of Man. Available at: http://www.ncbi.nlm.nih.gov/omim. Accessed January 24, 2012

Online Metabolic and Molecular Bases of Inherited Disease. Available at: http://www.ommbid.com/. Accessed January 24, 2012

Thousand Genomes Project. Available at: http://www.1000genomes.org. Accessed January 24, 2012

Important Advances that Contributed to the Development of Genetics

(This list represents a selection and should not be considered complete; apologies to all authors not included.)

1665 Cells described and named (*Robert Hooke*)

1827 Human egg cell described (*Karl Ernst von Baer*)

1839 Cells recognized as the basis of living organisms (*Schleiden, Schwann*)

1859 Concept and facts of evolution (*Charles Darwin*)

1865 Rules of inheritance by distinct "factors" acting dominantly or recessively (*Gregor Mendel*)

1869 "Nuclein": a new acidic, phosphorus-containing, long molecule (*F. Miescher*)

1874 Monozygotic and dizygotic twins distinguished (*C. Dareste*)

1876 "Nature and nurture" (*F. Galton*)

1879 Chromosomes in mitosis (*W. Flemming*)

1883 Quantitative aspects of heredity (*F. Galton*)

1888 Term "chromosome" (*W. Waldeyer*)

1889 Term "nucleic acid" (*R. Altmann*)

1892 Term "virus" (*R. Ivanowski*)

1897 Enzymes discovered (*E. Büchner*)

1900 Mendel's discovery recognized (*H. de Vries, E. Tschermak, K. Correns*, independently)
ABO blood group system (*Landsteiner*)

1901 Term "Mutation" coined (*H. DeVries*)

1902 Some diseases in man inherited according to Mendelian rules (*W. Bateson, A. Garrod*)
Sex chromosomes (*McClung*)
Chromosomes and Mendel's factors are related (*W. Sutton*)
Individuality of chromosomes (*T. Boveri*)

1906 Term "genetics" proposed (*W. Bateson*)

1907 Amphibian spinal cord culture (*Harrison*)

1908 Population genetics (*G. H. Hardy, W. Weinberg*)

1909 Inborn errors of metabolism (*A. Garrod*)
Terms "gene," "genotype," "phenotype" proposed (*W. Johannsen*)
Chiasma formation during meiosis (*Janssens*)
First inbred mouse strain DBA (*C. Little*)

1910 Beginning of *Drosophila* genetics (*T. H. Morgan*)
First *Drosophila* mutation (white-eyed)

1911 Sarcoma virus (*Peyton Rous*)

1912 Crossing-over (*T. H. Morgan and E. Cattell*)
Genetic linkage (*T. H. Morgan and C. J. Lynch*)
First genetic map (*A. H. Sturtevant*)

1913 First long-term cell culture (*A. Carrel*)
Nondisjunction (*C. B. Bridges*)

1915 Genes located on chromosomes (chromosomal theory of inheritance) (*Morgan, Sturtevant, Muller, Bridges*)
Bithorax mutant (*C. B. Bridges*)
First genetic linkage in vertebrates (*J. B. S. Haldane, A. D. Sprunt, N. M. Haldane*)
Term "intersex" (*R. B. Goldschmidt*)

1917 Bacteriophage discovered (*F. d'Herrelle*)

1922 Characteristic phenotypes of different trisomies in the plant *Datura stramonium* (*F. Blakeslee*)

1923 Chromosome translocation in *Drosophila* (*C. B. Bridges*)

1924 Blood group genetics (*Bernstein*)
Statistical analysis of genetic traits (*R. A. Fisher*)

1926 Enzymes are proteins (*J. Sumner*)

1927 Mutations induced by X-rays (*H. J. Muller*)
Genetic drift (*S. Wright*)

1928 Euchromatin/heterochromatin (*E. Heitz*)
Genetic transformation in bacteria (*F. Griffith*)

1933 Pedigree analysis (*Haldane, Hogben, Fisher, Lenz, Bernstein*)
Polytene chromosomes (*Heitz and Bauer, Painter*)

1934 Term "aneuploidy" coined (*A. F. Blakeslee*)

1935 First cytogenetic map in *Drosophila* (*C. B. Bridges*)

1937 Mouse H2 gene locus (*P. Gorer*)
First human linkage group hemophilia A – colorblindness (*J. Bell and J. B. S. Haldane*)

1938 Telomere defined (*H. J. Muller*)

1940 Polymorphism (*E. B. Ford*)
Rhesus blood groups (*Landsteiner and Wiener*)

1941 Evolution through gene duplication (*E. B. Lewis*)
Genetic control of enzymatic biochemical reactions (*Beadle and Tatum*)
Mutations induced by mustard gas (*C. Auerbach and J. M. Robson*)

1942 Concept of epigenetics (*C. H. Waddington*)

1943 Mutations in bacteria (*S. E. Luria and M. Delbrück*)

1944 DNA as the material basis of genetic information (*Avery, MacLeod, McCarty*)
What is Life? The Physical Aspect of the Living Cell. An influential book (*E. Schrödinger*)

1946 Genetic recombination in bacteria (*Lederberg and Tatum*)

1947 Genetic recombination in viruses (*Delbrück and Bailey, Hershey*)

1949 Sickle cell anemia, a genetically determined molecular disease (*Neel, Pauling*)
Hemoglobin disorders prevalent in areas of malaria (*J. B. S. Haldane*)
X chromatin (*Barr and Bertram*)

1950 Defined relation of the four nucleotide bases (*E. Chargaff*)

1951 Mobile genetic elements in Indian corn, *Zea mays* (*B. McClintock*)
α-Helix and β-sheet in proteins (*L. Pauling and R. B. Corey*)

1952 Genes consist of DNA (*Hershey and Chase*)
Plasmids (*Lederberg*)
Transduction by phages (*Zinder and Lederberg*)
First enzyme defect in man (*Cori and Cori*)
First linkage group in man (*Mohr*)
Colchicine and hypotonic treatment in chromosomal analysis (*Hsu and Pomerat*)
Exogenous factors as a cause of congenital malformations (*J. Warkany*)

1953 DNA structure (*Watson and Crick, Franklin, Wilkins*)
Conjugation in bacteria (*W. Hayes, L. L. Cavalli, J. and E. Lederberg*, independently)
Non-Mendelian inheritance (*Ephrussi*)
Cell cycle (*Howard and Pelc*)
Dietary treatment of phenylketonuria (*Bickel*)

1954 DNA repair (*Muller*)
HLA system (*J. Dausset*)
Leukocyte drumsticks (*Davidson and Smith*)
Cells in Turner syndrome are X-chromatin negative (*P. Polani*)
Cholesterol biosynthesis (*K. Bloch*)

1955 First genetic map at the molecular level (*S. Benzer*)
First amino acid sequence of a protein, insulin (*F. Sanger*)
Lysosomes (*C. de Duve*)
Buccal smear (*Moore, Barr, Marberger*)
5-Bromouracil, an analogue of thymine, induces mutations in phages (*A. Pardee and R. Litman*)

1956 46 Chromosomes in man (*Tijo and Levan, Ford and Hamerton*)
Amino acid sequence of hemoglobin molecule (*V. Ingram*)
DNA synthesis in vitro (*S. Ochoa, A. Kornberg*)
Synaptonemal complex, the area of synapse in meiosis (*M. J. Moses, D. Fawcett*)
Genetic heterogeneity (*H. Harris, C. F. Fraser*)

1957 Genetic complementation (*Fincham*)
Genetic analysis of radiation effects in man (*Neel and Schull*)

1958 Semiconservative replication of DNA (*M. Meselson and F. W. Stahl*)
Somatic cell genetics (*G. Pontecorvo*)
Ribosomes (*Roberts, Dintzis*)
Cloning of single cells (*Sanford, Puck*)

1959 First chromosomal aberrations in man: trisomy 21 (*Lejeune, Gautier, Turpin*)
Turner syndrome, 45,XO (*C. E. Ford*)
Klinefelter syndrome: 47 XXY (*Jacobs and Strong*)
DNA polymerase (*A. Kornberg*)
Isoenzymes (*Vesell, Markert*)
Pharmacogenetics (*Motulsky, Vogel*)

1960 Phytohemagglutinin-stimulated lymphocyte cultures (*Nowell, Moorhead, Hungerford*)

1961 The genetic code is read in triplets (*Crick, Brenner, Barnett, Watts-Tobin*)
The genetic code determined (*Nirenberg, Mathaei, Ochoa*)
X-chromosome inactivation (*M. F. Lyon*, confirmed by *Beutler, Russell, Ohno*)
Gene regulation, concept of operon (*Jacob and Monod*)
Galactosemia in cell culture (*Krooth*)
Cell hybridization (*Barski, Ephrussi*)
Thalidomide embryopathy (*Lenz, McBride*)

1962 Philadelphia chromosome (*Nowell and Hungerford*)
Molecular characterization of immunoglobulins (*Edelman, Franklin*)
Identification of individual human chromosomes by ^{3}H-autoradiography (*J. German, O. J. Miller*)
Term "codon" for a triplet of (sequential) bases (*S. Brenner*)
Replicon (*Jacob and Brenner*)
Cell culture (*W. Szybalski and E. K. Szybalska*)
Xg, the first X-linked human blood group (*Mann, Race, Sanger*)
Screening for phenylketonuria (*Guthrie, Bickel*)

1963 Lysosomal storage diseases (*C. de Duve*)
First autosomal deletion syndrome (cri-du-chat syndrome) (*J. Lejeune*)

1964 Colinearity of gene and protein gene product (*C. Yanofsky*)
Excision repair (*Setlow*)
MLC test (*Bach and Hirschhorn, Bain and Lowenstein*)
Microlymphotoxicity test (*Terasaki and McClelland*)
Selective cell culture medium HAT (*J. Littlefield*)
Spontaneous chromosomal instability (*J. German, T. M. Schröder*)
Cell culture from amniotic fluid cells (*H. P. Klinger*)
Hereditary diseases studied in cell cultures (*Danes, Bearn, Krooth, Mellman*)
Population cytogenetics (*Court Brown*)
Fetal chromosomal aberrations in spontaneous abortions (*Carr, Benirschke*)

1965 Sequence of alanine transfer RNA from yeast (*R. W. Holley*)
Limited life span of cultured fibroblasts (*Hayflick, Moorhead*)
Crossing-over in human somatic cells (*J. German*)
Cell fusion with Sendai virus (*H. Harris and J. F. Watkins*)

1966 Genetic code complete
Catalogue of Mendelian phenotypes in man (*V. A. McKusick*)

1968 Restriction endonucleases (*H. O. Smith, Linn and Arber, Meselson and Yuan*)
Okazaki fragments in DNA synthesis (*R. T. Okazaki*)
HLA-D the strongest histocompatibility system (*Ceppellini, Amos*)
Repetitive DNA (*Britten and Kohne*)
Biochemical basis of the ABO blood group substances (*Watkins*)
DNA excision repair defect in xeroderma pigmentosum (*Cleaver*)
First assignment of an autosomal gene locus in man (*Donahue, McKusick*)
Synthesis of a gene in vitro (*H. G. Khorana*)
Neutral gene theory of molecular evolution (*M. Kimura*)

1970 Reverse transcriptase (*D. Baltimore, H. Temin*, independently)
Synteny, a new term to refer to all gene loci on the same chromosome (*Renwick*)
Enzyme defects in lysosomal storage diseases (*Neufeld, Dorfman*)
Individual chromosomal identification by specific banding stains (*Zech, Casperson, Lubs, Drets and Shaw, Schnedl, Evans*)
Y-chromatin (*Pearson, Bobrow, Vosa*)
Thymus transplantation for immune deficiency (*van Bekkum*)

1971 Two-hit theory in retinoblastoma (*A. G. Knudson*)

1972 High average heterozygosity (*Harris and Hopkinson, Lewontin*)
Association of HLA antigens and diseases

1973 Receptor defects in the etiology of genetic defects, genetic hyperlipidemia (*Brown, Goldstein, Motulsky*)
Demonstration of sister chromatid exchanges with BrdU (*S. A. Latt*)
Philadelphia chromosome as translocation (*J. D. Rowley*)

1974 Chromatin structure, nucleosome (*Kornberg, Olins and Olins*)
Dual recognition of foreign antigen and HLA antigen by T lymphocytes (*P. C. Doherty and R. M. Zinkernagel*)
Clone of a eukaryotic DNA segment mapped to a specific chromosome location (*D. S. Hogness*)

1975 Southern blot hybridization (*E. Southern*)
Monoclonal antibodies (*Köhler and Milstein*)
First protein-signal sequence identified (*G. Blobel*)
Model for promoter structure and function (*D. Pribnow*)
First transgenic mouse (*R. Jaenisch*)
Asilomar conference about recombinant DNA

1976 Overlapping genes in phage ΦX174 (*Barell, Air, Hutchinson*)
Loci for structural genes on each human chromosome known (*Baltimore Conference on Human Gene Mapping*)
First diagnosis using recombinant DNA technology (*W. Kan, M. S. Golbus, A. M. Dozy*)

1977 Genes contain coding and noncoding DNA segments (*R. J. Roberts, P. A. Sharp, independently*)
First recombinant DNA molecule that contains mammalian DNA
Methods to sequence DNA (*F. Sanger, Maxam and Gilbert*)
Sequence of phage ΦX174 (*F. Sanger*)
X-ray diffraction analysis of nucleosomes (*Finch and coworkers*)

1978 Terms "exon" and "intron" for coding and noncoding parts of eukaryotic genes (*W. Gilbert*)
β-Globulin gene structure (*Leder, Weissmann, Tilghman and others*)
Mechanisms of transposition in bacteria
Production of somatostatin with recombinant DNA
Introduction of "chromosome walking" to find genes
First genetic diagnosis using restriction enzymes (*Y. H. Kan and A. M. Dozy*)
DNA tandem repeats in telomeres (*E. H. Blackburn and J. G. Gall*)

1979 Small nuclear ribonucleoproteins ("snurps") (*M. R. Lerner and J. A. Steitz*)
Alternative genetic code in mitochondrial DNA (*B. G. Barell, A. T. Bankier, J. Drouin*)
p53 protein (*D. P. Lane, A. Levine, L. Crawford, L. Old*)

1980 Restriction fragment length polymorphism for mapping (*D. Botstein and coworkers*)
Genes for embryonic development in *Drosophila* studied by mutational screen (*C. Nüsslein-Volhard and E. Wieschaus*)
First transgenic mice by injection of cloned DNA (*J. W. Gordon*)
Transformation of cultured mammalian cells by injection of DNA (*M. R. Capecchi*)
Structure of 16S ribosomal ribonucleoprotein (*C. Woese*)

1981 Sequence of a mitochondrial genome (*S. Anderson, S. G. Barrell, A. T. Bankier*)

1982 Tumor suppressor genes (*H. P. Klinger*)
Prions (proteinaceous infectious particles) as cause of central nervous system diseases (kuru, scrapie, Creutzfeldt–Jakob disease) (*S. B. Prusiner*)
Insulin made by recombinant DNA marketed (Eli Lilly)

1983 Cellular oncogenes (*H. E. Varmus and others*)
HIV virus (*L. Montagnier, R. Gallo*)
Molecular basis of chronic myelocytic leukemia (*C. R. Bartram, D. Bootsma and coworkers*)
First recombinant RNA molecule (*E. A. Miele, D. R. Mills, F. R. Kramer*) Bithorax

complex of *Drosophila* sequenced (*W. Bender*)

1984 Identification of the T cell receptor (*Tonegawa*)
Homeobox (Hox) genes in *Drosophila* and mice (*W. McGinnis*) Localization of the gene for Huntington disease (*Gusella*)
Description of *Helicobacter pylori* (*B. Marshall and R. Warren*)

1985 Polymerase chain reaction (*K. B. Mullis, R. K. Saiki*)
Hypervariable DNA segments as "genetic fingerprints" (*A. Jeffreys*)
Hemophilia A gene cloned (*J. Gietschier*)
Sequencing of the HIV-1 virus
Linkage analysis of the gene for cystic fibrosis (*H. Eiberg and others*)
Isolation of telomerase from *Tetrahymena* (*C. W. Greider and E. H. Blackburn*)
Isolation of a zinc finger protein from *Xenopus* oocytes (*J. R. Miller, A. D. McLachlin, A. Klug*)
Insertion of DNA by homologous recombination (*O. Smithies*)
Genomic imprinting in the mouse (*B. Cattanach*)

1986 First cloning of human genes
Human visual pigment genes characterized (*J. Nathans, D. Thomas, D. S. Hogness*)
RNA as catalytic enzyme (*T. Cech*)
First identification of a human gene based on its chromosomal location (positional cloning) (*B. Royer-Pokora and coworkers*)

1987 Fine structure of an HLA molecule (*Björkman, Strominger and coworkers*)
Knockout mouse (*M. Capecchi*)
A genetic map of the human genome (*H. Donis-Keller and coworkers*)
Mitochondrial DNA and human evolution (*R. L. Cann, M. Stoneking, A. C. Wilson*)

1988 Start of the Human Genome Project
Molecular structure of telomeres at the ends of chromosomes (*E. H. Blackburn and others*)
Cloning of the gene for Duchenne muscular dystrophy (*L. M. Kunkel and others*)
Mutations in human mitochondrial DNA (*D. C. Wallace*)
Transposable DNA as rare cause of hemophilia A (*H. H. Kazazian*)
Successful gene therapy in vitro

1989 Identification of the gene causing cystic fibrosis (*L.-C. Tsui and others*)
Microdissection and cloning of a defined region of a human chromosome (*Lüdecke, Senger, Claussen, Horsthemke*)

1990 Mutations in the *p53* gene as cause of Li-Fraumeni syndrome (*D. Malkin*)
Mutations in the gene wrinkled seed used by Mendel (*M. K. Bhattacharyya*)
A defective gene as cause of inherited breast cancer (*Mary-Claire King*)

1991 Odorant receptor multigene family (*Buck and Axel*)
Complete sequence of a yeast chromosome
Increasing use of microsatellites as polymorphic DNA markers
Trinucleotide repeat expansion as a new class of human pathogenic mutations

1992 High-density map of DNA markers on human chromosomes
X chromosome inactivation center identified
p53 Knockout mouse (*O. Smithies*)

1993 Gene for Huntington disease cloned (*M. E. MacDonald*)
Developmental mutations in zebra fish (*M. C. Mullins and C. Nüsslein-Volhard*)

1994 First physical map of the human genome in high resolution
Mutations in fibroblast growth factor receptor genes as cause of achondroplasia and other human diseases (*M. Muenke*)
Identification of genes for hereditary breast cancer

1995 Cloning of the *BLM* (Bloom syndrome) gene (*N. A. Ellis, J. Groden, J. German and coworkers*)
First genome sequence of a free-living bacterium, *Haemophilus influenzae* (*R. D. Fleischmann, J. C. Venter and coworkers*)
Master gene of the vertebrate eye, *sey*

(small-eye) (*G. Halder, P. Callaerts, W. J. Gehring*)
STS map of the human genome (*T. J. Hudson and coworkers*)

1996 Yeast genome sequenced (*A. Goffeau and coworkers*)
Mouse genome map with more than 7000 markers (*E. S. Lander*)

1997 Sequence of *E. coli* (*F. R. Blattner and coworkers*), *Helicobacter pylori* (*J. F. Tomb*)
Neanderthal mitochondrial DNA sequences (*M. Krings, S. Pääbo and coworkers*)
Mammal ("Dolly, the sheep") cloned by transfer of an adult cell nucleus into an enucleated oocyte (*I. Wilmut*)

1998 RNA interference (RNAi) (*A. Fire and coworkers*)
Nematode *C. elegans* genome sequenced
Human embryonic stem cells (*Thomson and Gearhart*)

1999 First human chromosome (22) sequenced
Ribosome crystal structure

2000 *Drosophila* genome sequenced (*M. D. Adams*)
First complete genome sequence of a plant pathogen (*Xylella fastidiosa*)
Arabidopsis thaliana, the first plant genome sequenced

2001 First draft of the complete sequence of the human genome (*F. H. Collins, J. C. Venter and coworkers*)

2002 Genome sequence of the mouse (*R. H. Waterston and coworkers*)
Sequence of the genome of rice, *Oryza sativa* (*J. Yu, S. A. Goff and coworkers*)
Sequence of the genomes of malaria parasite, *Plasmodium falciparum*, and its vector, *Anopheles gambiae*
Earliest hominid, *Sahelanthropos tchadiensis* (*M. Brunet*)

2003 International HapMap Project and ENCODE launched
Sequence of the human Y chromosome (*H. Skaletsky, D. C. Page and coworkers*)
Homo sapiens idaltù, the oldest anatomically modern man from pleistocene 154–160 years ago (*T. D. White and coworkers*)

2004 Genome sequence of the Brown Norway rat
A new small-bodied hominin from Flores island, Indonesia (*P. Brown and coworkers*)

2005 Massive parallel DNA sequencing methods ("Next generation sequencing") introduced
Genome sequence of the chimpanzee (*R. H. Waterston, E. S. Lander, R. K. Watson and coworkers*)
1.58 million human single-nucleotide polymorphisms mapped (*D. A. Hinds, D. R. Cox and coworkers*)
Human haplotype map
Sequence of the human X chromosome (*M. T. Ross and coworkers*)
Inactivation profile of the human X chromosome (*L. Carrel and H. F. Willard*)

2006 All human chromosomes sequenced

2007 Genome-wide studies applied to find predisposing factors for certain diseases
Genomic disorders recognized

2008 Synthetic bacterial genome (*C. Venter and coworkers*)
Sequencing of individual human genomes

2009 Whole genome analysis by microarrays
Cancer genomes sequenced
Ardipithecus ramidus defines new stages in human evolution (*T. White and others*)

2010 Exome sequencing
Neanderthal genome sequence
Induced pluripotent stem cells (iPS)

2011 Genome structural variation (*E. E. Eichler and coworkers*)
Chromothripsis, a catastrophic event in oncogenesis

References for the Chronology

In addition to personal notes, dates are based on the following main sources:

Dunn LC. A Short History of Genetics. New York: McGraw-Hill; 1965

King RC, Stansfield WD. A Dictionary of Genetics. 7th ed. Oxford: Oxford University Press; 2006

Lander ES, Weinberg RA. Genomics: journey to the center of biology. Science 2000;287:1777–1782

McKusick VA. Human genetics: the last 35 years, the present, and the future. Am J Hum Genet 1992;50: 663–670

Stent GS, ed. James D. Watson. The Double Helix: A Personal Account of the Discovery of the Structure of DNA. London: Weidenfeld & Nicolson; 1981

Sturtevant AH. A History of Genetics. New York: Cold Spring Harbor Press; 2001

The New Encyclopaedia Britannica. 15th ed. Chicago: Encyclopaedia Britannica; 2010

Vogel F, Motulsky AG. Human Genetics: Problems and Approaches. 3rd ed. Heidelberg: Springer-Verlag; 1997

Whitehouse HLK. Towards an Understanding of the Mechanism of Heredity. 3rd ed. London: Edward Arnold; 1973

Fundamentals

Phylogenetic Tree of Living Organisms

A phylogenetic tree attempts to show inferred evolutionary relationships of living organisms. The first example of such a tree was presented by Lamarck in 1809. It is the only figure in the *Origin of Species*, in which Charles Darwin wrote in 1859, "Probably all of organic beings which have ever lived on this Earth have descended from some primordial form." There is overall agreement that the earth is a little more than 4.5 billion years old, and that early forms of life date back about 3.5 billion years.

A. Three primary branches of the tree of life

The formal evolutionary hierarchy of groups of organisms proceeds from the largest to the smallest groups: domain—kingdom—phylum—order–class–family–genus–species. Living organisms are grouped according to the type of cells they consist of, either *prokaryotic* cells or *eukaryotic* cells. A third group of living organisms was recognized in the late 1960s, the Archaea (also called archaebacteria). They are assigned to two classes: Euryarchaeota and Crenarchaeota.
Archaea can live without molecular oxygen at high temperatures (70–110°C, *thermophiles*) or at low temperatures (*psychrophiles*), in water with high concentrations of sodium chloride (*halophiles*) or sulfur (*sulfothermophiles*), in a highly alkaline environment (pH as high as 11.5, alkaliphiles), in acid conditions with pH near zero (acidophiles), or a combination of such adverse conditions that would boil or dissolve ordinary bacteria. It is assumed that prokaryotes predate eukaryotes, and that two preexisting prokaryotes contributed their genomes to the first eukaryotic genome. Eukaryotes consist of several kingdoms, including animals, fungi, plants, algae, protozoa, and others. The three domains have a presumed common progenitor, called the last universal common ancestor.

B. Phylogeny of metazoa (animals)

The phylogeny of metazoa differs, depending on whether it is based on the traditional interpretation or on molecular evidence as revealed mainly by ribosomal RNA sequence comparisons.

C. Mammalian phylogeny

Mammals arose about 100 million years ago in the late Mesozoic period of the Earth. The time scale is only approximate. Of the 4629 known mammalian species, 4356 are placentals, and these fall into 12 orders. The first five placental orders according to their number of species are rodents (2015), followed by bats (925), insectivores (385), carnivores (271), and primates (233). For many of these DNA sequence data exist. This has resulted in some rearrangements of the phylogeny. (Figures modified from Klein & Takahata, 2001.)

Further Reading

Alberts B, et al. Molecular Biology of the Cell, 6th ed. New York: Garland Science; 2008

Allers T, Mevarech M. Archaeal genetics – the third way. Nat Rev Genet 2005;6:58–73

Dereeper A, et al. Phylogeny.fr: robust phylogenetic analysis for the non-specialist. Nucl Acids Res 2008;36(Web Server Issue):W465–469. Available at: http://www.phylogeny.fr/. Accessed January 24, 2012

Delsuc F, Brinkmann H, Philippe H. Phylogenomics and the reconstruction of the tree of life. Nat Rev Genet 2005;6:361–375

Delsuc F, et al. Tunicates and not cephalochordates are the closest living relatives of vertebrates. Nature 2006;439:965–968

Hazen RM. Genesis: the Scientific Quest for Life's Origins. Washington: Joseph Henry Press; 2005

Klein J, Takahata N. Where Do We Come From? The Molecular Evidence for Human Descent. Berlin, Heidelberg: Springer-Verlag; 2001

Lamarck JB. In: Elliot H, ed. Zoological Philosophy: An Exposition with Regard to the Natural History of Animals. London: Macmillan; 1914:179 [reprinted by the University of Chicago Press, 1984]

Murphy WJ, et al. Molecular phylogenetics and the origins of placental mammals. Nature 2001;409:614–618

Rivera MC, Lake JA. The ring of life provides evidence for a genome fusion origin of eukaryotes. Nature 2004;431:152–155

Rokas A. Genomics. Genomics and the tree of life. Science 2006;313:1897–1899

Woese CR. Interpreting the universal phylogenetic tree. Proc Natl Acad Sci U S A 2000;97:8392–8396

Woese CR. On the evolution of cells. Proc Natl Acad Sci U S A 2002;99:8742–8747

Woese CR. A new biology for a new century. Microbiol Mol Biol Rev 2004;68:173–186

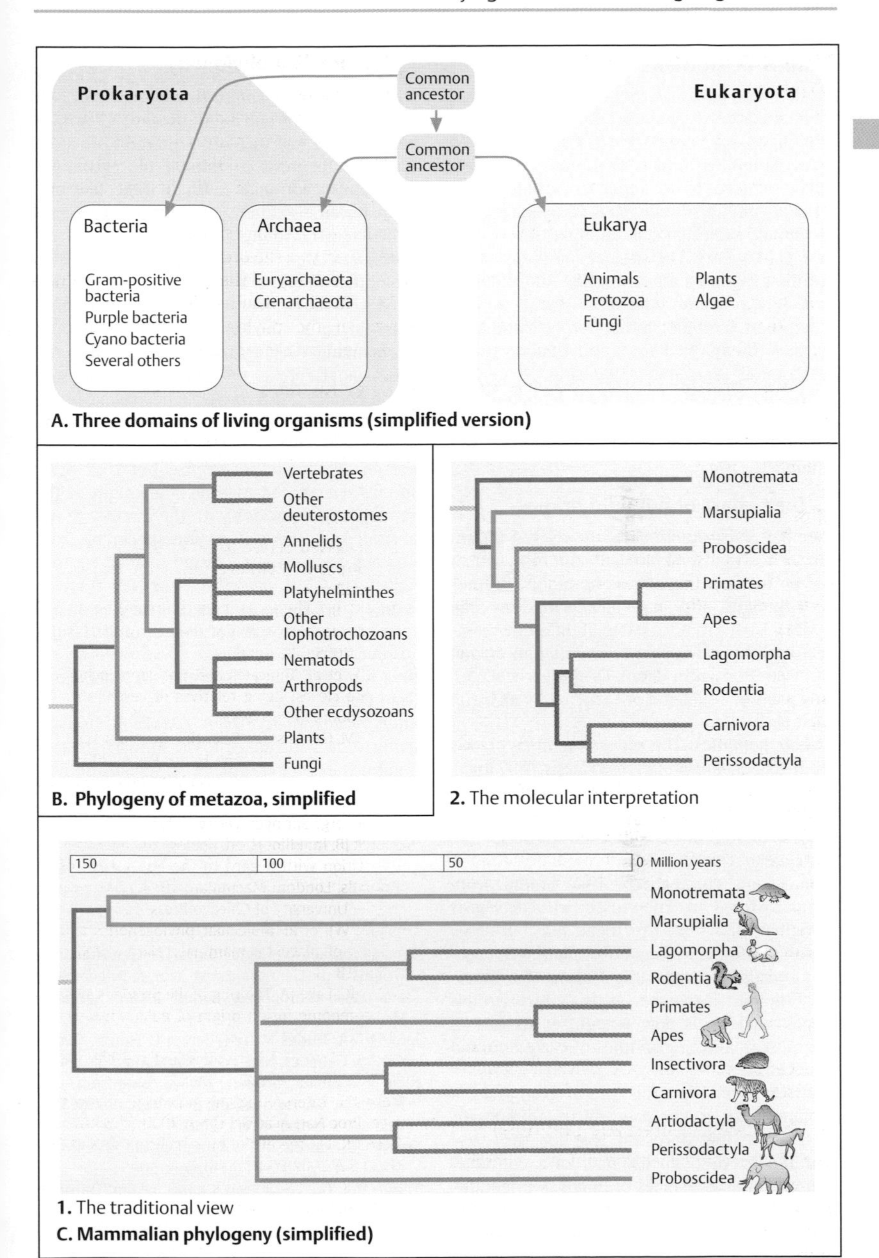

A. Three domains of living organisms (simplified version)

B. Phylogeny of metazoa, simplified

2. The molecular interpretation

1. The traditional view

C. Mammalian phylogeny (simplified)

Origins of Humans

Human and chimpanzee lineages separated from a common ancestor about 6–7 million years ago (mya). Several extinct species related in evolution to human evolution are collectively referred to as *hominids*, although the term *hominin* also is used. A complex pattern of human evolution has emerged during the past 10–15 years. The sites of the main hominid finds in Africa are: along the Rift Valley in East Africa in the Afar and Hadar regions (north-east Ethiopia), the region on both sides of Lake Turkana in Kenya, and Olduvai Gorge and Laetoli (Tanzania); two sites in Central Africa (Bar-el-Ghazal and Toros-Menalla, both in Chad); and several sites in South Africa (Sterkfontein, Kromdraai, Swartkrans, and Taung).

A. Time chart of hominid origins

Four hominid evolutionary age-related phases can be distinguished: archaic hominids, transient forms including robust hominids, premodern humans, and anatomically modern man (AMM). Each group consists of different members with a relationship (shown by an arrow) often not known in detail. Of the early stages, only parts of a skeleton or teeth are available in most cases.

Archaic hominids. The oldest member of this group is *Sahelanthropus tchadensis* (6–7 mya), found in Central Africa 2500 km west of the Rift Valley. The brain is of chimpanzee size (360–370 cm^3), but the face is relatively flat, and enamel thickness is intermediate between human and chimpanzee. Two genera from about 5–6 mya are known: *Orrorin tugenensis* ("original man from the Tugen hills," Baringo region of central Kenya) and *Ardipithecus ramidus kadabba* (from Middle Awash, Afar, Ethiopia). The fourth member of this group, *Ardipithecus ramidus*, is now recognized as one of the most revealing examples of early hominid evolution.

Transient hominids. This group comprises a subfamily of Hominidae, *Australopithecinae*, with a possible ancestral role for the early hominin species. Bipedal gait with concomitant anatomical changes of hands and feet, reduction of tooth size, and progressive development of the brain are main characteristics of this group. The member of this group discovered first, by Dart in 1924 ("Taung child") in South Africa, was *Australopithecus africanus* with an apparent ability to walk upright. Best known of this group is *A. afarensis* (3–4 mya) with upright gait and other hominid features. "Lucy" is its most prominent representative. The robust hominids with a large jaw and teeth became extinct and are not considered to be ancestral to the genus *Homo*.

Premodern Man. The major representative of premodern man is *Homo erectus*. It originated in Africa about 1.9 mya, and is the first earliest hominid to be found outside Africa in Asia ("Java man," "Peking man," 1.8 mya) and the Caucasus (Georgia, 1.6–1 mya). The status of *Homo habilis* (2.4–1 mya), "handy man," in the Olduvai Gorge, is controversial. Recent data from Koobi Fora indicate that *H. habilis* was not anagenetic to *H. erectus*, but that these species overlapped in time. The origin of the genus *Homo* coincides with the distinctive use of stone tools. Early *H. erectus* specimens from Africa are sometimes referred to as *H. ergaster* (2.3–1.4 mya). *H. heidelbergensis* (0.6–0.1 mya) is an extinct *Homo* species that may be an ancestor of both *H. neanderthalensis* and *H. sapiens* in Europe. *H. antecessor* (0.8 mya) is an extinct human species discovered 10 years ago in Gran Dolina, Spain.

Anatomically Modern Man (AMM). All humans living today, including fossil humans resembling us, are *Homo sapiens*. *H. sapiens* derived from *H. erectus* in Africa about 200–100 000 years ago. AMM left Africa about 50 000 years ago and migrated to all continents at different times. (Data based on Wood, 2005; and Stringer & Andrews, 2005.)

Further Reading

Grine FE, Leakey RE, Flagle JG, eds. The First Humans – Origins of the Genus *Homo*. Vertebrate Paleobiology and Paleoanthropology Series. Berlin: Springer-Verlag; 2009

Jobling MA, Hurles M, Tyler-Smith C. Human Evolutionary Genetics. New York, NY: Garland; 2004

Klein J, Takahata N. Where do we come from? The Molecular Evidence of Human Descent. Berlin, Heidelberg: Springer-Verlag; 2002

Stringer C, Andrews P. The Complete World of Evolution. London: Thames & Hudson; 2005

Tattersall I. The Fossil Trait. Oxford: Oxford University Press; 1995

Wood B. Human Evolution. Oxford: Oxford University Press; 2005

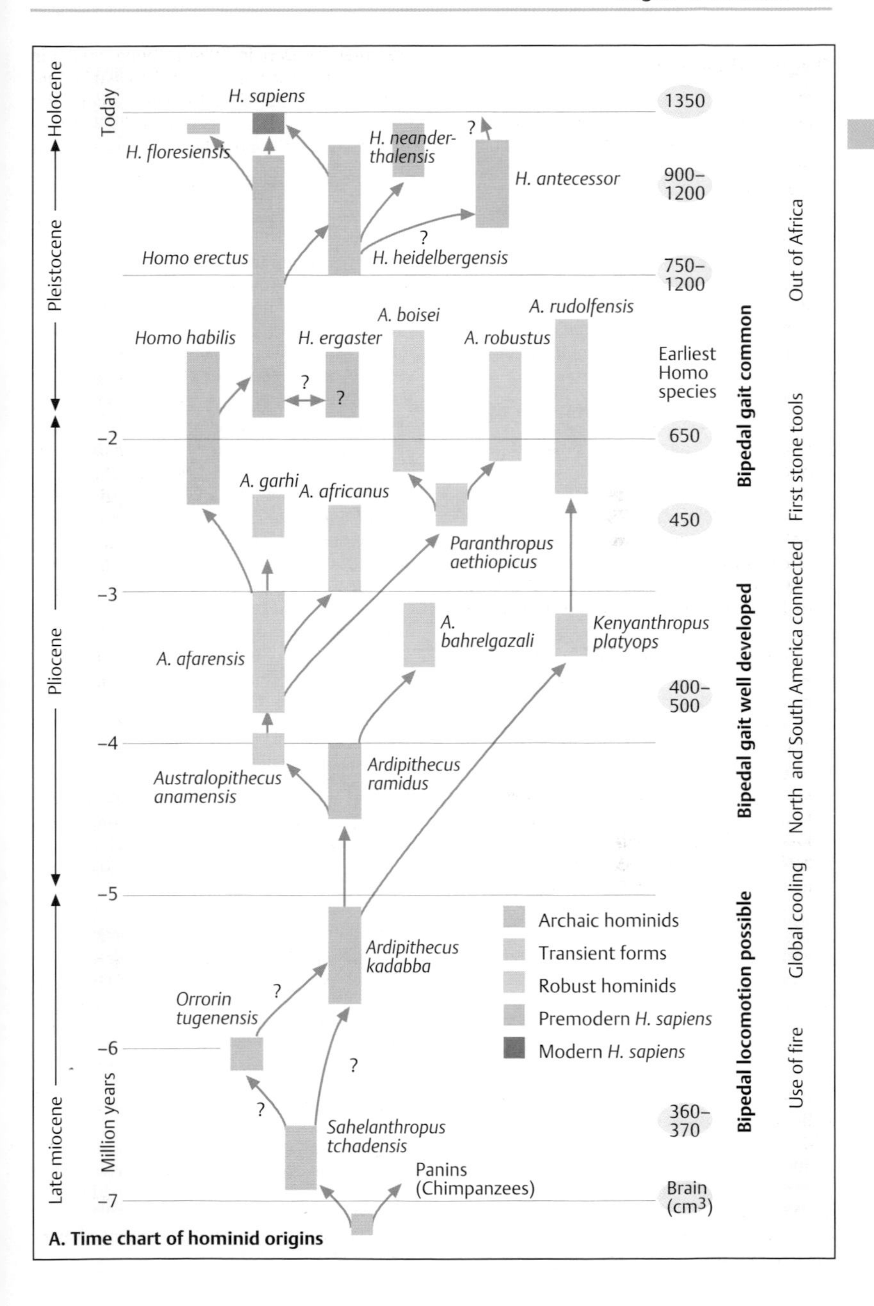

A. Time chart of hominid origins

Out of Africa: Toward Modern Humans

Early forms of humans emerged in Africa about 160 000–200 000 years ago. From there, they migrated out of Africa into other continents.

When modern humans arrived in Europe about 50 000 years ago, they met an older population that had been living there for about 400 000 years, the Neandertals. For the most part, they occupied different sites and developed different albeit overlapping forms of early culture, referred to as *Mousterian* for the Neandertals and *Acheulian* for the early humans, named after sites in France.

A. Origins of humans in relation to Neandertals

Modern humans and Neandertals share a common ancestor about 800 000 years ago. The population split occurred 270 000–440 000 years ago. For about 14 000 years, Neandertals and modern humans co-existed in regions extending from south-east Europe into Siberia. Neandertals became extinct about 28 000 years ago.

About 4% of European and Asian genomes share nuclear DNA with Neandertals, but African populations do not. A distinct third group has been discovered from the Denisovan cave in Siberia (Reich et al., 2010). This population contributed about 4–6% of its DNA to present-day Melanesians.

Mitochondrial DNA (mtDNA) shows that Neandertals share a common mtDNA ancestor with mtDNA of modern human about 500 000 years ago. (Figure adapted from Noonan et al., 2006.)

B. Dispersal of modern humans out of Africa

For the exodus out of Africa about 60 000 years ago (60 K), northern and southern routes have been recognized. Asia was reached first (50–70 K), then Australia and Europe somewhat later (40 K). The Americas were reached last (about 13 K). The dispersal from Africa coincides with growth of local populations and scarcity of resources. The climate must have been a main influence on the human populations that arrived in Europe. The use of tools developed rapidly, but cultural artifacts without practical use were also fabricated. The oldest tools for producing clothing, bone needles, are about 40 000 years old. Agriculture and domesticated animals were introduced about 12 K. (Figure adapted from Jones, 2007, contributed by A. Charkravarti, Johns Hopkins University, Baltimore, Maryland, USA.)

C. Tools and art of early modern humans

Modern humans in the Upper Paleolithic age (17–33 K) were sophisticated makers of tools and art. Four stages of development are distinguished: Magdalenian (17 K), Solutrean (21 K), Gravettian (27 K), and Aurignacian/Châtelperronian (33 K). The examples show a 26 000-year-old bone needle (**1**), and 33 000-year-old flutes made out of swan bone (Conard et al., 2009) (**2**).

Medical relevance

The field of Evolutionary Medicine deals with scientific and practical questions concerning the evolutionary background of many diseases in modern humans. Examples of such disorders are obesity, arteriosclerosis, hypertension, coronary heart disease, autoimmune diseases, and others (Gluckman et al., 2009).

Further Reading

Gluckman P, Beedle A, Hanson M. Principles of Evolutionary Medicine. Oxford: Oxford University Press; 2009

Green RE, et al. A draft sequence of the Neandertal genome. Science 2010;328:710–722

Green RE, et al. A complete Neandertal mitochondrial genome sequence determined by high-throughput sequencing. Cell 2008;134:416–426

Hublin JJ. Out of Africa: modern human origins special feature: the origin of Neandertals. Proc Natl Acad Sci USA 2009;106(38):16022–16027

Jones D. Going global: How humans conquered the world. New Sci 2007;27 October:36–41

Mellars P. Why did modern human populations disperse from Africa ca. 60,000 years ago? A new model. Proc Natl Acad Sci USA 2006;103:9381–9386

Noonan JP. Neanderthal genomics and the evolution of modern humans. Genome Res 2010;20:547–553

Noonan JP, et al. Sequencing and analysis of Neanderthal genomic DNA. Science 2006;314:1113–1118

Reich D, et al. Genetic history of an archaic hominin group from Denisova Cave in Siberia. Nature 2010;468:1053–1060

Stringer CB, Andrews P. Genetic and fossil evidence for the origin of modern humans. Science 1988;239(4845):1263–1268

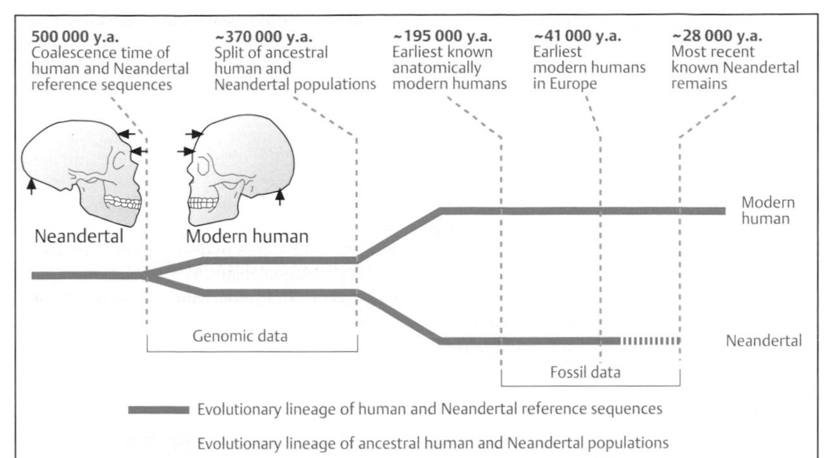

A. Origins of humans in relation to Neandertals

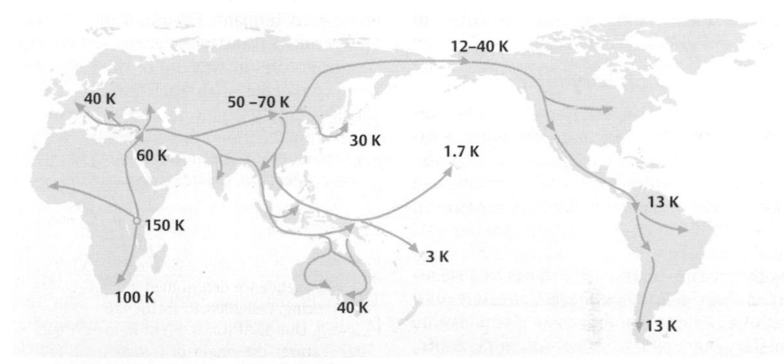

B. Dispersal of modern humans out of Africa

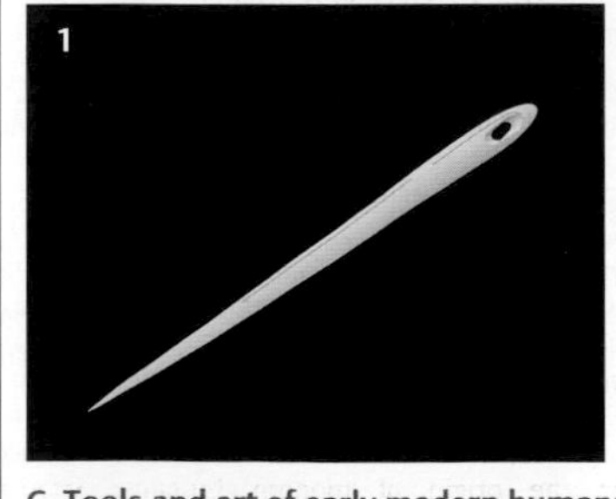

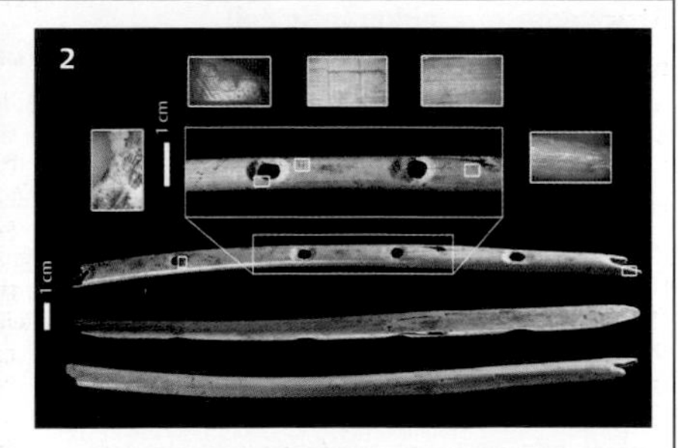

C. Tools and art of early modern humans

The Cell and its Components

Cells are the smallest organized structural units of living organisms. Surrounded by a membrane, they are able to carry out a wide variety of functions during a limited life span. Each cell originates from another living cell, as postulated by R. Virchow in 1855 ("*omnis cellula e cellula*"). Three basic types of cells exist: (i) *prokaryotic cells*, which carry their functional information in a circular genome without a nucleus; (ii) the bacteria (or eubacteria) and Archea (or archebacteria); and (iii) *eukaryotic cells*, which contain their genome in individual chromosomes in a nucleus and have a well-organized internal structure. Robert Hooke introduced the word *cell* in 1665 for the tiny cavities in cork, as they reminded him of the small rooms in which monks sleep. Cells were recognized as the "elementary particles of organisms," animals, and plants by Mathias Schleiden and Theodor Schwann in 1839. Today we understand many of the biological processes of cells at the molecular level.

A. Scheme of a prokaryotic cell

Prokaryotic cells (bacteria) are typically rod-shaped or spherical, a few micrometers in diameter, and without a nucleus or special internal structures. Within a cell wall consisting of a bilayered cell membrane, bacteria contain on average 1000–5000 genes tightly packed in a circular molecule of DNA. In addition, they usually contain small circular DNA molecules named *plasmids*. These replicate independently of the main chromosome and generally contain genes, which confer antibiotic resistance.

B. Scheme of a eukaryotic cell

A eukaryotic cell consists of cytoplasm and a nucleus. It is enclosed by a plasma membrane. The eukaryotic cell nucleus contains the genetic information. The cytoplasm contains a complex system of inner membranes that form discrete structures (organelles). These are the mitochondria (in which important energy-delivering chemical reactions take place), the endoplasmic reticulum (a series of membranes in which important molecules are formed), the Golgi apparatus (for transport functions), lysosomes (in which some proteins are broken down), and peroxisomes (for formation or degradation of certain molecules). Animal cells (**1**) and plant cells (**2**) share several features, but differ in important structures. A plant cell contains chloroplasts for photosynthesis. Plant cells are surrounded by a rigid wall of cellulose and other polymeric molecules, and they contain vacuoles for water, ions, sugar, nitrogen-containing compounds, or waste products. Vacuoles are permeable to water but not to the other substances enclosed within them.

C. Plasma membrane of the cell

Cells are surrounded by plasma membranes. These are water-resistant membranes composed of bipartite molecules of fatty acids. These molecules are phospholipids arranged in a double layer (*bilayer*). The plasma membrane contains numerous molecules that traverse the lipid bilayer once or many times to perform special functions. Cells communicate with each other by means of a broad repertoire of molecular signals. Different types of membrane proteins can be distinguished: (i) transmembrane proteins used as channels to transport molecules into or out of the cell, (ii) proteins connected with each other to provide stability, (iii) receptor molecules involved in signal transduction, (iv) molecules with enzyme function to catalyze internal chemical reactions in response to an external signal, and (v) gap junctions in specialized cells forming pores between adjacent cells. Gap junction proteins are composed of connexins. They allow the passage of molecules as large as 1.2 nm in diameter. Cells contain four major families of organic molecules: carbohydrates (sugars), fatty acids, amino acids, and nucleotides (see pp. 36–42). (Figure adepted from Alberts et al., 2008.)

Further Reading

Alberts B, et al. Essential Cell Biology. An Introduction to the Molecular Biology of the Cell. New York: Garland Publishing; 1998

Alberts B, et al. Molecular Biology of the Cell. 5th ed. New York: Garland Science; 2008

de Duve C. A Guided Tour of the Living Cell. Vols 1 and 2. New York: Scientific American Books; 1984

Lodish H, et al. Molecular Cell Biology. 6th ed. New York: W. H. Freeman; 2007

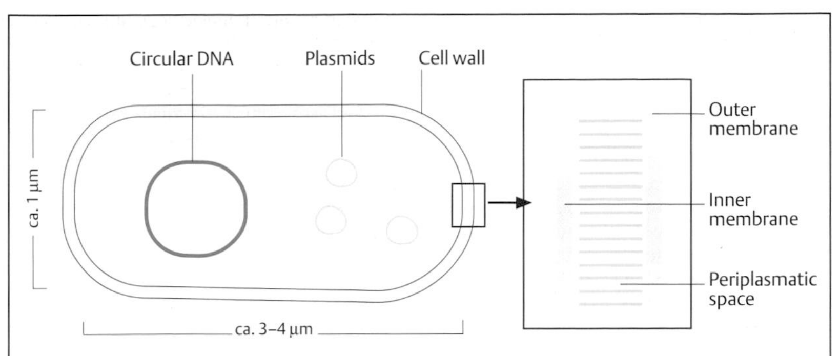

A. Scheme of a prokaryotic cell

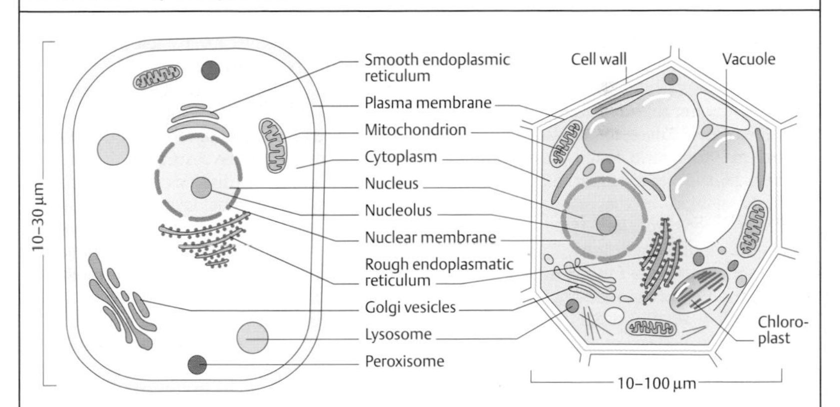

1. Animal cell

2. Plant cell

B. Scheme of an eukaryotic cell

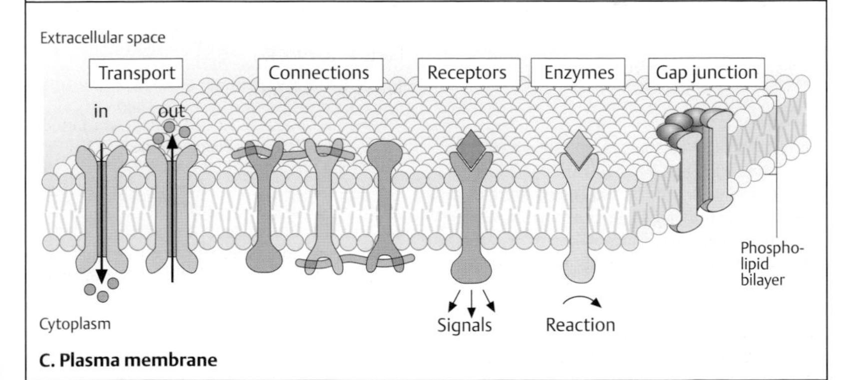

C. Plasma membrane

Some Types of Chemical Bonds

Chemical bonds form between molecules and allow building of complex structures. Each atom can establish chemical bonds with another in a defined way. Strong forces of attraction are present in *covalent* bonds, when two atoms share one pair of electrons. Weak forces of attraction occur in *noncovalent* bonds. They play a major role in many biomolecules such as carbohydrates, lipids, nucleic acids, and proteins. Four major types of noncovalent interactions are distinguished: hydrogen bonds, ionic interactions, van der Waals interactions, and hydrophobic effects.

Close to 99% of the weight of a living cell is composed of just four elements: carbon (C), hydrogen (H), nitrogen (N), and oxygen (O). Almost 50% of the atoms are hydrogen atoms, about 25% are carbon, and 25% are oxygen. Apart from water (about 70% of the weight of the cell) almost all components are carbon compounds. Carbon, a small atom with four electrons in its outer shell, is the central chemical building block of the living world. It can form four strong covalent bonds with other atoms. But most importantly, carbon atoms can combine with each other to build chains and rings, and thus large complex molecules with specific biological properties.

A. Functional groups with hydrogen (H), oxygen (O), and carbon (C)

Four simple combinations of these atoms occur frequently in biologically important molecules: hydroxyl (—OH; alcohols), methyl ($-CH_3$), carboxyl (—COOH), and carbonyl (C=O; aldehydes and ketones) groups. They impart characteristic chemical properties to the molecules, including possibilities to form compounds.

B. Acids and esters

Many biological substances contain a C=O bond with weak acidic or basic (alkaline) properties. The degree of acidity is expressed by the pH value, which indicates the concentration of H^+ ions in a solution, ranging from 10^{-1} mol/L (pH 1, strongly acidic) to 10^{-14} mol/L (pH 14, strongly alkaline). Pure water contains 10^{-7} moles H^+ per liter (pH 7.0). An ester is formed when an acid reacts with an alcohol. Esters are frequently found in lipids and phosphate compounds.

C. Carbon–nitrogen bonds (C—N)

C—N bonds occur in many biologically important molecules: amino groups, amines, and amides, especially in proteins. Of paramount significance are the amino acids (see. p. 42), the building blocks of proteins. Proteins have specific roles in the functioning of an organism.

D. Phosphate compounds

Ionized phosphate compounds play an essential biological role. HPO_4^{2-} is a stable inorganic phosphate ion from ionized phosphoric acid. A phosphate ion and a free hydroxyl group can form a phosphate ester. Phosphate compounds play an important role in energy-rich molecules and numerous macromolecules because they can store energy.

E. Sulfur groups

Sulfur often joins biological molecules together, especially when two sulfhydryl groups (—SH) react to form a disulfide bridge (—S—S—). Sulfur is a component of two amino acids (cysteine and methionine), and of some polysaccharides and sugars. Disulfide bridges play an important role in many complex molecules, serving to stabilize and maintain particular three-dimensional structures.

Further Reading

Alberts B, et al. Molecular Biology of the Cell. 5th ed. New York: Garland Science; 2008

Berg JM, Tymoczko JL, Stryer L. Biochemistry. 7th ed. New York: W. H. Freeman; 2011

Koolman J, Roehm KH. Color Atlas of Biochemistry. 2nd ed. Stuttgart: Thieme; 2005

Lodish H, et al. Molecular Cell Biology. 6th ed. New York: W. H. Freeman; 2007

Pauling L. The Nature of the Chemical Bond. 3rd ed. Ithaca: Cornell University Press; 1960

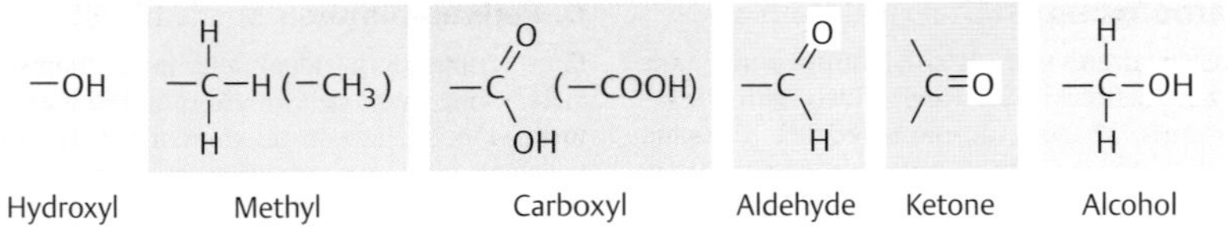

—OH Hydroxyl

$(-CH_3)$ Methyl

$(-COOH)$ Carboxyl

Aldehyde

Ketone

Alcohol

A. Functional groups with hydrogen (H), oxygen (O), and carbon (C)

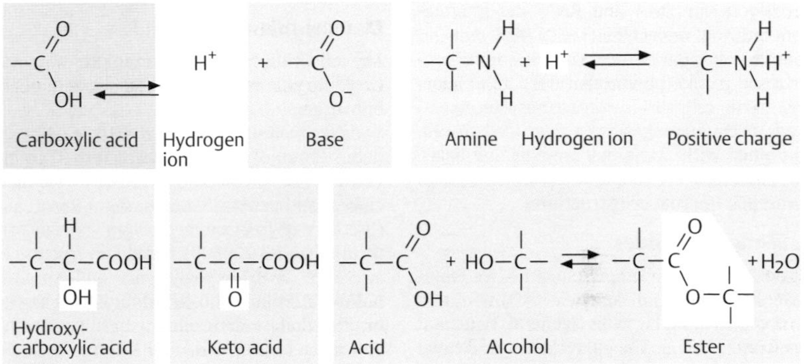

Carboxylic acid ⇌ H^+ Hydrogen ion + O^- Base

Amine + H^+ Hydrogen ion ⇌ Positive charge

Hydroxy-carboxylic acid

Keto acid

Acid + Alcohol ⇌ Ester $+ H_2O$

B. Acids and esters

Acid + Amine → Amide $+ H_2O$

α-C Atom

Amino-group

Side chain

Amino acid

$-COOH$; ^+H_3N ; COO^-

Amino acids are ionized in aqueous solutions at pH 7

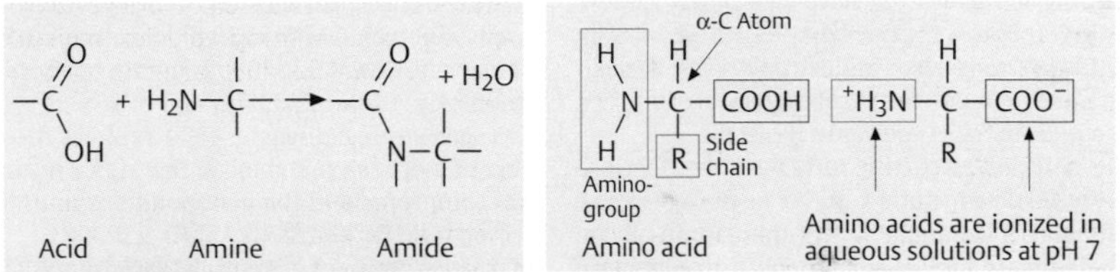

C. Carbon-nitrogen bonds (C–N)

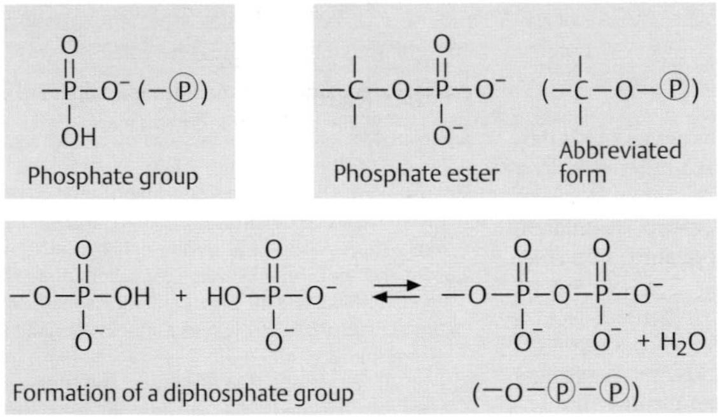

$(-Ⓟ)$ Phosphate group

Phosphate ester

$(-C-O-Ⓟ)$ Abbreviated form

Formation of a diphosphate group $+ H_2O$

$(-O-Ⓟ-Ⓟ)$

D. Phosphate compounds

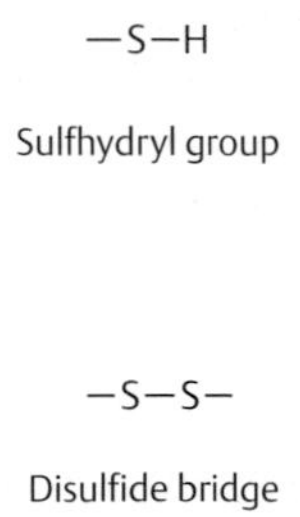

—S—H

Sulfhydryl group

—S—S—

Disulfide bridge

E. Sulfur groups

Carbohydrates

Carbohydrates are carbonyl compounds (aldehydes, ketones) that occur widely in living organisms as part of biomolecules. Carbohydrates are one of the most important classes of biomolecules. Their main functions can be classified into three groups: (i) to deliver and store energy; (ii) to help provide the basic framework for DNA and RNA, the information-carrying molecules (see p. 46); and (iii) to form structural elements of cell walls of bacteria and plants (polysaccharides). In addition, they form cell surface structures (receptors) used in conducting signals from cell to cell. Combined with numerous proteins and lipids, carbohydrates are important components of numerous internal cell structures.

A. Monosaccharides

Monosaccharides (simple sugars) are aldehydes (—C=O, —H) or ketones (C=O) with two or more hydroxyl groups (general structural formula: CH_2O). The aldehyde or ketone group can react with one of the hydroxyl groups to form a ring. This is the usual configuration of sugars that have five or six carbon atoms (pentoses and hexoses). The carbon atoms are numbered sequentially. The D (dextro) and L (levo) forms of sugars are mirror-image isomers of the same molecule.
The naturally occurring forms are the D forms. These further include β and α forms as stereoisomers. In the cyclic forms, the carbon atoms of sugars are not on a plane, but three-dimensionally take the shape of a chair or a boat. The arrangement of the hydroxyl groups can differ, so that stereoisomers such as mannose or galactose are formed.

B. Disaccharides

These are compounds of two monosaccharides. Sugars can be linked with each other. Usually a bond is formed between a hydroxyl group of one sugar and a hydroxyl group of another. Sucrose and lactose are frequently occurring disaccharides.

C. Derivatives of sugars

Sugar derivatives are formed by replacing certain hydroxyl groups by other groups.

D. Polysaccharides

Short (oligosaccharides) and long chains of sugars and sugar derivatives (polysaccharides) form essential structural elements of the cell. Complex oligosaccharides with bonds to proteins or lipids are part of cell surface structures, e.g., blood group antigens. Others are part of the extracellular matrix of cells.

Medical relevance

The following diseases are examples of human hereditary disorders in the metabolism of carbohydrates.
Diabetes mellitus (OMIM 125850): a heterogeneous group of disorders characterized by elevated levels of blood glucose, with complex clinical and genetic features (see p. 264).
Disorders of fructose metabolism: benign fructosuria (OMIM 229800), hereditary fructose intolerance with hypoglycemia and vomiting (OMIM 229600), and hereditary fructose-1,6-bisphosphatase deficiency with hypoglycemia, apnea, lactic acidosis, and often with lethal outcome in newborn infants (OMIM 229700).
Galactose metabolism: inherited disorders with acute galactose toxicity and long-term effects (galactosemia, OMIM 230400; galactokinase deficiency, OMIM 230200; galactose epimerase deficiency, OMIM 230350), and others.
Glycogen storage diseases: eight types of disorders of glycogen metabolism that differ in clinical symptoms and the genes and enzymes involved (OMIM 232200, 232210–232800).
In a large group of genetically determined disorders, complex polysaccharides cannot be degraded owing to reduced or absent enzyme function as storages diseases (mucopolysaccharidoses, mucolipidoses) (see p. 280).

Further Reading

Koolman J, Roehm KH. Color Atlas of Biochemistry. 2nd ed. Stuttgart: Thieme; 2005

McKusick VA. Online Mendelian Inheritance in Man (OMIM). 12th ed. Baltimore: Johns Hopkins Univ Press; 1998. Available at: /www.ncbi.nlm.nih.gov/

Ramachandran TS, ed. Disorders of Carbohydrate Metabolism. Available at: http://emedicine.medscape.com/article/1183033-overview. Accessed January 25, 2012

Scriver CR, et al., eds. The Metabolic and Molecular Bases of Inherited Disease. 8th ed. New York: McGraw-Hill; 2001

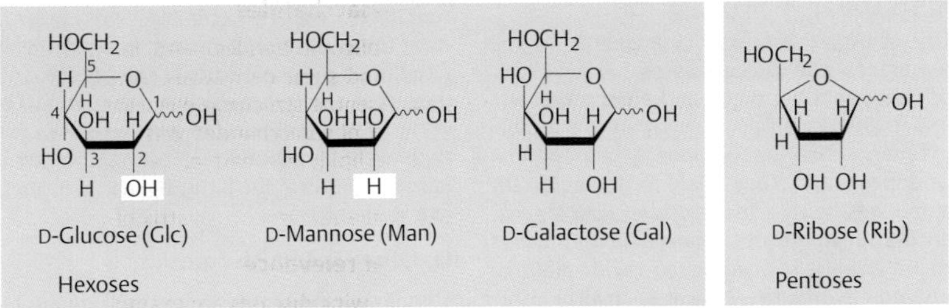

D-Glucose (Glc)

D-Mannose (Man)

D-Galactose (Gal)

Hexoses

D-Ribose (Rib)

Pentoses

A. Monosaccharides

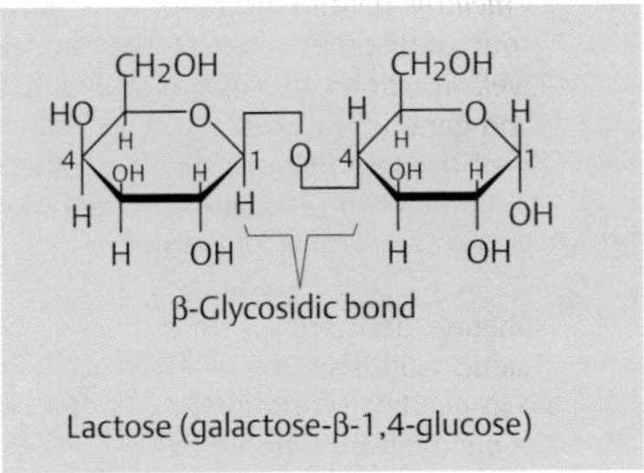

β-Glycosidic bond

Lactose (galactose-β-1,4-glucose)

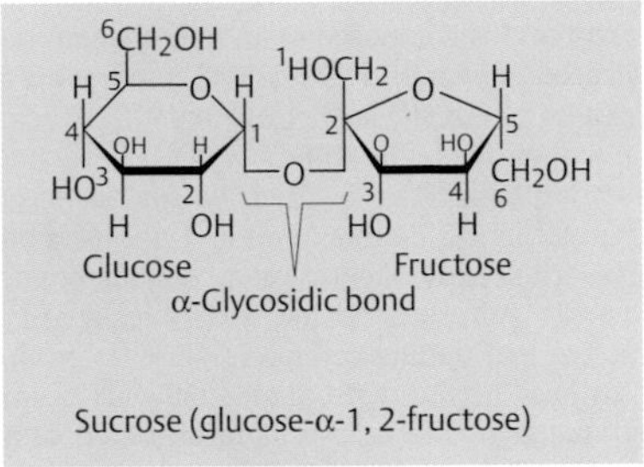

Glucose

Fructose

α-Glycosidic bond

Sucrose (glucose-α-1, 2-fructose)

B. Disaccharides

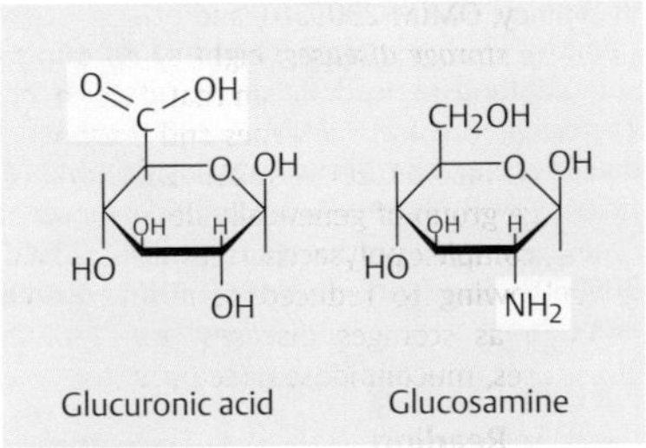

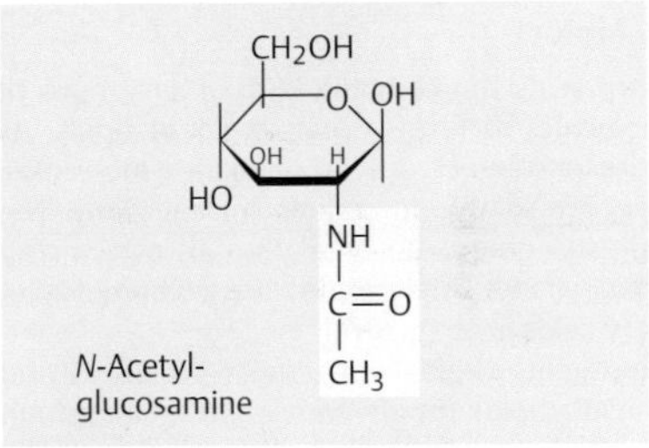

Glucuronic acid

Glucosamine

N-Acetyl-glucosamine

C. Sugar derivatives

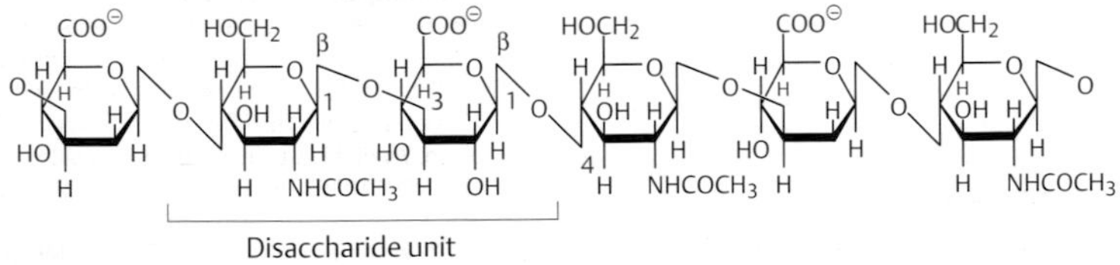

Disaccharide unit

D. Polysaccharides

Lipids (Fatty Acids)

Fatty acids are essential components of cell membranes and precursors of biomolecules, such as steroid hormones and other molecules used in signal transduction. Lipids are important energy-carrying components of food (dosage-dependent). They also form important compounds with carbohydrates (glycolipids) and phosphate groups (phospholipids). Lipids can be classified according to their ability to undergo hydrolytic cleavage (hydrolyzable and nonhydrolyzable).

A. Fatty acids

A fatty acid is composed of an unbranched hydrocarbon chain of 4–24 carbon atoms with a terminal carboxylic acid group (**1**). A fatty acid is polar, with a hydrophilic (–COOH) head and a hydrophobic tail ($-CH_3$). Saturated fatty acids without a double bond and unsaturated fatty acids with one or more double bonds exist (**2**). Palmitic acid has a chemically very reactive hydrophilic carboxyl (–COOH) group at its head, and a hydrophobic tail that is not very reactive. Oleic acid is an unsaturated fatty acid with a double bond between carbon atoms 8 and 9. Linoleic acid and arachidonic acid are essential in human nutrition.

B. Lipids

Fatty acids can combine with other groups of molecules to form various types of lipids. As water-insoluble (hydrophobic) molecules, they are soluble in organic solvents only. The carboxyl group can enter into an ester or an amide bond. Triglycerides are compounds of fatty acids with glycerol.

Glycolipids (lipids with sugar residues) and phospholipids (lipids with a phosphate group attached to an alcohol derivative) are the structural bases of important macromolecules. Sphingolipids are an important group of molecules in biological membranes. Here, sphingosine, instead of glycerol, is the fatty-acid-binding molecule. Sphingomyelin and gangliosides contain sphingosine. Gangliosides make up 6% of the central nervous system lipids. They are degraded by a series of enzymes.

C. Lipid aggregates

Owing to their bipolar properties, fatty acids can form lipid aggregates in water. The hydrophilic ends are attracted to their aqueous surroundings; the hydrophobic ends protrude from the surface of the water and form a surface film. If completely under the surface, lipids may form a micelle, which is compact and dry within. Phospholipids and glycolipids can form two-layered membranes (lipid membrane bilayers). These are the basic structural elements of cell membranes.

D. Other lipids: steroids

Steroids are small molecules consisting of four different rings of carbon atoms. Cholesterol is the precursor of five major classes of steroid hormones: prostagens, glucocorticoids, mineralocorticoids, androgens, and estrogens. Each of these hormone classes is responsible for important biological functions, such as maintenance of pregnancy, fat and protein metabolism, maintenance of blood volume and blood pressure, and sexual development.

Medical relevance

Several groups of disorders of lipoprotein and lipid metabolism exist. Important examples are familial hypercholesterolemia (OMIM 143890, see p. 272), hyperlipoproteinemia (OMIM 238600), dysbetalipoproteinemia (OMIM 107741), and high-density lipoprotein binding protein (OMIM 142695).

Genetically determined disorders of ganglioside catabolism lead to severe diseases, e.g., Tay–Sachs disease (OMIM 272800) caused by defective degradation of ganglioside GM2 (deficiency of β-*N*-acetylhexosaminidase), several types of gangliosidoses (OMIM 230500, 305650), Sandhoff disease (OMIM 268800), and others.

Further Reading

Gilbert-Barness E, Barness L. Metabolic Diseases. Foundations of Clinical Management, Genetics, and Pathology. Natick: Eaton Publishing; 2000

Koolman J, Roehm KH. Color Atlas of Biochemistry. 2nd ed. Stuttgart: Thieme; 2005

McKusick VA. Online Mendelian Inheritance in Man (OMIM). 12th ed. Baltimore: Johns Hopkins Univ Press; 1998. Available at: www.ncbi.nlm.nih.gov/sites/entrez?db=omim

Scriver CR, et al., eds. The Metabolic and Molecular Bases of Inherited Disease. 8th ed. New York: McGraw-Hill; 2001

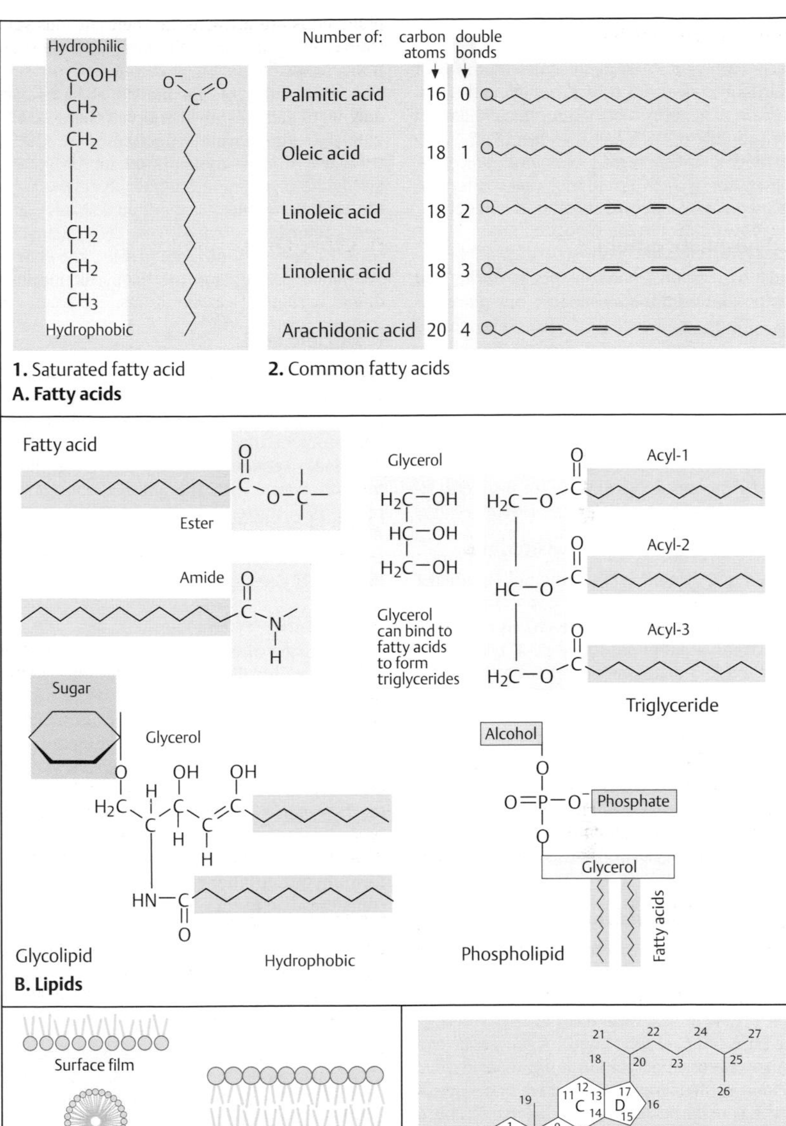

1. Saturated fatty acid

2. Common fatty acids

A. Fatty acids

Glycolipid

B. Lipids

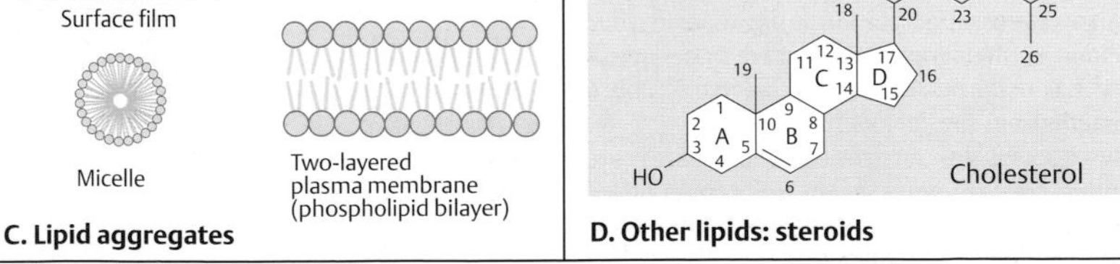

C. Lipid aggregates

D. Other lipids: steroids

Nucleotides and Nucleic Acids

Nucleic acids are macromolecules that as DNA and RNA are central to the storage and transmission of genetic information. Nucleotides are the subunits of DNA and RNA (see p. 48). They participate in numerous biological processes, convey energy, are part of essential coenzymes, and regulate numerous metabolic functions.

A. Phosphate groups

Phosphate groups occur in nucleic acids and nucleotides as monophosphates (one phosphorus atom), diphosphates (two phosphorus atoms), or triphosphates (three phosphorus atoms).

B. Sugar residues

The carbohydrate residues in nucleotides are usually derived from either ribose as β-D-ribose (in ribonucleic acid, RNA) or β-D-deoxyribose (in deoxyribonucleic acid, DNA).

C. Nucleotide bases of pyrimidine

Cytosine (C), thymine (T), and uracil (U) are the three pyrimidine nucleotide bases. They differ from each other in their side chains ($-NH_2$ on the carbon atom in position 4 [C-4] of cytosine, $-CH_3$ on C-5 in thymine, and O on C-4 in uracil). In addition, cytosine has a double bond between N-3 and C-4.

D. Nucleotide bases of purine

Adenine (A) and guanine (G) are the two nucleotide bases of purine. They differ in their side chains and in having a double bond between N-1 and C-6 (present in adenine, absent in guanine).

E. Nucleosides and nucleotides

A *nucleoside* is a compound of a sugar residue (ribose or deoxyribose) and a nucleotide base. The bond is between the carbon atom in position 1 (C-1) of the sugar and a nitrogen atom of the base (*N*-glycosidic bond). A *nucleotide* is a compound of a five-carbon-atom sugar residue (ribose or deoxyribose) attached to a nucleotide base (pyrimidine or purine base) and a phosphate group.

The nucleosides of the various bases are grouped as ribonucleosides or deoxyribonucleosides, e.g., adenosine or deoxyadenosine, guanosine or deoxyguanosine, cytidine or deoxycytidine. Uridine occurs only as a ribonucleoside. Thymidine occurs only as a deoxyribonucleoside.

The nucleotides of the individual bases are: adenylate (adenosine monophosphate, AMP), guanylate (guanosine monophosphate, GMP), uridylate (uridine monophosphate, UMP), and cytidylate (cytosine monophosphate, CMP) for the ribonucleotides (5′ monophosphates); and deoxyadenylate (dAMP), deoxyguanylate (dGMP), deoxythymidylate (dTMP), and deoxycytidylate (dCMP) for the deoxyribonucleotides.

F. Nucleic acid

A nucleic acid consists of a series of nucleotides. A phosphodiester bridge between the 3′ carbon atom of one nucleotide and the 5′ carbon atom of the next joins two nucleotides. The linear sequence is usually given in the 5′ to 3′ direction with the abbreviations of the respective nucleotide bases. For instance, ATCG would signify the sequence adenine (A), thymine (T), cytosine (C), and guanine (G) in the 5′ to 3′ direction.

Medical relevance

The following diseases are examples of human hereditary disorders in purine and pyrimidine metabolism.

Hyperuricemia and gout: a group of disorders resulting from genetically determined excessive synthesis of purine precursors (OMIM 240000).

Lesch–Nyhan syndrome: a variable, usually severe infantile X-chromosomal disease with marked neurological manifestations resulting from hypoxanthine–guanine phosphoribosyltransferase deficiency (OMIM 308000).

Adenosine deaminase deficiency: a heterogeneous group of disorders resulting in severe infantile immunodeficiency. Different autosomal recessive and X-chromosomal types exist (OMIM 102700).

Further Reading

Gilbert-Barness E, Barness L. Metabolic Diseases. Foundations of Clinical Management, Genetics, and Pathology. Natick: Eaton Publishing; 2000

Koolman J, Roehm KH. Color Atlas of Biochemistry. 2nd ed. Stuttgart: Thieme; 2005

Scriver CR, et al., eds. The Metabolic and Molecular Bases of Inherited Disease. 8th ed. New York: McGraw-Hill; 2001

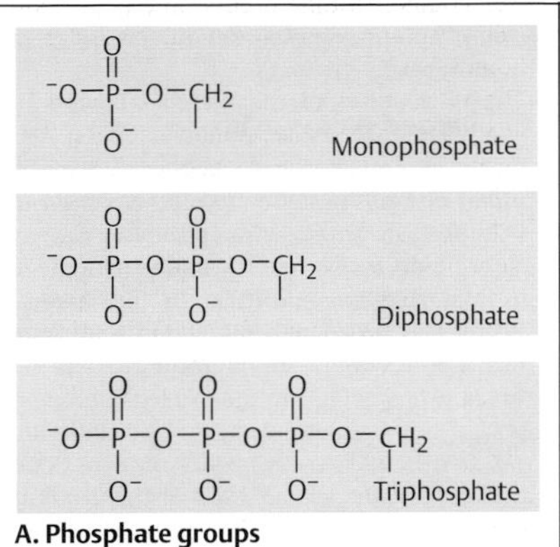

A. Phosphate groups

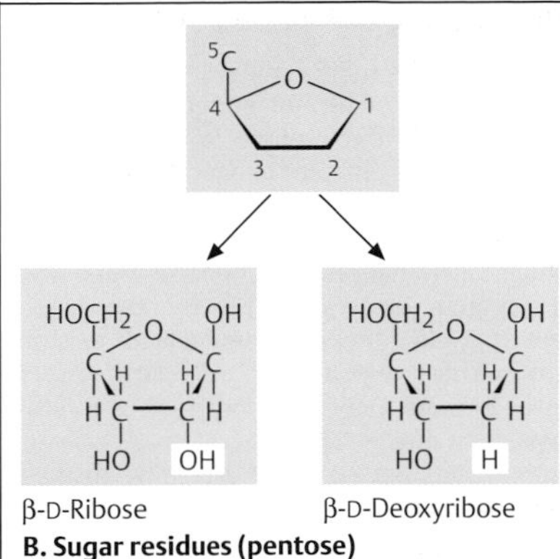

B. Sugar residues (pentose)

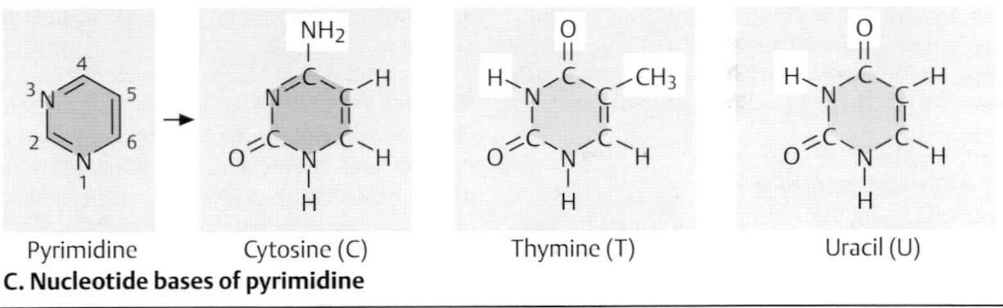

C. Nucleotide bases of pyrimidine

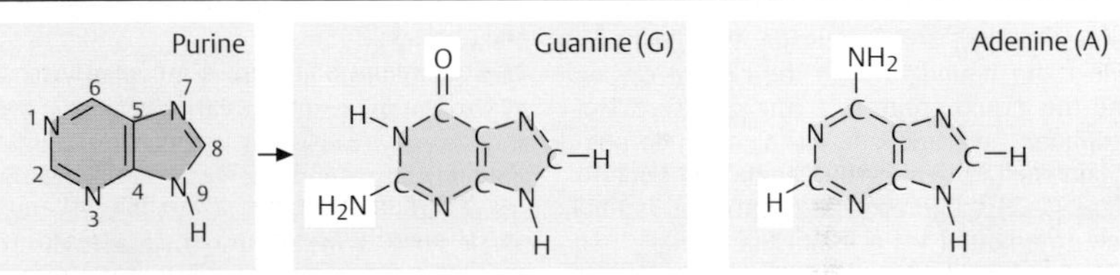

D. Nucleotide bases of purine

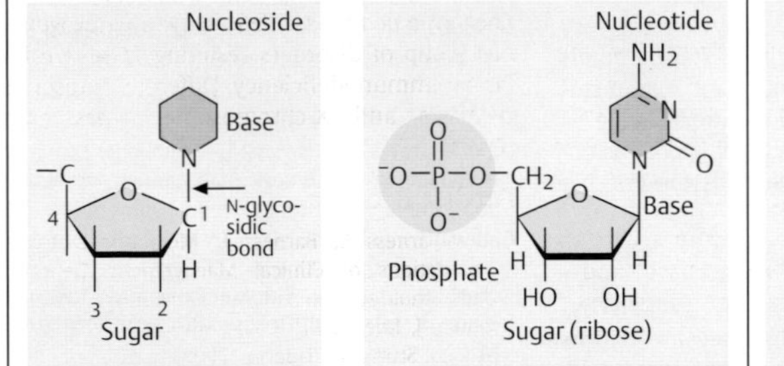

E. Nucleosides and nucleotides

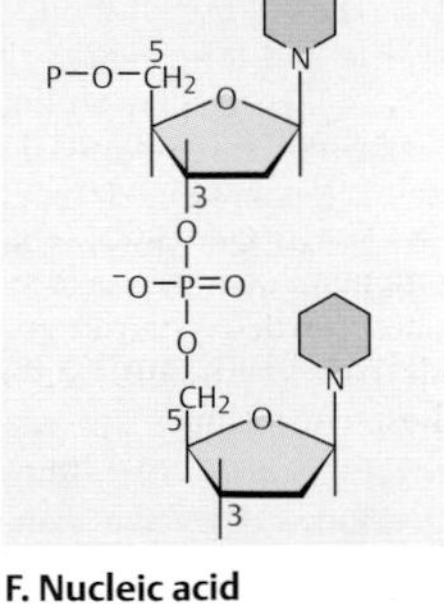

F. Nucleic acid

Amino Acids

Amino acids (2-aminocarboxylic acids) are the basic structural units of proteins. An amino acid consists of a central α carbon atom bonded to four different chemical groups: one bond to an amino group ($-NH_2$). The side chain (R) is the major determinant of the individual functional property of each amino acid in a protein. The α carbon is asymmetric except in glycine; therefore, amino acids exist in two mirror-image forms: D (dextro) and L (levo) isomers. Only the L forms occur in proteins, with rare exceptions.

Amino acids are classified according to their side chains and chemical reactivity. Each amino acid has its own three-letter and one-letter abbreviations. Essential amino acids for humans are valine (Val), leucine (Leu), isoleucine (Ile), phenylalanine (Phe), tryptophan (Trp), methionine (Met), threonine (Thr), and lysine (Lys). These have to be supplied by food intake.

A. Aliphatic amino acids

Aliphatic amino acids have an aliphatic side chain, e.g., glycine has a hydrogen atom (—H) and alanine a methyl group ($-CH_3$), or a larger, hydrophobic (water-repellent) side chain as in valine, leucine, and isoleucine. Proline has a side chain bound to both the central carbon and the amino group in a ring structure. Hydrophobic aromatic side chains occur in phenylalanine (a phenyl group bound via a methylene [$-CH_2-$] group) and tryptophan (an indole ring bound via a methylene group). Two hydrophobic amino acids contain sulfur (S) atoms: cysteine with a sulfhydryl group (—SH) and methionine with a thioether ($-S-CH_3$). The sulfhydryl group in cysteine is very reactive and forms stabilizing disulfide bonds (—S—S—). These play an important role in stabilizing the three-dimensional forms of proteins. Selenocysteine is a cysteine analogue occurring in a few proteins, such as the enzyme glutathione peroxidase.

B. Hydrophilic amino acids

Serine, threonine, and tyrosine contain hydroxyl groups (—OH). Thus, they are hydrolyzed forms of glycine, alanine, and phenylalanine. The hydroxyl groups make them hydrophilic and reactive. Both asparagine and glutamine contain an amino group and an amide group. At physiological pH their side chains are negatively charged.

C. Charged amino acids

These amino acids have either two ionized amino groups (basic) or two carboxyl groups (acidic). Basic amino acids (positively charged) are arginine, lysine, and histidine. Histidine has an imidazole ring and it can be uncharged or positively charged, depending on its surroundings. It is frequently found in the reactive centers of proteins (e.g., in the oxygen-binding region of hemoglobin). Aspartic acid and glutamic acid have two carboxyl groups (—COOH) and are usually acidic. Seven of the 20 amino acids have slightly ionizable side chains, making them highly reactive (asparagine, glutamic acid, histidine, cysteine, tyrosine, lysine, and arginine).

Medical relevance

Glycine, phenylalanine, tyrosine, histidine, proline, and lysine, and the branched chain amino acids valine, leucine, and isoleucine, are predominantly involved in various genetic diseases showing toxic metabolic symptoms that occur when their plasma concentration is too high or too low (disorders of amino acid metabolism).

Phenylketonuria: disorders of phenylalanine hydroxylation result in variable clinical signs and severity, caused by a spectrum of mutations in the responsible gene (OMIM 261600, see p. 266).

Maple syrup urine disease: a variable disorder caused by deficiency of branched chain α-keto acid dehydrogenase, which leads to accumulation of valine, leucine, and isoleucine (OMIM 248600). The classic severe form results in severe neurological damage to the infant.

Further Reading

Gilbert-Barness E, Barness L. Metabolic Diseases. Foundations of Clinical Management, Genetics, and Pathology. Natick: Eaton Publishing; 2000

Koolman J, Roehm KH. Color Atlas of Biochemistry. 2nd ed. Stuttgart: Thieme; 2005

Scriver CR, et al., eds. The Metabolic and Molecular Bases of Inherited Disease. 8th ed. New York: McGraw-Hill; 2001

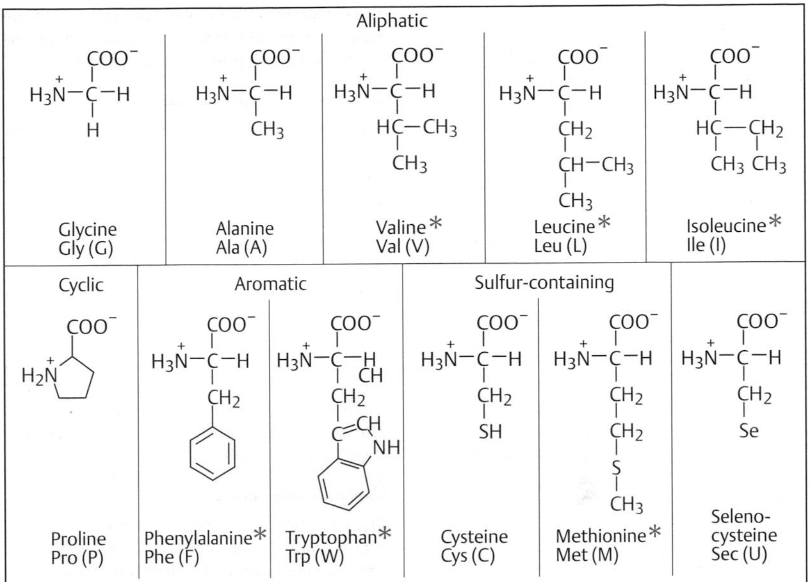

Aliphatic

COO^- $H_3\overset{+}{N}-C-H$ H	COO^- $H_3\overset{+}{N}-C-H$ CH_3	COO^- $H_3\overset{+}{N}-C-H$ $HC-CH_3$ CH_3	COO^- $H_3\overset{+}{N}-C-H$ CH_2 $CH-CH_3$ CH_3	COO^- $H_3\overset{+}{N}-C-H$ $HC-CH_2$ CH_3 CH_3
Glycine Gly (G)	Alanine Ala (A)	Valine * Val (V)	Leucine * Leu (L)	Isoleucine * Ile (I)

Cyclic	Aromatic		Sulfur-containing		
COO^- $H_2\overset{+}{N}$	COO^- $H_3\overset{+}{N}-C-H$ CH_2	COO^- $H_3\overset{+}{N}-C-H$ CH_2 C=CH NH	COO^- $H_3\overset{+}{N}-C-H$ CH_2 SH	COO^- $H_3\overset{+}{N}-C-H$ CH_2 CH_2 S CH_3	COO^- $H_3\overset{+}{N}-C-H$ CH_2 Se
Proline Pro (P)	Phenylalanine * Phe (F)	Tryptophan * Trp (W)	Cysteine Cys (C)	Methionine * Met (M)	Seleno- cysteine Sec (U)

A. Neutral amino acids, nonpolar side chains

* Essential amino acids for humans

		Aromatic		
COO^- $H_3\overset{+}{N}-C-H$ CH_2OH	COO^- $H_3\overset{+}{N}-C-H$ CHOH CH_3	COO^- $H_3\overset{+}{N}-C-H$ CH_2 OH	COO^- $H_3\overset{+}{N}-C-H$ CH_2 $H_2N-C=O$	COO^- $H_3\overset{+}{N}-C-H$ CH_2 CH_2 $H_2N-C=O$
Serine Ser (S)	Threonine * Thr (T)	Tyrosine Tyr (Y)	Asparagine Asn (N)	Glutamine Gln (Q)

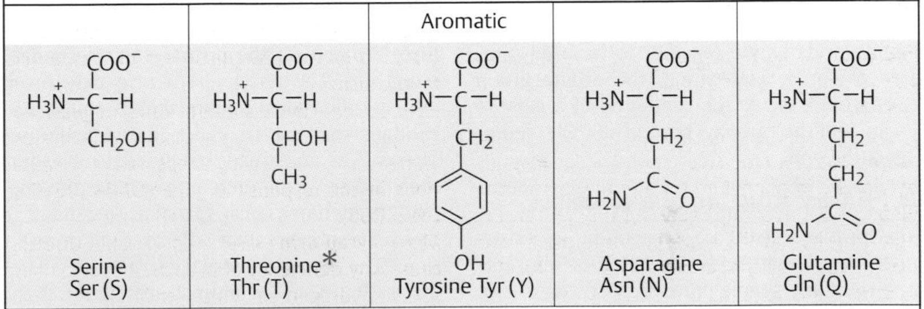

B. Hydrophilic amino acids, polar side chains

* Essential amino acids for humans

1. Basic (positively charged)			2. Acid (negatively charged)	
COO^- $H_3\overset{+}{N}-C-H$ CH_2 CH_2 CH_2 NH $H_2N-C=\overset{+}{N}H_2$	COO^- $H_3\overset{+}{N}-C-H$ CH_2 CH_2 CH_2 $CH_2-\overset{+}{N}H_3$	COO^- $H_3\overset{+}{N}-C-H$ CH_2 N N H	COO^- $H_3\overset{+}{N}-C-H$ CH_2 COO^-	COO^- $H_3\overset{+}{N}-C-H$ CH_2 CH_2 COO^-
Arginine Arg (R)	Lysine Lys (K) *	Histidine His (H)	Aspartic acid Asp (D)	Glutamic acid Glu (E)

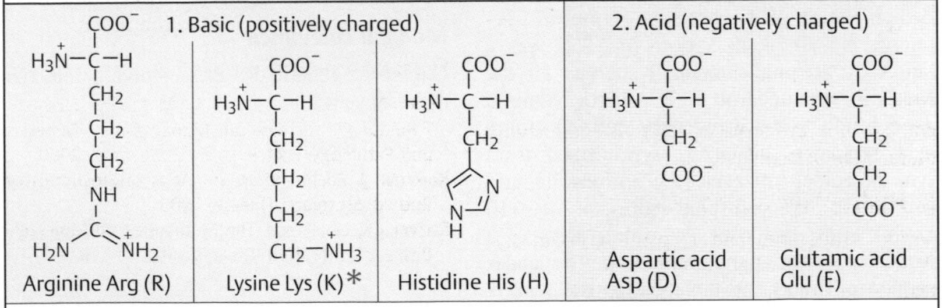

C. Charged amino acids

* Essential amino acids for humans

Proteins

Proteins are linear macromolecules (polypeptides) consisting of amino acids joined by peptide bonds, and they are arranged in a complex three-dimensional structure that is specific for each protein. Proteins are involved in all chemical processes in living organisms. As enzymes, they drive chemical reactions that in living cells would not occur spontaneously. They serve to transport small molecules, ions, or metals, and have important functions in cell division during growth and in cell and tissue differentiation. Proteins control the coordination of movements by regulating muscle cells and the production and transmission of impulses within and between nerve cells; they control blood homeostasis (blood clotting) and immune defense. They have mechanical functions in skin, bone, blood vessels, and other areas.

A. Peptide bonds

Amino acids are easily joined together owing to their dipolar ionization (zwitterions). The carboxyl group of one amino acid bonds to the amino group of the next (a peptide bond, sometimes also referred to as an amide bond). Amino acids bound together form a polypeptide chain. By convention, the amino group ($-NH_2$) represents the beginning of a peptide chain, and the carboxyl group ($-COOH$) represents the end.

B. Primary structure of a protein

The primary structure of protein is its sequence of amino acids. Insulin is an example of a relatively simple protein consisting of two polypeptide chains: an A chain of 21 amino acids and a B chain of 30 amino acids. The determination of its complete amino acid sequence by Frederick Sanger in 1955 was a landmark accomplishment. It showed for the first time that a protein, in genetic terms a gene product, has a precisely defined amino acid sequence. Insulin is synthesized from two precursor molecules: preproinsulin and proinsulin. Preproinsulin consists of 110 amino acids, including 24 amino acids of a leader sequence at the amino end. The leader sequence directs the molecule to the correct site within the cell, where it is removed to yield proinsulin with 86 amino acids. From this, a connecting (C) peptide is removed (amino acid numbers 31–65). This yields the two chains: B (amino acids 1–30) and A (amino acids 1–21). The A and B chains are connected by two disulfide bridges, which join the cysteines in positions 7 and 19 of the B chain with positions 6 and 20 of the A chain, respectively. The A chain contains a disulfide bridge between positions 7 and 11. The linear sequence of the amino acids yields information about the function and evolutionary origin of a protein. The positions of the disulfide bridges reflect the spatial arrangements (*the secondary structure*) of the amino acids.

C. Further protein structures

The secondary, tertiary, and quaternary structures of a protein are further levels of folding. The secondary structure of a protein refers to regions with a defined spatial arrangement. Two basic units of global proteins are alpha helix (α helix) formation and a flat sheet (β-pleated sheet).
The tertiary structure of a protein is the complete three-dimensional structure that is required for its biochemical and biological function. All functional proteins assume a well-defined three-dimensional structure. This structure is based on the primary and secondary structures. The tertiary structure may result in a specific spatial relationship of amino acid residues that are far apart in the linear sequence. The quaternary structure involves further folding of the protein, resulting in a specific three-dimensional spatial arrangement of different subunits that affects their interactions. The correct quaternary structure ensures proper function. The ultimate resolution is the detailed crystal atomic structure in three dimensions at a resolution of 3–10 Å.

Medical relevance

Numerous genetic diseases involve a defective or absent protein.

Further Reading

Berg JM, Tymoczko JL, Stryer L. Biochemistry. 7th ed. New York: W. H. Freeman; 2011

Koolman J, Röhm K-H. Color Atlas of Biochemistry. 2nd ed. Stuttgart: Thieme; 2005

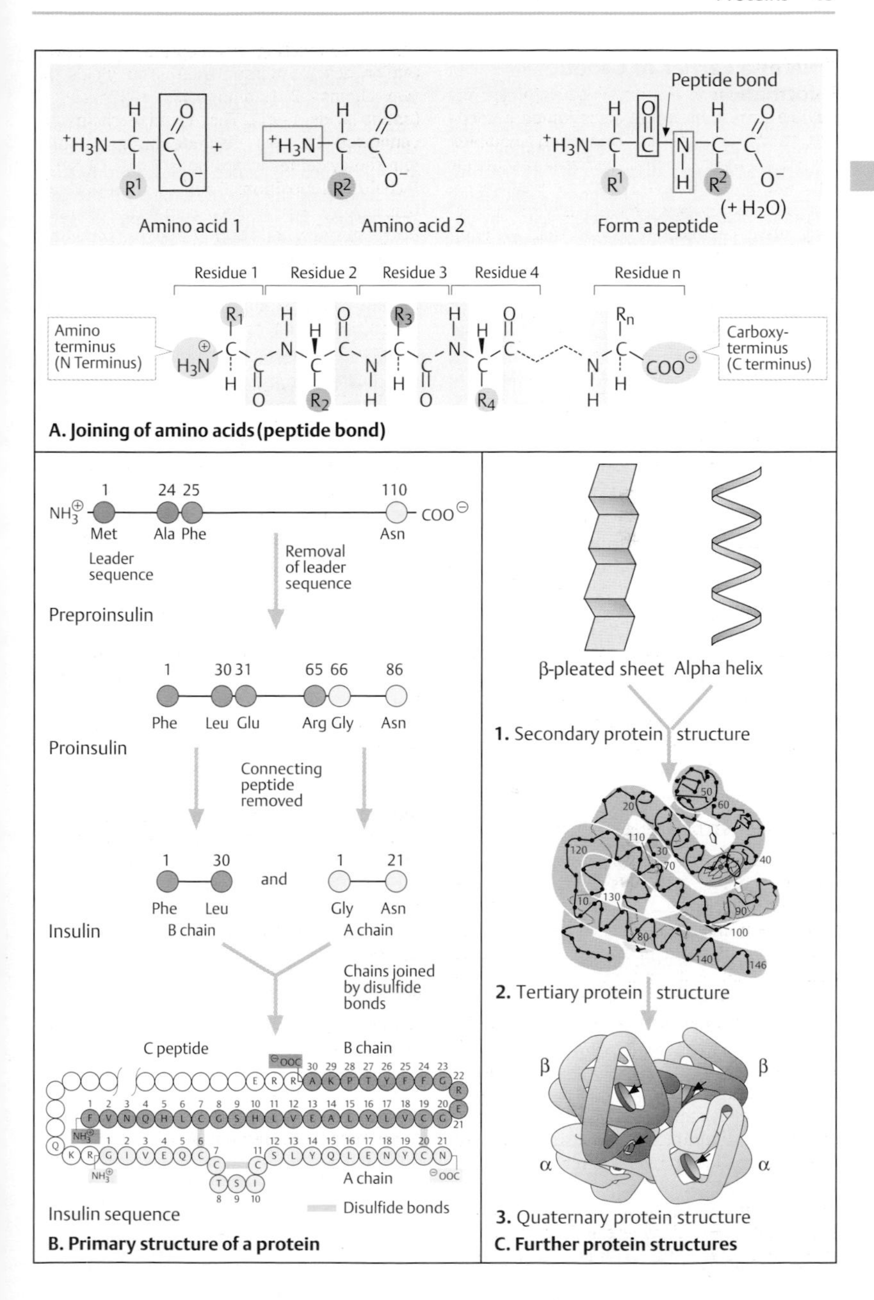

A. Joining of amino acids (peptide bond)

B. Primary structure of a protein

C. Further protein structures

DNA as a Carrier of Genetic Information

It took a surprisingly long time to realize that DNA is the carrier of genetic information. Friedrich Miescher's discovery in 1869 of "nuclein," a new, acidic, phosphorus-containing substance, later named "nucleic acid" by Richard Altmann in 1889, did not reveal its biological role.
By 1900, the purine and pyrimidine bases were known, and 20 years later the two kinds of nucleic acid, RNA and DNA, were distinguished. An incidental but precise observation in 1928 by Fred Griffith, and decisive experiments in 1944 by Oswald Avery, indicated that DNA is the carrier of genetic information.

A. Griffith's observation

In 1928, the English microbiologist Fred Griffith made a remarkable observation while investigating various strains of pneumococcus bacteria (*Streptococcus pneumoniae*, a cause of inflammation of the lungs, pneumonia). He determined that mice injected with strain S (smooth) died (**1**). On the other hand, animals injected with strain R (rough) survived (**2**). When he inactivated the lethal S strain by heat, the animals again survived (**3**). Surprisingly, a mixture of the nonlethal R strain and the heat-inactivated S strain had the same lethal effect as the original S strain (**4**). When he found normal living pneumococci of the S strain in the blood of the animals, he concluded that cells of the R strain must have changed into cells of the S strain. This is called *bacterial transformation*. For some time, this surprising result could not be explained and was met with skepticism. Its relevance for genetics was not apparent. (Figure adapted from Stent & Calendar, 1978.)

B. The transforming principle is DNA

Griffith's findings formed the basis for investigations by Oswald Avery, C. M. MacLeod, and M. McCarty in 1944 at the Rockefeller Institute in New York. They determined that the chemical basis of the transforming principle was DNA. From cultures of an S strain (**1**) they produced an extract of lysed cells (cell-free extract, **2**). After all the proteins, lipids, and polysaccharides had been removed, the extract retained the ability to transform pneumococci of the R strain into pneumococci of the S strain (transforming principle, **3**).
Avery and coworkers determined that this was caused by DNA alone. Thus, the DNA explained Griffith's observation. The DNA of the bacterial chromosomes remained intact after heating. The section of the chromosome with the gene responsible for capsule formation (S gene) could be released from the destroyed S cells and taken up by some R cells in subsequent cultures. After the S gene was incorporated into its DNA, an R cell was transformed into an S cell (**4**). This observation is based on the ability of bacteria to take up foreign DNA, which alters (*transforms*) some of their genetic attributes. (Figure adapted from Stent & Calendar, 1978.)

C. DNA transmits genetic information

The final evidence that DNA transmits genetic information was provided by Hershey & Chase (1952). They labeled the capsular protein of bacteriophages with radioactive sulfur (^{35}S) and the DNA with radioactive phosphorus (^{32}P). When bacteria were infected with the labeled bacteriophage, only ^{32}P (DNA) entered the cells, and not ^{35}S (capsular protein). The subsequent formation of new, complete phage particles in the cell proved that DNA was the exclusive carrier of the genetic information needed to form new phage particles.

Further Reading

Avery OT, Macleod CM, McCarty M. Studies on the chemical nature of the substance inducing transformation of pneumococcal types: Induction of transformation by a desoxyrobonucleic acid fraction isolated from pneumococcus type III. J Exp Med 1944;79:137–158

Griffith F. The Significance of Pneumoccocal Types. J Hyg (Lond) 1928;27:113–159

Hershey AD, Chase M. Independent functions of viral protein and nucleic acid in growth of bacteriophage. J Gen Physiol 1952;36:39–56

Judson MF. The Eighth Day of Creation. Makers of the Revolution in Biology. Expanded Edition. New York: Cold Spring Harbor Laboratory Press; 1996

McCarty M. The Transforming Principle. Discovering that Genes Are Made of DNA. London: W. W. Norton & Co; 1986

Stent GS, Calendar R. Molecular Genetics. An Introductory Narrative. 2nd ed. San Francisco: W. H. Freeman; 1978

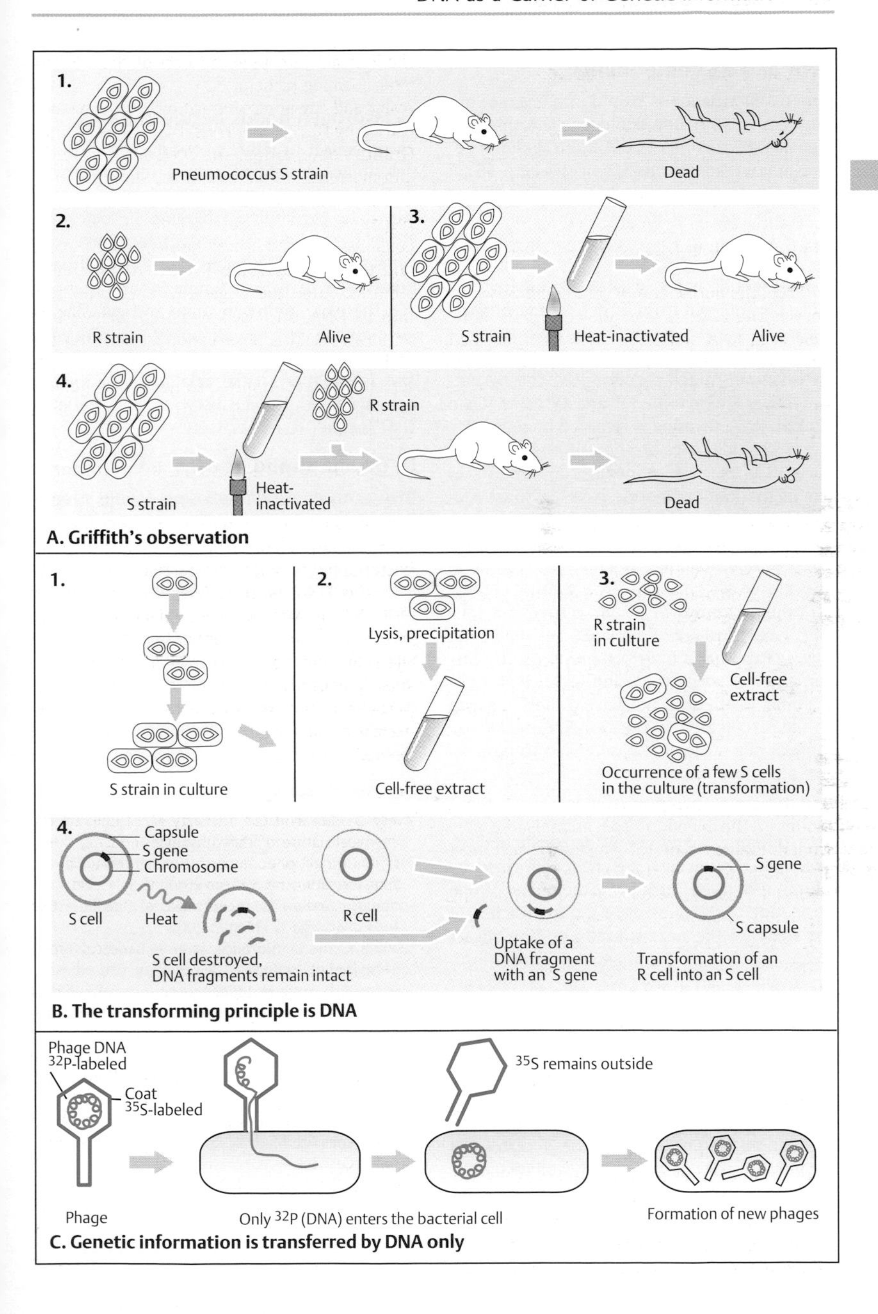
1.
Pneumococcus S strain
Dead
2.
R strain
Alive
3.
S strain
Heat-inactivated
Alive
4.
R strain
S strain
Heat-inactivated
Dead
A. Griffith's observation
1.
S strain in culture
2.
Lysis, precipitation
Cell-free extract
3.
R strain in culture
Cell-free extract
Occurrence of a few S cells in the culture (transformation)
4.
Capsule
S gene
Chromosome
S cell
Heat
S cell destroyed, DNA fragments remain intact
R cell
Uptake of a DNA fragment with an S gene
S gene
S capsule
Transformation of an R cell into an S cell
B. The transforming principle is DNA
Phage DNA ^{32}P-labeled
Coat ^{35}S-labeled
^{35}S remains outside
Phage
Only ^{32}P (DNA) enters the bacterial cell
Formation of new phages
C. Genetic information is transferred by DNA only

DNA and its Components

Genetic information is stored in a chain of nucleotides (p. 40). They determine the three-dimensional structure of DNA, from which it derives its functional consequences (see p. 56). A second type of nucleic acid contains ribose (RNA, ribonucleic acid).

A. Nucleotide bases

The nucleotide bases in DNA are heterocyclic molecules derived from either purine or pyrimidine. Five bases occur in the two types of nucleic acids DNA and RNA. The purine bases are adenine (A) and guanine (G). The pyrimidine bases are thymine (T) and cytosine (C) in DNA; in RNA, thymine is replaced by uracil (U). The nucleotide bases are part of a subunit, the nucleotide, of DNA. A nucleotide consists of one of the four nucleotide bases, a sugar (deoxyribose), and a phosphate group. The nitrogen atom in position 9 of a purine or in position 1 of a pyrimidine is bound to the carbon in position 1 of the sugar (*N*-glycosidic bond). RNA differs from DNA in two respects: it contains ribose instead of deoxyribose (unlike the latter, ribose has a hydroxyl group on the carbon atom at position 2), and uracil instead of thymine. Uracil does not have a methyl group on the carbon atom at position 5. A nucleoside consists of a nucleotide base linked to position 1 of a pentose sugar.

B. DNA nucleotide chain

DNA is a linear polymer of deoxyribonucleotide units. The nucleotide chain is formed by joining a hydroxyl group on the sugar of one nucleotide to the phosphate group attached to the sugar of the next nucleotide. The sugars linked together by the phosphate groups form the invariant part of the DNA. The variable part is in the sequence of the nucleotide bases A, T, C, and G.

A DNA nucleotide chain is polar. The polarity results from the way the sugars are attached to each other. The phosphate group at position C-5 (the 5′ carbon) of one sugar joins to the hydroxyl group at position C-3 (the 3′ carbon) of the next sugar by means of a phosphate diester bridge. Thus, one end of the chain has a 5′ phosphate group free and the other end has a 3′ hydroxyl group free (5′ end and 3′ end, respectively). By convention, the sequence of nucleotide bases is written in the 5′ to 3′ direction.

C. Hydrogen bonds between bases

The chemical structure of the nucleotide bases determines a defined spatial relationship. A purine (adenine or guanine) always lies opposite to a pyrimidine (thymine or cytosine). Three hydrogen-bond bridges form between cytosine (C) and guanine (G). Two hydrogen bonds form between adenine (A) and thymine (T). Therefore, either guanine and cytosine or adenine and thymine are posed opposite each other, forming complementary base pairs G–C and A–T. Other spatial relationships are not possible. The distance between two bases is 2.90 and 3.00 Å, respectively.

D. Double strand of DNA

DNA consists of two opposing double strands in a double helix (see p. 50). As a result of the spatial relationships of the nucleotide bases, a cytosine will always lie opposite to a guanine and a thymine opposite to an adenine. The sequence of the nucleotide bases on one strand of DNA (in the 5′ to 3′ direction) is complementary to the nucleotide base sequence (or simply the base sequence) of the other strand in the 3′ to 5′ direction. The specificity of base pairing is the most important structural characteristic of DNA.

Further Reading

Alberts B, et al. Molecular Biology of the Cell. 5th ed. New York: Garland Science; 2008

Berg JM, Tymoczko JL, Stryer L. Biochemistry. 7th ed. New York: W. H. Freeman; 2011

Koolman J, Roehm KH. Color Atlas of Biochemistry. 2nd ed. Stuttgart: Thieme; 2005

Lewin B. Genes IX. Sudbury: Jones & Barlett; 2008

Lodish H, et al. Molecular Cell Biology. 6th ed. New York: W. H. Freeman; 2007

Olinski R, Jurgowiak M, Zaremba T. Uracil in DNA—its biological significance. Mutat Res 2010;705:239–245

Strachan T, Read A. Human Molecular Genetics. 4th ed. New York: Garland Science; 2011

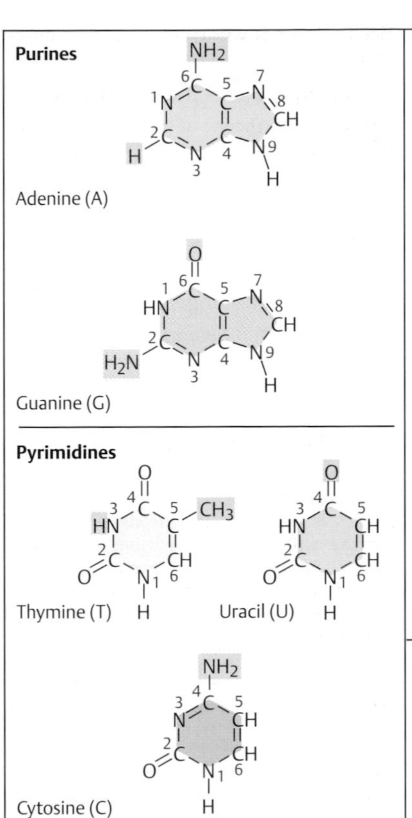

A. Nucleotide bases

5' end

O⁻
O=P–O–CH2 O base
O⁻
Sugars
H

O⁻
O=P–O–CH2 O base
O⁻
H

Phosphates
O⁻
O=P–O–CH2 O base
O⁻
H

O⁻
O=P–O–CH2 O base
O⁻

OH H 3' end

B. DNA nucleotide chain

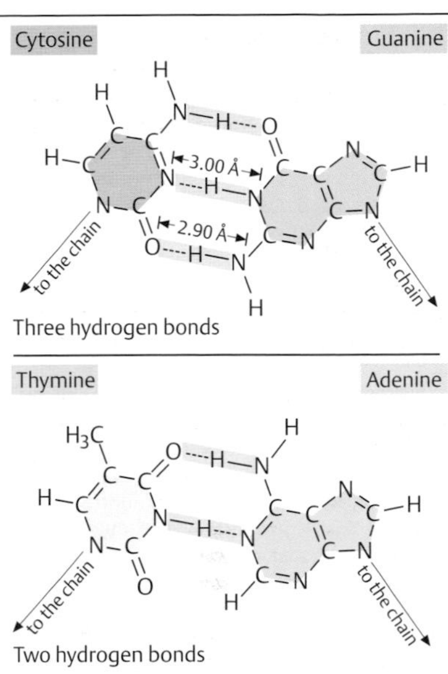

C. Hydrogen bonds between bases

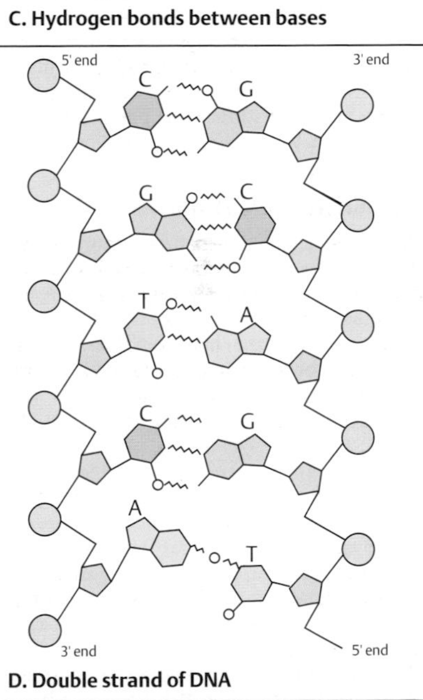

D. Double strand of DNA

DNA Structure

The elucidation of the structure of DNA as a double helix in 1953 by Watson and Crick is a landmark discovery, considered to be the cornerstone of modern genetics. The novel features of this structure explain two fundamental genetic mechanisms: storage of genetic information in a linear, readable pattern, and replication of genetic information to ensure its faithful transmission from generation to generation.

A. DNA double helix

DNA consists of two complementary polynucleotide chains wound around each other as a double helix along a common axis. The nucleotide bases lie inside the molecule in pairs, (adenine opposite to thymine, A–T, and guanine opposite to cytosine, G–C). The sugar–phosphate backbone lies on the outside.

B. Replication

The complementary arrangement of the base pairs along each strand allows that one strand can serve as a template to generate a new strand after the double helix has been opened. From one single strand a new complementary DNA molecule can be generated. DNA replication is semiconservative. This means that one completely new complementary strand is copied from the existing one.

C. Denaturation and renaturation

Although each of the noncovalent hydrogen bonds between the nucleotide base pairs is weak, DNA is stable at physiological temperatures because the two strands are held together by the many hydrogen bonds along the several hundred million nucleotides. The two strands separate by unwinding when exposed in solution to increasing temperature or weak chemical reagents (e.g., alkali, formamide, or urea). This reversible process is called *denaturation* or melting. The melting temperature (T_m) depends on many factors, including the proportion of G–C pairs, because the three hydrogen bonds between G and C are more stable than the two bonds between A–T pairs. The resulting single-stranded DNA molecules form random coils and do not maintain their helical structure. Lowering the temperature, increasing the ion concentration, or neutralizing the pH will reassociate the two strands into a double helix (*renaturation*), but only if they are complementary. If identical, they hybridize rapidly; if closely related, they hybridize slowly; if unrelated, they do not hybridize. The ability of single-stranded DNA to hybridize to another indicates that both strands are complementary. Hybridization of single-stranded DNA or RNA is widely used to explore whether DNA or RNA fragments contain complementary sequences.

D. Transmission of genetic information

The sequence of the nucleotide base pairs (A–T and G–C) is decoded and read in two steps called *transcription* and *translation*. In transcription, the sequence of one of the two DNA strands is converted into a complementary sequence of bases in a similar molecule, known as messenger RNA (mRNA). One DNA strand, the one in the 3′ to 5′ direction (coding strand), serves as the template. In translation, this sequence is converted into a defined sequence of amino acids. A triplet sequence, three base pairs (a codon), encodes one of the 20 amino acids. Beginning with a defined start point (methionine), the nucleotide sequence of the mRNA molecule is translated into a corresponding sequence of amino acids. DNA and RNA differ with respect to one nucleotide: RNA contains uracil (U) instead of the structurally related thymine (T, see p. 40). The mechanisms and biochemical reactions of transcription and translation are complex (not shown here).

Further Reading

Watson JD, Crick FHC. Molecular structure of nucleic acids; a structure for deoxyribose nucleic acid. Nature 1953;171:737–738

Watson JD, Crick FHC. Genetic implications of the structure of deoxyribonucleic acid. Nature 1953;171:964–967

Wilkins MFH, Stokes AR, Wilson HR. Molecular structure of deoxypentose nucleic acids. Nature 1953;171:738–740

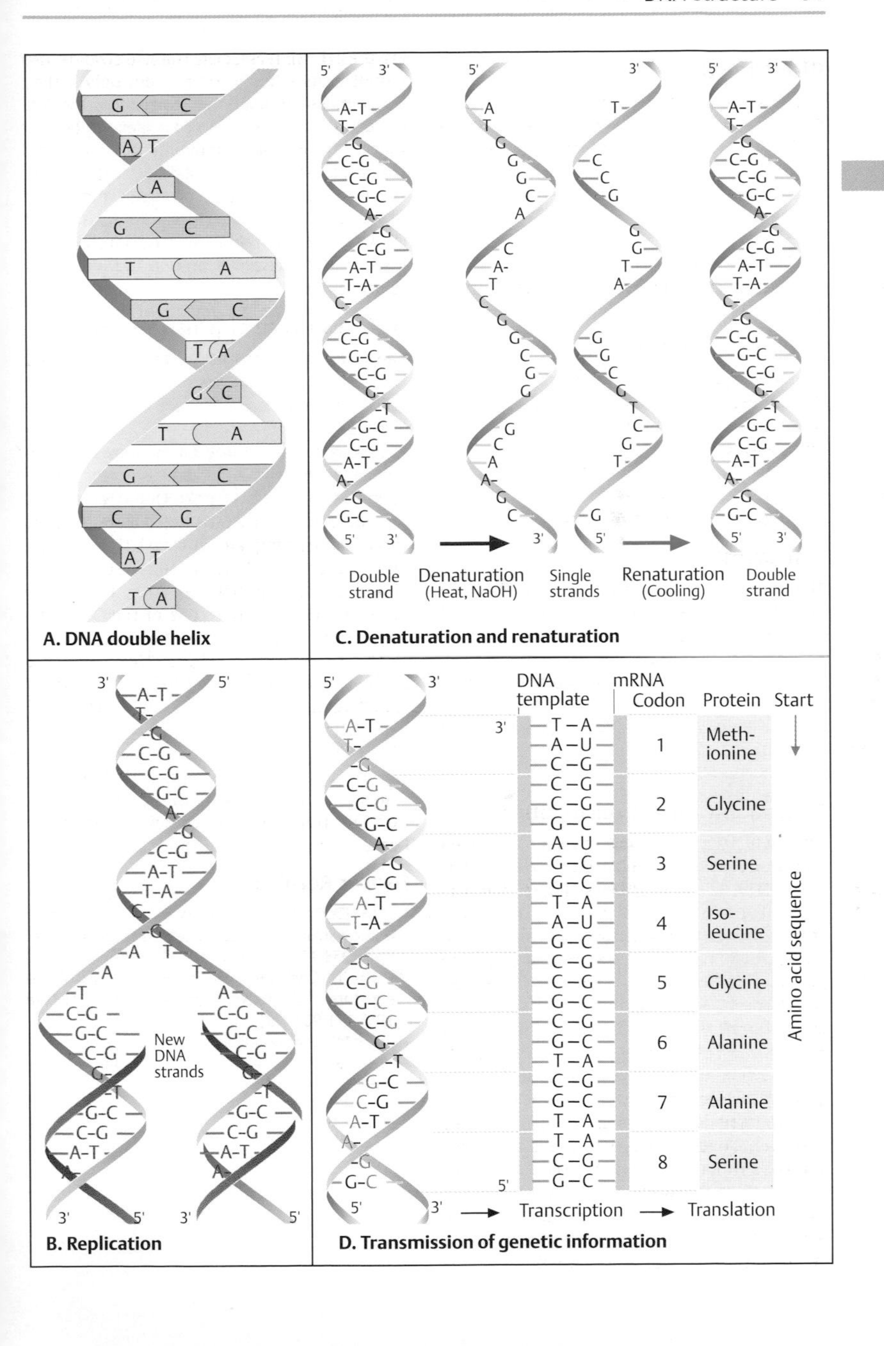

A. DNA double helix

B. Replication

C. Denaturation and renaturation

D. Transmission of genetic information

Alternative DNA Structures

The two strands of DNA wound around each other in the double helix can assume different conformations: the usual B-DNA (as shown on the preceding page), and A-DNA and Z-DNA. All forms produce a minor and a major groove on the outside.

A. Major and minor grooves in B-DNA

B-DNA forms two grooves on the outside: one large (major groove) and one small (minor groove). These result from differences in the base pairing of adenine–thymine (A–T) and guanine–cytosine (G–C). The purine and pyrimidine rings are 0.34 nm apart. The distance from one complete turn to the next is 3.4 nm.

B. Physical dimensions of the double helix (B form)

The usual B form of the DNA double helix has a diameter of 20 Å (2×10^{-6} mm) with 10.5 base pairs (bp) per turn. A helical turn repeats itself at intervals of 3.6 nm (36 Å). Bases on the same strand are 0.36 nm (3.6 Å) apart.

C. Three forms of DNA

The original classic form, determined by Watson and Crick in 1953, is B-DNA, a right-handed helix. In very low humidity, the B structure can change to another conformation, the A form. RNA/DNA and RNA/RNA helices assume this form in cells and in vitro. The A form is compact and has 11 bases per turn instead of 10.5, as in the B form. A-DNA exists only in the dehydrated state and differs from the B form by a 20-degree rotation of the perpendicular axis of the helix. A-DNA has a deep major groove and a flat minor groove.

The Z form is left-handed instead of right-handed. Its helix has a higher-energy form. This leads to a greater distance (0.77 nm) between the base pairs than in B-DNA and a zigzag form of the sugar–phosphate skeleton when viewed from the side (thus the designation Z-DNA). A segment of B-DNA consisting of GC pairs can be converted into Z-DNA when the bases are rotated by 180 degrees. Normally, Z-DNA is thermodynamically relatively unstable. However, transition to Z-DNA is facilitated when cytosine is methylated in position 5 (C-5). The modification of DNA by methylation of cytosine is frequent in certain regions of the DNA of eukaryotes.

Z-DNA has biological roles. Sequences favoring the formation of Z-DNA occur frequently near the promoter region, where Z-DNA stimulates transcription (Ha et al., 2005). Four families of specific Z-DNA-binding proteins of defined three-dimensional structure interact with Z-DNA. These are the editing enzyme ADAR1, an interferon-inducible protein DLM-1, an ortholog of the interferon-induced protein kinase R, and the N-terminal domain of the pox virulence factor protein E3L. The E3L protein is required for pathogenicity in mice. Its sequence is similar to other members of the family of Z-DNA binding proteins (Kim et al., 2003; Ha et al., 2005).

In addition to the three forms shown here, a triple-stranded DNA structure forms when synthetic polymers of poly(A) and polydeoxy (U) are mixed in a test tube (Figure adapted from Koolman & Röhm, 2005.)

Further Reading

Bacolla A, et al. Breakpoints of gross deletions coincide with non-B DNA conformations. Proc Natl Acad Sci USA 2004;101:14162–14167

Berg JM, Tymoczko JL, Stryer L. Biochemistry. 7th ed. New York: W. H. Freeman; 2011

Ha SC, et al. Crystal structure of a junction between B-DNA and Z-DNA reveals two extruded bases. Nature 2005;437:1183–1186

Kim Y-G, et al. A role for Z-DNA binding in vaccinia virus pathogenesis. Proc Natl Acad Sci USA 2003;100:6974–6979

Koolman J, Röhm KH. Color Atlas of Biochemistry. 2nd ed. Stuttgart: Thieme; 2005

Lodish H, et al. Molecular Cell Biology. 6th ed. New York: W. H. Freeman; 2007

Rich A, Zhang S. Timeline: Z-DNA: the long road to biological function. Nat Rev Genet 2003;4:566–572

Rich A, Nordheim A, Wang AH. The chemistry and biology of left-handed Z-DNA. Annu Rev Biochem 1984;53:791–846

Wang AH-J, et al. Molecular structure of a left-handed double helical DNA fragment at atomic resolution. Nature 1979;282:680–686

Watson JD, et al. Molecular Biology of the Gene. 6th ed. New York: Cold Spring Harbor Laboratory Press; 2008

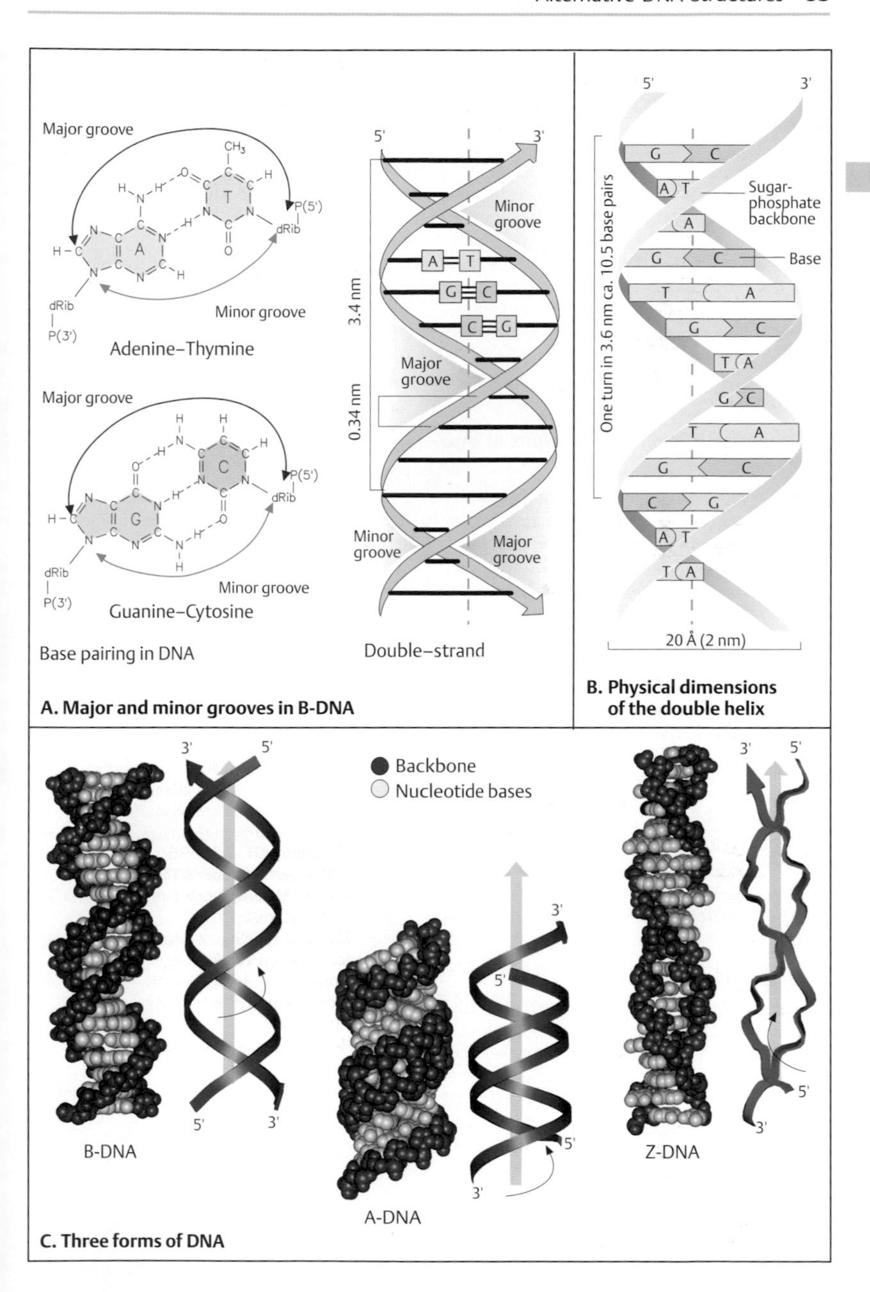

A. Major and minor grooves in B-DNA

B. Physical dimensions of the double helix

C. Three forms of DNA

DNA Replication

DNA replication refers to the process of copying each DNA strand into a new complementary strand. This occurs at every cell division and ensures that genetic information is transmitted to both daughter cells. DNA replication requires the highly coordinated action of many proteins in a complex called the replisome. Precision and speed are required. The two new DNA chains are assembled at a rate of about 1000 nucleotides per second in *Escherichia coli* (*E. coli*). During replication, each preexisting strand of DNA serves as a template for the formation of a new strand. This is referred to as semiconservative replication.

A. Prokaryotic replication begins at one site

In prokaryote cells, replication begins at a defined point in the ring-shaped bacterial chromosome, the origin of replication (**1**). From here, new DNA is formed in both directions. Replication can be visualized by autoradiography after the newly replicated DNA has incorporated tritium (^3H)-labeled thymidine (**2**).

B. Eukaryotic replication begins at several sites

In eukaryotic cells, DNA synthesis occurs during a defined phase of the cell cycle (S phase). Replication of eukaryotic DNA begins at numerous sites (replicons) (**1**). At each site the parental strands are unwound by helicases. Replication proceeds in both directions from each replicon until neighboring replicons fuse (**2**) and the entire DNA is duplicated (**3**). The electron micrograph (**4**) shows replicons at three sites (arrows).

C. DNA replication fork

During DNA replication, a characteristic structure forms at the site of the opened double helix where the new strands are synthesized; this structure is the replication fork (**1**). It moves along the double helix as the parental strands are unwound by helicases. Before this, an enzyme called topoisomerase I binds to DNA at random sites and relieves torsional stress by breaking a phosphodiester bond in one strand (making a nick). Each of the preexisting strands serves as a template for the synthesis of a new strand. However, the replication fork is asymmetric. One daughter DNA stand is synthesized continuously, and this is called the leading strand, whereas the other strand lags slightly behind (the lagging strand). New DNA is synthesized in the 5′ to 3′ direction only, because at the 3′ end nucleotides can be attached continuously to the leading strand (**2**). At the 5′ end (the lagging strand), this is not possible.

Replication requires a primer of complementary RNA at the start site. At the leading strand, synthesis of new DNA proceeds continuously from a single primer in the 5′ to 3′ direction. In contrast, at the lagging strand DNA is synthesized in small segments of 1000–2000 bases (Okazaki fragments) in the opposite direction. Each fragment requires its own primer. Subsequently, the primers are removed and replaced by DNA chain growth from the neighboring fragment, and the gaps are closed by DNA ligase. The enzyme responsible for DNA synthesis, DNA polymerase III, is complex and comprises several subunits. There are different enzymes for the leading and lagging strands in eukaryotes. A mismatch proofreading and repair system eliminates replication errors.

(Figure in 1 adapted from Alberts, 2003.)

Further Reading

Alberts B. DNA replication and recombination. Nature 2003;421:431–435

Alberts B, et al. Molecular Biology of the Cell. 5th ed. New York: Garland Science; 2008

Cairns J. The bacterial chromosome and its manner of replication as seen by autoradiography. J Mol Biol 1963;6:208–213

Chagin VO, Stear JH, Cardoso MC. Organization of DNA replication. Cold Spring Harb Perspect Biol 2011;2:a000737

Lodish H, et al. Molecular Cell Biology. 6th ed. New York: W. H. Freeman; 2007

Marx J. How DNA replication originates. Science 1995;270:1585–1587

Meselson M, Stahl FW. The replication of DNA in Escherichia coli. Proc Natl Acad Sci USA 1958;44: 671–682

Watson JD, et al. Molecular Biology of the Gene. 6th ed. New York: Cold Spring Harbor Laboratory Press; 2008

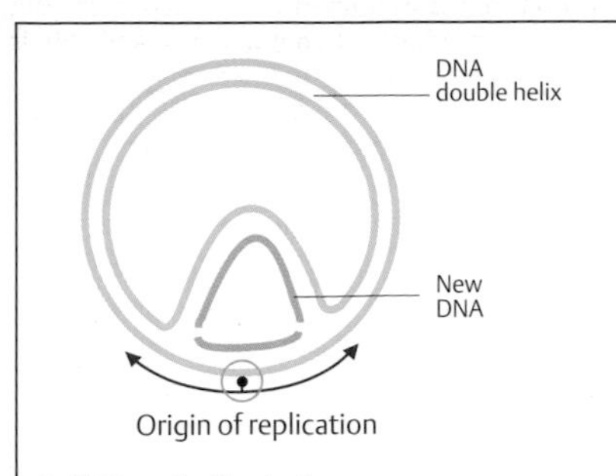

1. DNA replication in the bacterial chromosome

new DNA

2. Prokaryotic replication in an autoradiogram in *E. coli* (J. Cairns)

A. Prokaryotic replication begins at one site

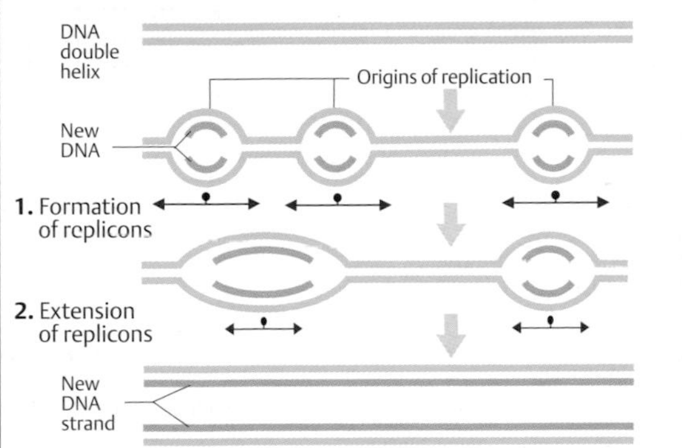

3. Replication completed

4. Eukaryotic replication in the EM (D. S. Hogness)

B. Eukaryotic replication begins at several sites

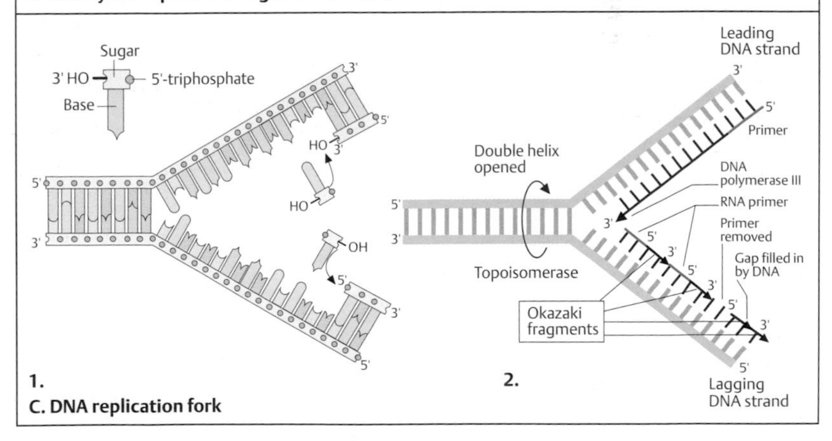

1.

2.

C. DNA replication fork

The Flow of Genetic Information: Transcription and Translation

The information contained in the nucleotide sequence of a gene is converted into useful biological function in two major steps: *transcription* and *translation*. This flow of genetic information is unidirectional. First, the information of the coding sequences of a gene is transcribed into an intermediary RNA molecule, which is synthesized in sequences that are precisely complementary to those of the coding strand of DNA (transcription). Second, the sequence information in the messenger RNA molecule (mRNA) is translated into a corresponding sequence of amino acids (translation). In eukaryotes RNA is not used directly for translation. It first is processed into an mRNA molecule before it can be used. In prokaryotes mRNA is directly transcribed from DNA.

A. Transcription

Here the nucleotide sequence of one strand of DNA is transcribed into a complementary molecule of RNA (messenger RNA, mRNA). The DNA helix is opened by a complex set of proteins (different DNA helicases). The DNA strand in the 3′ to 5′ direction (coding strand) serves as the template for the transcription of DNA into RNA by RNA polymerase. RNA is synthesized in the 5′ to 3′ direction. This strand is the RNA sense strand. RNA transcribed under experimental conditions from the opposing DNA strand is called antisense RNA. It inhibits regular transcription.

B. Translation

Translation refers to the process by which the nucleotide sequence of the mRNA is used to construct a chain of amino acids (a polypeptide chain, see p. 44) in the sequence encoded in the DNA. Translation occurs in a reading frame, which is defined at the start of translation (start codon, AUG). Translation involves two further types of RNA molecule in addition to mRNA: transfer RNA (tRNA) and ribosomal RNA (rRNA). tRNA deciphers the codons. Each amino acid has its own set of tRNAs, which bind the amino acid and carry it to the end of the growing polypeptide chain. Each tRNA has a region, called an anticodon, that is complementary to a codon of the mRNA. The figure shows codons 1, 2, 3, and 4 of the mRNA that have been translated into the amino acid sequence methionine (Met), glycine (Gly), serine (Ser), and isoleucine (Ile). Glycine and alanine are added next in this example.

C. Stages of translation

Translation is accomplished in three steps. First, at initiation (**1**) an initiation complex comprising mRNA, a ribosome, and tRNA is formed. This requires several initiation factors (IF1, IF2, IF3, etc., not shown). Then elongation follows (**2**): a further amino acid, determined by the next codon, is attached. A three-phase elongation cycle develops, with codon recognition, peptide binding of the next amino acid residue, and movement (translocation) of the ribosome three nucleotides further in the 3′ direction of the mRNA. Translation ends with termination (**3**), when one of three mRNA stop codons (UAA, UGA, or UAG) is reached.

Translation (protein synthesis) in eukaryotes occurs outside the cell nucleus in ribosomes in the cytoplasm (see p. 32). The biochemical processes of the stages shown here have been greatly simplified.

D. Structure of transfer RNA (tRNA)

Transfer RNA has a characteristic, cloverleaf-like structure, illustrated here by yeast phenylalanine tRNA (**1**). It has three single-stranded loop regions and four double-stranded "stem" regions. The three-dimensional structure (**2**) is complex, but various functional areas can be differentiated, such as the recognition site (anticodon) for the mRNA codon and the binding site for the respective amino acid (acceptor stem) on the 3′ end (acceptor end). (Figure adapted from Lodish et al., 2007.)

Further Reading

Alberts B, et al. Molecular Biology of the Cell. 5th ed. New York: Garland Science; 2008

Brenner S, Jacob F, Meselson M. An unstable intermediate carrying information from genes to ribosomes for protein synthesis. Nature 1961;190:576–581

Ibba M, Söll D. Quality control mechanisms during translation. Science 1999;286:1893–1897

Lodish H, et al. Molecular Cell Biology. 6th ed. New York: W. H. Freeman; 2007

Watson JD, et al. Molecular Biology of the Gene. 6th ed. New York: Cold Spring Harbor Laboratory Press; 2008

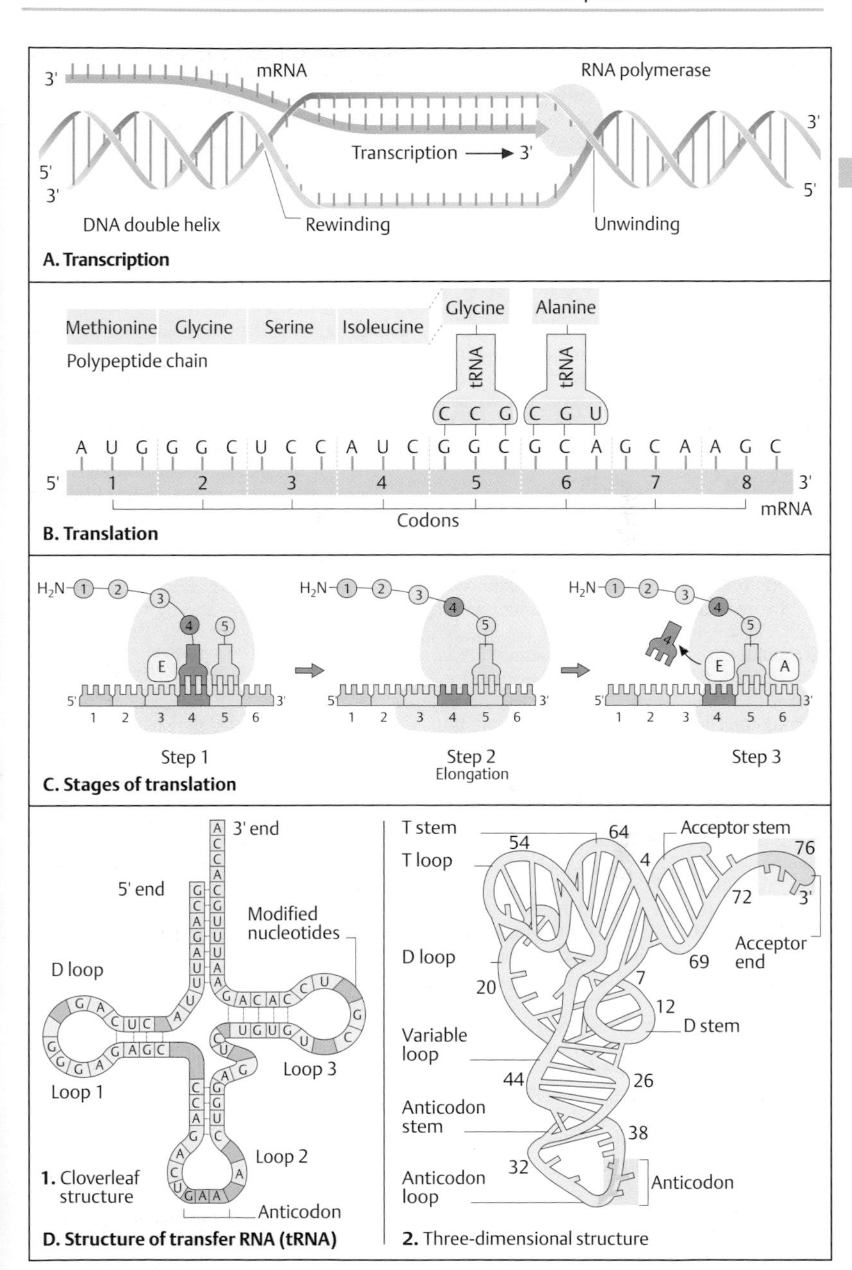
mRNA
RNA polymerase
3'
5'
3'
3'
5'
Transcription → 3'
DNA double helix
Rewinding
Unwinding
A. Transcription
Methionine
Glycine
Serine
Isoleucine
Glycine
Alanine
Polypeptide chain
tRNA
tRNA
C C G
C G U
A U G G G C U C C A U C G G C G C A G C A A G C
5'
1 2 3 4 5 6 7 8
3'
mRNA
Codons
B. Translation
H_2N
E
A
Step 1
Step 2
Elongation
Step 3
C. Stages of translation
3' end
5' end
Modified nucleotides
D loop
Loop 1
Loop 3
Loop 2
Anticodon
1. Cloverleaf structure
D. Structure of transfer RNA (tRNA)
T stem
T loop
Acceptor stem
D loop
Variable loop
Anticodon stem
Anticodon loop
Acceptor end
D stem
Anticodon
2. Three-dimensional structure

Genetic Code

The genetic code is the set of biological rules by which DNA nucleotide base pair sequences are translated into corresponding sequences of amino acids. The genetic code is a triplet code. Each code word (codon) for an amino acid consists of a sequence of three nucleotide base pairs. The genetic code is called degenerate because several amino acids are encoded by more than one codon (on average three, but up to six codons for leucine, serine, and arginine). The genetic code also specifies the beginning (start codon) and the end (stop codon) of the coding region. The genetic code is almost universal. Most organisms (bacteria and viruses, animals, and plants) use the same code, except for mitochondria and chloroplasts in plants.

A. Genetic code in mRNA

The first, second, and third nucleotide bases in the 5′ to 3′ direction determine the codon for one of the 20 amino acids. Each codon corresponds to one amino acid. The genetic code usually is written in the code language of mRNA, using uracil (U) instead of thymine (T) in DNA. An example for the redundancy of the genetic code is the amino acid phenylalanine, which is encoded by UUU and UUC. The six codons that define the amino acid serine are UCU, UCC, UCA, UCG, AGU, and AGC. The third position (at the 3′ end of the triplet) of a codon is the most variable. The only amino acids that are encoded by a single codon are methionine (AUG) and tryptophan (UGG). The start codon is AUG (methionine). Stop codons are UAA, UAG, and UGA.

The genetic code was elucidated in 1966 by analyzing how triplets transmit information from the genes to proteins. mRNA added to bacteria could be directly converted into a corresponding protein.

Some deviations from the universal genetic code are listed in the table below.

B. Abbreviated code

To save space, long sequences of amino acids are usually written using abbreviations, with each amino acid designated by just one letter, as shown.

C. Overlapping reading frames

A reading frame is a nucleotide sequence from the start codon to the stop codon. The interval between a start and a stop codon containing genetic information is called an open reading frame (ORF). An ORF does not contain a further stop codon. Normal reading frames do not overlap. Of the possible three reading frames A, B, or C, only one is correct (ORF A). The reading frames B and C are interrupted by a stop codon after three and five codons, respectively, and cannot serve to code sequences.

Further Reading

Alberts B, et al. Molecular Biology of the Cell. 5th ed. New York: Garland Science; 2008

Crick FHC, et al. General nature of the genetic code for proteins. Nature 1961;192:1227–1232

Lodish H, et al. Molecular Cell Biology. 6th ed. New York: W. H. Freeman; 2007

Rosenthal N. DNA and the genetic code. N Engl J Med 1994;331:39–41

Strachan T, Read A. Human Molecular Genetics. 4th ed. New York: Garland Science; 2011

Singer M, Berg P. Genes and Genomes: A Changing Perspective. Oxford, London: Blackwell Scientific; 1991

Deviations from the universal genetic code

Codon	Universal	Deviation	Occurrence
UGA	Stop	Tryptophan	*Mycoplasma*, mitochondria of some species
CUG	Leucine	Threonine	Mitochondria in yeast
UAA, UAG	Stop	Glycine	*Acetabularia*, *Tetrahymena*, *Paramecium*
UGA	Stop	Cysteine	*Euplotes*

Data from Lodish et al., (2007), p. 121.

Nucleotide base					
First	Second				Third
	Uracil (U)	Cytosine (C)	Adenine (A)	Guanine (G)	
Uracil (U)	F Phenylalanine (Phe)	S Serine (Ser)	Y Tyrosine (Tyr)	C Cysteine (Cys)	U
	F Phenylalanine (Phe)	S Serine (Ser)	Y Tyrosine (Tyr)	C Cysteine (Cys)	C
	L Leucine (Leu)	S Serine (Ser)	Stop Codon	Stop Codon	A
	L Leucine (Leu)	S Serine (Ser)	Stop Codon	W Tryptophan (Trp)	G
Cytosine (C)	L Leucine (Leu)	P Proline (Pro)	H Histidine (His)	R Arginine (Arg)	U
	L Leucine (Leu)	P Proline (Pro)	H Histidine (His)	R Arginine (Arg)	C
	L Leucine (Leu)	P Proline (Pro)	Q Glutamine (Gln)	R Arginine (Arg)	A
	L Leucine (Leu)	P Proline (Pro)	Q Glutamine (Gln)	R Arginine (Arg)	G
Adenine (A)	I Isoleucine (Ile)	T Threonine (Thr)	N Asparagine (Asn)	S Serine (Ser)	U
	I Isoleucine (Ile)	T Threonine (Thr)	N Asparagine (Asn)	S Serine (Ser)	C
	I Isoleucine (Ile)	T Threonine (Thr)	K Lysine (Lys)	R Arginine (Arg)	A
	Start (Methionine)	T Threonine (Thr)	K Lysine (Lys)	R Arginine (Arg)	G
Guanine (G)	V Valine (Val)	A Alanine (Ala)	D Aspartic acid (Asp)	G Glycine (Gly)	U
	V Valine (Val)	A Alanine (Ala)	D Aspartic acid (Asp)	G Glycine (Gly)	C
	V Valine (Val)	A Alanine (Ala)	E Glutamic acid (Glu)	G Glycine (Gly)	A
	V Valine (Val)	A Alanine (Ala)	E Glutamic acid (Glu)	G Glycine (Gly)	G

A. Genetic code for all amino acids in mRNA

Amino acid	Codons
Start	AUG
Stop	UAA, UAG, UGA
A (Ala)	GCU, GCC, GCG, GCA
C (Cys)	UGU, UGC
D (Asp)	GAU, GAC
E (Glu)	GAG, GAA
F (Phe)	UUU, UUC
G (Gly)	GGU, GGC, GGG, GGA
H (His)	CAU, CAC
I (Ile)	AUU, AUC, AUA
K (Lys)	AAG, AAA
L (Leu)	CUU, CUC, CUG, CUA, UUG, UUA
M (Met)	AUG
N (Asn)	AAU, AAC
P (Pro)	CCU, CCC, CCG, CCA
Q (Gln)	CAG, CAA
R (Arg)	CGU, CGC, CGG, CAA, AGG, AGA
S (Ser)	UCU, UCC, UCG, UCA, AGU, AGC
T (Thr)	ACU, ACC, ACG, ACA
V	GUU, GUC, GUG, GUA
W (Trp)	UGG
Y (Tyr)	UAU, UAC
B (Asx)	Asn or Asp
Z (Glx)	Gln or Glu

B. Abbreviated code

A — GCA– AAU– AAG– GUA– GAC– CAU – ; – Ala – Asn – Lys – Val – Asp – His – : ORF not interrupted

B — GGC– AAA– UAA (Stop) – GGU– AGA– CCA– UAC – : ORF interrupted by stop codon

C — UGG– CAA– AUA– AGG– UAG (Stop) – ACC– AUC – : ORF interrupted by stop codon

C. Overlapping reading frames

Eukaryotic Gene Structure

In eukaryotic genes, the coding sequences are interrupted by noncoding sequences of variable length. The coding sequences are called *exons*, and the noncoding sequences are called *introns*, two terms introduced in 1978 by W. Gilbert. The introns are removed before translation can begin. This process is called RNA processing. Some introns contain sequences necessary for the regulation of gene activity.

A. Exons and introns

In 1977, it was found unexpectedly that the DNA of a eukaryotic gene is longer than its corresponding mRNA. When mRNA is hybridized to its complementary single-stranded DNA, loops of single-stranded DNA remain, as shown in the electron micrograph (**1**). mRNA hybridizes only with certain sections of the single-stranded DNA because it is shorter than its corresponding DNA coding strand (**2**). Seven loops (A–G) and eight hybridizing sections are shown (1–7, and the leading section L). Of the total 7700 DNA base pairs of this gene (**3**), only 1825 hybridize with mRNA. Each hybridizing segment is an exon, here a total of seven. The single-stranded segments that do not hybridize correspond to the introns. The size and arrangement of exons and introns are characteristic for every eukaryotic gene (exon/intron structure). (Electron micrograph from Chambon, 1981.)

B. Intervening DNA sequences (introns)

In prokaryotes, DNA is colinear with mRNA and contains no introns (**1**). In eukaryotes, mature mRNA is complementary to only certain sections of DNA because the latter contains introns (**2**). (Figure adapted from Stryer, 1995.)

C. Basic eukaryotic gene structure and transcript processing

Exons and introns are numbered in the 5′ to 3′ direction of the coding strand. Both exons and introns are transcribed into a precursor RNA (primary transcript). The first and the last exons usually contain sequences that are not translated. These are called the 5′ untranslated region (5′ UTR) of exon 1, and the 3′ UTR at the 3′ end of the last exon. After transcription, RNA processing begins by removing the noncoding segments (introns) from the primary transcript. Then the exons are connected by a process called *splicing*. Splicing must be very precise to avoid an undesirable change of the correct reading frame. Introns almost always start with the nucleotides GT in the 5′ to 3′ strand (GU in RNA) and end with AG. The sequences at the 5′ end of the intron beginning with GT are called the splice donor site. The splice acceptor site is at the 3′ end of the intron. Mature mRNA is modified at its 5′ end by adding a stabilizing structure called a *cap*, and at its 3′ end by adding many adenines (polyadenylation).

D. Splicing pathway in GU–AG introns

RNA splicing is a complex process mediated by a large RNA-containing protein called a *spliceosome*. This consists of five types of small nuclear RNA molecules (snRNA) and more than 50 proteins (small nuclear riboprotein particles). Schematically, the basic mechanism of splicing involves autocatalytic cleavage at the 5′ end of the intron, resulting in lariat formation. This is an intermediate circular structure formed by connecting the 5′ terminus (UG) to a base (A) within the intron. This site is called the branch site. In the next stage, cleavage at the 3′ site releases the intron in lariat form. At the same time the right exon is ligated (spliced) to the left exon. The lariat is debranched to yield a linear intron, and this is rapidly degraded. The branch site identifies the 3′ end for precise cleavage at the splice acceptor site, and it lies 18–40 nucleotides upstream (in the 5′ direction) of the 3′ splice site. (Figure adapted from Strachan & Read, 2011.)

Further Reading

Berg JM, Tymoczko JL, Stryer L. Biochemistry. 4th ed. New York: W. H. Freeman; 1995

Chambon P. Split genes. Sci Am 1981;244(5):60–71

Lewin B. Genes IX. Sudbury: Jones & Bartlett; 2008

Strachan T, Read AP. Human Molecular Genetics. 4th ed. New York: Garland Science; 2011

Watson JD, et al. Molecular Biology of the Gene. 6th ed. New York: Cold Spring Harbor Laboratory Press; 2008

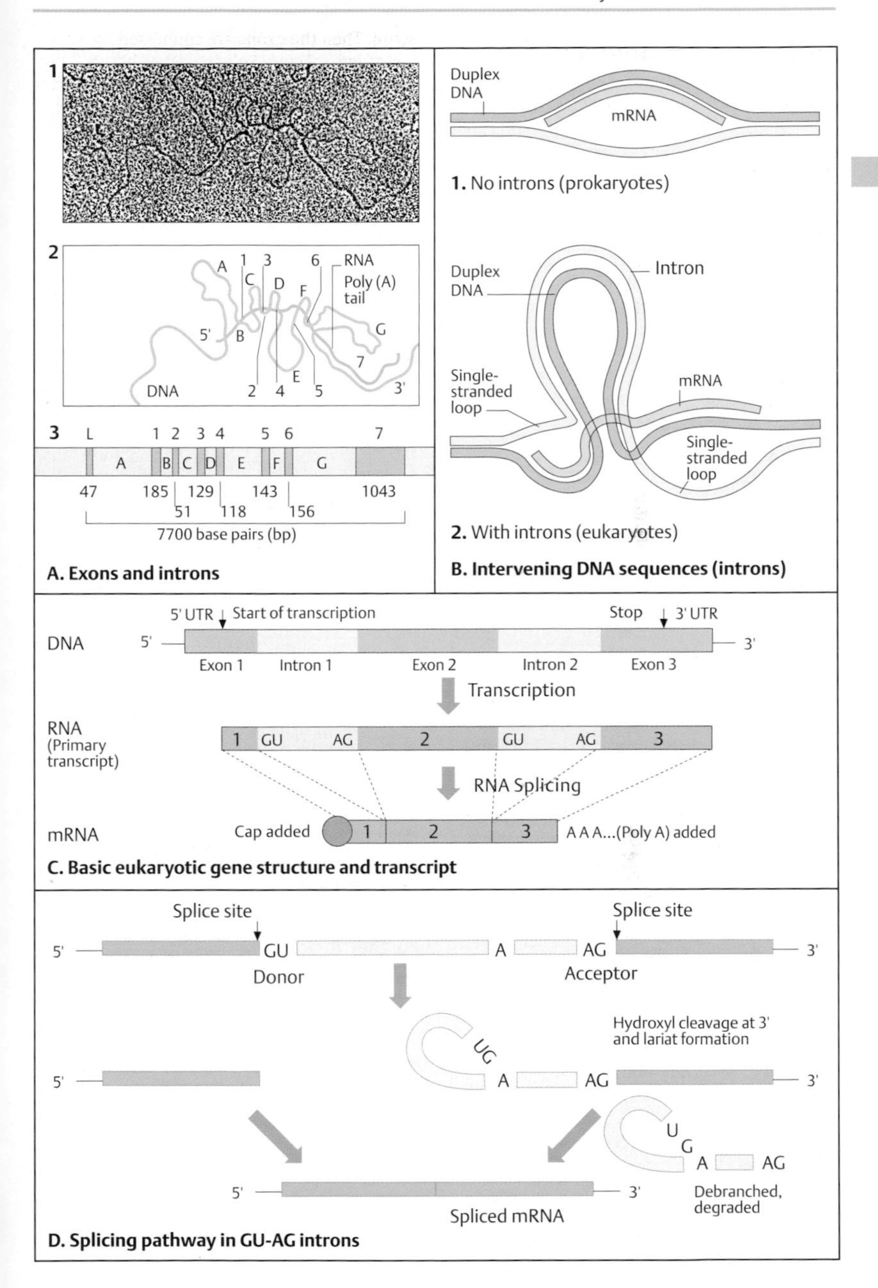

A. Exons and introns

B. Intervening DNA sequences (introns)

C. Basic eukaryotic gene structure and transcript

D. Splicing pathway in GU-AG introns

Restriction Enzymes

DNA can be cleaved into relatively small fragments in a reproducible manner by enzymes that cleave DNA at specific sites. Each enzyme cleaves only when an enzyme-specific nucleotide sequence occurs (its recognition sequence or restriction site). These enzymes are restriction endonucleases (called restriction enzymes), and they are present in different bacteria. About 400 different restriction endonucleases have been derived from various bacteria. Restriction endonucleases protect bacteria from invading foreign DNA. The sizes of the DNA fragments produced by a particular enzyme depend on the distribution of the restriction sites.

A. DNA cleavage by restriction nucleases

Many restriction enzymes cleave DNA asymmetrically. This leaves a single-stranded part of the DNA at both ends of a DNA fragment. For example the recognition site of the common restriction enzyme *Eco*RI, derived from the bacterium *Escherichia coli* restriction enzyme I, is 5′-GAATTC-3′ (**1**). It cleaves asymmetrically producing fragments with single-stranded ends (**2**). On one fragment, the single-stranded 3′ to 5′ end has four nucleotides (3′-TTAA) overhanging, and on the other fragment the 5′ to 3′ overhang is AATT-3′. Such an asymmetric cleavage pattern is called palindromic because it reads the same in opposite directions. Some restriction enzymes have a symmetric recognition site (**3**) and produce blunted ends (**4**); an example is *Hae*III (5′-CGCG-3′). The ends of fragments with single-stranded overhangs can be easily connected.

B. Examples of restriction enzymes

Restriction enzymes can be classified according to the type of ends they produce: (**1**) 5′ overhangs (e.g., *Eco*RI, see above); (**2**) 3′ overhangs (e.g., *Pst*I); (**3**) blunted ends (e.g., *Alu*I, *Hae*III [see above], *Hpa*I); or (**4**) nonpalindromic ends (*Mln*I). Some have a bipartite recognition sequence with different numbers of nucleotides at the single-stranded ends (e.g., *Bst*I).

In *Hind*II it suffices that the two middle nucleotides are a pyrimidine and a purine (GTPyPuAC), and it does not matter whether the former is thymine (T) or cytosine (C) or whether the latter is adenine (A) or guanine (G). Such recognition sites occur frequently and produce many relatively small fragments. Rare cutters recognize long sites of 10 and more nucleotides. Consequently they produce large fragments, which are useful for many purposes. Some enzymes have cutting sites with limited specificity.

C. Determining of the location of restriction sites

Since the fragment sizes reflect the relative positions of the cleavage sites, they can be used to characterize a DNA segment (restriction map). For example, if a DNA segment of 10 kilobases (kb) is cleaved by two enzymes, A and B, and the result is three fragments of 2 kb, 3 kb, and 5 kb, then the relative locations of the cleavage sites can be determined by using enzyme A and enzyme B. If enzyme A yields two fragments of 3 kb and 7 kb, and enzyme B yields two fragments of 2 kb and 8 kb, then the two recognition sites of enzymes A and B must be located 5 kb apart. The recognition site for A must be 3 kb from the left end end, and that for B 2 kb from the right end (red arrows).

D. Restriction map

In the example shown, a DNA segment is characterized by the distribution of the recognition sites for enzymes E (*Eco*RI) and H (*Hind*III). The individual sites are separated by intervals defined by the size of the fragments after digestion with the enzyme. Restriction mapping is of considerable importance in medical genetics and evolutionary research.

Further Reading

Alberts B. et al. Molecular Biology of the Cell. 5th ed. New York: Garland Science; 2008

Brown TA. Genomes. 3rd ed. New York: Garland Science; 2007

Strachan T, Read AP. Human Molecular Genetics. 4th ed. New York: Garland Science; 2011

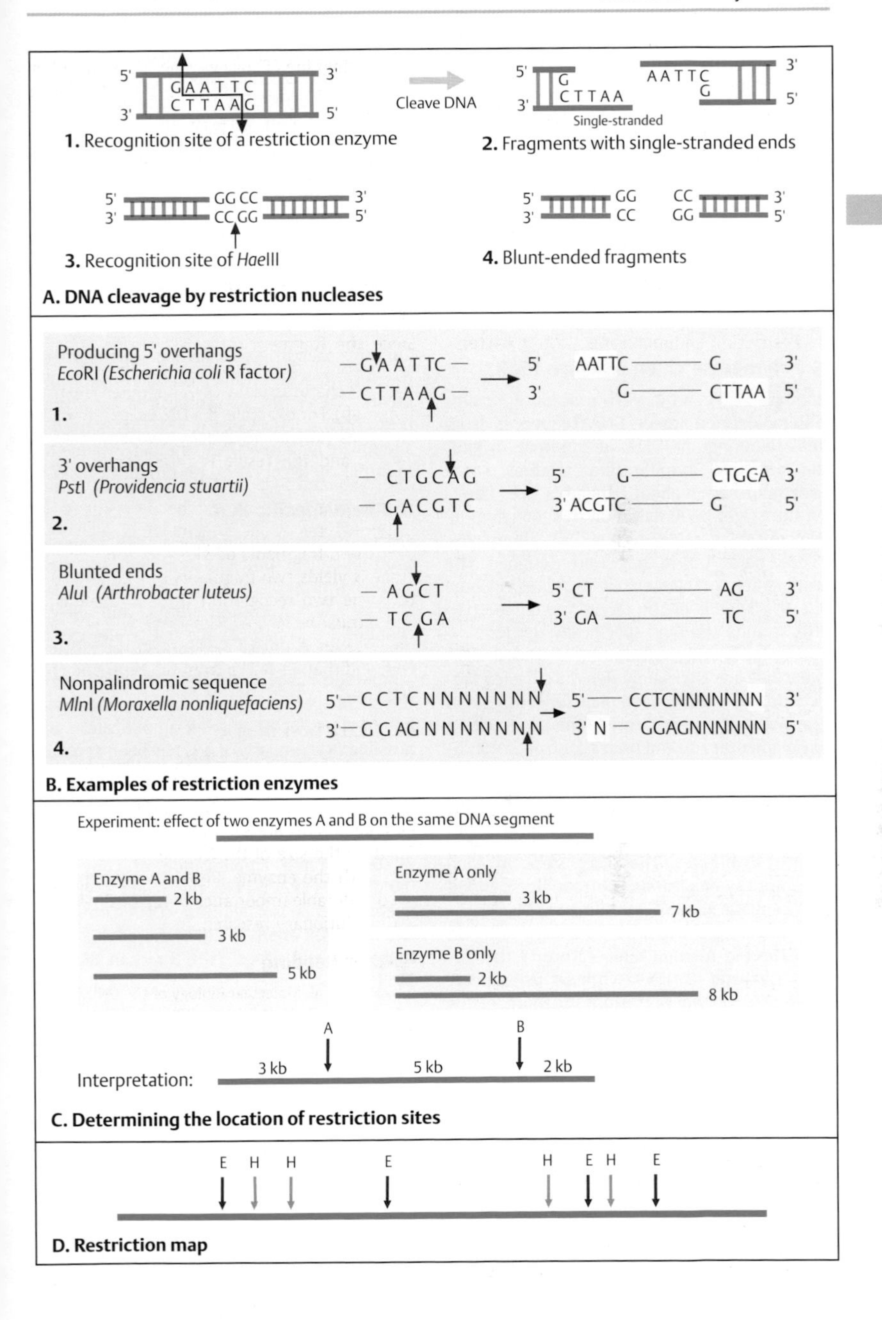
5'
GAATTC
CTTAAG
3'
3'
5'
Cleave DNA
G
CTTAA
AATTC
G
Single-stranded
1. Recognition site of a restriction enzyme
2. Fragments with single-stranded ends
GG CC
CC GG
3. Recognition site of HaeIII
4. Blunt-ended fragments
A. DNA cleavage by restriction nucleases
Producing 5' overhangs
EcoRI (Escherichia coli R factor)
— G A A T TC —
— C T T A A G —
AATTC G
G CTTAA
1.
3' overhangs
PstI (Providencia stuartii)
— C T G C A G
— G A C G T C
G CTGCA
3' ACGTC G
2.
Blunted ends
AluI (Arthrobacter luteus)
— A G C T
— T C G A
5' CT AG 3'
3' GA TC 5'
3.
Nonpalindromic sequence
MlnI (Moraxella nonliquefaciens)
5'— C C T C N N N N N N N N
3'— G G AG N N N N N N N N
5'— CCTCNNNNNNN 3'
3' N — GGAGNNNNNN 5'
4.
B. Examples of restriction enzymes
Experiment: effect of two enzymes A and B on the same DNA segment
Enzyme A and B
2 kb
3 kb
5 kb
Enzyme A only
3 kb
7 kb
Enzyme B only
2 kb
8 kb
A
B
3 kb
5 kb
2 kb
Interpretation:
C. Determining the location of restriction sites
E H H E H E H E
D. Restriction map

DNA Amplification by Polymerase Chain Reaction (PCR)

The introduction, in 1985, of cell-free methods for multiplying DNA fragments of defined origin from a complex mixture has greatly facilitated the molecular analysis of genes. The polymerase chain reaction (PCR) is a cell-free, rapid, sensitive, automated method for multiplying DNA fragments from minute amounts of DNA. It is so sensitive that contamination with undesired DNA occurs easily, and this requires special precautions.

A. Polymerase chain reaction (PCR)

Standard PCR is an in vitro procedure for amplifying defined target DNA sequences from small amounts of DNA of different origins. The selective amplification requires some prior information about DNA sequences flanking the target DNA. Based on this information, two oligonucleotide primers of about 15–25 base pairs (bp) in length are designed. The primers are complementary to sequences outside the 3′ ends of the target site on the two DNA strands and bind specifically to these.

During each PCR cycle double-stranded DNA molecules are alternately denatured, each single strand is used as a template for synthesis of a new strand, and renatured (annealed) with a complementary strand under controlled conditions. PCR is a chain reaction of about 25–35 cycles. Each cycle, involving three precisely time-controlled and temperature-controlled reactions in automated thermal cycler, takes about 1–5 minutes. The three steps in each cycle are (**1**) denaturation of double-stranded DNA at about 93–95°C for human DNA, (**2**) primer annealing at about 50–70°C depending on the expected melting temperature of the duplex DNA, and (**3**) DNA synthesis using heat-stable DNA polymerase (from microorganisms living in hot springs; for example, *Thermophilus aquaticus*, *Taq* polymerase), typically at about 70–75°C. At each subsequent cycle the template (shown in blue) and the DNA newly synthesized during the preceding cycle (shown in red) act as templates for another round of synthesis. The first cycle results in newly synthesized DNA of varied lengths (shown with an arrow) at the 3′ ends because synthesis continues beyond the target sequences. However, DNA strands of fixed length at both ends rapidly outnumber those of variable length because synthesis cannot proceed past the terminus of the primer at the opposite template DNA. At the end, at least 10^5 copies of the specific target sequence are present. This can be visualized as a distinct band of a specific size after using gel electrophoresis. In addition to the standard reaction, a wide variety of PCR-based methods have been developed for different purposes. (Figure adapted from Alberts et al., 2008.)

B. Reverse PCR (RT-PCR)

This approach utilizes mRNA as starting material. After the first primer is attached, complementary new DNA is synthesized by reverse transcription (cDNA, see p. 72). This is used as a template for a new DNA strand. Subsequently multiple copies of cDNA are produced by PCR.

C. Allele-specific PCR

This is designed to amplify a DNA sequence from one allele only and to exclude the other allele. For example, if allele 1 contains an A–T base pair at a particular site (**1**) and allele 2 contains a C–G pair (**2**), the two alleles can be distinguished by allele-specific PCR (**3** and **5**). If a mutation has changed the T to C in allele 1, the allele-specific oligonucleotide primer does not bind perfectly, making amplification impossible (**4**). Similarly, if a C has been replaced by an A in allele 2, allele-specific amplification is not possible (**6**). Reverse transcriptase PCR (RT-PCR) can be used when the known exon sequences are widely separated within a gene. By using rapid amplification of copy DNA (*cDNA*) ends (RACE-PCR), the 5′ and 3′ end sequences can be isolated from cDNA. Other variations of PCR are Alu-PCR, anchored PCR, real-time PCR, and others (see Strachan & Read, 2011, p. 182 ff).

Further Reading

Alberts B, et al. Molecular Biology of the Cell. 5th ed. New York: Garland Science; 2008

Brown TA. Genomes. 3rd ed. New York: Garland Science; 2007

Lodish H, et al. Molecular Cell Biology. 6th ed. New York: W. H. Freeman; 2007

Strachan T, Read AP. Human Molecular Genetics. 4th ed. New York: Garland Science; 2011

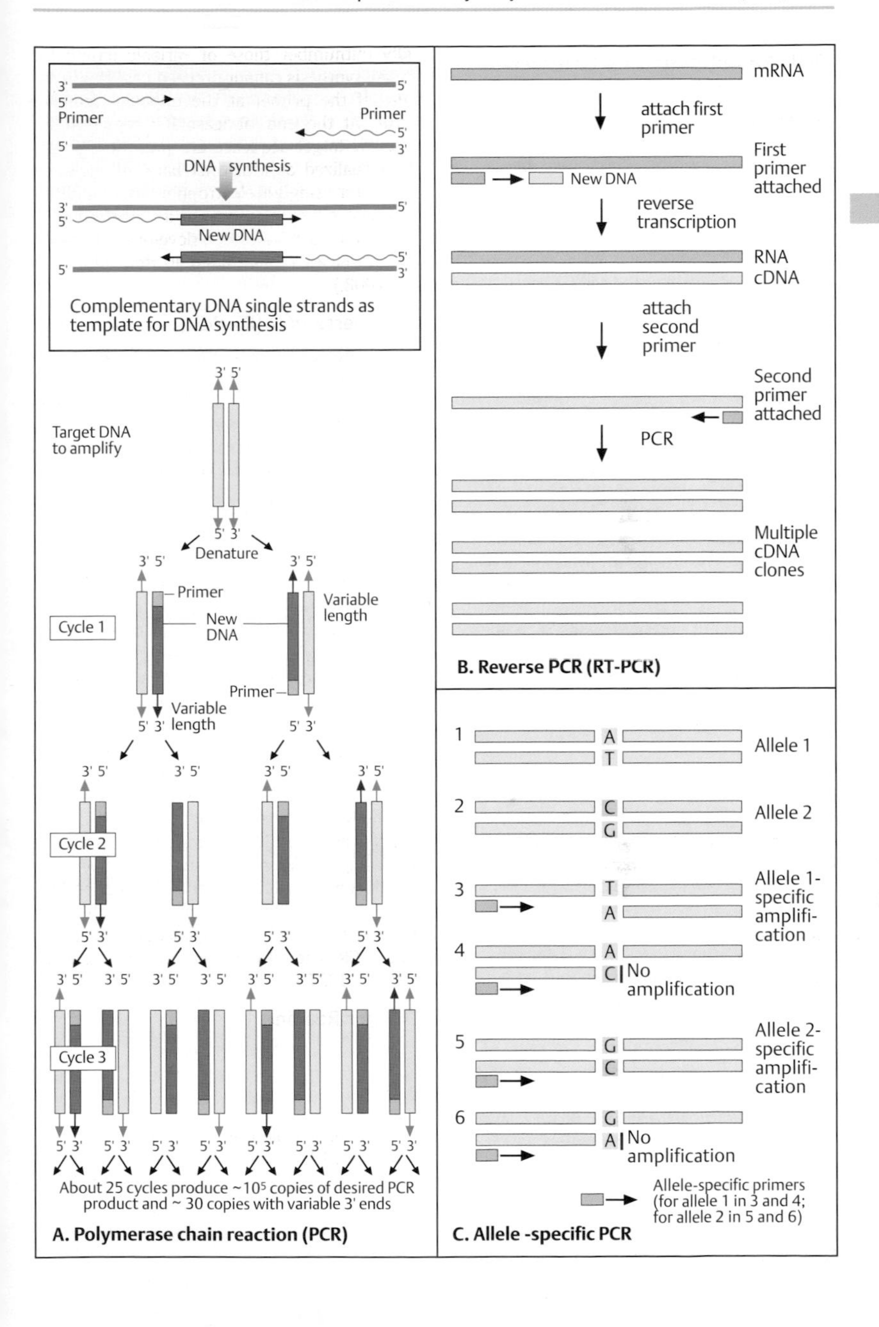

3'
5'
5'
Primer
Primer
5'
5'
3'
DNA synthesis
3'
5'
5'
New DNA
5'
5'
3'
Complementary DNA single strands as template for DNA synthesis
Target DNA to amplify
Denature
Primer
New DNA
Variable length
Cycle 1
Primer
Variable length
Cycle 2
Cycle 3
About 25 cycles produce ~10⁵ copies of desired PCR product and ~ 30 copies with variable 3' ends
A. Polymerase chain reaction (PCR)
mRNA
attach first primer
First primer attached
New DNA
reverse transcription
RNA
cDNA
attach second primer
Second primer attached
PCR
Multiple cDNA clones
B. Reverse PCR (RT-PCR)
Allele 1
Allele 2
Allele 1-specific amplification
No amplification
Allele 2-specific amplification
No amplification
Allele-specific primers (for allele 1 in 3 and 4; for allele 2 in 5 and 6)
C. Allele -specific PCR

DNA Sequencing

The introduction of two basic methods for DNA sequencing in 1977 represented a milestone in genetic analysis. These are the chemical cleavage method developed by A. M. Maxam and W. Gilbert, and an enzymatic method developed by F. Sanger and coworkers. The chemical cleavage method is no longer used (see 3rd edition, p. 62), whereas the Sanger principle was eventually applied in automated sequencing.

A. Dideoxy DNA sequencing

This method rests on the principle that DNA synthesis is terminated when a dideoxynucleotide (ddATP, ddTTP, ddGTP, ddCTP) is used instead of a normal deoxynucleotide (dATP, dTTP, dGTP, dCTP). A dideoxynucleotide (ddNTP) is an analogue of the normal dNTP. It differs by lack of a hydroxyl group at the 3′ carbon position. When a ddNTP is incorporated during DNA synthesis, no bond between its 3′ position and the next nucleotide is possible.

DNA synthesis is initiated using a primer of about 20 RNA nucleotides in four parallel reactions containing all four dNTPs and one of the four ddNTPs labeled with ^{32}P in the phosphate groups. When ddATP is incorporated into the new DNA chain, the further synthesis will not be possible beyond any site containing a T in the template. By setting the concentration of ddATP low enough, the chain will be terminated at all sites after an A is incorporated. The resulting fragments share the same 5′ end, but differ in length at their 3′ end, each ending with an A. Thus, each DNA fragment ends with an A at a defined position.

Similarly, in the other three reactions the growing DNA chain is terminated at a G by ddGTP, at a C by ddCTP, and at a T by ddTTP.

By aligning the ends of the fragments according to size, the nucleotides at the ends of the fragments will yield their sequence accurately. Initially this was done by gel electrophoresis (not shown, see 3rd edition, p. 63), later by automated sequencing as shown in part B.

B. Automated DNA sequencing

Automated DNA sequencing is based on fluorescence labeling of DNA and suitable detection systems. The direct fluorescent labels are fluorophores. These are molecules that emit a distinct fluorescent color when exposed to ultraviolet light of a specific wavelength: fluorescein, pale green when exposed to a wavelength of 494 nm; rhodamine, red at 555 nm; and aminomethylcoumarin acetic acid blue at 399 nm. In addition, a combination of different fluorophores can be used to produce a fourth color. Thus, each of the four bases can be labeled distinctly. Several variant techniques for sequencing have been devised.

Automated DNA sequencing involves four fluorophores, one for each of the four nucleotide bases. The resulting fluorescent signal is recorded at a fixed point when DNA passes through a capillary containing an electrophoretic gel. Each ddNTP is labeled with a different color, e.g., ddATP green, ddCTP blue, ddGTP black, and ddTTP red (**1**). (The actual colors for each nucleotide may be different.) All chains terminated at an adenine (A) will yield a green signal, all chains terminated at a cytosine (C) will yield a blue signal, and so on. The sequencing reactions (**2**) are performed in sequencing capillaries (**3**). The electrophoretic migration of the ddNTP-labeled chains in the gel in the capillary pass in front of a laser beam focused on a fixed position (**4**). The sequence is electronically read and recorded, and is visualized as peaks in one of the four colors, representing the alternating nucleotides in their sequence positions (**5**). (Parts 1–4 of the figure are adapted from Brown, 2007; and Strachan and Read, 2011.)

Further Reading

Alberts B, et al. Molecular Biology of the Cell. 5th ed. New York: Garland Science; 2008

Brown TA. Genomes. 3rd ed. New York: Garland Science; 2007

Lodish H, et al. Molecular Cell Biology. 6th ed. New York: W. H. Freeman; 2007

Pettersson E, Lundeberg J, Ahmadian A. Generations of sequencing technologies. Genomics 2009;93:105–111

Sanger F, Nicklen S, Coulson AR. DNA sequencing with chain-terminating inhibitors. Proc Natl Acad Sci USA 1977;74:5463–5467

Strachan T, Read AP. Human Molecular Genetics. 4th ed. New York: Garland Science; 2011

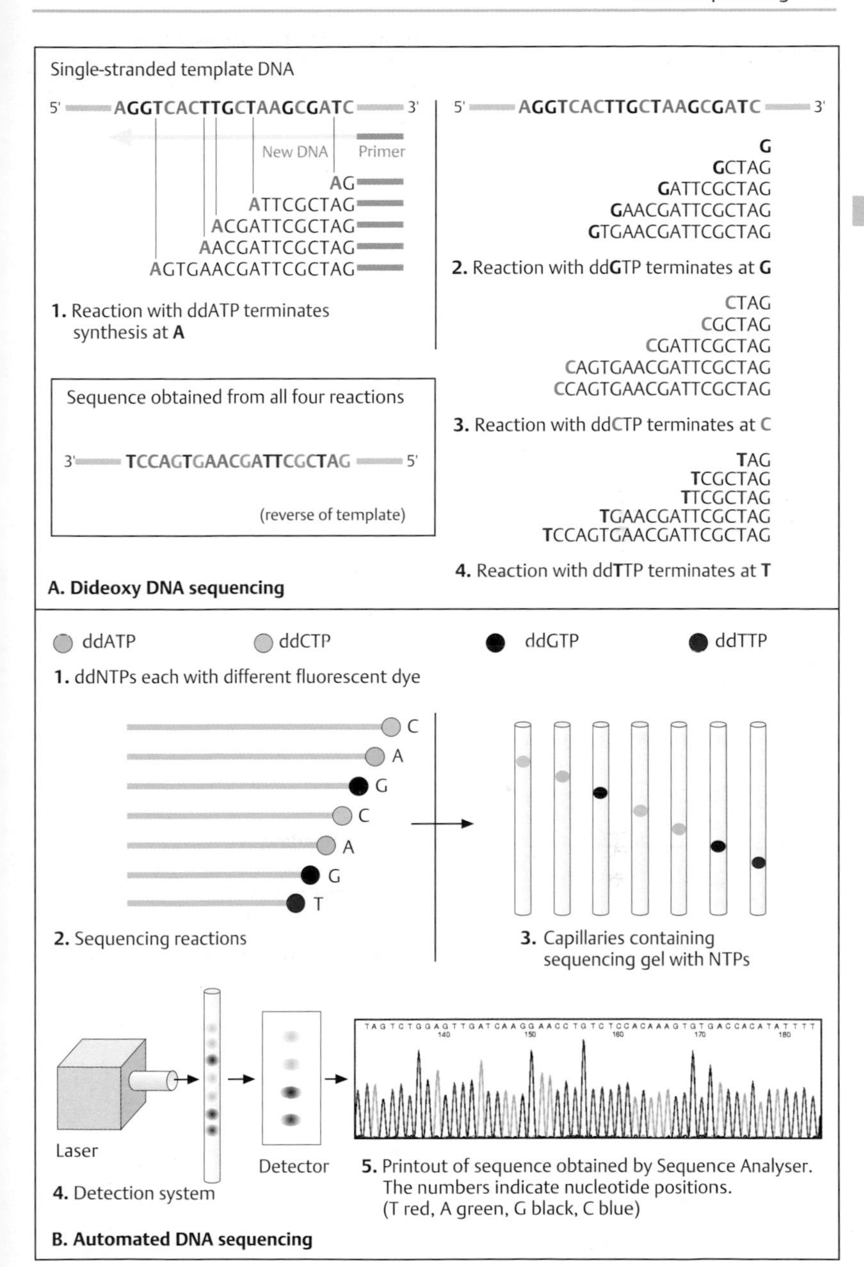

A. Dideoxy DNA sequencing

B. Automated DNA sequencing

Parallel DNA Sequencing

Since 2005 several completely new DNA sequencing technologies have been introduced. They are referred to as next-generation sequencing (NGS), although parallel sequencing is a more accurate term. These approaches differ fundamentally from dideoxy sequencing according to the Sanger principle (previous plate). They consist of a combination of template preparation, sequencing by DNA synthesis without electrophoresis, high-resolution imaging, sequence tag assembly, and genome alignment methods. Millions of sequences read in parallel yield whole-genome sequences in a relatively short time and at reduced costs (see table in Appendix, p. 409). The methods differ and yield different read length and sequenced output. About 1–50 giga bases (Gb, 10^9) can be generated in a matter of days. The principle of one selected example is presented here.

A. Selection of ligated DNA fragments

Double-stranded DNA is first fragmented (**1**) and, after end repair, overhangs of adenine (A) molecules are added (**2**). A unique adapter sequence is ligated to each fragment (**3**). This creates a library with a selection of ligated DNA (**4**).

B. Cluster generation and polony array formation

Adapter-modified single-stranded DNA fragments are attached to a fluid channel surface within a glass slide, called a flow cell. Solid-phase bridge amplification is achieved with an immobilized template, dNTPs, and DNA polymerase. This results in multiple (100–200 million) spatially separated template clusters. Following denaturation into single-stranded DNA, each member of this population of templates will undergo an individual sequencing reaction. This starts at a free end, to which a universal sequencing primer can be annealed to initiate the sequencing reaction. Attached to an array, polymerase-driven colonies of growing DNA strands (termed *polonies*) result.

C. Sequencing by synthesis and imaging

Sequencing is achieved by continuous DNA synthesis rather than dideoxy chain termination as in the Sanger principle. Each synthesis reaction is visualized by four colors, each representing one of the four nucleotides synthesized. From the template of the sequence CGAGGC, the synthesized sequence is generated (GCTCC shown).

D. Overview

Four principle phases are: DNA fragmentation (**1**), in vitro adapter ligation (**2**), generation of polony array (**3**), and cycle array sequencing (**4**).

(Figures derived from a poster that was kindly provided by Illumina, Inc., San Diego, California, USA. The text was read by Professor Thomas Langmann, Regensburg.)

Further Reading

Bentley DR, et al. Accurate whole human genome sequencing using reversible terminator chemistry. Nature 2008;456:53–59

Mardis ER. A decade's perspective on DNA sequencing technology. Nature 2011;470:198–203

Margulies M, et al. Genome sequencing in microfabricated high-density picolitre reactors. Nature 2005;437:376–380

Metzker ML. Sequencing technologies - the next generation. Nat Rev Genet 2010;11:31–46

Pettersson E, Lundeberg J, Ahmadian A. Generations of sequencing technologies. Genomics 2009;93: 105–111

Shendure J, et al. Accurate multiplex polony sequencing of an evolved bacterial genome. Science 2005;309:1728–1732

Strachan T, Read A. Human Molecular Genetics. 4th ed. New York: Garland Science; 2011

Voelkerding KV, Dames SA, Durtschi JD. Next-generation sequencing: from basic research to diagnostics. Clin Chem 2009;55:641–658

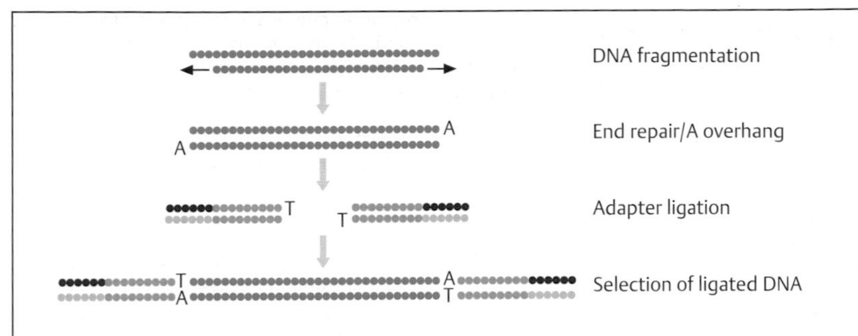

A. Selection of ligated DNA fragments

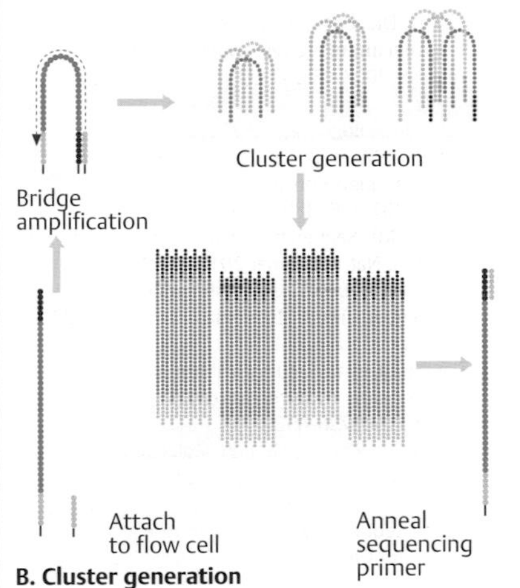

B. Cluster generation

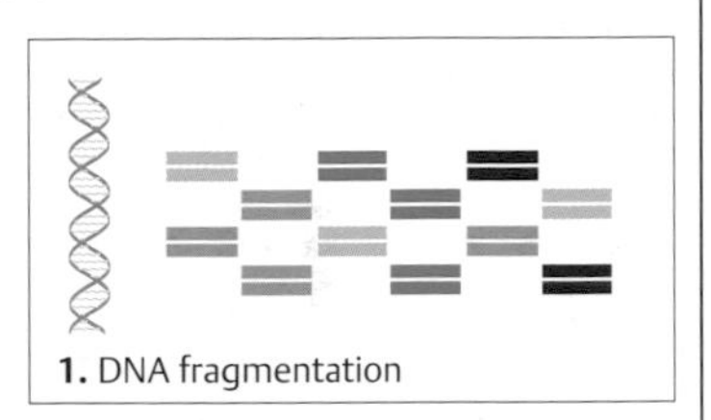

1. DNA fragmentation

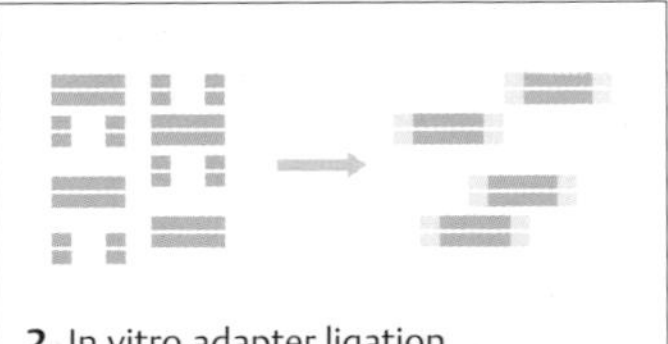

2. In vitro adapter ligation

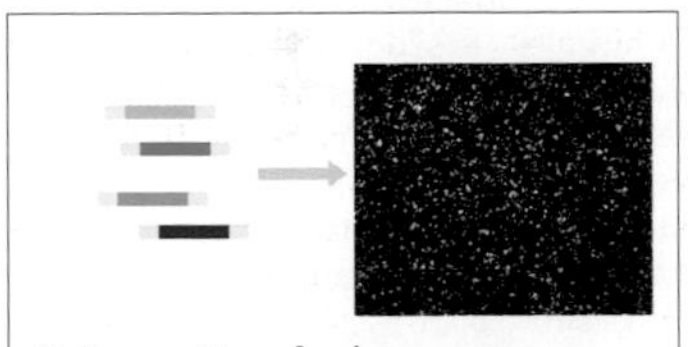

3. Generation of polony array

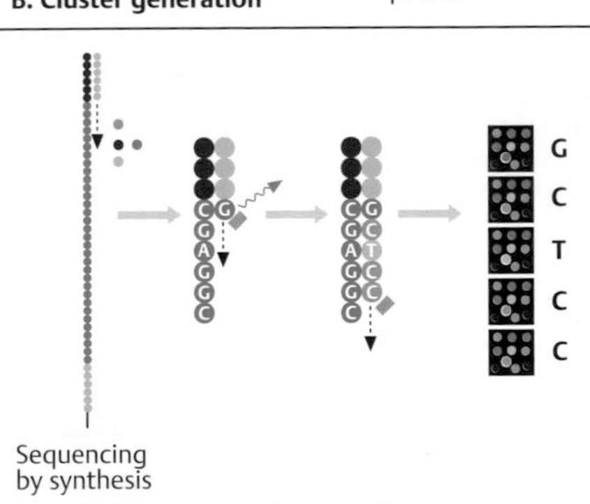

C. Sequencing by synthesis and imaging

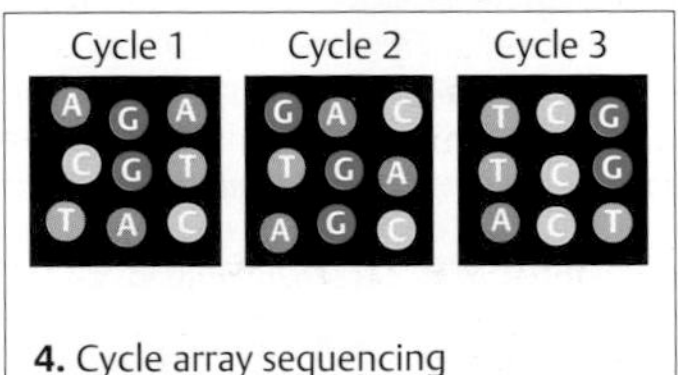

4. Cycle array sequencing

D. Overview

DNA Cloning

Cell-based DNA cloning selectively amplifies specific DNA sequences in bacterial cells. The correct DNA fragments are identified by hybridization of complementary single-stranded DNA. A short segment of single-stranded DNA (a DNA probe) originating from the sequence to be studied will hybridize to its complementary single-stranded sequences. This serves to identify and select the fragment under study. After the hybridized sequence has been separated from other DNA, it can be cloned.

A. Principle of cell-based DNA cloning

Cell-based DNA cloning requires a series of consecutive steps. First, a collection of different DNA fragments (labeled 1, 2, and 3) are obtained from the desired DNA (target DNA) by cleaving it with a restriction enzyme (see p. 62) (**1**). Fragments with short single-stranded ends resulting from restriction enzyme cleavage are ligated to DNA fragments containing the origin of replication (OR) of a replicon, enabling them to replicate, and a selectable marker, e.g., a DNA sequence containing an antibiotic resistance gene (**2**). The recombinant DNA molecules are transferred into host cells (bacterial or yeast cells), where they can replicate independently of the host cell genome (**3**). Host cells that have incorporated the fragment to be cloned will be transformed (change a genetic property; here fragment 1, a brown circle containing the number 1). Usually the host cell takes up only one (although occasionally more than one) foreign DNA molecule. The host cells transformed by recombinant (foreign) DNA are grown in culture, where they multiply (propagation, **4**). Selective growth of transformed cells containing the desired DNA fragment produces multiple copies of the fragment (**5**). The resulting recombinant DNA clones form a homogeneous population (**6**). They are used to build a large collection of cloned DNA fragments, called a clone library (**7**) (see pp. 74). In cell-based cloning, the replicon-containing DNA molecules are referred to as vector molecules. (Figure adapted from Strachan & Read, 2011.)

B. A vector for DNA cloning

Cell-based cloning requires a DNA sequence (a vector molecule) able to replicate in a cell. Many different DNA vector systems exist for cloning DNA fragments of different sizes. Plasmid vectors are used to clone small fragments. The experiment is designed so that the plasmid incorporating the fragment to be cloned confers antibiotic resistance to its bacterial host, which will be grown in culture medium containing the antibiotic.

One of the first vectors to be developed was the plasmid vector pBR322 (Bolivar et al., 1977). It is small, 4363 base pairs (bp), and has an origin of replication and two genes encoding antibiotic resistance, against ampicillin and tetracycline (**1**). Since it was constructed by ligating restriction fragments from three *Escherichia coli* plasmids that occur naturally, it contains recognition sites for seven restriction enzymes, as shown. Resistance to two antibiotics is used to provide selectable markers. They can serve to distinguish cells containing recombinant from those containing nonrecombinant pBR322 plasmids. Bacteria that have been exposed to pBR322 are plated on a medium containing both antibiotics. Only bacteria containing plasmids that have taken up new DNA at the *Bam*HI recognition site lose the tetracycline resistance, because it is inserted at the site for this resistance (insertional activation) (**2**). Cells containing the nonrecombinant pBR322 plasmid remain resistant to tetracycline. If the enzyme *Pst*I is used to incorporate a fragment, ampicillin resistance is lost (the bacterium becomes ampicillin sensitive), but tetracycline resistance is retained. Thus, with the help of replica plating, recombinant plasmids containing the DNA fragment to be cloned can be distinguished from nonrecombinant plasmids by altered antibiotic resistance. Cloning in plasmids (bacteria) has become less important since yeast artificial chromosomes (YACs) have become available for cloning relatively large DNA fragments.

Further Reading

Bolivar F, et al. Construction and characterization of new cloning vehicles. II. A multipurpose cloning system. Gene 1977;2:95–113

Brown TA. Genomes. 3rd ed. New York: Garland Science; 2007

Strachan T, Read AP. Human Molecular Genetics. 4th ed. New York: Garland Science; 2011

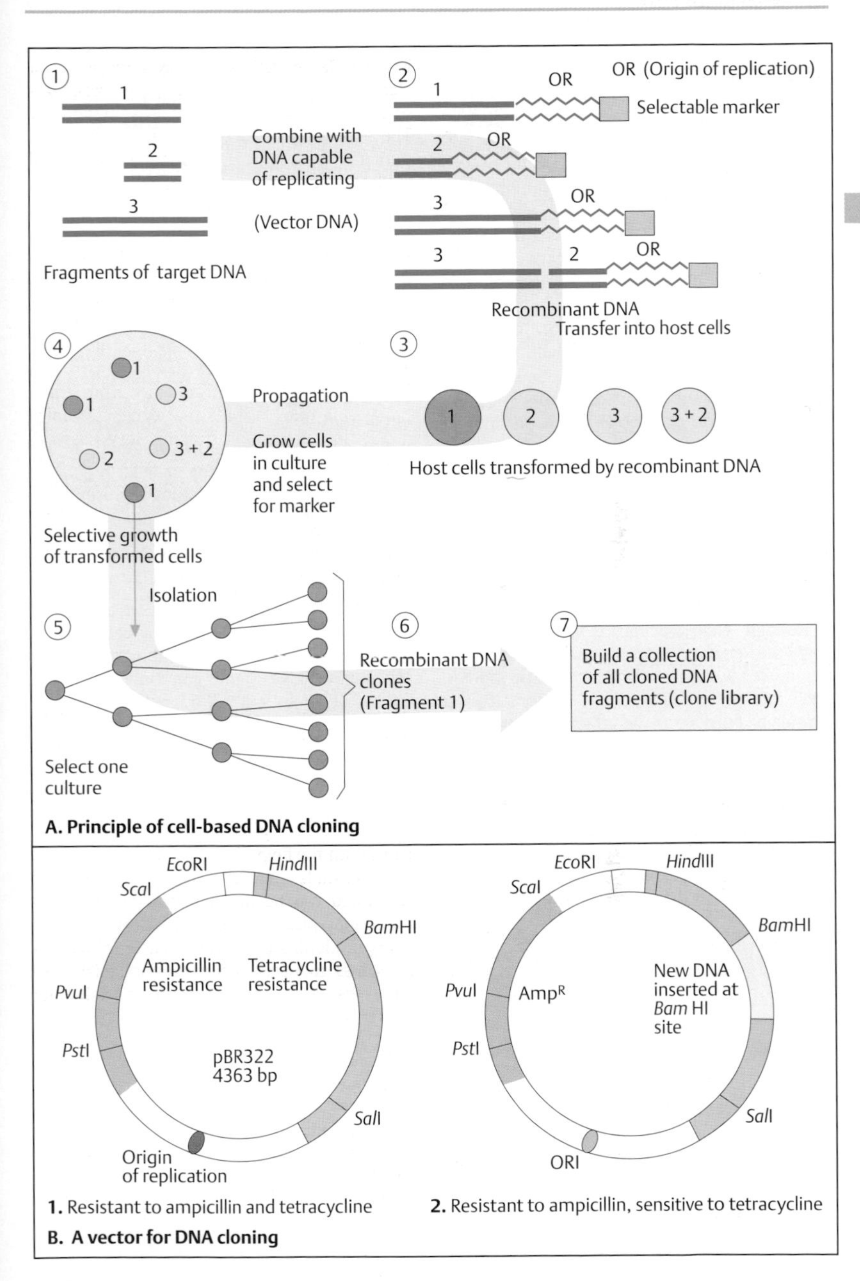

A. Principle of cell-based DNA cloning

1. Resistant to ampicillin and tetracycline

2. Resistant to ampicillin, sensitive to tetracycline

B. A vector for DNA cloning

cDNA Cloning

A DNA copy of RNA is called complementary DNA (cDNA). It is single stranded and derived from a coding DNA segment of an active (expressed) gene. It can be synthesized by reverse transcription and cloned. A collection of cDNA clones is called a cDNA library (see p. 74). Its advantage is that it corresponds to coding parts of the gene. Its disadvantage is that it does not yield information about the exon/intron structure of a gene. From the cDNA sequence, essential inferences can be made about a gene and its gene product. It can be used as a probe (cDNA probe) to recognize structural rearrangements of a gene. Thus, the preparation and cloning of cDNA are important.

A. Preparation of cDNA

cDNA is prepared from mRNA by reverse transcriptase. This is an enzyme system present in retroviruses. A tissue is chosen in which the respective gene is transcribed, and mRNA is produced in sufficient quantities. First, mRNA is isolated. Then a primer is attached to it so that the enzyme reverse transcriptase can begin to synthesize cDNA from the mRNA. Since mRNA contains poly(A) at its 3′ end, a primer of poly(T) can be attached. From here, the enzyme reverse transcriptase can start forming cDNA in the 5′ to 3′ direction. The RNA is then removed by ribonuclease. The cDNA serves as a template for the formation of a new strand of DNA, making the cDNA double stranded. This requires the enzyme DNA polymerase. The result is a double strand of DNA, one strand of which is complementary to the original mRNA. To this DNA, single sequences (linkers) are attached that are complementary to the single-stranded ends produced by the restriction enzyme to be used. The same enzyme is used to cleave the DNA of the cloning vector, e.g., a plasmid into which the cDNA is incorporated for cloning (see p. 70).

B. Cloning vectors

The cell-based cloning of DNA fragments of different sizes requires a suitable vector system. Plasmid vectors are used to clone small cDNA fragments in bacteria. Their main disadvantage is that only 5–10 kb of foreign DNA can be cloned. A plasmid cloning vector that has taken up a DNA fragment to be cloned (recombinant vector), e.g., pUC8 with 2.7 kb DNA, has to be distinguished from one that has not. First, ampicillin resistance (*Amp*⁺) serves to distinguish bacteria that have taken up plasmids from those that have not. Several unique restriction sites in the plasmid DNA segment, where a DNA fragment might be inserted, serve as markers along with a marker gene, such as the *lacZ* gene encoding β-galactosidase. β-Galactosidase cleaves an artificial sugar (5-bromo-4-chloro-3-indolyl-β-D-galactopyranoside) similar to lactose, which is the natural substrate for this enzyme, into two sugar components, one of which is blue. Colonies with active β-galactosidase appear blue, while those with inactive β-galactosidase are white. The uptake of a DNA fragment by the plasmid vector disrupts the gene for β-galactosidase. Thus, all white colonies represent bacteria that contain the recombinant plasmid with a cDNA fragment.

C. cDNA cloning

The white colonies are grown in a medium containing ampicillin. Only those bacteria that have incorporated the recombinant plasmid are ampicillin-resistant, and only the white colonies contain a cDNA fragment. By further propagation of these bacteria, the cDNA fragments can be cloned until there is enough material to be studied. Subsequently, a clone library can be constructed. (Figure adapted from Lodish et al., 2007.)

Further Reading

Brown TA. Genomes. 3rd ed. New York: Garland Science; 2007

Lodish H. Molecular Cell Biology. 6th ed. New York: WH Freeman; 2007

Watson JD, et al. Molecular Biology of the Gene. 6th ed. New York: Cold Spring Harbor Laboratory Press; 2008

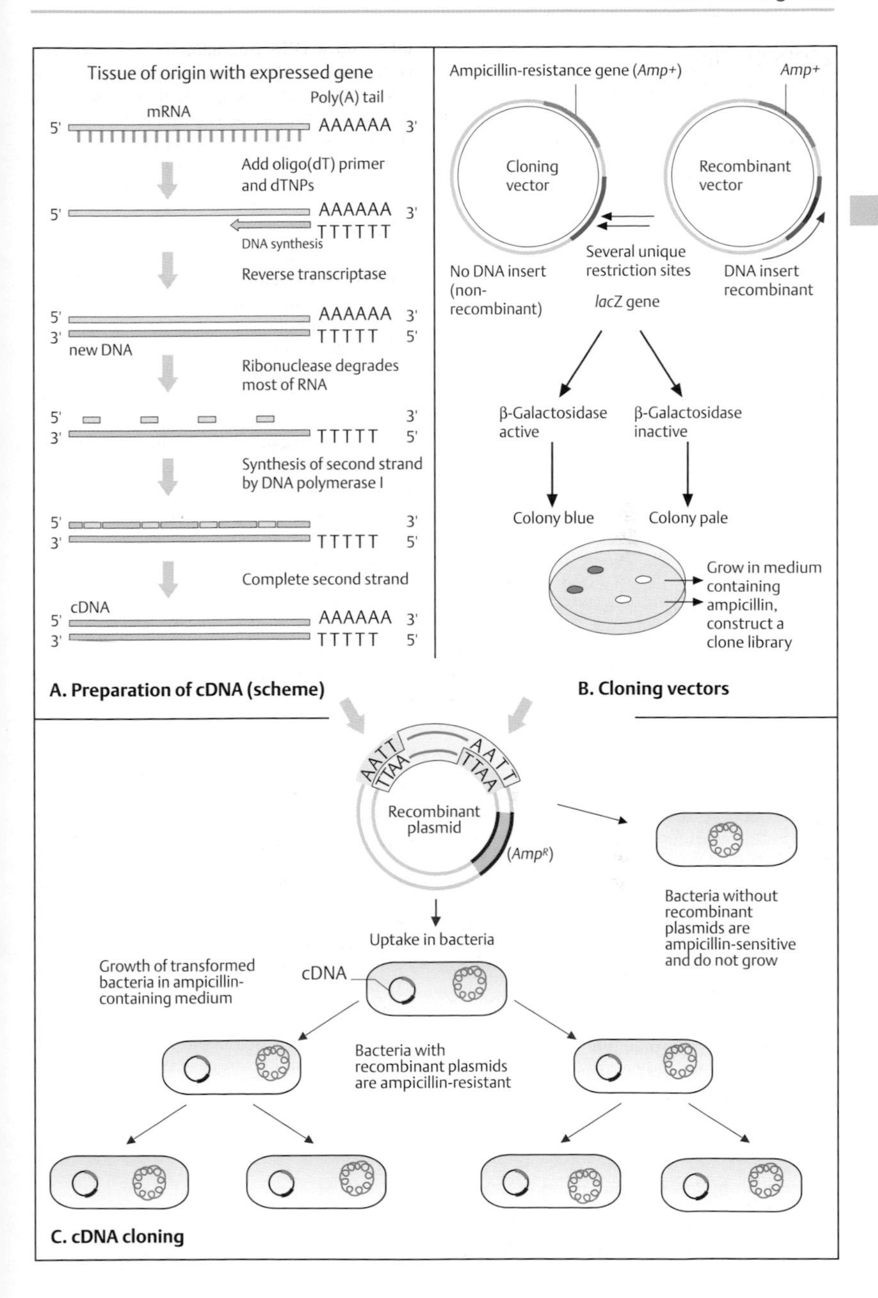

A. Preparation of cDNA (scheme)

B. Cloning vectors

C. cDNA cloning

DNA Libraries

A DNA library is a random collection of individual DNA fragments that in their entirety represent part or all of the genome of an organism. Two types of DNA libraries exist: (i) genomic DNA libraries and (ii) cDNA libraries. The first is a collection of DNA clones containing all unique nucleotide sequences of the genome. A sufficient number of clones must be present so that every segment of the genome is represented at least once. A cDNA library is a collection of cDNA that is representative of all mRNA molecules produced by the organism.

A. Genomic DNA library

Clones of genomic DNA are copies of DNA fragments from all of the chromosomes (**1**). They contain coding and noncoding sequences. Restriction enzymes cleave the genomic DNA into many fragments. Four fragments containing two genes, A and B, are shown schematically (**2**). Fragments of the genes and their surrounding DNA are incorporated into vectors, e.g., into phage DNA, and are cloned in bacteria or yeast cells. In eukaryotes, a genomic library will contain hundreds of thousands to more than a million of individual DNA clones. A screening procedure is required to find a particular gene (see **C**).

B. cDNA library

Since a cDNA library consists only of coding DNA sequences, it is smaller than a genomic library. The starting material is usually total RNA from a specific tissue or a particular developmental stage of embryogenesis. A limitation is that mRNA can be obtained only from cells in which the respective gene is transcribed, i.e., in which mRNA is produced (**1**).
Unlike a genomic library, which is complete and contains coding and noncoding DNA, a cDNA library contains only coding DNA. This specificity offers considerable advantages over genomic DNA. However, it requires that mRNA is available and it does not yield information about the structure of the gene. In eukaryotes, the primary transcript RNA undergoes splicing to form mRNA (**2**; see p. 60). cDNA is derived from mRNA by reverse transcription (**3**; see p. 72). The subsequent steps, i.e., incorporation into a vector and replication in bacteria, correspond to those of the procedure for producing a genomic library. The sequence of amino acids in a protein can be determined from cloned and sequenced cDNA. Furthermore, large amounts of a protein can be produced when the cloned gene is expressed in bacteria or yeast cells (*proteome library*).

C. Screening of a DNA library

To identify clones carrying a gene, parts of a gene, or another DNA region of interest, a screening procedure is required. Two principal approaches can be used: (i) detection using oligonucleotide probes that bind to the sequences of interest, and (ii) detection based on protein produced by the gene in question. Screening with oligonucleotide probes depends on the principle of hybridization. Single-stranded DNA or RNA molecules will specifically hybridize to their complementary single-stranded sequences (see p. 50).
In the membrane-based assay, colonies of cultured bacteria (**1**), some of which will have taken up a recombinant vector, are transferred to a filter paper or a membrane (**2**), lysed, and their DNA is denatured (made single stranded). Labeled complementary single-stranded DNAs or RNAs are used as a probe. The probe will hybridize with only complementary DNA or RNA (**4**). After hybridization, a signal appears on the membrane (**5**). DNA complementary to the labeled probe is located here; its exact position in the culture corresponds to that of the signal on the membrane (**5**). A sample is taken from the corresponding area of the culture (**6**). Bacteria from such colonies that have taken up the vectors are grown on agar in a Petri dish to produce multiple copies (clone) of the desired DNA fragments.

Further Reading

Lodish H, et al. Molecular Cell Biology. 6th ed. New York: W. H. Freeman; 2007
Strachan T, Read AP. Human Molecular Genetics. 4th ed. New York: Garland Science; 2011
Watson JD, et al. Recombinant DNA. 2nd ed. New York: Scientific American Books; 1992

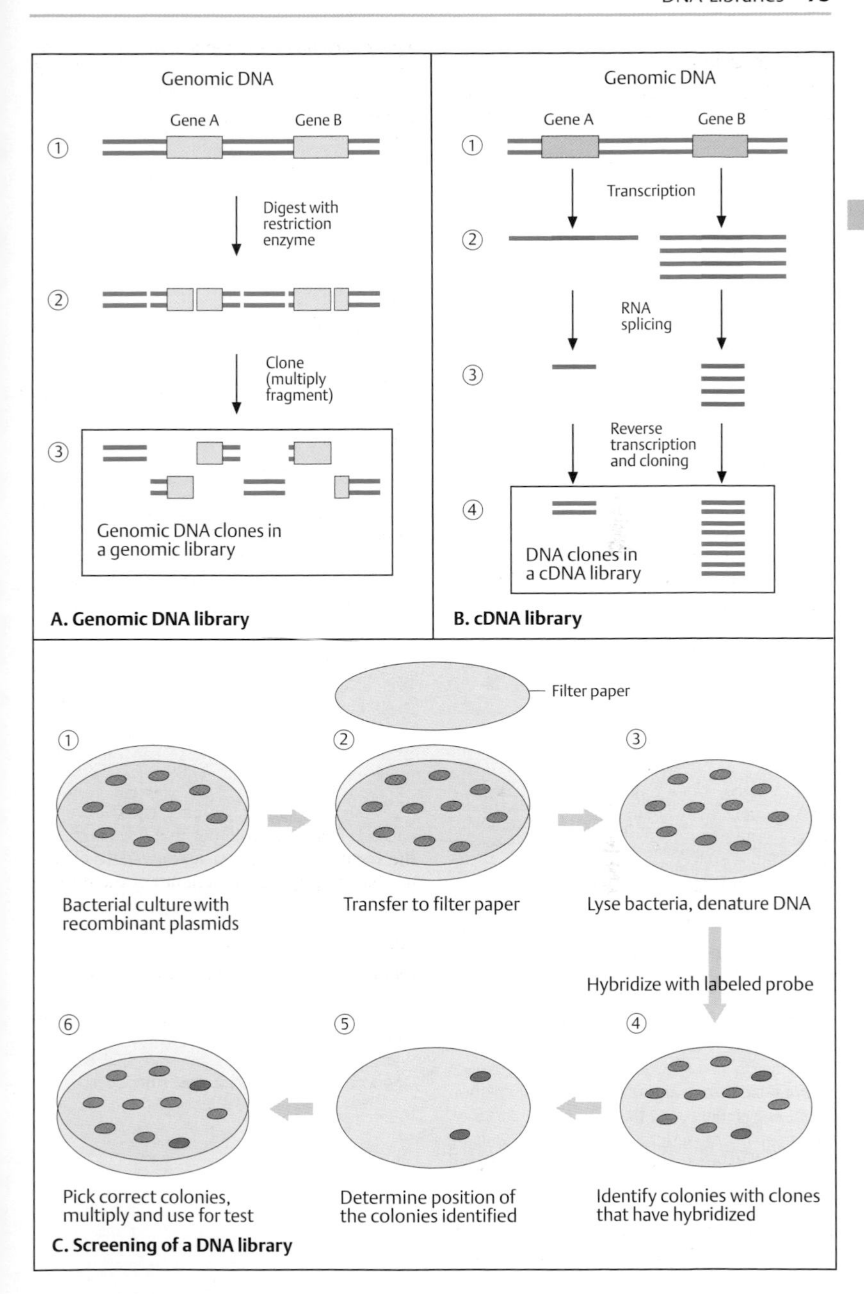

A. Genomic DNA library

B. cDNA library

C. Screening of a DNA library

Southern Blot Hybridization

Southern blot refers to a sensitive method of detecting one or more specific DNA fragments within a complex, random mixture of other DNA fragments. The procedure is named after E. M. Southern, who developed this method in 1975. A corresponding method for assaying RNA is called Northern blot hybridization (a word play on Southern). Immunoblotting (Western blot) detects proteins by an antibody-based procedure. Since each restriction enzyme cleaves DNA only at its specific recognition sequence, its recognition sites are distributed unevenly and the DNA fragments differ in size. This method today is rarely employed.

A. Principle of Southern blot hybridization

Total DNA is extracted from white blood cells or other cells (**1**). The DNA is isolated and digested with a restriction enzyme (**2**). The fragments are sorted by size in a gel (usually agarose) in an electric field, by using electrophoresis (**3**). The smallest fragments migrate fastest from the cathode to the anode; the largest fragments migrate slowest. Next, the blot is performed: the fragments contained in the gel are transferred to a nitrocellulose or nylon membrane (**4**). The DNA is denatured (made single stranded) with alkali and fixed to the membrane by moderate heating (80°C) or ultraviolet cross-linkage. The sample is incubated with a probe of single-stranded DNA (genomic DNA or cDNA) complementary to the region of the gene to be studied (**5**). The probe hybridizes solely with the complementary fragment being sought, and not with others. Since the probe is radiolabeled, the fragment being sought induces a signal on an X-ray film placed on the membrane. Here it becomes visible as a black band on the film after development (autoradiogram) (**6**).

B. Restriction fragment length polymorphism (RFLP)

Differences in individual DNA sequences result in differences in the location of the recognition sequences of restriction enzymes. This is apparent from differences in sizes of DNA fragments obtained by Southern hybridization (restriction fragment length polymorphism, RFLP). RFLP results when an individual difference of this type creates or eliminates a restriction enzyme recognition site. If an additional site is created, two smaller-than-usual fragments appear in the Southern blot hybridization. If a site is eliminated, one larger-than-usual fragment appears instead of two smaller ones.

An example is shown for a 5-kb (5000 base pairs) stretch of DNA. On the left it contains a restriction recognition site in the middle that is not present at the right. The first allele, with the polymorphic site, is arbitrarily called allele 1; the one at the right without this additional site is called allele 2.

Southern blot analysis distinguishes the two alleles. On the left, the restriction enzyme cleaves the 5-kb stretch into two fragments. A probe bridging the site will hybridize to both fragments: one 3 kb and the other 2 kb. On the left, a single 5-kb fragment will result. Thus, three possibilities, called the genotypes, can be distinguished: (i) two copies of allele 1, homozygous 1–1; (ii) one allele 1 and one allele 2, heterozygous; and (iii) two alleles 2, homozygous 2–2 (for explanations of the terms homozygous and heterozygous, see pp. 116–124).

This approach can be used for indirect detection of a disease-causing mutation. The RFLP itself is unrelated to the mutation. It simply distinguishes DNA fragments of different sizes from the same region. These can be used as markers within a family to determine who is likely to carry a disease-causing mutation and who is not.

Further Reading

Botstein D, et al. Construction of a genetic linkage map in man using restriction fragment length polymorphisms. Am J Hum Genet 1980;32:314–331

Brown TA. Genomes. 3rd ed. New York: Garland Science; 2007

Kan YW, Dozy AM. Antenatal diagnosis of sickle-cell anaemia by D.N.A. analysis of amniotic-fluid cells. Lancet 1978;2:910–912

Strachan T, Read AP. Human Molecular Genetics. 4th ed. New York: Garland Science; 2011

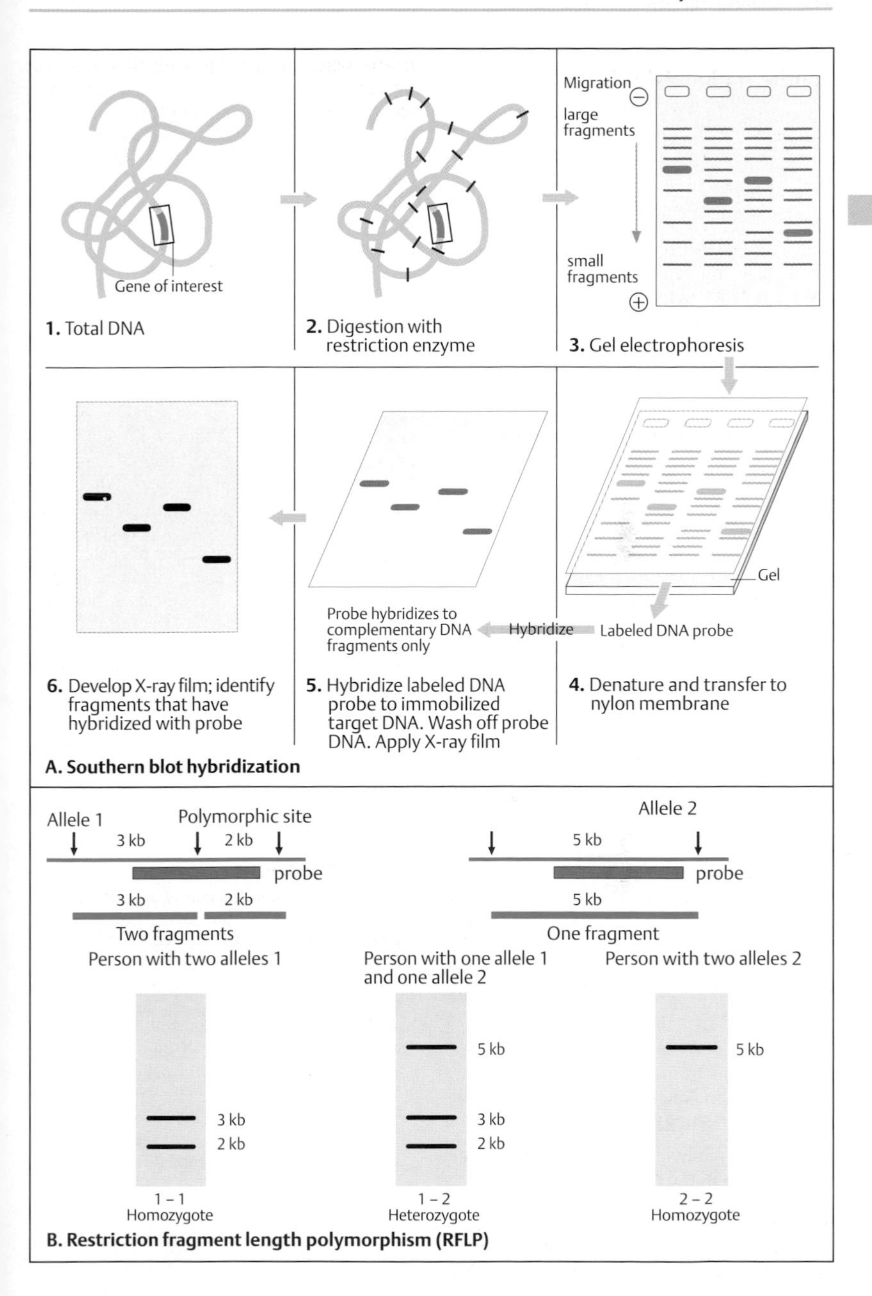

A. Southern blot hybridization

B. Restriction fragment length polymorphism (RFLP)

DNA Polymorphism

Each individual genome differs from any other by differences in the DNA nucleotide sequence (DNA polymorphism). On average, about one in 300 nucleotides is polymorphic. Three basic types of DNA polymorphisms occur: (i) a difference in a single nucleotide (single nucleotide polymorphism, SNP); (ii) small scale variants involving the presence or absence of repeated nucleotides (microsatellites); and (iii) tandem repeats of 10–50 nucleotides (minisatellites). The term satellite is derived from extra bands present in addition to the bulk of DNA when subjected to density gradient sedimentation in an ultracentrifuge.

A. Single nucleotide polymorphism (SNP)

Each SNP can be identified by a unique identifier, such as rs312476 (where rs stands for reference SNP, followed by a six-digit unique serial number). At any particular site, an SNP involving for example adenine (A) and guanine (G) can occur in one of three combinations: AA, AG, or GG. SNPs can be detected by methods based on a polymerase chain reaction (PCR; see p. 64), which does not require gel electrophoresis. (Figure adapted from Cichon et al., 2002.)

B. SNP, microsatellite, minisatellite

The three types of common DNA polymorphisms are shown: SNP, microsatellites, and minisatellites. Microsatellites are variable blocks of short tandem repeats (STRs) of nucleotide sequences. For example CA repeats can occur with three repeats (5′-CACACA-3′), five repeats, (5′-CACACACACA-3′), six repeats, etc. Each repeat defines an allele by its number of repeats, e.g., three and four.
Minisatellites (also called variable number of tandem repeats, VNTRs) consist of repeat units of 20–500 base pairs (bp). The size differences resulting from the number of repeats are determined by PCR. These allelic variants differing in the number of tandemly repeated short nucleotide sequences usually occur in noncoding DNA. They are referred to as repetitive DNA (see Part II, Genomics). (Figure adapted from Cichon et al., 2002.)

C. Genetic variability along a stretch of 100 000 bp

Along a typical stretch of DNA, most variability is represented by SNPs. Minisatellites are quite unevenly distributed and vary in density. (Figure adapted from Cichon et al., 2002.)

D. CEPH family

The inheritance patterns of DNA polymorphisms are best recognized in a collection of three-generation families with at least eight children in the third generation. DNA from such families has been collected by the Centre d'Étude du Polymorphisme Humain (CEPH) in Paris, now called the Centre Jean Dausset, after the founder. Immortalized cell lines are stored from each family. A CEPH family consists of four grandparents, the two parents, and eight children. The schematic figure shows the RFLP patterns of a family with four grandparents, two parents, and eight offspring. The four alleles present at a given locus analyzed by Southern blot are designated A, B, C, and D. Starting with the grandparents, the inheritance of each allele through the parents to the grandchildren can be traced. Of the four grandparents, three are heterozygous (AB, CD, BC) and one is homozygous (CC). Since the parents are heterozygous for different alleles (AD father, and BC mother), all eight children are heterozygous: BD, AB, AC, or CD.

Further Reading

Brown TA. Genomes. 3rd ed. New York: Garland Science; 2007

Cichon S, et al. Variabilität im menschlichen Genom. Dtsch Arztebl 2002;99:A3091–A3101

Dausset J, et al. Centre d'etude du polymorphisme humain (CEPH): collaborative genetic mapping of the human genome. Genomics 1990;6:575–577

Feuk L, Carson AR, Scherer SW. Structural variation in the human genome. Nat Rev Genet 2006;7:85–97 (with online links to databases)

Hinds DA, et al. Whole-genome patterns of common DNA variation in three human populations. Science 2005;307:1072–1079

Lewin B. Genes IX. Sudbury: Bartlett & Jones; 2008

Strachan T, Read AP. Human Molecular Genetics. 4th ed. New York: Garland Science; 2011

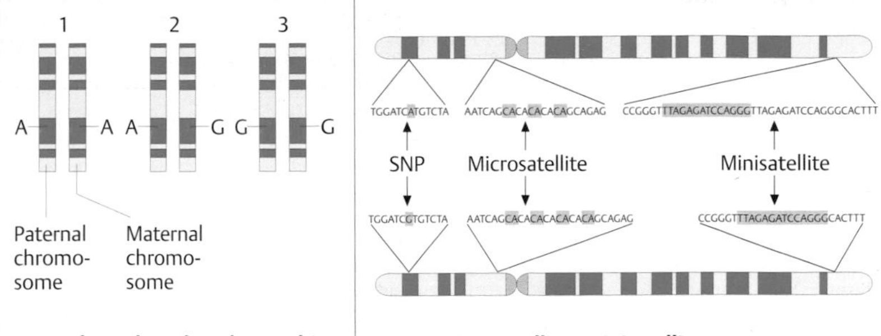

A. Single nucleotide polymorphism

B. SNP, microsatellite, minisatellite

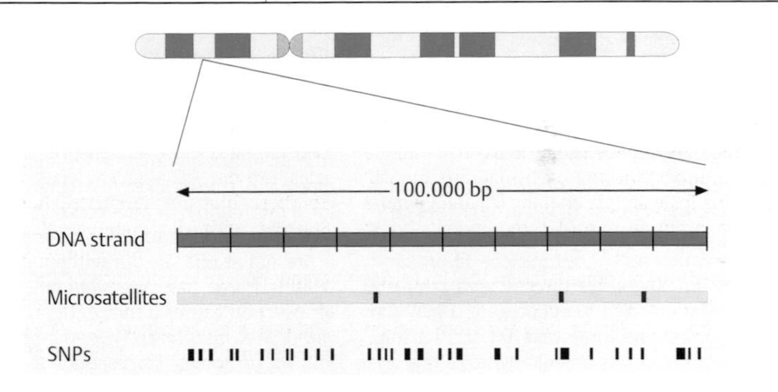

C. Genetic variability along a stretch of 100 000 bp

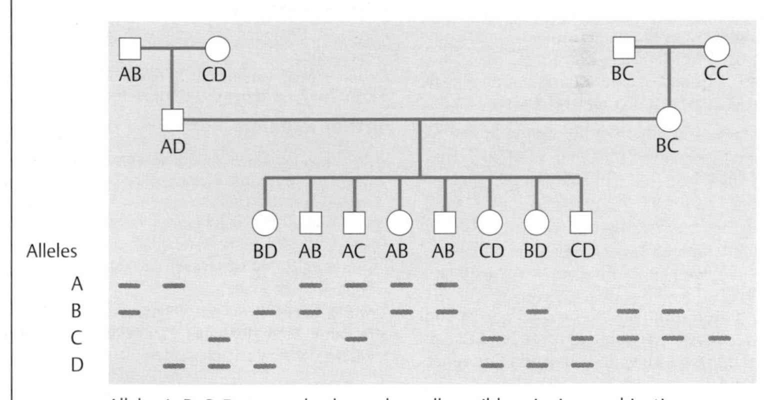

Alleles A, B, C, D at a marker locus show all possible pairwise combinations

D. CEPH family

Genes and Mutation

Mutation, a term introduced by H. de Vries in 1901, refers to a process by which a structural alteration changes the biological function of a gene. The discovery that mutations also occur in bacteria and other microorganisms paved the way to understanding how genes and mutations are related.

When it was recognized that changes (mutations) in genes occur not only spontaneously (T. H. Morgan in 1910), but can also be induced by X-rays (H. J. Muller in 1927), the mutation theory of heredity became a cornerstone of early genetics. The study of mutations is important for several reasons. Mutations cause diseases, including all forms of cancer; and without mutations, well-organized forms of life would not have evolved.

A. Transcription and translation in prokaryotes and eukaryotes

Transcription differs in unicellular organisms without a nucleus, such as bacteria (prokaryotes, **1**), and multicellular organisms (eukaryotes, **2**), which have a cell nucleus. In prokaryotes, the mRNA serves directly as a template for translation. The sequences of DNA and mRNA correspond in a strict 1:1 relationship, i.e., they are colinear. In eukaryotic cells a primary transcript of RNA is first formed. The mature mRNA is formed by removing noncoding sections from the primary transcript before it leaves the nucleus to act as a template for the synthesis of a polypeptide (RNA processing, p. 60).

B. Mutations have a defined site

The systematic analysis of mutations in microorganisms provided the first evidence that coding DNA and its corresponding polypeptides are colinear. Yanofsky et al. (1964) showed that the position of the mutation in the *Escherichia coli* (*E. coli*) gene encoding the protein tryptophan synthetase A corresponds to the position of the resulting change in the sequence of amino acids. Mutations are shown at four positions. At position 22, phenylalanine (Phe) is replaced by leucine (Leu); at position 49, glutamic acid (Glu) by glutamine (Gln); at position 177, Leu by arginine (Arg). Each mutation has a defined position. Whether it leads to incorporation of a different amino acid depends on how the corresponding codon has been altered. Different mutations at one position (one codon) in different DNA molecules are possible: two different mutations were observed at position 211: glycine (Gly) to arginine (Arg), and Gly to glutamic acid (Glu). Normally (in the wild-type), codon 211 is GGA and codes for glycine. A mutation of GGA to AGA leads to a codon for arginine; a mutation to GAA leads to a codon for glutamic acid.

C. Basic types of mutations

Three different types of mutation, i.e., changes from the usual or so-called wild-types, involving single nucleotides (point mutations) can be distinguished: (i) *substitution* (exchange of one nucleotide base for another, altering a codon); (ii) *deletion* (loss of one or more bases); and (iii) *insertion* (addition of one or more bases). Two types of substitution are distinguished: *transition* (exchange of one purine for another purine or of one pyrimidine for another pyrimidine) and *transversion* (exchange of a purine for a pyrimidine, or vice versa). A substitution may alter a codon so that a wrong amino acid is present at this site but without changing the reading frame (*missense mutation*). A deletion or insertion causes a shift of the reading frame (*frameshift mutation*). Thus, the sequence that follows no longer corresponds to the normal sequence of codons. No functional gene product is produced (*nonsense mutation*).

D. Different mutations at the same site

Different mutations may occur at the same site. In the example in **B** (position 211), glycine is replaced by either arginine or glutamic acid.

Further Reading

Alberts B, et al. Essential Cell Biology. An Introduction to the Molecular Biology of the Cell. New York: Garland Science; 1998

Alberts B, et al. Molecular Biology of the Cell. 5th ed. New York: Garland Science; 2008

Lodish H, et al. Molecular Cell Biology. 6th ed. New York: W. H. Freeman; 2007

Watson JD, et al. Molecular Biology of the Gene. 6th ed. New York: Cold Spring Harbor Laboratory Press; 2008

Yanofsky C, et al. On the colinearity of gene structure and protein structure. Proc Natl Acad Sci USA 1964;51:266–272

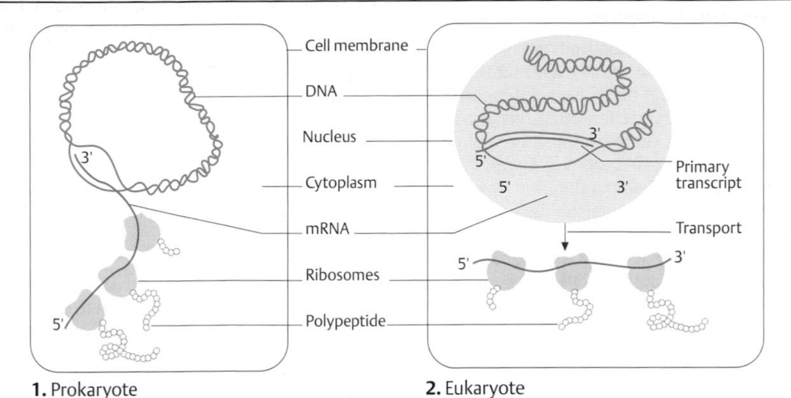

A. Transcription and translation in prokaryotes and eukaryotes

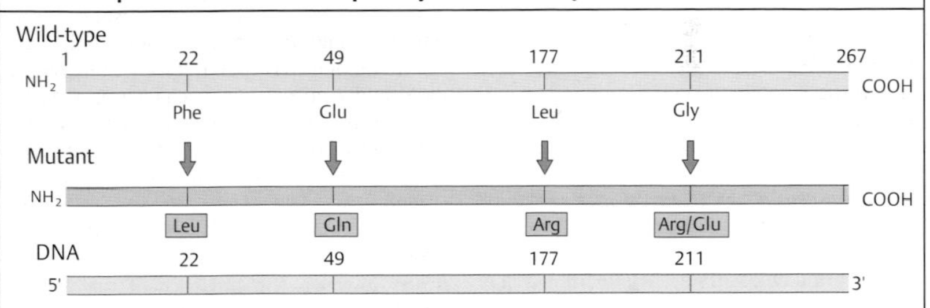

B. Mutations have a defined site

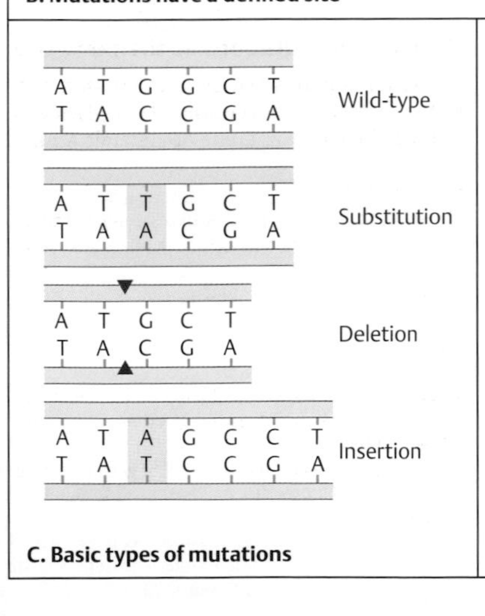

C. Basic types of mutations

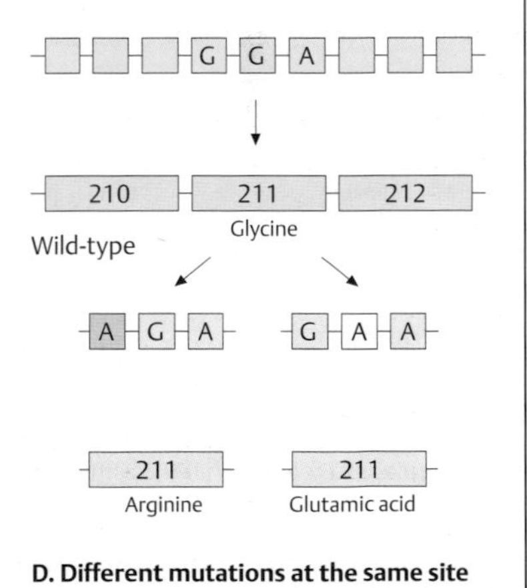

D. Different mutations at the same site

Mutations Due to Base Modifications

A mutation can result from a chemical or a physical event that leads to modification of a nucleotide base. If this affects the base-pairing pattern, it will interfere with replication or transcription. A chemical substance able to induce such changes is called a *mutagen*. Mutagens cause mutations in different ways. Spontaneous oxidation, hydrolysis, uncontrolled methylation, alkylation, and ultraviolet (UV) irradiation result in alterations that modify nucleotide bases. DNA-reactive chemicals change nucleotide bases into different chemical structures or remove a base.

A. Deamination and methylation

Cytosine, adenine, and guanine each contain an amino group. When this group is removed (deamination), a modified base with a different base-pairing pattern is the result. Nitrous acid typically removes the amino group. Deamination of cytosine removes the amino group at position 4 (**1**). The resulting molecule is uracil (**2**), and this pairs with adenine rather than guanine. Normally this change is efficiently repaired by uracil-DNA glycosylase. Methylation of the carbon atom at position 5 of cytosine results in 5-methylcytosine, which contains a methyl group at position 5 (**3**). Deamination of 5-methylcytosine will result in a change to thymine, containing an oxygen molecule at position 4 instead of an amino group (**4**) Methylation occurs at a rate of about 100 bases per cell per day (Alberts et al., 2008). This mutation will not be corrected because thymine is a natural base. Adenine (**5**) can be deaminated at position 6 to form hypoxanthine, which contains an oxygen molecule in this position instead of an amino group (**6**), and which pairs with cytosine instead of thymine. The resulting change after DNA replication is a cytosine instead of a thymine in the mutant strand.

B. Depurination

About 5000 purine bases (adenine and guanine) are lost per day from DNA in each cell (depurination) owing to thermal fluctuations (Alberts et al., 2008). Depurination of DNA involves hydrolytic cleavage of the *N*-glycosyl linkage of deoxyribose to the guanine nitrogen in position 9. This leaves a depurinated sugar. The loss of a base pair will lead to a deletion after the next replication if not repaired in time (see DNA Repair, p. 86).

C. Alkylation of guanine

Alkylation is the introduction of a methyl or an ethyl group into a molecule. The alkylation of guanine involves reaction with the ketone group at position 6 to form 6-methylguanine. This cannot form a hydrogen bond and thus is unable to pair with cytosine. Instead, it will pair with thymine. Thus, after the next replication the opposite cytosine (C) is replaced by a thymine (T) in the mutant daughter molecule. As a result, this molecule contains an abnormal GT pair instead of GC. Important alkylating agents are ethylnitrosourea (ENU), ethylmethane sulfonate (EMS), dimethylnitrosamine, and *N*-methyl-*N*-nitro-*N*-nitrosoguanidine.

D. Base analogues

Base analogues are purines or pyrimidines that are similar enough to the regular DNA nucleotide bases to be incorporated into the new strand during replication. 5-Bromodeoxyuridine (5-BrdU) is an analogue of thymine. It contains a bromine atom instead of the methyl group in position 5. Thus, it can be incorporated into the new DNA strand during replication. However, the presence of the bromine atom causes ambiguous and often incorrect base pairing.

E. UV-light-induced thymine dimers

Ultraviolet irradiation at 260 nm wavelength induces covalent bonds between adjacent thymine residues at carbon positions 5 and 6. If located within a gene, this will interfere with replication and transcription unless repaired. Another important type of UV-induced change is a photoproduct consisting of a covalent bond between the carbons in positions 4 and 6 of two adjacent nucleotides, the 4–6 photoproduct (not shown).

(Figures redrawn from Lewin, 2008.)

Further Reading

Alberts B, et al. Molecular Biology of the Cell. 5th ed. New York: Garland Science; 2008

Brown TA. Genomes. 3rd ed. New York: Garland Science; 2007

Lewin B. Genes IX. Sudbury: Bartlett & Jones; 2008

Strachan T, Read AP. Human Molecular Genetics. 4th ed. New York: Garland Science; 2011

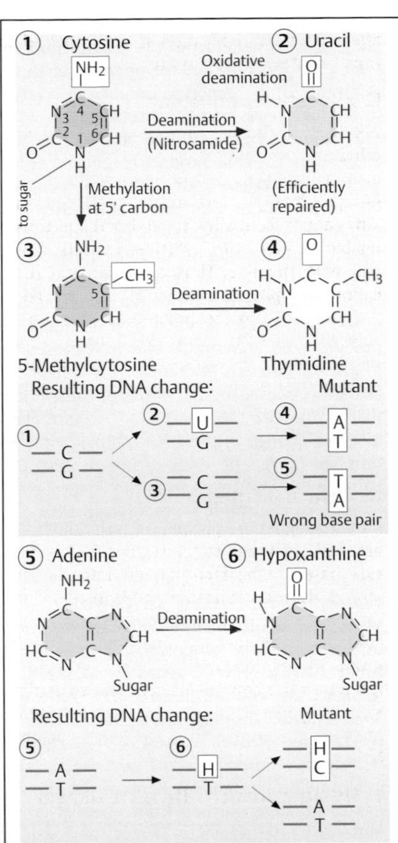

A. Deamination and methylation

Depurination

Deletion

Normal

Loss of base leads to deletion after next replication

B. Depurination

Guanine

O^6-Methyl-guanine

no hydrogen bond

Alkylation

Pairs with cytosine

Pairs with thymine

Resulting DNA change:

Mutant

C. Alkylation of guanine

Thymine

Adenine

Sugar

Change to base analogue

BrdU

Guanine

D. Base analogues

Ultraviolet irradiation forms thymine dimers with covalent bonds, distorts DNA. Corrected by excision repair

Thymine

Thymine

Thymine dimer

DNA sugar-phosphate backbone

E. UV-light-induced thymine dimers

Mutations Due to Errors in Replication

Errors occur during replication. If an error goes undetected or is not repaired, it will result in different types of mutations. Errors in replication occur at a rate of about 1 in 10^5. Proofreading and repair system (see p. 86) mechanisms reduce this rate to about 1 in 10^7–10^9. An important mechanism is replication slippage in regions with repeated nucleotide sequences.

A. Consequences of errors in replication

When an error in replication occurs before the next cell division, it might, for example, result in a cytosine (C) being incorporated instead of an adenine (A) at the fifth base pair, as shown here. If the error remains undetected, the next (second) division will result in a mutant molecule containing a CG instead of an AT pair at this position (blue field). This mutation will be perpetuated to all daughter cells. (Figure redrawn from Brown, 2007.)

B. Replication slippage

A different class of mutations does not involve an alteration of individual nucleotides, but results from incorrect alignment between allelic and nonallelic DNA sequences during replication. When the template strand contains short tandem repeats, e.g., CA repeats as in microsatellites (see DNA polymorphisms), the newly replicated strand and the template strand may shift their positions relative to each other (microsatellite instability). With replication or polymerase slippage, leading to incorrect pairing of repeats, some repeats are copied twice and others not at all, depending on the direction of the shift. Thus, one can distinguish forward slippage and backward slippage in relation to the newly replicated strand. Backward slippage of the new strand results in the addition (insertion) of nucleotides to the new strand. Forward slippage of the new DNA strand results in the loss (deletion) of nucleotides from the new DNA. (Figure redrawn from Brown, 2007.)

C. Functional consequences of mutations

Aside from their molecular type, mutations can be classified according to their functional consequences (molecular pathology). A principal goal is to understand the relationship between the genotype and the phenotype. This is referred to as *genotype/phenotype relation* (often incorrectly named correlation).

The main class is loss of function of a normal allele. Here the gene product is reduced or has no function. When both alleles are necessary for normal function, but one is inactivated by a mutation, the result is called *haploinsufficiency*. The opposite is an undesirable functional effect of a new gene product resulting from a mutation; this is called a *dominant negative effect*. Overexpression of a normal gene product with undesirable effects is caused by a gain-of-function mutation. An epigenetic change is caused by an alteration other than that of the DNA sequence, and is quite commonly a change in the DNA methylation pattern (see Epigenetic Modifications, p. 190). Dynamic mutations result from the abnormal expansion of nucleotide repeats (see p. 92).

Medical relevance

Mutations cause more than 3000 known individually defined diseases (see OMIM).

Microsatellite instability is a characteristic feature of hereditary nonpolyposis cancer of the colon (HNPCC). HNPCC genes are localized on human chromosomes at 2p22-p21, 2p16 and 3p21.3, 14q23.3, and others (OMIM 120435, 609310, and others). About 15% of all colorectal, gastric, and endometrial carcinomas show microsatellite instability. Replication slippage has to be distinguished from unequal crossing-over during meiosis. This is the result of recombination between adjacent sequences of homologous chromosomes

Further Reading

Brown TA. Genomes. 3rd ed. New York: Garland Science; 2007

Lewin B. Genes IX. Sudbury: Bartlett & Jones; 2008

OMIM. Online Mendelian Inheritance of Man. Available at: http://www.ncbi.nlm.nih.gov/omim

Pray L. DNA replication and causes of mutation. Nature Education 2008;1(1). Available at: www.nature.com/scitable/topicpage/dna-replication-and-causes-of-mutation-409. Accessed January 25, 2012

Strachan T, Read AP. Human Molecular Genetics. 4th ed. New York: Garland Science; 2011

Vogel F, Rathenberg R. Spontaneous mutation in man. Adv Hum Genet 1975;5:223–318

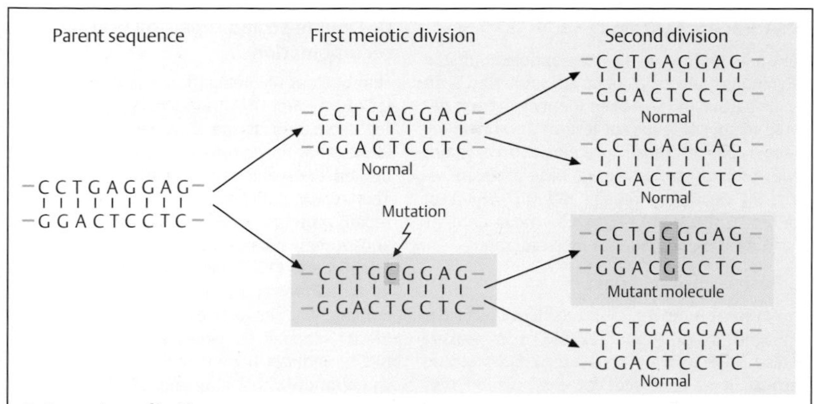

A. Errors in replication

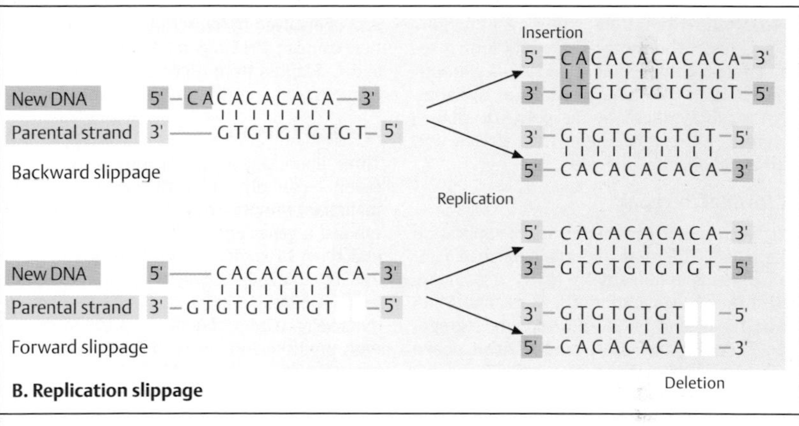

B. Replication slippage

	Two alleles	Genetic mechanisms involved
1. Normal		
2. Loss of function		Type of change causing loss of function:
Haploinsufficiency		• Frameshift resulting from deletion /insertion • Premature termination codon
Dominant negative effect		• Nonsense-mediated mRNA decay • Alteration of splice site • Interference with normal gene product
3. Gain of function		Undesirable overexpression of an allele
4. Epigenetic change		Change in DNA methylation pattern
5. Dynamic mutations		Expansion of nucleotide triplet repeats

C. Functional consequences of mutations

DNA Repair Systems

DNA damage can result from endogenous and exogenous sources. At least six multistep DNA repair pathways have been identified, each devoted to specific types of lesions and involving many types of repair proteins. Three major types of DNA repair are: (i) base-excision repair, (ii) mismatch repair, and (iii) repair of double-strand breaks. Double-strand damage is a common consequence of γ-radiation.

A. Base-excision repair

This system removes 27–29 nucleotides of the damaged strand (about 12 or 13 in prokaryotes). It recognizes a damaged DNA strand because it is distorted, for example by thymine–thymine dimers induced by ultraviolet light. Three proteins, XPA, XPB, and XPC in human cells (UvrA, UvrB, and UvrC in prokaryotes), detect the damaged site and form a repair protein complex. XP-C and XP-G endonucleases cleave the damaged strand at two sites. DNA repair synthesis by the poly(ADP-ribose) polymerase 1 restores the missing stretch, and a DNA ligase closes the gap.

B. Mismatch repair

Mismatch repair corrects errors of replication (see p. 84). The most important mismatch repair proteins in humans are MSH1, 2, 3, and 6, MLH1 (homologous to MutH, MutL, and MutS in bacteria), and PMS2 (postmeiotic segregation 2). MSH2 and MLH1 bind to mismatched base pairs, while others cleave DNA and remove the strand with erroneous bases. DNA polymerase III replaces the damaged strand.

C. Double-strand repair by non-nomologous end joining (NHEJ)

This type of repair joins the two ends of the DNA following a double-strand break. The ends bind to specific dimeric proteins of 80 and 90 kDa (Ku80 and Ku90). Additional repair proteins form a repair complex (Artemis, a DNA double-strand repair/V(D)J recombination protein that is mutated in human severe combined immune deficiency; a DNA-dependent protein kinase [DNA-PK]; and XRCC4 [X-ray repair complementing defective repair]). In this type of repair, bases may be lost or added. Several polymerases bypass lesions during replication, and these accumulate as a result.

D. Double-strand repair by homologous recombination

Homologous recombination acts through a series of complex DNA transactions, as shown for recombination at the DNA level (p. 88). The identical sister chromatid is used to align the broken ends and insert missing information. This repair pathway requires several central repair proteins: ATM, BRCA1, BRCA2, RAD51, and others, such as the Fanconi anemia proteins (see p. 312). ATM, a member of a protein kinase family, is activated in response to DNA damage (**1**). Its active form phosphorylates BRCA1 at specific sites (**2**). Phosphorylated BRCA1 induces homologous recombination in cooperation with BRCA2 and RAD51, the mammalian homolog of the *Escherichia coli* (*E. coli*) RecA repair protein (**3**). Phosphorylated BRCA1 is also involved in transcription and transcription-coupled DNA repair (**4**). (Figures in A, B, and C adapted from Alberts et al., 2008; in D adapted from Ventikaraman, 1999.)

Medical relevance

Three important heterogeneous groups of excision repair diseases with increased risk for malignant tumors (see p. 320) result from mutations in genes encoding repair proteins: several types of xeroderma pigmentosum (OMIM 278700–278780), Cockayne syndrome types A and B (OMIM 216400, 133540), and trichothiodystrophy (OMIM 601675). Other disorders with predisposition to cancer are ataxia-telangiectasia (see p. 318) and hereditary predisposition to breast cancer (BRCA1 and BRCA2, see p. 310).

Further Reading

Alberts B, et al. Molecular Biology of the Cell. 5th ed. Garland Science; New York; 2008

Cleaver JE, Lam ET, Revet I. Disorders of nucleotide excision repair: the genetic and molecular basis of heterogeneity. Nat Rev Genet 2009;10:756–768

Hoeijmakers JHJ. DNA damage, aging, and cancer. N Engl J Med 2009;361:1475–1485

O'Driscoll M, Jeggo PA. The role of double-strand break repair - insights from human genetics. Nat Rev Genet 2006;7:45–54

Venkitaraman AR. Breast cancer genes and DNA repair. Science 1999;286:1100–1102

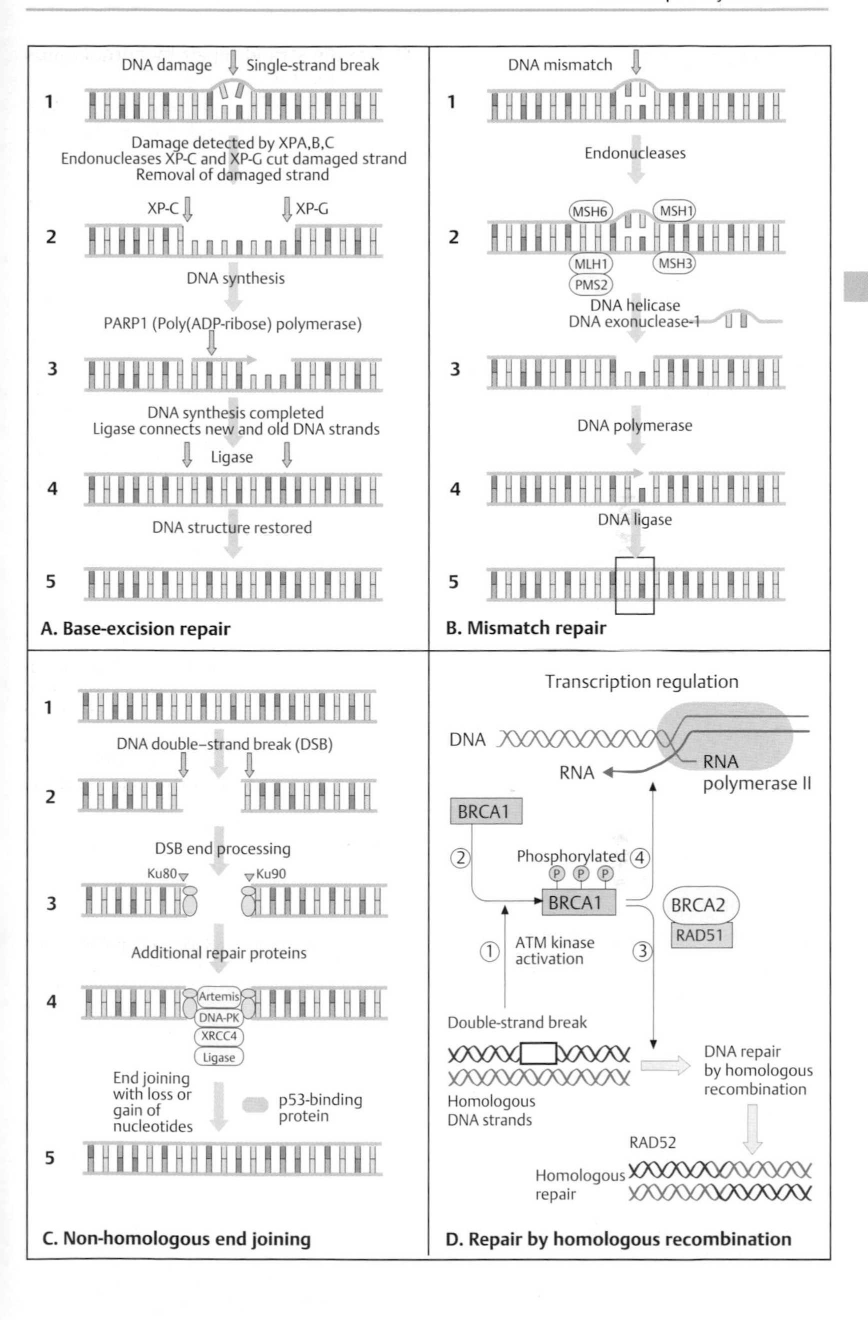
DNA damage
Single-strand break
1
Damage detected by XPA,B,C
Endonucleases XP-C and XP-G cut damaged strand
Removal of damaged strand
XP-C
XP-G
2
DNA synthesis
PARP1 (Poly(ADP-ribose) polymerase)
3
DNA synthesis completed
Ligase connects new and old DNA strands
Ligase
4
DNA structure restored
5
A. Base-excision repair
DNA mismatch
1
Endonucleases
MSH6
MSH1
2
MLH1
MSH3
PMS2
DNA helicase
DNA exonuclease-1
3
DNA polymerase
4
DNA ligase
5
B. Mismatch repair
1
DNA double–strand break (DSB)
2
DSB end processing
Ku80
Ku90
3
Additional repair proteins
4
Artemis
DNA-PK
XRCC4
Ligase
End joining
with loss or
gain of
nucleotides
p53-binding
protein
5
C. Non-homologous end joining
Transcription regulation
DNA
RNA
RNA
polymerase II
BRCA1
2
Phosphorylated
4
P P P
BRCA1
BRCA2
RAD51
1
ATM kinase
activation
3
Double-strand break
DNA repair
by homologous
recombination
Homologous
DNA strands
RAD52
Homologous
repair
D. Repair by homologous recombination

Recombination

Genetic recombination is an exchange between two homologous DNA molecules. Recombination provides the means to restructure genetic information. It confers an evolutionary advantage by helping to eliminate unfavorable mutations, maintain and spread favorable mutations, and endow each individual with a unique set of genetic information.

Recombination must occur between precisely corresponding sequences (homologous recombination) to ensure that not one base pair is lost or added. Two types of recombination can be distinguished: (i) generalized or homologous recombination, which occurs at meiosis in eukaryotes (see p. 102); and (ii) site-specific (specialized) recombination. A third process, transposition, utilizes recombination to insert one DNA sequence into another without regard to sequence homology (see p. 90). The examples here show homologous recombination, which is a complex biochemical reaction between two duplexes of DNA. The enzymes that are required to break and rejoin the DNA strands are not described. Two general models are described: recombination initiated from single-strand DNA breaks and recombination initiated from double-strand breaks.

A. Single-strand breaks

This model for homologous recombination assumes that the process starts with breaks at corresponding positions at each of one of the strands of homologous DNA (same sequences of different parental origin, shown in blue and red) (**1**). A nick is made in each molecule by a single-strand-breaking enzyme (endonuclease) at corresponding sites (**2**). This allows the free ends of one nicked strand to join with the free ends of the other nicked strand from the other molecule, and allows a single-strand exchange between the two duplex molecules at the recombination joint (**3**). The recombination joint moves along the duplex, a process called branch migration (**4**). This ensures a sufficient distance for a second nick at another site in each of the two strands (**5**). After the two other strands have joined and gaps have been sealed (**6**), a reciprocal recombinant molecule is generated (**7**).

Recombination involving DNA duplexes requires topological changes, i.e., either the molecules have to be free to rotate or the restraint has to be relieved in some other way. This structure is called a Holliday structure (not shown), first described in 1964. This model has an unresolved difficulty: how is it ensured that the single-strand nicks shown in step **2** occur at precisely the same position in the two double helix DNA molecules?

B. Double-strand breaks

The current model for recombination is based on initial double-strand breaks in one of the two homologous DNA molecules (**1**). Both strands are cleaved by an endonuclease. The break is enlarged to a gap by an exonuclease. It removes the new 5′ ends of the strands at the break and leaves 3′ single-stranded ends (**2**). One free 3′ end recombines with a homologous strand of the other molecule, generating a D loop (displacement) (**3**). This consists of a displaced strand from the "donor" duplex. The D loop is extended by repair synthesis from the 3′ end (**4**). The displaced strand anneals to the single-stranded complementary homologous sequences of the recipient strand and closes the gap (**5**). DNA repair synthesis from the other 3′ end closes the remaining gap (**6**). In contrast to the single-strand exchange model, the reciprocal double-strand breaks result in heteroduplex DNA in the entire region that has undergone recombination (**7**).

Double-strand breaks occur in meiosis (see p. 102) and DNA repair (see p. 86). A disadvantage of this model is the temporary loss of information in the gaps after the initial cleavage. However, the ability to retrieve this information by resynthesis from the other duplex avoids permanent loss.

(Figures adapted from Lewin, 2008.)

Further Reading

Alberts B, et al. Molecular Biology of the Cell. 5th ed. New York: Garland Science; 2008

Brown TA. Genomes. 3rd ed. New York: Garland Science; 2007

Holliday R. A mechanism for gene conversion in fungi. Genet Res 1964;5:282–304

Kanaar R, Hoeijmakers JH. Genetic recombination. From competition to collaboration. Nature 1998;391:335–337, 337–338

Lewin B. Genes IX. Sudbury: Bartlett & Jones; 2008

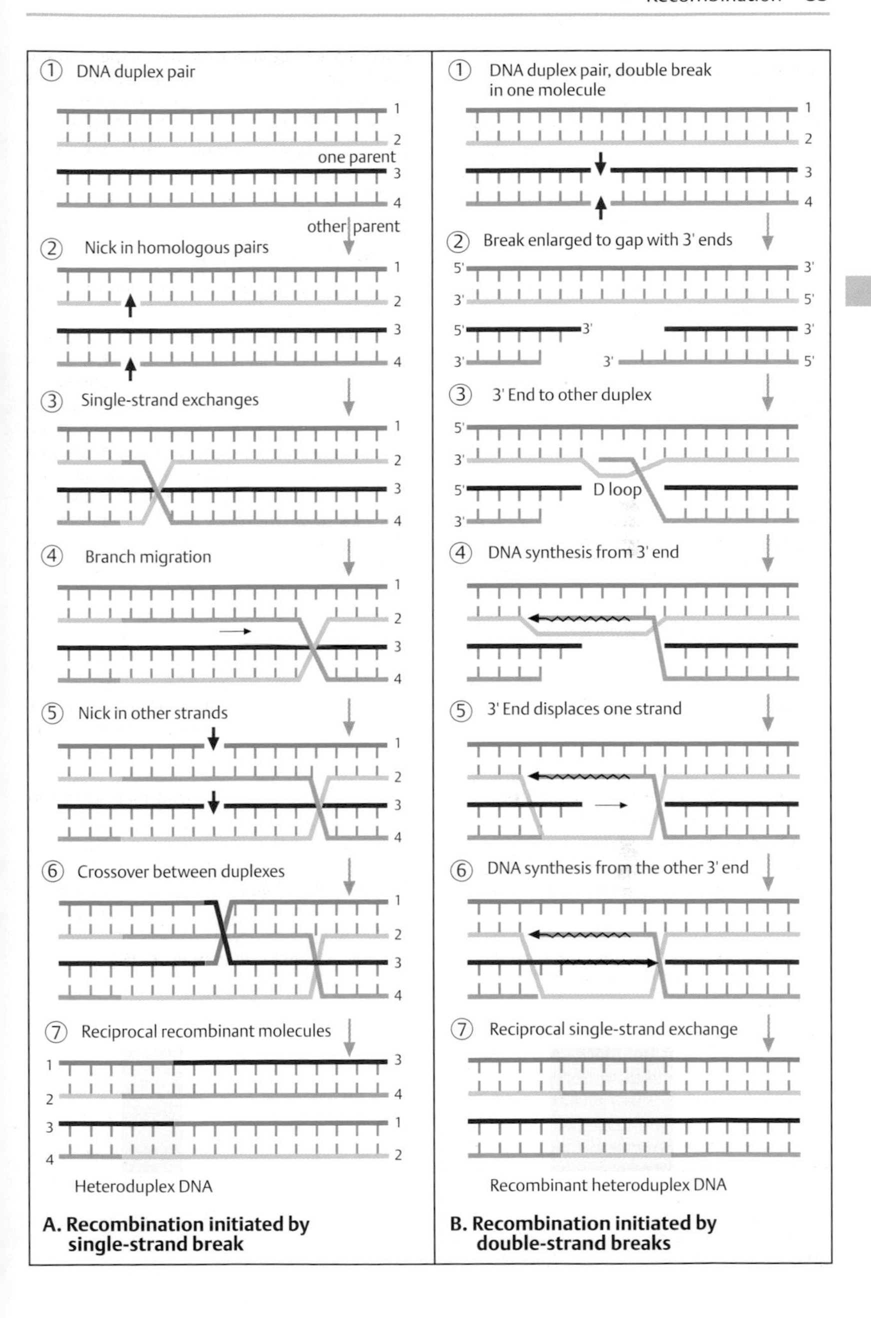

A. Recombination initiated by single-strand break

B. Recombination initiated by double-strand breaks

Transposition

Transposition refers to a DNA sequence that inserts itself at a new location in the genome. Transposons are a major source of genetic variation. They play an important role in the evolution of genomes. Tranposition utilizes recombination, but does not result in an exchange. Instead, a transposon moves directly from one site of the genome to another without an intermediary, such as plasmid or phage DNA. This results in rearrangements that create new sequences and change the functions of target sequences. In some cases, they cause disease when inserted into a functioning gene. Three examples of the different classes of transposons are presented.

A. Insertion sequences (IS) and transposons (Tn)

The host DNA contains a target site of about 4–10 base pairs (bp) (**1**). The selection of the target site of the host DNA is either random or selective for particular sites. The insertion sequence (IS) consists of about 700–1500 bp, depending on the particular class. It contains a transposase gene encoding the enzyme responsible for transposition of mobile sequences. It is flanked by inverted repeats of about 9 bp at both ends. This is a characteristic feature of IS transposition. The IS inserts itself at the target site by means of the transposase activity (**2**). Transposons (Tn) may contain other genes, such as those for antibiotic resistance, and have direct (**3**) or inverted (**4**) repeats at either end. Direct repeats are identical or closely related sequences oriented in the same direction. Inverted repeats are oriented in opposite directions.

B. Replicative and nonreplicative transposition

In replicative transposition (**1**) the donor transposon remains in place and creates a new copy of itself, and this inserts into a recipient site elsewhere. This mechanism leads to an increase in the number of copies of the transposon in the genome. It involves two enzymatic activities: a transposase, acting on the ends of the original transposon, and resolvase, acting on the duplicated copies. In nonreplicative transposition (**2**) the transposing element itself moves as a physical entity directly to another site.

C. Transposition of retroelements

Retrotransposition requires synthesis of an RNA copy of the inserted retroelement. Retroviruses, including the human immunodeficiency virus and RNA tumor viruses, are important retroelements. The first step in retrotransposition is the synthesis of an RNA copy of the inserted retroelement, followed by reverse transcription up to the polyadenylation sequence in the 3′ long terminal repeat (LTR). Three important classes of mammalian transposons that undergo or have undergone retrotransposition through an RNA intermediary are shown. Endogenous retroviruses (**1**) are sequences that resemble retroviruses but cannot infect new cells and are restricted to one genome. Nonviral retrotransposons (**2**) lack LTRs and usually other parts of retroviruses. Both types contain reverse transcriptase and are therefore capable of independent transposition. Processed pseudogenes (**3**) or retropseudogenes lack reverse transcriptase and cannot transpose independently. They contain two groups: low copy number of processed pseudogenes transcribed by RNA polymerase II; and high copy number of mammalian SINE sequences, such as human *Alu* and the mouse B1 repeat families. One in 600 mutations is estimated to arise from retrotransposon-mediated insertion (Prak and Kazazian, 2000).
(Figures adapted from Lewin, 2008, and Brown, 2007.)

Further Reading

Brown TA. Genomes. 3rd ed. New York: Garland Science; 2007

Lewin B. Genes IX. Sudbury: Bartlett & Jones; 2008

Lodish H, et al. Molecular Cell Biology. 6th ed. New York: W. H. Freeman; 2007

Mills RE, et al. Which transposable elements are active in the human genome? Trends Genet 2007;23:183–191

Prak ET, Kazazian HH. Mobile elements and the human genome. Nature Rev Genet 2000;1:134–144

Strachan T, Read AP. Human Molecular Genetics. 4th ed. New York: Garland Science; 2011

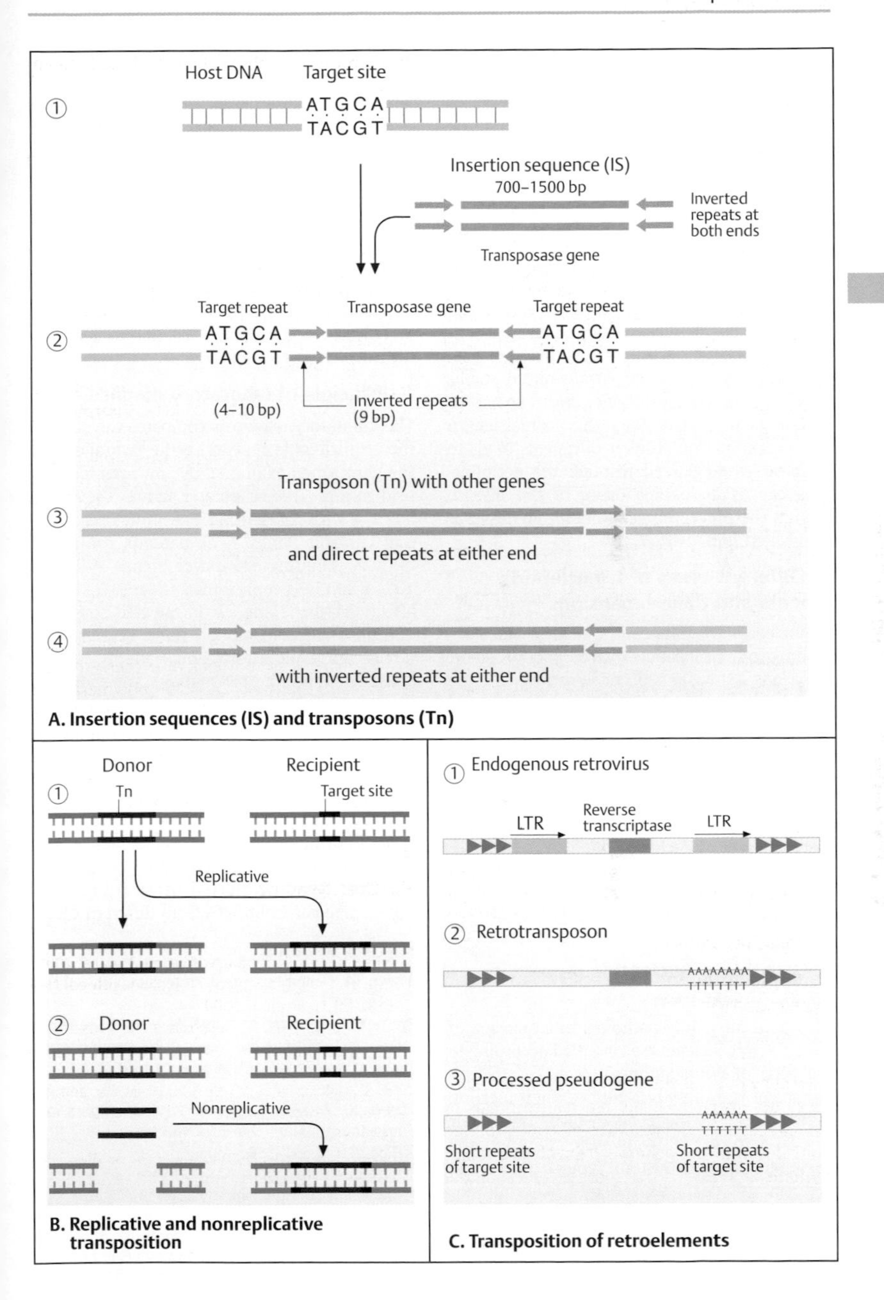

A. Insertion sequences (IS) and transposons (Tn)

B. Replicative and nonreplicative transposition

C. Transposition of retroelements

Trinucleotide Repeat Expansion

A new class of mutations was discovered in 1991. These are pathogenic microsatellite variants that may become unstable (*unstable mutations*). The human genome contains numerous tandem repeats of three nucleotides (trinucleotides or triplets), or more than three nucleotides. These can expand abnormally within or near certain genes, interfere with gene expression, and cause disease depending on the gene involved (*trinucleotide expansion disorders*). Normally triplets usually occur in groups of 5–35 repeats, but this varies depending on the site. Although usually transmitted stably, they can become unstable and expand to pathological lengths. Once the normal length has expanded, the number of repeats tends to increase when passed through the germline. This causes an earlier onset of the disease than in the preceding generations, an observation called *anticipation*.

A. Different types of trinucleotide repeats and their expansions

Trinucleotide repeats can be distinguished according to their location with respect to a gene. Very long expansions occur in introns (**1**). The increase in the number of these repeats can be drastic, up to 1000 or more repeats. The first stages of expansion do not usually lead to clinical signs of a disease, but they do predispose to increased expansion of the repeat in the offspring of a carrier (premutation). Within the coding regions, in exons, expansions are more moderate (**2**). However, their effect is dramatic, as seen in several severe neurological diseases, because they result in expanded glutamine tracts.

B. Unstable trinucleotide repeats in different diseases

Disorders caused by pathological expansion of trinucleotide repeats are classified according to the type of trinucleotide repeat, i.e., the sequence of the three nucleotides, their location with respect to the gene involved, and their clinical features. All involve the central or the peripheral nervous system. Type I trinucleotide diseases are characterized by CAG trinucleotide expansions within the coding regions of different genes. The triplet CAG encoding glutamine usually has about 20 CAG repeats. Thus, about 20 glutamine residues occur in the gene product. In the disease state, the number of glutamine residues is greatly increased in the protein. Hence, the diseases are collectively referred to as polyglutamine disorders.

Type II trinucleotide diseases are characterized by expansions of CTG, GAA, GCC, or CGG trinucleotides within a noncoding region of the gene involved at the 5′ UTR (untranslated region of exon 1 [CGG in fragile X syndrome type A, FRAXA]), at the 3′ end (CGG in FRAXE; CTG in myotonic dystrophy), or in an intron (GAA in Friedreich ataxia). A brief review of these disorders is given on p. 364.

C. Principle of laboratory diagnosis

The laboratory diagnosis compares the sizes of the trinucleotide repeats in the two alleles of the gene when examined by Southern blot hybridization. The schematic figure shows 11 lanes, each representing one individual: normal controls (lanes 1–3), patients confirmed to have Huntington chorea (lanes 4–7 and 10). A family is represented in lanes 7–11: an affected father (lane 7), an affected son (lane 10), an unaffected mother (lane 11), and two unaffected children (a son, lane 8; and a daughter, lane 9). Size markers are shown at the left. Each lane represents a polyacrylamide gel and the (CAG) repeat of the Huntington locus amplified by polymerase chain reaction is shown as a band of defined size. The two alleles are shown for each individual. In the affected individuals, the band representing the abnormal allele is located above the threshold in the expanded region (in practice the bands may be blurred because the exact repeat size varies in DNA from different cells).

Further Reading

Brown TA. Genomes. 3rd ed. New York: Garland Science; 2007

Kremer EJ, Pritchard M, Lynch M, et al. Mapping of DNA instability at the fragile X to a trinucleotide repeat sequence p(CCG)n. Science 1991;252:1711–1714

McIvor EI, Polak U, Napierala M. New insights into repeat instability: role of RNA.DNA hybrids. RNA Biol 2010;7:551–558

Oberle I, et al. Instability of a 550-base pair DNA segment and abnormal methylation in fragile X syndrome. Science 1991;252:1097–1102

Rosenberg RN. DNA-triplet repeats and neurologic disease. N Engl J Med 1996;335:1222–1224

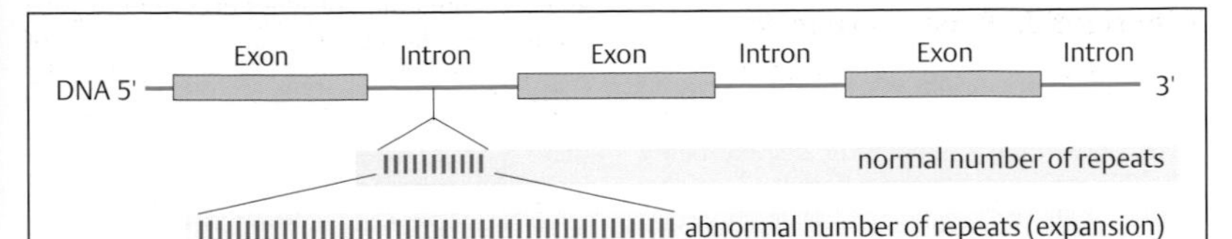

1. Very large expansions of repeats outside coding sequences

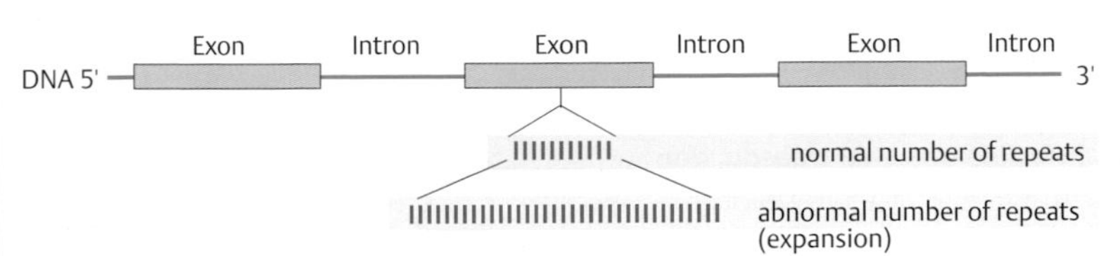

2. Modest expansion of CAG repeats within coding sequences

A. Different types of trinucleotide repeat expansion

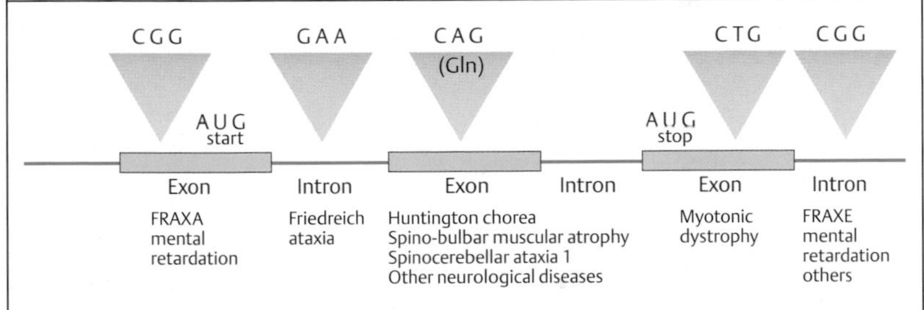

B. Unstable trinucleotide repeats in different diseases

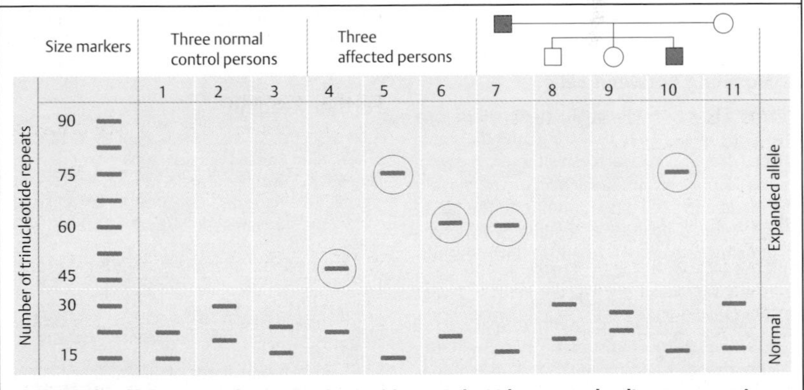

C. Principle of laboratory diagnosis of unstable trinucleotide repeats leading to expansion

Cell Communication

Multicellular organisms use extensive systems of cell-to-cell communication for processes such as growth (e.g., embryonic development), cell differentiation into the different types, regulation of the cell cycle, and other important functions. Cells communicate via a vast variety of extracellular signal proteins that mediate cell-specific responses, some over a long distance. These proteins include extracellular signal molecules, cell surface receptors, intracellular receptors, and intracellular signal molecules that transmit signals.

A. Principle of signal transduction

The transduction of a signal elicits a cell-specific effect involving several types of interacting proteins. A cell membrane-bound receptor, consisting of an extracellular and an intracellular portion (called domains), responds to a signal molecule. The specificity of the response is achieved by the binding of specific signal molecule (called a ligand) to the extracellular domain of the receptor. This in turn activates a series of further signaling proteins downstream. Generally, one activated protein activates the next by a specific biochemical reaction, called a signaling cascade. Only two are shown here (designated signaling proteins 1 and 2), but quite often many more are involved. The final steps of a signaling cascade reach the target protein and elicit the desired cellular response. Extracellular signal molecules typically act at very low concentrations, at about 10^{-8} molar concentration. (Figure adapted from Alberts et al., 2008.)

B. Signaling between cells

Different types of cell signals exist. Often they form a signal pathway that transmits the signal through several relays to the target. A signal molecule may remain attached to the surface of the signaling cell when binding to the target cell (**1**, *contact-dependent signaling*). This type of signaling is common in embryonic development and in the immune system. In (**2**), *paracrine signaling*, the signal molecule is secreted and released into the extracellular space. Here it acts over a short distance on cells in the neighborhood. Many growth and differentiation factors act in this manner. Some important long-distance signaling is mediated via hormones in the blood circulation (**3**, *endocrine signaling*). In this type of signaling, specialized cells, called *endocrine cells*, secrete a substance, called a *hormone*, into the bloodstream. From there it can reach the target cells at a distance in another part of the body. *Synaptic signaling* refers to nerve cells or the junction of nerve and muscle cells (**4**). In this case, a specialized cell, a nerve cell (neuron), sends electrical impulses along a cellular extension, the *axon*, which can be quite long. At the end of the axon, a chemical signal, called a neurotransmitter, is secreted at the junction (the synapse) between the signaling cell (the neuron) and the postsynaptic target cell. In some cases the same types of signaling molecules are used in paracrine, endocrine, and synaptic signaling, but in different contexts of selectivity. (Figure adapted from Lodish et al., 2007.)

The term hormone is derived from a Greek word meaning to spur on. It was first used in 1904 by William Bayliss and Ernest Starling to describe the action of a secreted molecule (Stryer, 1995, p. 342). Five major classes of hormones can be defined: (i) amino acid derivatives (e.g., catecholamine, dopamine, thyroxine); (ii) small neuropeptides (e.g., thyrotropin-releasing hormone, somatostatin, vasopressin); (iii) proteins (e.g., insulin, luteinizing hormone); (iv) steroid hormones, derived from cholesterol (e.g., cortisol, sex hormones); and (v) vitamin derivatives (e.g., retinoids [vitamin A], peptide growth hormones).

Medical relevance

Mutations in the genes encoding proteins involved in signal transduction cause a vast array of human genetic disorders (Jameson, 2011).

Further Reading

Alberts B, et al. Molecular Biology of the Cell. 5th ed. New York: Garland Science; 2008

Jameson JL. Principles of endocrinology. In: Longo DL, et al., eds. Harrison's Principles of Internal Medicine. 18th ed. New York: McGraw-Hill; 2012:2866–2875

Lewin B. Genes IX. Sudbury: Bartlett & Jones; 2008

Lodish H, et al. Molecular Cell Biology. 6th ed. New York: W. H. Freeman; 2007

Mapping Cellular Signaling. Special Issue Science 2002;296:1557–1752

Stryer L. Biochemistry. 4th ed. New York: W. H. Freeman; 1995.

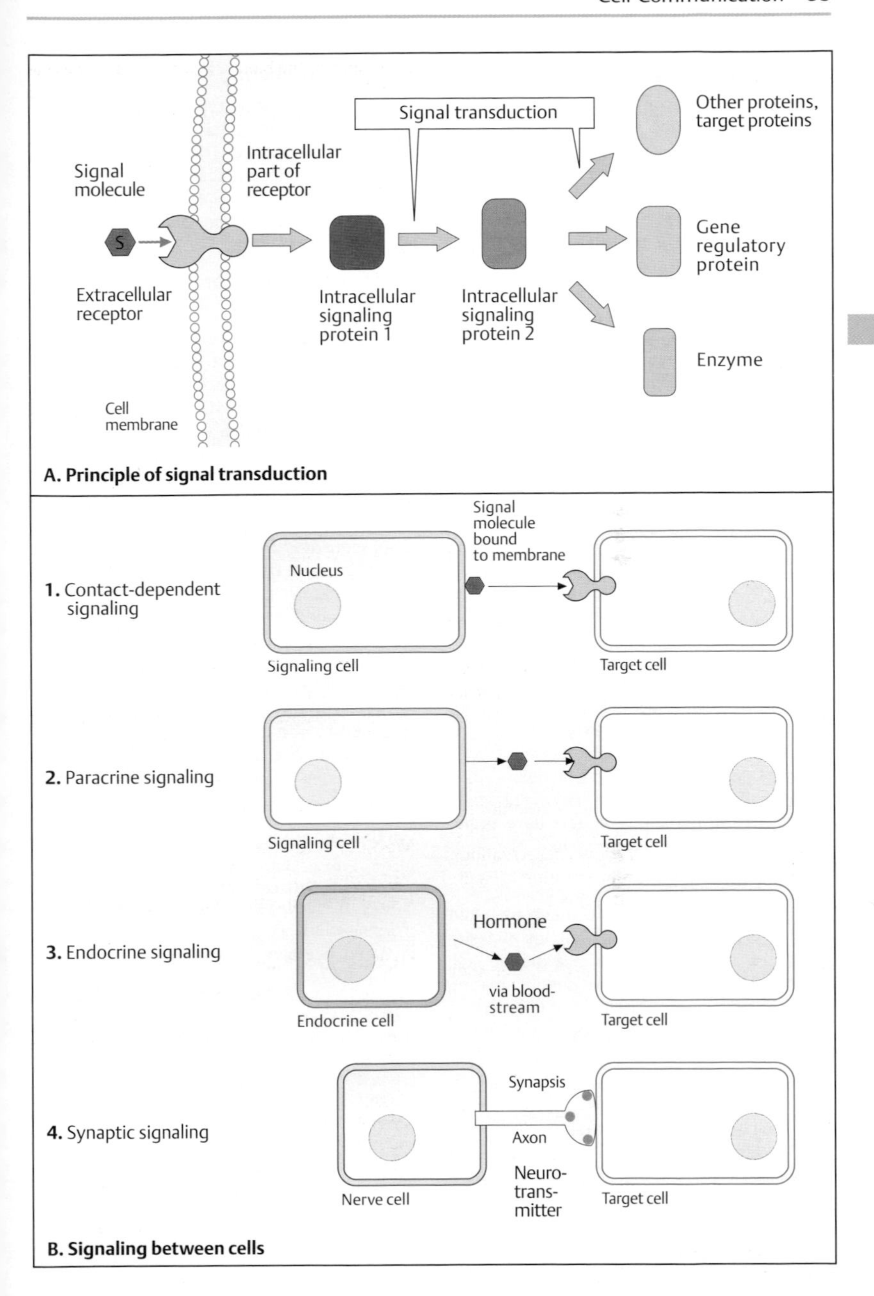

A. Principle of signal transduction

B. Signaling between cells

Haploid and Diploid Yeast Cells

Budding yeast, the common baker's yeast (*Saccharomyces cerevisiae*), is an important model organism that is used widely in genetics. It is easy to grow and manipulate. It can differentiate into three distinct cell types: two haploid and one diploid. It is a single-celled eukaryotic fungus with a genome of individual linear chromosomes enclosed in a nucleus and with cytoplasmic organelles such as endoplasmic reticulum, Golgi apparatus, mitochondria, peroxisomes, and a vacuole analogous to a lysosome. About 40 different types of yeast are known. Baker's yeast consists of oval cells about 5 µm diameter. Under good nutritional conditions, a cell can divide by budding every 90 minutes. Fission yeast, *Schizosaccharomyces pombe*, has rod-shaped cells that divide by elongation at the ends.

The haploid genome of *S. cerevisiae* contains about 6200 genes in 1.4×10^7 DNA base pairs distributed in 16 chromosomes (Goffeau et al., 1996). The genes are involved in the following functions: cell structure, 250 genes (4%); DNA metabolism, 175 genes (3%); transcription and translation, 750 genes (13%); energy production and storage, 175 genes (3%); biochemical metabolism, 650 genes (11%); and transport, 250 genes (4%). The *S. cerevisiae* genome is very compact compared with other eukaryotic genomes, with about one gene every 2 kb. Nearly half of the human proteins known to be defective in hereditary disease have amino acid similarities to a yeast protein.

A. Yeast life cycle

The life cycle of yeast passes through a haploid or a diploid phase. Haploid cells of opposite types can fuse (mate) to form a diploid cell. Haploid cells are of one of two possible mating types, called **a** and α. The mating is mediated by a small secreted polypeptide called a pheromone or mating factor. A cell-surface receptor recognizes the pheromone secreted by cells of the opposite type, i.e., **a** cell receptors bind only α factor, and α cell receptors bind only **a** factor. Mating and subsequent mitotic divisions occur in favorable conditions for growth. In starvation conditions, a diploid yeast cell undergoes meiosis and forms four haploid spores (*sporulation*): two of type **a**, and two of type α.

B. Switch of mating type

The switch of mating type (mating-type conversion) is initiated by a double-strand break in the DNA at the *MAT* locus (recipient) and may involve the boundary to either of the flanking donor loci (*HMR* or *HML*). This is mediated by a HO endonuclease through site-specific DNA cleavage.

C. Cassette model for mating-type switch

Mating-type switch is regulated at three gene loci near the centromere (cen) of chromosome III of *S. cerevisiae*. The central locus is *MAT* (mating-type locus), which is flanked by loci *HML*α (left) and *HMR***a** (right). Only the *MAT* locus is active and transcribed into mRNA. Transcription factors regulate other genes responsible for the **a** or the α phenotype. The *HML*α and *HMR***a** loci are repressed (silenced). DNA sequences from either the *HML*α or the *HML***a** locus are transferred into the *MAT* locus once during each cell generation by a specific recombination event called gene conversion. The presence of *HMR*α sequences at the *MAT* locus determines the phenotype of the **a** cell. When *HML*α sequences are transferred (switched to an α cassette), the phenotype is switched to α. A gene is repressed when it is placed near the yeast mating-type silencer. (Figure in A–C adapted from Lodish et al., 2007.)

Further Reading

Alberts B, et al. Molecular Biology of the Cell. 5th ed. New York: Garland Science; 2008

Botstein D, Chervitz SA, Cherry JM. Yeast as a model organism. Science 1997;277:1259–1260

Brown TA. Genomes. 3rd ed. New York: Garland Science; 2007

Goffeau A, et al. Life with 6000 genes. Science 1996;274:546, 563–567

Haber JE. A locus control region regulates yeast recombination. Trends Genet 1998;14:317–321

Lewin B. Genes IX. Sudbury: Bartlett & Jones; 2008

Lodish H, et al. Molecular Cell Biology. 6th ed. New York: W. H. Freeman; 2007

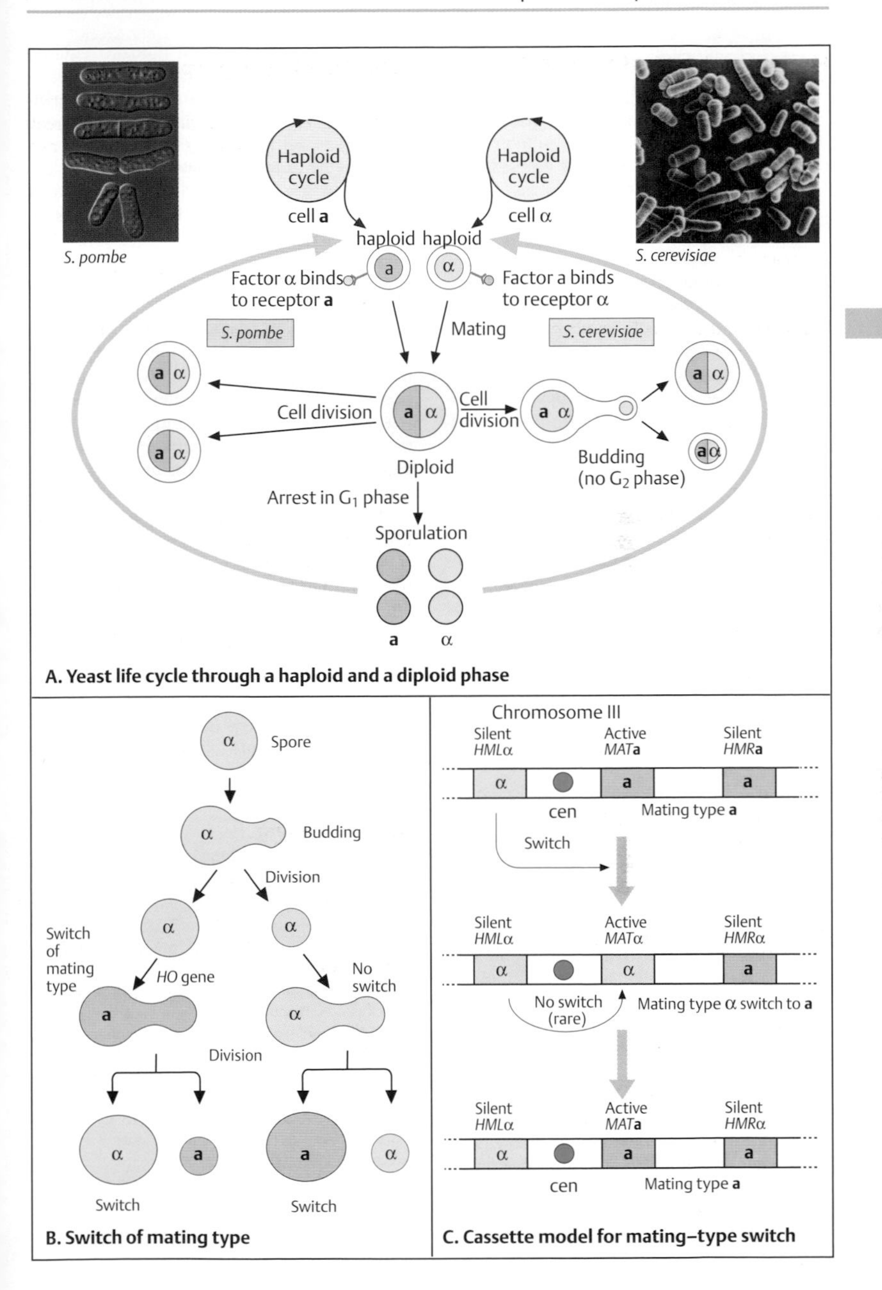

A. Yeast life cycle through a haploid and a diploid phase

B. Switch of mating type

C. Cassette model for mating–type switch

Mating Type Determination in Yeast Cells and Yeast Two-Hybrid System

The asymmetric cell division in yeast resembles the generation of many different cell types in different tissues during embryogenesis of multicellular organisms.

A. Regulation of cell-type specificity in yeast

Each of the three *S. cerevisiae* cell types expresses cell-specific genes. The different resulting combinations of DNA-binding proteins determine the specific cell types. Haploid cells of an **a** type produce a regulatory protein a1. Haploid cells of the α type produce two regulatory proteins, α1 and α2, with different effects. Protein α1 has no effect on cells of the a type, but has an effect on haploid cells (see below). Protein α1 is a transcription factor that activates haploid α-specific genes, whereas protein α2 suppresses a-cell-specific genes. Together they activate haploid-specific genes corresponding to their own type. Following fusion of a cell of the a type with an α cell, the combination of the three proteins α1, α2, and a1 in a diploid cell (**a**/α) follows a completely different pattern: α2 inactivates genes of the a type, but in combination with α1 it now also inactivates all haploid-specific genes. Some haploid cells can switch repeatedly between the two types.

The principle is that each of the three cell types is determined by a cell-specific set of transcription factors acting as activators or suppressors, depending on the regulatory sequences to which they bind. These regulatory proteins, are encoded in the *MAT* locus. Cells of type a express a-specific genes only. In diploid (**a**/α) cells, haploid-specific genes are suppressed. This stimulates transcription of the a-specific genes, but it does not bind too efficiently to the α-specific upstream regulatory sequences when α1 protein is absent.

B. Yeast two-hybrid system

Yeast cells can be used to investigate whether different proteins interact with each other or a protein interacts with DNA. This approach uses a yeast vector for producing a DNA-binding protein and a downstream reporter gene. The two-hybrid (Y2H) method rests on observing whether two different proteins, each hybridized to a different protein domain required for transcription factor activity, are able to interact and thereby reassemble the transcription factor. When this occurs, a reporter gene is activated. Neither of the two hybrid proteins alone is able to activate transcription. Hybrid 1 consists of protein X, the protein of interest (the "bait"), attached to a transcription factor DNA-binding domain (BD). This fusion protein alone cannot activate the reporter gene because it lacks a transcription factor activation domain (AD). Hybrid 2, consisting of a transcription factor AD and an interacting protein, protein Y (the "prey"), lacks the BD. Therefore, hybrid 2 alone also cannot activate transcription of the reporter gene. Different (prey) proteins expressed from cDNAs in vectors can be tested. Fusion genes encoding either hybrid 1 or hybrid 2 are produced using standard recombinant DNA methods. Cells are cotransfected with the genes. Only cells producing the correct hybrids, i.e., those in which the X and Y proteins interact and thereby reconnect AD and BD to form an active transcription factor, can initiate transcription of the reporter gene. This can be observed as a color change or by growth in selective medium.

Using several modifications of Y2H multiple interactions can de determined ("interactome").

(Figures adapted from Oliver, 2000; and Frank Kaiser, University of Lübeck, Germany, personal communication.)

Further Reading

Fields S, Song O. A novel genetic system to detect protein-protein interactions. Nature 1989;340: 245–246

Lemmens I, Lievens S, Tavernier J. Strategies towards high-quality binary protein interactome maps. J Proteomics 2010;73:1415–1420

Lodish H, et al. Molecular Cell Biology. 6th ed. New York: W. H. Freeman; 2007

Oliver S. Guilt-by-association goes global. Nature 2000;403:601–603

Strachan T, Read AP. Human Molecular Genetics. 4th ed. New York: Garland Science; 2011

Uetz P, et al. A comprehensive analysis of protein–protein interaction in *Saccharomyces cerevisiae.* Nature 2000;403:623–627

Yu H, et al. High-quality binary protein interaction map of the yeast interactome network. Science 2008;322:104–110

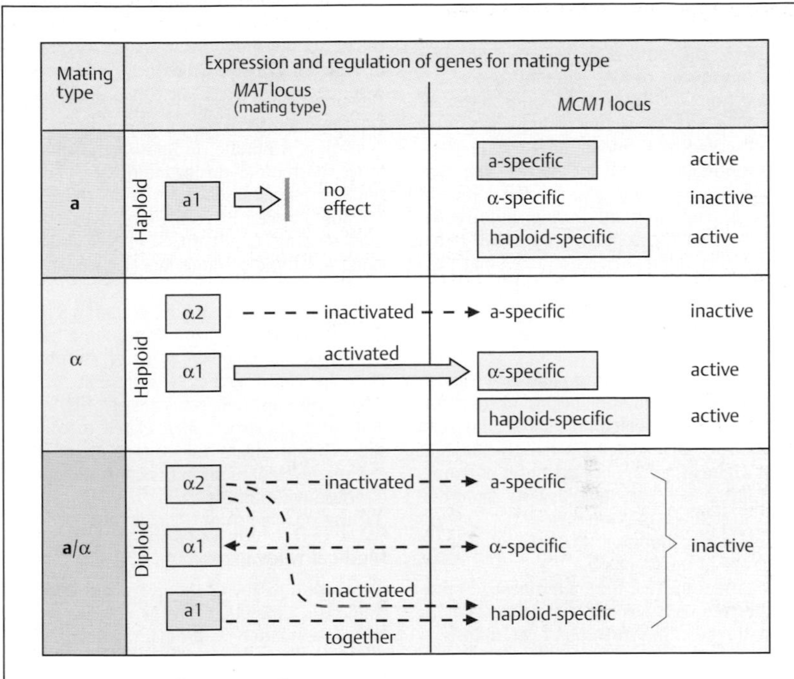

A. Regulation of cell-type specificity in yeast

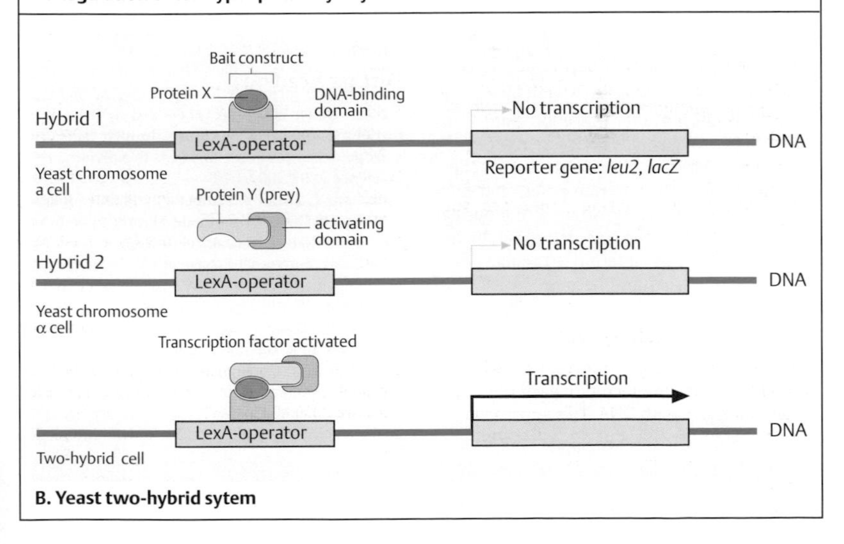

B. Yeast two-hybrid sytem

Cell Division: Mitosis

Mitosis is the process of cell division. This term, introduced by W. Flemming in 1882, is derived from the Greek word mitos, a thread. Threadlike structures in dividing cells were first observed by Flemming in 1879. In 1884, E. Strasburger coined the terms *prophase*, *metaphase*, and *anaphase* for the different stages of cell division. During mitosis the newly replicated chromosomes are distributed equally into different parts of a cell and then separated into their own cell. This results in two genetically identical daughter cells.

A. Mitosis

A cell divides in consecutive phases. In eukaryotic cells, each cell division begins with a phase of DNA synthesis, which lasts about 8 hours (*S phase*). This is followed by the G_2 *phase* of about 4 hours (gap 2) until the onset of *mitosis* (M). Mitosis in eukaryotic cells lasts about one hour. It follows the interphase (G_1) of extremely varied duration. Cells that no longer divide are in the G_0 phase.
During the transition from interphase to mitosis, the chromosomes become visible as elongated threads; this first phase of mitosis is called *prophase*. The chromosomes contract during late prophase to become thicker and shorter (chromosomal condensation). In late prophase, the nuclear membrane disappears and *metaphase* begins. The chromosomes become arranged on the equatorial plate, but homologous chromosomes do not pair. In late metaphase during the transition into *anaphase*, the chromosomes divide at the centromere region. The two chromatids of each chromosome migrate to opposite poles, and *telophase* begins with the formation of the nuclear membranes. Finally, the cytoplasm also divides (cytokinesis). In early interphase the individual chromosomal structures become invisible in the cell nucleus.

B. Metaphase chromosome

In 1888, Waldeyer coined the term chromosome for the stainable threadlike structures visible during mitosis. A metaphase chromosome consists of two chromatids (*sister chromatids*) and the centromere, which holds them together. The regions at both ends of the chromosome are the telomeres. The point of attachment to the mitotic spindle fibers is the kinetochore.

C. Role of condensins

The progressive compaction of chromosomes entering mitosis is called chromosome condensation. A mitotic chromosome is about 50 times shorter than during interphase. Chromosomes are condensed by proteins called condensins. Condensins consist of five subunits (not shown). Condensins can be visualized along a mitotic chromosome (Figure 31.18 in Lewin, 2008).
When chromosomes are duplicated in S phase, the two copies of each chromosome remain tightly bound together as sister chromatids. They are held together by multiunit proteins called cohesins. Cohesins consist of four subunits and are structurally related to condensins. They regulate the separation of sister chromatids. Mutations in fission yeast cohesins interfere with mitosis.
(Figures adapted from Uhlmann, 2002).

Medical relevance

A mutation in any of the five genes encoding condensin subunits results in a severe growth and malformation syndrome: the Roberts syndrome (OMIM 268300; Vega et al., 2005).

Further Reading

Karsenti E, Vernos I. The mitotic spindle: a self-made machine. Science 2001;294:543–547
Lewin B. Genes IX. Sudbury: Bartlett & Jones; 2008
Nurse P. The incredible life and times of biological cells. Science 2000;289:1711–1716
Rieder CL, Khodjakov A. Mitosis through the microscope: advances in seeing inside live dividing cells. Science 2003;300:91–96
Tsukahara T, Tanno Y, Watanabe Y. Phosphorylation of the CPC by Cdk1 promotes chromosome bi-orientation. Nature 2010;467:719–723
Uhlmann F, Lottspeich F, Nasmyth K. Sister-chromatid separation at anaphase onset is promoted by cleavage of the cohesin subunit Scc1. Nature 1999;400:37–42
Vega H, et al. Roberts syndrome is caused by mutations in ESCO2, a human homolog of yeast ECO1 that is essential for the establishment of sister chromatid cohesion. Nat Genet 2005;37:468–470

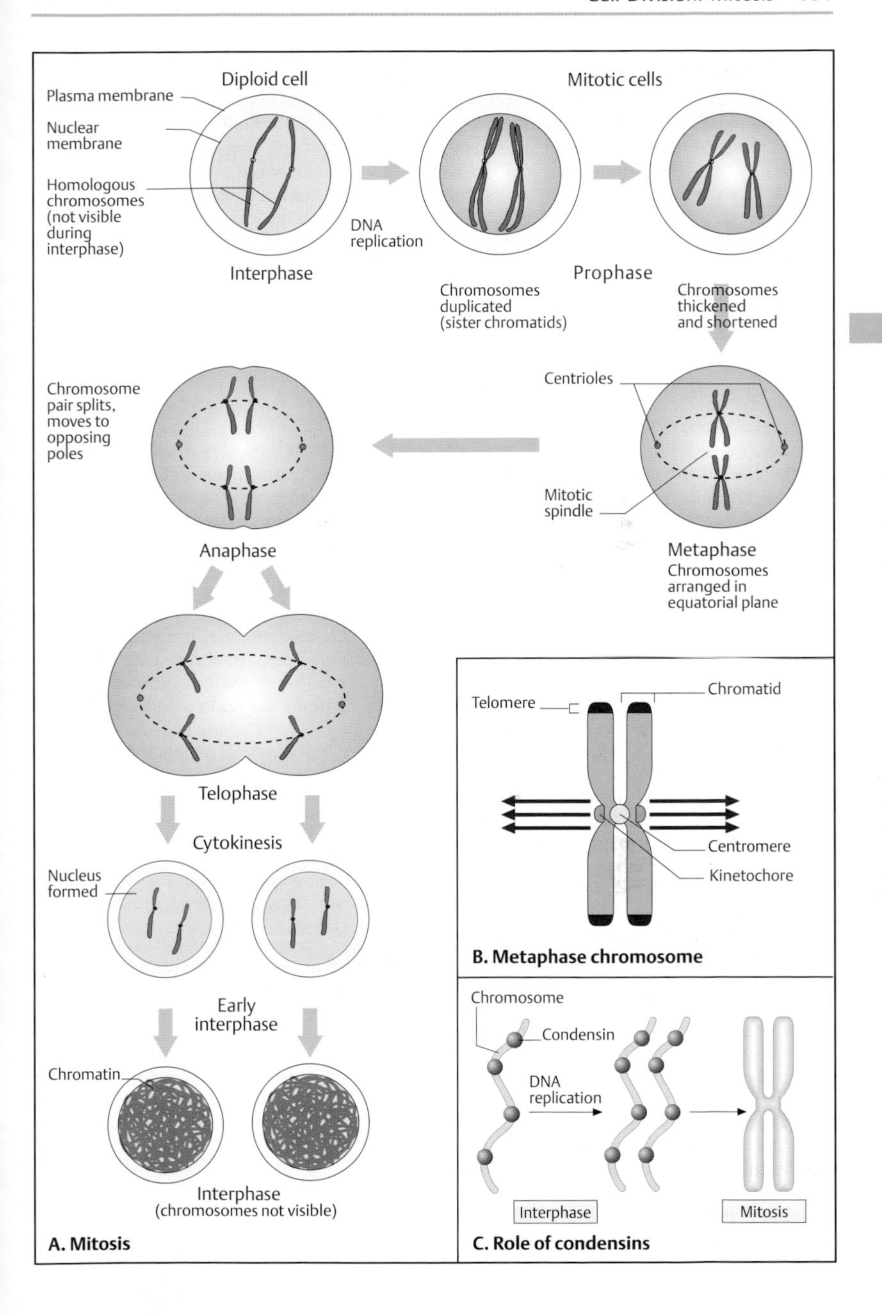

A. Mitosis

B. Metaphase chromosome

C. Role of condensins

Meiosis in Germ Cells

Meiosis is a special type of cell division that in most eukaryotes produces haploid germ cells (eggs and sperm cells). This term was introduced by Strasburger in 1884 and derived from a Greek work meaning diminution (or maturation division). Meiosis consists of two nuclear divisions but only one round of DNA replication. As a result, the four daughter cells are haploid, i.e., they contain only one chromosome of each pair.

Meiosis differs fundamentally from mitosis in genetic and cytological respects. First, homologous chromosomes pair at prophase of the first division. Second, exchanges between homologous chromosomes (*crossing-over*) occur regularly. As a result all new chromosomes consist of segments of both maternal and paternal origin. The process of creating new combinations of genetic information is called *genetic recombination*. Third, the chromosome complement is reduced to half during the first cell division, meiosis I.

Meiosis is a complex cellular and biochemical process. The cytologically observable course of events and the genetic consequences do not correspond exactly in time. A genetic process occurring in one phase usually becomes visible cytologically at a later phase.

A. Meiosis I

A gamete-producing cell goes through two cell divisions at meiosis: meiosis I and meiosis II. The relevant genetic events, genetic recombination by means of crossing-over and reduction to the haploid chromosome complement, occur in meiosis I. Meiosis begins with DNA replication. Initially the chromosomes in late interphase are visible only as threadlike structures. At the beginning of prophase I, the chromosomes are doubled. The pairing allows an exchange between homologous chromosomes (crossing-over), made possible by juxtapositioning homologous chromatids. At certain sites a chiasma forms. As a result of crossing-over, chromosome material of maternal and paternal origin is exchanged between two chromatids of homologous chromosomes. After the homologous chromosomes migrate to opposite poles, the cell enters anaphase I.

B. Meiosis II

Meiosis II consists of longitudinal division of the duplicated chromosomes (chromatids) and a further cell division. Each daughter cell is haploid, as it contains one chromosome of a pair only. On each chromosome, recombinant and nonrecombinant sections can be identified. The genetic events relevant to these changes have occurred in the prophase of meiosis I (see p. 104).

During meiosis several meiosis-specific proteins in the cohesion complex ensure correct homologous pairing and subsequent separation.

Medical relevance

The independent distribution of chromosomes (independent assortment) during meiosis explains the segregation (separation or splitting) of observable traits according to the rules of Mendelian inheritance (1:1 segregation, see p. 114).

Errors in the correct distribution of the chromosomes, called nondisjunction, result in gametes with an extra chromosome or a chromosome missing, and after fertilization the zygote will have either three homologous chromosomes (*trisomy*) or only one chromosome (*monosomy*). Both trisomy and monosomy result in embryonic developmental disturbances (see p. 382).

Further Reading

Carpenter ATC. Chiasma function. Cell 1994;77 (7):957–962

Kitajima TS, et al. Distinct cohesin complexes organize meiotic chromosome domains. Science 2003; 300:1152–1155

Moens PB, ed. Meiosis. New York: Academic Press; 1987

Page SL, Hawley RS. Chromosome choreography: the meiotic ballet. Science 2003;301:785–789

Petronczki M, Siomos MF, Nasmyth K. Un ménage à quatre: the molecular biology of chromosome segregation in meiosis. Cell 2003;112:423–440

Whitehouse LHK. Towards an Understanding of the Mechanism of Heredity. 3rd ed. London: Edward Arnold; 1973

Zickler D, Kleckner N. Meiotic chromosomes: integrating structure and function. Annu Rev Genet 1999;33:603–754

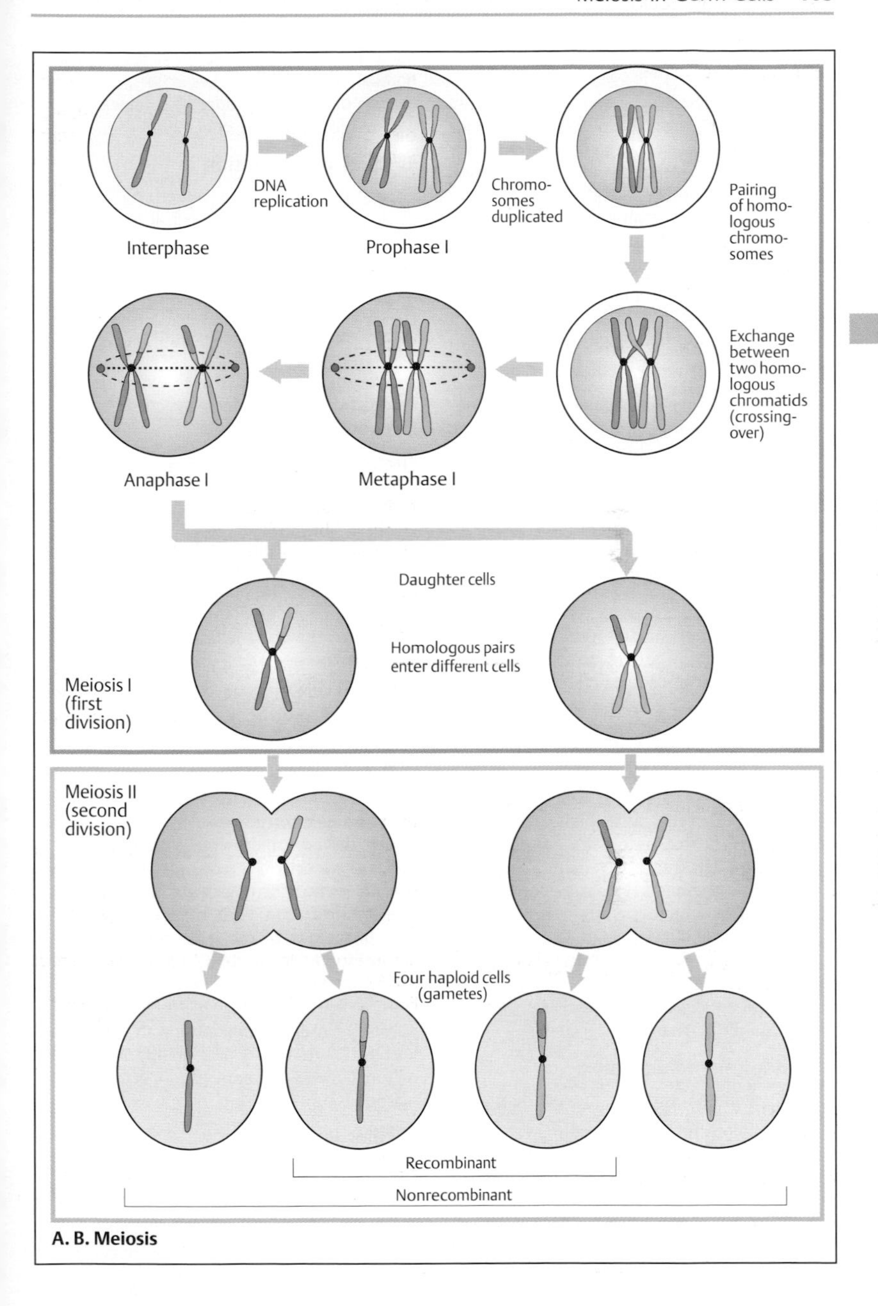
DNA replication
Chromosomes duplicated
Interphase
Prophase I
Pairing of homologous chromosomes
Exchange between two homologous chromatids (crossing-over)
Anaphase I
Metaphase I
Daughter cells
Homologous pairs enter different cells
Meiosis I (first division)
Meiosis II (second division)
Four haploid cells (gametes)
Recombinant
Nonrecombinant

A. B. Meiosis

Meiosis Prophase I

The decisive cytological and genetic events take place in the prophase of meiosis I. In prophase I, exchanges between homologous chromosomes occur regularly by crossing-over. Crossing-over, a term introduced by Morgan and Cattell in 1912, is an elaborate cytological process by which parts of chromosomes of maternal and paternal origin exchange stretches of DNA. This results in new combinations of chromosome segments (genetic recombination).

A. Prophase of meiosis I

The prophase of meiosis I passes through consecutive stages. The first is the *leptotene* stage. Here the chromosomes first become visible as fine threadlike structures (only one chromosome pair is shown). Next is *zygotene*: each chromosome is visible as a paired structure, the result of DNA replication prior to the beginning of prophase. Consequently, each chromosome has been doubled and consists of two identical chromatids (sister chromatids). These are held together at the centromere. Each chromatid contains a DNA double helix. Two homologous chromosomes that have paired are referred to as a bivalent. In the *pachytene* stage, the bivalents become thicker and shorter. In *diplotene*, the two homologous chromosomes separate, but remain attached to each other at a few points, each called a chiasma (see below). In the next phase, *diakinesis*, each of the chromosome pairs has separated further, although they still remain attached to each other at the ends. A chiasma corresponds to a region at which crossing-over has taken place previously. However, in late diakinesis, the chiasmata shift distally, called chiasma terminalization. The mechanisms of meiosis II correspond to those of mitosis. The different stages cannot be sharply separated.

B. Synaptonemal complex

The synaptonemal complex, independently observed in spermatocytes by D. Fawcett and M. J. Moses in 1956, is a complex structure formed during meiotic prophase I. It consists of two chromatids (1 and 2) of maternal origin (mat) and two chromatids (3 and 4) of paternal origin (pat). It initiates chiasma formation and is the prerequisite for crossing-over and subsequent recombination. Double-strand breaks in homologous chromosomes occur prior to formation of the synaptonemal complex. (Figure adapted from Alberts et al., 2008.)

C. Chiasma formation

Chiasma is the term introduced by F. A. Janssens in 1909 for the cytological manifestation of crossing-over during meiotic prophase I. A chiasma forms between one chromatid of a chromosome of maternal origin (chromatids 1 and 2 in the figure) and one chromatid of a chromosome of paternal origin (chromatids 3 and 4). Either of the two chromatids of one chromosome can cross over with one of the chromatids of the homologus chromosome (e. g., 1 and 3, 2 and 4, etc.).

D. Genetic recombination

Through crossing-over, new combinations of chromosome segments arise (recombination). As a result, recombinant and nonrecombinant chromosome segments can be differentiated. In the diagram, the areas A–E (shown in pink) of one chromosome and the corresponding areas a–e (shown in blue) of the homologous chromosome become a–b–C–D–E and A–B–c–d–e, respectively.

E. Pachytene and diakinesis under the microscope

During pachytene and diakinesis, the individual chromosomes can be readily visualized by light and electron microscopy. Here diakinesis under the light microscope (a) and pachytene under the electron microscope (b) are shown. (a) An extra chromosome 21 (red arrow) present in a man with trisomy 21 does not pair. In meiosis in the male, the X and the Y chromosomes form an XY body. Pairing of X and Y is limited to the extreme end of the short arms (see p. 240). (b) The thickened (duplicated) chromosomes and the XY bivalent are visible. (Figure kindly provided by Dr R. Johannisson, Lübeck, Germany [b]; [a] is from Johannisson et al., 1983.)

Further Reading

Johannison R, et al. Down's syndrome in the male. Reproductive pathology and meiotic studies. Hum Genet 1983;63:132–138

Lewin B. Genes IX. Sudbury: Bartlett & Jones; 2008

Miller OJ, Therman E. Human Chromosomes. 4th ed. New York: Springer-Verlag; 2001

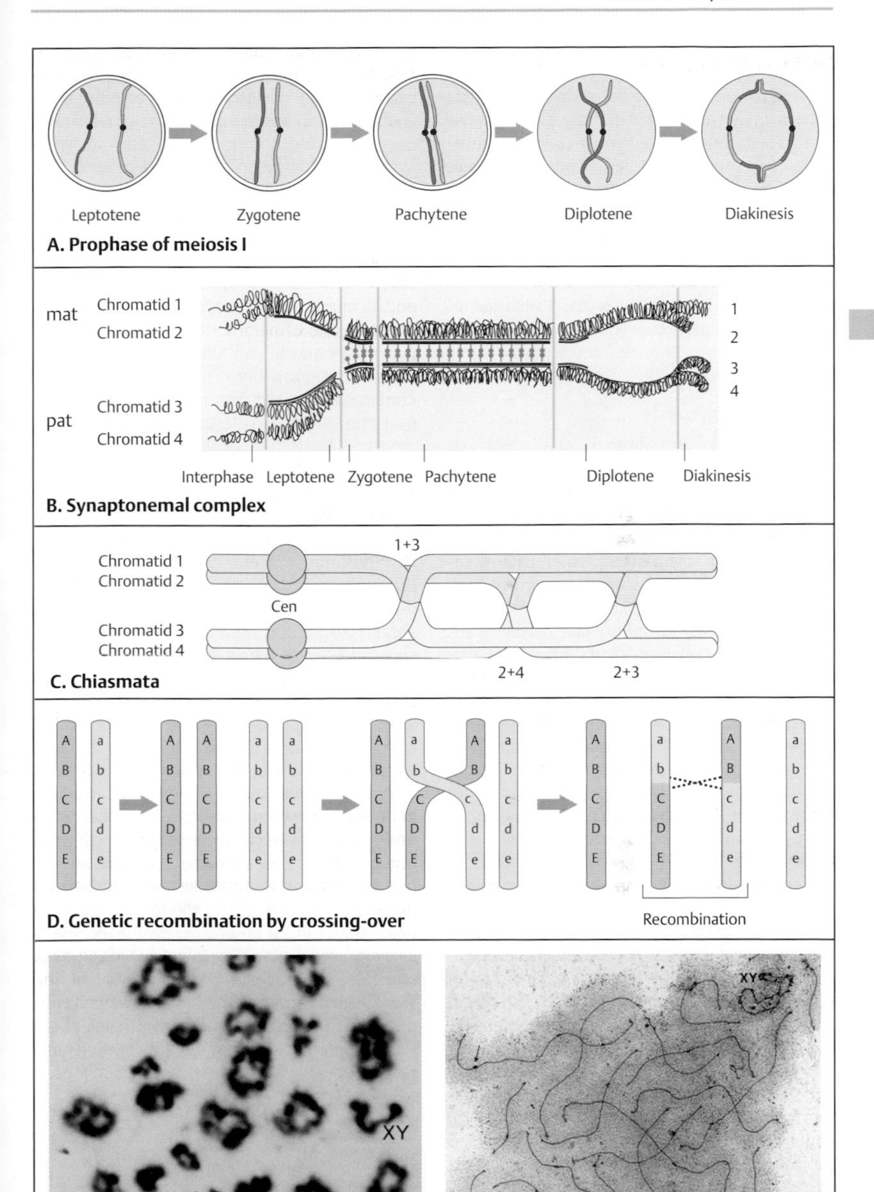

A. Prophase of meiosis I

B. Synaptonemal complex

C. Chiasmata

D. Genetic recombination by crossing-over

E. Diakinesis (a, light microscope) and early pachytene (b, electron microscope)

Formation of Gametes

Gametes (germ cells) are the cells that transmit genetic information from one generation to the next. They are produced in the gonads. In females the process is called *oogenesis* (formation of oocytes), and in males it is called *spermatogenesis* (formation of spermatozoa). Primordial germ cells migrate during early embryonic development from the genital ridge to the gonads. Here they increase in number by mitotic divisions. The actual formation of germ cells (gametogenesis) begins with meiosis. Gametogenesis in males and females differs in duration and results.

A. Spermatogenesis

Spermatogonia are diploid cells that go through mitotic divisions in the gonads of male animals. The primary spermatocytes result from the first meiotic division beginning at the onset of puberty. At the completion of meiosis I, one primary spermatocyte gives rise to two secondary spermatocytes. Each has a haploid set of duplicated chromosomes. In meiosis II, each secondary spermatocyte divides to form two spermatids. Thus, one primary spermatocyte forms four spermatids, each with a haploid chromosome complement. The spermatids differentiate into mature spermatozoa in about 6 weeks. Male spermatogenesis is a continuous process. In human males, the time required for a spermatogonium to develop into a sperm cell is about 90 days.

B. Oogenesis

Oogenesis is the formation of eggs (oocytes) in females. It differs from spermatogenesis in timing and result. During early embryogenesis the germ cells migrate from the genital ridge to the ovary, where they form oogonia by repeated mitoses. A primary oocyte results from the first meiotic division of an oogonium. In human females, meiosis I begins about 4 weeks before birth. Then meiosis I is arrested in a stage of prophase designated *dictyotene*. The primary oocyte persists in this stage until ovulation. Only then is meiosis I continued.

In primary oocytes the cytoplasm divides asymmetrically in both meiosis I and meiosis II. The result each time is two cells of unequal size. One cell is larger and will eventually form the egg; the other, smaller cell becomes a polar body, not a germ cell. When the secondary oocyte divides, the daughter cells again differ; one secondary oocyte and another polar body (polar body II) are the result. The polar bodies degenerate and do not develop. On rare occasions a polar body may become fertilized. This can give rise to an incompletely developed twin. In the secondary oocyte, each chromosome still consists of two sister chromatids. These do not separate until the next cell division (meiosis II). At ovulation the secondary oocyte is released from the ovary, and if fertilization occurs, meiosis is then completed. A polar body may be examined to assess whether a genetic abnormality is present in the fetus.

The maximum number of germ cells in the ovary of the human fetus at about the 5th month is 6.8×10^6. By the time of birth, this has been reduced to 2×10^6, and by puberty to about 200 000. Of these, about 400 eventually go through ovulation.

Medical relevance

Most new mutations occur during gametogenesis. The difference in time in the formation of gametes during oogenesis and spermatogenesis is reflected in the difference in germline cell divisions. The number of cell divisions in spermatogenesis and oogenesis differs considerably. On average about 380 chromosome replications have taken place in the progenitor cells of spermatozoa by age 30 years, and about 610 chromosome replications by age 40 years. Altogether, 25 times more cell divisions occur during spermatogenesis than during oogenesis (Crow, 2000). This probably accounts for the higher mutation rate in males, especially with increased paternal age. In the female, an average number of 22 mitotic cell divisions occurs before meiosis, resulting in a total of 23 chromosome replications.

Faulty distribution of the chromosomes (nondisjunction) during meiosis I or meiosis II is the cause of aberrations of the chromosome number (see p. 382).

Further Reading

Crow JF. The origins, patterns and implications of human spontaneous mutation. Nat Rev Genet 2000;1:40–47

Miller OJ, Therman E. Human Chromosomes. 4th ed. New York: Springer-Verlag; 2001

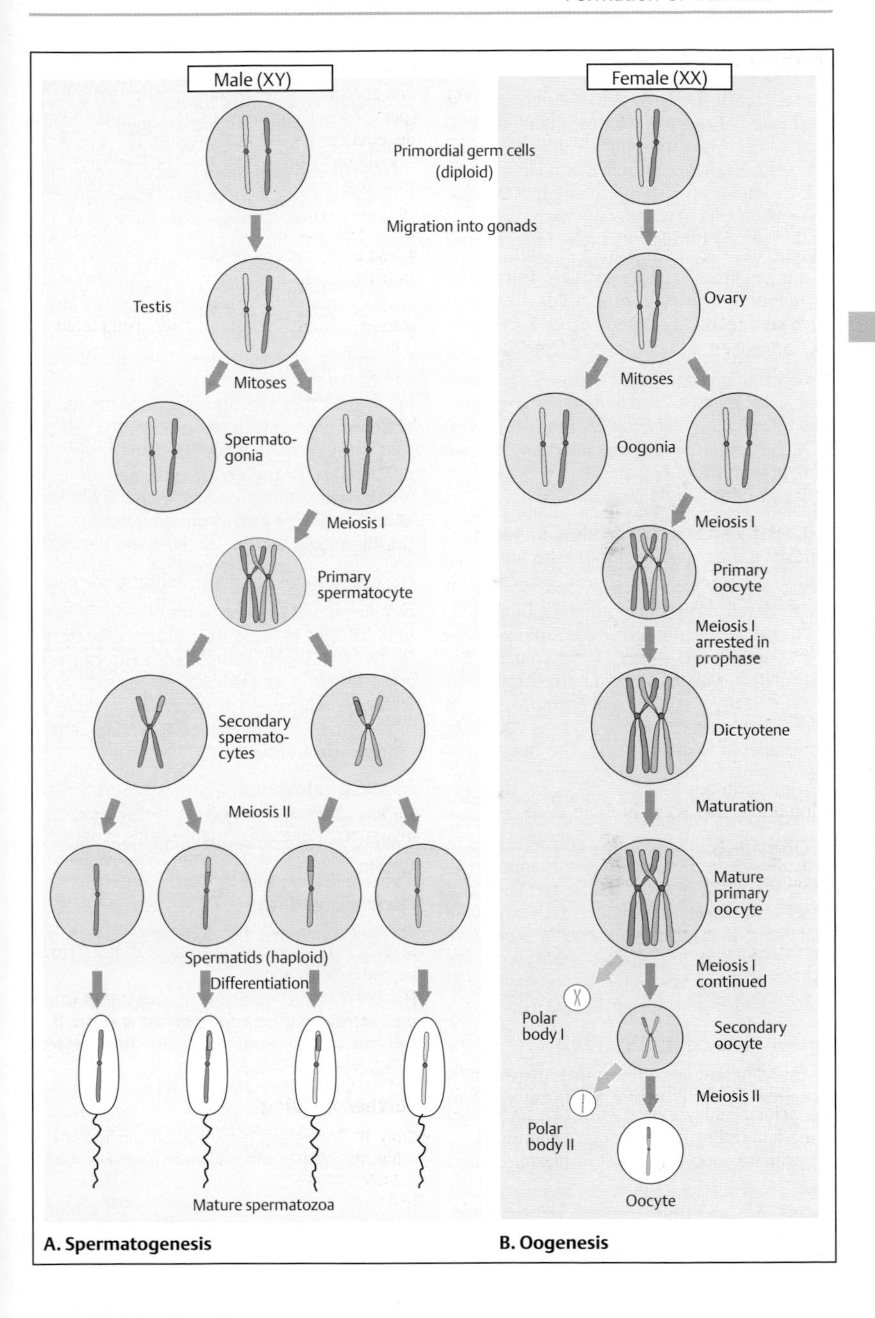

A. Spermatogenesis

B. Oogenesis

Cell Cycle Control

The cell cycle leads to cell division and the production of two daughter cells with an identical set of chromosomes. The cell cycle, as first defined by A. Howard and S. R. Pelc in 1953, has two main phases: interphase and mitosis. An elaborate set of controlling macromolecules guide the cell through its cycle. Here, control systems can detect and eliminate errors, and abandon defective cells.

A. Yeast cells

Genetic studies in yeast cells (see p. 96) have provided important insights into cell cycle control. Baker's yeast (*Saccharomyces cerevisiae*) and fission yeast (*Schizosaccharomyces pombe*) have cell cycle control mechanisms that are similar to those of higher eukaryotes. (Images obtained from Google, July 2005.)

B. Cell division cycle models in yeast

Mitotic division in budding yeast (baker's yeast) results in one large and one small daughter cell. Since a microtubule mitotic spindle forms very early during the S phase, there is practically no G_2 phase (**1**). In contrast, fission yeast (*S. pombe*) forms a mitotic spindle at the end of the G_2 phase, and then proceeds to mitosis to form two daughter cells of equal size (**2**). Unlike in vertebrate cells, the nuclear envelope remains intact during mitosis. An important regulator of yeast cell division is cdc2 protein (cell division cycle 2). Absence of cdc2 activity (cdc2 mutants) in *S. pombe* results in cycle delay and prevents entry into mitosis (**3**). Thus, the result is a cell that is too large and only has one nucleus. Increased activity of cdc2 (dominant mutant *cdc*^{D}) results in premature mitosis and cells that are too small (wee phenotype, from the Scottish word for small). (Figure adapted from Lodish et al., 2007.)

C. Cell cycle control systems

Several different growth factors initiate cell division. Other systems detect DNA or cell damage and induce various repair systems (p. 86). The eukaryotic cell cycle is driven by cell cycle "engines," which are a set of interacting proteins known as the cyclin-dependent kinases (Cdks). An important member of this family of proteins is cdc2 (also called Cdk1). Other proteins act as rate-limiting steps in cell cycle progression and are able to induce cell cycle arrest at defined stages (checkpoints). The cell is induced to progress through G_1 by growth factors (mitogens) acting through receptors that transmit signals to proceed toward the S phase. Other proteins can induce G_1 arrest. The p53 protein (p. 306) plays a leading role in detecting DNA damage and subsequent cell cycle arrest.

In early G_1 phase, cdc2 is inactive. It is activated in late G_1 by associating with G_1 cyclins, such as cyclin E. Once the cell has passed the G_1 check point, cyclin E is degraded and the cell enters the S phase. This is initiated, among many other activities, by cyclin A binding to Cdk2 and phosphorylation of the retinoblastoma protein (RB; see p. 314). The cell can pass through the mitosis checkpoint provided no damage is present. Association of Cdc2 (Cdk1) with mitotic cyclins A and B activates and forms the mitosis-promoting factor (MPF). During mitosis, cyclins A and B are degraded, and an anaphase-promoting complex is formed (details not shown). When mitosis is completed, cdc2 is inactivated by the S phase inhibitor Sic1 in yeast. At the same time the RB (p. 312) is dephosphorylated. Cells can progress to the next cell cycle stage only when feedback controls have ensured the integrity of the genome. (The figure omits many important protein transactions.)

Medical relevance

Mutations in one of the many cell cycle controlling genes may result in different types of cancer.

Further Reading

Hartwell LH, Weinert TA. Checkpoints: controls that ensure the order of cell cycle events. Science 1989;246:629–634

Howard A, Pelc S. Synthesis of deoxyribonucleic acid in normal and irradiated cells and its relation to chromosome breakage. Heredity 1953;6 (Suppl.):261–273

Lodish H, et al. Molecular Cell Biology. 6th ed. New York: W. H. Freeman; 2007

Nurse P. A long twentieth century of the cell cycle and beyond. Cell 2000;100:71–78

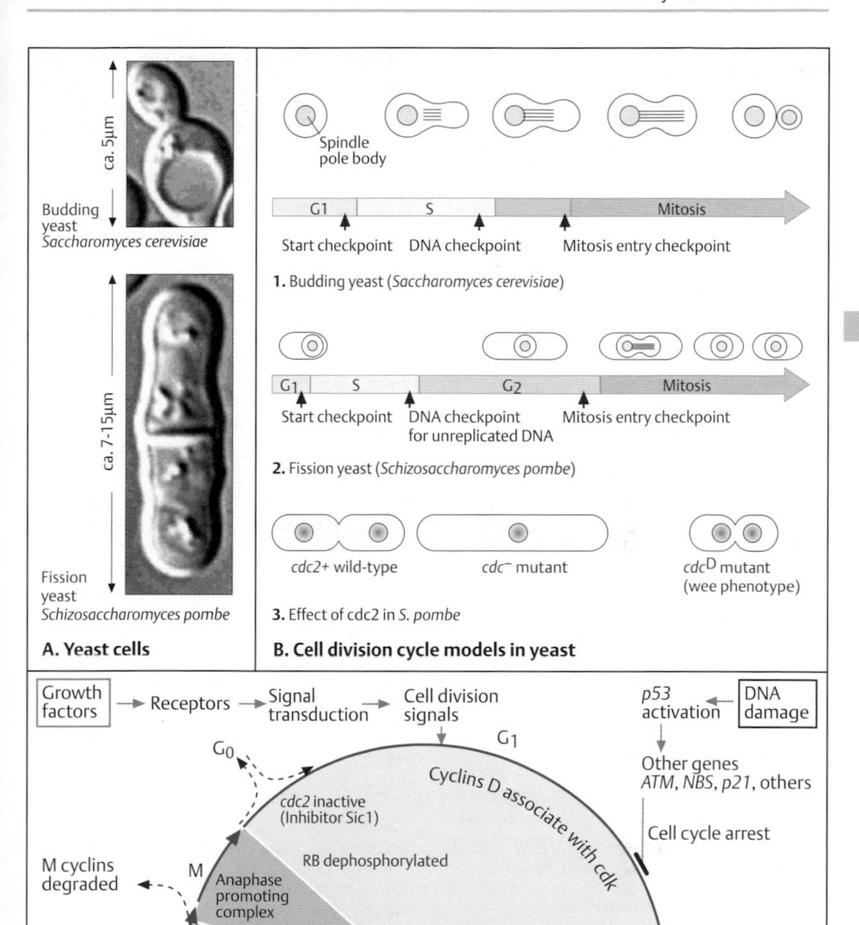

A. Yeast cells

B. Cell division cycle models in yeast

C. Cell cycle control systems

Programmed Cell Death

At certain stages in the development of a multicellular organism, some cells must die. This well regulated process is called *apoptosis* (programmed cell death, derived from the Greek *apo* meaning from/off/without, and *ptosis* meaning fall), as suggested by Kerr in 1972. The importance of this biological phenomenon was first realized in studies of a tiny worm, the soil nematode *Caenorhabditis elegans* (see p. 212). Apoptosis is regulated in several apoptosis pathways by many proteins, which either trigger or prevent apoptosis. Apoptosis can be triggered from outside the cell (extrinsic pathway) or from within the cell (intrinsic pathway).

A. Importance of apoptosis

Apoptosis occurs mainly during development. For example, the digits in the developing mammalian embryo are sculptured by apoptosis (**1**). The paws (hands) start out as spade-like structures. The formation of digits requires that cells between them die (here shown as bright green dots on the left). More staggering is the amount of apoptosis in the developing vertebrate nervous system. Normally up to half of the nerve cells die soon after they have been formed. In the embryos of mice that lack an important gene regulating apoptosis (the gene encoding caspase 9, see below), neurons proliferate excessively and the brain protrudes above the face (**2**). (**1**, illustration modified from Alberts et al., 2008; **2**, from Gilbert, 2010, according to Kuida et al., 1998.)

B. Cellular events in apoptosis

The first visible signs of apoptosis are condensation of chromatin and shrinking of the cell. The cell membrane shrivels (membrane blebbing), and the cell begins to disintegrate (nuclear segmentation, DNA fragmentation). An apoptotic body of cell remnants forms, and this eventually dissolves in a process called lysis. (Figure modified from that produced by Dr. A. J. Cann, Microbiology, Leicester University, and displayed on Google Images, 22 March, 2005.)

C. Regulation of apoptosis

Specialized cysteine-containing aspartate proteinases, called caspases, activate or inactivate each other in a defined sequence. Binding of a ligand, Fas, of a cytotoxic T cell (see section on immune system) to the Fas receptor (also called CD95) activates the intracellular adaptor protein FADD (Fas-associated death domain). This binds to and activates procaspase 8 into active caspase 8. Caspase 8 causes release of cytochrome *c* in mitochondria (see p. 244) and activates several different effector caspases.

The mouse and human genomes contain 13 caspase genes (1–12 and 14; see Appendix, Table p. 410). Human caspases 3 and 6–10 are involved in apoptosis; the others are involved in inflammation. Caspase 8 also serves as a selective signal transducer for nuclear factor κB (NF-κB) during the early genetic response to an antigen. Other regulators of apoptosis are members of the Bcl-2 family. (Figure from Koolman & Röhm, 2005.)

Medical relevance

Mutations in genes regulating apoptosis cause diseases, e.g., B cell lymphoma (OMIM 151430) caused by a mutation in the *BCL2* gene.

Further Reading

Alberts B, et al. Molecular Biology of the Cell. 5th ed. New York: Garland Science; 2008

Danial NN, Korsmeyer SJ. Cell death: critical control points. Cell 2004;116:205–219

Gilbert SF. Developmental Biology. 9th ed. Sunderland: Sinauer; 2010

Hengartner MO. The biochemistry of apoptosis. Nature 2000;407:770–776

Hotchkiss RS, et al. Cell death. N Engl J Med 2009;361:1570–1583

Kerr JF, Wyllie AH, Currie AR. Apoptosis: a basic biological phenomenon with wide-ranging implications in tissue kinetics. Br J Cancer 1972;26:239–257

Koolman J, Roehm KH. Color Atlas of Biochemistry. 2nd ed. Thieme: Stuttgart–New York: 2005

Kuida K, et al. Reduced apoptosis and cytochromosome c-mediated caspase activation in mice lacking caspase 9. Cell 1998;94:325–337

Nagata S. DNA degradation in development and programmed cell death. Annu Rev Immunol 2005;23:853–875

Su H, et al. Requirement for caspase-8 in NF-kappaB activation by antigen receptor. Science 2005;307: 1465–1468

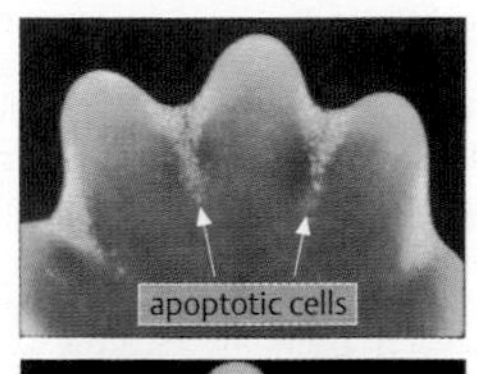

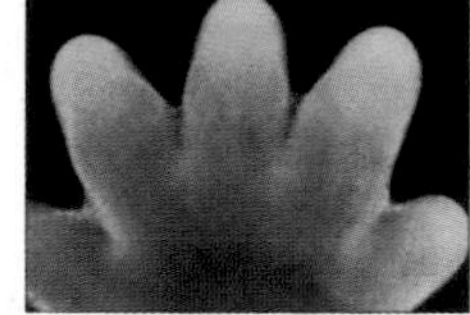

1. Apoptosis in the paw of a mouse embryo

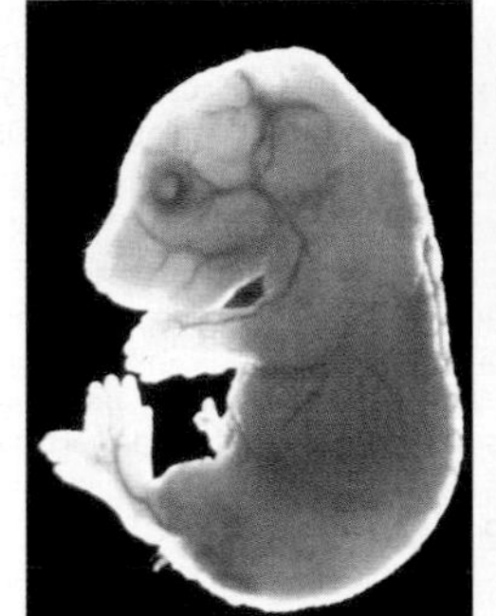

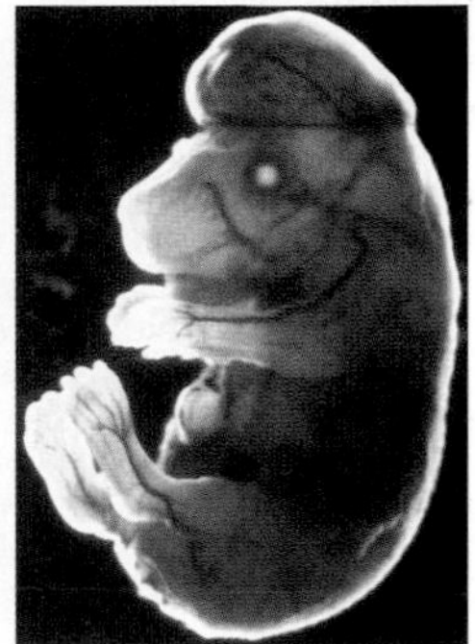

2. Disruption of brain development (right) in a mouse embryo lacking caspase 9

A. Importance of apoptosis

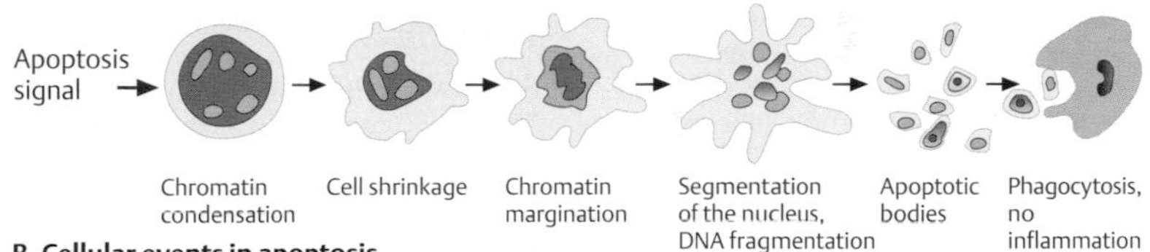

B. Cellular events in apoptosis

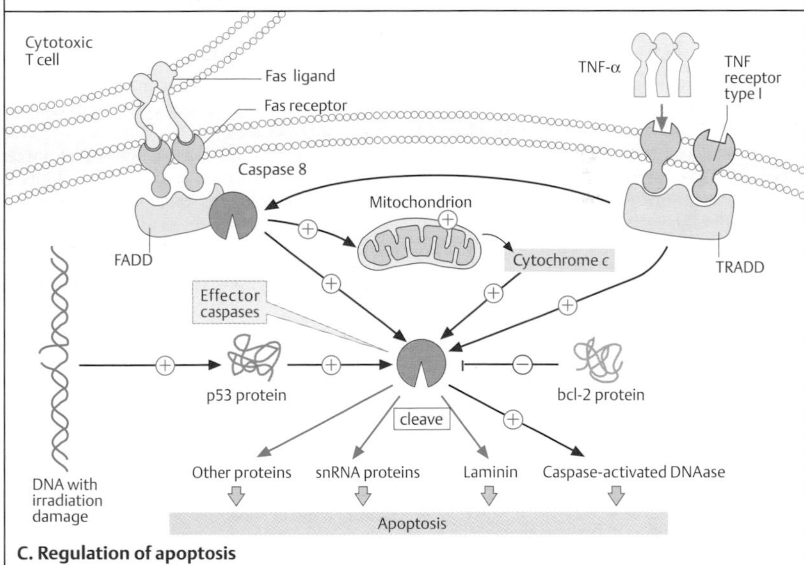

C. Regulation of apoptosis

Cultured Cells

Cells of animals and plants can live and multiply outside the body in a tissue-culture dish as a cell culture at 37°C in a medium containing vitamins, sugar, serum (containing numerous growth factors and hormones), the nine essential amino acids for vertebrates, and usually glutamine and cysteine. Cultured cells usually represent one particular cell type, are generally easy to handle, and can, under certain conditions, be grown into colonies of identical cells (cell cloning). Use of cell cultures began in 1940 when W. Earle established a permanent mouse cell strain, and T. T. Puck and coworkers grew clones of human cells in vitro. In 1965, this led to the development of *somatic cell genetics.*

Cultured cells have a finite life span (Hayflick, 1997). Human cells have a capacity for about 30 doublings until they reach a state called senescence. Cells derived from adult tissues have a shorter lifespan than those derived from fetal tissues. In contrast, cells that have acquired certain genetic changes can grow indefinitely. This is referred to as oncogenic transformation or transformed for short.

Embryonic stem cells can be cultured, and they give rise to different cell types of the body.

A. Skin fibroblast culture

The predominant cell type that grows from a piece of mammalian tissue in culture is the fibroblast. To initiate a culture, a small piece of skin (2 × 4 mm) is obtained in sterile conditions and cut into smaller pieces, which are placed in a culture dish. The pieces attach to the bottom of the dish (adhesion culture due to anchorage dependency of the cells) and after about 8–14 days, cells begin to grow out from each piece and multiply. When the bottom of the culture vessel is covered with a dense layer of cells, they stop dividing because of contact inhibition (this is lost in tumor cells). When transferred into new culture vessels (subcultures), the cells will resume growing until they again become confluent. With a series of subcultures, several million cells can be obtained for a given study. Cultured cells are highly sensitive to increased temperature and do not survive above about 39°C, whereas in special conditions they can be stored alive in vials kept in liquid nitrogen at −196°C. They can be thawed after many years or even decades, and cultured again.

B. Hybrid cells

Cells in culture can be induced to fuse by exposure to polyethylene glycol or Sendai virus. If parental cells from different species are fused, interspecific (from different animal species) hybrid cells can be derived. The hybrid cells can be distinguished from the parental cells by using parental cells deficient in thymidine kinase (TK^-) or hypoxanthine phosphoribosyltransferase ($HPRT^-$). When cell cultures of parental type A (TK^-, **1**) and type B ($HPRT^-$, **2**) are cultured together (co-cultivation, **3**), some cells fuse (**4**). In a selective medium containing hypoxanthine, aminopterin, and thymidine (HAT medium; Littlefield, 1964) only fused cells with a nucleus from each of the parental cells (**1** and **2**) can grow. Cells that have not fused cannot grow in HAT medium (**5**). The reason for this is that TK^- cells cannot synthesize thymidine monophosphate; $HPRT^-$ cells cannot synthesize purine nucleoside monophosphates. The fused cells complement each other. The two nuclei of fused cells (heterokaryon) will also fuse (**6**). This forms a hybrid cell (**7**). Hybrid cells are used for many purposes.

C. Radiation hybrids

A radiation hybrid is a rodent cell containing small fragments of chromosomes from another organism (McCarthy, 1996). When human cells are irradiated with lethal roentgen doses of 3–8 Gy, the chromosomes break into small pieces (**1**) and the cells cannot divide in culture. However, if these cells are fused with nonirradiated rodent cells (**2**), some human chromosome fragments will be integrated into the rodent chromosomes (**3**). Those cells which contain human DNA can be identified by human chromosome-specific probes.

Further Reading

Alberts B, et al. Molecular Biology of the Cell. 5th ed. New York: Garland Science; 2008

Hayflick L. Mortality and immortality at the cellular level. A review. Biochemistry (Mosc) 1997;62: 1180–1190

Lodish H, et al. Molecular Cell Biology. 6th ed. New York: W. H. Freeman; 2007

McCarthy LC. Whole genome radiation hybrid mapping. Trends Genet 1996;12:491–493

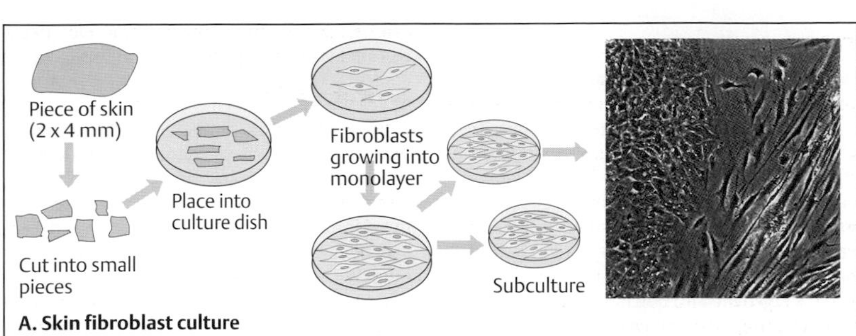

A. Skin fibroblast culture

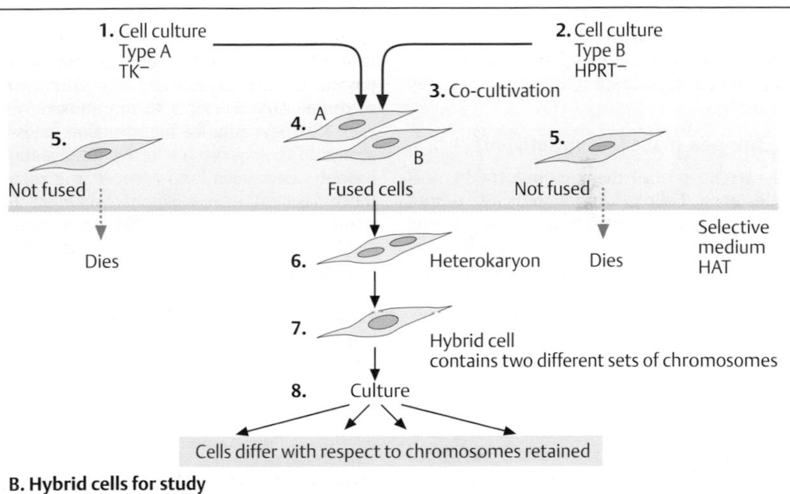

B. Hybrid cells for study

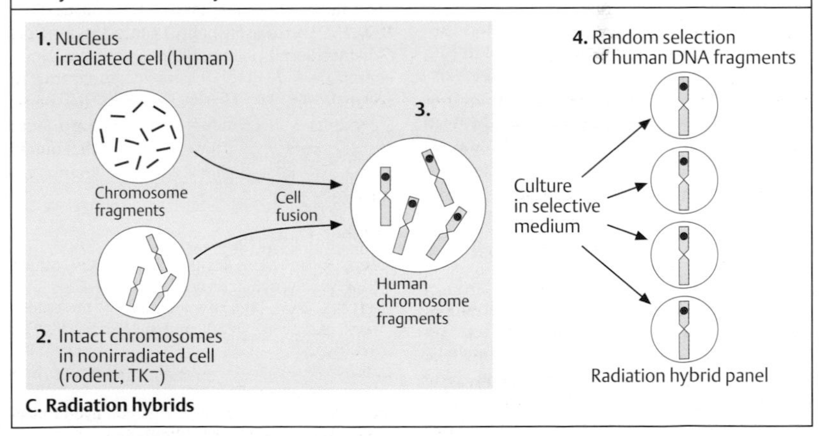

C. Radiation hybrids

The Mendelian Traits

A scientific foundation for the rules underlying inheritance was laid in 1865, but recognized only 35 years later. The Augustinian monk Gregor Mendel presented remarkable observations at the Naturgeschichtliche Vereinigung von Brünn (The Natural History Society of Brünn/Brno), published in 1866. In this work, entitled *Versuche über Pflanzenhybriden* (Experiments with Hybrid Plants) Mendel observed that certain traits in the garden pea (*Pisum sativum*) are inherited according to regular patterns and independently of one another. It was not until 1900 that H. de Vries, C. Correns, and E. Tschermak independently recognized the importance of Mendel's discovery for biology.

A. The pea plant (*Pisum sativum*)

The garden pea normally reproduces by self-fertilization. Pollen from the anther falls onto the stigma of the same blossom. However, one can easily cross-fertilize (cross-pollinate) pea plants. The plant (left) consists of a stem, leaves, blossoms, and seedpods. In the blossom (right), the female and male reproductive organs are visible. The female pistil comprises stigma, style, and ovule. The male organ is the stamen, comprising the anther and filament. For cross-fertilization, Mendel opened a blossom and removed the anther to avoid self-fertilization. Then he transferred pollen from another plant to the receptive stigma directly.

B. The traits observed by Mendel

Mendel observed seven characteristic traits: (**1**) height of the plants, (**2**) location of the blossoms on the stem of the plant, (**3**) the color of the pods, (**4**) the form of the pods, (**5**) the form of the seeds, (**6**) the color of the seeds, and (**7**) the color of the seed coat. These traits were transmitted in defined proportions to the next plant generation.

Deviations from the Mendelian pattern of inheritance

Mendelian traits can deviate from the proportions described on the next page. *Epistasis* is a nonreciprocal interaction between nonallelic genes. As a result, the effect of one gene masks the expression of alleles of another gene. In 1902, W. Bateson described this phenomenon in the recessive gene *apterous* (*ap*) in *Drosophila*. Homozygotes are wingless, but, in addition, other genes affecting wing morphology, such as curled wing, are masked (the *ap* gene is epistatic to curled wing). The Bombay blood group is another example (Bhende et al., 1952; Race and Sanger, 1975).

Meiotic drive refers to the preferential transmission of one allele over others. As a result, one trait occurs in offspring much more frequently than others. Striking examples are the t complex of the mouse (about 99% instead of 50% of offspring of heterozygous t/+ male mice are also heterozygous) and segregation disorder in *Drosophila*.

In 1993, a mouse population in Siberia was described in which 85% and 65% of offspring were heterozygous for a chromosomal inversion. Homozygosity for the inversion leads to reduced fitness and is a selective disadvantage. Possibly, deviations from Mendelian laws are more frequent than previously assumed. *Genomic imprinting* (see p. 194) is a further cause of deviation from the Mendelian pattern of inheritance.

Further Reading

Bhende YM, et al. A "new" blood group character related to the ABO system. Lancet 1952;1:903–904

Brink RA, Styles ED. Heritage from Mendel. Madison: University of Wisconsin Press; 1967

Corcos AF, Monaghan FV. Gregor Mendel's Experiments on Plant Hybrids. New Brunswick: Rutgers University Press; 1993

Griffith AJF, et al. An Introduction to Genetic Analysis. 7th ed. New York: W. H. Freeman; 2000

Mendel G. Versuche über Pflanzenhybriden. Verh naturf Ver Brünn 1866;4:3–47

Pomiankowski A, Hurst LD. Evolutionary genetics. Siberian mice upset Mendel. Nature 1993;363:396–397

Race RR, Sanger R. Blood Groups in Man. 6th ed. Oxford: Blackwell; 1975

Vogel F, Motulsky AG. Human Genetics. Problems and Approaches. 3rd ed. Heidelberg: Springer-Verlag; 1997

Weiling F. Johann Gregor Mendel: Der Mensch und Forscher. II Teil. Der Ablauf der Pisum Versuche nach der Darstellung. Med Genetik 1993;2:208–222

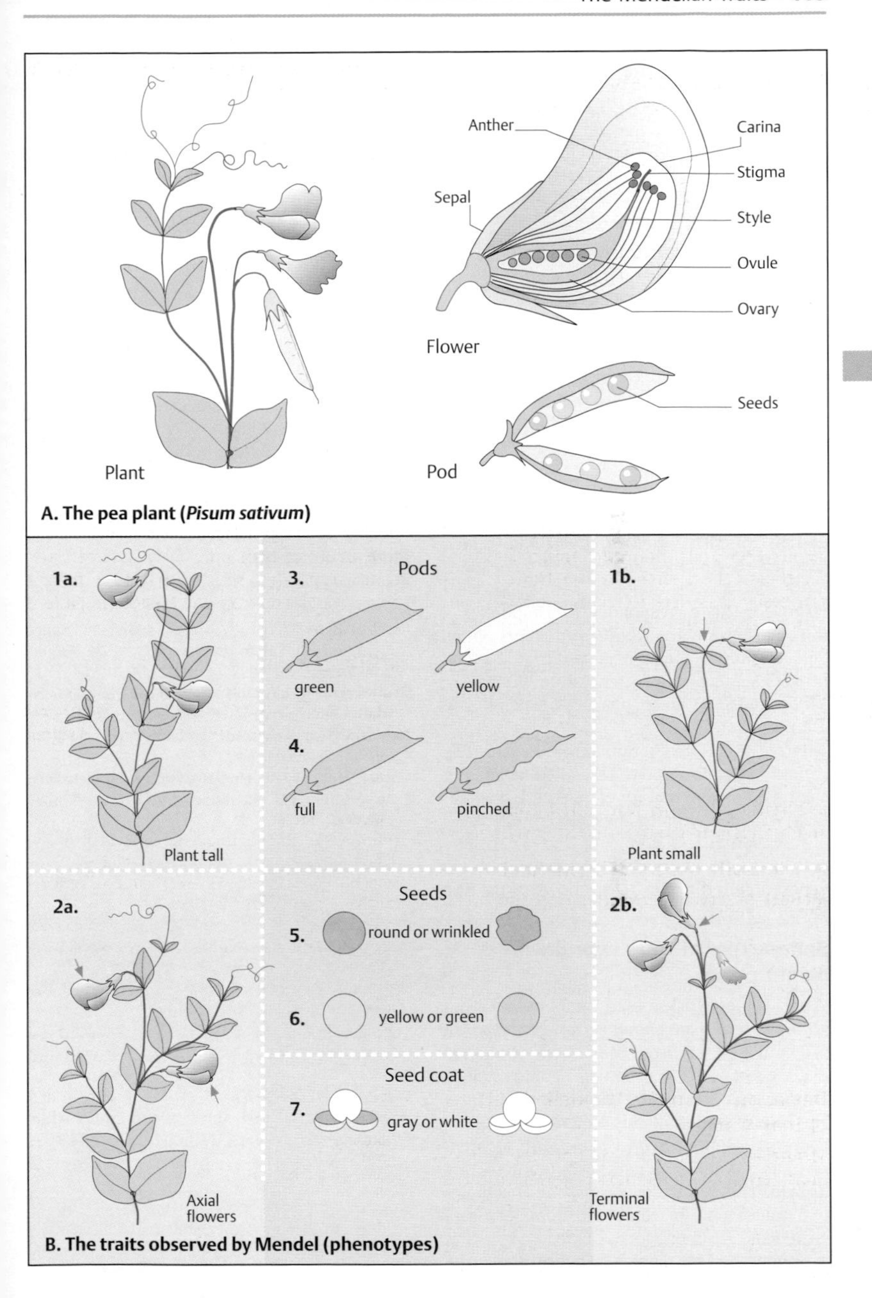

A. The pea plant (*Pisum sativum*)

B. The traits observed by Mendel (phenotypes)

Transmission to the Next Generation

Mendel observed that the traits of the pea plant (*Pisum sativum*) (described on the previous page) were transmitted to the next generation according to a defined pattern. He provided a meaningful biological interpretation, later called Mendel's laws of inheritance.

A. Segregation of dominant and recessive traits

In two different experiments, Mendel observed the shape (smooth or wrinkled) and the color (yellow or green) of the seeds. When he crossed the plants of the parental generation P, i.e., smooth and wrinkled or yellow and green, he observed that in the first filial (daughter) generation, F_1, all seeds were smooth and yellow.

In the next generation, F_2, which arose by self-fertilization, the traits observed in the P generation (smooth, wrinkled, green and yellow) reappeared. Among 7324 seeds of one experiment, 5474 were smooth and 1850 were wrinkled. This corresponded to a ratio of 3:1. In the experiment with different colors (green versus yellow), Mendel observed that in a total of 8023 seeds of the F_2 generation, 6022 were yellow and 2001 green, again corresponding to a ratio of 3:1.

The trait the F_1 generation showed exclusively (round or yellow), Mendel called *dominant*; the trait that did not appear in the F_1 generation (wrinkled or green) he called *recessive*. His observation that a dominant and a recessive pair of traits occurring (segregating) in the F_2 generation in the ratio 3:1 is known as the first law of Mendel.

B. Backcross of an F_1 hybrid with a parent plant

When Mendel backcrossed the F_1 hybrid plant with a parent plant showing the recessive trait (**1**), both traits occurred in the next generation in a ratio of 1:1 (106 round and 102 wrinkled). This is called the second law of Mendel.

The interpretation of this experiment (**2**), the backcross of an F_1 hybrid plant with a parent plant, is that different germ cells (gametes) are formed. The F_1 hybrid plant (round) contains two traits: one for round (R, dominant over wrinkled, r) and one for wrinkled (r, recessive to round, R). This plant is a hybrid (*heterozygous*) and therefore can form two types of gametes (R and r).

In contrast, the other plant is *homozygous* for wrinkled (r). It can form only one type of gamete (r, wrinkled). Half of the offspring of the heterozygous plant receive the dominant trait (R, round), the other half the recessive trait (r, wrinkled). The resulting distribution of the observed traits is a ratio of 1:1, or 50% each.

The observed trait is called the *phenotype* (the observed appearance of a particular characteristic). The composition of the two factors (genes) R and r, (Rr) or (rr), is called the *genotype*. The alternative forms of a trait (here, round and wrinkled) are called *alleles*. They are the result of different genetic information at one given gene locus.

If the alleles are different, the genotype is heterozygous; if they are the same, it is homozygous (this statement is always in reference to a single, given gene locus).

Further Reading

Brink RA, Styles ED. Heritage from Mendel. Madison: University of Wisconsin Press; 1967

Corcos AF, Monaghan FV. Gregor Mendel's Experiments on Plant Hybrids. New Brunswick: Rutgers University Press; 1993

Griffith AJF, et al. An Introduction to Genetic Analysis. 9th ed. New York: W. H. Freeman; 2007

Mendel G. Versuche über Pflanzenhybriden. Verh naturf Ver Brünn 1866;4:3–47

Vogel F, Motulsky AG. Human Genetics. Problems and Approaches. 3rd ed. Heidelberg: Springer-Verlag; 1997

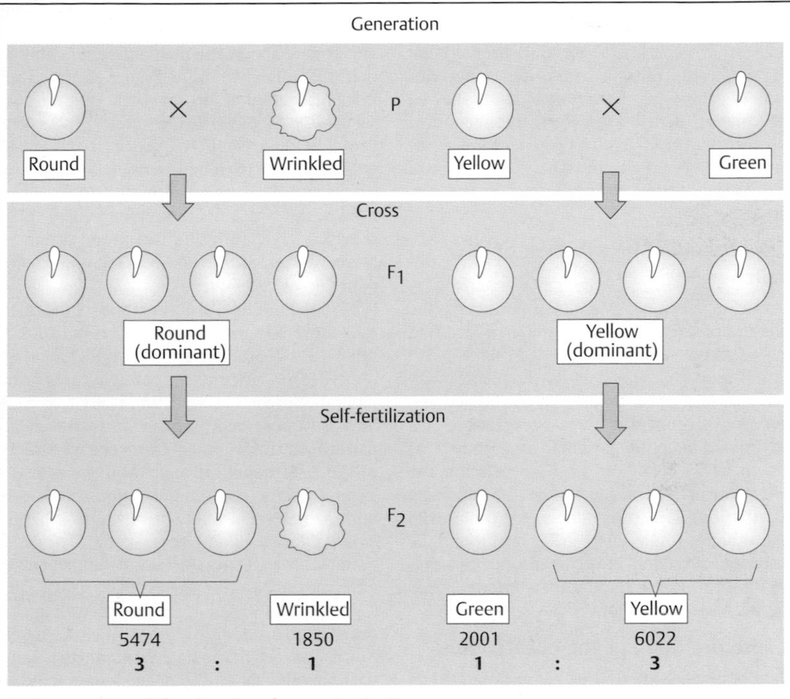

A. Segregation of dominant and recessive traits

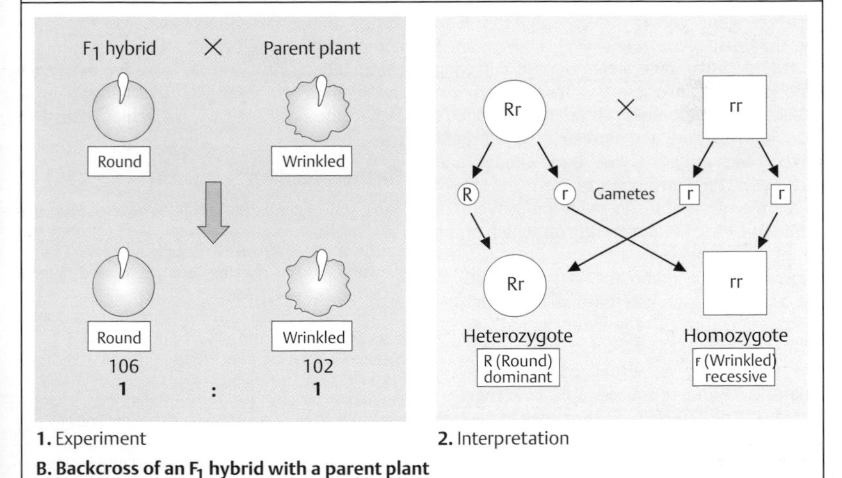

1. Experiment **2.** Interpretation

B. Backcross of an F_1 hybrid with a parent plant

Independent Distribution

In a further experiment, Mendel observed that two different traits are inherited independently of each other. Each pair of traits showed the same 3:1 distribution of the dominant over the recessive trait in the F_2 generation as he had observed previously. The segregation of two pairs of traits again followed certain patterns.

A. Independent distribution of two traits

In one experiment, Mendel investigated the crossing of the trait pair round/wrinkled and yellow/green. When he crossed plants with round and yellow seeds with plants with wrinkled and green seeds, only round and yellow seeds occurred in the F_1 generation. This corresponded to the original experiments as shown. Of 556 plants in the F_2 generation, the two pairs of traits occurred in the following distribution: 315 seeds yellow and round, 108 yellow and wrinkled, 101 green and round, 32 green and wrinkled. This corresponds to a segregation ratio of 9:3:3:1. This is referred to as the third Mendelian law.

B. Interpretation of the observations

Mendel's observations can be summarized as follows: If we assign the capital letter **G** to the dominant gene *yellow*, a lowercase **g** to the recessive gene *green*, the capital letter **R** to the dominant gene *round*, and a lowercase **r** to the recessive gene *wrinkled*, the following nine genotypes of these two traits can occur: **GGRR**, **GGRr**, **GgRR**, **GgRr** (all *yellow* and *round*); **GGrr**, **Ggrr** (*yellow* and *wrinkled*); **ggRR**, **ggRr** (*green* and *round*); and **ggrr** (*green* and *wrinkled*). The distribution of the traits shown in A is the result of the formation of gametes of different types, i.e., depending on which of the genes they contain.

The ratio of the dominant trait yellow (**G**) to the recessive trait green (**g**) is 12:4, or 3:1. Also, the ratio of dominant round (**R**) to wrinkled (**r**) seeds is 12:4, i.e., 3:1.

The results can be visualized in a diagram called the Punnett square. This is a checkerboard way of determining the types of zygotes produced when two gametes with a defined genotype fuse. It was first published in a book entitled *Mendelism* in its 2nd edition by Punnett (1910).

This square shows the nine different genotypes that can be formed in the zygote after fertilization. Altogether there are 9/16 yellow round seeds (**GRGR**, **GRGr**, **GrGR**, **GRgR**, **gRGR**, **GRgr**, **GrgR**, **gRGr**, **grGR**), 3/16 green round (**gRgR**, **gRgr**, **grgR**), 3/16 yellow wrinkled (**GrGr**, **Grgr**, **grGr**), and 1/16 green wrinkled seeds (**grgr**). Each of the two traits (dominant yellow versus recessive green, or dominant round versus recessive wrinkled) occurs in a 3:1 ratio (dominant versus recessive).

Why were Mendel's observations fundamentally new and completely different from all other 19th-century attempts to understand heredity? First, Mendel simplified the experimental approach by selecting traits that could be easily observed; second, he assessed the pattern of transmission from one generation to the next quantitatively; third, he provided a biologically meaningful interpretation by noting that each pair of traits was inherited independently of the other pairs of traits in predictable pattern. This was a fundamentally new insight into the process of heredity. Since it distinctly diverged from the prevailing concepts about heredity at the time, its significance was not immediately recognized. Today we know that genetically determined traits are independently inherited (segregation) only when they are located on different chromosomes or are far enough apart on the same chromosome to be separated each time by recombination. This was the case for the genes investigated by Mendel; these have been cloned and have had their molecular structures characterized.

Further Reading

Brink RA, Styles ED. Heritage from Mendel. Madison: University of Wisconsin Press; 1967

Corcos AF, Monaghan FV. Gregor Mendel's Experiments on Plant Hybrids. New Brunswick: Rutgers University Press; 1993

Griffith AJF, et al. An Introduction to Genetic Analysis. 9th ed. New York: W. H. Freeman; 2007

Mendel G. Versuche über Pflanzenhybriden. Verh naturf Ver Brünn 1866;4:3–47

Punnett RC. Mendelism. 2nd ed. Cambridge: Bowes & Bowes; 1910 (3rd ed. Cosimo Classics, 2007)

Vogel F, Motulsky AG. Human Genetics. Problems and Approaches. 3rd ed. Heidelberg: Springer-Verlag; 1997

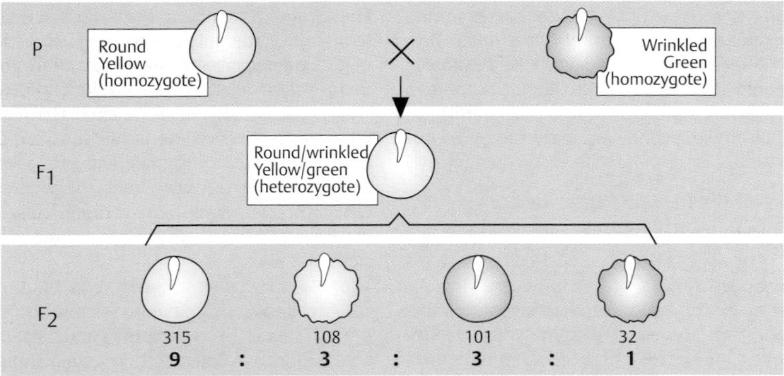

P

Round
Yellow
(homozygote)

Wrinkled
Green
(homozygote)

F1

Round/wrinkled
Yellow/green
(heterozygote)

F2

315 108 101 32

9 : 3 : 3 : 1

A. Independent segregation of two traits

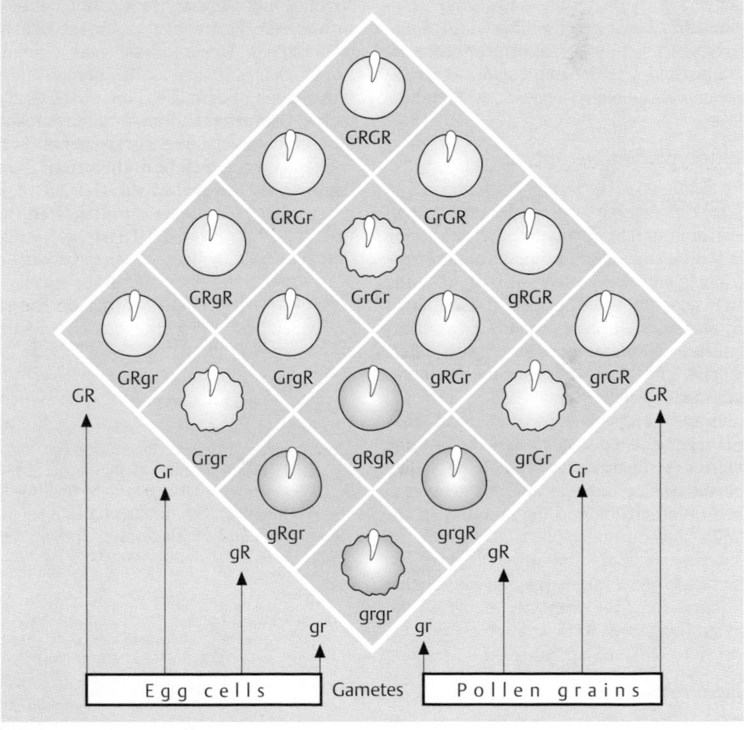

GRGR

GRGr GrGR

GRgR GrGr gRGR

GRgr GrgR gRGr grGR

Grgr gRgR grGr

gRgr grgR

grgr

GR Gr gR gr

GR Gr gR gr

Egg cells Gametes Pollen grains

B. Interpretation of the observation

Phenotype and Genotype

Formal genetic analysis in humans examines the genetic relationship of individuals presented in a pedigree (pedigree analysis). An observed *phenotype* could be a disease, a blood group, a protein variant, a laboratory result, or any other attribute determined by observation.

A. Symbols in a pedigree drawing

The symbols shown here represent a common way of drawing a pedigree. Males are shown as squares, females as circles. Individuals of unknown sex (e.g., because of inadequate information) are shown as diamonds. In medical genetics, the degree of reliability in determining the phenotype, e.g., presence or absence of a disease, should be stated. In each case it must be stated which phenotype (e.g., which disease) is being dealt with. Established diagnoses (data complete), possible diagnoses (data incomplete), and questionable diagnoses (statements or data dubious) should be differentiated.

B. Genotype and phenotype

The definitions of genotype and phenotype refer to a given *gene locus*. Different forms of genetic information at one and the same gene locus are called *alleles*. In diploid organisms—all animals and many plants—there are three possible genotypes with respect to two alleles at any one locus: (1) *homozygous* for two identical alleles, (2) *heterozygous* for the two different alleles, and (3) *homozygous* for the other two identical alleles.

If they can be recognized in the heterozygous state, they are called *dominant*. If they can be recognized in the homozygous state only, they are *recessive*. The concepts dominant and recessive are an attribute of the accuracy in observation and do not apply at the molecular level. If the two alleles can both be recognized in the heterozygous state, they are designated *codominant* (e.g., the alleles A and B of the blood group system ABO; O is recessive to A and B).

Medical relevance

The Mendelian pattern of inheritance provides the foundation for genetic counseling of patients with monogenic diseases. *Genetic counseling* is a communication process relating to all aspects related to the occurrence of a genetic disorder, in particular the diagnosis and assessment of the potential occurrence of a genetically determined disease in a family and in more distant relatives. The individual affected with a disease, who first attracted attention to a particular pedigree, is called the index patient (or proposita if female and propositus if male). The person who seeks information is called the consultand. The index patient and the consultand are very often different persons.

The goal of genetic counseling is to provide comprehensive information about the expected course of the disease, medical care, and possible treatments or an explanation, for why treatment is not possible. Genetic counseling includes a review of possible decisions about family planning as a consequence of a genetic risk. Professional confidentiality must be observed. The counselor makes no decisions. The increasing availability of information about a disease based on a DNA test (predictive DNA testing) prior to disease manifestation requires the utmost care in establishing whether it is in the interest of a given individual to have a test performed.

Further Reading

Griffith AJF, et al. An Introduction to Genetic Analysis. 9th ed. New York: W. H. Freeman; 2007

Harper PS. Practical Genetic Counselling. 7th ed. London: Edward Arnold; 2010

Jameson JL, Kopp P. Principles of human genetics In: Longo DL, et al., eds. Harrison's Principles of Internal Medicine. 18th ed. New York: McGraw-Hill; 2012:486–509

Rimoin DL, et al., eds. Emery and Rimoin's Principles and Practice of Medical Genetics. 6th ed. Philadelphia: Elsevier-Churchill Livingstone; 2013 (in press)

Vogel F, Motulsky AG. Human Genetics. Problems and Approaches. 3rd ed. Heidelberg: Springer-Verlag; 1997

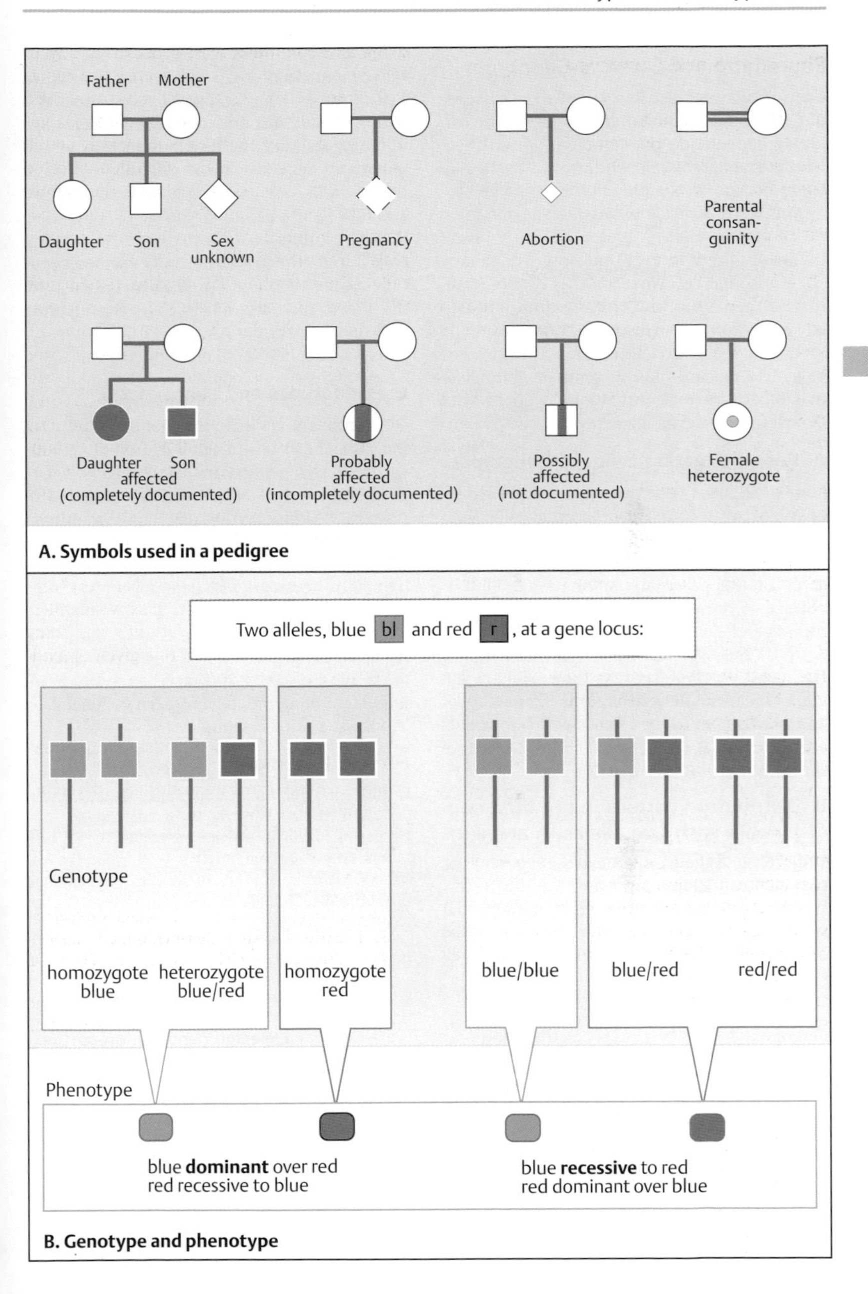

A. Symbols used in a pedigree

B. Genotype and phenotype

Segregation of Parental Genotypes

The segregation (distribution) of the genotypes of the parents (parental genotypes) in the offspring depends on the combination of the alleles in the parents. Depending on the effect of the genotype on the phenotype in the heterozygous state, an allele is classified as dominant or recessive. Hence, there are three basic modes of inheritance: (1) autosomal dominant, (2) autosomal recessive, and (3) X-chromosomal. For genes on the X chromosome, it is usually not important to distinguish dominant and recessive (see below). Since the Y chromosome bears few disease-causing genes, Y-chromosomal inheritance can usually be ignored with regard to monogenic diseases.

A. Possible mating types of genotypes

For a gene locus with two alleles, there are six possible combinations of parental genotypes (**1–6**). Here two alleles, blue (bl) and red (r), are shown, with blue dominant over red. In three of the parental combinations (**1**, **3**, **4**) neither of the parents is homozygous for the recessive allele (red). In three parental combinations (**2**, **5**, **6**), one or both parents manifest the recessive allele because they are homozygous. The distribution patterns of genotypes and phenotypes in the offspring of the parents are shown in B. In these examples, the sex of the parents is interchangeable.

B. Distribution pattern in the offspring of parents with two alleles: A and a

With three of the parental mating types for the two alleles **A** (dominant over **a**) and **a** (recessive to **A**), there are three combinations that lead to segregation (separation during meiosis) of allelic genes. These correspond to the parental combinations 1, 2, and 3 shown in A. In mating types 1 and 2, one of the parents is a heterozygote (**Aa**) and the other parent is a homozygote (**aa**). The distribution of observed genotypes expected in the offspring is 1:1; in other words, 50% (0.50) are **Aa** heterozygotes and 50% (0.50), **aa** homozygotes.
If both parents are heterozygous **Aa** (mating type 3 in A), the proportions of expected genotypes of the offspring (**AA**, **Aa**, **aa**) occur in a ratio of 1:2:1. In each case, 25% (0.25) of the offspring will be homozygous **AA**, 50% (0.50) heterozygous **Aa**, and 25% (0.25) homozygous **aa.**

C. Phenotypes and genotypes

One dominant allele (in the first pedigree, **A**, in the father) can be expected in 50% of the offspring. If both parents are heterozygous, 25% of the offspring will be homozygous **aa**. If both parents are homozygous, one for the dominant allele **A**, the other for the recessive allele **a**, then all offspring are obligate heterozygotes (i.e., must necessarily be heterozygotes).

Medical relevance

An attempt to determine the mode of inheritance of a disorder (pedigree analysis) is an important fundamental approach in genetic diagnostics and counseling.

Further Reading

Griffith AJF, et al. An Introduction to Genetic Analysis. 9th ed. New York: W. H. Freeman; 2007
Harper PS. Practical Genetic Counselling. 7th ed. London: Edward Arnold; 2010
Vogel F, Motulsky AG. Human Genetics. Problems and Approaches. 3rd ed. Heidelberg: Springer-Verlag; 1997

Expected distribution of genotypes in the offspring of different parental genotype combinations

Parents	Offspring	Expected genotype proportion
AA × AA	AA	1 (100%)
AA × Aa	AA, Aa	1:1 (each 50%)
Aa × Aa	AA, Aa, aa	1:2:1 (25%, 50%, 25%)
AA × aa	Aa	1 (100%)
Aa × aa	Aa, aa	1:1 (50% each)
aa × aa	aa	1 (100%)

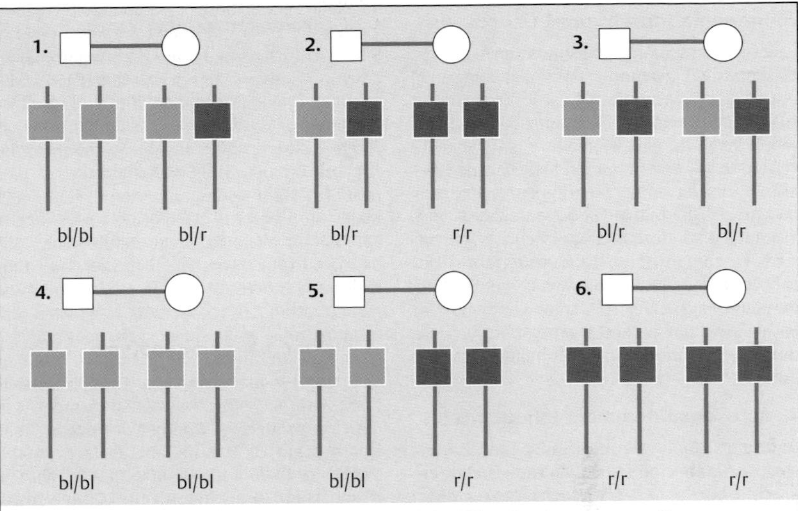

A. Possible mating types of genotypes for two alleles (blue dominant over red)

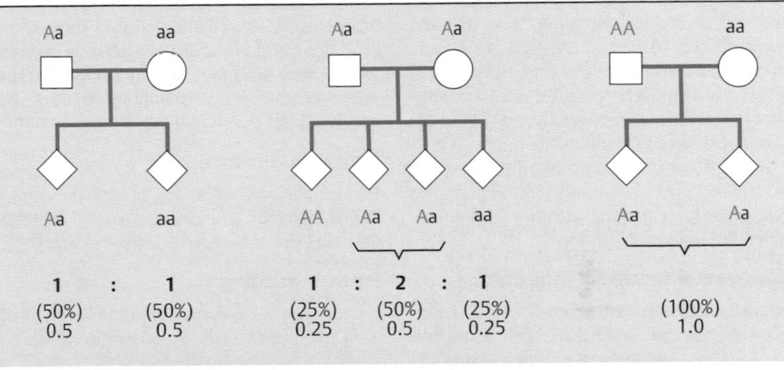

B. Expected distribution of genotypes in offspring of parents with two alleles, A and a

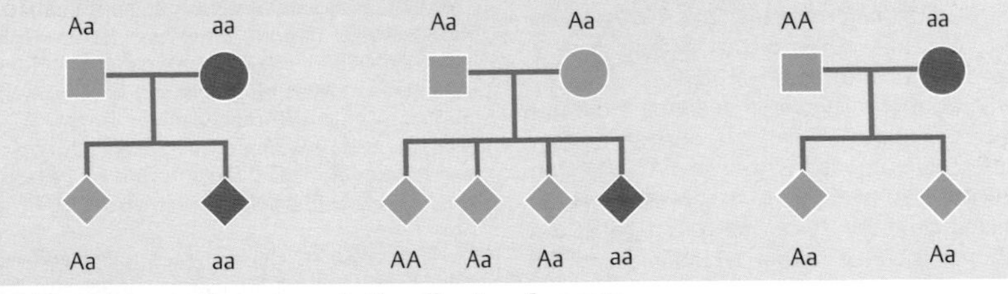

C. Phenotypes and genotypes in the offspring of parents with a dominant allele A and a recessive allele a

Monogenic Inheritance

Four types of monogenic inheritance can be distinguished: autosomal dominant, autosomal recessive, X-chromosomal, and Y-chromosomal. Since the human Y chromosome contains only a small number of genes (p. 240) that are related to sex determination and spermatogenesis, Y-chromosomal inheritance can usually be ignored. In human pedigree analysis it is customary to designate consecutive generations by roman numerals, starting with I for the first generation about which some information is available, and arabic numerals for each individual within a generation, or nonoverlapping numbers for computer calculations.

A. Autosomal dominant inheritance

Autosomal dominant inheritance is characterized by: (i) affected individuals are directly related in one or more successive generations; (ii) both males and females are affected in a 1:1 ratio; (iii) the expected proportion of affected and unaffected offspring of an affected individual is 1:1 (0.50, or 50%) each. An important consideration in autosomal disorders is whether a new mutation is present in a patient without affected parents (Pedigrees 2 and 3). In some autosomal dominant disorders, a carrier of the mutation does not manifest the disease. This is called *nonpenetrance.* The degree of manifestation can vary within a family. This is called *variable expressivity.*

B. Autosomal recessive inheritance

Autosomal recessive inheritance is characterized by: (i) affected siblings of both sexes in a ratio of 1:1, (ii) nonaffected parents. The genotypes of children of heterozygous patients are distributed in a ratio 1:2:1 (25% homozygous normal, 50% heterozygous, 25% homozygous affected). Thus, heterozygous parents have a risk of 25% of affected offspring.

In the examples shown, in pedigree **1** the unaffected parents (II-3 and II-4) must be heterozygous; the same holds true for I-1 and I-2 of pedigree **2**. In pedigree **3**, the ancestors of the affected child (IV-2) can be traced back to a common ancestor (either I-1 and I-2) of the parents (III-1 and III-2), who are first cousins (parental consanguinity, blood relationship). The double line in the pedigree indicates the parental consanguinity.

C. X-chromosomal inheritance

Since males have only one X chromosome, all daughters always inherit the father's X chromosome in addition to one of the two X chromosomes of their mother. Sons inherit the paternal Y chromosome and one of the maternal X chromosomes. Thus, the characteristic pattern of X-chromosomal inheritance is: (i) only males are affected; (ii) affected males inherit the mutant allele from the mother only; (iii) no male-to-male transmission. Therefore, a female who is heterozygous for an X-chromosomal mutation has a 50% risk of an affected son. Heterozygous females may show mild manfestations of an X-linked disease as a result of incomplete X-inactivation (p. 196). In isolated cases with an X-chromosososomal disorder it is difficult to decide whether it results from transmission of the mutant allele from the mother or from a new mutation. Therefore, it is important to distinguish these three situations.

Females with an affected son and an affected brother or with two affected sons must be heterozygous and are said to be *obligate heterozygotes.* Those who may or may not be heterozygous are *facultative heterozygotes* (e.g., III-5 and IV-2).

Medical relevance

Human diseases following one of these inheritance patterns are referred to as *Mendelian disorders* (for a complete list see OMIM).

Further Reading

Griffith AJF, et al. An Introduction to Genetic Analysis. 9th ed. New York: W. H. Freeman; 2007

Harper PS. Practical Genetic Counselling. 7th ed. London: Edward Arnold; 2010

Jameson JL, Kopp P. Principles of human genetics In: Longo DL, et al., eds. Harrison's Principles of Internal Medicine. 18th ed. New York: McGraw-Hill; 2012:486–509

OMIM. Online Mendelian Inheritance of Man. Available at: http://www.ncbi.nlm.nih.gov/omim

Rimoin DL, et al., eds. Emery and Rimoin's Principles and Practice of Medical Genetics. 6th ed. Philadelphia: Elsevier-Churchill Livingstone; 2013 (in press)

Vogel F, Motulsky AG. Human Genetics. Problems and Approaches. 3rd ed. Heidelberg: Springer-Verlag; 1997

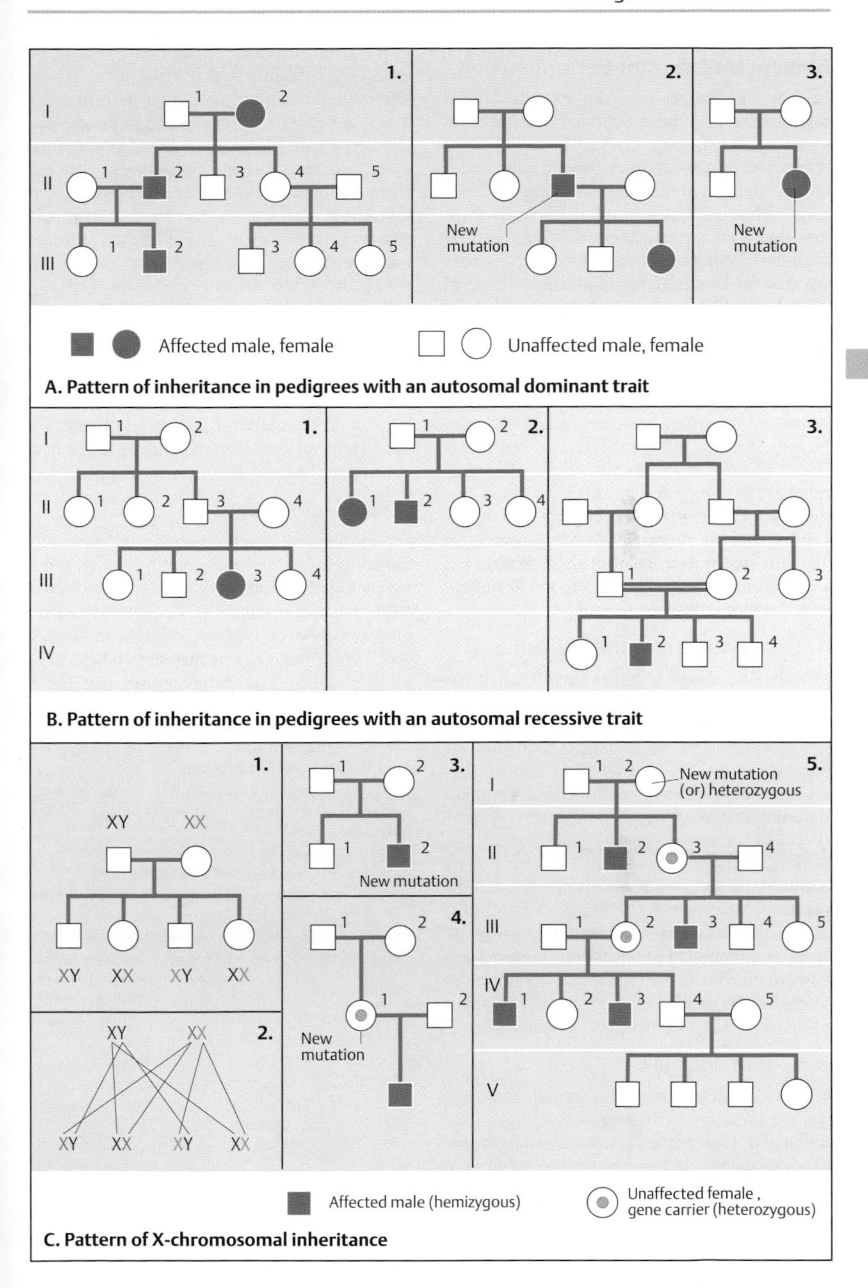

A. Pattern of inheritance in pedigrees with an autosomal dominant trait

B. Pattern of inheritance in pedigrees with an autosomal recessive trait

C. Pattern of X-chromosomal inheritance

Genetic Linkage and Recombination

Genetic linkage refers to an exception from segregation and independent assortment of genetic traits.

It is based on the fact that two or more genes located near each other on the same chromosome often are transmitted together. Linkage was first reported in 1902 by Correns, and established by Bateson and coworkers in genes for colored flowers and long pollen shape of sweet peas. The ratio of being transmitted together versus being separated allows an estimate of their distance from each other.

Linkage relates to gene loci, not to specific alleles. Gene loci located on the same chromosome and inherited together are called a *haplotype*. Alleles at different loci that are inherited together more frequently or less frequently than expected by their individual frequencies are said to show *linkage disequilibrium* (p. 128). *Synteny* (H. J. Renwick in 1971) refers to gene loci being located on the same chromosome without regard to linkage or the distance between them.

A. Detection of recombination

Whether neighboring genes on the same parental chromosome remain together or become separated depends on whether crossing-over between the two gene loci **A** and **B**, having the respective alleles **Aa** and **Bb**, has occurred or not during meiosis. If no recombination took place, this part of the chromosome is called nonrecombinant. If recombination took place, it is recombinant (**1**). The genetic result (**2**) for two neighboring gene loci **A** and **B** on the same chromosome is one of two possibilities: not recombinant (gametes correspond to parental genotype) or recombinant (new combination). The two possibilities can be differentiated only when the parental genotype is informative for both gene loci (**Aa** and **Bb**).

B. Genetic linkage

The inheritance pattern of a mutation causing a disease may not be recognizable, e.g., in autosomal recessive or X-chromosomal inheritance. However, if the disease locus is linked to one or several loci with polymorphic alleles, its presence or absence can be indirectly assessed. Two pedigrees are shown, one without recombination (**1**) and with recombination (2).

In the first example (**1**) the father and three children (red symbols in the pedigree) are affected (the children are shown as diamonds, their genders ignored). Since all three affected children share the marker allele **A** with each other and their father, this indicates that the allele **B** at the disease locus must be the mutant allele. The unaffected children share the paternal marker locus **a**. This would indicate that the disease locus **b** is not the mutant allele. Thus, recombination has not occurred.

The situation differs if recombination has occurred, as in two individuals shown (**2**). An affected individual has inherited alleles **a** and **B** from the father, instead of **A** and **B**; an unaffected individual has inherited allele **A** and allele **b.** This situation can only be observed if the father (affected parent) is heterozygous at the marker locus (**Aa**).

Medical relevance

DNA analysis of polymorphic marker loci linked to a disease locus can be used within a family to determine whether unaffected family members have inherited the mutant allele known to be present in an affected family member (indirect DNA genetic analysis).

Further Reading

Bateson W, Saunders ER, Punnett RC. Experimental studies in the physiology of heredity. Reports Evolut Comm Royal Soc 1906;3:1–53

Correns C. Scheinbare Ausnahmen von der Mendelschen Spaltungsregel für Bastarde. Ber deutsch bot Ges 1902;20:97–172

Griffith AJF, et al. An Introduction to Genetic Analysis. 9th ed. New York: W. H. Freeman; 2007

Harper PS. Practical Genetic Counselling. 7th ed. London: Edward Arnold; 2010

Vogel F, Motulsky AG. Human Genetics. Problems and Approaches. 3rd ed. Heidelberg: Springer-Verlag; 1997

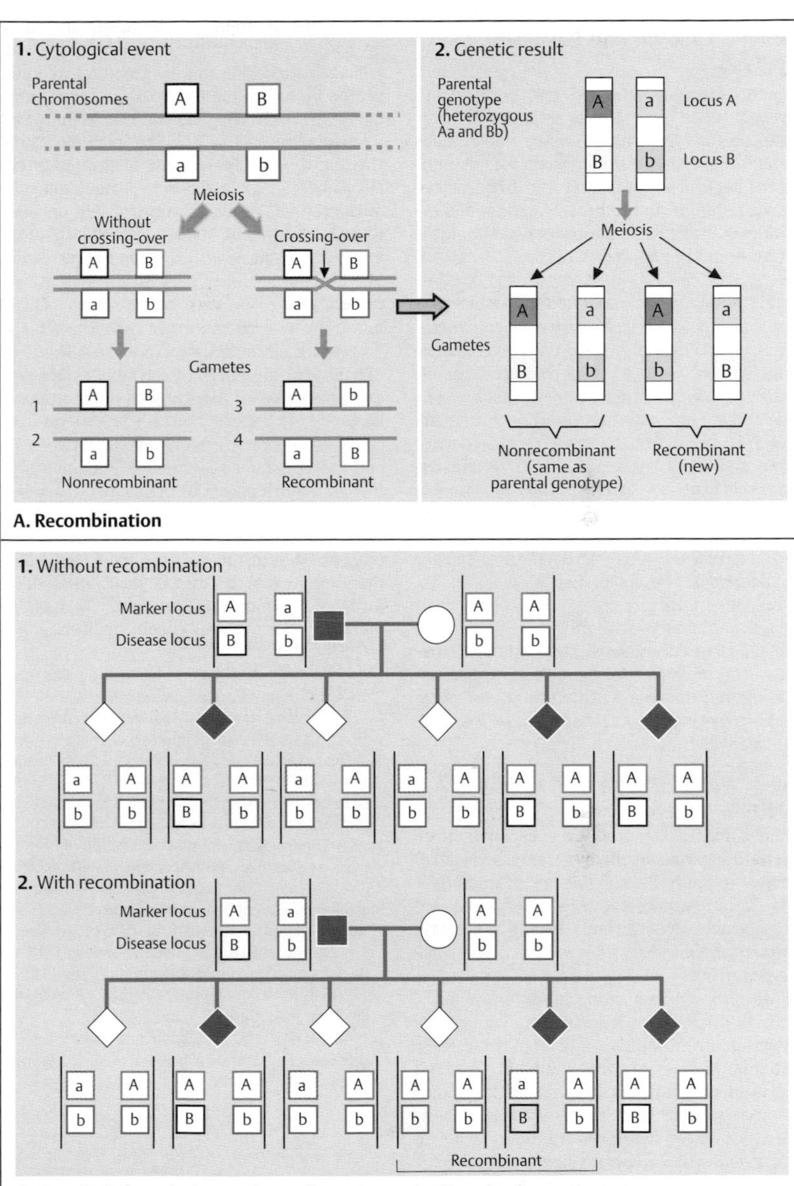

A. Recombination

B. Genetic linkage between two adjacent gene loci (marker locus, A, and autosomal dominant mutation, B)

Genetic Linkage and Association Analysis

Genetic linkage estimates the distance between two or more loci or adjacent DNA regions on the same chromosome. It determines either the physical distance, expressed in number of base pairs, or simply the frequency of recombination. In contrast, association expresses a statistical relation between specific alleles and a phenotype.

A. LOD score in linkage analysis

The four LOD score curves are a graphic representation of linkage analysis (a, no recombination; b, Θ of 0.15, i.e. 15% recombination; c, unlikely; d, inconclusive). Genetic linkage analysis determines how often two or more gene loci are transmitted together compared with how often they are separated by recombination. This ratio is called *recombination fraction*, designated by the Greek letter *theta* (Θ). It expresses the genetic distance. Loci that are not linked have a Θ value of 0.5. If Θ is zero, the loci are identical. Linkage of two gene loci is assumed when the probability of linkage divided by the probability of no linkage is equal to or greater than the ratio of 1000:1 (10^3:1). The logarithm of this ratio, the odds, is called the LOD score. The four LOD score curves are a graphic represention of linkage analysis. (Figure adapted from Emery, 1986.)

B. Use of haplotypes in association analysis

A deviation of the frequency of a particular allele in a population from the expected frequency is referred to as *linkage disequilibrium* (LD). It can be detected by a set of associated polymorphic DNA variants (single nucleotide polymorphisms [SNPs], see p. 78), called a *haplotype*.

Three SNP markers along a short stretch of DNA (**1**) from four unrelated inividuals are shown (chromosomes 1–4). The three SNPs differ by their polymorphic alleles (A/G, T/C, C/G, respectively). This defines four haplotypes (**2**). Genotyping just three SNPs (TagSNPs) identifies the haplotypes (**3**). For example ACG defines haplotype 2. (Figure from The International HapMap Consortium, 2003.)

C. Linkage disequilibrium (LD)

Linkage disequilibrium is graphically represented by haplotype blocks. This is a pairwise comparison of all SNPs in the DNA region studied with each other. The example shows this for 10 SNPs (rs7190220 at the top at the left to rs9931018 at the right). The first square at the top left shows the relationship between 39 and 40 (SNPs rs7190220 and rs178023041). The bottom square displays pair 39 and 49, etc. The LD between each pair is shown by a color reflecting the strength of LD. A red square without a number indicates complete linkage, whereas a white square points to absence of linkage. The numbers in the squares are percentages between pairs of loci. In the example shown, 42 (SNP rs11863156) to 49 (rs9931018) form a block of LD between 86 and 100 (bright red square without a number). If such a block occurs in higher frequency in a cohort of patients compared with controls, one may assume that this region contributes to susceptibility to the disease under study. (Figure kindly provided by Tea Berulava, Institute for Human Genetics Essen, Essen, Germany.)

Further Reading

Emery AEH. Methodology in Medical Genetics. 2nd ed. Edinburgh: Churchill Livingstone; 1986

Evseeva I, et al. Linkage disequilibrium and age of HLA region SNPs in relation to classic HLA gene alleles within Europe. Eur J Hum Genet 2010;18:924–932

International HapMap Consortium. The International HapMap Project. Nature 2003;426:789–796

Lander ES, Kruglyak L. Genetic dissection of complex traits: guidelines for interpreting and reporting linkage results. Nat Genet 1995;11:241–247

Morton NE. Sequential tests for the detection of linkage. Am J Hum Genet 1955;7:277–318

Ott J. Analysis of Human Genetic Linkage. 3rd ed. Baltimore: Johns Hopkins University Press; 1999

Strachan T, Read AP. Human Molecular Genetics. 4th ed. New York: Garland Science; 2011

Terwilliger J, Ott J. Handbook for Human Genetic Linkage. Baltimore: Johns Hopkins University Press; 1994

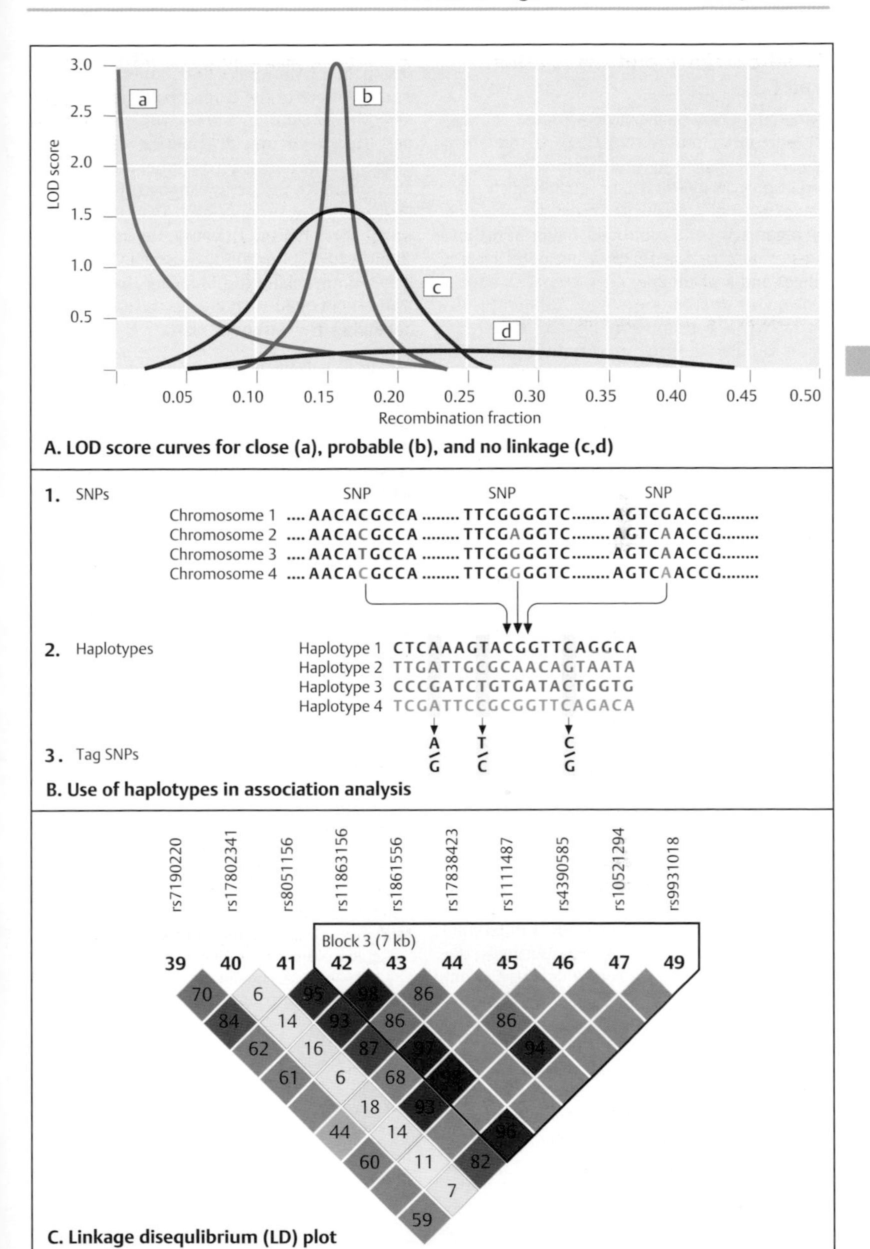

A. LOD score curves for close (a), probable (b), and no linkage (c,d)

B. Use of haplotypes in association analysis

C. Linkage disequlibrium (LD) plot

Quantitative Differences in Genetic Traits

Many phenotypes are not determined at a single gene locus, but by two, three or more loci. This is referred to as multifactorial or multigenic (polygenic) inheritance. Often such phenotypes are of a quantitative nature, obtained by measurements, e.g., body height. Furthermore, interaction with environmental factors may occur. Quantitative characteristics have a continuous distribution among different individuals. The transmission of the underlying genes cannot be recognized individually. The term *quantitative genetics* was introduced by Francis Galton (1822–1911) in 1883. The underlying genetic variation (genotypic variation) is interchangeably called polygenic (many genes), multigenic (several genes), or multifactorial (many factors).

A. Length of the corolla in *Nicotiana longiflora*

The plant *Nicotiana longiflora* is an example of how the genetic transmission of a quantitative trait can be analyzed experimentally. When parent plants with average corolla lengths of 40 cm and 90 cm are crossed, the next generation (F_1) shows a corolla length distribution that is longer than that of the short parent and shorter than that of the tall parent. In the next generation (F_2) the distribution spreads at both ends toward long and short. If plants from the short, middle, and long varieties are crossed, the mean distribution in the next generation (F_3) corresponds to the corolla length of the parental plants. They are short (shown on the left), long (right), or average (in the middle). This can be explained by a difference in the distribution of genes contributing to the variation in the trait. (Figure adapted from Ayala and Kiger, 1984.)

B. Influence of the number of gene loci on a quantitative trait

Four situations with 1, 2, 3, and 4 loci and two allele pairs A and a are shown for the parental generation P. Two phenotypes are shown (yellow bars). The hypothetical genotypes aa and AA, aabb and AABB, etc. appear uniform in the F_1 generation. In the F_2 generation, the number of distinct groups depends on the number of genes involved. With four loci, the difference in size between the groups is small. It begins to resemble a continuous distribution, as shown in the bottom of the diagram on the right. The smooth distribution curve corresponds to a bell-shaped normal distribution or Gaussian curve. The variance of the phenotype (V_P) is the sum of the genetic variance (V_G) and the environmental variance (V_E). The ratio V_G/V_T is called heritability. However, in many cases these types of variance cannot be distinguished, especially in humans. A locus contributing to a quantitative characteristic is called a quantitative trait locus (QTL).

Medical relevance

Gene loci that are part of a quantitative trait are called QTLs. The analysis of QTLs is important in the genetic analysis of complex diseases (multifactorial diseases). Examples of common quantitative genetic characteristics in humans are height, weight, eye and skin color, blood pressure, plasma concentrations of glucose and lipids, intellectual abilities, behavioral patterns, and others.

Further Reading

Ayala FJ, Kiger JA. Modern Genetics. 2nd ed. Menlo Park, CA: Benjamin/Cummings; 1984

Falconer DS, Mackay TFC. Introduction to Quantitative Genetics. 4th ed. London: Longman; 1996

Griffith AJF, et al. An Introduction to Genetic Analysis. 9th ed. New York: W. H. Freeman; 2007

King R, Rotter J, Motulsky AG, eds. The Genetic Basis of Common Disorders. 2nd ed. Oxford; Oxford University Press; 2002

Mackay TFC. The genetic architecture of quantitative traits. Annu Rev Genet 2001;35:303–339

Vogel F, Motulsky AG. Human Genetics. Problems and Approaches. 3rd ed. Heidelberg: Springer-Verlag; 1997

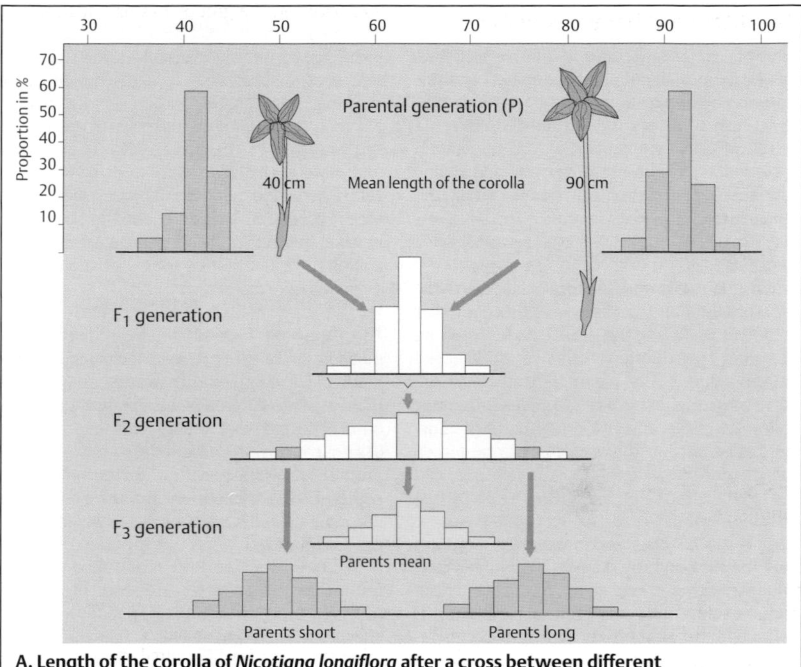

A. Length of the corolla of *Nicotiana longiflora* after a cross between different types of parental plants

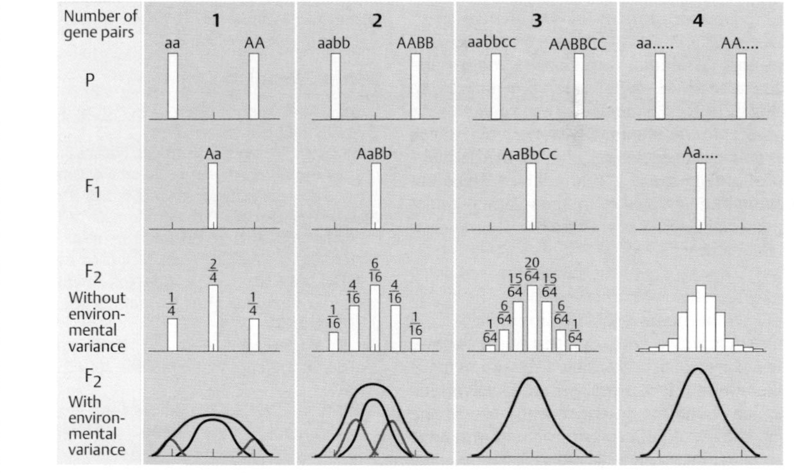

B. Distribution of frequency in the F_2 generation with a different number of gene loci

Polygenic Threshold Model

Polygenic inheritance of quantitative traits occurs in all animal and plant species. Its analysis requires the application of statistical methods. These help to assess the difference between a sample of measurements and the population from which the sample is derived and define how much confidence can be placed in the conclusions.

A. Normal distribution

When quantitative data from a large sample are plotted along the abscissa and the number of individuals along the ordinate, the resulting frequency distribution forms a bell-shaped Gaussian curve. The mean ($\bar{X}$) intersects the curve at its highest point and divides the area under the curve into two equal parts (**1**). Further perpendicular intersections can be placed one standard deviation (s) to the left ($-1s$) and to the right ($+1s$) of the mean to yield two additional areas: c to the left and d to the right. Areas a and b each comprise 34.13% of the total area under the curve (**2**). Further partitioning with perpendiculars to the abscissa at two and three standard deviations ($-2s$ and $-3s$ to the left and $+2s$ and $+3s$ to the right) results in further subsections (**3**).

The mean of a sample ($\bar{X}$) is determined by the sum of individual measurements (Σx) divided by the number of individuals (n) (equation 1). Many measurements yield the frequency f_x of observed individuals (equation 2). The population variance (σ^2) defines the variability of the population. It is expressed as the square of the sum (Σ) of individual measurements (x) minus the population mean (μ), divided by the number of individuals in the population (N) (equation 3). The variance of the population cannot be determined directly. Therefore, the variance of the sample (s^2) has to be estimated (equation 4). The square of the sum of the individual measurements (x) minus the mean ($\bar{X}$) is divided by the number of measurements (n). A correction factor $n/n-1$ is introduced because the number of independent measurements is $n-1$, resulting in a simplified equation 5. The standard deviation (s, equation 6), the square root of the sample variance (s^2), rests on a large number of successive samples. (Figure adapted from Burns & Bottino, 1989.)

B. Polygenic threshold model

Some continuously variable characteristics represent susceptibility to a disease manifest at a certain threshold. According to the threshold model, populations differ with respect to susceptibility to the disease (**1**). The liability to disease in the offspring of affected individuals (first-degree relatives, **2**) or in second-degree relatives (**3**) is closer to the threshold than in unrelated individuals from the general population.

C. Sex differences in threshold

The threshold beyond which a disease will manifest may differ between males and females (**1**). In families with at least one affected individual, a difference in the proportion of other affected individuals will be observed (**2**). In contrast to expectation, however, the proportion of affected first-degree relatives is higher if the index patient is female than if he is a male (**3**). This seeming paradox, known as the Carter effect after the medical geneticist Cedric O. Carter (London), is assumed to result from a greater or lesser influence of disease-causing genetic factors in the gender less often affected in the population.

Medical relevance

The genetic analysis of complex diseases (multifactorial, multigenic) is based in part on the polygenic threshold model.

Further Reading

Burns GW, Bottino PJ. The Science of Genetics. 6th ed. London: Macmillan; 1989

Comings DE. Polygenic disorders. Nature Encyclopedia of the Human Genome. Volume 4; London: Nature Publishing Group; 2003:589–595

Falconer DS, Mackay TFC. Introduction to Quantitative Genetics. 4th ed. London: Longman; 1996

Fraser FC. The William Allan Memorial Award Address: evolution of a palatable multifactorial threshold model. Am J Hum Genet 1980;32(6):796–813

Glazier AM, Nadeau JH, Aitman TJ. Finding genes that underlie complex traits. Science 2002;298:2345–2349

Hill WG, Mackay TF. D. S. Falconer and Introduction to quantitative genetics. Genetics 2004;167:1529–1536

King R, Rotter J, Motulsky AG, eds. The Genetic Basis of Common Disorders. 2nd ed. Oxford; Oxford University Press; 2002

1.

a b

$\bar{x}$

2.

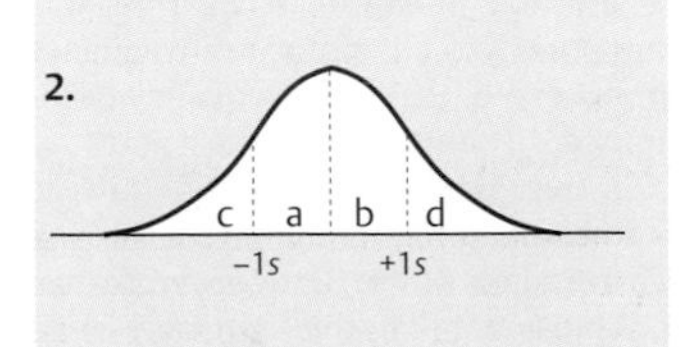

3.

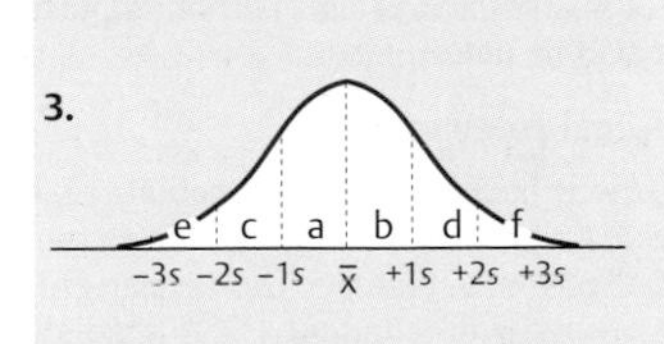

1. $$\bar{x} = \frac{\sum x}{n} \quad \text{(Mean)}$$

2. $$\bar{x} = \frac{\sum fx}{n}$$

3. $$\sigma^2 = \frac{\sum (x - \mu)^2}{N} \quad \text{Population variance}$$

4. $$s^2 = \left(\frac{\sum (x - \bar{x})^2}{n}\right)\left(\frac{n}{n-1}\right)$$

5. $$s^2 = \frac{\sum (x - \bar{x})^2}{(n-1)}$$

6. $$s = \sqrt{\frac{\sum f(x - \bar{x})^2}{n-1}}$$

(Standard deviation)

A. Normal distribution of a trait

1.

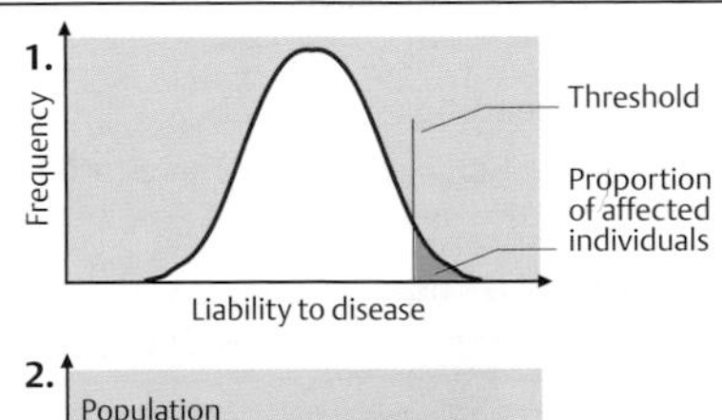

2.

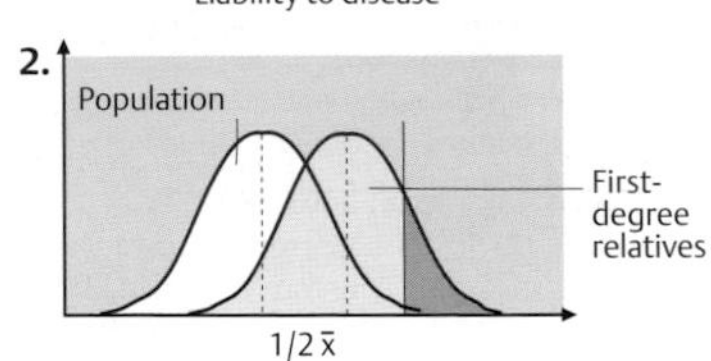

3.

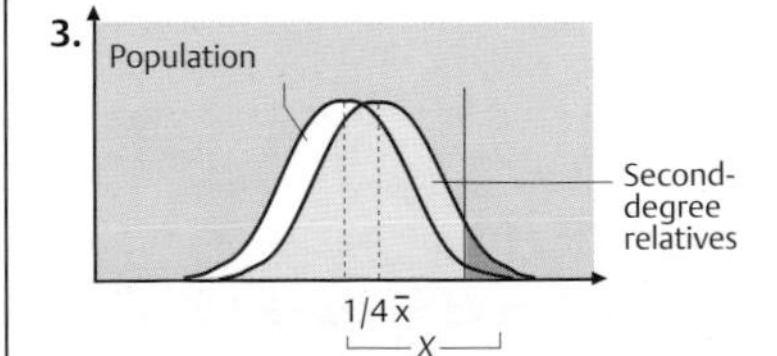

B. Polygenic threshold model

1.

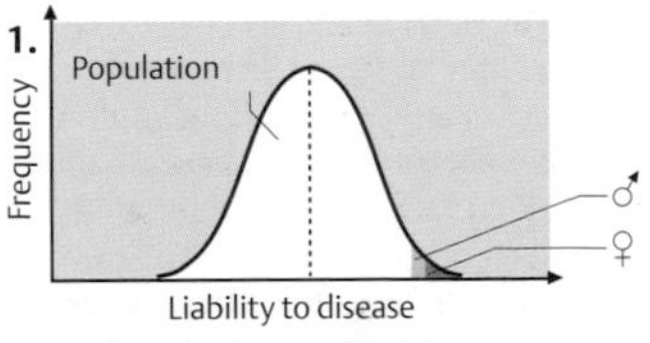

2.

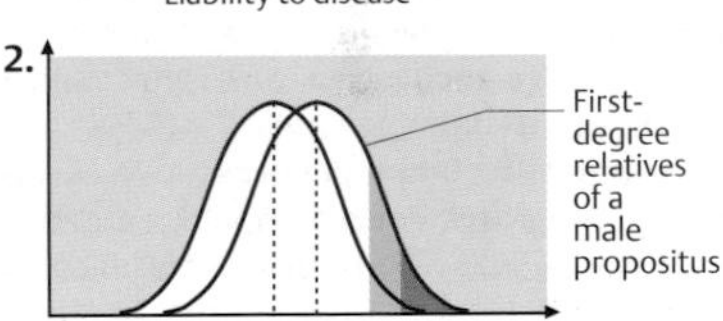

3.

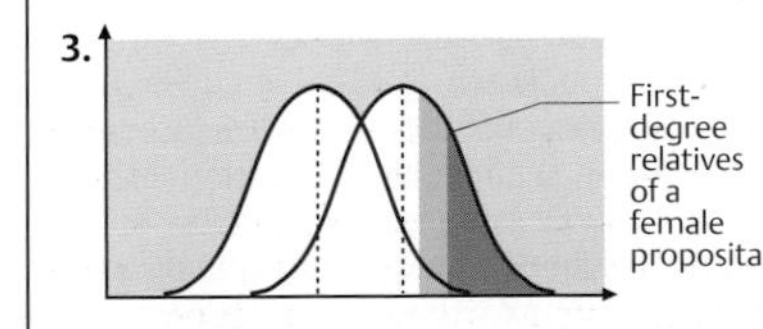

C. Sex differences in thresholds

Allelic Distribution in a Population

Population genetics is the scientific study of the genetic composition of populations. A principal goal is to estimate the frequency of alleles at different gene loci in natural populations (*allele frequency*, also called *gene frequency*). From this, conclusions can be drawn about possible selective influences that might explain differences observed. A population can be characterized on the basis of the frequency of alleles at various gene loci.

A. Frequency of genotypes in the children of parents with various genotypes

With regard to an allele pair A (dominant) and a (recessive), six types of parental genotype matings are possible (**1–6**). Each of these has an expected distribution of genotypes in the offspring according to the Mendelian laws, as indicated in the figure. This pattern will only be observed if all genotypes can participate in mating and are not prohibited by a severe disease. The frequency with which each mating type occurs depends on the frequencies of the alleles in the population.

B. Allele frequency

The allele frequency designates the proportion of a given allele at a given locus in a population. If an allele accounts for 20% of all alleles present (at a given locus) in the population, its frequency is 0.20. The allele frequency determines the frequencies of the individual genotypes in a population.
For example, for a gene locus with two possible alleles **A** and **a**, three genotypes are possible: **AA**, **Aa**, or **aa.** The frequency of the two alleles together (p the frequency of **A** and q the frequency of **a**) is 1.0 (100%). If two alleles **A** and **a** are equally frequent (each 0.5), they have the frequencies of $p = 0.5$ for allele **A**, and $q = 0.5$ for allele **a** (**1**). Thus, the equation $p + q = 1$ defines the population at this locus. The frequency distribution of the two alleles in a population follows a simple binomial relationship: $(p + q)^2 = 1$. Accordingly, the distribution of genotypes in the population corresponds to $p^2 + 2pq + q^2 = 1.0$. The expression p^2 corresponds to the frequency of the genotype **AA**; the expression $2pq$ corresponds to the frequency of the heterozygotes **Aa**; and q^2 corresponds to the frequency of the homozygotes **aa.**
When the frequency of an allele is known, the frequency of the genotype in the population can be determined. For instance, if the frequency p of allele **A** is 0.6 (60%), then the frequency q of allele **a** is 0.4 (40%, derived from $q = 1 - p$ or $1 - 0.6$). Thus, the frequency of the genotype **AA** is 0.36; that of **Aa** is $2 \times 0.24 = 0.48$; and that of **aa** is 0.16 (**2**).
And conversely, if genotype frequency has been observed, the allele frequency can be determined. If only the homozygotes **aa** are known (e.g., they can be identified because of an autosomal recessive inherited disease), then q^2 corresponds to the frequency of the disorder. From $p = 1 - q$, the frequency of heterozygotes ($2pq$) and of normal homozygotes (p^2) can also be determined.

Medical relevance

In genetic counseling, the probability of an unaffected sibling of an individual with an autosomal recessive disorder can be derived from the genotype distribution 1:2:1. Since the unaffected sibling cannot be homozygous for the mutant allele, the probability of heterozygosity would be ⅔ (66.66%), derived from the remaining ratio of 2:1.

Further Reading

Cavalli-Sforza LL, Bodmer WF. The Genetics of Human Populations. San Francisco: W. H. Freeman; 1971

Cavalli-Sforza LL, Menozzi P, Piazza A. The History and Geography of Human Genes. Princeton: Princeton University Press; 1994

Eriksson AW, et al., eds. Population Structure and Genetic Disorders. London: Academic Press; 1980

Hamilton MB. Population Genetics. Chichester: John Wiley & Sons; 2009

Kimura M, Ohta T. Theoretical Aspects of Population Genetics. Princeton: Princeton University Press; 1971

Kruglyak L. Prospects for whole-genome linkage disequilibrium mapping of common disease genes. Nat Genet 1999;22:139–144

Terwilliger J, Ott J. Handbook for Human Genetic Linkage. Baltimore: Johns Hopkins University Press; 1994

Vogel F, Motulsky AG. Human Genetics. Problems and Approaches. 3rd ed. Heidelberg: Springer-Verlag; 1997

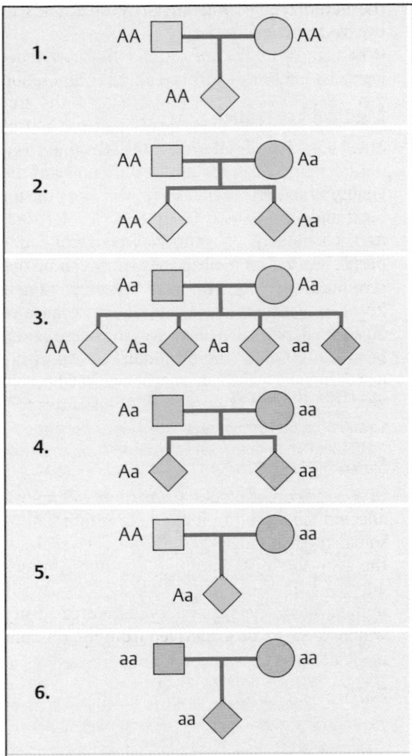

	Genotype of parents	Genotype of offspring
1. AA × AA → AA	AA and AA	1.0 AA
2. AA × Aa → AA, Aa	AA and Aa	0.50 AA 0.50 Aa
3. Aa × Aa → AA, Aa, Aa, aa	Aa and Aa	0.25 AA 0.50 Aa 0.25 aa
4. Aa × aa → Aa, aa	Aa and aa	0.50 Aa 0.50 aa
5. AA × aa → Aa	AA and aa	1.0 Aa
6. aa × aa → aa	aa and aa	1.0 aa

A. Expected frequency of genotypes in children of parents with different genotypes

1.

Parents	0.5 A	0.5 a
0.5 A	AA 0.25	Aa 0.25
0.5 a	Aa 0.25	aa 0.25

Offspring

$p = 0.50$ (Frequency of A)
$q = 0.50$ (Frequency of a)

2.

	A = 0.60	a = 0.40	
A 0.6	0.36 AA	0.24 Aa	p
a 0.4	0.24 Aa	0.16 aa	q
	p	q	

$$p^2 + 2pq + q^2 = 1$$
$$0.36 + 0.48 + 0.16 = 1.0$$
(AA) (Aa) (aa)

B. Allele frequency

Hardy–Weinberg Equilibrium Principle

The Mendelian segregation under random mating results in an equilibrium distribution of genotypes in every new generation. Thus, genetic variation is maintained, unless external influences interfere with random mating. This is the Hardy–Weinberg equilibrium principle, first formulated independently by the English mathematician G. F. Hardy and the German physician W. Weinberg in 1908.

A. Constant allele frequency

An autosomal recessive allele (allele **a**) that leads to a severe disorder in the homozygous state remains undetectable in the heterozygous state in a population. Only the homozygotes (**aa**) can be recognized because of their disease. The frequency of affected individuals (homozygotes **aa**) depends on the frequency of allele **a** (corresponding to q). The frequency of the three genotypes is determined by the binomial relationship $(p + q)^2 = 1$, where p represents the frequency of allele **A,** and q represents the frequency of allele **a** (see p. 134). The homozygous alleles (**aa**) eliminated in one generation by illness are replaced by new mutations. This results in an equilibrium between elimination due to illness and frequency of the mutation.

B. Some factors influencing the allele frequency

The Hardy–Weinberg equilibrium principle is valid only in certain conditions: First, it applies only if there is no selection for one genotype. Selection for heterozygotes will increase the frequency of the allele with a selective advantage (see p. 138). Second, nonrandom matings (assortative mating) will change the allele frequency (proportion of p and q). Third, a change in the rate of mutations will increase the frequency of the allele resulting from mutations. Fourth, in a small population random fluctuation may change the frequency. This is called *genetic drift.*

Other causes of a change in allele frequency may occur. If a population experiences a drastic reduction in size, followed by a subsequent increase in the number of individuals, an allele that was previously rare in this population may by chance subsequently become relatively common as the population expands again. This is called a *founder effect.*

(Photograph from *Coney Island 1938*, by Weegee, also known as Arthur Fellig.)

Medical relevance

The frequency of heterozygotes for an autosomal recessive disorder can be derived from the Hardy–Weinberg equation ($p + q = 1$). If, for example, the disease frequency is 1:10000 (0.01%), then $q^2 = 1/10000$ would yield $q = 1/100$. Since $2pq$ corresponds to the heterozygote frequency, this would be 1:50, since $2pq$ is $2 \times 1 \times 1/100$. A founder effect in some rare autosomal recessive disorders has frequently been observed in closed human populations.

Further Reading

Cavalli-Sforza LL, Bodmer WF. The Genetics of Human Populations. San Francisco: W. H. Freeman; 1971

Cavalli-Sforza LL, Menozzi P, Piazza A. The History and Geography of Human Genes. Princeton: Princeton University Press; 1994

Croucher PJP. Linkage Disequlibrium. Nature Encyclopedia of the Human Genome. Vol 3. London: Nature Publishing Goup; 2003:727–728

Eriksson AW, et al., eds. Population Structure and Genetic Disorders. London: Academic Press; 1980

Hardy GH. Mendelian proportions in a mixed population. Science 1908;28:49–50

Jorde LB. Linkage disequilibrium and the search for complex disease genes. Genome Res 2000;10: 1435–1444

Kimura M, Ohta T. Theoretical Aspects of Population Genetics. Princeton: Princeton University Press; 1971

Kruglyak L. Prospects for whole-genome linkage disequilibrium mapping of common disease genes. Nat Genet 1999;22:139–144

Vogel F, Motulsky AG. Human Genetics. Problems and Approaches. 3rd ed. Heidelberg: Springer-Verlag; 1997

Weinberg W. Über den Nachweis der Vererbung des Menschen. Jahreshefte Verein vaterländ Naturk Württemberg 1908;64:368-382

Wigginton JE, Cutler DJ, Abecasis GR. A note on exact tests of Hardy-Weinberg equilibrium. Am J Hum Genet 2005;76:887–893

Zöllner S, von Haeseler A. Population History and Linkage Equilibrium. Nature Encyclopedia of the Human Genome. Vol 4. London: Nature Publishing Goup; 2003:628–637

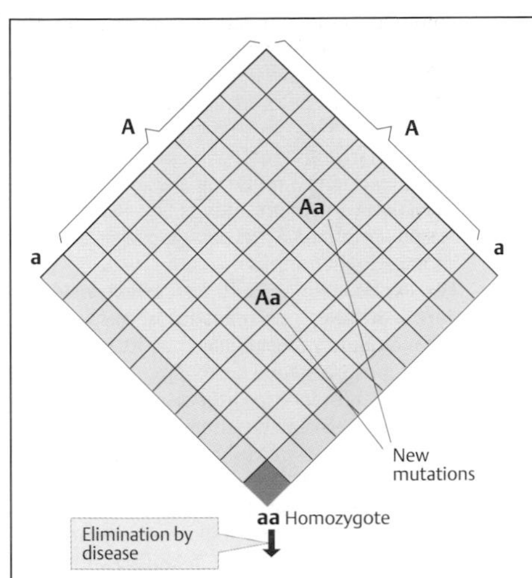

Alleles **a** are eliminated from the population owing to severe illness in homozygotes **aa**, but are replaced by new mutation. Equilibrium results

Genotypes		Frequency
Homozygote	**AA**	p^2
Heterozygote	**Aa**	$2pq$
Homozygote	**aa**	q^2

A. Constant allele frequency (Hardy–Weinberg equilibrium)

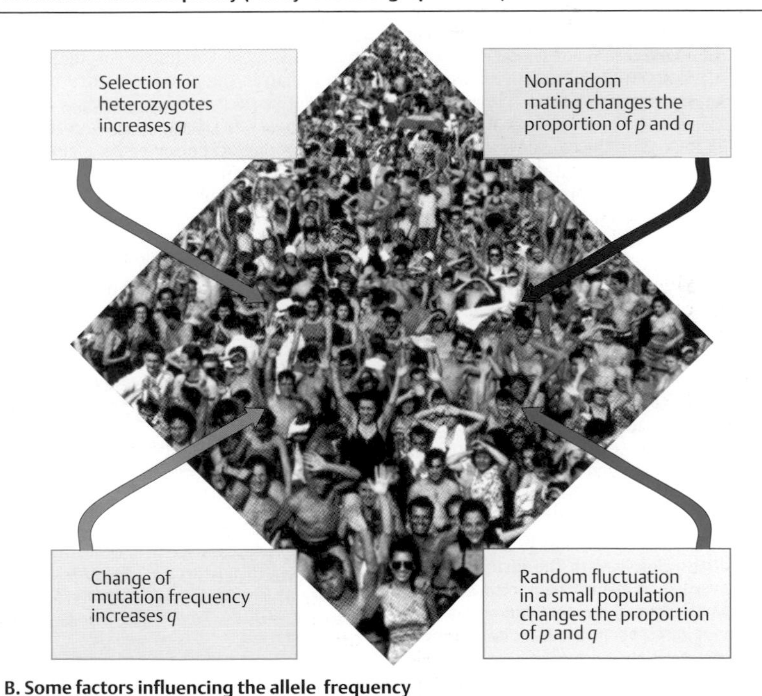

B. Some factors influencing the allele frequency

Geographical Differences in Allelic Distribution

Human populations are genetically considerably diverse. This is a result of human evolution and migration of human ancestors to different parts of the world. The genetic diversity is defined by differences in the frequency of DNA nucleotide sequences and the frequency of certain alleles at various gene loci. Certain mutant alleles may differ widely in frequency. Therefore, autosomal recessive diseases that are common in one population may be rare in others. This may either be the result of a selective advantage of an allele or be due to a random founder effect in small populations.

A. Different frequencies

Finland represents an example of a small population in which several rare recessive diseases occur at a much higher frequency than elsewhere. The most likely explanation is a combination of founder effect and genetic isolation. Three diseases that are clustered in different regions of Finland are (**1**) congenital flat cornea (cornea plana 2, OMIM 217300) in the western part of the country;(**2**) the Finnish type of congenital nephrosis, a severe renal disorder (OMIM 256300) in the southwestern part; and (**3**) diastrophic skeletal dysplasia (OMIM 222600) in the southeastern region. The recessive mutant alleles in these diseases are identical by descent. They must have arisen independently in the different regions because the grandparents live predominantly there. There is no known basis for a selective advantage of these mutant alleles in the heterozygotes to explain their frequencies. The different distributions merely reflect the places and relative points in time of the mutations. Similar examples are found in many other regions of the world and in other populations. (Figure adapted from Norio, 2003.)

B. Malaria and hemoglobin disorders

The distribution of the parasitic disease malaria overlaps closely with the distribution of different types of hemoglobin disorder (see p. 344). Malaria is common in tropical and subtropical regions (**1**). It used to be common around the Mediterranean Sea, but here the frequency of malaria has been reduced as a result of successful prevention. In the same regions with endemic malaria, several types of hemoglobin disorder are prevalent (**2**, **3**). Typical examples are sickle cell anemia (OMIM 141900) and different types of thalassemia (OMIM 187550). In 1954, A. C. Allison proposed that individuals who are heterozygous for the sickle cell mutation are less susceptible to malaria infection. This is the first and the best example of a heterozygous selective advantage known in man.

The sickle cell mutant has arisen independently at least four times in different regions and has become established as a result of selective advantage for heterozygotes.

Another red blood cell disease confers an advantage against malaria infection on heterozygotes: glucose-6-phosphate dehydrogenase deficiency (OMIM 305900), an X-chromosomal anemia disorder leading to severe anemia in hemizygous males, whereas heterozygous females are normal and relatively protected against malaria.

The selective advantage of heterozygotes for these genetically determined diseases is based on less favorable conditions for the malaria parasite than in the blood of normal homozygotes. The protection of heterozygotes occurs at the expense of affected homozygotes who suffer from one of the severe hemoglobin disorders. The benefits are at the population level. More than 400 million people are infected with malaria each year in Africa, Asia, and South America; about 1–3 million die each year, mainly in sub-Saharan Africa. Hopefully, preventive measures can be introduced in endemic areas.

Further Reading

Cavalli-Sforza LL, Menozzi P, Piazza A. The History and Geography of Human Genes. Princeton: Princeton University Press; 1994

Marshall E. Malaria. A renewed assault on an old and deadly foe. Science 2000;290:428–430

Norio R. Diseases of Finland and Scandinavia. In: Rothschild HR, ed. Biocultural Aspects of Disease. New York: Academic Press; 1981

Norio R. The Finnish Disease Heritage III: the individual diseases. Hum Genet 2003;112:470–526

Turnpenny P, Ellard S. Emery's Elements of Medical Genetics. 14th ed. Edinburgh: Elsevier-Churchill Livingstone; 2011

Weatherall DJ, Clegg JB. Thalassemia—a global public health problem. Nat Med 1996;2:847–849

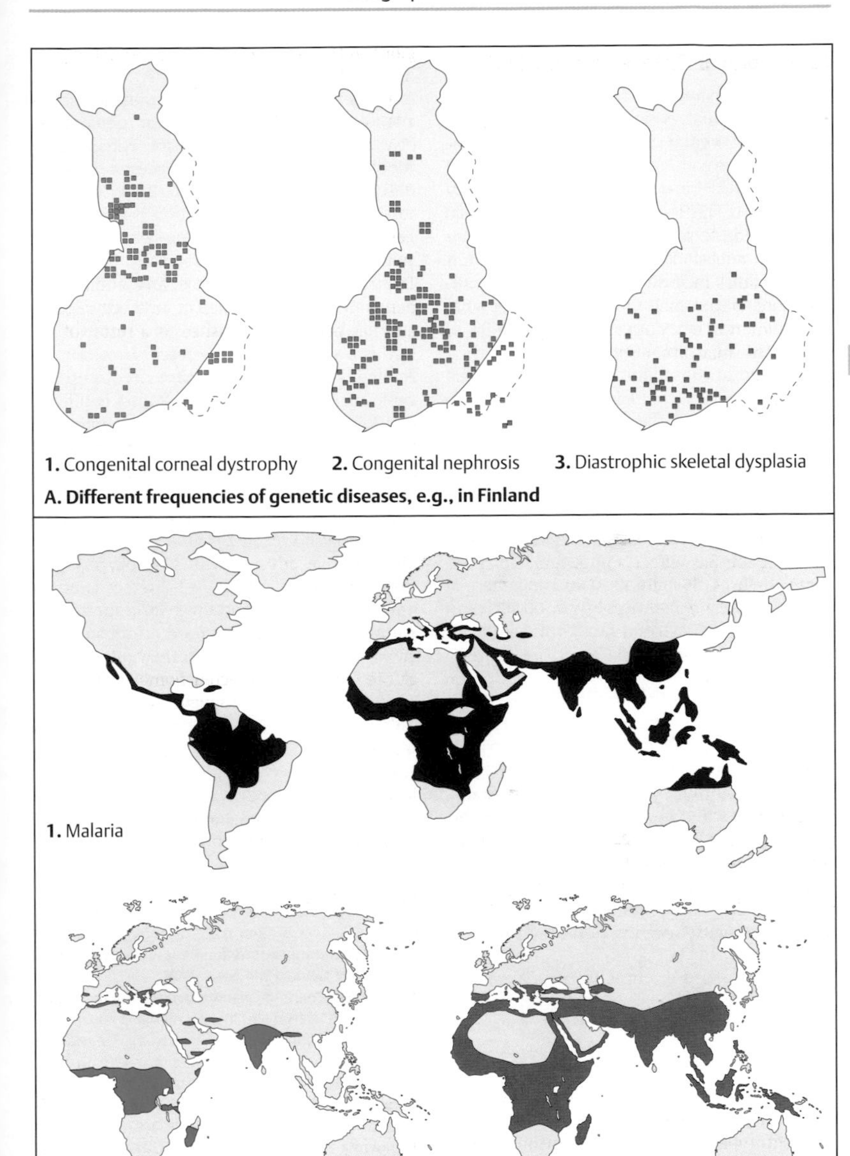

A. Different frequencies of genetic diseases, e.g., in Finland

B. Distribution of malaria and hemoglobin diseases

Inbreeding

Inbreeding in humans refers to matings between individuals who are closely related to each other. Parental consanguinity refers to parents who have at least one common ancestor (being of the "same blood") in the past four generations. This increases the chance that an allele will become homozygous in a descendant. As both alleles will be identical, this is called *identity by descent*. Although consanguineous marriage is widespread in some populations, where it may account for 25%–40% of marriages, the usual rate is about 1%–2%.

The degree of consanguinity can be expressed by two measures: coefficient of kinship or relationship (Φ) and the inbreeding coefficient (*F*). Φ is the probability that an allele from A is identical by common descent with the allele from B at the same locus. The inbreeding coefficient expresses the probability that two alleles will be homozygous in offspring of consanguineous parents. *F* of an individual is the same as Φ of its parents. The coefficient of relationship (*r*) is the proportion of alleles in any two individuals that are identical by descent.

A. Simple types of consanguinity

A mating between brother and sister or father and daughter is called incest (**1**). For two alleles descending from both parents, here designated as A and B, the probability of transmission to each of two offspring, C and D, respectively, is 0.5 ($^1/_2$). The siblings (C and D) share half of their genes, corresponding to a coefficient of relationship of $^1/_2$. The chance of homozygosity by descent at a given locus in their offspring is $^1/_4$ ($^1/_2$ each from C and D to E).

First cousins share $^1/_8$ of their genes (**2**), second cousins share $^1/_{32}$. The chance of homozygosity by descent in the offspring of first cousins is $^1/_{16}$. An uncle–niece union (**3**) has a coefficient of relationship of $r = {^1/_4}$ because they share $^1/_4$ of their genes. The chance of homozygosity by descent in their offspring is ⅛. The possibility that an unrelated individual E transmits a mutant allele at this locus can usually be disregarded.

B. Identity by descent (IBD)

Identity by descent refers to regions in the genome that are identical because they are derived from a common ancestor. For example, for two alleles A and B in ancestor I, and C and D in ancestor II, the probability of transmission to the next generation is 0.5 ($^1/_2$) for each. The same applies to the next generation with individuals III and IV. Finally, the probability of transmission from III to V, and IV to V, is again 0.5 or $^1/_2$ each. The number of steps (the probability of transmission) from I to III, and from I to IV, is $(^1/_2)^2$ each. However, for individual V to be homozygous requires that both alleles are transmitted from I to III and to IV. This corresponds to $(½)^4$. For individual V to be homozygous by IBD the probabilities $(^1/_2)^4$ have to be added and multiplied by the probability of ($^1/_2$), which yields $^1/_{16}$. The resulting probability of homozygosity at a given locus is the inbreeding coefficient, *F*.

Medical relevance

Consanguinity is a frequent reason for genetic counseling. First cousins have a probability of homozygosity by descent in their offspring (*F*) of $^1/_{16}$. The risk of a harmful allele reaching their offspring in the homozygous state is $^1/_{64}$. The total risk is $^1/_{32}$ because the other common ancestor (both grandparents) also has to be taken into account. Although at first this risk of 3.125% seems high, it is not when compared with the risk in the general population. The overall risk for a newborn to have a disorder of any kind is estimated to be 1%–2%.

Further Reading

Bittles AH, Neel JV. The costs of human inbreeding and their implications for variations at the DNA level. Nat Genet 1994;8:117–121

Griffith AJF, et al. An Introduction to Genetic Analysis. 9th ed. New York: W. H. Freeman; 2007

Harper PS. Practical Genetic Counselling. 7th ed. London: Edward Arnold; 2010

Jaber L, Halpern GJ, Shohat M. The impact of consanguinity worldwide. Community Genet 1998;1:12–17

Turnpenny P, Ellard S. Emery's Elements of Medical Genetics. 14th ed. Edinburgh: Elsevier-Churchill Livingstone; 2011

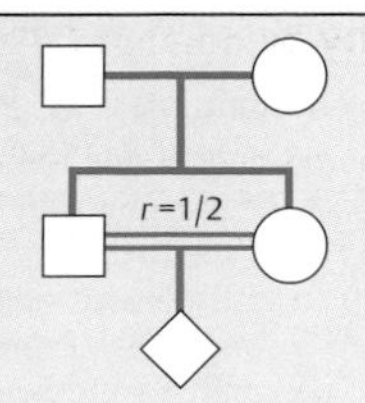

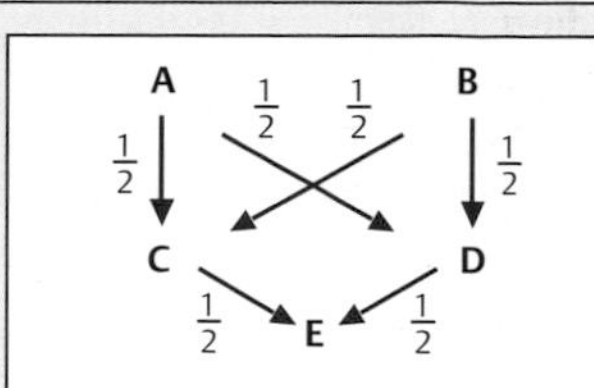

1. Brother/sister mating

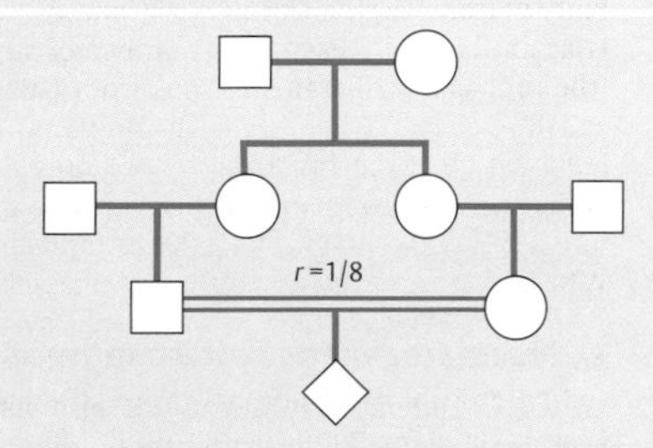

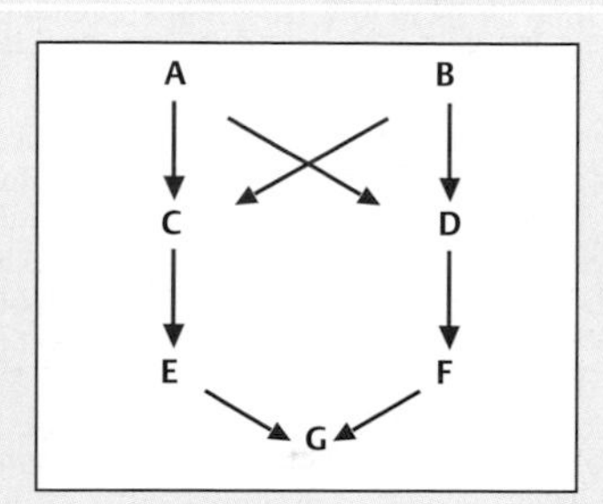

2. First cousins

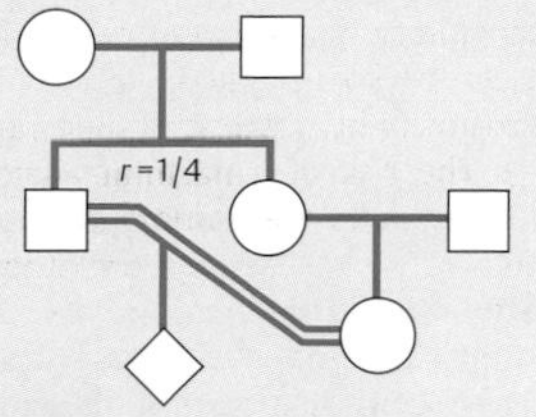

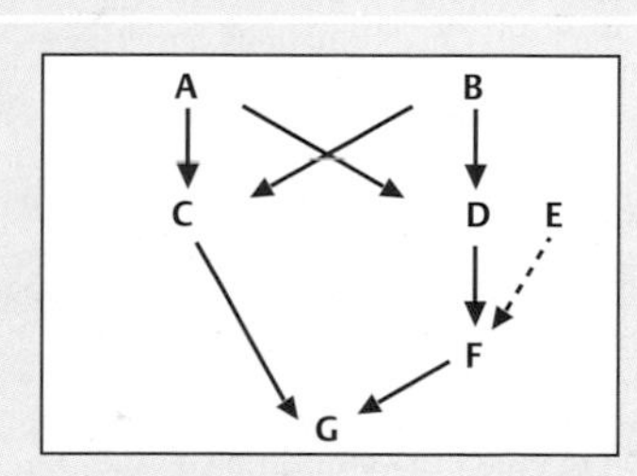

3. Uncle/niece

A. Simple types of consanguinity

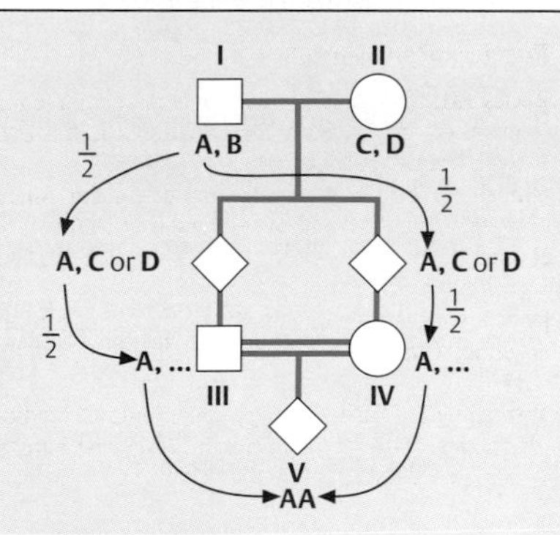

I ⟶ III $= \left(\frac{1}{2}\right)^2$

I ⟶ IV $= \left(\frac{1}{2}\right)^2$

I ⟶ III and IV $= \left(\frac{1}{2}\right)^4$

III and IV ⟶ V $= \left(\frac{1}{2}\right)^2$

IBD: $\left[\left(\frac{1}{2}\right)^4 + \left(\frac{1}{2}\right)^4\right] \times \frac{1}{2} = \frac{1}{16}$

Probability of descent from a common ancestor

B. Identity by descent (IBD)

Twins

Twin, triplet, or quadruple pregnancies occur regularly in many species of animals. In humans, twinning is detected during pregnancy by ultrasonography in about 1 of 40 pregnancies, but in live births in 1 of 80. This difference results from the early intrauterine death of one twin and its subsequent resorption. The frequency of twinning varies widely from about 6 in 1000 births in Asia, 10–20 in 1000 in Europe, and 40 in 1000 in Africa. Twins can arise from a single fertilized egg (genetically identical monozygotic twins, MZ) or from two different eggs (dizygotic twins, DZ), a distinction first proposed by C. Dareste in 1874. The rate of monozygotic twins at birth is relatively constant. F. Galton initiated research on twins in 1876. Multiple systematic comparisons of MZ and DZ twins have attempted to disentangle the contribution of genetic and environmental factors in the etiology of diseases, disease susceptibility, intellectual and behavioral attributes, congenital malformations, etc. However, the resulting data remain controversial in many respects.

A. Types of human monozygotic twins

Monozygotic twins arise during very early stages of embryonic development by splitting of the inner cell mass of the early embryo. Three stages in the timing of splitting can be distinguished: (**1**) after trophoblast formation, resulting in twins with individual amnions but a common chorion; (**2**) after amnion formation, resulting in twins in a single chorion and amnion; and (**3**) before formation of the trophoblast, resulting in twins each with its own chorion and amnion.
About 66% of MZ twins have one chorion and two amnions, suggesting a split after formation of the chorion at day 5, but before formation of the amnion at day 9. About 33% of MZ twins have two complete separate chorions and an individual amnion. (Figure adapted from Gilbert, 2010, and Goerke, 2002.)

B. Pathological conditions in twins

Conjoined twins result from late splitting after formation of the amnion as monochorionic, monoamniotic twins. Such twins are at risk of becoming conjoined (so-called Siamese twins). A relatively frequent form of incomplete separation is thoracopagus, in which the twins are joined to various extents at the thoracic region (**1**). Dizygotic twins may be affected by erroneous blood supply. If one twin receives insufficient blood due to a shunt in the blood circulation, this twin might be retarded in growth or die (**2**). Especially severe malformations may result from incompletely formed organs, e.g., absence of the heart in one twin (acardius) (**3**).

C. Concordance rates in twins

When twins show the same trait, they are said to be concordant for that trait; when they differ, they are discordant. Comparing the rate of concordance in monozygotic and dizygotic twins may reflect the relative contribution of genetic factors in the etiology of complex traits.

D. Pharmacogenetic pattern in twins

Dizygotic and monozygotic twins also differ biochemically. As a result of genetic differences, many chemical substances used in therapy are metabolized or excreted at different rates, owing to different activities of corresponding enzymes. Phenylbutazone is excreted at the same rate in identical twins, whereas the rates of excretion differ between dizygotic twins or among siblings (Vesell, 1978).

Further Reading

Boomsma D, Busjahn A, Peltonen L. Classical twin studies and beyond. Nat Rev Genet 2002;3:872–882

Bouchard TJ Jr, et al. Sources of human psychological differences: the Minnesota Study of Twins Reared Apart. Science 1990;250:223–228

Gilbert SF. Developmental Biology. 9th ed. Sunderland: Sinauer; 2010

Goerke K. Taschenatlas der Geburtshilfe. Stuttgart: Thieme; 2002

Hall JG. Twinning. Lancet 2003;362:735–743

MacGregor AJ, et al. Twins. Novel uses to study complex traits and genetic diseases. Trends Genet 2000;16:131–134

Phelan MC, Hall JG. Twins. In: Stevenson RE, Hall JG, eds. Human Malformations and Related Anomalies. 2nd ed. Oxford: Oxford University Press; 2006:1377–1411

Vesell ES. Twin studies in pharmacogenetics. Hum Genet Suppl 1978; 1(1, Suppl) 19–30

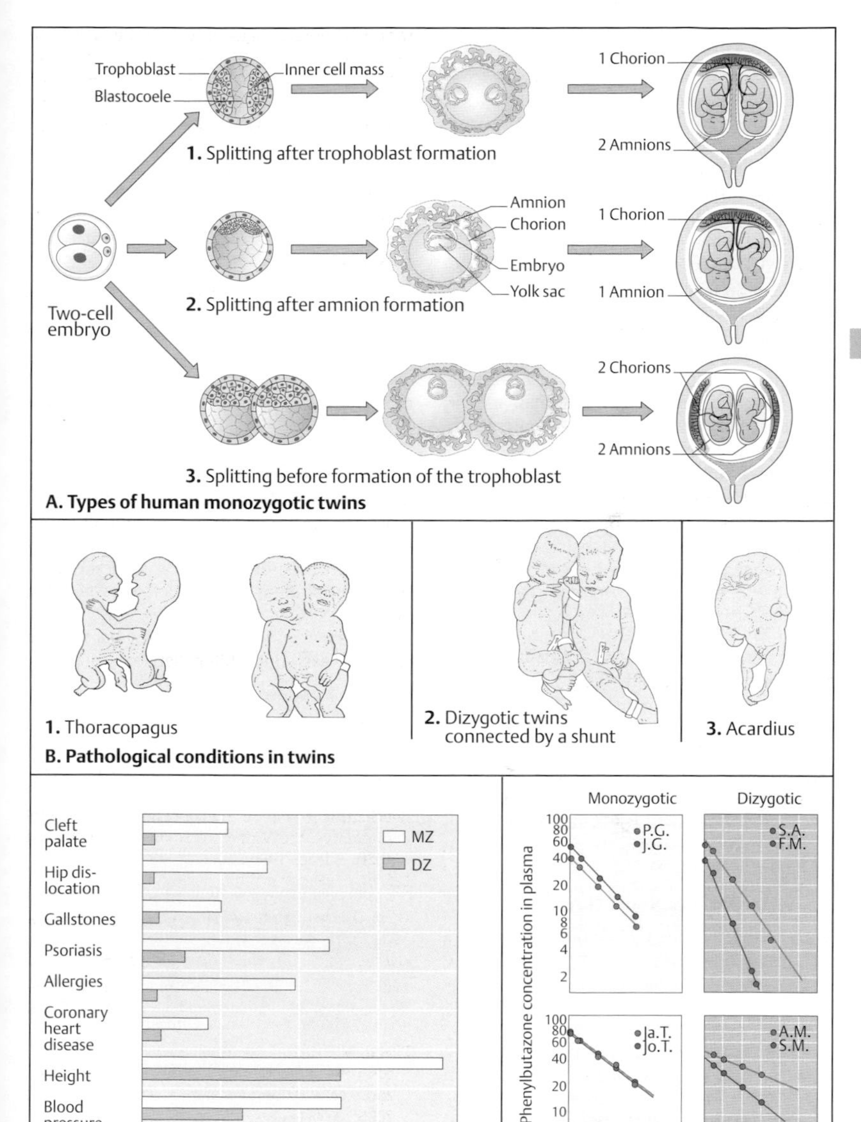

A. Types of human monozygotic twins

B. Pathological conditions in twins

C. Concordance of some traits in monozygotic (MZ) and dizygotic (DZ) twins

D. Pharmacogenetic pattern in twins

Chromosomes in Metaphase

In eukaryotes, the cellular genetic material (DNA and associated molecules) is packaged into chromosomes. Each chromosome contains a continuous thread of DNA double helix. The term chromosome, derived from Greek for "colored thread," was introduced by W. Waldeyer in 1888. In prokaryotes, the single chromosome usually has a circular structure. In eukaryotes, the chromosomes are located in the nucleus at defined positions. As first noted by W. Flemming in 1879, chromosomes can be visualized under a light microscope during mitosis as separate, individual structures. In 1956, Tjio and Levan in Lund, and Ford and Hamerton in Oxford, established that man has 46 chromosomes, not 48 as had been previously assumed. Each organism has a defined number of chromosomes with distinct morphological appearances (the karyotype).

A. The metaphase chromosomes of man

A human metaphase is shown here at about 2800-fold magnification. Each chromosome pair is defined by its length, position of the centromere, and size and arrangement of the transverse light and dark bands (banding pattern). Parts of a chromosome also can be identified by its banding pattern. In a typical metaphase preparation, about 400–550 distinct bands can be distinguished. In prometaphase, the chromosomes are longer than in metaphase and show more bands. Thus, for certain purposes chromosomes are also studied in prometaphase.

B. Types of metaphase chromosome

Each chromosome is classified as *submetacentric*, *metacentric*, or *acrocentric* according to the location of its centromere. This appears as a constriction and is the point of attachment of the spindle during mitosis. The centromere divides a submetacentric chromosome into a short arm (p arm, derived from the French *petit*) and a long arm (q, next letter after p). In metacentric chromosomes, the short and long arms are about the same length. The short arm (p) of acrocentric chromosomes consists of a gene-free dense appendage called a satellite (not to be confused with satellite DNA) at the end of a stalk.

C. Karyogram

The karyogram is a systematic visual representation of the karyotype. It displays all chromosomes in homologous pairs, one from the mother and one from the father, arranged according to their relative lengths and the positions of their centromeres. Chromosomes are numbered as 22 pairs of chromosomes (autosomes, pairs 1–22) and in addition either two X chromosomes, in females, or an X and a Y chromosome, in males (karyotype 46,XX or 46, XY, respectively). The 22 pairs of autosomes in man are divided into seven groups (A–G). The karyotypes of different species of animals show differences and similarities depending on their evolutionary relationships.

Medical relevance

The analysis of chromosomes in metaphase (or under certain circumstances in prometaphase) is an important tool of diagnostics, because about one in 400 newborns carries an aberration of the normal number or structure of a chromosome. This may result in developmental disturbances, depending on the chromosome involved (see Chromosomal Location of Human Genetic diseases, p. 398).

Further Reading

Caspersson T, Zech L, Johansson C. Differential binding of alkylating fluorochromes in human chromosomes. Exp Cell Res 1970;60:315–319

Dutrillaux B, Lejeune J. Sur une nouvelle technique d'analyse du caryotype humain. CR Acad Sci Paris D 1971;272:2638–2640

Ford CE, Hamerton JL. The chromosomes of man. Nature 1956;178:1020–1023

Lewin B. Genes IX. Sudbury: Bartlett & Jones; 2008

Miller OJ, Therman E. Human Chromosomes. 4th ed. New York: Springer-Verlag; 2001

Riddihough G. Chromosomes through space and time. Science 2003;301:779–802

Shaffer LG, Slovak ML, Campbell LJ, eds. An International System for Human Cytogenetic Nomenclature. Basel: Karger; 2009

Tjio JH, Levan A. The chromosome number of man. Hereditas 1956;42:1–6

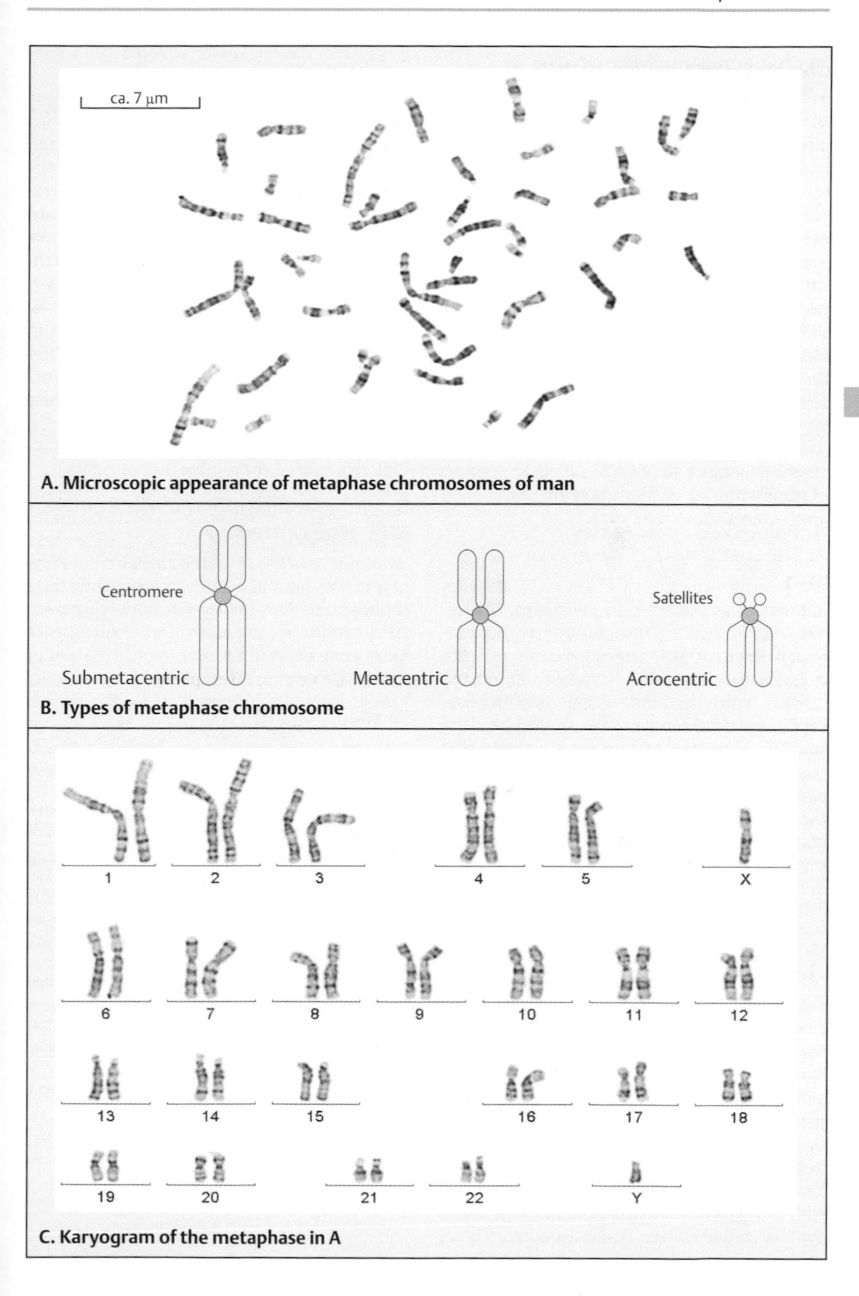

A. Microscopic appearance of metaphase chromosomes of man

B. Types of metaphase chromosome

C. Karyogram of the metaphase in A

Visible Functional Structures of Chromosomes

In certain cells and tissues of some insects and amphibians, chromosomal structures can be observed that relate to their function. Polytene ("many threads") chromosomes, formed as a result of repeated DNA synthesis without cell division, have a distinct pattern of chromosome banding readily visible under the light microscope. These chromosomes, first observed in cells of insect salivary glands (*Drosophila melanogaster* and *Chironomus)* by E. G. Balbiani in 1881, show regions of temporary localized enlargement called Balbiani rings.

In the oocytes of some animals, fine loops protrude from the chromosomes during the diplotene phase of meiosis (see p. 102). Because of their appearance they were called lampbrush chromosomes by W. Flemming in 1882.

A. Polytene chromosomes

In the salivary glands of *Drosophila* larvae, chromosomes are greatly enlarged in diameter and length as polytene chromosomes. This results from 10 cycles of replication without division into daughter chromosomes. Thus, there are about 1024 (2^{10}) identical chromatid strands, which lie strictly side by side. The *Drosophila* genome contains about 5000 bands. A micrographic detail of a polytene chromosome from a *Drosophila* salivary gland shows the characteristic banding pattern (shown at the bottom of **A**). The dark bands are the result of chromatin condensation in the large interphase polytene chromosomes. (Figure from Alberts et al., 2008; and modified from Painter, 1934.)

B. Functional stages in polytene chromosomes

Expansions (puffs) appear and recede at defined positions in temporal stages in the polytene chromosomes (**1**). Chromosome puffs are decondensed, expanded segments that represent regions containing genes that are being transcribed. The locations and durations of the puffs reflect different stages of larval development. The incorporation of radioactively labeled RNA (**2**) has been used to demonstrate that RNA synthesis occurs in these regions as a sign of gene activity (transcription). (Figure from Alberts et al., 2008, adapted from Ashburner et al., 1974.)

C. Lampbrush chromosomes in oocytes

Lampbrush chromosomes are greatly extended chromosome bivalents that occur during the diplotene phase of meiosis in oocytes of certain amphibians. A meiotic bivalent of two pairs of sister chromatids can be seen under the light microscope (**1**). They are held together at points of chiasma formation. The interpretation is that loops of a paired chromosome form mirror-image structures (**2**). Lampbrush chromosomes in the newt *Notophthalmus viridens* are unusually large compared with their mitotic chromosomes, about 400–800 µm long compared with 15–20 µm at most during later stages of meiosis. (Microphotograph by J. G. Gall, reproduced from Alberts et al., 2008).

D. Visible transcription of ribosomal RNA gene clusters

Ribosomes are the sites of translation. Tandem repeats of ribosomal RNA (rRNA) genes transcribed in the nucleolus of an amphibian, *Triturus viridiscens*, are shown here. Along each gene many rRNA molecules are synthesized by RNA polymerase I. The growing RNA molecules extend from a backbone of DNA, the shorter ones being at the start site of transcription and the longer ones having been completed. (Photograph by O. L. Miller and B. A. Hamkalo, reproduced from Griffiths et al., 2007.)

Further Reading

Alberts B, et al. Molecular Biology of the Cell. 5th ed. New York: Garland Science; 2008

Ashburner M, Chihara C, Meltzer P, Richards G. Temporal control of puffing activity in polytene chromosomes. Cold Spring Harb Symp Quant Biol 1974;38:655–662

Callan HG. The Croonian Lecture, 1981. Lampbrush chromosomes. Proc R Soc Lond B Biol Sci 1982; 214:417–448

Gall JG. On the submicroscopic structure of chromosomes. Brookhaven Symp Biol 1956;8:17–32

Griffith AJF, et al. An Introduction to Genetic Analysis. 9th ed. New York: W. H. Freeman; 2007

Lewin B. Genes IX. Sudbury: Bartlett & Jones; 2008

Miller OL Jr. The visualization of genes in action. Sci Am 1973;228:34–42

Painter TS. Salivary chromosomes and the attack on the gene. J Hered 1934;25:465–476

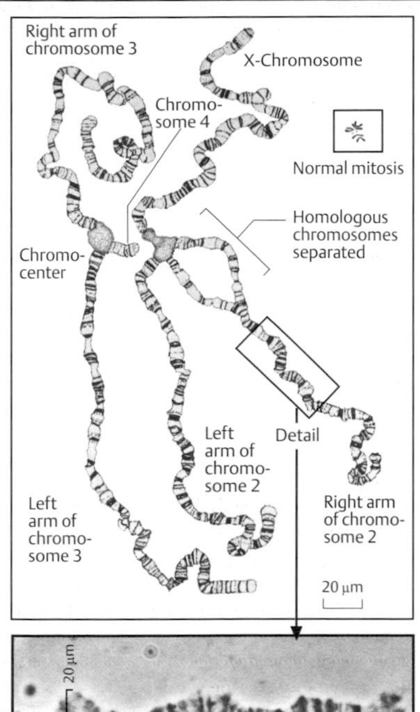

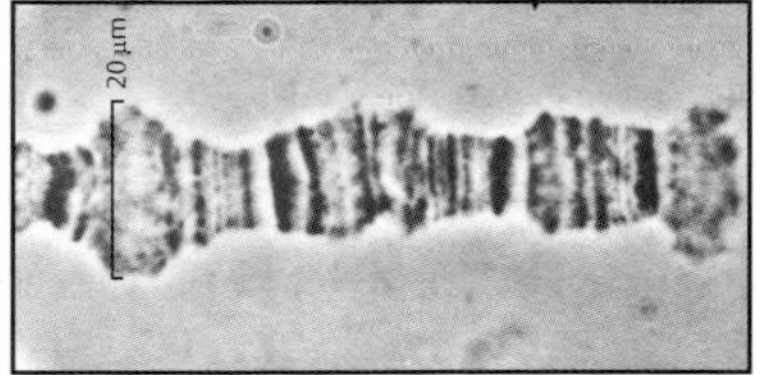

A. Polytene chromosomes in salivary glands of *Drosophila* larvae

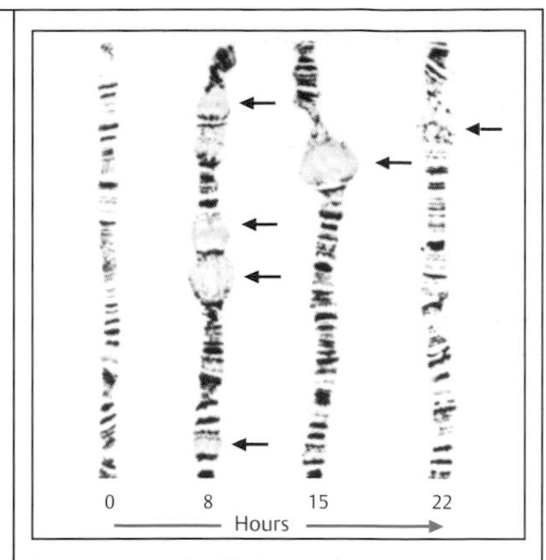

1. Formation of puffs (arrows)

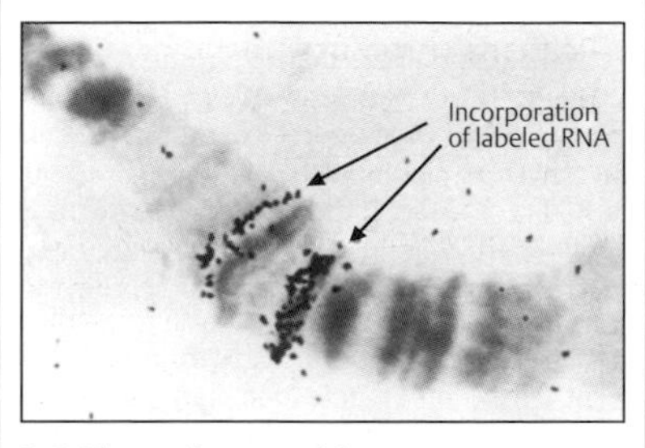

2. Evidence of gene activity

B. Functional stages in polytene chromosomes

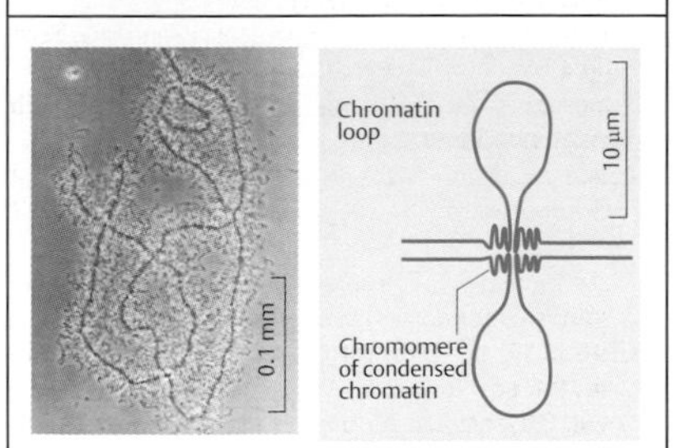

1. Lampbrush chromosome

2. Schematic section of a chromatin loop

C. Chromosome structure in amphibian oocytes ("lampbrush chromosomes")

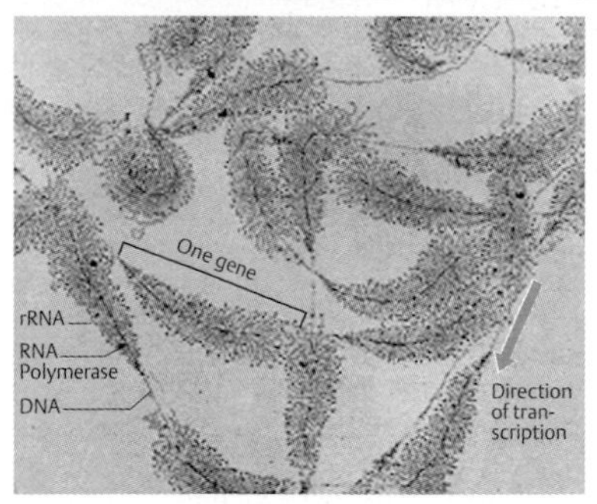

D. Visible transcription of ribosomal RNA gene clusters

Chromosome Organization

The structural chromosome organization differs between interphase and mitosis. Chromosomes are visible as separate structures during mitosis only; during interphase they appear as a tangled mass called chromatin. The density of chromatin varies, referred to as *heterochromatin* and *euchromatin* (E. Heitz, 1928). These relate to the overall activity of genes: euchromatin to active genes, heterochromatin to inactive genes.

A. A histone-depleted chromosome under the electron microscope

A protein scaffold of dense fibers surrounds each metaphase chromosome. This is best seen in histone-depleted chromosomes (for histone proteins see p. 152). Without histone proteins, a DNA chromosomal skeleton becomes visible under the electron microscope (**1**). A histone-depleted chromosome is visible in the center as a dark scaffold surrounded by a halo of DNA (darkly stained threads). A higher magnification (**2**) shows that DNA is a single continuous thread. (Photographs by Paulson & Laemmli, 1977).

B. Levels of chromosome organization

From the chromosome to its DNA strand, different levels of organization can be distinguished. The total length of haploid DNA in a dividing human cell is about 1 m. During mitosis, this has to fit into 23 chromosomes of about 3–7 m each. When a portion of a chromosome arm corresponding to 10% of that chromosome is magnified 10-fold, it might be seen to contain about 40 genes, depending on the chromosome segment chosen (eight are shown here, **2**). Magnifying a tenth of that section another 10-fold (**3**) would yield a region containing three or four genes on average. A further 10-fold magnification shows a single gene (**4**) with its exon/intron structure. The last level (**5**) would be that of the nucleotide sequence. (Figure adapted from Alberts et al., 2008.)

C. Heterochromatin and euchromatin

In 1928, Emil Heitz observed that certain parts of the chromosomes of a moss (*Pellia epiphylla*) remain thickened and deeply stained during interphase. He named these structures *heterochromatin*. Those parts that were less densely stained and became invisible during late telophase and subsequent interphase he called *euchromatin*. Subsequent studies showed that heterochromatin consists of regions with few or no active genes, whereas euchromatin corresponds to regions with active genes. (Figure from Heitz, 1928.)

D. Constitutive heterochromatin at the centromeres (C bands)

The centromeres of eukaryotic chromosomes contain repetitive DNA, called α-satellites. They are specific for each chromosome. These sequences are visible in the centromeric region (constitutive heterochromatin). This can be specifically stained (C bands). The heterochromatin of the centromeres of chromosomes 1, 9, and 16, and of the long arm of the Y chromosome, differs in length in different human individuals (chromosomal polymorphism). (Photograph from Verma & Babu, 1989.)

Further Reading

Brown SW. Heterochromatin. Science 1966;151:417–425

Grewal SIS, Jia S. Heterochromatin revisited. Nat Rev Genet 2007;8:35–46

Heitz E. Das Heterochromatin der Moose. I. Jahrb Wiss Bot 1928;69:762–818

Lewin B. Genes IX. Sudbury: Bartlett & Jones; 2008

Passarge E. Emil Heitz and the concept of heterochromatin: longitudinal chromosome differentiation was recognized fifty years ago. Am J Hum Genet 1979;31:106–115

Paulson JR, Laemmli UK. The structure of histone-depleted metaphase chromosomes. Cell 1977;12:817–828

Sumner A. Chromosomes: Organization and Function. Malden: Blackwell; 2003

Verma AS, Babu A. Human Chromosomes. New York: Pergamon Press; 1989

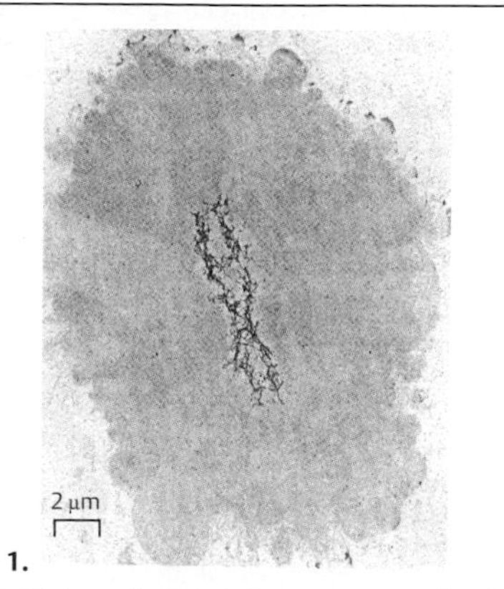

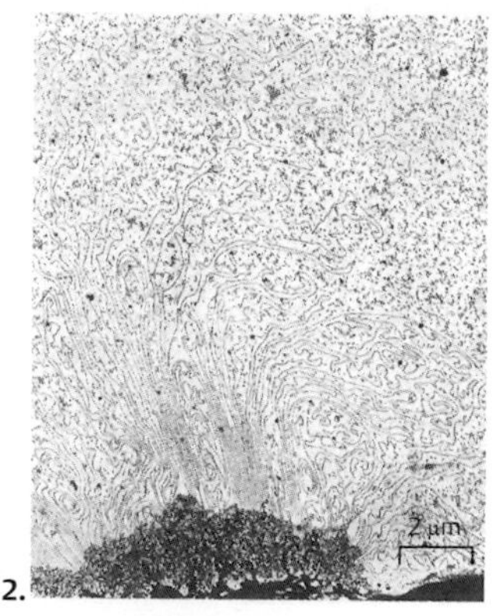

A. Histone-depleted chromosome under the electron microscope

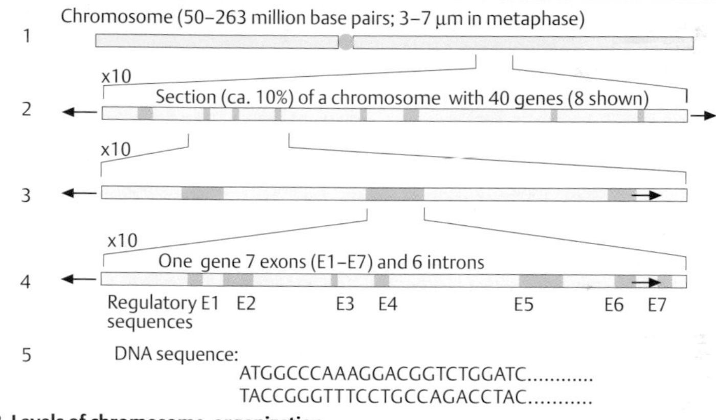

B. Levels of chromosome organization

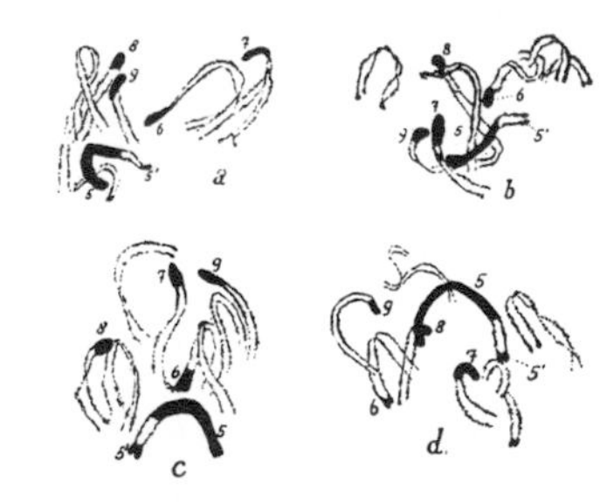

C. Heterochromatin and euchromatin

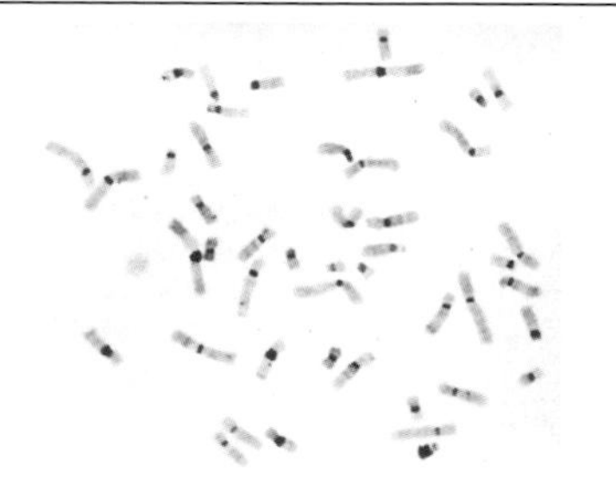

D. Constitutive heterochromatin (C bands) at the centromeres

Functional Elements of Chromosomes

The eukaryotic chromosome can be viewed as a segregation device (Lewin, 2008). This requires three types of structures: (i) centromeric sequences (CEN); (ii) autonomous replicating sequences (ARS, origin of replication); and (iii) telomeric sequences (TEL). The individual contributions of these three types of functional chromosomal elements have been demonstrated in mutant yeast cells (*Saccharomyces cerevisiae*, baker's yeast). Using this information yeast artificial chromosomes (YACs) can be constructed.

A. Basic features of a eukaryotic chromosome

The *centromere* and the two *telomeres*, one at each end, are distinct hallmarks of a chromosome. The centromere attaches a chromosome to the spindle at mitosis. Centromeric DNA contains repetitive DNA called α-satellite DNA, the most abundant type being long tandem repeats of a 170-base-pair (bp) monomeric sequence. The total length of centromeric DNA ranges from about 300 to 5000 kb. Cloned α-satellite fragments hybridize specifically to individual chromosomes.
The subtelomeric sequences, located proximal to the telomeric sequences, contain sequence homologies shared among subsets of other chromosomes (p. 156).

B. Autonomous replicating sequences (ARS)

ARS are required for initiation of eukaryotic DNA replication. Mutant yeast cells that cannot synthesize the amino acid leucine (Leu) can be transformed with cloned plasmids that contain the gene for leucine synthesis. However, such cells still cannot grow in culture medium lacking leucine because they cannot replicate DNA (**1**). Replication can be restored if ARS are transferred along with the *Leu* gene, because plasmid DNA containing the *Leu* gene and ARS is able to replicate (**2**). However, only about 5%–20% of daughter cells receive the plasmid DNA and can grow on Leu^- media.

C. Centromeric sequences (CEN)

CEN are required for correct distribution of replicated chromosomes during mitosis. This is shown by adding yeast CEN sequences to the plasmid in addition to the *Leu* gene and ARS. In this case nearly 90% of progeny can grow on Leu^- medium (**1**) because normal mitotic segregation takes place (**2**). Thus, CEN sequences are necessary for normal distribution of the chromosomes at mitosis (see p. 100).

D. Telomeric sequences (TEL)

When Leu^- yeast cells are transfected with a plasmid that, in addition to the *Leu* gene, contains ARS and CEN sequences (**1**) but that is linear instead of circular, as in **C** and **D**, the cells fail to grow in Leu^- medium (**2**). However, if telomere sequences (TEL) are attached to both ends of the plasmid (**3**) before it is incorporated into the cells (**4**), normal growth in Leu^- medium takes place (**5**). In this case the linearized plasmid behaves as a normal chromosome.
(Figures in A–C adapted from Lodish et al., 2007.)

Medical relevance

Losses or rearrangements of subtelomeric sequences are important causes of human developmental disturbances, and are found in about 5% of patients with mental retardation (De Vries et al., 2003; Knight & Flint, 1999).

Further Reading

Clarke L, Carbon J. The structure and function of yeast centromeres. Annu Rev Genet 1985;19:29–55

De Vries BBA, et al. Telomeres: a diagnosis at the end of the chromosomes. J Med Genet 2003;40:385–398

Knight SJL, Flint J. Perfect endings: a review of subtelomeric probes and their use in clinical diagnosis. J Med Genet 2000;37:401–409

Lodish H, et al. Molecular Cell Biology. 6th ed. New York: W. H. Freeman; 2007

Murray AW, Szostak JW. Construction of artificial chromosomes in yeast. Nature 1983;305:189–193

Schlessinger D. Yeast artificial chromosomes: tools for mapping and analysis of complex genomes. Trends Genet 1990;6:248–258, 255–258

Schueler MG, et al. Genomic and genetic definition of a functional human centromere. Science 2001;294:109–115

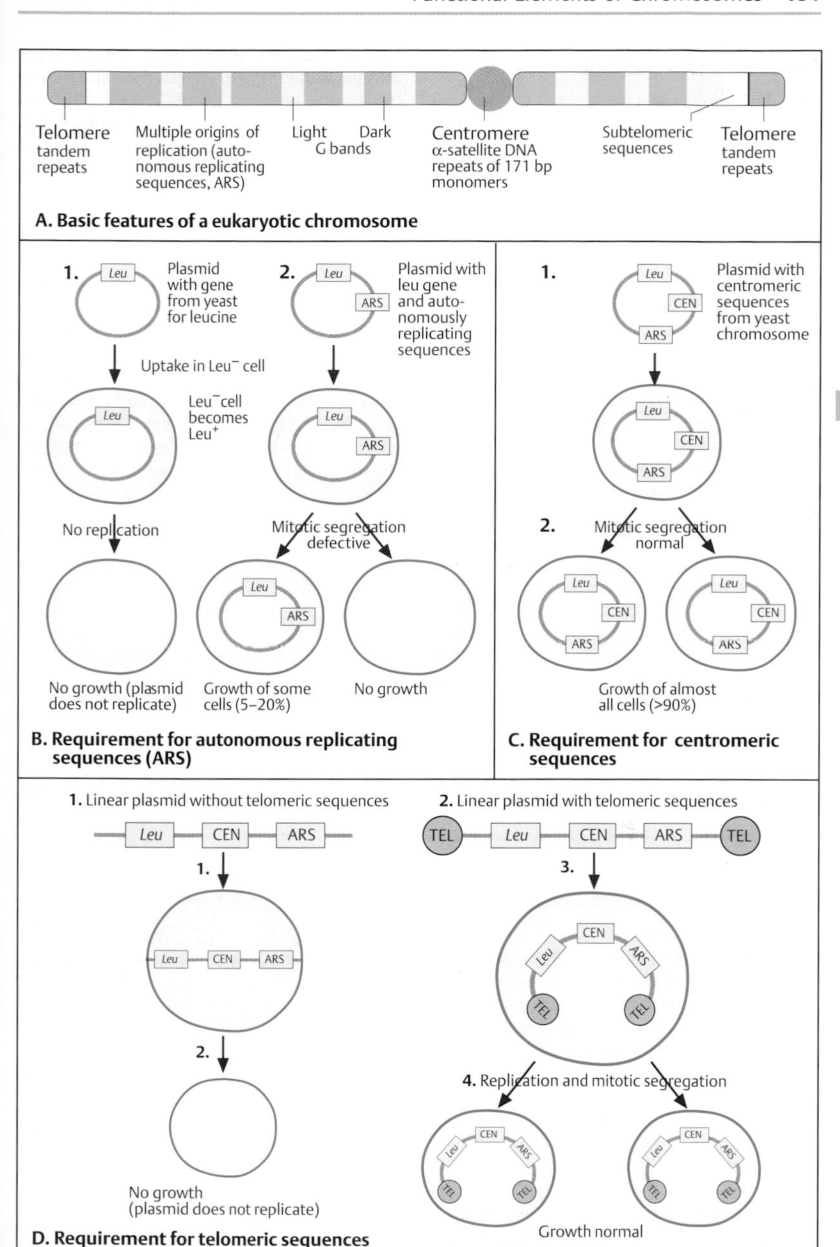

A. Basic features of a eukaryotic chromosome

B. Requirement for autonomous replicating sequences (ARS)

C. Requirement for centromeric sequences

D. Requirement for telomeric sequences

DNA and Nucleosomes

The basic subunit of chromatin is the nucleosome, first delineated by R. D. Kornberg, and by Olins and Olins in 1974. It consists of eight small basic proteins known as the histone proteins or histones. The DNA double helix is wound around each nucleosome. This allows a tight packing of DNA, achieving a packing ratio of about 1:10 000.

A. The nucleosome, the basic unit of DNA packing

A nucleosome consists of eight core histone molecules (octamer): two copies each of H2A, H2B, H3, and H4 (**1**), and about 150 bp DNA wrapped around it. The histone octamer forms a disc-shaped core about 140–150 bp (146 bp in humans) of DNA are wrapped 1.65 times in left-handed turns (a left-handed superhelix, **2**) around the histone core to form a nucleosome about 11 nm diameter and 6 nm high (the illustrations in **A** are highly schematic). The DNA enters and leaves the nucleosome at points close to each other. Linker DNA of variable length between 8 and 114 bp links two adjacent nucleosomes with each other. It is associated with histone H1. Each nucleosome is separated from the other by 50–70 bp linker DNA, which yields a repeat length of 157–240 bp. For transcription and repair, the tight association of histones and DNA has to be loosened (see p. 192). H4 and H3 belong to the most conserved proteins in evolution; H2A and H2B are present in all eukaryotes, but their sequence varies between species.

B. Three-dimensional structure of a nucleosome

The ribbon diagram of the nucleosome shown from above, based on the X-ray structure at a high resolution of 2.8 Å, shows DNA wrapped around its histone core. One strand of DNA is shown in green, the other in brown. The histones are shown in different colors as indicated. (Photograph from Luger et al., 1997, with kind permission from T. J. Richmond.)

C. Chromatin structures

Chromatin occurs in a condensed (tightly folded), a less condensed (partially folded), and an extended, unfolded form. When extracted from cell nuclei in isotonic buffers, most chromatin appears as fibers of about 30 nm diameter. The corresponding electron microscopy photographs obtained by different techniques show the condensed (folded) chromatin as compact 300–500 Å structures (top), a 250-Å fiber when partially folded (middle), and as a "beads on a string" 100-Å chromatin fiber (bottom). (Figure adapted from Lodish et al., 2007; electron micrographs are from Thoma et al., 1979).

D. Chromatin segments

The chromatin structures shown correspond to the third level of organization, the packing of the 30-nm fiber. This yields an overall packing ratio of 1000 in euchromatin (about the same in mitotic chromosomes) and 10 000-fold in heterochromatin in both interphase and mitosis. (Figure adapted from Alberts et al., 2008.)

Medical relevance

Faulty changes in chromatin structure cause different disorders and some forms of cancer (see Part III).

Further Reading

Alberts B, et al. Molecular Biology of the Cell. 5th ed. New York: Garland Science; 2008

Kornberg RD, Lorch Y. Twenty-five years of the nucleosome, fundamental particle of the eukaryote chromosome. Cell 1999;98:285–294

Khorasanizadeh S. The nucleosome: from genomic organization to genomic regulation. Cell 2004;116: 259–272

Lewin B. Genes IX. Sudbury: Bartlett & Jones; 2008

Lodish H, et al. Molecular Cell Biology. 6th ed. New York: W. H. Freeman; 2007

Luger K, Mäder AW, Richmond RK, Sargent DF, Richmond TJ. Crystal structure of the nucleosome core particle at 2.8 Å resolution. Nature 1997;389:251–260

Mellor J. Dynamic nucleosomes and gene transcription. Trends Genet 2006;22:320–329

Olins AL, Olins DE. Spheroid chromatin units (v bodies). Science 1974;183:330–332

Richmond TJ, Davey CA. The structure of DNA in the nucleosome core. Nature 2003;423:145–150

Schalch T, Duda S, Sargent DF, Richmond TJ. X-ray structure of a tetranucleosome and its implications for the chromatin fibre. Nature 2005;436:138–141

Thoma F, Koller T, Klug A. Involvement of histone H1 in the organization of the nucleosome and of the salt-dependent superstructures of chromatin. J Cell Biol 1979;83(2 Pt 1):403–427

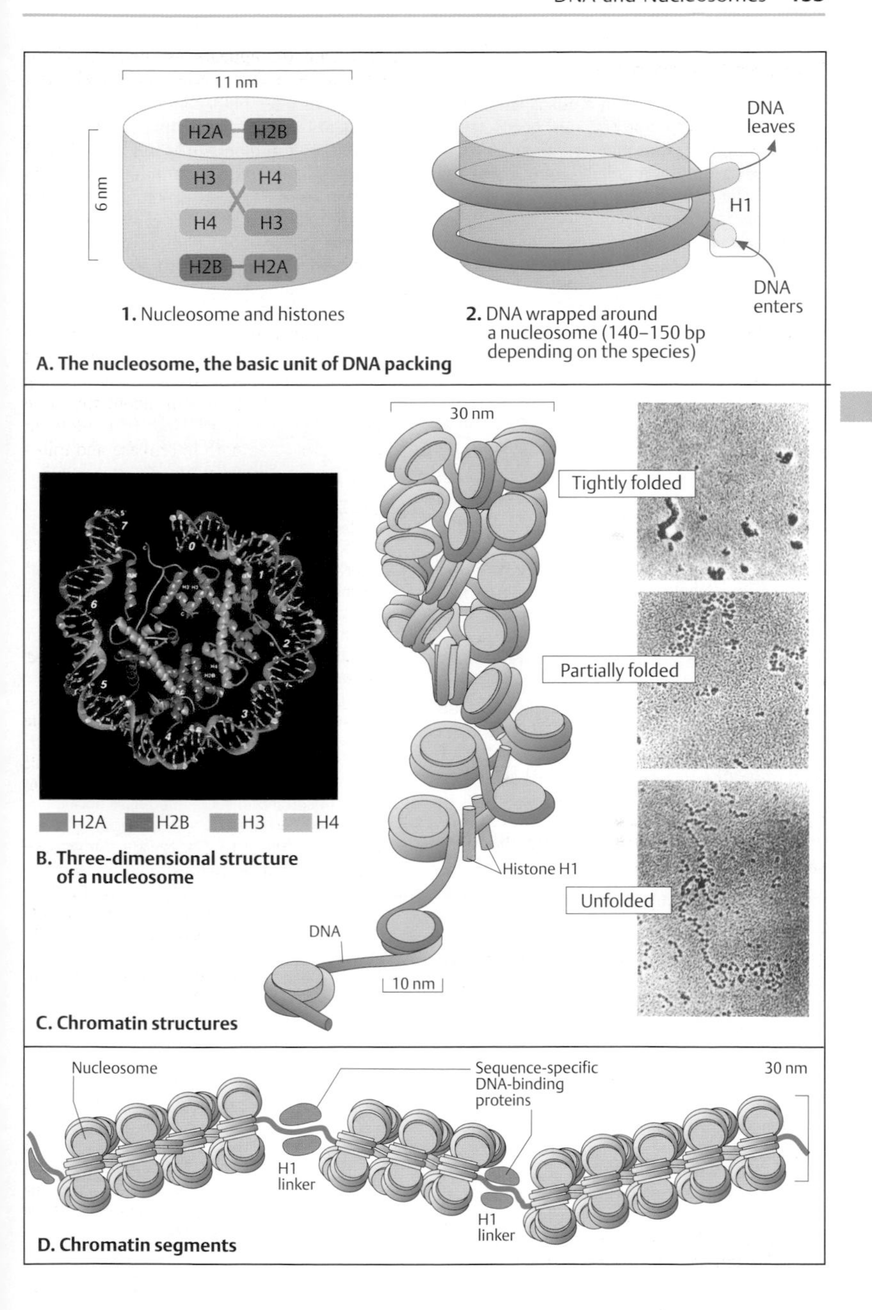

1. Nucleosome and histones

2. DNA wrapped around a nucleosome (140–150 bp depending on the species)

A. The nucleosome, the basic unit of DNA packing

B. Three-dimensional structure of a nucleosome

C. Chromatin structures

D. Chromatin segments

DNA in Chromosomes

The total length of DNA in a eukaryotic cell is about 1 m. During mitosis, this needs to be packed into chromosomes of about 3–7 µm long. The packing occurs in a highly organized manner. The change from interphase to mitotic chromosomes requires a class of proteins called *condensins*. These use energy from ATP hydrolysis to coil each interphase chromosome into a mitotic chromosome. Two of the five subunits of condensins interact with ATP and DNA. The cell cycle protein cdc2 (see Cell Cycle Control, p. 108) is required for both interphase and mitotic condensation.

A. Model for packing DNA in chromatin

Chromosomal DNA is folded and packed in six successive levels of hierarchical organization in a metaphase chromosome. These levels are shown schematically from top to bottom. First, a condensed section loops out of a metaphase chromosome. A higher magnification of the section shows a slightly extended part in a scaffold-associated region with loops of DNA attached to a scaffold. These loops correspond to the 30-nm chromatin fiber of packed nucleosomes shown at the next level below. The 11-nm "beads-on-a-string" form of chromatin follows. A short region of DNA double helix (five turns) marks the molecular level of DNA. (Figure modified from Alberts et al., 2008, and Lodish et al., 2008.)

B. Chromosome territories during interphase

Individual chromosomes occupy particular territories in an interphase nucleus. The work of T. Cremer and coworkers (Bolzer et al., 2005) has shown that small chromosomes are located preferentially toward the center of fibroblast nuclei, whereas large chromosomes are positioned preferentially toward the nuclear rim. Measurements along the optical axes of the chromosome territories of human chromosomes 18 and 19 suggest that the gene-poor chromosome 18 is closer to the top or bottom of the nuclear envelope than chromosome 19. This agrees with observations that a layer of Alu- and gene-poor chromatin lies close to the nuclear envelope, whereas chromatin that is rich in Alu sequences and genes has been found preferentially in the interior of the nucleus. Probably complex genetic and epigenetic mechanisms act at different levels to establish, maintain, or alter higher-order chromatin arrangements as required for proper nuclear functions. The numbers in (**1**) on the left indicate different degrees of chromosome decondensation (Monto Carlo relaxation steps 200, 1000, and 400000; see Bolzer et al., 2005). (Images from Bolzer et al., 2005; courtesy of Professor Thomas Cremer, Institute of Anthropology and Human Genetics, Munich, Germany.)

Further Reading

Alberts B, et al. Molecular Biology of the Cell. 5th ed. New York: Garland Science; 2008

Aono N, et al. Cdn2 has dual roles in mitotic condensation and interphase. Nature 2002;417:197–202

Bolzer A, et al. Three-dimensional maps of all chromosomes in human male fibroblast nuclei and prometaphase rosettes. PLoS Biol 2005;3:e157

Cremer T, Cremer C. Chromosome territories, nuclear architecture and gene regulation in mammalian cells. Nat Rev Genet 2001;2:292–301

Gilbert N, et al. Chromatin architecture of the human genome: gene-rich domains are enriched in open chromatin fibers. Cell 2004;118:555–566

Hagstrom KA, Meyer BJ. Condensin and cohesin: more than chromosome compactor and glue. Nat Rev Genet 2003;4:520–534

Lodish H, et al. Molecular Cell Biology. 6th ed. WH Freeman, New York, 2007

The Dynamic Chromosome. Science 2003;301:779–802

Sun HB, Shen J, Yokota H. Size-dependent positioning of human chromosomes in interphase nuclei. Biophys J 2000;79:184–190

Tyler-Smith C, Willard HF. Mammalian chromosome structure. Curr Opin Genet Dev 1993;3:390–397

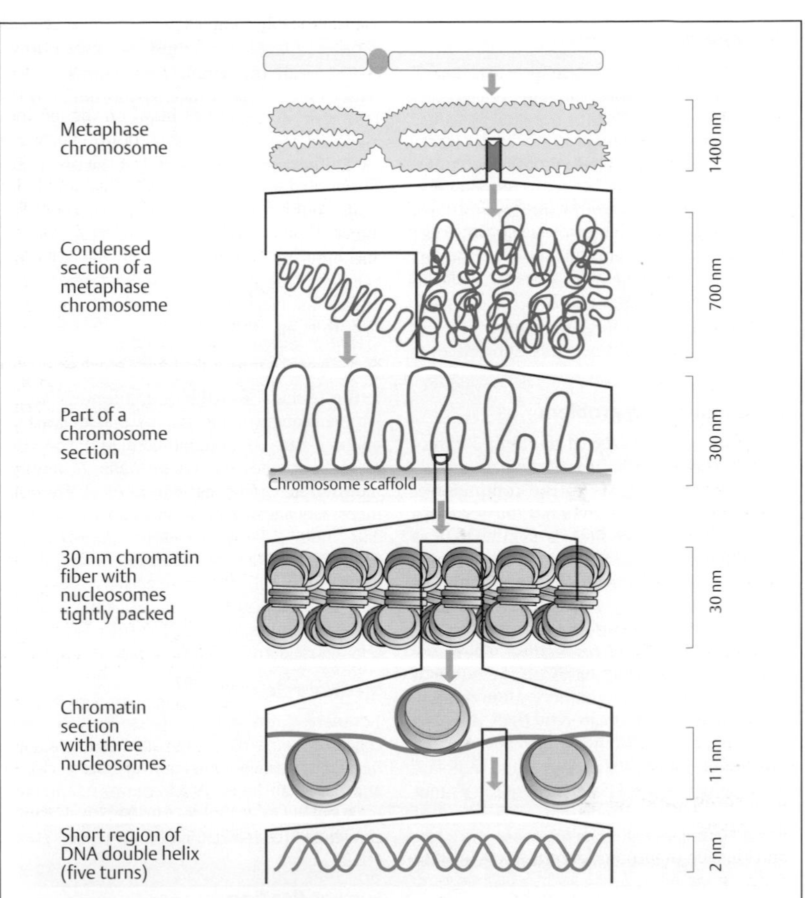

A. Model for packing of DNA in chromatin

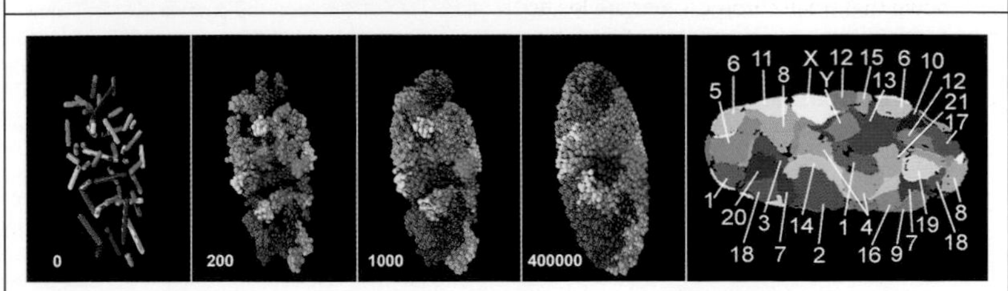

1. Chromosome position in metaphase and interphase

2. Territories for each chromosome at interphase

B. Chromosome territories during interphase

The Telomere

The telomere is a special structure that "seals" both ends of a chromosome. This is achieved by telomeric DNA sequence repeats. Mammalian telomeric DNA sequences are tandemly repeated along approximately 60 kb. One strand is G-rich, and the other is C-rich. All telomeric sequences can be written in the general form $C_n(A/T)_m$, with $n > 1$ and $m = 1–4$ (Lewin, 2008, p. 748). The TTAGGG repeat region in humans is 10–15 kb long, and in mice it is 25–50 kb (Blasco, 2005). At each cell division, somatic cells lose nucleotides from their telomeres. As a result the chromosome ends shorten over time.

A. The replication problem

Since DNA is synthesized in the 5′ to 3′ direction only, the two templates of the parent molecule differ with respect to the continuity of synthesis. From the 3′ to 5′ strand template (leading strand), new DNA is synthesized as one continuous 5′ to 3′ strand from a single short RNA primer (see DNA Replication, p. 54). However, from the 5′ to 3′ template strand (lagging strand), DNA must be synthesized in the opposite direction. This requires a new primer every few hundred bases on the parental strand (the Okazaki fragments), After the primer is removed and replaced by DNA, the gaps are closed by a DNA ligase. However, 8–12 bases at the end of the lagging strand template cannot be synthesized because here the primer cannot be attached. Hence, at each round of replication before cell division, these 8–12 nucleotides will be lost at the chromosome ends.

B. G-rich repetitive sequences

DNA at the telomeres consists of G-rich tandem sequences (5′-TTAGGG-3′ in vertebrates, 5′-TGTGGG-3′ in yeast, 5′-TTGGGG-3′ in protozoa). The G-strand overhang is important for telomeric protection by formation of a duplex loop (see **C**).

C. DNA duplex loop formation

Telomeres are formed by a ribonucleoprotein enzyme (telomerase). It forms a loop of telomeric DNA, which stabilizes the chromosome ends. Telomerase belongs to a family of modified reverse transcriptases (TERT, telomerase reverse transcriptase) composed of an RNA molecule (about 450 nucleotides) with a well-defined secondary structure. The RNA nucleotides provide the template for adding nucleotides to the 3′ end of the chromosome.

Telomeric duplex DNA forms a loop that is mediated by two related proteins, TRF1 and TRF2 (telomeric repeat-binding factor 1 and 2), binding to double-stranded telomere repeats. These are part of the six-subunit telomere-specific protein shelterin. The loop is anchored by the insertion of the G-strand overhang (see **B**) into a proximal segment of duplex telomeric DNA. (Figure adapted from Griffith et al., 1999.)

D. General structure of a telomere

In the terminal 6–10 kb of a chromosome, telomere sequences and telomere-associated sequences can be differentiated (**1**). The telomere-associated sequences contain autonomously replicating sequences (ARS). The telomere sequences themselves consist of about 250–1500 G-rich repeats (approximately 9 kb). They are highly conserved among different species (**2**). Telomerase activity is essential for survival in protozoans and yeast. In vertebrates, it occurs mainly in germ cells, and no telomerase activity is found in somatic tissues.

Medical relevance

Telomere shortening is a cause for cell senescence (growing old). High telomerase activity has been observed in immortal cells in culture and malignant cells. A telomerase component is defective in a severe disease: dyskeratosis congenita (OMIM 305000; Mitchell et al., 1999).

Further Reading

Blasco MA. Telomeres and human disease: ageing, cancer and beyond. Nat Rev Genet 2005;6:611–622

de Lange T. How telomeres solve the end-protection problem. Science 2009;326:948–952

Griffith JD, et al. Mammalian telomeres end in a large duplex loop. Cell 1999;97:503–514

Lewin B. Genes IX. Sudbury: Bartlett & Jones; 2008

Mitchell JR, Wood E, Collins K. A telomerase component is defective in the human disease dyskeratosis congenita. Nature 1999;402:551–555

Sahin E, et al. Telomere dysfunction induces metabolic and mitochondrial compromise. Nature 2011;470:359–365

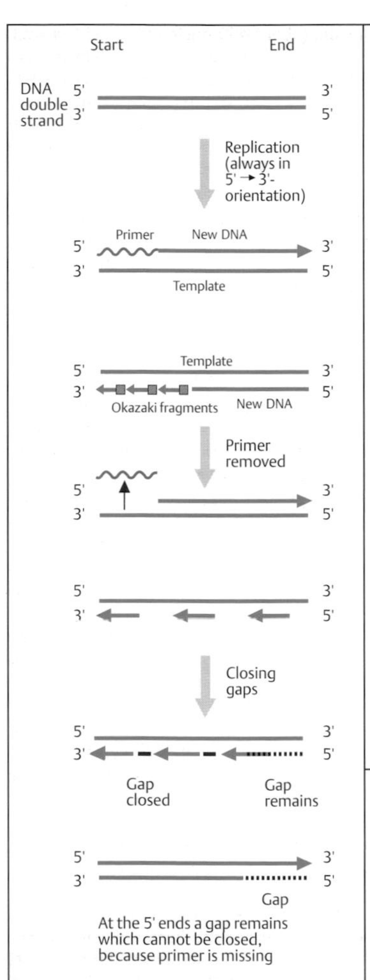

A. The replication problem at the end of linear DNA

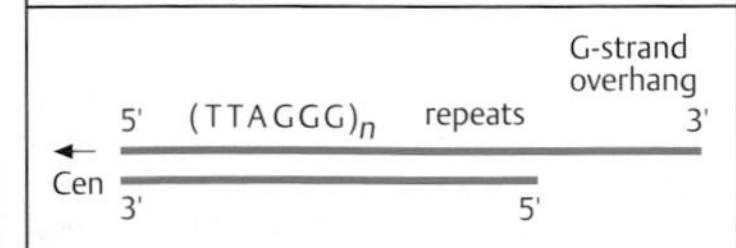

B. Telomeric DNA has G-rich repetitive repeats at the 3' end

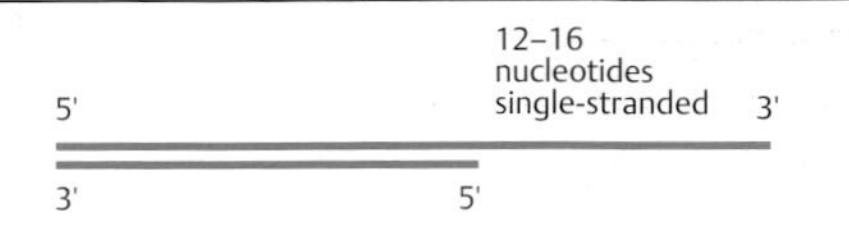

1. Single-stranded telomere end

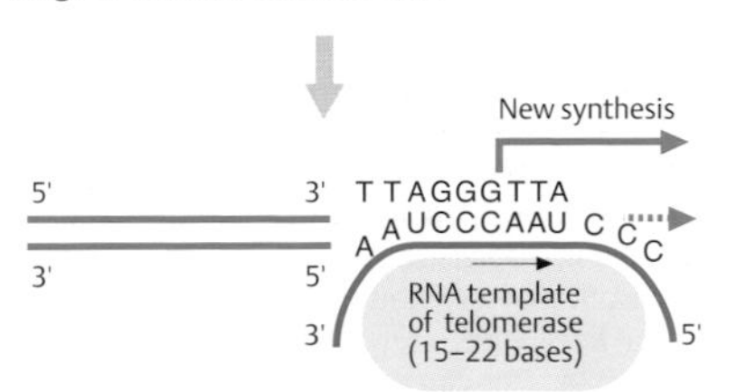

2. Addition of nucleotides to 3' end

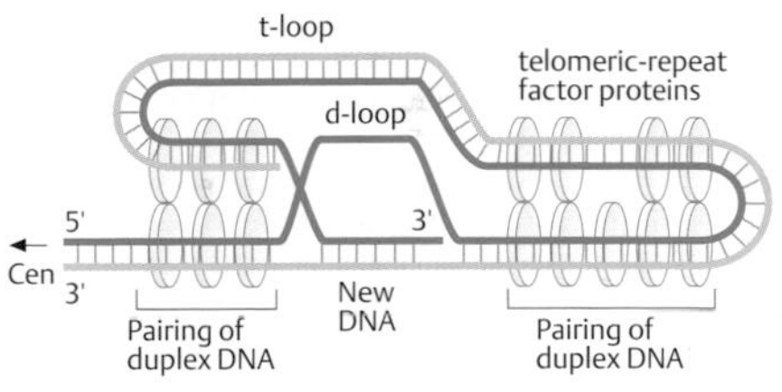

3. DNA duplex loop formation (t-loop)

C. DNA duplex loop formation in the mammalian telomere

1.

9 kb

Area for genes	Telomere-associated sequences	Telomeric sequences
Cen	ARS ARS	(TTAGGG)$_n$
		n = 250–1500

2. Examples of telomeric repeats

Protozoa	e.g., *Tetrahymena* microchromosomes 5' – TTGGGG – 3'
Yeasts	e.g., *Saccharomyces* 5' – TGTGGGG – 3'
Vertebrates	5' – TTAGGG – 3'
General	5' – $(T/A)_{1-4}(G)_{1-8}$ – 3' (telomere to the right)

D. General structure of a telomere

The Banding Patterns of Human Chromosomes

With special staining techniques, a reproducible, chromosome-specific pattern of transverse intensively and less intensively stained regions (bands) can be visualized. This allows breakpoints and genes to be assigned to defined regions along a chromosome. When Tjio and Levan, and Ford and Hamerton, independently in 1956, determined the chromosome number of man to be 46, many of the chromosome pairs could not be distinguished. The complete set of chromosomes of an individual or a species is the *karyotype*, a term introduced by Levitsky in 1924. The chromosomes are arranged in homologous pairs in a *karyogram*. The most frequently used type of banding is *G-banding* (also referred to as GTG banding). This name is derived from the Giemsa stain, which is applied to the chromosomal preparation after it has been treated with the proteolytic enzyme trypsin. Other types of band are Q bands (quinacrine-induced), R bands (reverse of G), C bands (centromeric constitutive heterochromatin), T bands (telomeric), and others (see International System of Chromosome Nomenclature [ISCN], 2009; and Appendix, Table 3, p. 410).

A. Banding patterns and sizes of human chromosomes 1–12

A band is defined as part of a chromosome that can be clearly distinguished from its adjacent band and shows staining of different intensity. This figure schematically shows the G-banding pattern of human chromosomes 1–12 (for chromosomes 13–22, and the X and Y chromosomes, see p. 161). At standard resolution, about 450 individual bands can be distinguished per haploid set of human chromosomes. With variant techniques a higher resolution (800 bands) can be achieved.

Each chromosome is divided into regions along the short (p) arm and the long (q) arm. The regions are numbered from the centromere to the telomeres. For example, the proximal short arm of chromosome 1 begins (next to the centromere) with region 1, containing a dark and a light band, followed by regions 2 and 3. Within each region, bands are also numbered proximally to distally (toward the telomere). The number of each band is stated directly after the number of the region. For example, bands 2 and 3 of region 2 are designated 22 and 23, respectively; bands 1–6 of region 3 are designated 31, 32, 33, 34, 35, and 36, respectively. Each band is designated according to its chromosome, chromosome arm, region, and band. Thus, 1p23 indicates region 2, band 3 of the short arm of chromosome 1. Additional bands that become visible with higher resolution are indicated by a decimal: if one additional band can be distinguished in 1p21, this region is designated as 1p2,1p21.2, 1p21.3, etc. This system is outlined in detail in ISCN, 2009. The size of each chromosome in millions of base pairs (Mb) is given below each chromosome.

Medical relevance

The banding pattern is used to define the location of structural chromosomal aberrations and the approximate location of gene loci on a chromosome (a precise location can be given by the number of DNA bases along the sequence of a chromosome). For example, 1p22.1–1p21.1 corresponds to the location of base pair numbers 94700000–107200000 (UCSC Genome Bioinformatics).

Further Reading

Bickmore WA, Craig J. Chromosome Bands: Patterns in the Genome. New York: Chapman & Hall; 1997

Ford CE, Hamerton JL. The chromosomes of man. Nature 1956;178:1020–1023

Gersen SL, Keagle MB, eds. The Principles of Clinical Cytogenetics. 2nd ed. Totowa, NJ: Humana Press; 2005

McKinley Gardner RJ, Sutherland GR. Chromosome Abnormalities and Genetic Counseling. 3rd ed. Oxford: Oxford University Press; 2004

Miller OJ, Therman E. Human Chromosomes. 4th ed. New York: Springer-Verlag; 2001

Philip AGS, Polani PE. Historical perspectives: Chromosomal Abnormalities and Clinical Syndromes. NeoReviews, Aug 2004;5:e315–e320. Available at: http://NeoReviews.org. Accessed January 26, 2012

Shaffer LG, Slovak ML, Campbell LJ, eds. ISCN. An International System for Human Cytogenetic Nomenclature. Basel: Karger; 2009

Tjio JH, Levan A. The chromosome number of man. Hereditas 1956;42:1–6

UCSC Genome Bioinformatics. Available at: http://www.genome.ucsc.edu. Accessed January 26, 2012

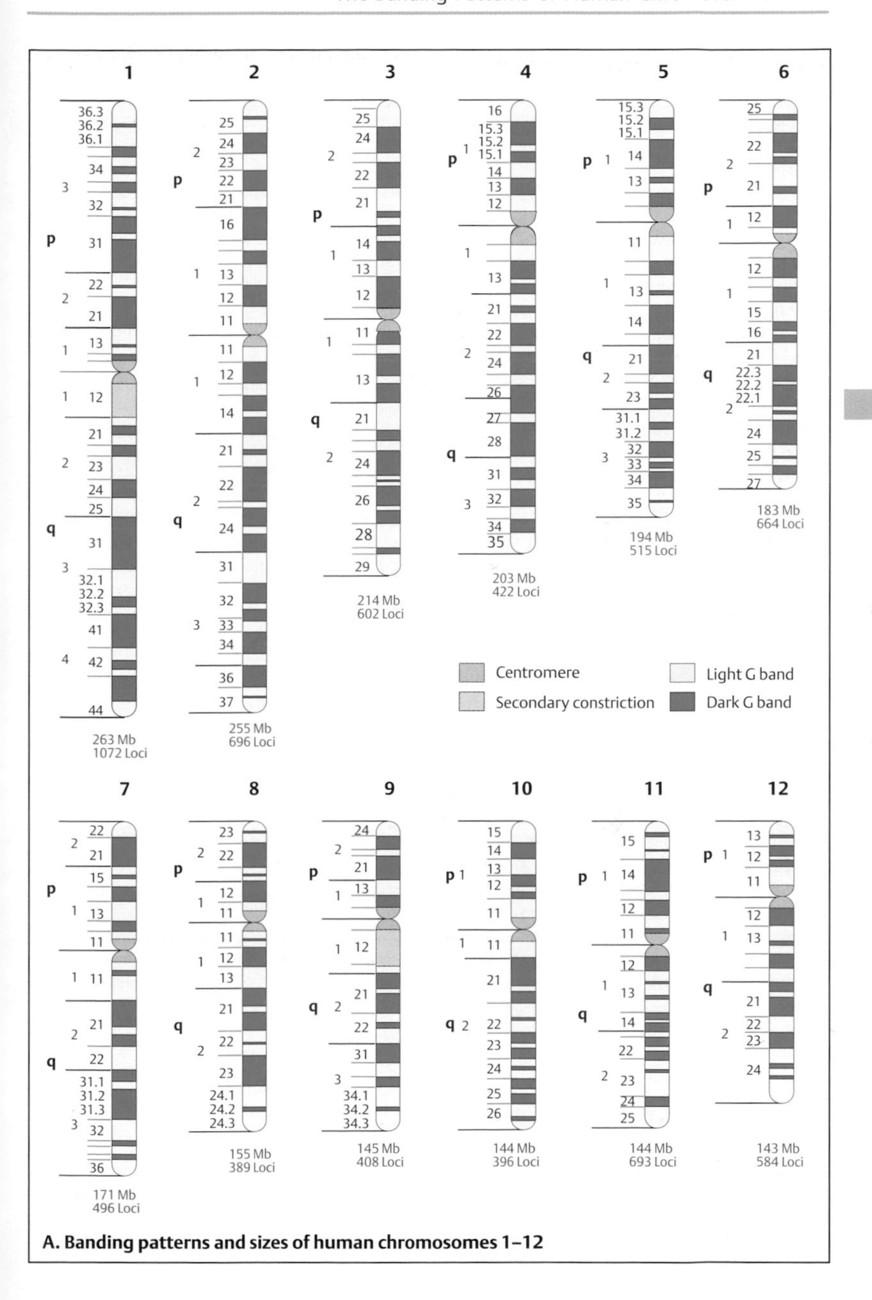

A. Banding patterns and sizes of human chromosomes 1–12

Karyotype of Man and Mouse

The 22 chromosome pairs of man are arranged in seven groups (A–G): group A consists of three pairs, chromosomes 1–3, which are large submetacentric chromosomes; group B (chromosomes 4–5) consists of two pairs of fairly long submetacentric chromosomes, which are slightly smaller than group A; members of group C (chromosomes 6–12) are all submetacentric; members of group D (chromosomes 13–15) are all acrocentric; members of group E (chromosomes 16–18) are all submetacentric; members of group F (chromosomes 19–20) are all metacentric; and members of Group G (chromosomes 21–22) are all small acrocentric chromosomes. In addition, the X and the Ychromosomes do not belong to a group.

A. Banding patterns of human chromosomes 13–22, X, and Y

The schematic representation of the human karyotype is continued from the previous page.

B. Karyotype of the mouse

The mouse (*Mus musculus*) has 20 chromosome pairs, each with a specific banding pattern (**1**). All chromosomes except the X chromosome are acrocentric, with the centromere at the very end.

Variant strains of mice have a karyotype with metacentric chromosomes. These are the result of centric fusion between two acrocentric chromosomes. The example shows fused chromosome pairs 4 and 2 (4/2), 8 and 3, 7 and 6, 13 and 5, 12 and 10, 14 and 9, 18 and 11, 17 and 16. Chromosome pairs 1, 15, and 19 are unchanged, as in the standard karyotype. The fused chromosomes are present in all mice of this particular population. They represent an example of chromosome evolution by fusion. (Photographs: **1** from Traut, 1991; **2** kindly provided by Dr H. Winking, Medical University of Lübeck, Lübeck, Germany.)

Human chromosome nomenclature

An elaborate system of nomenclature has been developed. It is updated from time to time, with the last update in 2009. This is used to designate normal and abnormal chromosome findings. The most important examples are listed in the Appendix, Table 4. For details, the reader is referred to the International System for Human Cytogenetic Nomenclature (ISCN 2009).

Further Reading

Gersen SL, Keagle MB, eds. The Principles of Clinical Cytogenetics. 2nd ed. Totowa: Humana Press; 2005

Miller OJ, Therman E. Human Chromosomes. 4th ed. New York: Springer-Verlag; 2001

Philip AGS, Polani PE. Historical perspectives: Chromosomal Abnormalities and Clinical Syndromes. NeoReviews, Aug 2004;5:e315–e320. Available at: http://NeoReviews.org. Accessed January 26, 2012

Schwartz S, Hassold T. Chromosome disorders. In: Longo DL, et al., eds. Principles and practice of Internal Medicine. 18th ed. New York: McGraw Hill; 2012:509–518

Shaffer LG, Slovak ML, Campbell LJ, eds. ISCN. An International System for Human Cytogenetic Nomenclature. Basel: Karger; 2009

Traut W. Chromosomen. Klassische und molekulare Cytogenetik. Heidelberg: Springer-Verlag; 1991

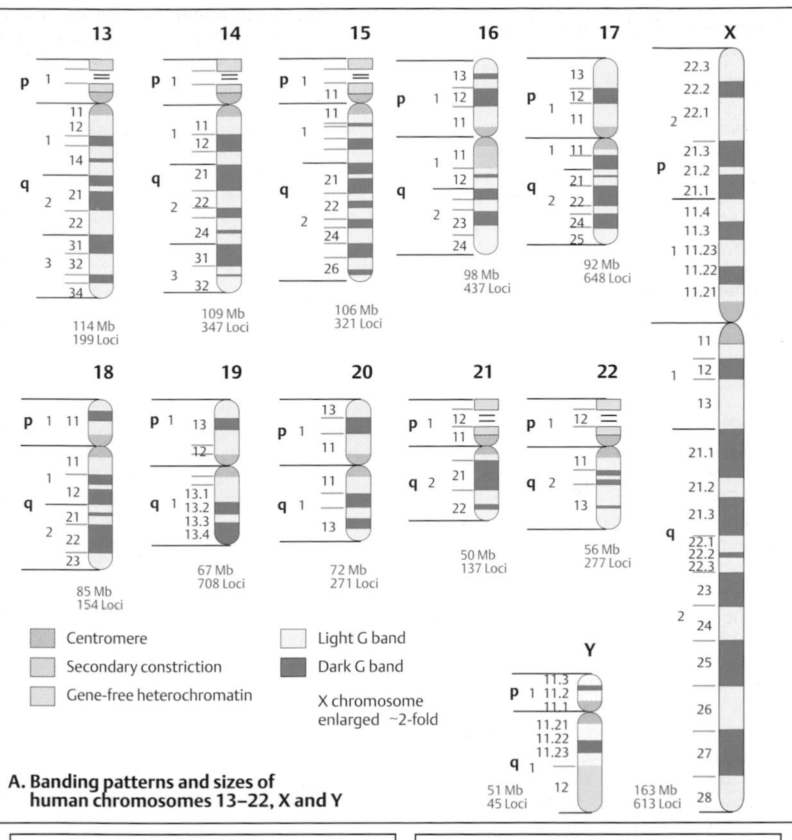

A. Banding patterns and sizes of human chromosomes 13–22, X and Y

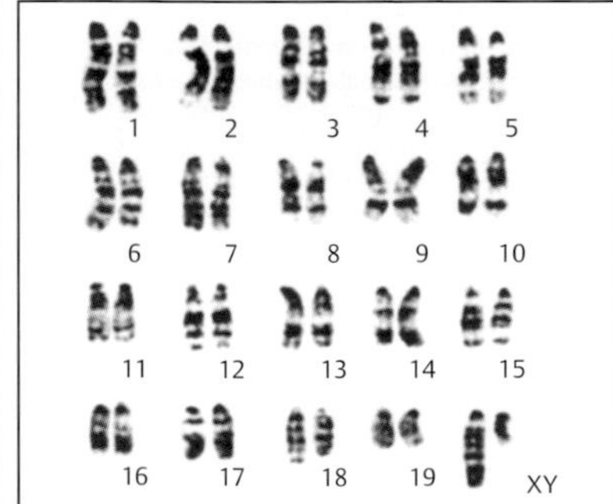

1. Standard

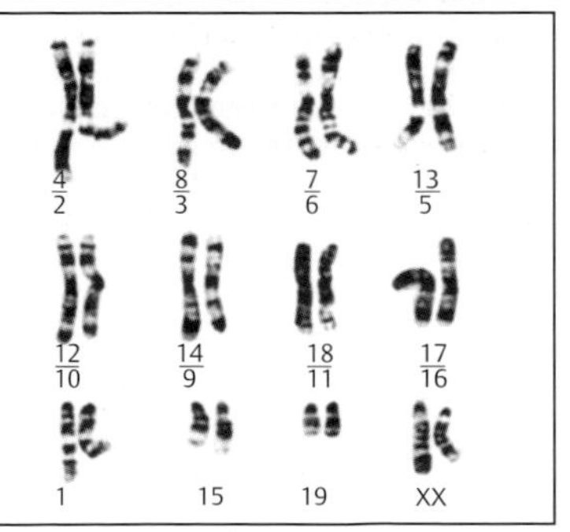

2. Variants in a population with fused chromosomes

B. Karyotype of the mouse (*Mus musculus*)

Preparation of Metaphase Chromosomes for Analysis

Chromosomal analysis requires cells in metaphase (or occasionally prometaphase). These are usually obtained from lymphocyte cultures, but can also be obtained from cultured skin fibroblasts, cultured amniocytes from amniotic fluid, cultured chorionic cells from chorionic villae, or directly from bone marrow cells. Peripheral blood lymphocytes grow in a suspension culture after they have been stimulated to divide by phytohemagglutinin. Their lifespan is limited to a few cell divisions. However, by exposing the culture to Epstein–Barr virus they can be transformed into a lymphoblastoid cell line with permanent growth potential (p. 112).

A. Chromosome analysis from blood

Five principal steps are required: (1) lymphocyte culture; (2) harvest of metaphase chromosomes; (3) chromosome preparation; (4) staining with special dyes that stain chromosomes (and chromatin); and (5) analysis by microscopy, which is now assisted by computer analysis.

Heparinized whole blood is the predominant source of a lymphocyte culture. The proportion of heparin is about 1/10 of the blood volume. A sample of about 0.5 mL peripheral blood is needed. Heparin must be added to prevent blood clotting. Following stimulation of cell division by phytohemagglutinin, a lymphocyte culture requires about 72 hours at 37°C for two cell divisions. Cells reaching mitosis are arrested in metaphase by adding a colchicine derivative (colcemid) for 2 hours prior to harvest. Colcemid arrests mitosis in metaphase. The culture is then terminated and the cells in metaphase are harvested (**1**).

At harvest, the culture solution is centrifuged (**2**). Hypotonic potassium chloride solution (KCl, 0.075 M) is added for 20 minutes (**3**), after which a fixative solution (3:1 mixture of methyl alcohol and glacial acetic acid) is added (**4**). Usually the fixative is changed between four and six times, with subsequent centrifugation. The fixed cells are taken up in a pipette, dropped onto a clean, wet, fat-free glass slide suitable for microscopic analysis, and air-dried (**5**). The preparation is treated according to the type of bands desired (**6**), stained (**7**), and the slide is covered with a cover glass (**8**).

Suitable metaphases are located under the microscope at approximately 100-fold magnification, and are subsequently examined at about 1250-fold magnification (**9**). More than one cell has to be analyzed. Depending on the purpose of the analysis, between 5 and 100 metaphases (usually 10–20) are examined. Some of the metaphases are photographed under the microscope and subsequently can be cut out from the photograph (karyotyping). In this way a karyogram can be obtained from the photograph of a metaphase. The time needed for a chromosome analysis varies depending on the problem, but it is usually 3–4 hours. Analysis and karyotyping time can be considerably shortened by computer procedures. The karyogram can be obtained by computer-assisted analysis.

Additional methods for chromosomal analysis. A variety of methods exist that do not use metaphase chromosomes for analysis. Examples are comparative genomic hybridization, CGH (see p. 230), interphase fluorescence in-situ hybridization (FISH; see p. 164), fiber FISH, chromosome sorting by flow cytometry, and others.

Medical relevance

Chromosomal analysis is an important tool in the diagnosis of certain genetic disorders. Under given circumstances, metaphase analysis can be replaced by molecular techniques, e.g., array CGH (p. 230).

Further Reading

Arakaki DT, Sparkes RS. Microtechnique for culturing leukocytes from whole blood. Cytogenetics 1963; 85:57–6014099759

Gosden JR, ed. Chromosome Analysis protocols. Methods in Molecular Biology. Volume 29. Totowa: Humana Press; 1994

Miller OJ, Therman E. Human Chromosomes. 4th ed. New York: Springer-Verlag; 2001

Moorhead PS, et al. Chromosome preparations of leukocytes cultured from human peripheral blood. Exp Cell Res 1960;20:613–616

Schwarzacher HG, Wolf U, Passarge E, eds. Methods in Human Cytogenetics. Berlin: Springer-Verlag; 1974

Sumner T. Chromosomes. Organization and Function. Oxford: Blackwell; 2003

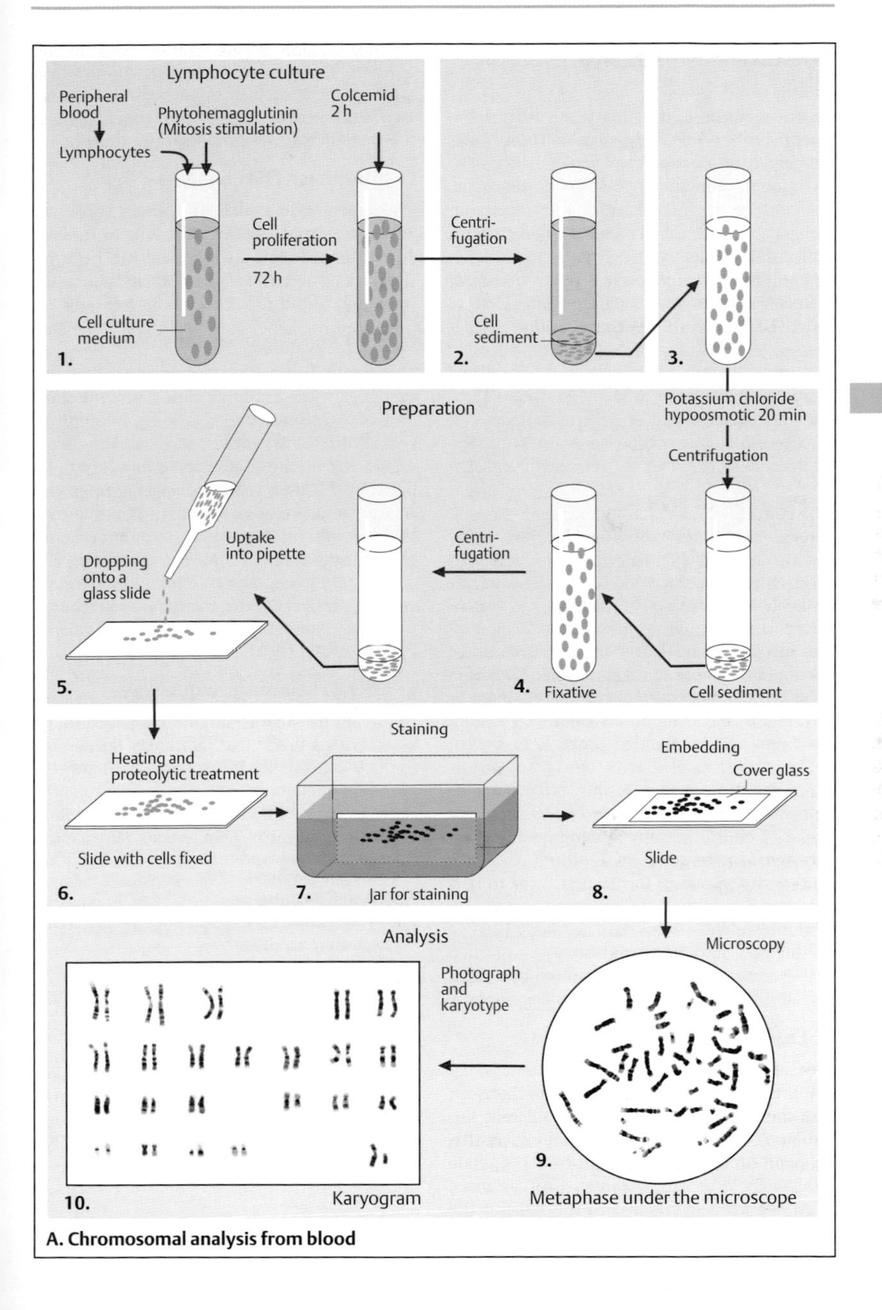

A. Chromosomal analysis from blood

Fluorescence In-Situ Hybridization (FISH)

FISH was introduced by Mary Lou Pardue and Joseph Gall at Yale University in 1969. It has developed into a variety of molecular genetic techniques that are applied to chromosome preparations in metaphase or interphase nuclei, an approach called *molecular cytogenetics*. Conventional metaphase chromosomal analysis can detect the loss or gain of chromosomal material of about 4 million base pairs (4 Mb) or more. FISH analysis can detect much smaller rearrangements. The principle rests on using metaphase or interphase cells and hybridizing a labeled DNA probe to single-stranded chromosomal DNA at one or several defined sites directly on a microscope slide (in situ). The hybridization results in a site-specific signal.

A. Principle of FISH

Direct and indirect nonisotopic labeling can be distinguished. In *direct labeling* the fluorescent label, a modified nucleotide (often 2′-deoxyuridine 5′-triphosphate) containing a fluorophore, is directly incorporated into DNA. *Indirect labeling* requires labeling the DNA probe with a fluorophore to make the signal visible.
In indirect nonisotopic labeling, metaphase or interphase cells fixed on a slide are denatured (**1a**) into single-stranded DNA (**2**). A DNA probe (**1b**) is labeled with biotin and hybridized in situ to its specific site on the chromosome (**3**). This site is visualized by fluorescence in dark-field microscopy by binding a fluorescent-dye-labeled antibody (for biotin this antibody is streptavidin) to the biotin (**4**). This is the primary antibody. To enhance the intensity of fluorescence, a secondary antibody (here, a biotinylated anti-avidin antibody) is attached (**5**). The resulting signal is amplified by attaching additional labeled antibodies (**6**).

B. Example of FISH in metaphase

Here all chromosomes are stained dark blue by DAPI (4′,6-diamidino-2-phenylindole) except five chromosomes carrying a fluorescent signal. Two copies of chromosome 3 are identified by two red signals. A green fluorescent probe (D354559; Vysis Inc., Downers Grove, Illinois, USA) has been hybridized to the ends of the short arms (3p). A further green signal is visible over the long arm of one chromosome 16. These three green signals indicate the presence of three chromosomal segments of the long arm of chromosome 3 (*partial trisomy 3q*). Two chromosomes 21 at the upper right of the metaphase are nonspecifically stained.

C. Interphase FISH analysis

Two green signals identify the chromosomes 22. A red signal identifies the long arm (22q). The upper cell (**1**) is normal, whereas the lower one lacks a red signal (**2**). This indicates loss (deletion) of the 22q chromosomal region.

D. FISH analysis of a translocation

This shows a reciprocal translocation between the long arm of a chromosome 8 and the short arm of a chromosome 4, yielding two derivative chromosomes: der4 and der8. The region embracing the breakpoint on 8 at q24 was labeled by a 170-kb yeast artificial chromosome. As a result, three signals are visible: one over the normal chromosome 8, one over the altered chromosome 8 (der8), and one over the altered chromosome 4 (der4). Centromere-specific probes identify the four chromosomes involved. (Photograph courtesy of H. J. Lüdecke, Essen, Germany.)

E. FISH of telomere sequences

In this human metaphase all telomeres are labeled with a probe that hybridizes specifically to the telomere sequences. Two signals are visible at each telomere, one over each chromatid. (Photographs with kind permission from Robert M. Moyzes, Los Alamos Laboratory, New Mexico, USA.)

Further Reading

Liehr T, ed. Fluorescence in Situ Hybridization (FISH). Berlin: Springer-Verlag; 2010

Protocol Online: Cytogenetics: In Situ Hybridization. Available at: http://www.protocol-online.org/prot/Genetics___Genomics/Cytogenetics/In_Situ_Hybridization/index.html. Accessed January 26, 2012

Ried T, Schröck E, Ning Y, Wienberg J. Chromosome painting: a useful art. Hum Mol Genet 1998;7: 1619–1626

Speicher MR, Carter NP. The new cytogenetics: blurring the boundaries with molecular biology. Nat Rev Genet 2005;6:782–792

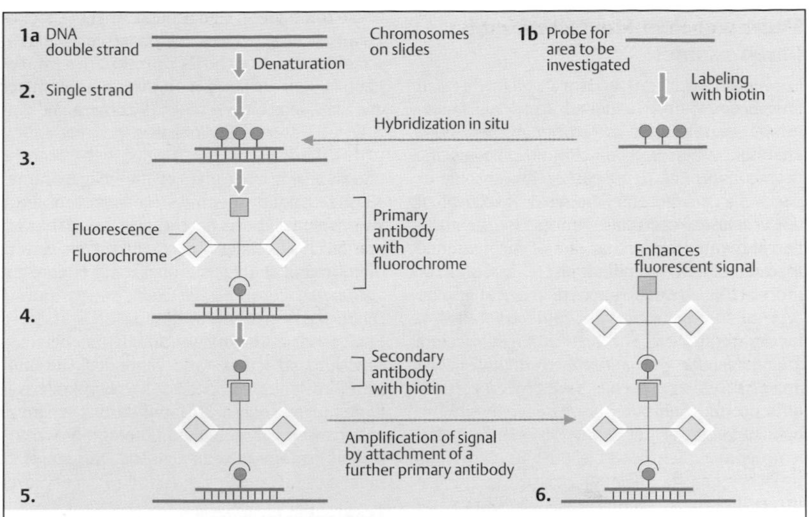

A. Principle of fluorescence in situ hybridization

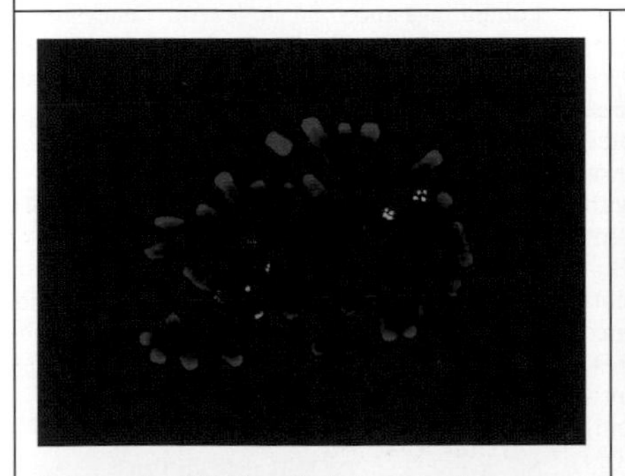

B. Example of FISH analysis in metaphase

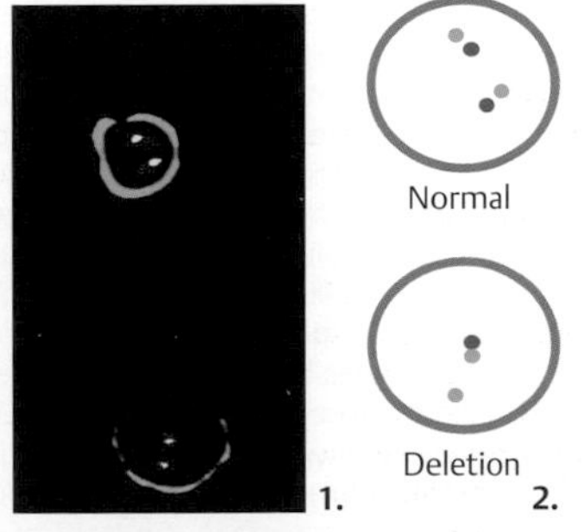

C. Interphase FISH analysis

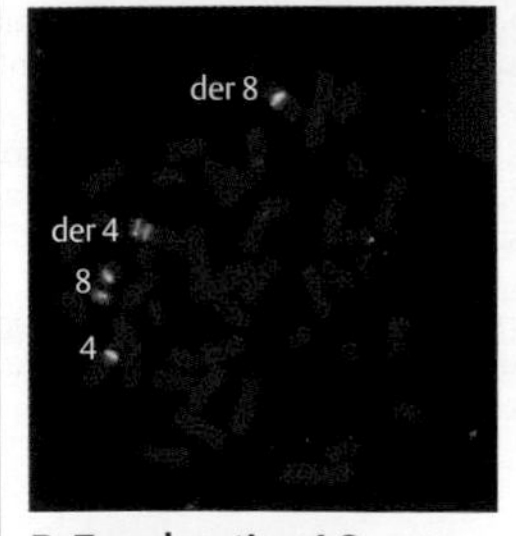

D. Translocation 4;8

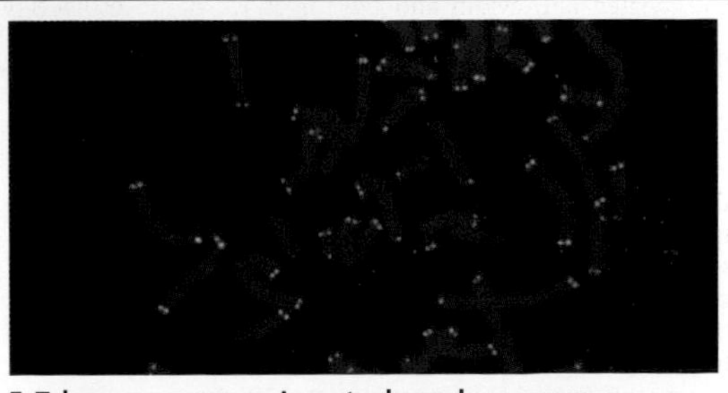

E. Telomere sequences in metaphase chromosomes

Multicolor FISH Identification of Chromosomes

Computer-based methods can identify each chromosome and to detect small rearrangements that escape conventional metaphase analysis. When sets of chromosome-specific probes hybridize to an entire chromosome or parts of a chromosome, the result is referred to as *chromosome painting*. To increase the number of chromosomes that can be differentiated at one time, a combination of labels (DNA probes that are labeled with more than one type of fluorochrome) and different ratios of labels are applied. The resulting mixed colors are detectable by automated digital image analysis. Two approaches have proved particularly useful: multiplex fluorescence in-situ hybridization (Speicher et al., 1996) and spectral karyotyping (Schröck et al., 1996). Other approaches and modifications exist, e.g., the use of artificially extended DNA or chromatin fibers.

A. Multicolor FISH

Multicolor FISH (M-FISH; Speicher et al., 1996) uses sets of chromosome-specific DNA probes, which are hybridized to denatured metaphase chromosomes. Each probe is contained in a yeast artificial chromosome and labeled with a combination of DNA-binding fluorescent dyes specific for each chromosome. Five different fluorophores can produce 24 different colors for image analysis by epifluorescence microscopy using a charge-coupled device (CCD) camera. An image of each chromosome in a pseudocolor is displayed by appropriate computer software. (Photograph courtesy of Drs Sabine Uhrig and Michael Speicher, Graz, Austria.)

B. Spectral karyotyping

Spectral karyotyping (SKY; Schröck et al., 1996) combines optical microscopy, Fourier spectroscopy, and CCD imaging.

First, all 24 human chromosomes (chromosomes 1–22, X, and Y) are separated by flow-sorting and labeled with at least one, but possibly as many as five fluorochrome combinations (SKY probes). Following DNA denaturation, the SKY probe is hybridized to a metaphase chromosome preparation at 37°C for 24–72 hours.

An interferometer and a CCD camera (SpectraCube) are coupled to an epifluorescence microscope to visualize the SKY probes. The emitted light is sent through a Sagnac interferometer and focused onto the CCD camera. In this way, an interferogram based on the optical path difference of the divided light beam is measured for each pixel of the image. The resulting unique spectral "signature" of each chromosome based on the raw spectral information is displayed as an RGB (red, green, blue) image of the metaphase and is used for automated chromosome classification. During this process, a discrete false color is assigned to all pixels of the image with identical spectra, revealing structural and numerical chromosome aberrations. Spectral karyotyping has a wide range of diagnostic applications in clinical and cancer cytogenetics. (Photographs courtesy of Professor Evelin Schröck, Dresden, Germany.)

Medical relevance

Computer-assisted cytogenetic analysis is widely used to identify structural chromosomal aberrations below the resolution power of light microscopy.

Further Reading

Chrombios. Molecular Cytogenetics. Multicolor FISH. Available at: http://www.chrombios.com/cms/website.php. Accessed January 26, 2012

Liehr T, ed. Fluorescence in Situ Hybridization (FISH). Heidelberg: Springer-Verlag; 2010

Liehr T, et al. Multicolor FISH probe sets and their applications. Histol Histopathol 2004;19:229–237

Miller OJ, Therman E. Human Chromosomes. 4th ed. New York: Springer-Verlag; 2001

NCI and NCBI SKY/M-FISH and CGH Database. Available at: (http://www.ncbi.nlm.nih.gov/projects/sky/. Accessed January 26, 2012

Ried T, et al. Chromosome painting: a useful art. Hum Mol Genet 1998;7:1619–1626

Schröck E, et al. Multicolor spectral karyotyping of human chromosomes. Science 1996;273:494–497

Speicher MR, Gwyn Ballard S, Ward DC. Karyotyping human chromosomes by combinatorial multi-fluor FISH. Nat Genet 1996;12:368–375

Uhrig S, et al. Multiplex-FISH for pre- and postnatal diagnostic applications. Am J Hum Genet 1999;65:448–462

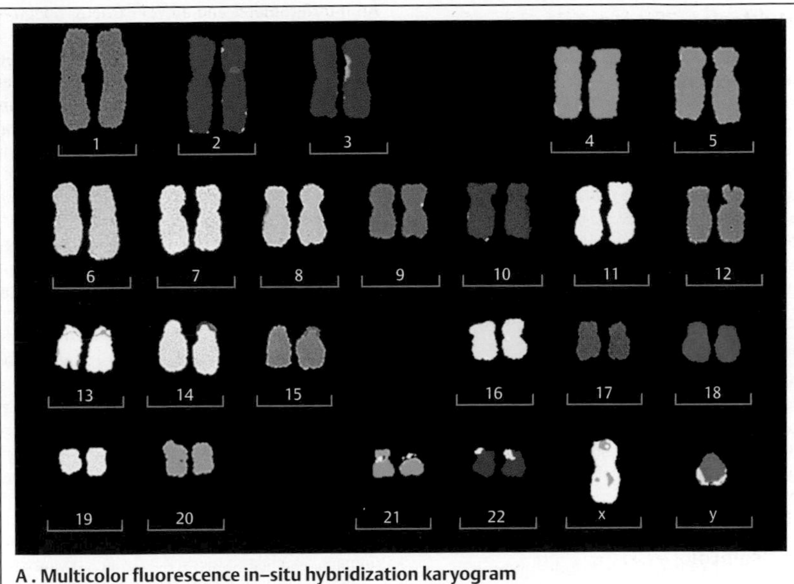

A. Multicolor fluorescence in-situ hybridization karyogram

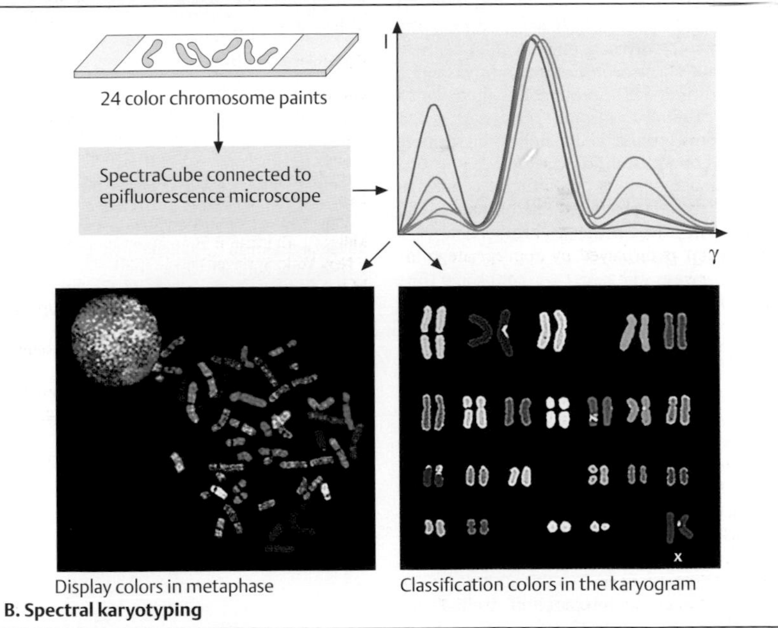

Display colors in metaphase

Classification colors in the karyogram

B. Spectral karyotyping

Aneuploidy

Aneuploidy is a deviation of the normal chromosome number. It leads to loss or gain of one or several individual chromosomes from the diploid set. A loss is called *monosomy*, a gain is called *trisomy*. Both result from nondisjunction at meiosis (see p. 102). Nondisjunction was discovered 1913 in *Drosophila melanogaster* by C. B. Bridges, who coined the term. The frequency of nondisjunction in humans is influenced by the age of the mother at the time of conception.

A. Nondisjunction in meiosis I or meiosis II

Nondisjunction may occur during either meiosis I or meiosis II. During meiosis I, one daughter cell will receive two chromosomes instead of one, whereas the other daughter cell receives none (in the figure opposite only one pair of chromosomes is shown). As a result, gametes will be formed carrying either two chromosomes (disomy) or none (nullisomy), instead of one each. If nondisjunction occurs during meiosis II, the first meiotic division (see p. 104) is normal. However, during the second meiotic division, one daughter cell will receive two chromosomes (becoming disomic), and the other will receive none (nullisomic). After fertilization, a disomic gamete gives rise to a trisomic zygote, and a nullisomic gamete gives rise to a monosomic zygote.

B. Common types of aneuploidy

Three states of aneuploidy can be distinguished: (i) trisomy (three chromosomes instead of two in one pair), (ii) monosomy (one chromosome instead of two), (iii) triploidy and tetraploidy (all chromosomes present in triplicate or quadruplicate). Triploidy does not result from nondisjunction at meiosis, but from one of a variety of other processes. Two sperm cells may have penetrated the egg (*dispermy*); the egg or sperm may have an unreduced chromosome set as a result of restitution in the first or second meiotic division; or the second polar body may have reunited with the haploid egg nucleus. With dispermy, two of the three sets of chromosomes will be of paternal origin, resulting in 69,XYY, 69,XXY, or 69, XXX. Dispermy is the cause of triploidy in about 66% of cases; fertilization of a haploid egg by a diploid sperm is the cause of triploidy in 24% of cases (failure at meiosis I); and a diploid egg is the cause of triploidy in 10% of cases. Triploidy is one of the most frequent chromosomal aberrations in man, causing 17% of spontaneous abortions (see p. 384). Tetraploidy is rarer than triploidy (for details, see Miller & Therman, 2001).

C. Autosomal trisomies in humans

In humans, only three autosomal trisomies occur in liveborn infants: trisomy 13, trisomy 18, and trisomy 21 at 1 in 650 (see p. 382).

D. Additional X or Y chromosome

An additional X or Y chromosome occurs in about 1 in 800 newborns. The most common are three X chromosomes (XXX), XXY and XYY.

Further Reading

Bridges CB. Nondisjunction of the sex chromosomes of Drosophila. J Exp Zool 1913;15:587–606

Gersen SL, Keagle MB, eds. The Principles of Clinical Cytogenetics. 2nd ed. Totowa: Humana Press; 2005

Grant R, McKinlay J, Sutherland GR. Chromosome Anormalities and Genetic Counseling. 3rd ed. Oxford: Oxford University Press; 2004

Harper P. Practical Genetic Counseling. 7th ed. London: Edward Arnold; 2010

Hassold TJ, Jacobs PA. Trisomy in man. Annu Rev Genet 1984;18:69–97

Jacobs PA, Hassold TJ. The origin of numerical chromosome abnormalities. Adv Genet 1995;33:101–133

Jacobs PA, Angell RR, Buchanan IM, Hassold TJ, Matsuyama AM, Manuel B. The origin of human triploids. Ann Hum Genet 1978;42:49–57

Miller OJ, Therman E. Human Chromosomes. 4th ed. New York: Springer-Verlag; 2001

Rooney DE, Czepulski BH, eds. Human Cytogenetics. A Practical Approach. 2nd ed. Oxford: Oxford University Press; 2001

Tolmie JL, MacFaden U. Cinical genetics of common autosomal trisomies. In: Rimoin DL, et al. Principles and Practice of Medical Genetics. 5th ed. Philadelphia: Churchill Livingstone-Elsevier; 2007: 1015–1037

Schinzel A. Catalogue of Unbalanced Chromosome Aberrations in Man. 2nd ed. Berlin: De Gruyter; 2001

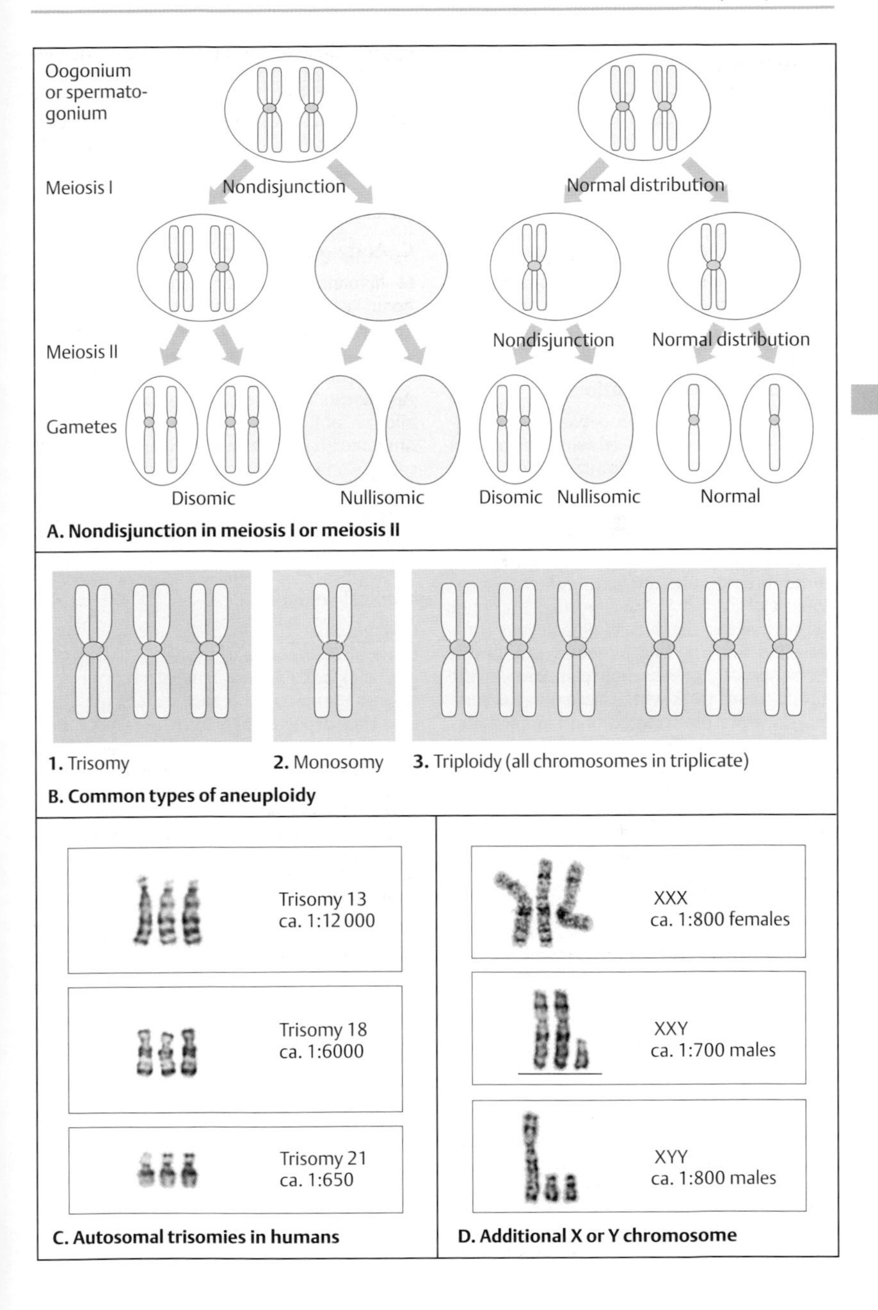

A. Nondisjunction in meiosis I or meiosis II

B. Common types of aneuploidy

C. Autosomal trisomies in humans

D. Additional X or Y chromosome

Chromosome Translocation

In cytogenetics, translocation refers to a structural change involving a break in the chromosome involved. It results in a change in position of a part of (or a whole) chromosome from its usual location to another, without changing the total number of chromosomes or genes. Usually a translocation is a reciprocal event, one part being exchanged with another.
A special type of translocation, the fusion of two acrocentric chromosomes, is called a Robertsonian translocation after W. R. B. Robertson who first observed it in insects in 1911.

A. Reciprocal translocation

A reciprocal translocation between two chromosomes is called balanced when it does not interrupt the structure and function of a gene. However, phenotypically normal carriers of a reciprocal translocation are prone to an increased risk of a chromosomal aberration in the offspring as a result of meiotic segregation of the chromosomes involved in the translocation.
During meiosis, the normal homologous chromosomes and the chromosomes involved in the reciprocal translocation pair. Each of the chromosomes pairs with its homologous partner. However, the translocation chromosomes can only pair by forming a quadriradial configuration. Disregarding crossing-over and secondary nondisjunction, several outcomes are possible. The simplest is segregation of the two normal chromosomes to one gamete and the two translocation chromosomes to the other (*alternate segregation*). This results in chromosomally balanced gametes.
However, if two neighboring (adjacent) chromosomes are distributed (segregate) to the same gamete, the latter will be chromosomally unbalanced. It will either contain a chromosomal segment twice (duplication) or lack it (deficiency). There are two types of adjacent segregation. In *adjacent-1* segregation, the gametes receive a normal chromosome and a translocation chromosome with opposite centromeres. Two types of unbalanced gametes can result. In the rare *adjacent-2* segregation, homologous centromeres go to the same pole. This produces two generally more extreme types of unbalanced gamete (1:3 and 0:4 segregation).

B. Centric fusion of acrocentric chromosomes

Robertsonian translocations (centric fusion) involve either a pair of homologous or two nonhomologous acrocentric chromosomes. In homologous fusion, only disomic and nullisomic gametes are produced. Fusion of nonhomologous acrocentric chromosomes is much more common. Chromosome 14 and chromosome 21 (**1**) are most frequently involved in centric fusion (in about 1 in 1000 newborns). When the long arm of a chromosome 21 (21q) and the long arm of a chromosome 14 (14q) fuse to form a chromosome 14q21q (**2**), four types of gamete can be formed: normal, balanced, disomic, or nullisomic for chromosome 21 (**3**). After fertilization, the corresponding zygotes contain only one chromosome 21 (nonviable monosomy 21), a normal chromosome complement, a balanced chromosome complement with the fused chromosome, or three chromosomes 21 (trisomy 21).

Medical relevance

Chromosomal translocations are a common cause of chromosomal imbalance in offspring and an increased risk familial occurrence in siblings. Some translocations are more common than expected by chance (breakpoint hotspots). Nearly 90% of translocations with a breakpoint in chromosome 11q23 also involve chromosome 22q11. A similarity of DNA sequence resulting from genomic duplication events may predispose to such aberrations (see Part II, Genomic Disorders, p. 242).

Further Reading

Gersen SL, Keagle MB, eds. The Principles of Clinical Cytogenetics. 2nd ed. Totowa: Humana Press; 2005
Harper P. Practical Genetic Counseling. 7th ed. London: Edward Arnold; 2010
Miller OJ, Therman E. Human Chromosomes. 4th ed. New York: Springer-Verlag; 2001
Rooney DE, Czepulski BH, eds. Human Cytogenetics. A Practical Approach. 2nd ed. Oxford: Oxford University Press; 2001
Schinzel A. Catalogue of Unbalanced Chromosome Aberrations in Man. 2nd ed. Berlin: De Gruyter; 2001

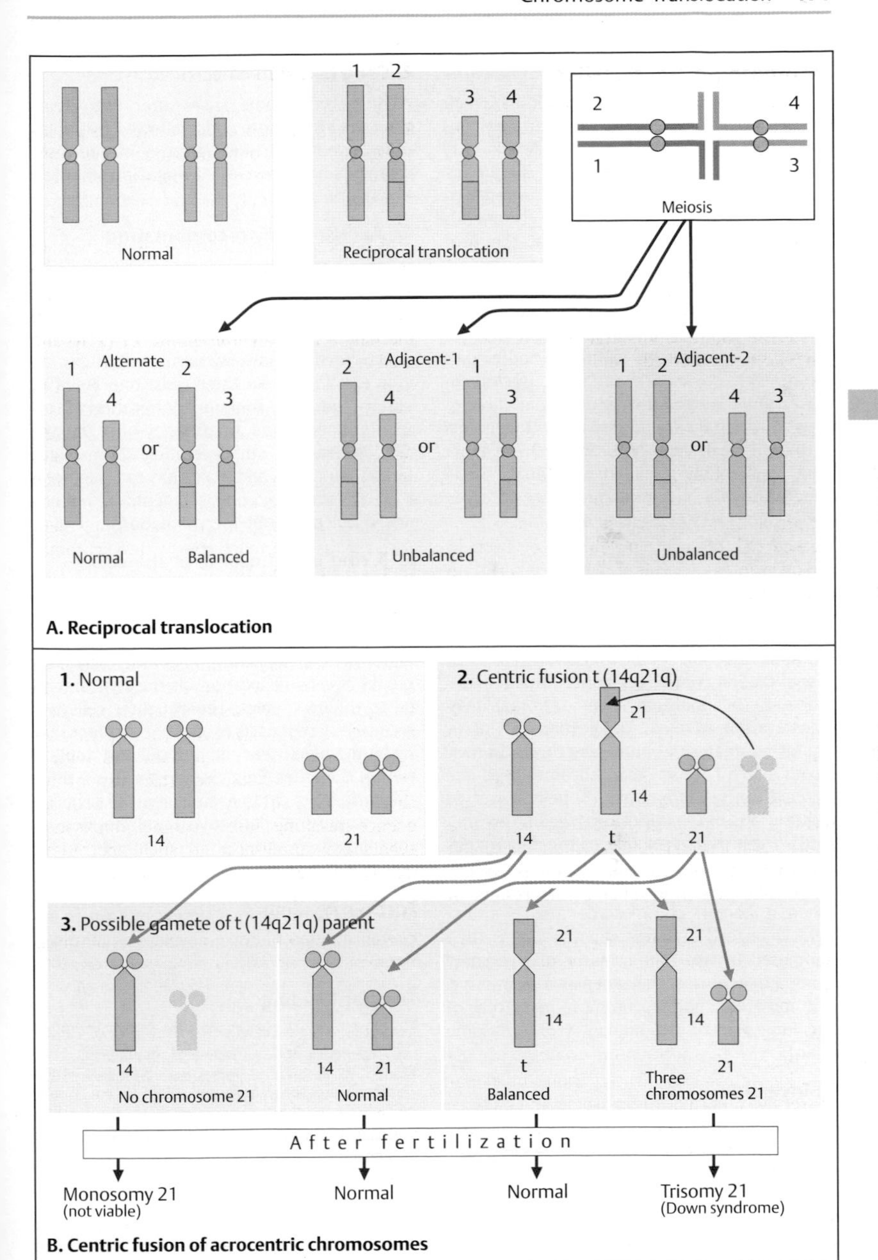
Normal
1
2
3
4
Reciprocal translocation
Meiosis
Alternate
or
Normal
Balanced
Adjacent-1
Unbalanced
Adjacent-2
Unbalanced
A. Reciprocal translocation
1. Normal
14
21
2. Centric fusion t (14q21q)
t
3. Possible gamete of t (14q21q) parent
No chromosome 21
Normal
Balanced
Three chromosomes 21
After fertilization
Monosomy 21
(not viable)
Normal
Normal
Trisomy 21
(Down syndrome)
B. Centric fusion of acrocentric chromosomes

Structural Chromosomal Aberrations

Structural chromosome abnormalities result from a break or breaks that disrupt the continuity of a chromosome. The break may occur at any stage of the cell cycle. If it occurs during the G_1 stage and remains unrepaired during S phase, it will involve both chromatids as a *chromosome break*, visible at the next metaphase stage. When the break occurs during G_2, it will be present in one chromatid of the next metaphase (*chromatid break*).

The principal types of structural chromosomal aberration as observed in human individuals are *deletion, duplication, inversion, isochromosome*, and the special case of a *ring chromosome*. Structural chromosomal rearrangements occur with a frequency of about 0.7–2.4 per 1000 in mentally impaired individuals. Small supernumerary chromosomes are observed about once in 2500 prenatal diagnoses.

A. Deletion, duplication, isochromosome

Chromosomal deletion (deficiency) arises from a single break with loss of the distal fragment (terminal deletion, **1**) or from two breaks and loss of the intervening segment (interstitial deletion, **2**). In molecular terms, terminal deletions are not terminal. Duplications (**3**) occur mainly as small supernumerary chromosomes. About half of these small chromosomes are derived from chromosome 15, being inverted duplications of the pericentric region (inv dup [15]). These represent one of the most common structural aberrations in man. An isochromosome (**4**) is an inverted duplication. It arises when a normal chromosome divides transversely instead of longitudinally, and is then composed of two long arms or of two short arms. In each case, the other arm is missing. The most common isochromosome is that of the long arm of the human X chromosome (i[Xq]).

B. Inversion

An inversion is a 180-degree change in direction of a chromosomal segment caused by a break at two different sites, followed by reunion of the inverted segment. Depending on whether the centromere is involved, a *pericentric* inversion (the centromere lies within the inverted segment) and a *paracentric* inversion can be differentiated.

C. Ring chromosome

A ring chromosome arises after two breaks with the loss of both ends, followed by joining of the two newly resulting ends. Since its distal segments have been lost, a ring chromosome is unbalanced.

D. Aneusomy by recombination

When a normal chromosome pairs with its inversion-carrying homolog during meiosis, a loop is created in the region of the inversion (**1**). When the inverted segment is relatively large, crossing-over may occur within this region (**2**). In the daughter cells, one resulting chromosome will contain a duplication of segments A and B and a deficiency of segment F (**3**), whereas the other resulting chromosome lacks segments A and B and has two segments F (**4**). These chromosome segments are not balanced (aneusomy by recombination).

E. A ring chromosome at meiosis

A ring chromosome at mitosis and meiosis is unstable and is frequently lost. If crossing-over occurs during meiosis, a dicentric ring is created. At the following anaphase, the ring breaks at variable locations as the centromeres go to different poles. The daughter cells will receive different parts of the ring chromosome, resulting in deficiency in one cell and duplication in the other. Ring chromosomes are often dicentric. Ring chromosomes tend to generate new variants of derivative chromosomes, all imbalanced.

Medical relevance

Structural chromosomal aberrations are an important group of human developmental disorders.

Further Reading

Gersen SL, Keagle MB, eds. The Principles of Clinical Cytogenetics. 2nd ed. Totowa: Humana Press; 2005

Madan K. Paracentric inversions: a review. Hum Genet 1995;96:503–515

Meltzer PS, Guan X-Y, Trent JM. Telomere capture stabilizes chromosome breakage. Nat Genet 1993;4:252–255

Miller OJ, Therman E. Human Chromosomes. 4th ed. New York: Springer-Verlag; 2001

Schinzel A. Catalogue of Unbalanced Chromosome Aberrations in Man. 2nd ed. Berlin: De Gruyter; 2001

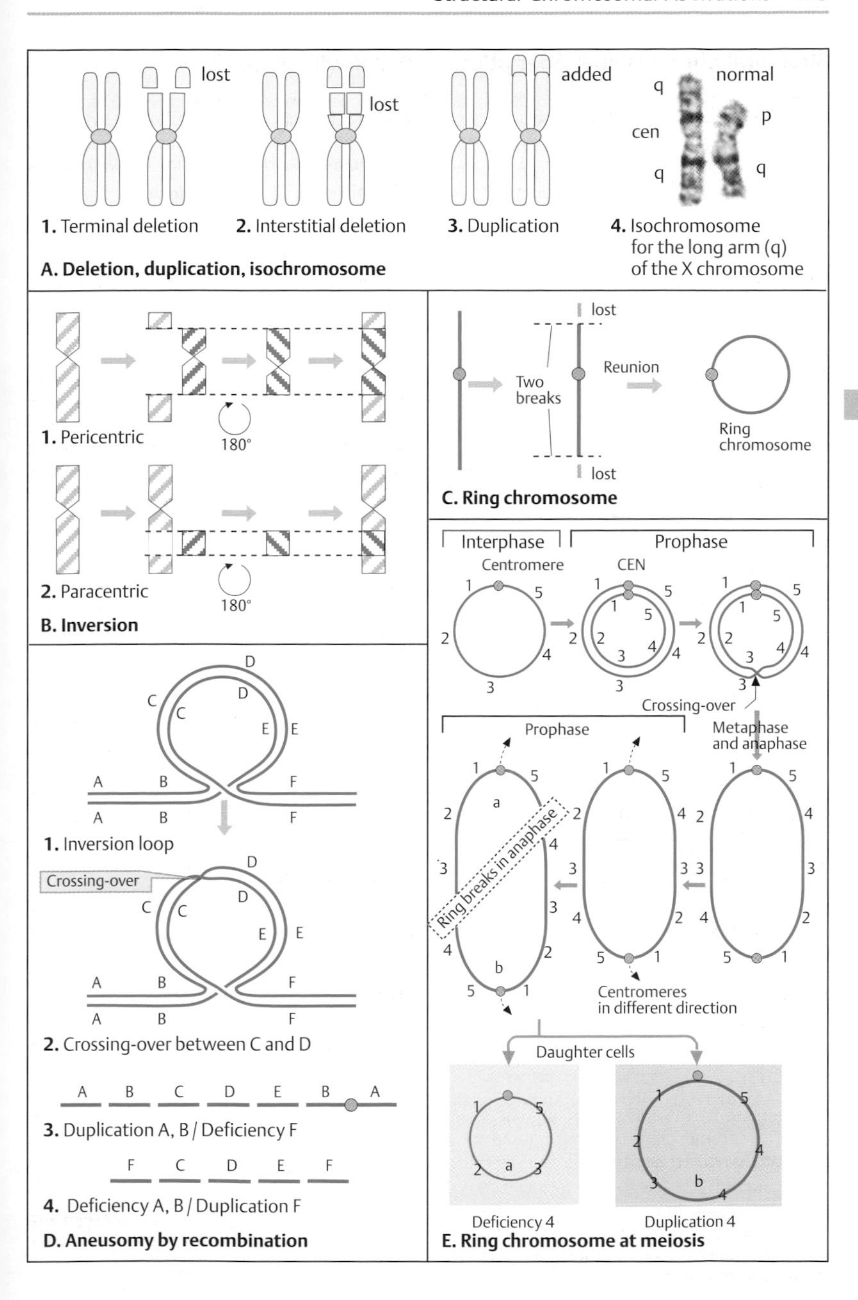

A. Deletion, duplication, isochromosome

B. Inversion

C. Ring chromosome

D. Aneusomy by recombination

E. Ring chromosome at meiosis

Ribosomal RNA and Protein Assembly

Ribosomes are large ribonucleoprotein particles that coordinate the interaction of messenger RNA (mRNA) and transfer RNA (tRNA) during protein synthesis. They consist of a small and a large subunit. The subunit contains tRNAs and proteins.

All proteins produced by a cell at any given time are called the *proteome*. The total number of different proteins required for eukaryotes is estimated to be in the range of 90 000. Ribosomes are the products of individual genes (ribosomal genes).

A. Structure and components of ribosomes

Ribosomes in prokaryotes and eukaryotes differ in size and composition, although their overall structure is similar. The size of a ribosome is expressed by its sedimentation coefficient (70S in prokaryotes; the sedimentation coefficient is a measure of the rate of sedimentation in an ultracentrifuge of a molecule suspended in a less dense solvent; it is measured in Svedberg units, S; values are not additive). The prokaryote small subunit is 30S and the large subunit is 50S in size. The prokaryote ribosome is composed of three different rRNA molecules (5S, 35S, and 165S in size), and 34 proteins. A bacterial cell contains about 20 000 ribosomes, which account for approximately 25% of its mass. The 30S subunit, containing a large 16S rRNA and 21 proteins, is the site where genetic information is decoded. It also has a proofreading mechanism. The 50S subunit provides peptidyltransferase activity. The entire ribosome has a molecular mass of 2.5 million daltons (MDa).

The eukaryotic ribosome is much larger (80S, 4.2 MDa), consisting of a 40S and a 60S subunit. The 60S subunit contains 5S, 5.8S, and 28S rRNAs (120, 160, and 4800 bases, respectively) in addition to 50 proteins. The 40S subunit has 18S rRNAs (1900 bases) and 33 proteins. Bacterial 30S and 50S ribosomal subunit structures have been recognized at 5 Å resolution.

B. From gene to protein

Transcription and processing of the primary transcript (splicing) takes place in the nucleus. RNA in the nucleus is bound to nuclear RNA-binding proteins for stabilization. The mature RNA is released from the nucleus into the cytoplasm, where it associates with ribosomes for translation into the corresponding sequence of amino acids.

C. Nucleolus and synthesis of ribosomes

The nucleolus is a morphologically and functionally specific region in the cell nucleus in which ribosomes are synthesized. In man, the rRNA genes (200 copies per haploid genome) are transcribed by RNA polymerase I to form 45S rRNA precursor molecules. After the 45S rRNA precursors have been produced, they are rapidly packaged with ribosomal proteins (from the cytoplasm). Before they are transferred from the nucleus to the cytoplasm, they are cleaved to form three of the four rRNA subunits. These are released into the cytoplasm with the separately synthesized 5S subunits. Here they form functional ribosomes. Two types of small RNAs have important functions. Small nuclear RNAs (snRNAs) are a family of RNA molecules that bind specifically with a small number of nuclear ribonucleoprotein particles (snRNP, pronounced "snurps"). These play important roles in the modification of RNA molecules after transcription (posttranscriptional modification). snRNAs base-pair with pre-mRNA and with each other during the splicing of RNA. Small nucleolar RNA molecules (snoRNA) assist in processing pre-rRNAs and in assembly of ribosomes.

(Figures based on Alberts et al., 2008, and Lodish et al., 2007.)

Medical relevance

Mutations in genes encoding small ribonucleoproteins cause different human disorders.

A variety of chemical compounds occurring naturally as poisons or synthetic products are used for cancer therapy by inhibition of transcription or translation.

Further Reading

Agalarov SC, et al. Structure of the S15,S6,S18-rRNA complex: assembly of the 30S ribosome central domain. Science 2000;288:107–113

Alberts B, et al. Molecular Biology of the Cell. 5th ed. New York: Garland Science; 2008

Berg JM, Tymoczko JL, Stryer L. Biochemistry. 7th ed. New York: W. H. Freeman; 2011

Wimberly BT, et al. Structure of the 30S ribosomal subunit. Nature 2000;407:327–339

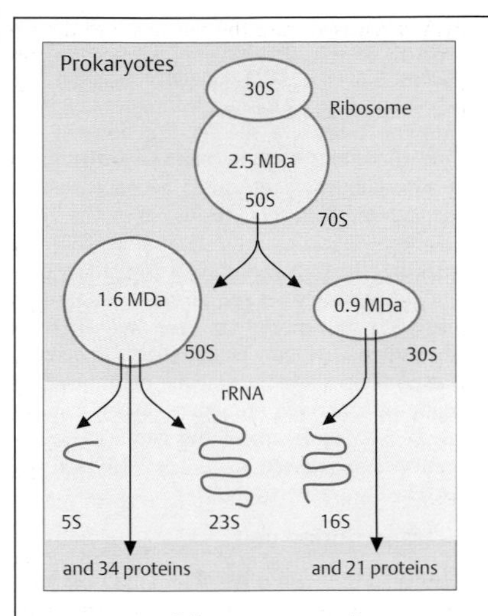

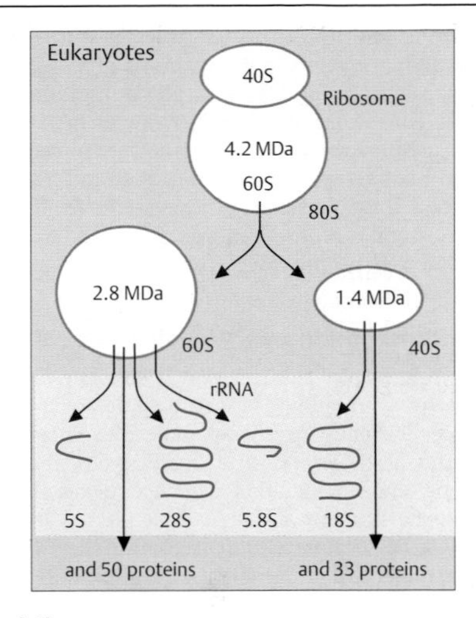

A. Overview of the structure and components of ribosomes

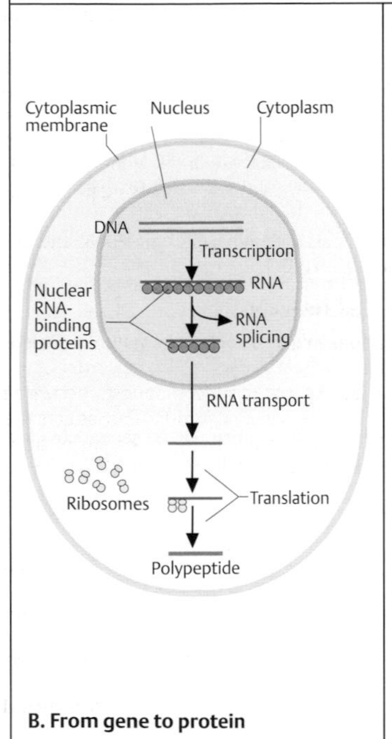

B. From gene to protein

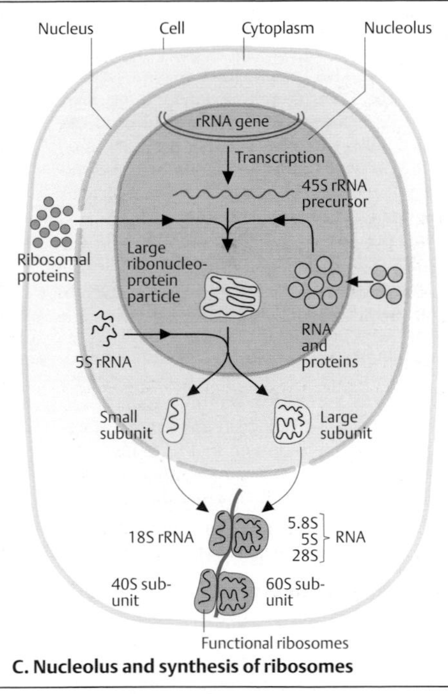

C. Nucleolus and synthesis of ribosomes

Transcription

Transcription refers to the synthesis of an RNA molecule from complementary DNA as a template. For this purpose, the DNA double helix is transiently opened, and the template strand is used to direct synthesis of an RNA strand. This process starts and ends at a defined site. The RNA strand is synthesized from the 5′ end to 3′ end by RNA polymerase II. Transcription occurs in four stages.

A. Transcription by RNA polymerase II

The first step of transcription is *template recognition* with binding of RNA polymerase II to a specific sequence of DNA molecules known as the promoter (**1**). The double helix then opens, and an initiation complex makes the template strand available for base pairing. *Initiation* (**2**) begins with synthesis of the first RNA molecules at the initiation complex. RNA polymerase remains at the promoter while it synthesizes the first nine nucleotide bonds. Initiation requires several other proteins, which are collectively referred to as activators and transcription factors. *Elongation* (**3**) begins when the RNA polymerase enzyme moves along the DNA, thereby extending the RNA chain. As it moves, DNA helicases unwind the DNA double helix. The DNA that has been transcribed rewinds into the double helix behind the polymerase. At *termination* (**4**), the RNA polymerase is removed from the DNA. At this point, the formation of the unstable primary transcript is completed. Since it is unstable, it is immediately translated in prokaryotes and modified (processed) in eukaryotes (see p. 60). All processes are mediated by the complex interaction of a variety of enzymes (not shown).

B. Polymerase-binding site

The polymerase-binding site defines the starting point of transcription. At the termination site it is closed again (rewinding). Bacterial RNA polymerase binds to a specific region of about 60 bp of the DNA.

C. Promoter of transcription

A promoter is a DNA sequence that specifies the site of RNA polymerase binding from which transcription is initiated. The promoter is organized into several regions with sequence homology. In eukaryotes, transcription of protein-coding genes begins at multiple sites, often extending hundreds of base pairs upstream. A specific DNA sequence of about 4–8 base pairs (bp) is located approximately 25–35 bp upstream (in the 5′ direction) of the gene. Since this sequence is nearly the same in all organisms, it is called a consensus sequence. One such sequence is called the TATA box, because it contains the sequence TATA (or T)AA(or T)TA/G. It is highly conserved in evolution. In prokaryotes the promoter contains a consensus sequence of six base pairs, TATAAT (also called a Pribnow box after its discoverer), located 10 bp above the starting point. Another region of conserved sequences, TTGACA, is located 35 bp upstream of the gene. These sequences are referred to as the −10 box and the −35 box, respectively.

D. Transcription unit

A transcription unit is the segment of DNA between the sites of initiation and termination of transcription.

E. Determining transcription start sites

A segment of DNA suspected to contain a transcription start site can be analyzed using the nuclease S1 protection assay. Endonuclease S1, an enzyme present in the mold *Aspergillus oryzae*, cleaves single-stranded DNA and RNA, but not double-stranded DNA or RNA. Therefore, a DNA fragment that is denatured and mixed with total RNA from cells containing the gene to be analyzed will hybridize with cognate RNA and form a DNA/RNA heteroduplex. Following digestion with S1, all single-stranded DNA will be removed, allowing the relevant DNA to be identified.

Medical relevance

Changes in regulatory DNA sequences are causes of human genetic disorders.

Further Reading

Alberts B, et al. Molecular Biology of the Cell. 5th ed. New York: Garland Science; 2008

Lewin B. Genes IX. Sudbury: Bartlett & Jones; 2008

Lodish H, et al. Molecular Cell Biology. 6th ed. New York: W. H. Freeman; 2007

Strachan T, Read AP. Human Molecular Genetics. 4th ed. New York: Garland Science; 2011

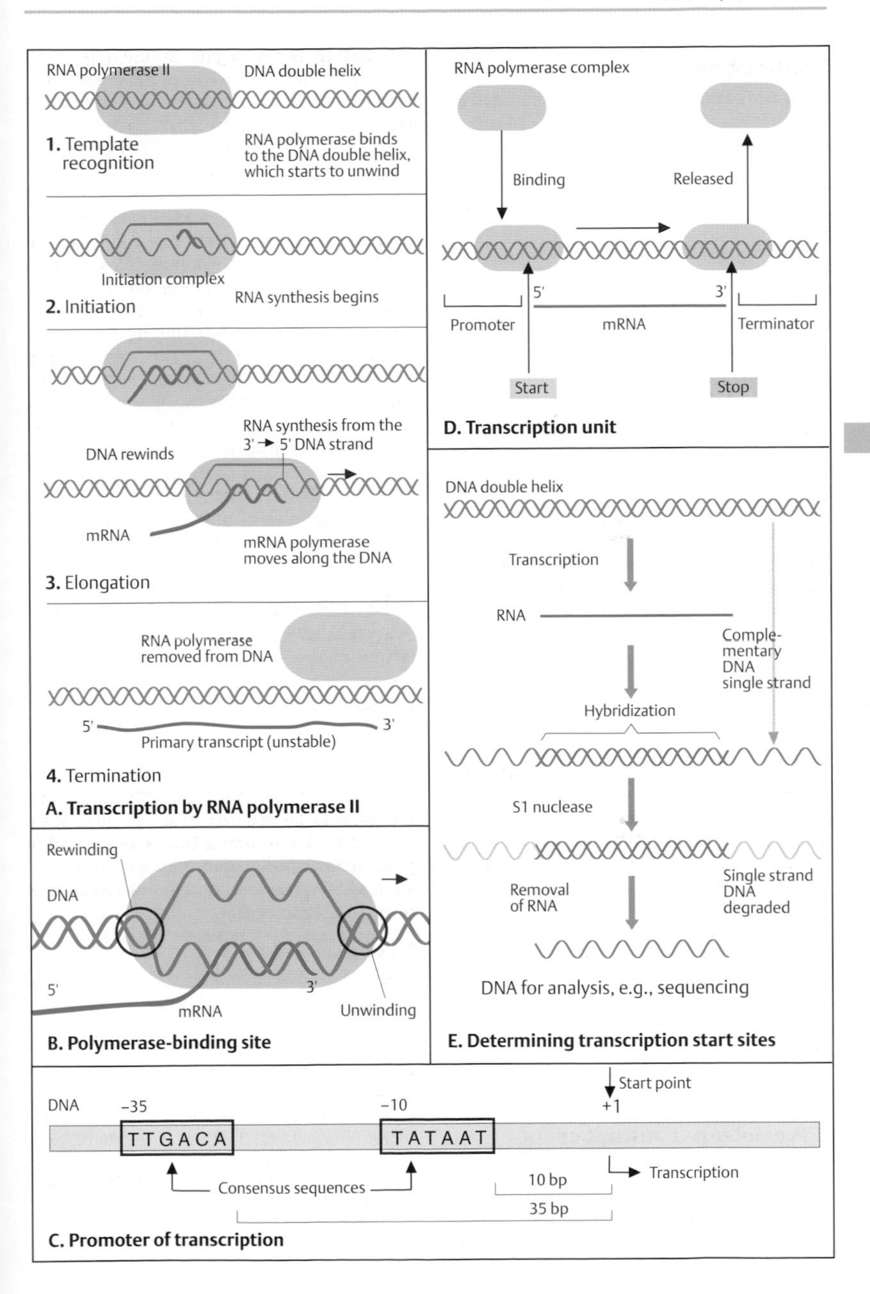

A. Transcription by RNA polymerase II

B. Polymerase-binding site

C. Promoter of transcription

D. Transcription unit

E. Determining transcription start sites

Basic Principles of Gene Control

Regulatory DNA sequences (promoters and enhancers) are regions that bind to DNA-specific proteins (transcription factors) and regulate gene activity. The binding of transcription factors at the promoter is the most important mechanism for regulating gene activity.

A. Consensus sequences at the promoter

Promoters are well conserved in evolution. Owing to their functional importance they do not tolerate mutational changes. Therefore their sequence remains constant across evolutionarily remote organisms. In prokaryotes, two important regulatory sequences are 35 and 10 nucleotide base pairs upstream (in the 5′ direction) of the starting point of transcription (Pribnow, 1975). The −10 site (TATA box) is 5′-TATAAT-3′. The −35 sequence is 5′-TATTGACA-3′ (**1**). Mutations at different sites upstream of a gene have different effects, depending on their precise location (**2**). (Figure adapted from Rosenberg & Court, 1979.)

B. Assembly of general transcription factors

In eukaryotes, the TATA box in the promoter region is located about 25–35 bp upstream of the transcription start site (**1**). Several other promoter-proximal elements (different transcription factors, TFs) are involved in regulating gene activity. General transcription factors associate in an ordered sequence. First, TFIID (transcription factor D for polymerase II) binds to the TATA region (**2**). The TATA box is recognized by a small, 30-kDa TATA-binding protein (TBP), which is part of one of the many subunits of TFIID (the bending of the DNA by TBP is not shown here). Then TFIIB binds to the complex (**3**). Subsequently, other transcription factors (TFIIH, followed by TFIIE) and RNA polymerase II (Pol II), escorted by TFIIF, join the complex and ensure that Pol II is attached to the promoter (**4**). Pol II is activated by phosphorylation and transcription can begin (**5**). Other activities of TFIIH involve a helicase and an ATPase. The site of phosphorylation is a polypeptide tail, in which the serine (S) and threonine (T) side chains are phosphorylated. (The figure is a simplified scheme adapted from Alberts et al., 2008, and Lewin, 2008).

C. RNA polymerase promoters

Eukaryotic cells contain three RNA polymerases (Pol I, Pol II, and Pol III). Each uses a different type of promoter. RNA polymerase II is a 550 kDa complex of 12 subunits. It requires a transcription factor complex (TFIID, see **B**) that binds to a single upstream promoter (**1**). RNA polymerase I has a bipartite promoter, one part 170–180 bp upstream (5′ direction) and the other from about 45 bp upstream to 20 bp downstream (3′ direction). The latter is called the core promoter (**2**). Pol I requires two ancillary factors: UPE1 (upstream promoter element 1) and SL1. RNA polymerase III uses either upstream promoters or two internal promoters downstream of the transcription start site (**3**). Three transcription factors are required with internal promoters: TFIIIA (a zinc finger protein, see p. 182), TFIIIB (a TBP and two other proteins), and TFIIIC (see Lewin, 2008, p. 615ff).

Pol I is located in the nucleolus and synthesizes ribosomal RNA. It accounts for about 50%–70% of the relative activity. Pol II and Pol III are located in the nucleoplasm (the part of the nucleus excluding the nucleolus). Pol II represents 20%–40% of cellular activity. It is responsible for the synthesis of heterogeneous nuclear RNA (hnRNA), the precursor of mRNA. Pol III is responsible for the synthesis of tRNAs and other small RNAs. It contributes only minor activity of about 10%. Each of the large eukaryotic RNA polymerases (500 kDa or more) has 8–14 subunits and is more complex than the single prokaryotic RNA polymerase. (Figure modified from Lewin, 2008.)

Further Reading

Alberts B, et al. Molecular Biology of the Cell. 5th ed. New York: Garland Science; 2008

Lewin B. Genes IX. Sudbury: Bartlett & Jones; 2008

Lodish H, et al. Molecular Cell Biology. 6th ed. New York: W. H. Freeman; 2007

Pribnow D. Nucleotide sequence of an RNA polymerase binding site at an early T7 promoter. Proc Natl Acad Sci U S A 1975;72:784–788

Rosenberg M, Court D. Regulatory sequences involved in the promotion and termination of RNA transcription. Annu Rev Genet 1979;13:319–353

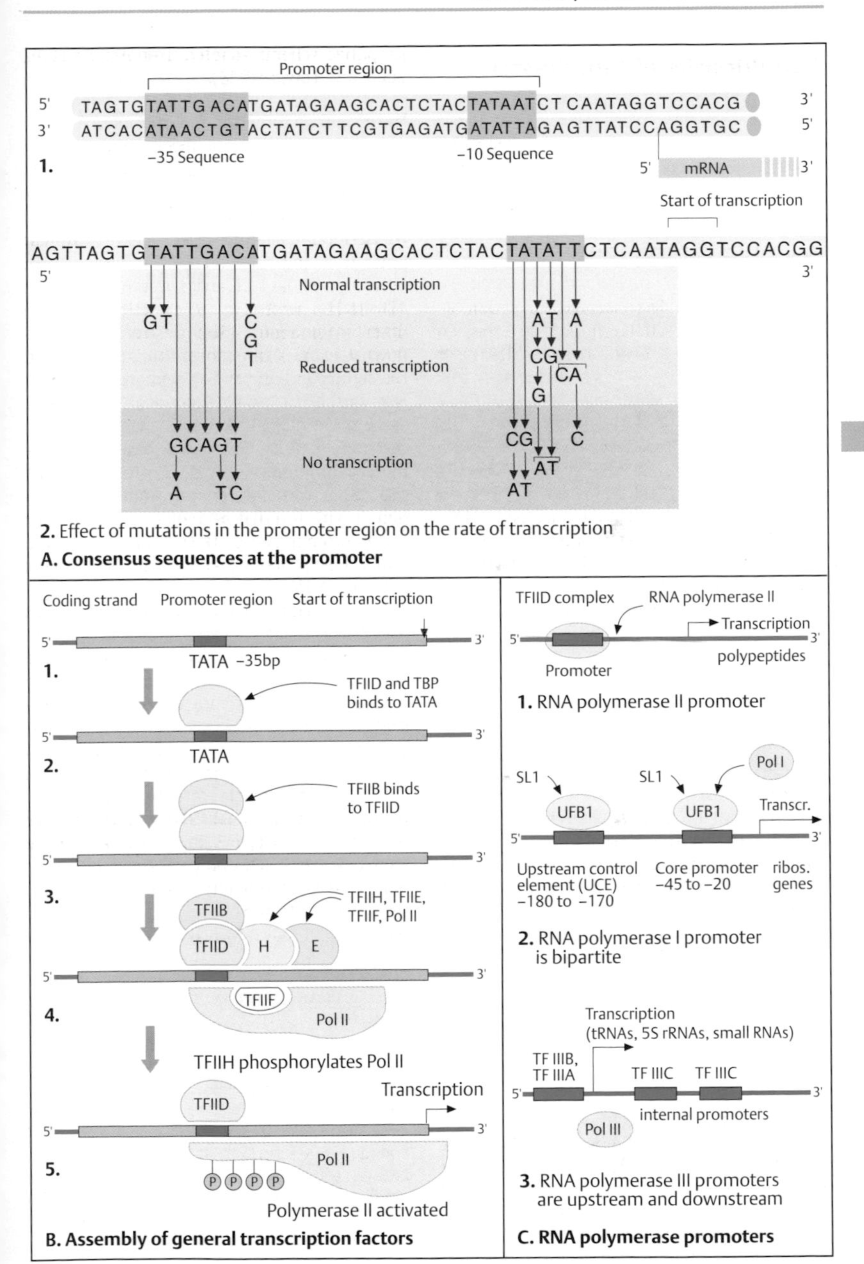
Promoter region
5' TAGTGTATTG ACATGATAGAAGCACTCTACTATAATCT CAATAGGTCCACG 3'
3' ATCACATAACTGTACTATCT TCGTGAGATGATATTAGAGTTATCCAGGTGC 5'
-35 Sequence
-10 Sequence
5' mRNA 3'
Start of transcription
1.
AGTTAGTGTATTGACATGATAGAAGCACTCTACTATATTCTCAATAGGTCCACGG
5'
3'
Normal transcription
Reduced transcription
No transcription
2. Effect of mutations in the promoter region on the rate of transcription
A. Consensus sequences at the promoter
Coding strand
Promoter region
Start of transcription
TATA -35bp
TFIID and TBP binds to TATA
TFIIB binds to TFIID
TFIIH, TFIIE, TFIIF, Pol II
TFIIB
TFIID
H
E
TFIIF
Pol II
TFIIH phosphorylates Pol II
Transcription
Polymerase II activated
B. Assembly of general transcription factors
TFIID complex
RNA polymerase II
Transcription
Promoter
polypeptides
1. RNA polymerase II promoter
SL1
Pol I
UFB1
Transcr.
Upstream control element (UCE) -180 to -170
Core promoter -45 to -20
ribos. genes
2. RNA polymerase I promoter is bipartite
Transcription (tRNAs, 5S rRNAs, small RNAs)
TF IIIB, TF IIIA
TF IIIC
internal promoters
Pol III
3. RNA polymerase III promoters are upstream and downstream
C. RNA polymerase promoters

Regulation of Gene Expression in Eukaryotes

The term *gene expression* refers to the entire process of decoding the genetic information of active genes. When genes are active (expressed) throughout the life of a cell or an organism, they show *constitutive expression*. Those genes that are transcribed only under certain circumstances, in specific cells or at specific times, show *conditional expression*.

A. Levels of control

Schematically, gene expression in eukaryotes can be regulated at four distinct levels. The first, and by far the most important, is primary control of the initiation of transcription. The next level, processing of the transcript to mature mRNA, can be regulated at the level of the primary RNA transcript. Different forms of mRNA are usually obtained from the same gene by alternative splicing (see **D**). A newly recognized form of gene expression control is RNA interference (see p. 186). Control is possible at the level of translation by mRNA editing (see **B**). Finally, at the protein level, posttranslational modifications can determine the activity of a protein.

B. RNA editing

RNA editing modifies genetic information at the RNA level. An important example is the gene encoding apolipoprotein (Apo) B-100, which is involved in lipid metabolism (OMIM 107730). It encodes a 512-kDa protein of 4536 amino acids. This is synthesized in the liver and secreted into the blood, where it transports lipids. Apo B-48 (250 kDa), a functionally related shorter form of the protein with 2152 amino acids, is synthesized in the intestine. An intestinal deaminase converts a cytosine in codon 2152 CAA (glutamine) to uracil (UAA). This change results in a stop codon (UAA) and thereby terminates translation at this site.

C. Long-range gene activation by an enhancer

The term enhancer refers to a DNA sequence that stimulates the initiation of transcription (see p. 176). Enhancers act at a distance from the gene. They may be located upstream or downstream on the same DNA strand (*cis*-acting) or on a different DNA strand (*trans*-acting). An enhancer effect is mediated by sequence-specific DNA-binding proteins. One model suggests that DNA forms a loop between an enhancer and the promoter. An activator protein bound to the enhancer, e.g., a steroid hormone, could then come into contact with the general transcription factor complex at the promoter. Enhancer elements provide tissue-specific or time-dependent regulation.

D. Alternative RNA splicing

Alternative splicing is an important mechanism for generating multiple protein isoforms from a single gene. The resulting proteins differ slightly in their amino acid sequence. This may result in small functional differences. Quite often these differences are restricted to certain tissues, as schematically shown here for the calcitonin gene (OMIM 114130). The primary transcript for the calcitonin gene contains six exons. They are spliced into two different types of mature mRNA. One type, calcitonin, consisting of exons 1–4 (excluding exons 5 and 6), is produced in the thyroid. The other, consisting of exons 1, 2, 3, 5, and 6 and excluding exon 4, encodes a calcitonin-like protein in the hypothalamus (calcitonin gene-related product, CGRP).

Alternative splicing clearly represents an evolutionary advantage because it allows for a high degree of functional flexibility.

Medical relevance

Human genetic disorders may result from defective splicing or DNA changes in regulatory sequences (promoters and enhancers).

Further Reading

Alberts B, et al. Molecular Biology of the Cell. 5th ed. New York: Garland Science; 2008

Graveley BR. Alternative splicing: increasing diversity in the proteomic world. Trends Genet 2001;17: 100–107

Lewin B. Genes IX. Sudbury: Bartlett & Jones; 2008

Lodish H, et al. Molecular Cell Biology. 6th ed. New York: W. H. Freeman; 2007

Modrek B, Lee C. A genomic view of alternative splicing. Nat Genet 2002;30:13–19

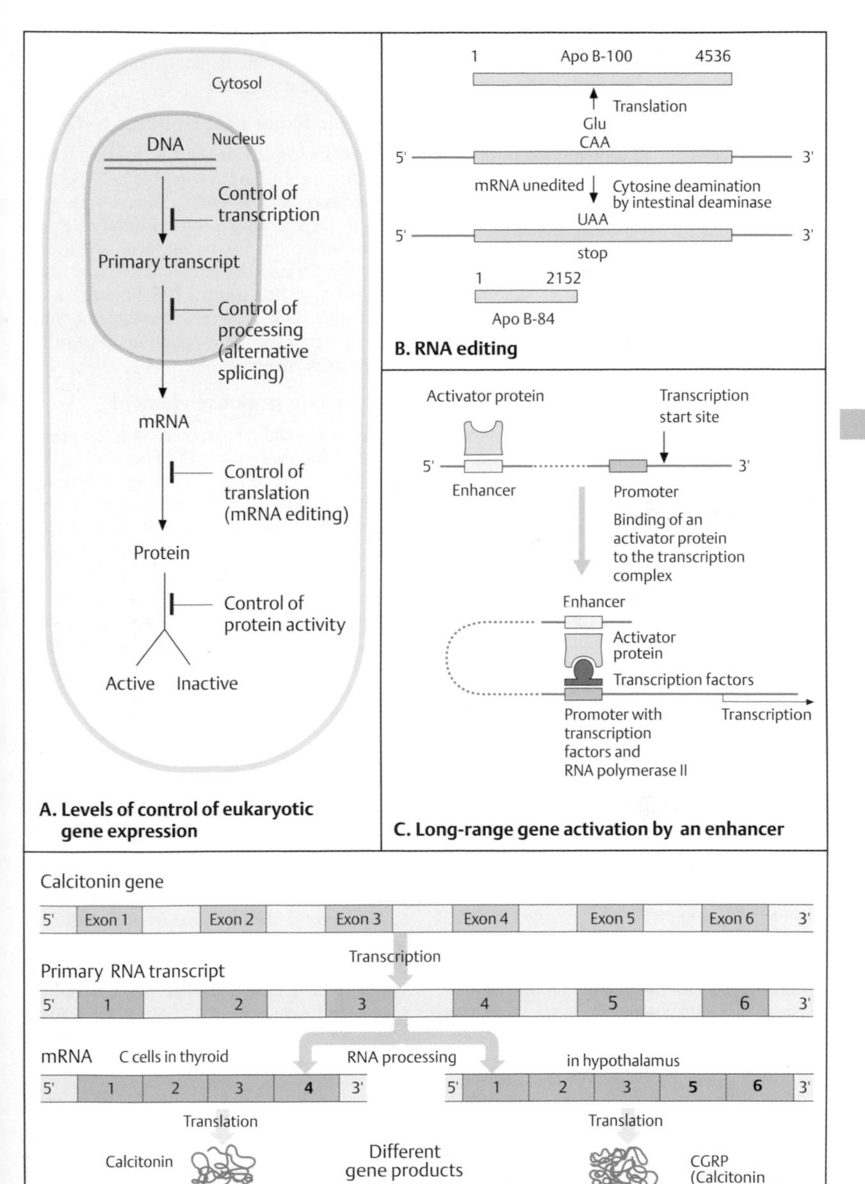

A. Levels of control of eukaryotic gene expression

B. RNA editing

C. Long-range gene activation by an enhancer

D. Alternative RNA splicing

DNA–Protein Interaction

Regulatory DNA sequences exert their control function by specific interaction with specific DNA-binding proteins. Regulatory proteins recognize specific DNA sequences by a precise fit of their surface to the DNA double helix.

A. Binding of a regulatory protein to DNA

Gene regulatory proteins recognize DNA sequence information without affecting the hydrogen bonds within the helix. Each base pair represents a distinctive pattern of hydrogen bond donors (shown in the rectangle in red) and hydrogen acceptors (shown in the rectangle in green). These proteins bind to the major groove of DNA. A single contact of an asparagine (Asn) residue of a gene-regulatory protein with a DNA base adenine (A) is shown here. A typical area of surface-to-surface contact involves 10–20 such interactions, resulting in high specificity. (Figure adapted from Alberts et al., 2008.)

B. Specific DNA–protein interaction

The α-helix of a DNA-binding regulatory protein recognizes specific DNA sequences. Many bacterial repressor proteins are dimeric, so that an α-helix from each dimer can insert itself into two adjacent major grooves of the DNA double helix (recognition or sequence-reading helix). This structural motif is called a *helix–turn–helix* motif, because two helices lie next to each other. The example shows the tight interaction of the bacteriophage 434 repressor protein with one side of the DNA molecule over a length of 1.5 turns. (Figure adapted from Lodish et al., 2008.)

C. Zinc finger motif

Several eukaryotic regulatory proteins harbor regions that fold around a central zinc atom (Zn^{2+}) to form a structural motif resembling a finger (hence, zinc finger). The basic zinc finger motif consists of a zinc atom connected to four amino acids of a polypeptide chain. The three-dimensional structure on the right consists of an antiparallel β-sheet (amino acids 1–10), an α-helix (amino acids 12–24), and the zinc connection. Four amino acids, two cysteines at positions 3 and 6 and two histidines at positions 19 and 23, are bonded to the zinc atom and hold the carboxy (COOH) end of the α-helix to one end of the β-sheet. (Figure adapted from Alberts et al., 2008.)

D. A zinc finger protein binds to DNA

The α-helix of each zinc finger can contact the major groove of the DNA double helix and establish specific and strong interactions over several turns in length. Extraordinary flexibility in gene control has been acquired during evolution through adjustments to the number of interacting zinc fingers. Zinc finger proteins serve important functions during embryonic development and differentiation. (Figure redrawn from Alberts et al., 2008.)

E. Hormone response element

Some DNA-binding proteins act as signal-transmitting molecules. The signal may be a hormone or a growth factor that activates an intracellular receptor. Steroid hormones enter target cells and bind to specific receptor proteins. Three views of the glucocorticoid receptor and its binding to DNA are shown here. The dimeric glucocorticoid receptor consists of two polypeptide chains (*hormone response element*, HRE). Each is stabilized by a zinc ion connected to four cysteine side chains (**1**). The skeletal model shows the binding of the dimeric protein to the DNA double helix (**2**). The space-filling model (**3**) shows how tightly the recognition helix of each dimer of this protein (brown above, red below) fits into two neighboring major grooves of DNA (shown in red and green). (Figures adapted from Stryer, 1995.)

Medical relevance

Defective function of transcription factors cause certain developmental disorders, e.g., Waardenburg syndrome type 1 (OMIM 193500) by mutation in the PAX3 transcription factor (Paired Box; OMIM 606597), Waardenburg syndrome type 2A (OMIM 193510) by mutations in MITF (microphthalmia-associated transcription factor; OMIM 156845), or cancer (see *TP53* gene, p. 306).

Further Reading

Alberts B, et al. Molecular Biology of the Cell. 5th ed. New York: Garland Science; 2008

Lodish H, et al. Molecular Cell Biology. 6th ed. New York: W. H. Freeman; 2007

Stryer L. Biochemistry. 4th ed. New York: W. H. Freeman; 1995

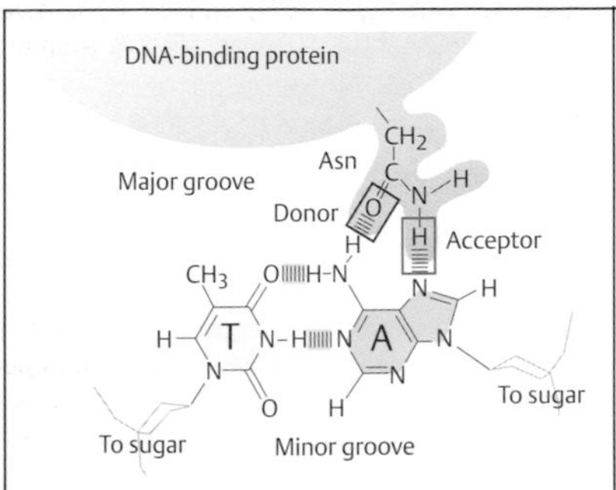

A. **Binding of a regulatory protein to DNA**

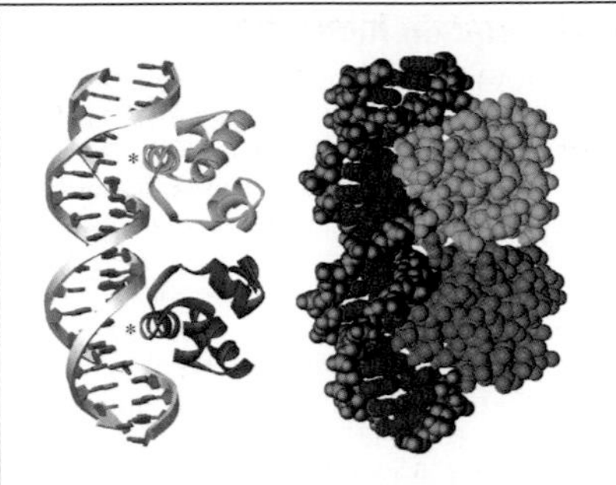

B. Interaction of a DNA-binding protein with DNA

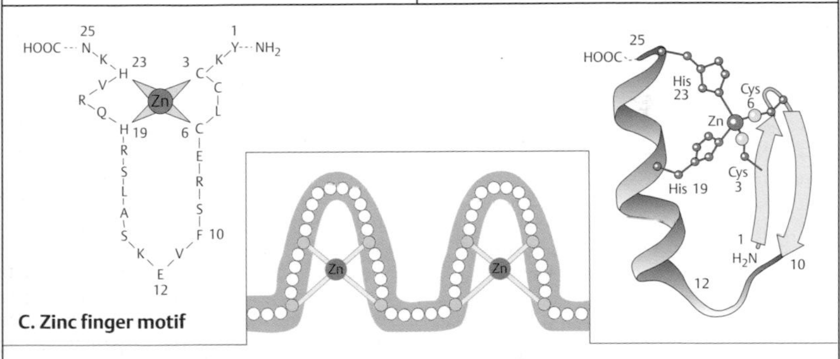

C. Zinc finger motif

D. A zinc finger protein binds to DNA

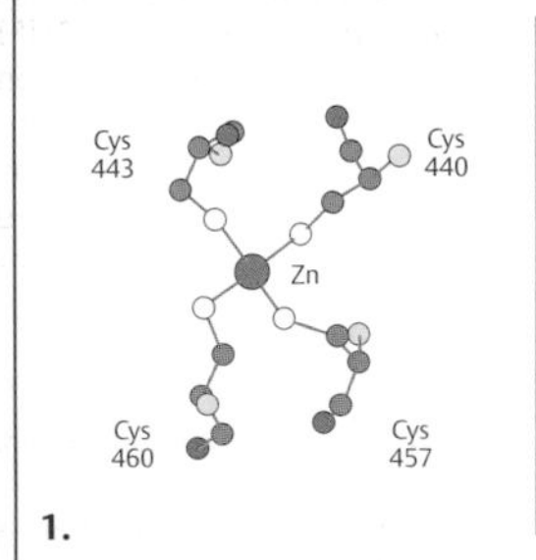

1.

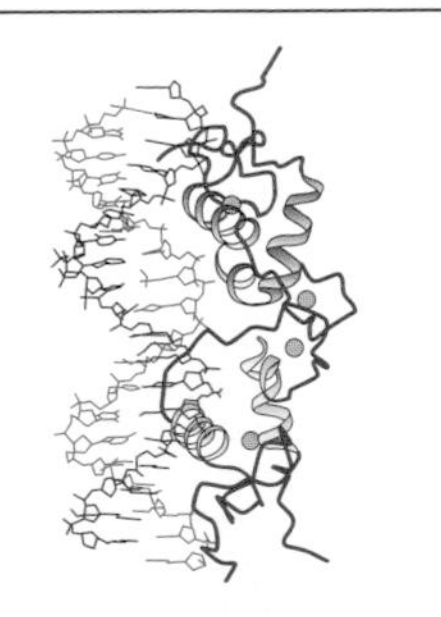

2.

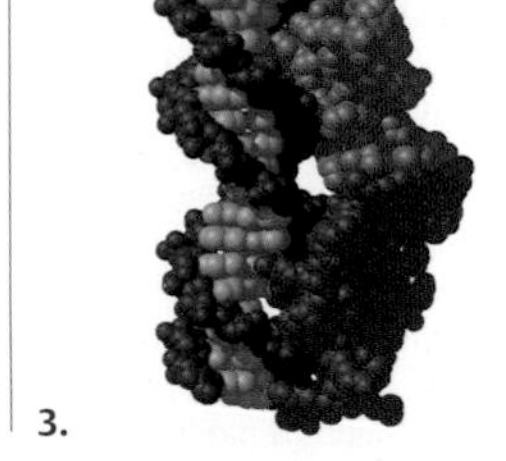

3.

E. Binding to a response element

Transcription Activation

The DNA-binding domains of eukaryotic transcription activators or repressors are characterized by different structural motifs. The specific DNA–protein binding usually involves noncovalent interactions between atoms in an α-helix of the binding domain of the protein and atoms on the outside of the major groove of the DNA double helix. About 2000 transcription factors are encoded by the human genome (Lodish et al., 2007). Examples of important classes of DNA-binding domains are (i) *homeodomain proteins* containing a 180-bp sequence that is highly conserved as a *homeobox* in evolution (see section on Genes in Embryonic Development, p. 210); (ii) *zinc finger proteins* (see previous page); (iii) *leucine-zipper proteins* (see below); and (iv) *basic helix–loop–helix* (bHLH) *proteins* (see p. 182).

A. Leucine zipper and bHLH protein

Most DNA-binding regulatory proteins have a dimeric structure, which allows a dual function: one part of the molecule recognizes specific DNA sequences, and the other stabilizes the molecule. One frequently occurring class of proteins has a characteristic structural motif called a *leucine zipper*. The name is derived from its basic structure. A typical leucine zipper protein consists of a dimer with a periodic repeat of leucine every seven residues. The leucine residues are aligned along one face of each α-helix and interact with the DNA at adjacent major grooves (**1**). In basic helix–loop–helix (bHLH) proteins the DNA-binding helices at the N-terminal near the DNA are separated by nonhelical loops (**2**). (Figure adapted from Lodish et al., 2007.)

B. Alternative combinations

Leucine zipper proteins and bHLH proteins often exist in alternative combinations of dimers consisting of two different monomers. This dramatically increases the number of combinatorial possibilities. In higher eukaryotes, leucine zipper proteins often mediate the effect of cyclic adenosine monophosphate (cAMP) on transcription. Genes under this type of control contain a cAMP response element (CRE), which is a palindromic 8-bp recognition sequence. A protein of 43 kDa binds to this target sequence. It is therefore known as the cAMP-response-element-binding protein (CREB). Leucine zipper proteins were first described in 1988 by Landschulz et al. (1988). (Figure based on Alberts et al., 2008.)

C. Activation by steroid hormone

Transcriptional enhancers are regulatory regions of DNA that increase the rate of transcription. Binding of a hormone receptor complex to a specific DNA sequence, a hormone response element, activates an enhancer. This activates the promoter, and transcription begins. Numerous genes in mammalian development are regulated by steroid-responsive transcription.

D. Detection of DNA–protein interactions

Different regulatory proteins bind at specific sites of DNA (transcription-control elements). An approach for detecting such sites is the DNase I footprinting assay (**1**). This is based on the observation that DNA is protected from digestion by nucleases at the protein-binding site, whereas DNA outside the binding site is digested by DNase I. The protected protein-binding sites of DNA are visible as missing bands ("footprint") after separation of the DNA fragments by gel electrophoresis. The electrophoretic mobility shift assay, or band-shift assay (**2**), is based on the principle that a DNA–protein complex retards the speed at which a fragment migrates in gel electrophoresis.

Medical relevance

Mutations in a bHLH protein coding gene, *TWIST* (OMIM 601622), cause acrocephaly-syndactyly syndrome type III (Saethre-Choten syndrome, OMIM 101400).

Further Reading

Alberts B, et al. Molecular Biology of the Cell. 5th ed. New York: Garland Science; 2008

Landschulz WH, et al. The leucine zipper: A hypothetical structure common to a new class of DNA binding proteins. Science 1988;240:1759–1764

Lodish H, et al. Molecular Cell Biology. 6th ed. New York: W. H. Freeman; 2007

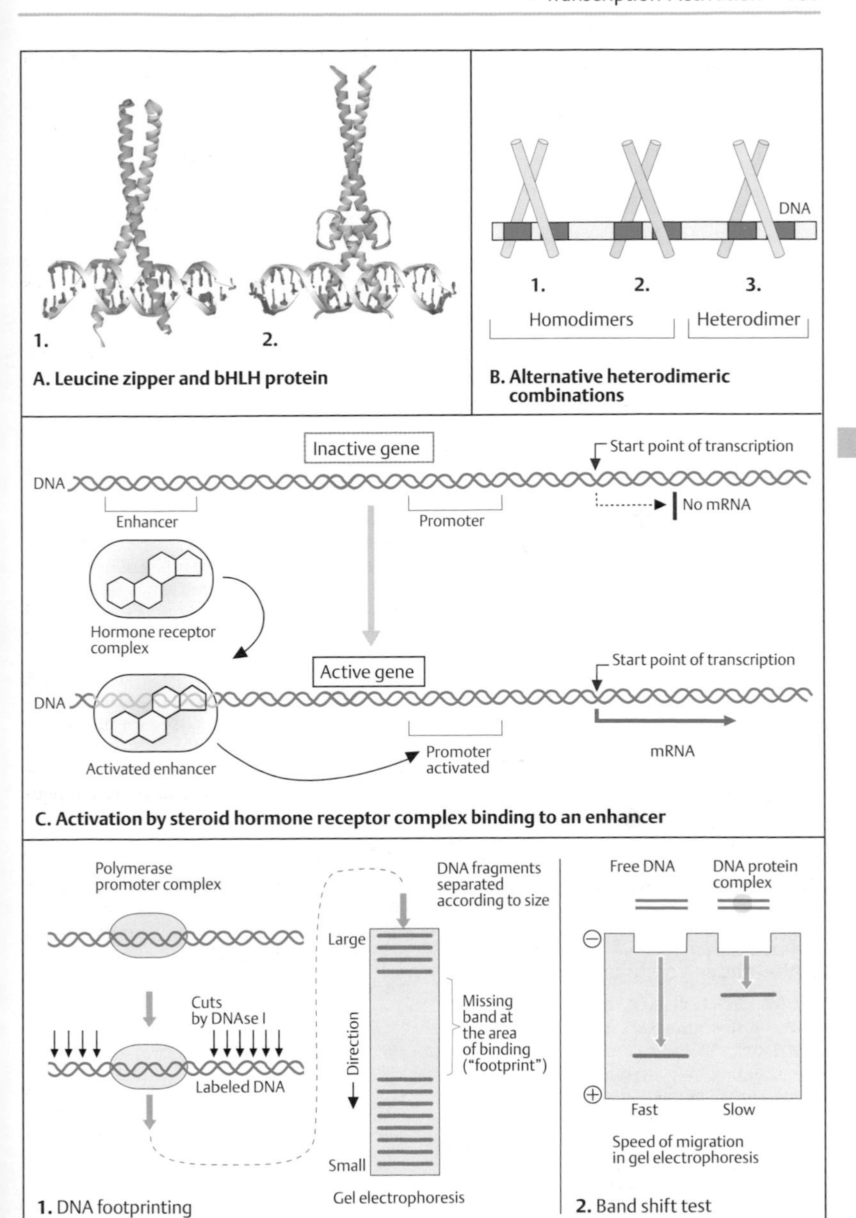
1.
2.
A. Leucine zipper and bHLH protein
DNA
1.
2.
3.
Homodimers
Heterodimer
B. Alternative heterodimeric combinations
Inactive gene
Start point of transcription
DNA
Enhancer
Promoter
No mRNA
Hormone receptor complex
Active gene
Start point of transcription
DNA
Activated enhancer
Promoter activated
mRNA
C. Activation by steroid hormone receptor complex binding to an enhancer
Polymerase promoter complex
DNA fragments separated according to size
Large
Cuts by DNAse I
Labeled DNA
Direction
Missing band at the area of binding ("footprint")
Small
Gel electrophoresis
1. DNA footprinting
Free DNA
DNA protein complex
Fast
Slow
Speed of migration in gel electrophoresis
2. Band shift test
D. Detection of DNA–protein interaction

RNA Interference (RNAi)

In addition to mRNA, tRNA, and rRNA, various small RNAs (microRNAs [miRNAs]) exist. These are 20–30 nucleotide sequences that are not translated into protein. Rather they assume a variety of biological regulatory functions. Often they interfere with mRNA translation (RNA interference [RNAi].) Different noncoding regulatory RNAs can be distiguished: small interfering RNAs (siRNAs), miRNAs, Piwi-associated RNAs (pRNAs), and longer noncoding RNAs. Small RNAs usually arise by fragmentation of longer precursors by distinctive pathways.

RNAi, discovered in 1998, selectively blocks transcription. It is regarded as a natural defense mechanism against endogenous parasites and exogenous pathogenic nucleic acids, and can be used for analyzing gene function. The human genome contains about 200–255 genes for miRNAs (Lim et al., 2003). miRNAs function as antisense regulators of other genes.

A. Small interfering RNA (siRNA)

RNAi is induced by siRNA. This typically consists of 21–23 nucleotides of double-stranded RNA with a 2-bp overhang at both ends, and with high specificity for the nucleotide sequence of the target molecule: an mRNA.

B. RNA-induced silencing complex

siRNA interacts with proteins that have helicase and RNA nuclease activity, resulting in an RNA-induced silencing complex (RISC). RISC is a sequence-specific nuclease with helicase and RNase activity to cleave RNA. The helicase unwinds the double-stranded RNA, and the RNA endonuclease cleaves it.

C. Posttranscriptional gene silencing

The long exogenous double-stranded RNA (dsRNA, **1**) is cleaved by a dsRNA-specific RNase III ribonuclease, called a dicer. The helicase activity of the RISC unwinds the RNA (**2**). The resulting single-stranded segment of the siRNA binds sequence-specifically to the target mRNA and silences gene expression (**3**). A specialized ribonuclease III (RNase III) in the RISC cleaves the neighboring single-stranded RNA (red arrows). The mRNA fragments resulting from the degradation are then rapidly degraded by cellular nucleases (**4**).

D. Degradation of dsRNA by dicer

A (biological) dicer is a complex molecule with endonuclease and helicase activity (RNase III helicase) that cleaves double-stranded RNA (**1**). The dicer promotes the formation of RISC, which binds to the dsRNA (**2**). The helicase activity unwinds the dsRNA, and the RNA endonuclease activity (RNase type III enzyme) cleaves the dsRNA (**3**) into small interterfering siRNA (**4**).

E. Functional effect of RNAi

RNAi can be used for intentional silencing of a selected gene to assess its normal function. Here RNAi has been shown to silence a gene in the developing worm *Caenorhabditis elegans* (see p. 214). Double-stranded RNAi targeted to a specific gene was injected into the gonad of an adult worm (**1**). Its effect is observed in the developing embryo (**2**). The degrading effect of the dsRNA is visualized by fluorescence in-situ hybridization of a labeled probe of mRNA from the target gene. The probe hybridizes to the normal, noninjected embryo (purple on the left, **2a**) but not to the injected embryo (**2b**), whose target gene mRNA has been destroyed. (**A–D**, figures modified from McManus & Sharp, 2002, and Kitabwalla & Ruprecht, 2002; **E** is from Lodish et al., 2007.)

Further Reading

Fire A, et al. Potent and specific genetic interference by double-stranded RNA in Caenorhabditis elegans. Nature 1998;391:806–811

Ghildiyal M, Zamore PD. Small silencing RNAs: an expanding universe. Nat Rev Genet 2009;10:94–108

Grosshans H, Filipowicz W. Molecular biology: the expanding world of small RNAs. Nature 2008;451:414–416

Kitabwalla M, Ruprecht RM. RNA interference—a new weapon against HIV and beyond. N Engl J Med 2002;347:1364–1367

Lim LP, et al. Vertebrate microRNA genes. Science 2003;299:1540

Lodish H, et al. Molecular Cell Biology. 6th ed. New York: W. H. Freeman; 2007

McManus MT, Sharp PA. Gene silencing in mammals by small interfering RNAs. Nat Rev Genet 2002;3:737–747

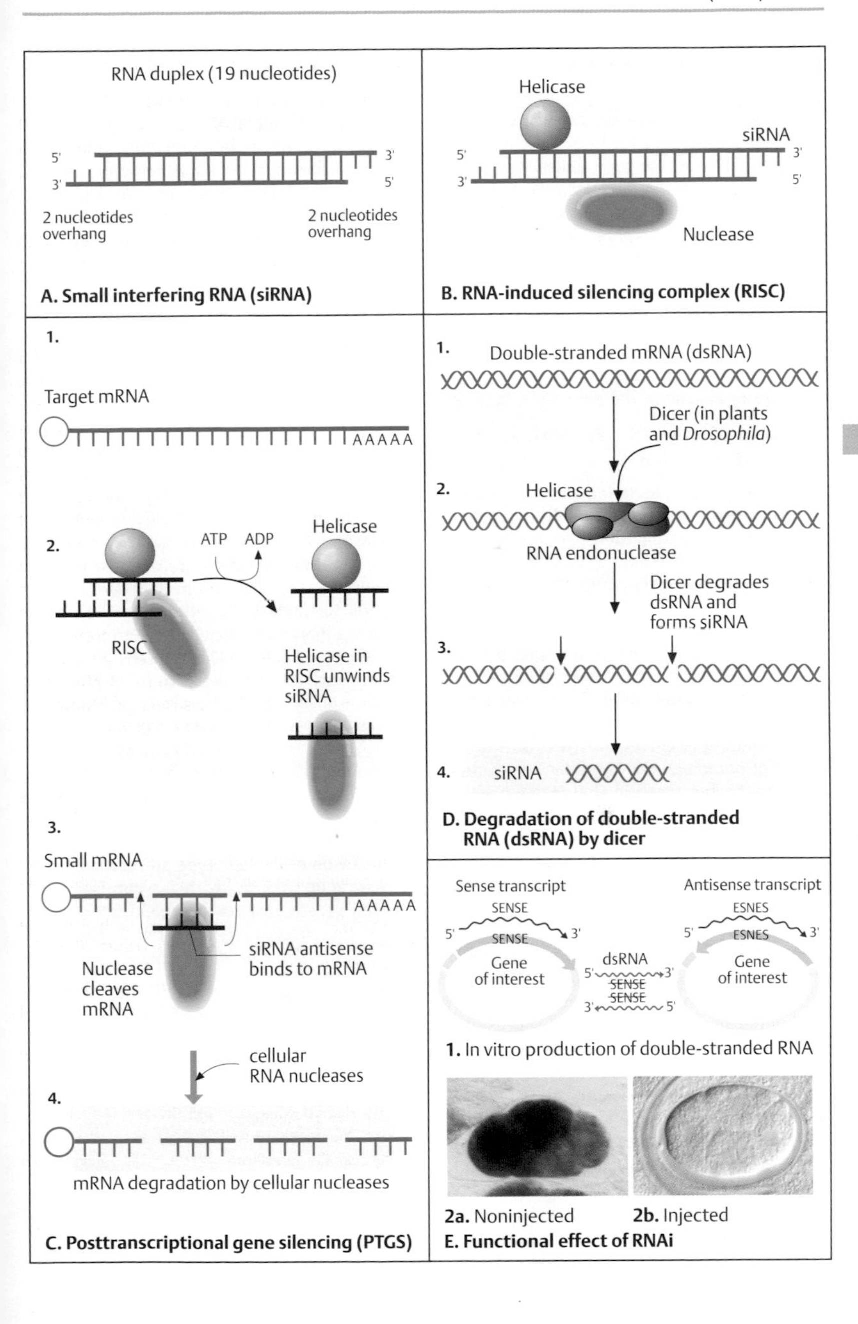
RNA duplex (19 nucleotides)
5'
3'
3'
5'
2 nucleotides overhang
2 nucleotides overhang
A. Small interfering RNA (siRNA)
Helicase
siRNA
5'
3'
3'
5'
Nuclease
B. RNA-induced silencing complex (RISC)
1.
Target mRNA
AAAAA
2.
ATP
ADP
Helicase
RISC
Helicase in RISC unwinds siRNA
3.
Small mRNA
AAAAA
Nuclease cleaves mRNA
siRNA antisense binds to mRNA
cellular RNA nucleases
4.
mRNA degradation by cellular nucleases
C. Posttranscriptional gene silencing (PTGS)
1. Double-stranded mRNA (dsRNA)
Dicer (in plants and Drosophila)
2. Helicase
RNA endonuclease
Dicer degrades dsRNA and forms siRNA
3.
4. siRNA
D. Degradation of double-stranded RNA (dsRNA) by dicer
Sense transcript
SENSE
5'
3'
SENSE
Gene of interest
dsRNA
5'
3'
SENSE
SENSE
3'
5'
Antisense transcript
ESNES
5'
3'
ESNES
Gene of interest
1. In vitro production of double-stranded RNA
2a. Noninjected
2b. Injected
E. Functional effect of RNAi

Targeted Gene Disruption

Targeted gene disruption refers to experimental inactivation of a gene to investigate its function in higher organisms. In a "knockout" animal, usually mice, the gene under study is inactivated in the germline by disrupting it (*gene knockout*). The effects can be studied at different embryonic stages and after birth. This knowledge can be utilized to understand the effects of mutations in homologous human genes as seen in human genetic diseases. A variant of knockout is known as gene *knock-in*. In this case, the targeting construct contains a normal gene that is introduced either in addition to or instead of the gene to be studied.

A. Preparation of ES cells with a knockout mutation

The target gene is disrupted (knocked-out) in yeast or embryonic stem (ES) cells by homologous recombination with an artificially produced nonfunctional allele. The isolation of ES cells with disrupted gene requires positive and negative selection. A bacterial gene conferring resistance to neomycin (*neo*^{R}) is introduced into the DNA of the artificial allele, partially cloned from the normal target gene (**1**,**2**). In addition, DNA containing the thymidine kinase gene (*tk*$^+$) from herpes simplex virus is added to the gene replacement construct outside the region of homology (**3**). The selective medium contains the positive and the negative selectable markers neomycin and ganciclovir (an analogue of the nucleotide guanine). Nonrecombinant cells and cells with nonhomologous recombination at random sites cannot grow in this medium. Nonrecombinant cells remain sensitive to neomycin, whereas recombinant cells are resistant (positive selection, not shown). The gene encoding thymidine kinase (*tk*$^+$) confers sensitivity to ganciclovir. Since nonhomologously recombinant ES cells contain the *tk*$^+$ gene at random sites, they are sensitive to ganciclovir and cannot grow in its presence (negative selection, **4**). Only cells that have undergone homologous recombination can survive, because they contain the gene for neomycin resistance (*neo*^{R}) and do not contain the *tk*$^+$ gene (**5**). (Figure redrawn from Lodish et al., 2007, p. 389.)

B. Transgenic mouse

In a transgenic animal, one copy of the target gene has been replaced by an altered gene in the germline. In the first step embryonic stem (ES) cells from a mouse blastocyst are isolated (**1**) after 3.5 days of gestation (of a total of 19.5 days) and transferred to a cell culture grown on a feeder layer of irradiated cells that are unable to divide (**2**). ES cells heterozygous for the knockout mutation are added (**3**). These ES cells are derived from a mouse that is homozygous for a different coat color (e.g., black) from that of the mouse that will develop from the blastocyst (e.g., white). The recombinant ES cells are injected into the recipient blastocyst (**4**). The early embryos are transplanted into a pseudopregnant mouse (**5**). Those offspring that have taken up ES-derived cells are chimeric, i.e., consisting of normal cells and cells with an altered gene. The transgenic mice can be recognized by black coat color spots on a white (or brown) background (**6**). The chimeric mice are then backcrossed to homozygous white mice (**7**). Black offspring from this mating are heterozygous for the disrupted (mutant) gene (**8**). By further breeding of the heterozygous mice (**9**), some of their offspring, the knockout mice, will be homozygous for the disrupted gene, as visualized by their new coat color. (Figure adapted from Alberts et al., 2008, p. 567, and Lodish et al., 2007, p. 390.)

Medical relevance

Comparison of the genotype and phenotype of a knockout mouse with a corresponding human genetic disease may yield information about the effects of a mutation, in particular during embryonic development.

Further Reading

Alberts B, et al. Molecular Biology of the Cell. 5th ed. New York: Garland Science; 2008

Capecchi MR. Targeted gene replacement. Sci Am 1994;270:52–59

Lodish H, et al. Molecular Cell Biology. 6th ed. New York: W. H. Freeman; 2007

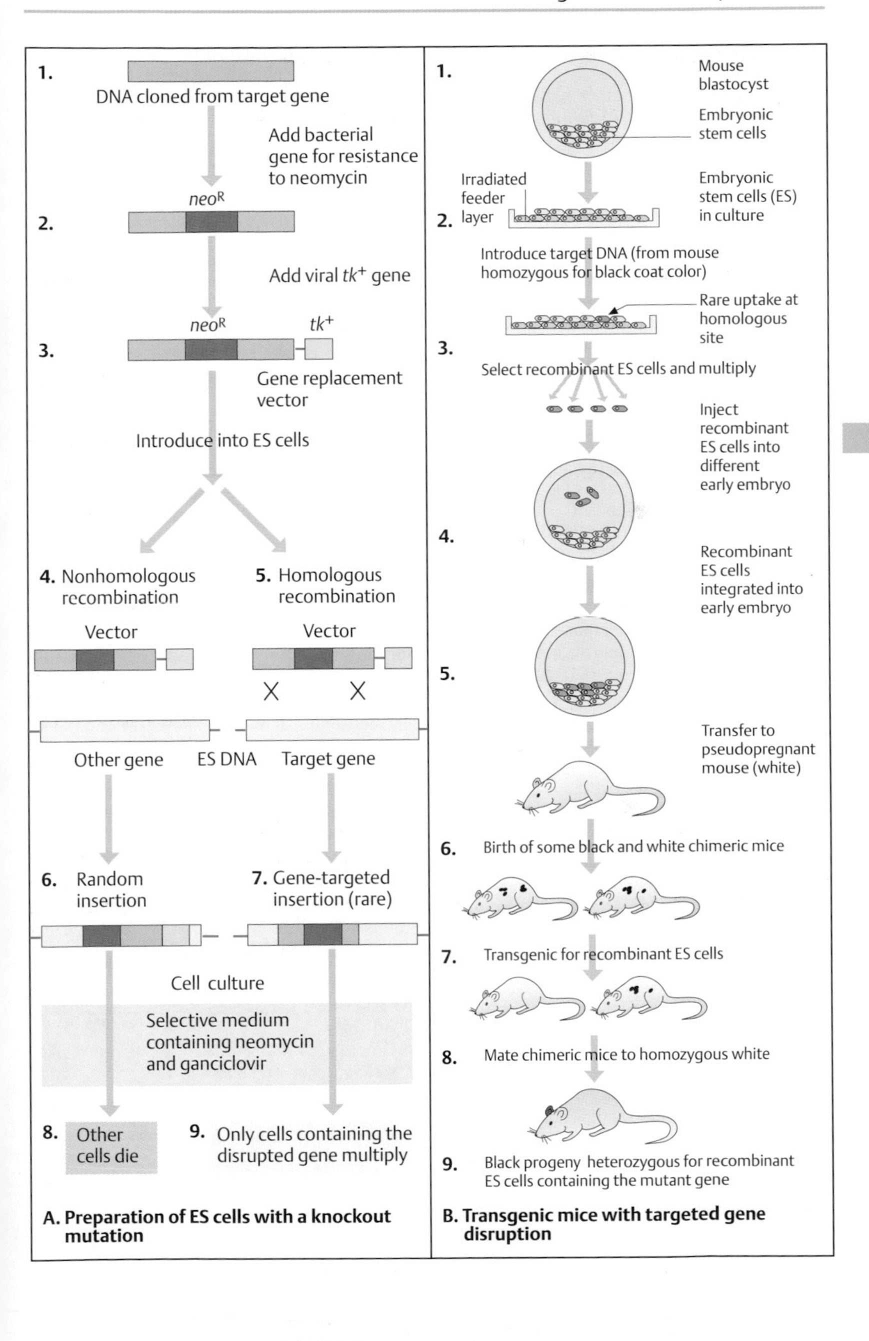

A. Preparation of ES cells with a knockout mutation

B. Transgenic mice with targeted gene disruption

DNA Methylation

An important mechanism of regulating gene function is chemical modification of DNA itself. In vertebrates, this is the addition of a methyl group cytosine at position 5 to form 5-methylcytosine (see p. 82).

Methylation is present in CG doublets, called CpG islands, present at the 5′ end of many genes. Up to 10% of cytosine residues are methylated in higher organisms. Mammalian cells contain enzymes that maintain DNA methylation and establish it in the new strand of DNA after replication. These are the DNA methyltransferases (DNMTs) and methylcytosine-binding proteins (MeCPs) binding to CpG islands. Two types of methyltransferase can be distinguished: maintenance methylation (DNMT1) and de novo methylation (DNMT3A and 3B).

A. Maintenance of DNA methylation

The enzyme maintenance methyltransferase adds methyl groups to the newly synthesized DNA strand. The methylated sites in the parental DNA (**1**) serve as templates for correct methylation of the two new strands after replication (**2**). This ensures that the previous methylation pattern is correctly maintained at the same sites as the parental DNA (**3**). The enzyme responsible for this is Dnmt1 (DNA methylase 1, DNMT1 in humans; OMIM 126375).

B. De novo DNA methylation

DNA methyltransferases 3A and 3B (DNMT3A, OMIM 602769; DNMTA3B, OMIM 602900) add methyl groups at new positions to both strands of DNA. Unmethylated DNA (**1**) is methylated by these enzymes in a site-specific and tissue-specific manner (**2**). Targeted homozygous disruption of the mouse *Dnmt3a* and *Dnmt3b* genes results in severe developmental defects.

C. Recognition of methylated DNA

Certain restriction enzymes do not cleave DNA when their recognition sequence is methylated (**1**). The enzyme *Hpa*II cleaves DNA only when its recognition sequence 5′-CCGG-3′ is not methylated (**2**). *Msp*I recognizes the same 5′-CCGG-3′ sequence irrespective of methylation and cleaves DNA at this site every time. This difference in cleavage pattern, resulting in DNA fragments of different sizes, serves to distinguish the methylation pattern of the DNA. Modern methods to identify methylated regions apply sodium bisulfite, which deaminates cytosine to uracil, whereas 5-methylcytosine remains unchanged.

D. Human *DNMT3B* gene and mutations

Mutations in the human gene *DNMT3B* encoding type 3B de novo methyltransferase causes ICF syndrome type 1 (*i*mmunodeficiency, *c*entromeric chromosomal instability, and *f*acial anomalies; OMIM 242860; Hansen et al., 1999; Xu et al., 1999). The centromeres of chromosomes 1, 9, and 16, where satellite DNA types 2 and 3 are located, are unstable. The human *DNMT3B* gene (**1**) consists of 23 exons spanning 47 kb. Six exons are subject to alternative splicing. The protein (**2**) has 845 amino acids with five DNA methyltransferase motifs (I, IV, V, IX, X) in the C-terminal region. The arrows point to six different mutations. One mutation (**3**) changes an A into a G in codon 809, i.e., GAC (Asp) to GGC (Gly), which results in a replacement of asparagine (Asn) by glycine (Gly). Both parents are heterozygous for this mutation. Multiradiate chromosomes with multiple p and q arms are typical in lymphocytes in ICF syndrome, here derived from chromosomes 1 and 16 shown in R banding (**4**). (Figure adapted from Xu et al., 1999.)

Further Reading

Hagleitner MM, et al. Clinical spectrum of immunodeficiency, centromeric instability and facial dysmorphism (ICF syndrome). J Med Genet 2008;45: 93–99

Hansen RS, et al. The DNMT3B DNA methyltransferase gene is mutated in the ICF immunodeficiency syndrome. Proc Natl Acad Sci U S A 1999;96: 14412–14417

Laird PW. Principles and challenges of genomewide DNA methylation analysis. Nat Rev Genet 2010;11: 191–203

Robertson KD. DNA methylation and human disease. Nat Rev Genet 2005;6:597–610

Xu GL, et al. Chromosome instability and immunodeficiency syndrome caused by mutations in a DNA methyltransferase gene. Nature 1999;402:187–191

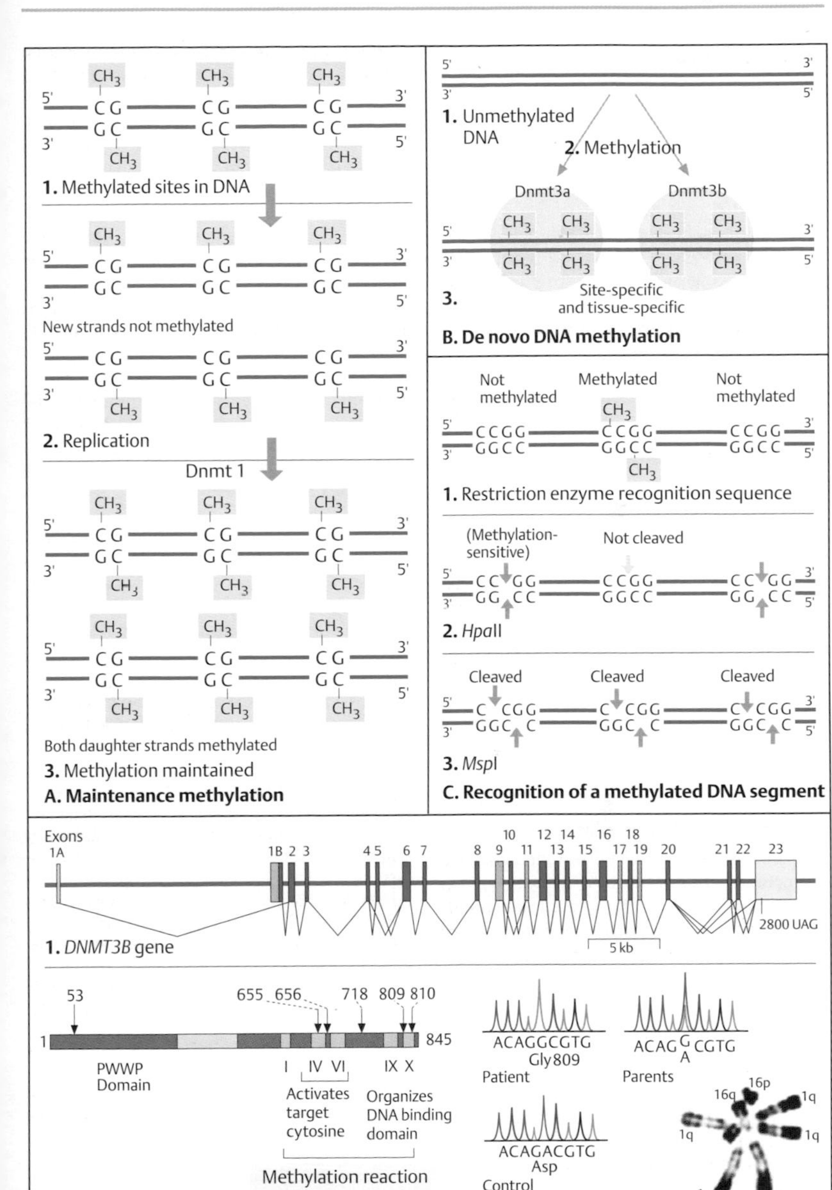
CH3
5'
CG
GC
3'
1. Methylated sites in DNA
New strands not methylated
2. Replication
Dnmt 1
Both daughter strands methylated
3. Methylation maintained
A. Maintenance methylation
1. Unmethylated DNA
2. Methylation
Dnmt3a
Dnmt3b
3.
Site-specific and tissue-specific
B. De novo DNA methylation
Not methylated
Methylated
CCGG
GGCC
1. Restriction enzyme recognition sequence
(Methylation-sensitive)
Not cleaved
2. HpaII
Cleaved
3. MspI
C. Recognition of a methylated DNA segment
Exons
1A
1B 2 3
4 5 6 7
8 9 10 11 12 13 14 15 16 17 18 19 20
21 22 23
2800 UAG
5 kb
1. DNMT3B gene
53
655 656 718 809 810
1
845
PWWP Domain
I IV VI IX X
Activates target cytosine
Organizes DNA binding domain
Methylation reaction
2. Protein and site of six mutations
ACAGGCGTG
Gly809
Patient
ACAG G/A CGTG
Parents
ACAGACGTG
Asp
Control
3. Mutation A809G
16q 16p 1q 1p
4. Chromosomes in ICF syndrome
D. Human DNMT3B gene and mutations

Reversible Changes in Chromatin Structure

Chromatin can assume one of two alternative stages: open (euchromatin) or closed (heterochromatin). These are associated with an active or inactive transcriptional state, respectively. In differentiated cells the transcriptional state is either permanent, but in certain regions it is reversible. This reversible change (chromatin remodeling) is a key event in gene regulation. The local structure of chromatin is an *epigenetic* state. Chromatin remodeling is an active, reversible process by which histones of the nucleosomes are displaced from the DNA molecule to make genes accessible for replication or transcription.

A. Histone modification

Changes in chromatin structure are initiated at the N-terminal tails of mainly histones H3 and H4. Here there are several sites for modification by methylation, acetylation (adding acetyl groups; $-NH-CH_3$), or phosphorylation. Examples are shown for histones H3 (**1**) and H4 (**2**). A typical inactive state in H3 is methylated lysine in position 4 (H3K4Me3), and an inactive state is methylated lysine at position 27 (H3K27Me3). The modifications are mediated by methylases and demethylases, acetylases and deacetylases, and phosphorylation kinases.

Active chromatin is acetylated at the lysine residues of H3 and H4 histones. Inactive chromatin is methylated at the position 9 lysine of H3 and at other lysine residues, and methylated at the cytosine residues of CpG islands. Lys-9 in H3 can be either methylated or acetylated. Thus, multiple modifications can occur and influence each other. Combinations of different signals are called the histone code. (Figure based on data in Strachan & Read, 2011, and Lewin, 2008.)

B. Histone acetylation and deacetylation

Acetylation of core histones occurs during DNA replication and activation of certain genes. Major targets are lysines in the N-terminal tails of H3 and H4. Active genes are acetylated directly in nucleosomal histones.

Acetylation is mediated by histone acetyltransferases (HATs), which are part of a large activating complex (**1**). The acetyl groups can be removed in a reversible process, deacetylation (**2**), mediated by deacetylases (HDACs), which are associated with repressors. Acetylases are associated with gene activators.

Deacetylation and methylation may be connected. Two methylcytosine-binding proteins, MeCP1 and MeCP2, selectively bind to methylated DNA. Transcriptional repression by the methyl-CpG-binding proteins 1 and 2 involves histone deacetylation in a multiprotein complex. (Figure adapted from Lodish et al., 2007, p. 475.)

C. Chromatin remodeling

Activator proteins can reverse the "gene off" state of condensed heterochromatin. These are specific DNA-binding control elements that are able to interact with multiprotein complexes. The activator proteins bind to a mediator protein. As a result, chromatin becomes decondensed and the gene assumes the "gene on" state. General transcription factors and RNA polymerase assemble at the promoter and initiate transcription. (Figure adapted from Lodish et al., 2007, p. 479.)

Medical relevance

Mutations of the *MECP2* gene on Xq28 cause Rett syndrome (OMIM 312750; Amir et al., 1999).

Further Reading

Amir RE, et al. Rett syndrome is caused by mutations in X-linked MECP2, encoding methyl-CpG-binding protein 2. Nat Genet 1999;23:185–188

Chahrour M, et al. MeCP2, a key contributor to neurological disease, activates and represses transcription. Science 2008;320:1224–1229

Jaenisch R, Bird A. Epigenetic regulation of gene expression: how the genome integrates intrinsic and environmental signals. Nat Genet 2003;33 (Suppl):245–254

Lewin B. Genes IX. Sudbury: Bartlett & Jones; 2008

Lodish H, et al. Molecular Cell Biology. 6th ed. New York: W. H. Freeman; 2007

Petronis A. Epigenetics as a unifying principle in the aetiology of complex traits and diseases. Nature 2010;465:721–727

Strachan T, Read AP. Human Molecular Genetics. 4th ed. New York: Garland Science; 2011

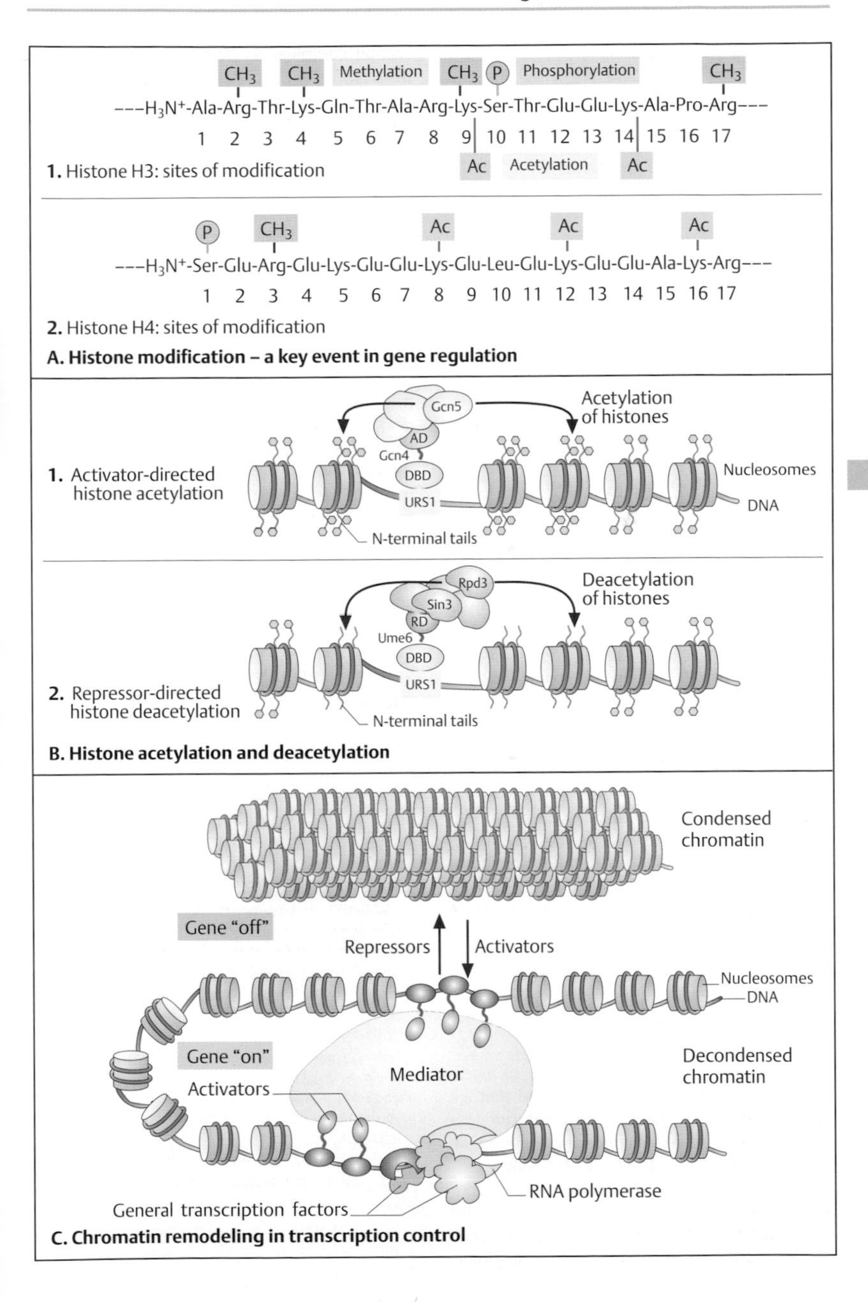

A. Histone modification – a key event in gene regulation

B. Histone acetylation and deacetylation

C. Chromatin remodeling in transcription control

Genomic Imprinting

Genomic imprinting is an epigenetic process leading to parent-of-origin specific difference in gene expression. It occurs in placental mammalian and marsupial genomes only. The inherited imprints are maintained during cell division, but are erased and re-established in the primordial germ cells according to the sex of the developing embryo. Imprinting is assumed to have evolved in mammals in response to intrauterine competition for resources (parental conflict hypothesis). Natural selection acts differently on maternal and paternal genomes. Presumably it favors a balance between preserving maternal resources and fetal growth. Currently, about 60 genes in the human genome are known to be imprinted.

A. The importance of two different parental genomes

Experimental observations in mice show that a contribution from both parental genomes is required for normal development. When from a normal zygote (**1**) the pronucleus of female origin is replaced by another male pronucleus (androgenote, **2**), it fails to complete preimplantation. The rare few that reach postimplantation develop abnormally and do not progress beyond the 12-somite stage. In contrast, a normal zygote develops normally (**3**). If the male pronucleus is replaced by another female pronucleus (gynogenote, **4**), about 85% of gynogenotes develop normally until preimplantation. However, subsequently the extra-embryonic membranes do not develop properly and the embryo dies at or before the 40-somite stage. (Figure adapted from Sapienza & Hall, 2001, and Barton et al., 1984.)

B. Requirement for a maternal and a paternal genome

A naturally occurring human androgenetic zygote is a hydatidiform mole (**1**). This is an abnormal placenta containing two sets of paternal chromosomes and none from the mother. Implantation takes place, but an embryo does not develop. The placental tissues develop many cysts (**2**). In a naturally occurring gynogenetic zygote, when only maternal chromosomes are present, an ovarian teratoma with many different types of fetal tissue develops (**3**). No placental tissue is formed. In triploidy, a relatively frequent fetal human chromosomal disorder (see p. 382), extreme hypoplasia of the placenta and fetus is observed when the additional chromosomal set is of maternal origin (**4**, 69,XXX). (Photographs kindly provided by Professor Helga Rehder, Vienna, Austria and Marburg, Germany.)

C. Imprint erasure, establishment, and maintenance

The epigenetic changes responsible for imprinting occur in early embryogenesis. The imprint pattern typically present in somatic cells according to parental origin of the chromosomes (**1**; P, paternal; M, maternal), is erased in primordial germ cells (**2**). During the formation of gametes, the sex-specific imprint patterns are re-established (**3**). Oocytes receive the maternal imprinting pattern (M) and spermatozoa receive the paternal imprinting pattern (P). Imprinted chromosomal regions of paternal origin (P) receive the paternal pattern; those of maternal origin (M) receive the maternal pattern. As a result, after fertilization the correct imprint pattern is present in the zygote (**4**) and is maintained through all subsequent cell divisions. Imprinted regions differ in their DNA methylation pattern and histone modifications.

Medical relevance

Failure in imprint erasure, establishment, or maintenance causes an important heterogeneous group of *imprinting diseases* (see p. 368).

Further Reading

Barton SC, Surani MAH, Norris ML. Role of paternal and maternal genomes in mouse development. Nature 1984;311:374–376

Constância M, Kelsey G, Reik W. Resourceful imprinting. Nature 2004;432:53–57

Horsthemke B. Mechanisms of imprint dysregulation. Am J Med Genet C Semin Med Genet 2010;154C:321–328

Moore T, Haig D. Genomic imprinting in mammalian development: a parental tug-of-war. Trends Genet 1991;7:45–49

Reik W, Walter J. Genomic imprinting: parental influence on the genome. Nat Rev Genet 2001;2:21–32

Sapienza C, Hall JG. Genetic imprinting in human disease. In: Scriver CR, et al., eds. The Metabolic and Molecular Bases of Inherited Disease. 8th ed. New York: McGraw-Hill; 2001. (www.ommbid.com.)

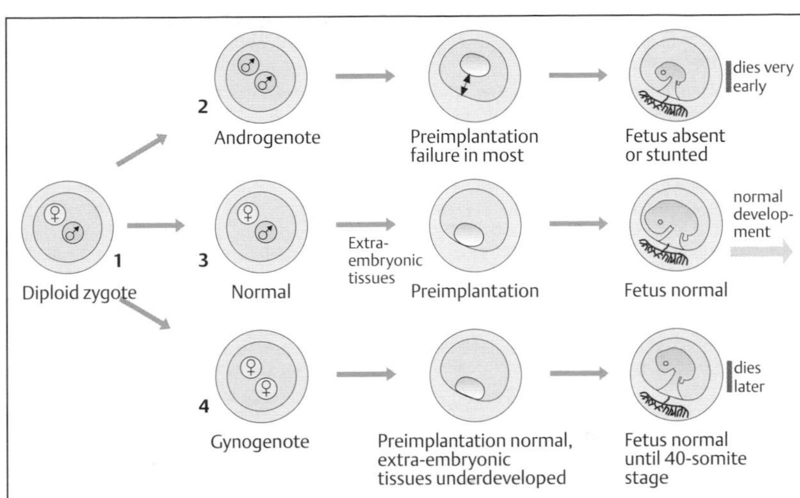

A. The importance of two different parental genomes

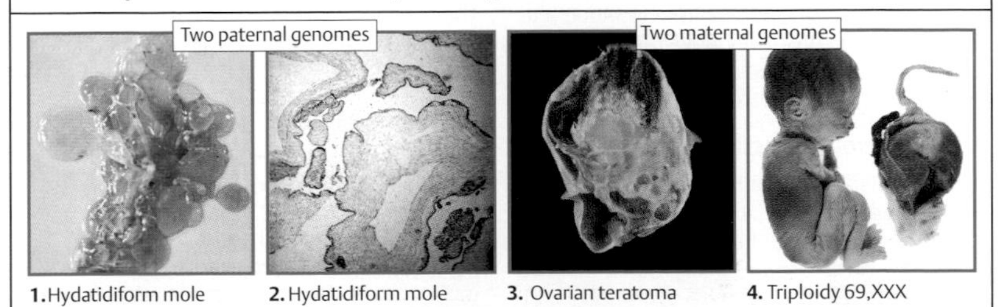

1. Hydatidiform mole **2.** Hydatidiform mole **3.** Ovarian teratoma **4.** Triploidy 69,XXX

B. Human embryonic development depends on presence of a maternal and a paternal genome

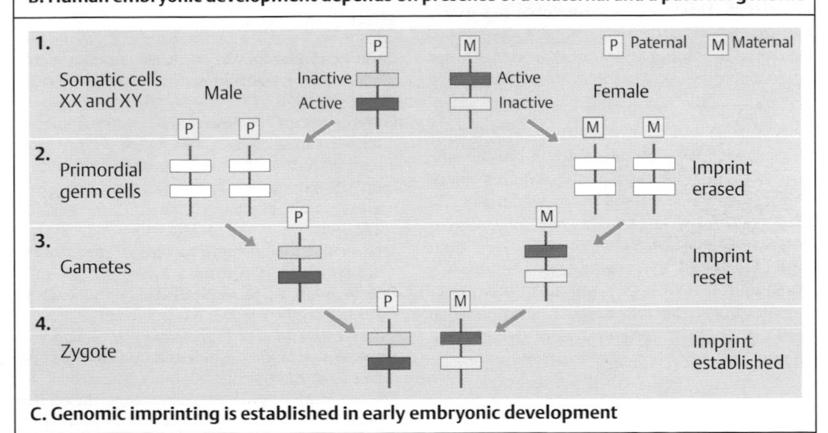

C. Genomic imprinting is established in early embryonic development

Mammalian X Chromosome Inactivation

In mammals, most genes on one of the two X chromosomes in every cell of females are inactivated at random. The result is *dosage compensation*, a term introduced in 1948 by H. J. Muller. The inactivation is initiated from an X inactivation center (Xic). This is a *cis*-acting region on the X chromosome. In early embryonic development, a noncoding RNA (Xist X-inactivation specific transcript) silences one of the X chromosomes in cells with two or more X chromosomes by methylation and histone modifications. Xist is the product of the *Xist* gene (OMIM 314670).

A. X chromatin

Small, darkly staining bodies were described by Barr & Bertram (1949) in the nerve cells of female cats (**1**, **3**), but not in males (**2**). In 1954, Davidson & Smith described similar structures as drumsticks in peripheral human blood leukocytes (**4**). X chromatin is visible as a dark density in the nuclei of oral mucosal cells (**5**). (Figure from Barr & Bertram, 1949.)

B. Scheme of X inactivation

The maternal *Xist* gene is expressed from the morula stage onwards. It follows random X inactivation involving either the maternal or the paternal X chromosome, stably transmitted to all daughter cells.

C. Mosaic pattern of expression

In 1961, Mary F. Lyon described a mosaic distribution pattern of X-linked coat colors in female mice as a manifestation of X inactivation (**1**). Fingerprints of human females heterozygous for X-linked hypohidrotic ectodermal dysplasia (OMIM 305100) show a mosaic pattern of areas with normal sweat pores (black points) and areas without in affected males (**2**). In cell cultures from females heterozygous for X-chromosomal HGPRT (hypoxanthine-guanine phosphoribosyltransferase) deficiency (OMIM 308000), the colonies are either $HGPRT^-$ or $HGPRT^+$ (**3**). (Figures: **1**, Thompson, 1965; **2**, Passarge & Fries, 1973; **3**, Migeon, 1971.)

D. X-inactivation profile

Genes in defined regions of the human X chromosome are not activated. An X-inactivation profile (Carrel & Willard, 2005) reveals that 458 (75%) genes are inactivated and 94 (15%) regularly escape inactivation. Surprisingly, 65 genes (10%) are inactivated in some females, but not in others. Thus, 25% of X-linked human genes are not regularly inactivated, and 10% exhibit an interindividual inactivation pattern. To the left (**a**) of the X chromosome, nine vertical lanes represent nine rodent/human cell hybrids. Genes expressed on the inactive X chromosome are shown in blue; silenced genes are in yellow. The right (**b**) illustrates the level of expression in the inactive X chromosome. (Figure kindly provided by Dr Laura Carrel, Hershey Medical Center, Pennsylvania, USA; from Carel & Willard, 2005.)

E. Evolutionary strata on the X

The human X chromosome harbors strata (S1–S5) of different evolutionary origin and time (see p. 240).

Medical relevance

Deviation from the random choice of either the maternal or the paternal X chromosome to be inactivated may lead to mild clinical manifestations in heterozygous females for an X-chromosomal mutation (OMIM 300087).

Further Reading

Barr ML, Bertram EG. A morphological distinction between neurones of the male and female, and the behaviour of the nucleolar satellite during accelerated nucleoprotein synthesis. Nature 1949;163:676–677

Carrel L, Willard HF. X-inactivation profile reveals extensive variability in X-linked gene expression in females. Nature 2005;434:400–404

Lyon MF. Gene action in the X-chromosome of the mouse (Mus musculus L.). Nature 1961;190:372–373

Migeon BR. Studies of skin fibroblasts from 10 families with HGPRT deficiency, with reference in X-chromosomal inactivation. Am J Hum Genet 1971;23:199–210

Passarge E, Fries E. X chromosome inactivation in X-linked hypohidrotic ectodermal dysplasia. Nat New Biol 1973;245:58–59

Thompson MW. Genetic implications of heteropyknosis of the X chromosome. Can J Genet Cytol 1965;52:202–213

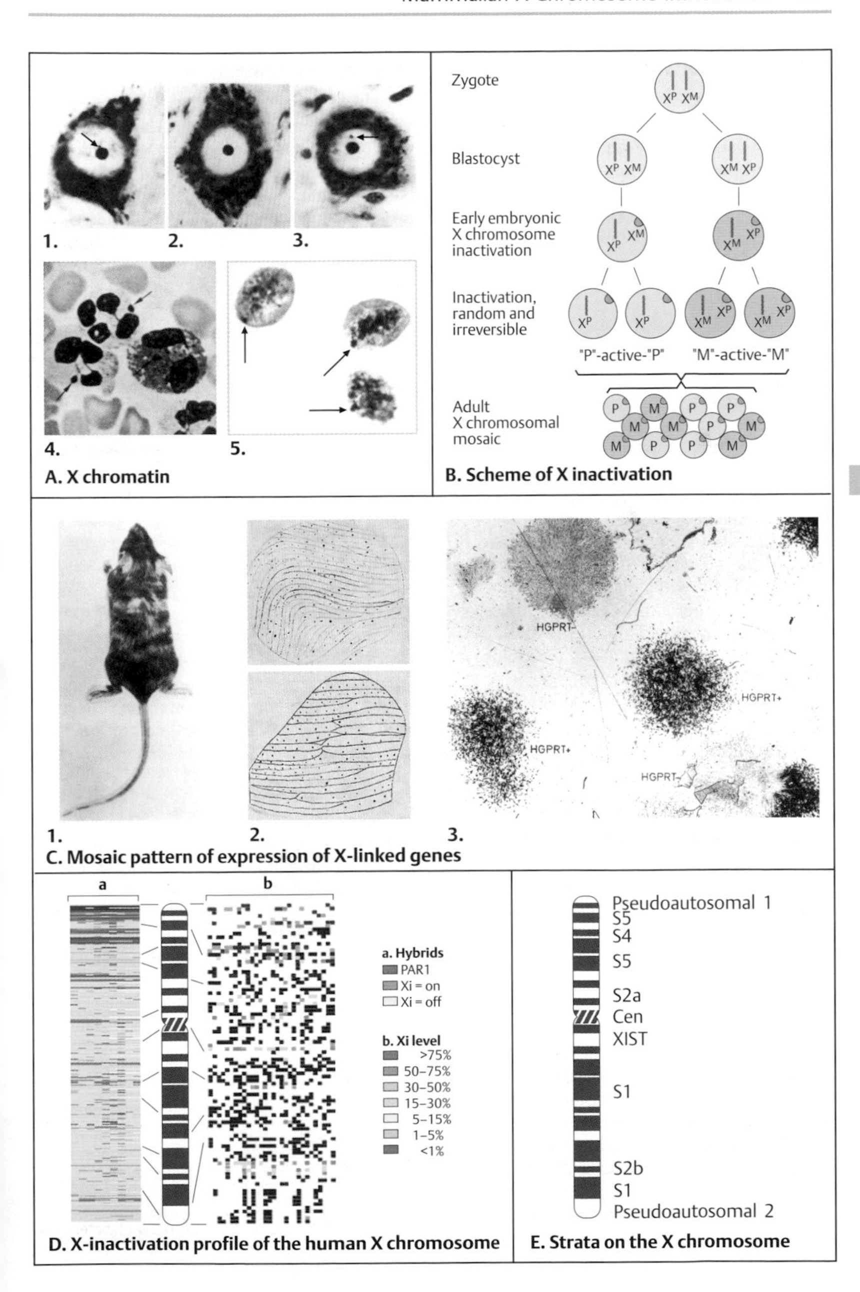

A. X chromatin

B. Scheme of X inactivation

C. Mosaic pattern of expression of X-linked genes

D. X-inactivation profile of the human X chromosome

E. Strata on the X chromosome

Cellular Signal Transduction

Multicellular organisms use a broad repertoire of signaling molecules for communication between and within cells (p. 94). The specific binding of an extracellular signaling molecule (ligand) to its receptor on a target cell triggers a specific response. This is followed by a series of mutually activating or inhibitory molecular events, called a *signal transduction pathway* (or signaling pathway).

A. Main intracellular functions controlling growth

Growth factors are important signaling molecules comprising a large group of secreted proteins (**1**). Each binds with high specificity to a cell surface receptor protein (**2**). This activates intracellular signal transduction proteins (**3**) and initiates a cascade of activations of responsive proteins (often by phosphorylation) that act as second messengers (**4**). Hormones are small signaling molecules (**5**) that arrive via the bloodstream. They enter the cell either by diffusion or by binding to a cell surface receptor (**6**). Some hormones bind to an intranuclear receptor (**7**). Activated transcription factors (**8**) together with cofactors initiate transcription (**9**). Prior to transcription, an elaborate system of DNA damage recognition and repair mechanisms (**10**) checks DNA integrity (cell cycle control, **11**). Cell division proceeds if faults in DNA structure have been repaired; if not, the cell is sacrificed by apoptosis (cell death, **12**). (Figure adapted from Lodish et al., 2007.)

B. Receptor tyrosine kinase family

Receptor tyrosine kinases (RTKs) are a major class of cell surface receptors. They consist of a single transmembrane protein with an extracellular N-terminal part, a transmembrane part (TM), and an intracellular C-terminal part. The intracellular part contains the tyrosine kinase domain. RTK ligands are growth factors that control a wide variety of functions involving growth and differentiation. RTK receptors share structural features but differ in function. The extracellular, ligand-binding domains of RTKs contain cysteine-rich regions. In certain RTKs the binding domains resemble immunoglobulin chains (immunoglobulin [Ig]-like domains), which are known for their ability to bind other molecules (see p. 284).

Medical relevance

Mutations in genes encoding RTKs may result in a proliferative signal in the absence of a growth factor and cause errors in embryonic development and differentiation (congenital malformations) or cancer. RTK mutations cause a group of important human diseases and malformation syndromes. The phenotypes due to the mutations differ according to the particular type of RTK involved and the type of mutation (see Appendix, Table 5, p. 412).

The nine members of the Receptor Tyrosine Kinase (RTK) family shown from left to right are:

EGFR (epidermal growth factor, OMIM 131550) and its ligands transmit signals in a variety of cellular functions, such as cell proliferation, differentiation, motility, and survival.

IR (insulin receptor, OMIM 147670, see pp. 44 and 264).

FGFR 1, 2, 3 (fibroblast growth factor receptor 1 —OMIM 136350; FGFR2—OMIM 176943; FGFR3—OMIM 134934, see p. 328), a family of receptors for fibroblast growth factors (FGF) involved in cell proliferation, differentiation, skeletal development, angiogenesis, and other functions.

PDGFR (platelet-derived growth factor receptors, type A—OMIM 173490, type B—OMIM 173410, and C—OMIM 608452) are receptors for platelet-derived factors (PDGF) required for angiogenesis and blood vessel functions.

RET (rearranged during transformation, OMIM 164761, see p. 340).

v-erbB (a viral oncogene, a homolog to the human ERBB2 gene, OMIM 164870).

Further Reading

Alberts B, et al. Molecular Biology of the Cell. 5th ed. New York: Garland Science; 2008

Brivanlou AH, Darnell JEJr. Signal transduction and the control of gene expression. Science 2002;295: 813–818

Lodish H, et al. Molecular Cell Biology. 6th ed. New York: W. H. Freeman; 2007

Robertson SC, Tynan JA, Donoghue DJ. RTK mutations and human syndromes: when good receptors turn bad. Trends Genet 2000;16:265–271

Tata JR. One hundred years of hormones. A new name sparked multidisciplinary research in endocrinology, which shed light on chemical communication in multicellular organisms. EMBO Reports 2005;6:490–496

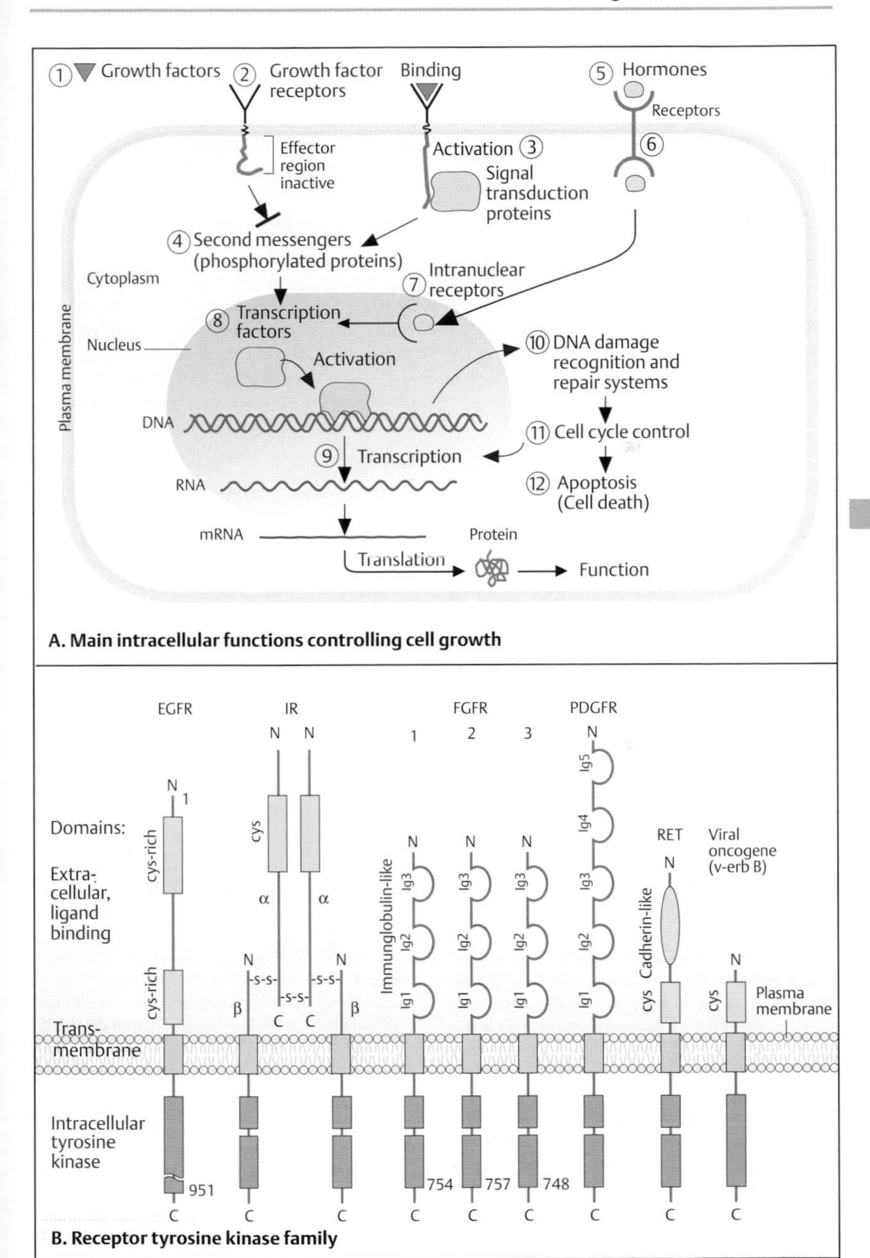

A. Main intracellular functions controlling cell growth

B. Receptor tyrosine kinase family

Heterotrimeric G Proteins

Heterotrimeric G proteins, or just G proteins, are a family of guanine-nucleotide-binding proteins involved in transmitting extracellular signals into the cell by means of a G-protein-coupled receptor.

Binding of its ligand to the receptor on the surface of a responding cell causes a conformational change in the structure of the receptor structure. This induces a cascade of intracellular activations and inhibitions of proteins, which transmit the signal to the nucleus. There it induces or inhibits the transcription of a target gene. Activation may be achieved by inhibiting an inhibitor.

A. Receptor tyrosine kinase (RTK)

The binding of a ligand activates a receptor tyrosine kinase (RTK, see p. 198) by dimerization of two neighboring transmembrane proteins. As a result, intracellular tyrosine residues are phosphorylated. This activates another target protein and further signal transduction events.

B. G-protein-coupled receptors

G proteins are trimeric guanine-nucleotide-binding proteins with subunits, α, β, and γ. They act as molecular switches between an inactive and a brief active state. In their inactive state, the α subunit is bound to guanosine diphosphate (GDP); in their active state, the α subunit is bound to guanosine triphosphate (GTP). G-protein-coupled receptors traverse the cell membrane seven times back and forth (therefore they are sometimes called serpentine receptors). Specific binding of a ligand activates the receptor by releasing the GDP from the α subunit. GDP is replaced by GTP. This change dissociates the trimer into two activated components: the α subunit and a β γ complex. The β γ complex thus released interacts with a target protein (effector protein), which is either an enzyme or an ion channel in the plasma membrane.

The α subunit is a GTPase and it rapidly hydrolyzes the bound GTP to GDP. As a result the α subunit and the β γ complex reunite, forming the inactive G protein. Therefore, the dissociated, active state of the α subunit and the β γ complex is short-lived. If the rapid reversal to the inactive state is delayed or impossible because of a toxin or a mutation, normal function is severely impaired.

The α subunits of mammalian G proteins form a large family of signaling molecules that bind to a wide variety of effector proteins. About 20 mammalian α subunits, 5 β subunits, and 12 γ subunits have been identified.

C. Cyclic AMP (cAMP)

The binding of the ligand to a signal transducing receptor molecule (the "first messenger") causes a brief increase or decrease in the concentration of low-molecular intracellular signaling molecules acting as second messengers. 3′,5′-cyclic adenosine monophosphate (cAMP) contains a phosphate group bound to the sugar 5′ and 3′ carbons in a cyclic structure.

D. Degradation of cyclic AMP

cAMP activates protein kinase A (PKA). Phosphodiesterase rapidly degrades cAMP into adenosine monophosphate (AMP). Active PKA phosphorylates more than 100 signaling proteins and transcription factors.

Medical relevance

Cholera toxin inhibits GTP hydrolysis and induces a permanently active state. In epithelial cells of the gastrointestinal tract this causes a massive efflux of water and chloride ions. Pertussis toxin (whooping cough) inhibits adenyl cyclase (inhibitory G protein, G_i) and prevents the α subunit of the inhibitory G protein from interacting with receptors.

Several endocrine disorders result from mutations of genes encoding G-protein-coupled receptors or the G proteins themselves. See Appendix, Table 5 (p. 412), for selected examples.

Further Reading

Alberts B, et al. Molecular Biology of the Cell. 5th ed. New York: Garland Science; 2008

Clapham DE. Mutations in G protein-linked receptors: novel insights on disease. Cell 1993;75: 1237–1239

Lodish H, et al. Molecular Cell Biology. 6th ed. New York: W. H. Freeman; 2007

Newley SE, van Aelst L. Guanine nucleotide-binding proteins. In: Epstein CJ, Erickson RP, Wynshaw-Boris A, eds. Inborn Errors of Development. The Molecular Basis of Clinical Disorders of Morphogenesis. 2nd ed. Oxford: Oxford University Press; 2008:1258–1276

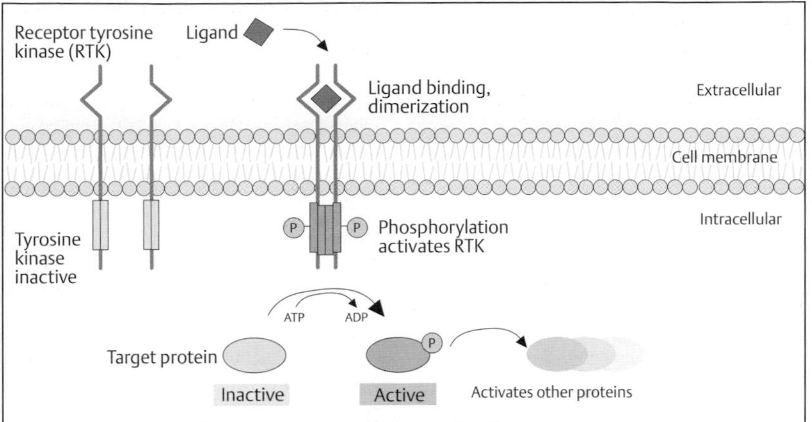

A. Signal transduction by a receptor tyrosine kinase (RTK)

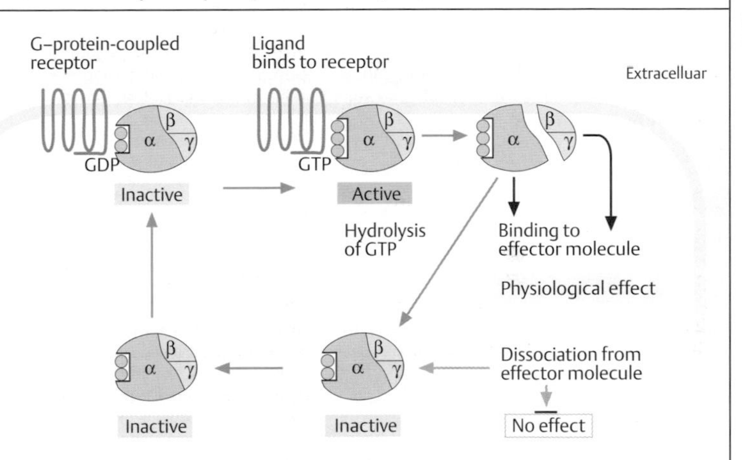

B. Signal transduction by a G–protein-coupled receptor

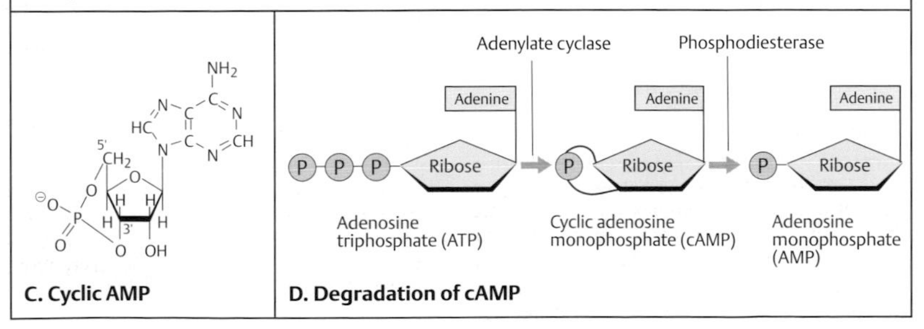

C. Cyclic AMP

D. Degradation of cAMP

TGFβ and Wnt/β-Catenin Signaling Pathways

Transforming growth factor β (TGFβ) and wingless (Wnt)/β-catenin signal transduction pathways are important examples of signaling systems involved in a variety of developmental processes and cellular functions.

A. The TGFβ signaling pathway

The TGFβ superfamily includes about 30 structurally related secreted polypeptides regulating many different biological processes involving growth and differentiation. They bind to dimeric transmembrane proteins with intracellular serine/threonine kinase domains as receptors. Binding induces a heterodimeric complex consisting of two copies each of receptors type II and type I (only one copy of each is shown). A constitutively phosphorylated receptor type II phosphorylates the GS domain, which is a conserved glycine/serine-rich sequence. This activates the kinase domain of receptor type I (TGFβ signaling). As a result, members of the Smad family of transcription factors become serine-phosphorylated and form a complex with a nonphosphorylated common Smad (Co-Smad). Smad 4 moves into the nucleus and binds with DNA-binding proteins to initiate transcription (Smad-dependent, canonical signaling). Non-Smad signaling pathways (non-canonical) may involve the RAS-MAPK pathway (see p. 342).
The TGFβ superfamily has important roles in embryonic development, differentiation, organogenesis, self-renewal and maintenance of stem cells in their undifferentiated stage, selection of cell differentiation lineage, and suppression of carcinogenesis. Phylogenetic evidence suggests that this family is one of the oldest signaling pathways, having arisen about 1.3 billion years ago, before the divergence of arthropods and vertebrates (see p. 26). (Figure redrawn from Descotte et al., 2008.)

B. Wnt/β-catenin signaling pathway

Wingless (*wg*) is a *Drosophila* segment-polarity mutant (see p. 208). The vertebrate ortholog is *Wnt* (derived from wingless and *int-1*, a mouse gene). It regulates β-catenin (in *Drosophila* called armadillo) by interacting with other proteins to activate target genes. The receptor for Wnt is a seven-transmembrane receptor named frizzled and a coreceptor, which is the low-density lipoprotein (LDL; see p. 272) protein receptor-related LRP6. Binding of Wnt to its receptor activates a protein called disheveled (Dsh). Dsh inhibits a protein kinase, glycogen synthase kinase-3 (GSK3), which normally prevents nuclear accumulation of β-catenin and γ-catenin by proteolysis. Inhibited GSK3 activity releases β-catenin from the APC protein (adenomatous polyposis coli, see p. 150). β-catenin then migrates into the nucleus, forms a complex with several other proteins (transcription factors LEF and TCF, and others), and regulates numerous target genes. In the absence of a Wnt signal, β-catenin is phosphorylated by GSK3, APC, and axin (a scaffolding protein), and then degraded (not shown). Of the at least three Wnt-activated pathways downstream of Dsh the main ("canonical") pathway is shown here. Different Wnt signals have roles in development, e.g., for gastrulation, brain development, limb patterning, and organogenesis. (Figure redrawn from Kategaya et al., 2008.)

Medical relevance

Dysregulated TGFβ signaling occurs in carcinogenesis and as a cause of Marfan syndrome (see p. 330). Mutations in APC occur in 80% of human cancers of the colon (see p. 308). *Wnt4* is a potential ovary-determining gene.

Further Reading

Brivanlou AH, Darnell JEJr. Signal transduction and the control of gene expression. Science 2002;295: 813–818

Descotte V, O'Connor MB, Jadrich J. An introduction to the transforming growth factor β (TGF-β) signaling pathway. In: Epstein CJ, Erickson RP, Wynshaw-Boris A, eds. Inborn Errors of Development. The Molecular Basis of Clinical Disorders of Morphogenesis. 2nd ed. Oxford: Oxford University Press; 2008:358–368

Mishra L, Derynck R, Mishra B. Transforming growth factor-β signaling in stem cells and cancer. Science 2005;310:68–71

Moon RT, et al. WNT and β-catenin signalling: diseases and therapies. Nat Rev Genet 2004;5:691–701

Sheldahl LC, Kategaya LS, Moon RT. The Wnt (Wingless-type) signaling pathway. In: Epstein CJ, Erickson RP, Wynshaw-Boris A, eds. Inborn Errors of Development. The Molecular Basis of Clinical Disorders of Morphogenesis, 2nd ed. Oxford: Oxford University Press; 2008:330–335

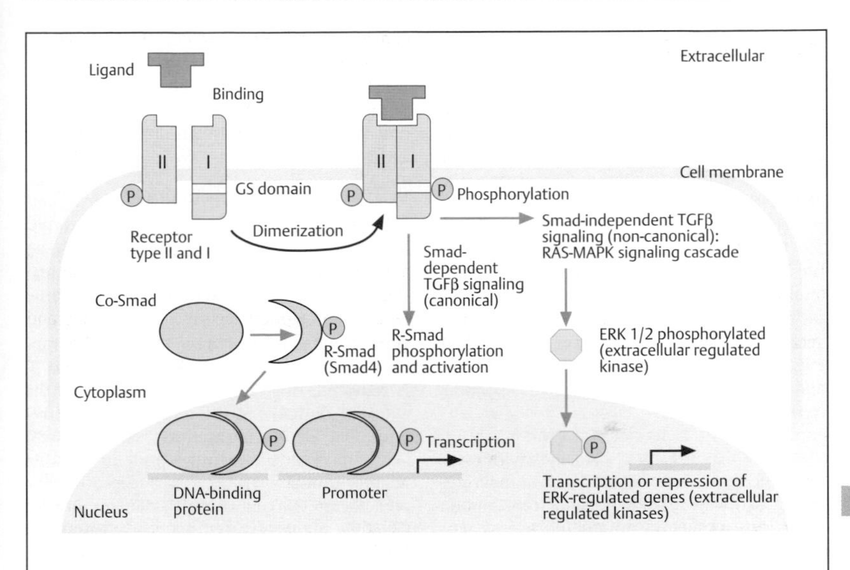

A. TGF β signaling pathway

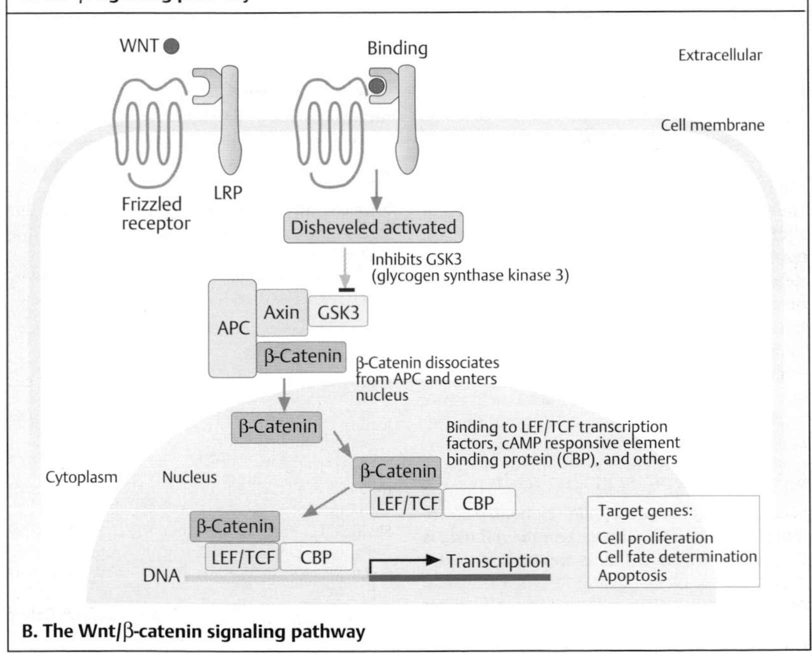

B. The Wnt/β-catenin signaling pathway

The Hedgehog and TNFα Signal Transduction Pathways

A family of secreted proteins (paracrine factors) called hedgehog conveys signals required in many developmental processes, in particular vertebrate limb and neural differentiation. The gene *hedgehog* (*hh*) in *Drosophila* is a segment polarity gene whose protein acts as a transcription factor with both activating and repressing functions in concert with other proteins. Mutant *hh* flies resemble hedgehogs owing to visible spikes on their surfaces. The genomes of vertebrates contain three hedgehog genes: *Sonic hedgehog, Desert hedgehog*, and *Indian hedgehog* see table in Appendix (p. 412).

Tumor necrosis factors (TNFs; OMIM 191160) are macrophage-secreted cytokines with cytotoxic effects. TNFα is involved in two main signaling pathways: (i) kinase and transcription factor activation pathway, and (ii) caspase activation in the cell death (apoptosis) pathway (see p. 110). TNFα belongs to a large family of 29 TNF receptors (TNFRs) and 18 TNF ligands in humans.

A. The hedgehog signaling pathways

More than 12 genes form a network rather than a one-directional pathway. The hedgehog (Hh) signal, secreted as a 45-kDa precursor protein, is cleaved into a 20-kDa N-terminal fragment and a 25-kDa C-terminal fragment. Hh is the ligand for a transmembrane receptor protein named patched (Ptch; PTCH in humans). Patched responds to ligand binding by negative control on another transmembrane protein called smoothened (Smo; SMO in humans; names derived from mutant *Drosophila* phenotypes).

Smoothened, a protein with seven hydrophobic membrane-spanning domains, acts as a hedgehog signal transducer. Without the signal, a microtubule-bound group of proteins (Costal 2, a kinesinlike protein, and Fused, a serine-threonine kinase) are attached to a 155-kDa effector protein Ci. Two other proteins, PKA (a protein kinase) and Slimb, cleave Ci into two fragments, one of which represses transcription of hedgehog-responsive genes. Binding of the Hh signal to Ptch suppresses the inhibitory function of Ptch on Smo. This inhibits the action of PKA and Slimb. As a result, Costal 2 and Fused are phosphorylated and they activate Ci. Activated Ci moves into the nucleus and initiates transcription together with coactivator CREB-binding protein (CBP) and other factors. (Figure adapted from Gilbert, 2010.)

B. TNFα signal transduction

The TNF ligands and their receptors are trimeric proteins. The extracellular moiety of TNFRα consists of between two and eight similar structural motifs of elongated shape stabilized by disulfide bridges.

Upon binding of TNFα to the receptor, the transcription factor NF-κB (nuclear factor kappa B) is activated. This causes rearrangements of the cytosolic domains of the receptor and the recruitment of an intracellular signal protein (RIP, receptor-interacting protein kinase), adaptor proteins, TRAF2 (TNF-receptor-associated factor 2), and a death-domain protein (TRADD). Signal-induced phosphorylation and degradation of an inhibitory κB (IκB) are the essential features of this pathway. NF-κB proteins control the expression of many genes involved in cell-specific differentiation, and inflammatory and apoptotic responses.

Medical relevance

Mutations and deletions in more than 10 human genes in the hedgehog gene network result in a group of malformation syndromes (Appendix, Table 6, p. 412). Mutations and polymorphisms in TNFα are associated with numerous disorders, e.g., rheumatoid arthritis, asthma, colitis ulcerosa, and others.

Further Reading

Gilbert SF. Developmental Biology. 9th ed. Sunderland, MA: Sinauer Publ.; 2010

Iannaccone P, et al. The sonic hedgehog pathway. In: Epstein CJ, et al., eds. Inborn Errors of Development. 2nd ed. Oxford: Oxford University Press; 2008:pp 263–279

Karin M. Nuclear factor-κB in cancer development and progress. Nature 2006;441:431–436

Lum L, Beachy PA. The Hedgehog response network: sensors, switches, and routers. Science 2004;304: 1755–1759

Schneider PL. The tumor necrosis factor signaling pathway. In: Epstein CJ, et al., eds. Inborn Errors of Development. 2nd ed. Oxford: Oxford University Press; 2008:433–441

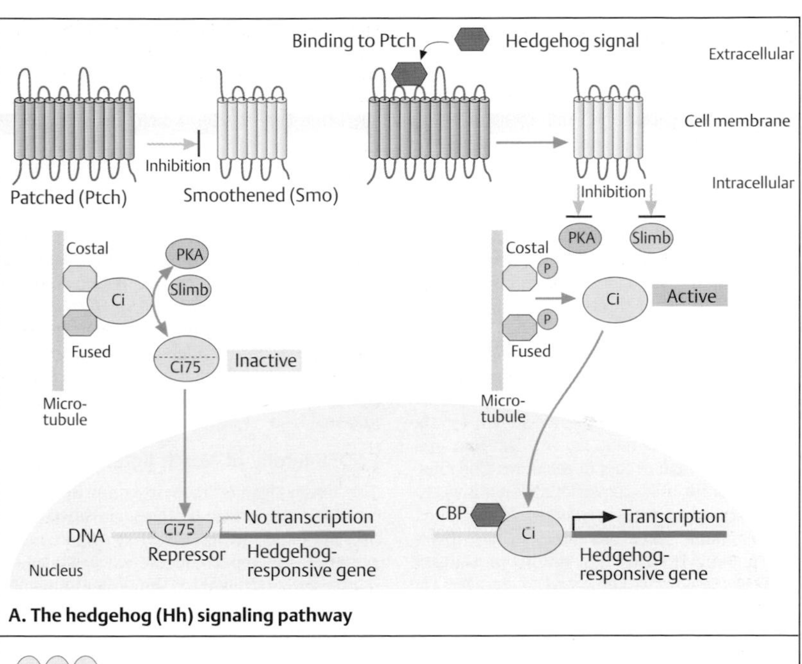

A. The hedgehog (Hh) signaling pathway

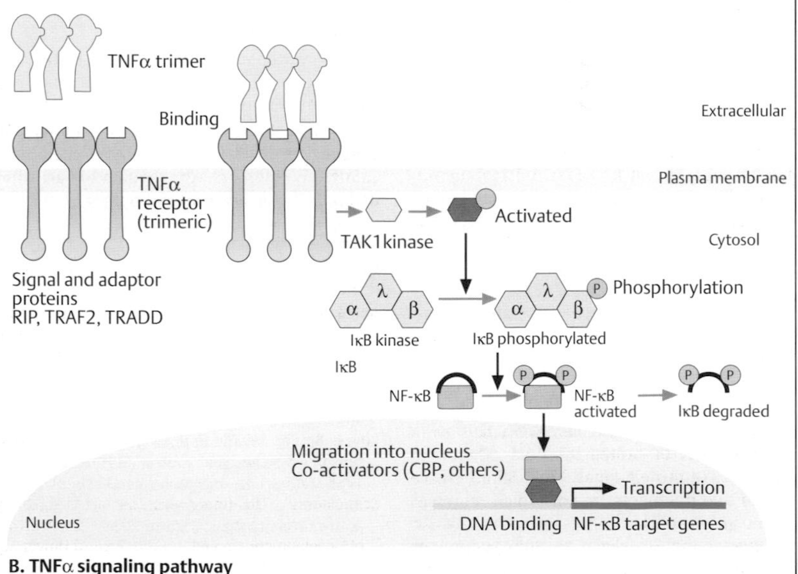

B. TNFα signaling pathway

The Notch/Delta Signaling Pathway

Notch signaling determines cell specification (e.g., neurogenesis, myogenesis, hematopoiesis), embryo patterning, and morphogenesis of various tissues in the developing vertebrate and invertebrate embryo. Notch signals are transmitted by direct interaction of a sending and a receiving cell. A receptor, Notch, and different types of ligands, called the DSL group (Delta, Serrate, Lag-2 ligands) mediate the signal. Notch activation involves regulated proteolytic cleavage at three sites of the receptor.

A. Transmission of a Notch signal

Both the ligand and the receptor are single-pass transmembrane proteins, and both require proteolytic cleavage (not shown for the ligand in the schematic figure). The first proteolytic cleavage occurs in the *trans* Golgi network by a furinlike convertase. Notch is *O*-glycosylated by a glycosyltranferase, fringe, which adds a single fucose to some serines, threonines, and hydroxylysines. Binding to its ligand induces the next two proteolytic cleavages. The Notch tail migrates to the nucleus and binds to the major effector of Notch signaling: the CSL protein. Binding with other gene regulatory proteins induces transcription of Notch-responsive target genes. (Figure based on Alberts et al., 2008.)

B. Lateral inhibition by Notch signaling

A characteristic function of Notch signaling is lateral inhibition in nerve cell development of *Drosophila*. Nerve cells arise as isolated single cells within a sheet of epithelial precursor cells. Cells determined to develop into nerve cells use the Notch signal pathways to inhibit the same cell fate in their neighboring cells. If this fails, a fatal excess of nerve cells causes embryonic death. (Figure based on Alberts et al., 2008.)

C. The *Drosophila* mutant Notch

Notch is a mutant phenotype in *Drosophila melanogaster*, described by T. H. Morgan in 1919. Female heterozygotes exhibit a notch of varied size in the distal part of their wings. Hemizygous males die during embryonic development from hypertrophy of the nervous system.

D. Notch family of receptors

Notch receptors constitute a large family of 50 surface proteins in mammals. The prototype is the *Drosophila* Notch (dNotch), consisting of 36 epidermal growth factor receptor (EGFR)-like tandem repeats in the extracellular domain. Other features are three cysteine (cys)-rich repeats, LNR (Lin-12/Notch-related region), and a signal peptide (SP) at the N-terminal extracellular part. The intracellular domain contains six ankyrin repeats (ANK), flanked by two nuclear localization signals (NLS). Humans have four slightly different types of Notch receptors: hNotch 1–4 (the smaller *C. elegans* notch receptors Lin-2 and Glp-1 are not shown). (Figure adapted from Miyamoto & Weinmeister, 2008.)

E. DSL family of Notch ligands

The Notch ligands likewise comprise a large family of cell surface proteins, classified either as Delta or Serrate ligands. They contain multiple EGFR-like repeats in the extracellular domains, an N-terminal DSL-binding motif, and a signal peptide (SP). (Figure adapted from Miyamoto & Weinmeister, 2008.)

Medical relevance

A wide variety of human diseases result from disrupted Notch signaling, for example: Alagille syndrome (OMIM 118450), which is caused by mutations in the gene encoding jagged1 or *NOTCH2*; spondylocostal dysplasia (OMIM 122600, 271529), resulting from mutations in the *deltalike 3* (*DLL3*) gene; and aortic valve disease (OMIM 109730), resulting from mutations in *hNotch1*.

Further Reading

Alberts B, et al. Molecular Biology of the Cell. 5th ed. New York: Garland Science; 2008

Krantz ID, Colliton RP, Genin A, et al. Spectrum and frequency of jagged1 (JAG1) mutations in Alagille syndrome patients and their families. Am J Hum Genet 1998;62:1361–1369

Miyamoto A, Weinmaster G. The notch signaling pathway. In: Epstein CJ, et al., eds. Inborn Errors of Development. The Molecular Basis of Clinical Disorders of Morphogenesis. 2nd ed. Oxford: Oxford University Press, Oxford; 2008:536–551

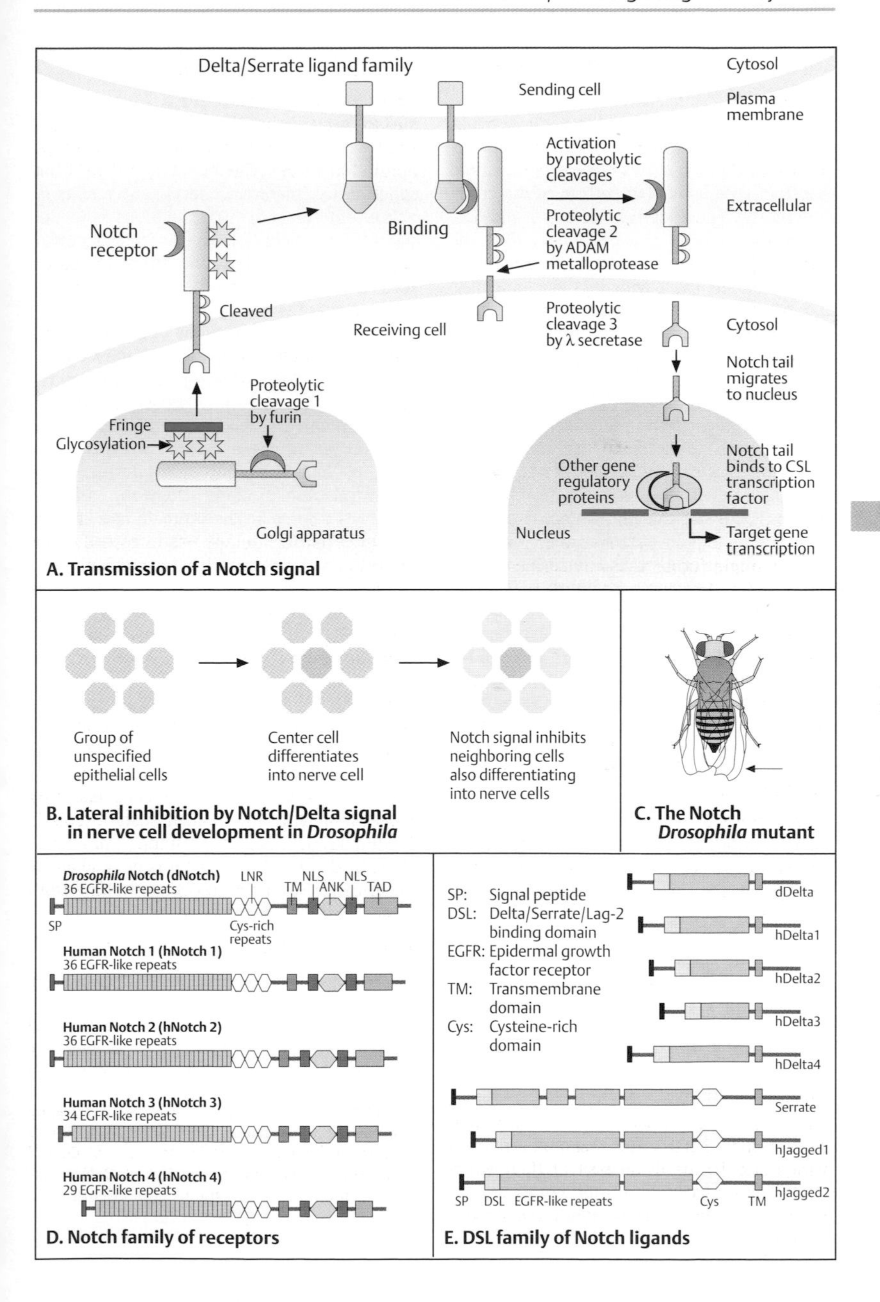

A. Transmission of a Notch signal

B. Lateral inhibition by Notch/Delta signal in nerve cell development in *Drosophila*

C. The Notch *Drosophila* mutant

D. Notch family of receptors

E. DSL family of Notch ligands

Embryonic Development Genes in *Drosophila melanogaster*

Different genes determine the early development of the *Drosophila* embryo. They are expressed at defined stages of development and determine the body plan pattern of the growing embryo. First, anterior–posterior and dorsal–ventral polarity is determined, then the segmental pattern of the embryo, and finally the head, the thorax with legs and antennae, and the abdomen. Homologous genes direct the developmental pattern of the mammalian embryo.

A. Life cycle of the fruit fly

The development from a fertilized egg (about 0.5 × 0.15 mm) to the adult fly (2 mm) takes 9 days. Nine nuclear divisions every 8 minutes without cell division leave the nuclei in a syncytium. Following the ninth nuclear division, 90 minutes after fertilization, the nuclei migrate to the periphery and form a *syncytial blastoderm*. After another four nuclear divisions, plasma membranes grow from the periphery and enclose each nucleus. This establishes the *cellular blastoderm* of about 6000 cells. Up to this stage, the embryo depends largely on maternal mRNA and proteins present before fertilization. The embryo passes through three defined larval stages before forming a pupa cocoon. After 5 days of metamorphosis, an adult fly emerges.

B. Segmental organization

The adult fly is organized into 14 segments: three segments (C1–3) form the head, three form the thorax (T1–3), and eight (A1–8) form the abdomen. Each segment has an anterior and a posterior compartment. Each of the initially formed 14 parasegments consists of the posterior compartment of the preceding segment and the anterior compartment of the following segment.

C. Embryonic developmental genes

Three classes of developmental genes exist. These are in hierarchical order of development: (i) *egg polarity* (*maternal effect*) genes defining anterior–posterior and dorsal–ventral polarity in the egg and early embryo; (ii) genes determining the cell fate to induce segmentation (*gap* genes and *pair-rule* genes); and (iii) genes determining the structure of body parts after the segmental boundaries are established (*homeotic selector* genes, see p. 210). The normal embryo shows the region for the head, three segments for the thorax, and eight for the abdomen (**1**). The mutation bicoid in an egg-polarity (maternal effect) gene results in lack of anterior parts (**2**). Mutations in one of about nine gap genes induce irregular segmentation, as in Krüppel (**3**) and Knirps (**4**). Characteristic mutant phenotypes of the eight pair-rule genes are even-skipped (**5**) and fushi tarazu (**6**) with loss of portions of alternate segments (fushi tarazu is Japanese for too few segments). More than 10 segment polarity genes determine the anterior–posterior polarity of each segment, e.g., the mutant gooseberry (**7**). Segment polarity genes encode proteins in the Wingless and Hedgehog signal transduction pathways (see p. 204). Homeotic selector genes (**8**) determine the ultimate fate of each segment. In the mutant antennapedia (*Ant*), the antenna is replaced by a leg (homeotic leg). The genes determining *Drosophila* embryonic development have been identified by embryonic fatal mutations induced in male flies with ethylmethane sulfonate (EMS). Males carrying random mutations were mated with females heterozygous for various mutations. The mutations have colorful names derived from the appearance of the mutant and are applied in the untranslated language of the discoverers.

(Figures adapted from Lawrence, 1992, and Nüsslein-Volhard & Wieschaus, 1980.)

Medical relevance

Mutations in several homologous developmental genes in humans cause disorders with congenital malformations, e.g., Waardenburg syndrome type 1 (OMIM 193500).

Further Reading

Gilbert SF. Developmental Biology. 9th ed. Sunderland, Mass: Sinauer; 2010

Lawrence PA. The Making of a Fly. The Genetics of Animal Design. Oxford: Blackwell Scientific; 1992

Nüsslein-Volhard C, Wieschaus E. Mutations affecting segment number and polarity in Drosophila. Nature 1980;287:795–801

Nüsslein-Volhard C, Frohnhöfer HG, Lehmann R. Determination of anteroposterior polarity in Drosophila. Science 1987;238:1675–1681

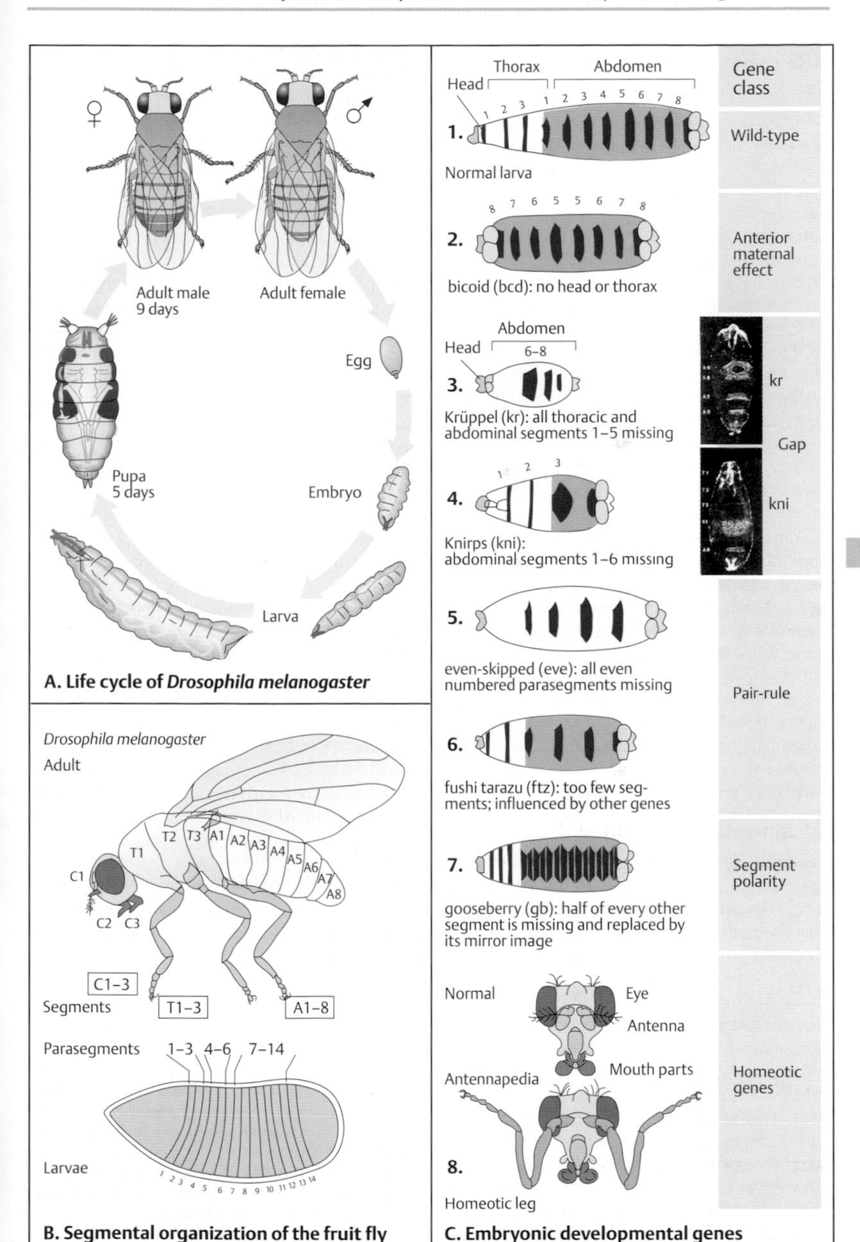

A. Life cycle of *Drosophila melanogaster*

B. Segmental organization of the fruit fly

C. Embryonic developmental genes

Hox Genes

Hox genes (OMIM 142950ff) form a group of evolutionary related genes that determine basic structures and their orientation during embryonic development, in particular segmental patterns. The name is derived from homeobox, a 180-nucleotide conserved sequence with important functions during embryonic development in *Drosophila* and higher organisms.

A. Regulatory hierarchy of pattern-determining genes

The five principal pattern-determining gene systems in *Drosophila* development act in hierarchical order: (i) *egg polarity* genes, (ii) *gap* genes, (iii) the *pair-rule* genes, (iv) *segment polarity* genes, and (v) *homeotic selector* (*Hox*) genes. Vertebrate Hox genes are homologous to homeotic *Drosophila* genes.

The anterior and posterior system (bicoid, anterior and nanos, posterior) is generated from localized mRNA molecules of maternal origin. The proteins diffuse in a decreasing gradient from the anterior and the posterior pole, respectively. The dorso–ventral axis is determined by a transmembrane receptor called Toll. A fourth signal is generated at both ends, a transmenbrane tyrosine kinase receptor called Torso. The three most important gap genes, Krüppel, Hunchback, and Knirps, determine local patterns. Gap genes induce the pair rule genes, such as even-skipped and fushi-tarazu. The segment polarity genes, expressed in parasegments, determine the correct anterior–posterior orientation of each individual segment. Homeotic selector genes determine the development of antennae, wings, legs, and other structures.

B. Homeotic selector genes

Homeotic selector genes form a complex called Hox genes (HOX in man). *Drosophila* has one set of Hox genes, mammals have four. Each gene belongs to the antennapedia cluster or the bithorax/ultrabithorax (*btx/ubx*) cluster. Five genes of the Antp complex are labial (*lb*), proboscis (*pb*), deformed (*dfd*), sex comb reduced (*scr*), and antennapedia (*antp*). The mammalian Hox genes share a single ancestral homeoselector gene complex. During the evolution of mammals this was duplicated twice, resulting in 39 genes occurring in four clusters, A–D, in man and mouse. In the process, some genes were lost, while a few genes were added. HOX genes are responsible for pattern formation along the body axis, and are involved in mammalian limb development. They are sequentially expressed in anterior–posterior orientation. The human chromosomal locations are 7p15 (*HOXA*), 17q21-q22 (*HOXB*), 12q13 (*HOXC*), and 2q31-q32 (*HOXD*). (Figure adapted from Alberts et al., 2008.)

C. The bithorax mutation

Mutations in the bithorax complex (*Bx-t*) induce the development of an additional thoracic segment with completely developed wings (Calvin Bridges, 1915). In 1978, E. B. Lewis recognized that the bithorax genes have evolved from a small number of ancestral genes by duplication and subsequent specialization into specific functions. (Photograph from Lawrence, 1992, after E. B. Lewis.)

D. The homeobox

Homeotic selector (Hox) genes encode gene regulatory proteins, e.g., transcription factors, and are highly conserved in evolution. The antennapedia gene contains a 180-bp conserved nucleotide sequence, the homeobox, from which the term Hox is derived. In the protein the corresponding 60 conserved amino acids constitute the homeodomain with four DNA-binding domains (I–IV), a common motif in transcription factors. Their expression depends on segment polarity genes.

Medical relevance

At least 20 human HOX genes are involved in various disorders. Mutations in *HOXD13* cause synpolydactyly (OMIM 142989).

Further Reading

Alberts B, et al. Molecular Biology of the Cell. 5th ed. New York: Garland Science; 2008

Garcia-Fernàndez J. The genesis and evolution of homeobox gene clusters. Nat Rev Genet 2005;6:881–892

Krumlauf R. Hox genes in vertebrate development. Cell 1994;78:191–201

Lawrence PA. The Making of a Fly. The Genetics of Animal Design. Oxford: Blackwell Scientific; 1992

Pearson JC, Lemons D, McGinnis W. Modulating Hox gene functions during animal body patterning. Nat Rev Genet 2005;6:893–904

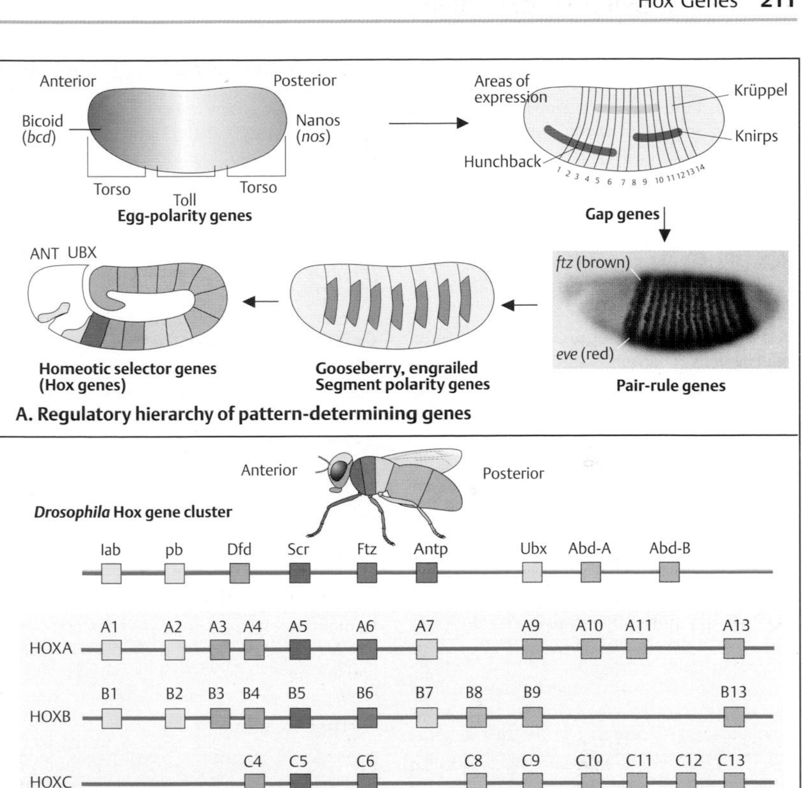

A. Regulatory hierarchy of pattern-determining genes

B. Homeotic selector genes

C. Bithorax mutation

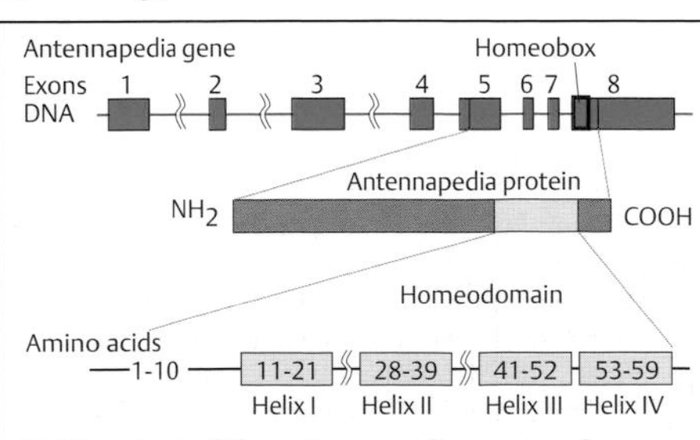

D. Structure of the antennapedia gene and homeobox

Zebrafish: A Translucent Vertebrate

Zebrafish *(Danio rerio)*, a tropical freshwater fish, is an important model organism for genetic and genomic studies. It is the first vertebrate to be studied by systematic genetic and genomic analysis. The zebrafish genome contains 18 000 known protein-coding genes and 1.5 billion base pairs. Developmental studies are facilitated by its translucent anatomy.

A. Embryo stages

In the optically clear embryo (pharyngula period), the main parts of the brain (forebrain, midbrain, and hindbrain) and the neural tube, somites, and floor plate are discernible 29 hours after fertilization. At 48 hours (hatching period), pigmentation begins, and the fins, eyes, brain, heart, and other structures become visible. At 5 days (swimming larva), the outline of a fish becomes apparent.

B. Induced mutagenesis

Random mutations induced in male parents were studied in thousands of offspring at various embryonic stages (genetic screen). Zebrafish adult males were exposed to 3 mM ethylnitrosourea (ENU) in an aqueous solution. The mutagenized males were crossed with wild-type females (P), resulting in the first generation (F_1) being heterozygous for mutations (m). Breeding the next generation (F_2) resulted in 50% carrying at least one mutation. Random matings of parents heterozygous for the same mutation resulted in 25% homozygous mutant offspring. Two examples are described below.

C. Skeletal phenotype of the fused somites (fss) mutation

In wild-type fish, the somite anlagen result in the formation of a normal segmental pattern of the vertebrae and muscles of the trunk and the tail (**1**). Five mutants with abnormal somite boundaries have been identified. Four show posterior somite defects and neuronal hyperplasia, while the fifth mutant, called fused somites (*fss*), completely lacks somite formation (**2**). Irregularly shaped spines grow ectopically at the wrong sites. The *fss* gene encodes a T-box transcription factor, Tbx24 (OMIM 607044), required for presomitic mesoderm maturation.

D. The no isthmus midbrain mutation

This mutation, in the *noi* gene, is an example of the more than 60 distinct mutant phenotypes affecting the central nervous system and spinal cord in zebrafish. Mutant *noi* embryos lack a conspicuous constriction at the boundary between midbrain and hindbrain. Normal 28-hour embryos (wild-type, **1**) show strong expression of the segment polarity gene engrailed between the midbrain and hindbrain. Normal eight-somite-stage embryos double stained for *eng* and *krox20* RNA, a marker for rhombomeres 3 and 5 in this region, express *eng* and *krox20* at the midbrain–hindbrain boundary (**2–3**), whereas *noi* mutants do not express *eng* (**4**). The *noi* mutation also eliminates production of wingless (Wnt1) protein in the posterior tectum at the border of the two brain regions in the 20-somite-stage embryo (**5**, **6**). (Figures adapted from Brand et al., 1996; van Eeden et al., 1996; Haffter et al., 1996.)

Medical relevance

Many human genes carrying disease-causing mutations have a homolog in the zebrafish genome.

Further Reading

Brand M, et al. Mutations in zebrafish genes affecting the formation of the boundary between midbrain and hindbrain. Development 1996;123:179–190

Dodd A, et al. Zebrafish: bridging the gap between development and disease. Hum Mol Genet 2000;9:2443–2449

van Eeden FJ, et al. Mutations affecting somite formation and patterning in the zebrafish, Danio rerio. Development 1996;123:153–164

Haffter P, et al. The identification of genes with unique and essential functions in the development of the zebrafish, Danio rerio. Development 1996;123:1–36

The *Danio rerio* Sequencing Project. Available at: http://www.sanger.ac.uk/Projects/D_rerio/

The Zebrafish Information Network. Available at: http://zfin.org/. Accessed January 27, 2012

Wylie C, ed. Development: Zebrafish Issue. Vol 123. Cambridge: Company of Biologists; 1996:1–481

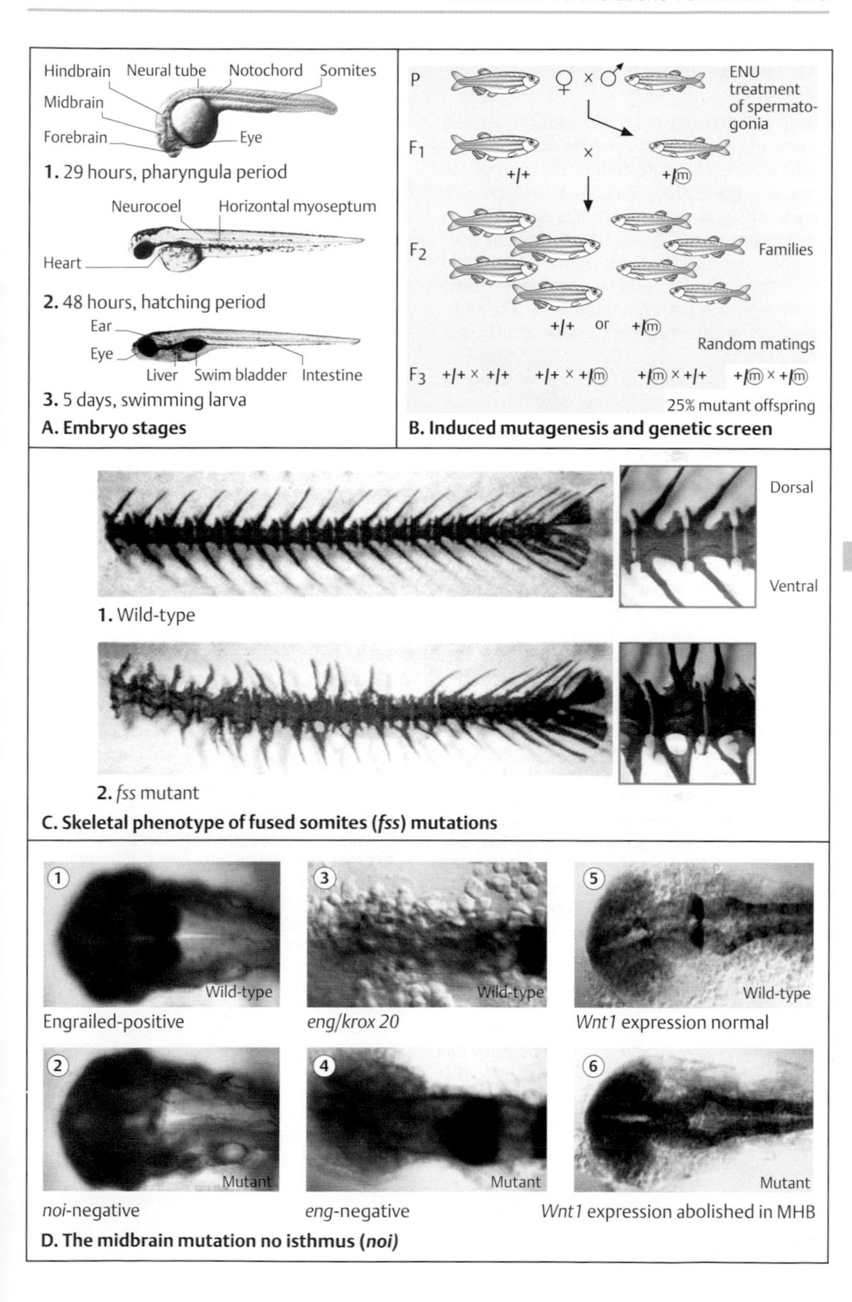

A. Embryo stages

B. Induced mutagenesis and genetic screen

C. Skeletal phenotype of fused somites (*fss*) mutations

D. The midbrain mutation no isthmus (*noi*)

Cell Lineage in a Nematode, *C. elegans*

Caenorhabditis elegans (*C. elegans*) is a small organism consisting of a precise number of somatic cells, each of which can be traced back to a founder cell. Sydney Brenner introduced this tiny worm as a model organism in 1965. Systematic genetic analysis of many mutant phenotypes has yielded important insights into the interaction of genetic, anatomical, and physiological traits in development. The complete cell lineage has been established in this organism.

The 97-Mb genome of *C. elegans* contains about 20 000 protein-coding genes and over 160 000 genes encoding untranslated RNA. About 32% of the coding sequences are homologous to sequences in man, and about 70% of known human proteins have homologies in *C. elegans*. The largest group of genes encodes transmembrane receptors (790), in particular chemoreceptors, zinc finger transcription factors (480), and proteins with protein-kinase domains. Interfering RNA (RNAi, see p. 186) has important functions in *C. elegans*.

A. *Caenorhabditis elegans*

C. elegans is a transparent worm, about 1 mm long, with a life cycle of about 3 days. Its basic structure is a bilaterally symmetric elongated body of nerves, muscles, skin, and intestines. It exists as one of two sexes: hermaphrodite or male. Hermaphrodites produce eggs and spermatozoa and can reproduce by self-fertilization. The adult hermaphrodite worm has 959 somatic cell nuclei; the adult male worm has 1031. In addition, there are 1000–2000 germ cells. (Figure adapted from Wood, 1988, after Sulston & Horvitz, 1977.)

B. Origin of individual cells

All tissues arise from six founder cells. At each cell division, genetically established rules determine the fate of the two daughter cells. Differentiated cells are derived from more than one founder cell, except for cells of the intestine and gonad. Of the 959 adult cells, 302 are nerve cells.

C. Developmental control genes

Many genes directing development have been identified by analysis of the mutations induced by ethylmethanesulfonate. The principal types of mutation cause cells to differentiate into incorrect cell types (e.g., Z instead of B), or other cells to divide too early or too late (division mutants).

D. Apoptosis in *C. elegans*

Programmed cell death (apoptosis) is a normal part of vertebrate and invertebrate development (see p. 110). During embryonic development of *C. elegans*, 131 of 947 nongonadal cells of the adult hermaphrodite undergo apoptosis at a defined time and branching point during different stages of development (**1**). Mutations in the *ced-9* gene induce apoptosis. Normally, *ced-9* suppresses apoptosis, and *ced-3* and *ced-4* are proapoptotic genes. Apoptosis does not occur in *ced-9*/*ced-3* double mutants because *ced-9* is upstream of *ced-3* in the apoptosis pathway. The photographs (**2**) show the death of a cell (cell P11.aap) over a time span of approximately 40 minutes. (Photographs from Wood, 1988, after Sulston & Horvitz, 1977).

Medical relevance

The human *BCL-2* gene, a homolog of *ced-9*, encodes an inner mitochondrial membrane protein that inhibits apoptosis in pro-B lymphocytes. Disruption of this gene causes follicular lymphoma, a B cell tumor (OMIM 151430).

Further Reading

Brenner S. The genetics of *Caenorhabditis elegans*. Genetics 1974;77:71–94

C. elegans Sequencing Consortium. Genome sequence of the nematode C. elegans: a platform for investigating biology. Science 1998;282:2012–2018

Culetto E, Sattelle DB. A role for *Caenorhabditis elegans* in understanding the function and interactions of human disease genes. Hum Mol Genet 2000;9:869–877

Stricklin SL, et al. *C.elegans* noncoding RNA genes. Available at: http://www.wormbook.org/chapters/www_noncodingRNA/noncodingRNA.html

Sulston JE, Horvitz HR. Post-embryonic cell lineages of the nematode, Caenorhabditis elegans. Dev Biol 1977;56:110–156

Wood WB, ed. The Nematode *Caenorhabditis elegans*. Monograph 17. New York: Cold Spring Harbor Laboratory Press; 1988

Wormbase. Available at: http://www.wormbase.org/

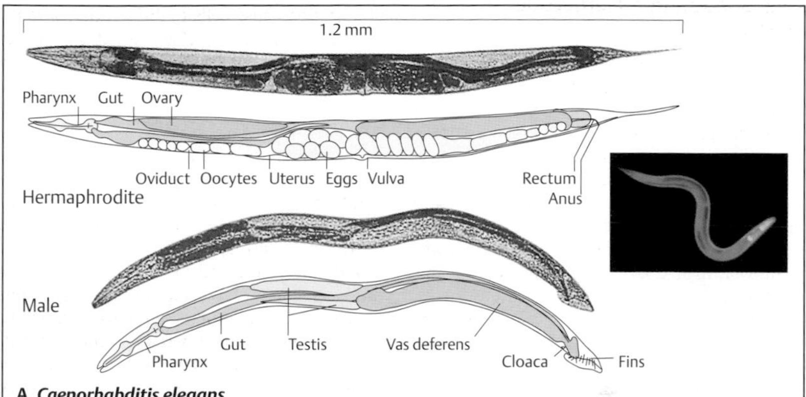

A. *Caenorhabditis elegans*

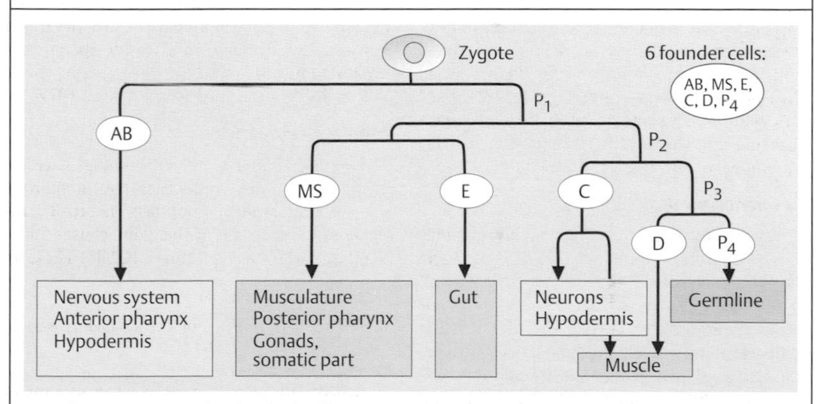

B. Embryonic origin of individual cells

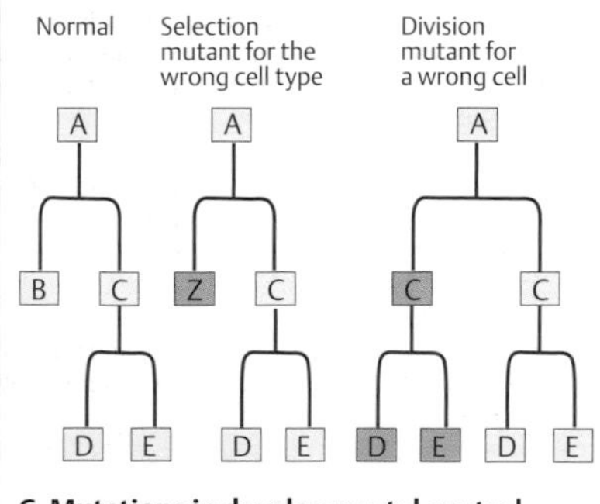

C. Mutations in developmental control genes

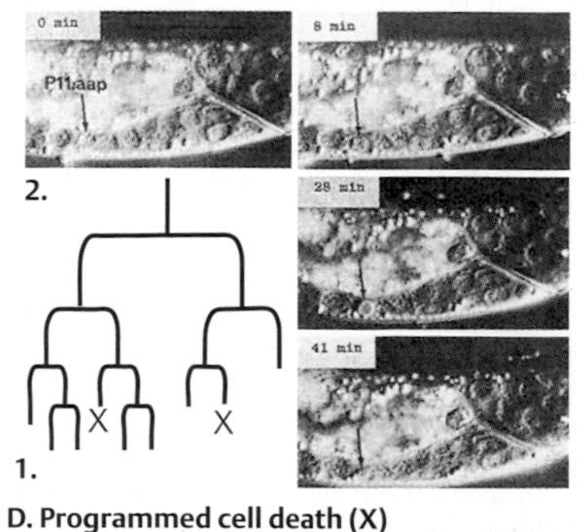

D. Programmed cell death (X)

Genomics

Genomics, the Study of the Organization of Genomes

Genomics is the scientific field dealing with all structural and functional aspects of the genome. A central goal is to determine entire DNA nucleotide sequences. Highly efficient methods of sequencing (p. 68) have enhanced knowledge in this field. The genomes of more than 4000 bacteria and viruses, 250 eukaryotes, and 15 extinct species have been sequenced. Related areas are concerned with all molecules involved in transcription and translation and their regulation (the *transcriptome*); all proteins that a cell or an organism is able to produce (the *proteome*); the functional analysis of all genes (*functional genomics*); and the assembly, storage, and management of data (*bioinformatics*). The term genomics was introduced in 1987 by V. A. McKusick and F. H. Ruddle, the founders of the journal *Genomics*.

A. Examples of organisms whose genomes have been sequenced

The DNA sequence of the human genome can be compared with that of an extinct hominid lineage, the Neandertal, that of the great apes, other primates, and many mammals and some marsupials.

Sources of images: **1**, A. Dürer, 1507. Adam und Eva. Museo Nacional del Prado, Madrid; **2**, J. Weissenbach, 2004; **3**, Weissenbach J. Genome sequencing: Differences with the relatives. Nature 2004;429:353–355; **4**, The Jackson Laboratory; **5**, Robert Geisler, Max Planck Institute for Developmental Biology, Tübingen, Germany. Available at: (http://zf-health.org/); **6**, Marco van Kerkhoven (http://www.kennislink.nl); **7**, M. Ashburner, University of Cambridge, UK (http://www.eurasnet); **8**, Dr Jordan Boyle (www.computescotland.com/c.elegans); **9**, Wikimedia Commons; **10**, Wikipedia/*E. coli*; **11**, The Arabidopsis Information Resource (TAIR).

Further Reading

Homo sapiens: Human Genome Resources NCBI. Available at: http://www.ncbi.nlm.nih.gov/projects/genome/guide/human/.

International Human Genome Sequencing Consortium. Finishing the euchromatic sequence of the human genome. Nature 2004;431:931–945

Lander ES. Initial impact of the sequencing of the human genome. Nature 2011;470:187–197

Nature: Human Genome Collection. Available at: http://www.nature.com/nature/supplements/collections/humangenome/.

UCSC Genome Bioinformatics. Available at: http://genome.ucsc.edu/. Accessed January 27, 2012

Neandertal: Green RE, Krause J, Briggs AW, et al. A draft sequence of the Neandertal genome. Science 2010;328:710–722

Chimpanzee: Waterston RH, et al; Chimpanzee Sequencing and Analysis Consortium. Initial sequence of the chimpanzee genome and comparison with the human genome. Nature 2005;437:69–87

Dog: Lindblad-Toh K, et al. Genome sequence, comparative analysis and haplotype structure of the domestic dog. Nature 2005;438:803–819

Mouse: Waterston RH, et al; Mouse Genome Sequencing Consortium. Initial sequencing and comparative analysis of the mouse genome. Nature 2002;420:520–562

Rat (not shown): Gibbs RA, et al; Rat Genome Sequencing Project Consortium. Genome sequence of the Brown Norway rat yields insights into mammalian evolution. Nature 2004;428:493–521

Zebrafish (*Brachydanio rerio*): Sanger Institute. Available at: http://www.sanger.ac.uk/resources/zebrafish/. Accessed January 27, 2012

Anopheles gambiae, the vector of the malaria parasite *Plasmodium falciparum*: Holt RA et al The genome sequence of the malaria mosquito Anopheles gambiae. Science 2002:298:129–149

***Drosophila melanogaster*:** Adams MD, Celniker SE, Holt RA, et al. The genome sequence of *Drosophila melanogaster*. Science 2000;287:2185–2195

***C. elegans*:** C. elegans Sequencing Consortium. Genome sequence of the nematode C. elegans: a platform for investigating biology. Science 1998;282:2012–2018

Yeast: Goffeau A, et al. Life with 6000 genes. Science 1996;274:546, 563–567

Bacterium *E. coli*: Blattner FR, et al. The complete genome sequence of Escherichia coli K-12. Science 1997;277:1453–1462

Plant: Arabidopsis Genome Initiative. Analysis of the genome sequence of the flowering plant Arabidopsis thaliana. Nature 2000;408:796–815

Sea Urchin: Special issue. Sea Urchin Genome. Science 2006;314:877–1032

1. *Homo sapiens*
3000 Mb ~22 000 genes

2. Chimpanzee
3000 Mb
1.2% differences to humans

3. *Canis domesticus*
24100 Mb, 19 300 genes

4. *Mus musculus*
2500 Mb ~25 000 genes

5. Zebrafish 1600 Mb
(70% sequenced) ~22 000 genes

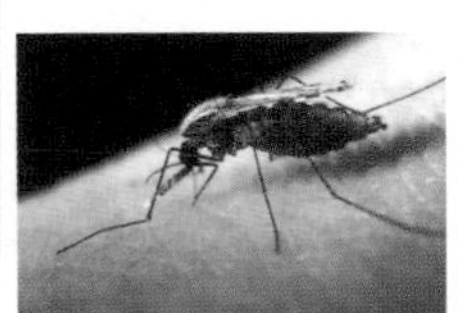

6. *Anopheles gambiae*
278 Mb, ~14 000 genes

7. *Drosophila melanogaster*
180 Mb, 13 600 genes

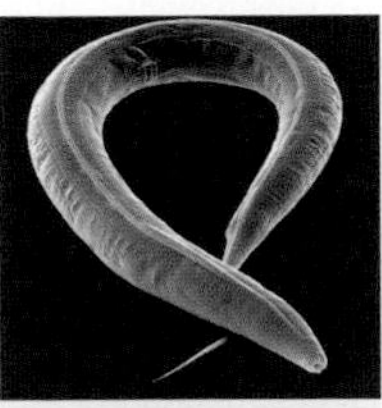

8. *C. elegans*
97 Mb, 19 000 genes

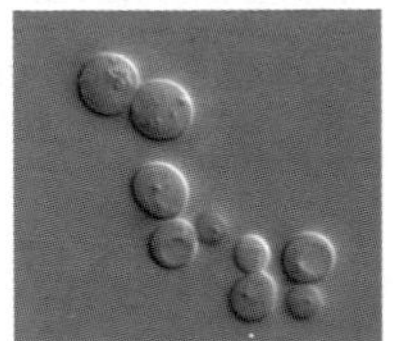

9. *S. cerevisiae*
12.1 Mb, 6300 genes

10. *E. coli* 4700 kb
4300 genes

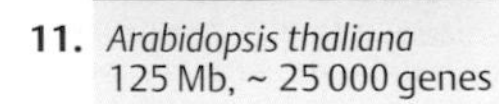

11. *Arabidopsis thaliana*
125 Mb, ~ 25 000 genes

A. Examples of organisms whose genomes have been sequenced

Identification of Disease-related Genes

Prior to the possibility of identifying a gene by complete sequencing of the entire genome, genes had to be identified by different approaches. The first step is to determine the exact chromosomal localization of the gene locus. Three principles applied in identifying a gene are outlined here.

A. Different approaches to identifying a disease-related gene

In principle, three approaches have proved useful: (i) *positional cloning*, (ii) *functional cloning*, and (iii) the *candidate gene* approach. The first and crucial step in all approaches is the clinical identification of the disease phenotype, i.e., the clinical diagnosis. The use of the McKusick catalog of human genes and phenotypes (McKusick, 1998) or its online system OMIM is indispensable for this purpose. It is important to consider the likely existence of genetic heterogeneity. In multigenic, complex inheritance it is usually not possible to identify a single gene as illustrated here.

Positional cloning starts with information about the chromosomal map position of the gene to be investigated. Commonly this information has been previously obtained by linkage analysis with neighboring loci. When the gene has been identified and isolated, it can be examined for mutations, which can then be related to impaired function. It has to be proved that a presumptive mutation is present in affected individuals only, and not in unaffected family members and normal controls.

Functional cloning requires prior knowledge of the function of the gene. As this information is rarely available at the outset, the utility of this approach is limited. It can be applied when a gene with a known function has been mapped previously and the clinical manifestations of the disease suggest a functional relationship.

The candidate gene approach utilizes independent paths of information. If a gene with a function relevant to the disorder is known and has been mapped, mutations of this gene can be sought in patients. If mutations are present in the candidate gene of patients, this gene is likely to be causally related to the disease.

B. Principal steps in gene identification

To identify a suspected human disease gene, clinical and family data together with blood samples for DNA have to be collected from affected and unaffected individuals. In monogenic diseases, the disorder will follow one of the three modes of inheritance: autosomal recessive, autosomal dominant, or X-chromosomal (**1**). A chromosome region likely to harbor a disease gene can be identified by one of several genetic mapping techniques, such as linkage analysis, or physical mapping using a chromosomal structural aberration such as a deletion or a translocation (**2**). The map position is refined by narrowing the region where the gene could be located to about 2–3 Mb (**3**). A contig map of overlapping DNA clones contained in a YAC or BAC (yeast or bacterial artificial chromosome) or a cosmid library is established from the region (**4**). This is further refined by a set of localized polymorphic DNA marker loci previously mapped to this region (**5**). Genes are identified by the presence of open reading frames (ORFs), transcripts, exons, and polyadenylated sites in this region, and are then isolated (**6**). Each gene in this region is subjected to mutational analysis (**7**). Genes without mutations in patients are excluded. When a mutation is found in one of the analyzed genes and a polymorphism is excluded, the correct gene has been identified (**8**). Now its exon/intron structure, size, and transcript can be determined. For confirmation, the expression pattern is analyzed and compared with that of homologous genes in other organisms. Finally, the DNA sequence of the entire gene can be determined.

Further Reading

Brown TA. Genomes. 3rd ed. New York: Garland Science; 2007

McKusick VA. Mendelian Inheritance in Man. Catalog of Human Genes and Genetic Disorders. 12th ed. Baltimore: Johns Hopkins University Press; 1998. Available at: http://www.ncbi.nlm.nih.gov/omim. Accessed January 24, 2012

Strachan T, Read AP. Human Molecular Genetics. 4th ed. New York: Garland Science; 2011

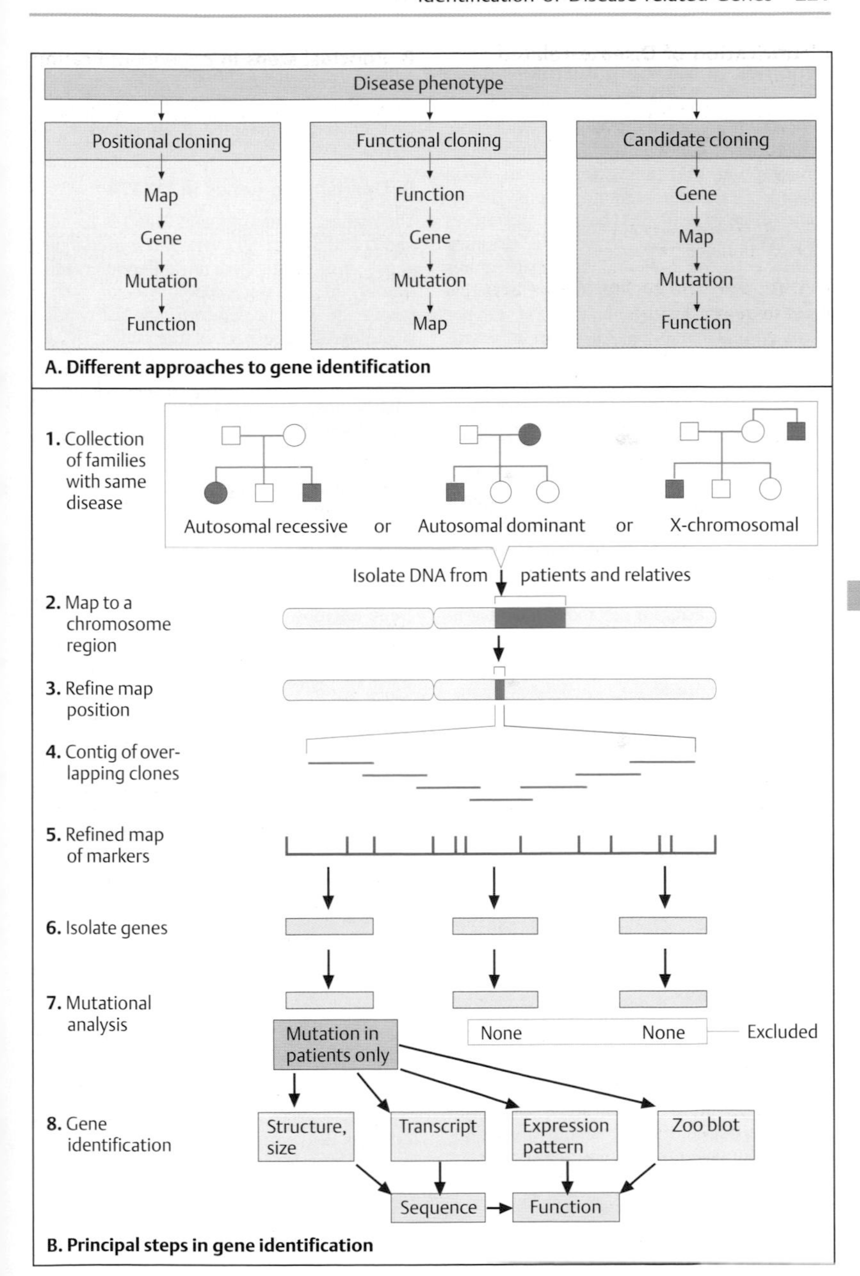

A. Different approaches to gene identification

B. Principal steps in gene identification

Genomes of Microorganisms

Bacterial genomes are small, ranging from 500 to 10 000 kb, and are tightly packed, with genes aligned nearly contiguously along the circular chromosome. The coding regions are small (average 1 kb) and contain no introns. The genomes of more than 4000 bacteria and viruses have been sequenced. They can be classified according to the sizes and principal features of their genomes. By selectively eliminating genes from a bacterial genome, it has been determined that about 265–350 genes are essential for survival under laboratory conditions.

Owing to their small size, bacterial genomes were the first to be completely sequenced, e.g., *Haemophilus influenzae* in 1995. Bacterial genomes can contain pseudogenes. About 1 in 20 of the coding regions of *Escherichia coli* K-12 is a pseudogene. The smallest bacterial genome is that of *Mycoplasma genitalium* (580 073 nucleotide base pairs, 483 genes). It is an obligate intracellular pathogen. Many genes encoding proteins for metabolic functions are absent. The limited capacity for metabolism in *M. genitalium* is compensated by the transport of life-supporting molecules from its extracellular environment into the cell.

The smallest free-existing bacterium is *Pelagibacter ubique* (1 308 759 bp, 1354 genes; Giovannoni et al., 2005). Its small, compact genome contains the complete biosynthetic pathways for all amino acids, in contrast to all other bacteria.

A. The genome of a small bacteriophage

Phage ΦX174, with 10 genes (A–J) contained in 5386 nucleotides in single-stranded DNA, was the very first organism to have its genome sequenced (Sanger et al., 1977). The genome of phage ΦX174 is so compact that several genes overlap. (Figure adapted from Sanger et al., 1977.)

B. Overlapping genes in ΦX174

The reading frames of genes A and B, B and C, and D and E partially overlap. The overlapping genes are transcribed in different reading frames: in the start codon ATG of gene E (above the sequences shown), the last two nucleotides (AT) are part of the codon TAT for tyrosine (Tyr) in gene D. Similarly, the stop codon TGA for the E gene is part of codons GTG (valine) and ATG (methionine) in the D gene. Thus, this small genome is used very efficiently.

C. Genome of *Escherichia coli*

This simplified figure shows the essential features of a bacterial genome. Functionally related genes usually cluster in operons (four of many shown). About half of the genes of *E. coli* are in operons.

Further Reading

Brown TA. Genomes. 3rd ed. New York: Garland Science; 2007

Fleischmann RD, et al. Whole-genome random sequencing and assembly of *Haemophilus influenzae* Rd. Science 1995;269:496–512

NCBI. Microbial Genomes. (www.ncbi.nlm.nih.gov/genomes/MICROBES/)

Ochman H, Davalos LM. The nature and dynamics of bacterial genomes. Science 2006;311:1730–1733

Sanger F, et al. Nucleotide sequence of bacteriophage phi X174 DNA. Nature 1977;265:687–695

US Department of Energy. Microbial Genomics. (http://microbialgenomics.energy.gov/)

Size and general contents of bacterial genomes

Main genomic features	Free-living	Facultative pathogen	Pathogen or obligate symbiont
Genome size	Large (5–10 Mb)	2–5 Mb	Small (0.5–1.5 Mb)
Genome stability	Stable or unstable	Unstable	Stable
Lateral gene transfer	Frequent	Frequent/rare	Rare or none
Number of pseudogenes	Few	Many	Rare
Population size	Large	Small	Small
Pathogenicity factors	Absent	Present	Present

Data from Ochman & Davalos (2006). Mb, one million base pairs.

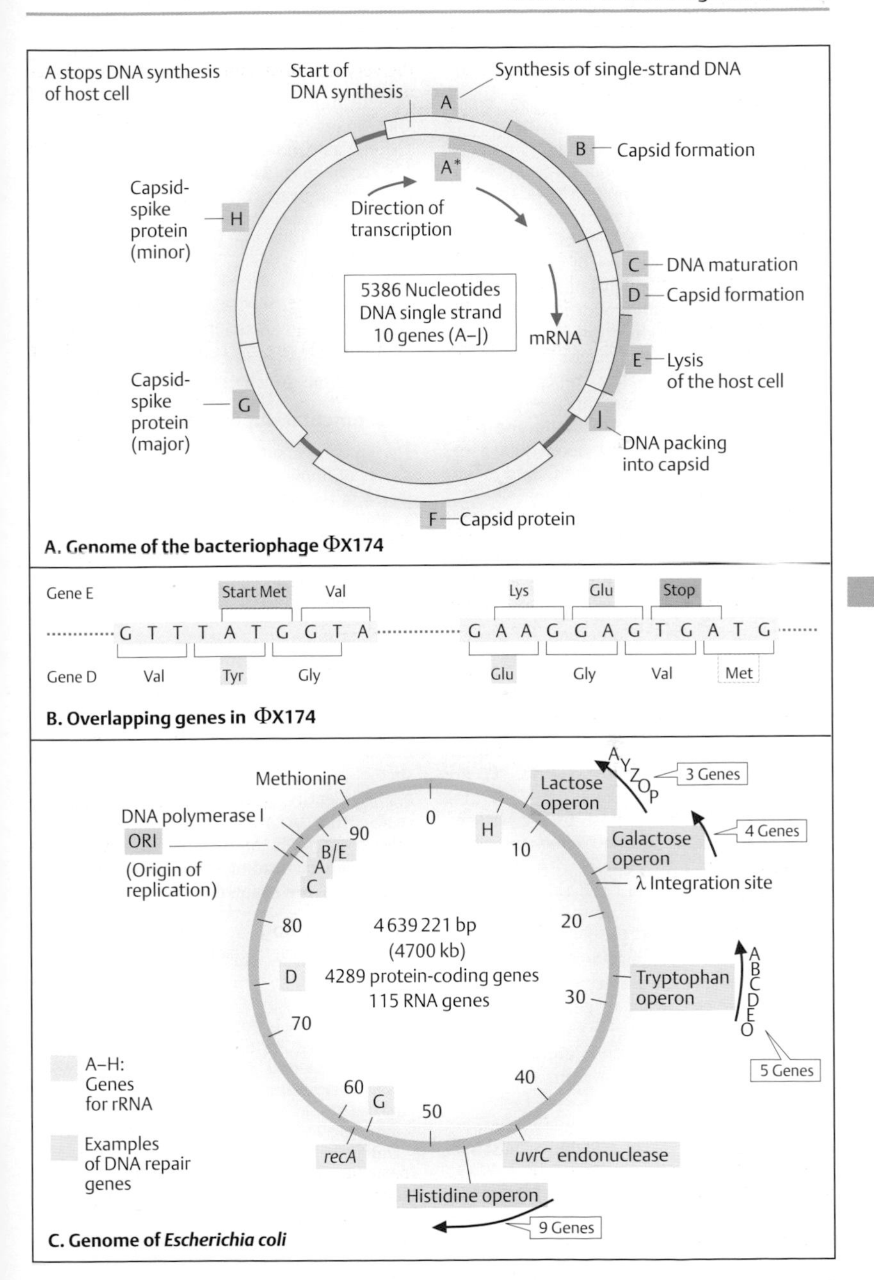

A. Genome of the bacteriophage ΦX174

B. Overlapping genes in ΦX174

C. Genome of *Escherichia coli*

Architecture of the Human Genome

The human genome contains 2.85 billion (2.85 Gb) nucleotides (approximately 3×10^9 bp). The about 22 000 protein-coding genes make up make up about 1.5% of the genome. About 6% of the human DNA sequence consists of conserved noncoding elements (CNEs). Basic features of the human genome are: a low average density of genes along a given stretch of DNA; many chromosomal segments that result from duplication events during evolution; a high proportion of interspersed, repetitive sequences; evidence of transposition events. The repeats derived from transposons arose during evolution mostly through RNA intermediates and make up about 40% of the genome. About 3000 genes encode a great variety of medium-sized to long regulatory RNAs.

A. Types of sequences

The types of sequences can be classified into sequences that are parts of genes or their regulatory parts (gene-related) and repetitive sequences outside genes (intergenic sequences). (Figure adapted from Brown, 2007.)

B. Interspersed repetitive DNA

Genome-wide repeats consist of mainly four types that are present at many sites. *Long interspersed repeat sequences* (LINEs, **1**) are mammalian retrotransposons that lack long terminal repeats in contrast to retroviruses. LINEs consist of repetitive sequences up to 6500 bp long that are adenine-rich at their 3′ ends. LINEs encode two open reading frames (ORF1 and ORF2), which are translated. In addition to a 5′ promoter (P) they have an internal promoter. Approximately 600 000 L1 elements are dispersed throughout the human genome. This can result in genetic disease if one is inserted into a gene (e.g., hemophilia A, see p. 254). LINE-2 and LINE-3 are inactive because reverse transcription from the 3′ end often fails to proceed to the 5′ end. LINE-1 accounts for 21% of the genome.

Short interspersed repeat sequences (SINEs, **2**) are repetitive segments of about 100–400 bp with tandem duplication of CG-rich segments separated by A-rich segments. They do not encode a protein and are not capable of autonomous insertion. The most abundant type of SINEs in the primate genome is the Alu family (Alu sequences) with about 1.2 million copies in humans (approximately 6% of the genome). One Alu repeat occurs about once every 3 kb in the human genome.

LTR retroposons (**3**) are flanked by long terminal direct repeats (LTRs) containing transcriptional regulatory elements. The autonomous retrotransposons contain *gag* and *pol* genes, which encode proteins required for retrotransposition.

DNA transposons (**4**) form different classes resembling bacterial transposons with inverted repeats and encoding a transposase. (Figure adapted from Strachan & Read, 2011.)

C. Segmental duplications

The human genome contains more than 1000 blocks of duplicated segments, which account for 5% of the genome. They are located mainly in the subtelomeric and pericentromeric regions or occur interspersed along each chromosome (segmental duplications).

They occur as blocks of 1–200 kb genomic sequences that are also present at another site in a chromosome (intrachromosomal) or another chromosome (interchromosomal). The three examples show segmental duplications present in the X chromosome and two autosomes: chromosome 20 and chromosome 4. Blocks shared with other chromosomes are shown as connecting lines. When their sequence identity exceeds 95% and their size exceeds 10 kb or more, unequal crossing-over may occur. This predisposes to duplication and deletion, leading to *genomic disorders* (p. 242). (Figure adapted from IHSG 2001.)

Further Reading

Brown TA. Genomes. 3rd ed. New York: Garland Science; 2007

Emanuel BS, Shaikh TH. Segmental duplications: an ‘expanding’ role in genomic instability and disease. Nat Rev Genet 2001;2:791–800

IHSG: International Human Sequencing Genome Consortium: Nature 2001;409:860–921

Kazazian HH Jr. Mobile DNA: Finding Treasure in Junk. Upper Saddle River: FT Press Science, Pearson Education; 2011

Nature Collections: Human Genome Collection. (www.nature.com/nature/supplements/collections/humangenome/)

Strachan T, Read AP. Human Molecular Genetics. 4th ed. New York: Garland Science; 2011

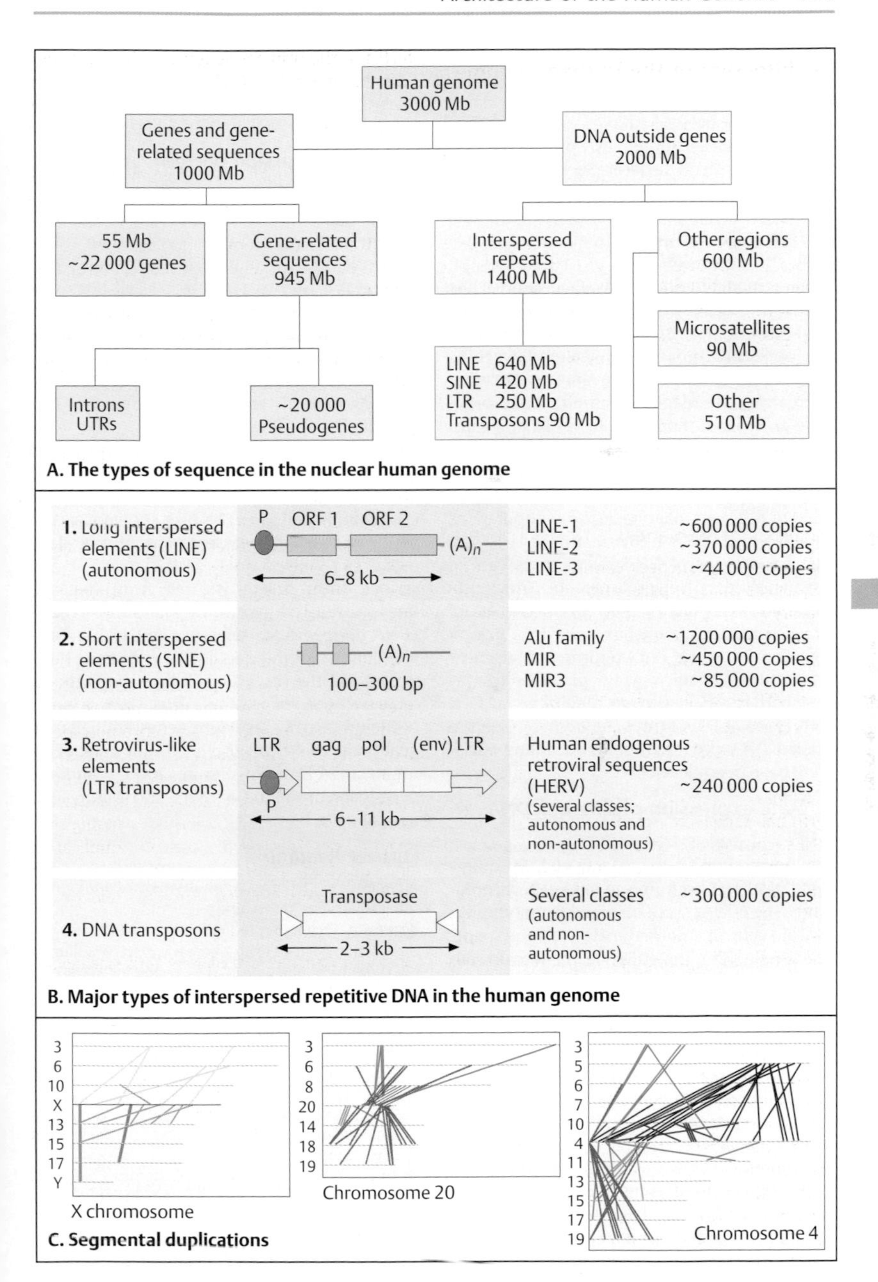

A. The types of sequence in the nuclear human genome

B. Major types of interspersed repetitive DNA in the human genome

C. Segmental duplications

Genome Analysis with DNA Microarrays

A microarray or DNA chip is an assembly of oligonucleotides or other DNA probes fixed on a small, fine grid of surfaces. Such high-density microarrays can be used to analyze complex RNA or DNA samples. Other specimens such as cDNA clones can also be used. With a microarray, the expression states of many genes can be examined simultaneously. One approach is to use complementary DNA (cDNA) prepared from mRNA (expression screening) or to recognize sequence variations in genes (screening for DNA variation). The advantages of using microarrays are manifold: simultaneous large-scale analysis of thousands of genes at a time, automation, small sample size, and easy handling. Several manufacturers offer highly efficient microarrays that contain more than 6 million different oligonucleotide populations on a $1.7\,cm^2$ ($1.28 \times 1.28\,cm$) surface.

Two basic types of DNA microarrays can be distinguished: (i) microarrays of previously prepared DNA clones or PCR products that are attached to the surface, arranged in a grid of high-density arrays in two-dimensional linear coordinates; (ii) microarrays of oligonucleotides synthesized in situ on a suitable surface. Both types of DNA arrays can be hybridized to labeled DNA probes in solution. Many variations have been developed.

A. Gene expression profile by cDNA array

This figure shows a microarray of 1500 different cDNAs derived from the human X chromosome. The cDNAs were obtained from lymphoblastoid cells of a normal male (XY) and a normal female (XX). The cDNAs of the female cells were labeled with the fluorophore Cy3 (red), and cDNAs of the male cells were labeled with Cy5 (green). The inactivation of most genes in one of the two X chromosomes in female cells leads to a 1:1 ratio of cDNAs from the expressed genes in the male and female X chromosomes. The result is a yellow signal at most sites because the superimposed red fluorescent (female) and green fluorescent (male) signals are present in equal amounts.

At seven sites, marked by yellow circles, the signal is red. Genes that have escaped inactivation on the inactive female X chromosome (see p. 240) are expressed at a double dosage. (Photograph kindly provided by Drs G. M. Wieczorek, U. Nuber, and H. H. Ropers, Max-Planck Institute for Molecular Genetics, Berlin.)

B. Gene expression patterns in human cancer cell lines

Microarrays can be used to analyze the pattern of gene expression in cancer cells, as an aid to diagnosis and the monitoring of therapy. The gene expression patterns in 60 cell lines derived from different human cancers are shown. Approximately 8000 genes have been analyzed by this procedure (Ross et al., 2000). A consistent relationship between gene expression patterns and tissue of origin was detectable.

A cell line dendrogram is shown, relating the patterns of gene expression to the tissue of origin of the cell lines as derived from 1161 cDNAs in 64 cell lines (**1**). Also shown is a colored microarray representation of the data using Cy3-labeled (red) cDNA reverse-transcribed from mRNA isolated from the cell lines compared with Cy5-labeled (green) cDNA derived from reference mRNA (**2**). Several clusters of red dots in the columns (1161 genes) and the rows (60 cell lines) indicate an increased gene expression. Thus, these regions of the microarray represent genes with altered gene expression patterns in tumor cells. (Figure adapted from Ross, et al., 2000, with kind permission from the authors and *Nature Genetics.*)

Further Reading

Affymetrix GeneChip. (www.affymetrix.com/jp/products_services/index.affx)

BeadArray Technology. A fundamentally different approach to high-density microarrays. (www.illumina.com/technology/beadarray_technology.ilmn)

Brown TA. Genomes. 3rd ed. New York: Garland Science; 2007

Gresham D, Dunham MJ, Botstein D. Comparing whole genomes using DNA microarrays. Nat Rev Genet 2008;9:291–302

Ross DT, et al. Systematic variation in gene expression patterns in human cancer cell lines. Nat Genet 2000;24:227–235

Strachan T, Read AP. Human Molecular Genetics. 4th ed. New York: Garland Science; 2011

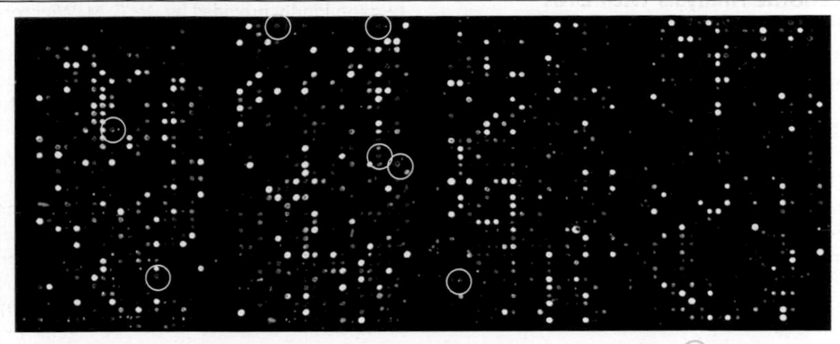

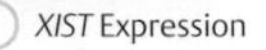

A. Gene expression profile by cDNA array

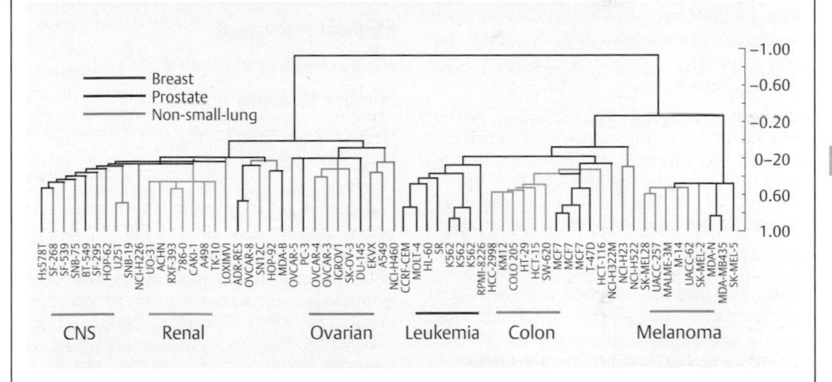

1. Cell line dentogram

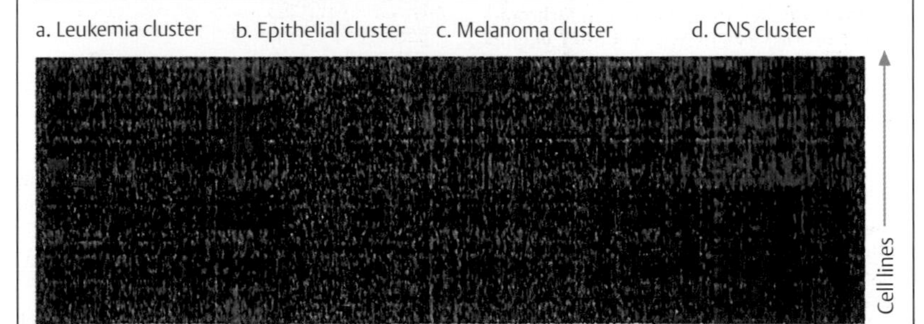

2. Microarray data

B. Gene expression patterns in human cancer cell lines

Genome Scan and Array CGH

A whole-genome scan and array-comparative genomic hybridization (CGH, see p. 230) aim at the analysis of the entire genome rather than selected groups of genes. Genome-wide studies have become a new approach in the analysis of linkage and association (see p. 128). A widely used whole genome scan is a single-nucleotide polymorphism (SNP) array. It detects polymorphic DNA sequence variants (SNPs) at multiple loci along all chromosomes in search for linkage or association with a region that might harbor gene loci involved in predisposing to or causing a disease. About 12 million variants have been identified in the human genome (SNP database). Array-comparative genomic hybridization (aCGH) combines the microarray methodology with genomic hybridization. The principle is illustrated schematically here.

A. Whole-genome scan

A whole-genome scan analyzes all chromosomes (only four are shown here schematically) for linkage of a disease locus with polymorphic markers along each chromosome (multipoint linkage analysis). A peak of the LOD score of nearly 4 over chromosome 2 indicates that linkage between a disease locus and two marker loci (shown in red and blue) may be present. A LOD score of 3 or above indicates a probability ratio for linkage/against linkage of 1000:1 or above.

More often, a genome-wide scan aims at detecting an association between polymorphic variants and a region containing susceptibility loci for complex diseases. An association will result in linkage disequilibrium (LD). LD is the result of a preferential segregation of certain alleles with a susceptibility region. Many different computer-based test procedures for analyzing linkage exist.

B. Array-comparative genomic hybridization (aCGH)

aCGH is a combination of microarray techniques and CGH (see p. 230). In aCGH, many mapped clones in a microarray are used instead of metaphase chromosomes. This greatly increases the detection rate for small deletions or duplications. The genome to be tested and a normal genome as reference are used as probes. They are differently labeled. In the example shown here, the DNA from an individual to be tested is labeled with the fluorophore Cy5, which induces green fluorescence. The DNA from another individual as control is labeled with Cy3, which induces red fluorescence.

If the amount of DNA in the test and the control sample is the same, a yellow signal will result. If there is an imbalance, the signal will be red if the amount of DNA is reduced due to a deletion. A green signal will indicate a duplication. Two sites with this type of imbalance are indicated by arrows. Array CGH is a fast and efficient means of detecting deletions and duplications on a genome-wide scale. (Figure kindly provided by Dr Evan E. Eichler, Seattle.)

Medical relevance

aCGH is widely used in diagnostics.

Further Reading

Eichler EE. Widening the spectrum of human genetic variation. Nat Genet 2006;38:9–11

Hinds DA, et al. Whole-genome patterns of common DNA variation in three human populations. Science 2005;307:1072–1079

International HapMap Project. (http://hapmap.ncbi.nlm.nih.gov/karyogram/gwas.html)

NCBI. SNP Database. Available at: http://www.ncbi.nlm.nih.gov/snp. Accessed January 27, 2012

Pinkel D, Segraves R, Sudar D, et al. High resolution analysis of DNA copy number variation using comparative genomic hybridization to microarrays. Nat Genet 1998;20:207–211

Thomas DC, Haile RW, Duggan D. Recent developments in genomewide association scans: a workshop summary and review. Am J Hum Genet 2005;77(3):337–345

Vissers LE, et al. Identification of disease genes by whole genome CGH arrays. Hum Mol Genet 2005;14 Spec No. 2:R215–R223

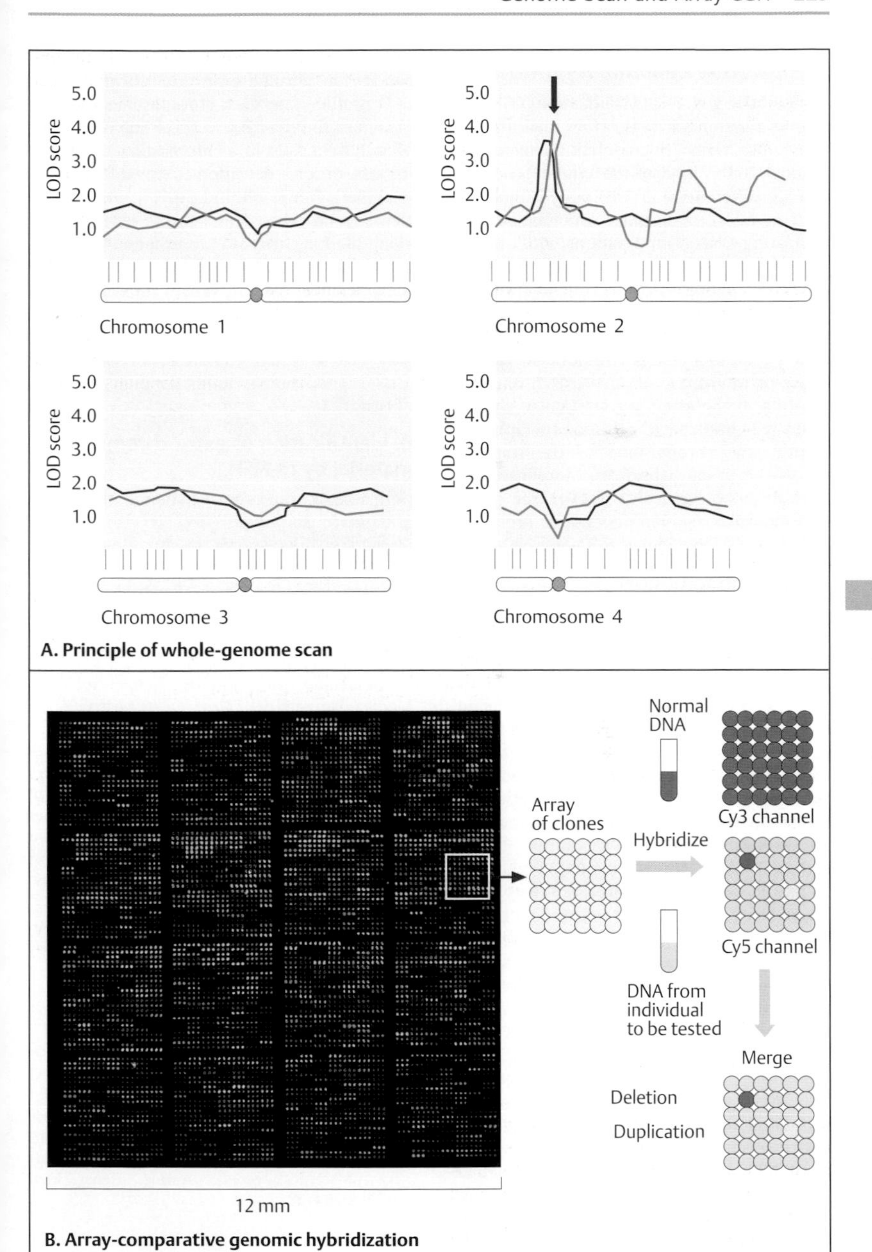

A. Principle of whole-genome scan

B. Array-comparative genomic hybridization

Comparative Genomic Hybridization

Comparative genomic hybridization (CGH) represents a systematic search for small deletions and duplications (microdeletions/microduplications) at the level of the whole genome. It detects differences in the copy number of DNA in all chromosomes simultaneously by comparing DNA from a patient with an unaffected control by competitive hybridization. It combines fluorescence in-situ hybridization and chromosome painting by using probes that hybridize to entire chromosomes. CGH is widely used to investigate tumor cells in metaphase or interphase. In forward chromosome painting, the probe for a particular chromosome is hybridized to a tumor metaphase to identify the chromosome of the tumor to which the probe hybridizes. Modifications of CGH are array-based CGH (aCGH, see p. 228) and multiplex ligation-dependent probe amplification (MLPA; Sellner & Taylor, 2004).

A. Application of CGH

CGH is useful to detect changes in the genome in tumor cells by comparing DNA isolated from a tumor sample with genomic DNA from a normal individual as a control. The figure shows a metaphase (top) from a patient with a hereditary form of renal cell carcinoma (von Hippel-Lindau syndrome; OMIM 193300). On the left, the metaphase is labeled green by fluorescein isothiocyanate (FITC), which in this case labels tumor DNA (**1**). In the middle the same metaphase is labeled red by rhodamine labeling control DNA (**2**). On the right (**3**) is an electronically blended composite of the fluorophore labels of green and red. Yellow fluorescence indicates a normal 1:1 green:red ratio without loss or gain of chromosomal material. Red fluorescence indicates chromosomal loss in the tumor, while green fluorescence indicates gain. The hybridization pattern is shown for all chromosomes in the CGH profile (**4**). The individual chromosomes are identified by a blue dye that stains AT-rich sequences (DAPI, 4′,6-diamidino-2-phenylindole), visualized by fluorescence microscopy. The metaphase is sequentially analyzed with a charge-coupled device (CCD) camera for red, green, and blue (DAPI counterstain) using different filter systems. The blue DAPI fluorescence is converted into a black-and-white image. This is enhanced to result in a banding pattern similar to G-bands for chromosome identification. The CGH profile scans each chromosome for a deviation to the red (first vertical line on the left of each field right to a chromosome) to search for loss, or for a deviation to green (third vertical line), which will indicate a gain. A deviation beyond the red line to the left is visible along the long arm of chromosome 4. This indicates chromosomal loss in 4q. A deviation to the green over the long arm of chromosome 10 indicates a gain of chromosomal material in 10q. The centromeres are shown as horizontal gray boxes. (Images courtesy of Drs Nicole McNeil and Thomas Ried, National Institutes of Health, USA.)

B. Identification of extra chromosomal material by M-FISH

Here, extra chromosomal material, invisible in a standard karyotype (left), is visualized by multicolor fluorescence in-situ hybridization (M-FISH, see p. 166). The multiplex FISH karyogram (right) shows a small extra band at the end of the long arm of a chromosome 1 (arrow). This extra band is derived from a chromosome 12. (Photographs kindly provided by Dr Sabine Uhrig and Dr Michael Speicher, Medical University of *Graz*, Graz, Austria.)

Further Reading

Chudoba I, et al. High resolution multicolor-banding: a new technique for refined FISH analysis of human chromosomes. Cytogenet Cell Genet 1999;84:156–160

NCBI. SKY/M-FISH & CGH Database. Available at: http://www.ncbi.nlm.nih.gov/projects/sky/. Accessed January 27, 2012

Sellner LN, Taylor GR. MLPA and MAPH: new techniques for detection of gene deletions. Hum Mutat 2004;23:413–419

Speicher MR, Carter NP. The new cytogenetics: blurring the boundaries with molecular biology. Nat Rev Genet 2005;6:782–792

Vissers LELM, et al. Identification of disease genes by whole genome CGH arrays. Hum Mol Genet 2005;14 Spec No. 2:R215–R223

Wong A, et al. Detection and calibration of microdeletions and microduplications by array-based comparative genomic hybridization and its applicability to clinical genetic testing. Genet Med 2005;7:264–271

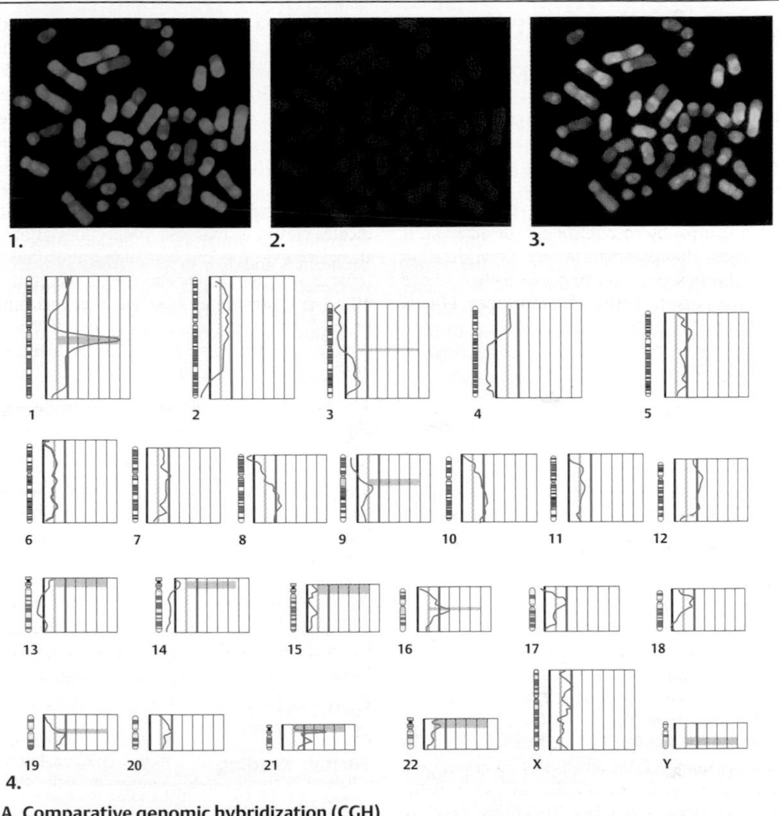

A. Comparative genomic hybridization (CGH)

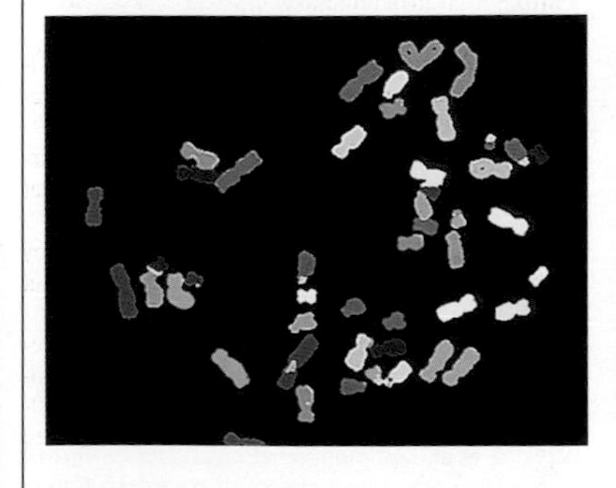

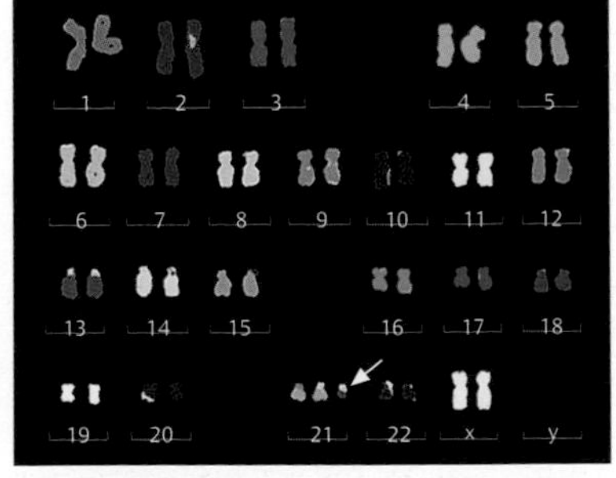

B. Identification of extra chromosomal material by M-FISH

Genome-Wide Association Study (GWAS)

GWAS is a whole-genome approach to search for chromosomal regions or known gene loci for an association with a disease or risk for disease. Hundreds of thousands of single-nucleotide polymorphisms (SNPs), the most common form of genomic variation, are tested in large populations of individuals affected and compared with large numbers of controls. In contrast to single gene disorders, the search by GWAS is aimed at complex diseases. Since the chromosomal location of each SNP is known, it is possible to relate its allelic variants to the trait under study. Nearly 600 GWASs involving 190 areas of investigation have been performed during the past 5 years.

About 12 million SNPs, each with a reference SNP number (rs), are recorded in the National Center for Biotechnology Information dbSNP database, with information on allelic forms, allele frequency, and other genomic information. Problems may arise from false-positive results in the huge number of statistical tests performed. High-throughput genotyping methods are employed to interrogate the entire human genome, unconstrained by prior hypotheses. GWA genotyping platforms on arrays or chips comprising 500 000 to 1 million SNPs are available.

A. Graphic representation of GWAS data

The association of SNPs with a chromosomal region or known genes is represented in a so-called Manhattan plot. This displays all human chromosomes with a graphic representation of the SNPs determined along each chromosome. It is shown typically as the $-\log_{10}$ of the P value (probability of the observed association arising by chance). Thus, $P = 0.01$ would be plotted as 2 or $P = 10^{-7}$ as 7 on the y-axis. In the example shown, 317 503 SNPs were studied in 1522 individuals with rheumatoid arthritis and 1850 controls. In addition to the major histocompatibility (MHC) region (here *HLA-DRb1*) on the short arm of chromosome 6 and the *PTPN22* gene on chromosome 1 known to be related to the immunological aspects of the disease, two additional genes with a possible etiological role were identified: *TRAF1* (encoding tumor necrosis factor receptor-associated factor-1) and *C5* (encoding complement component 5). (Figure adapted from Plenge et al., 2007.)

B. Identification of common variant in a complex disorder

In this example, a GWAS of two cohorts of 780 families (3101 subjects) and 1204 affected individuals with an autism spectrum disorder are compared with 6491 control subjects. Six SNPs between two genes, *CDH10* and *CDH9* (encoding neuronal cell-adhesion molecules cadherin 9 and 10), show a strong association of signals on the short arm of chromosome 5 (5p14.1) with a $-\log_{10}$ P value above 5 (**1**). Chromosome 5 is shown schematically (**2**) below the plot for all chromosomes. In (**3**), the enlarged region 5p14.1, the nucleotides 24 500 000–27 000 000, and the two genes and their 5′ to 3′ orientation are shown. SNPs are plotted with their P values (**4**). Genotyped SNPs are colored according to their correlation with one of the SNPs (rs4307059): red for above 0.5, yellow between 0.5 and 0.2, and white for below 0.2 (Figure adapted from Wang et al., 2009.)

Medical relevance

Examples disease gene or genome region identification by GWAS: Crohn disease, age-related macular degeneration, diabetes mellitus, neuropsychiatric diseases, Alzheimer disease, hypertension, and others.

Further Reading

Alkan C, Coe BP, Eichler EE. Genome structural variation discovery and genotyping. Nat Rev Genet 2011;12:363–376

NIH. Genome-Wide Association Studies (GWAS). (http://gwas.nih.gov/ and http://www.genome.gov/20019523)

Manolio TA. Genomewide association studies and assessment of the risk of disease. N Engl J Med 2010;363:166–176

NCBI. SNP Database. (http://www.ncbi.nlm.nih.gov/snp)

Pearson TA, Manolio TA. How to interpret a genome-wide association study. JAMA 2008;299:1335–1344

Plenge RM, et al. TRAF1-C5 as a risk locus for rheumatoid arthritis—a genomewide study. N Engl J Med 2007;357:1199–1209

Wang K, et al. Common genetic variants on 5p14.1 associate with autism spectrum disorders. Nature 2009;459:528–533

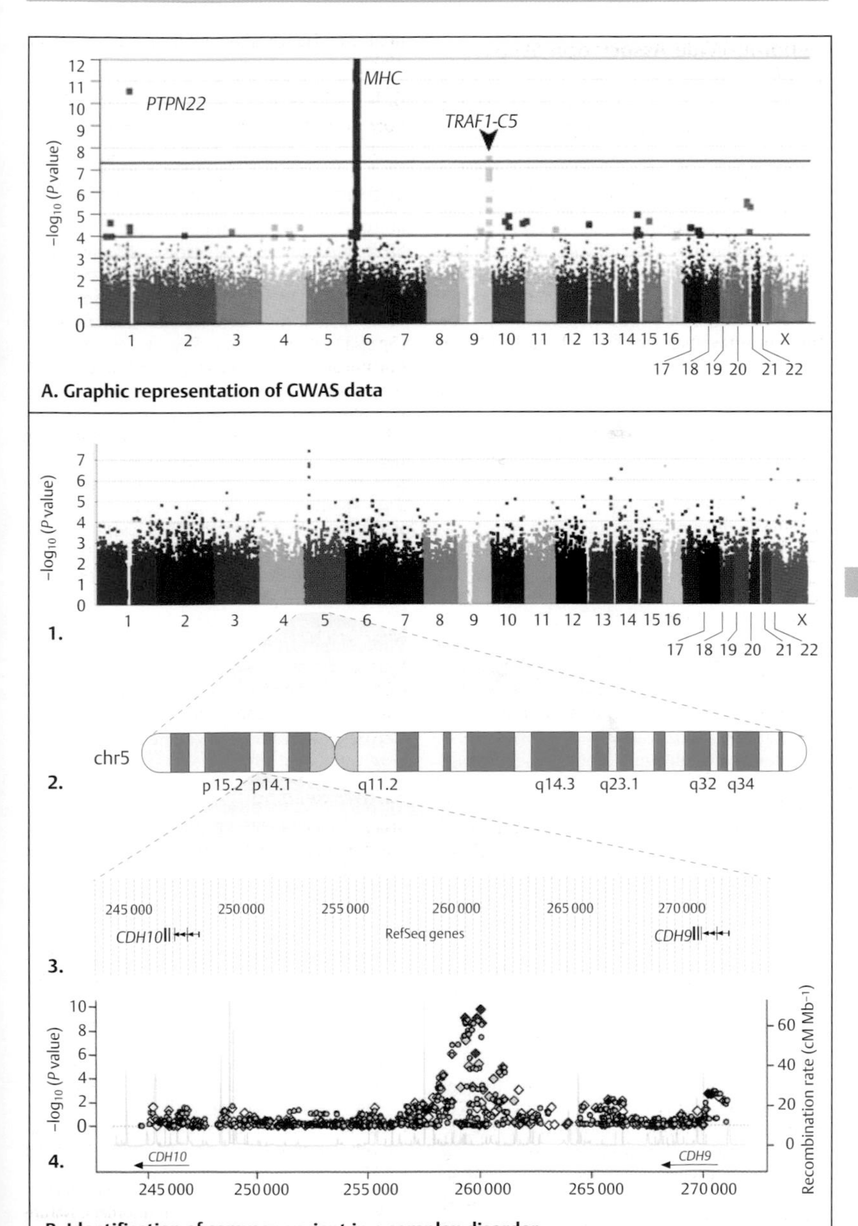

MHC
PTPN22
TRAF1-C5
−log10 (P value)
A. Graphic representation of GWAS data
chr5
p 15.2 p14.1
q11.2
q14.3
q23.1
q32 q34
CDH10
RefSeq genes
CDH9
Recombination rate (cM Mb−1)
B. Identification of common variant in a complex disorder

The Dynamic Genome: Mobile Genetic Elements

Mobile genetic elements are DNA sequences that are able to move from one site to another by transposition (p. 90). The term "dynamic genome" indicates that the genome of all living organisms is not static. Instead, it is flexible and subject to changes. This unusual phenomenon was first observed in the late 1940s by Barbara McClintock while investigating the genetics of Indian corn (maize, *Zea mays*). She found that certain genes were able to alter their position spontaneously. She named this phenomenon "jumping genes," and later *mobile genetic elements*. Although McClintock's observations were initially met with skepticism, she was awarded the Nobel Prize for this work in 1984 (Fox-Keller, 1983; McClintock, 1984). Transposable genetic elements or their remnants account for nearly half of the genome in mammals and up to 90% of the genome in some plants. Transposition provides mechanisms for genomes to acquire new sequences and rearrange existing ones during the course of evolution.

A. Stable and unstable mutations

McClintock (1953) determined that certain mutations in maize are unstable. A stable mutation at the C locus causes violet corn kernels (**1**) whereas unstable mutations cause fine pigment spots in individual kernels (variegation, **2**).

B. Effect of mutation and transposition

Normally a gene at the C locus produces a violet pigment of the aleurone in cells of Indian corn (**1**). This gene can be inactivated by insertion of a mobile element (*Ds*) into the gene, resulting in a colorless kernel (**2**). If *Ds* is removed by transposition, C-locus function is restored and small pigmented spots appear (**3**).

C. Insertion and removal of Ds

Activator-dissociation (*Ac/Ds*) is a system of controlling elements in maize. *Ac* is an inherently unstable autonomous element. It can activate another locus, dissociation (*Ds*), and cause a break in the chromosome (**1**). While *Ac* can move independently (*autonomous transposition*), *Ds* can move to another location in the chromosome only under the influence of *Ac* (*nonautonomous transposition*). The *Ac* locus is a 4.6-kb transposon; *Ds* is defective without a transposase gene (see p. 90). The C locus (**2**) is inactivated by the insertion of *Ds*. *Ds* can be removed under the influence of *Ac*. This restores normal function at the C locus.

D. Transposons in bacteria

Transposons are classified according to their effect and molecular structure: simple insertion sequences (IS) and the more complex transposons (Tn). A transposon contains additional genes, e.g., for antibiotic resistance in bacteria.

Transposition is a special type of recombination by which a DNA segment of about 750 bp to 10 kb is able to move from one position to another, either on the same or on another DNA molecule. The insertion occurs at an integration site (**1**) and requires a break (**2**) with subsequent integration (**3**). The sequences on either side of the integrated segment at the integration site are direct repeats. At both ends, each IS element or transposon carries inverted repeats whose lengths and base sequences are characteristic for different IS and Tn elements. (Photographs from Fedoroff, 1984.)

Medical relevance

Insertion of a transposon into a functional gene may be a rare cause of a genetic disease.

Further Reading

Fedoroff NV. Transposable genetic elements in maize. Sci Am June 1984;250:65–74

Fedoroff NV, Botstein D, eds. The Dynamic Genome: Barbara McClintock's Ideas in the Century of Genetics. New York: Cold Spring Harbor Laboratory Press; 1992

Fox-Keller E. A Feeling for the Organism: The Life and Work of Barbara McClintock. San Francisco: W. H. Freeman; 1983

Kazazian HH Jr. Mobile DNA: Finding Treasure in Junk. Upper Saddle River: FT Press Science, Pearson Education; 2011

Kazazian HH Jr. Mobile elements: drivers of genome evolution. Science 2004;303:1626–1632

Lewin B. Genes IX. Sudbury: Bartlett & Jones; 2008

McClintock B. Induction of Instability at Selected Loci in Maize. Genetics 1953;38:579–599

McClintock B. The significance of responses of the genome to challenge. Science 1984;226:792–801

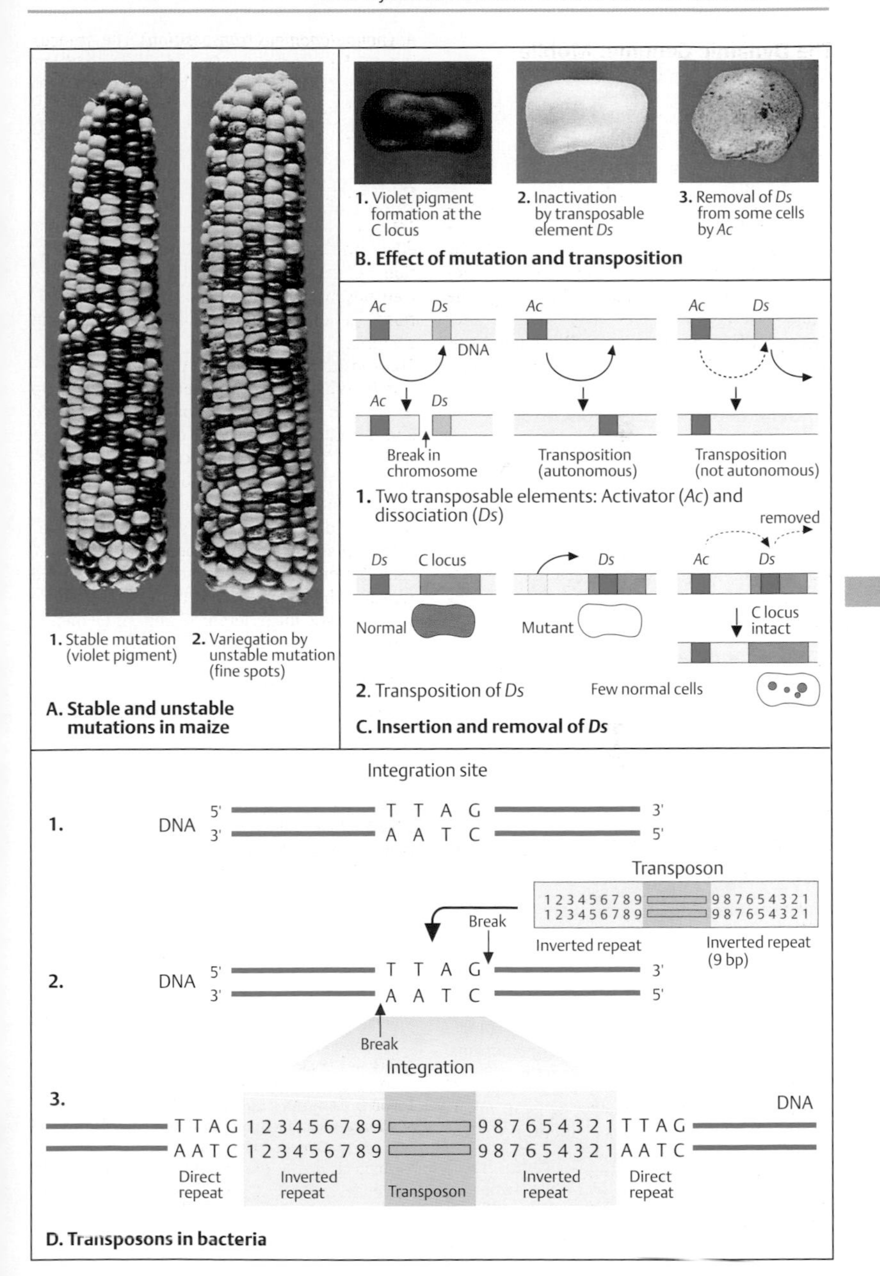

1. Stable mutation (violet pigment)
2. Variegation by unstable mutation (fine spots)
A. Stable and unstable mutations in maize
1. Violet pigment formation at the C locus
2. Inactivation by transposable element Ds
3. Removal of Ds from some cells by Ac
B. Effect of mutation and transposition
Ac
Ds
DNA
Break in chromosome
Transposition (autonomous)
Transposition (not autonomous)
1. Two transposable elements: Activator (Ac) and dissociation (Ds)
removed
C locus
Normal
Mutant
C locus intact
2. Transposition of Ds
Few normal cells
C. Insertion and removal of Ds
Integration site
1.
DNA
5'
3'
T T A G
A A T C
Transposon
123456789
987654321
Break
Inverted repeat
Inverted repeat (9 bp)
2.
Integration
3.
T T A G 1 2 3 4 5 6 7 8 9
9 8 7 6 5 4 3 2 1 T T A G
A A T C 1 2 3 4 5 6 7 8 9
9 8 7 6 5 4 3 2 1 A A T C
Direct repeat
Inverted repeat
Transposon
D. Transposons in bacteria

Evolution of Genes and Genomes

Genes and genomes existing today are the cumulative result of events that have taken place in the past. The classical theory of evolution, as formulated by Charles Darwin in 1859, states that (i) all living organisms today have descended from organisms living in the past; (ii) organisms that lived during earlier times differed from those living today; (iii) the changes were more or less gradual, with only small changes at a time; and (iv) the changes usually led to divergent organisms, with the number of ancestral types of organisms being smaller than the number of types today.

A. Gene evolution by duplication

Studies of the various genomes indicate that different types of duplication must have occurred: of individual genes or parts of genes (exons), subgenomic duplications, and rarely, duplications of the whole genome (Ohno, 1970). Duplication of a gene relieves the selective pressure on that gene. After a duplication event, the gene can accumulate mutations without compromising the original function, provided the duplicated gene has separate regulatory control. The duplication also can result in a genetic dead end. In this case, mutations inactivate the gene, leading to a pseudogene. Each human chromosome harbors almost as many pseudogenes as functional genes.

B. Gene evolution by exon shuffling

The exon/intron structure of eukaryotic genes provides great evolutionary versatility. New genes can be created by placing parts of existing genes into a new context, using functional properties in a new combination. This is referred to as exon or domain shuffling.

C. Evolution of chromosomes

Evolution also occurs by structural rearrangements of the genome at the chromosomal level. Related species, e.g., mammals, differ in the number of their chromosomes and chromosomal morphology, but not in the number of genes, which are often conserved to a remarkable degree. The human chromosome 2 appears to have evolved from the fusion of two primate chromosomes. The differences in chromosome 3 are much more subtle. The orang-utan chromosome 3 differs from that of man and other primates by a pericentric inversion. The banding patterns of all primate chromosomes are remarkably similar. This reflects their close evolutionary relationship. (Figure adapted from Yunis & Prakash, 1982.)

D. Molecular phylogenetics and evolutionary tree reconstruction

A phylogenetic tree can be reconstructed based on different types of evidence: fossils, differences in proteins, immunological data, DNA–DNA hybridization, and DNA sequence similarity. The number of events that have to have taken place to explain the diversity observed today is determined. In the path from an ancestral gene (**1**) two events are shown schematically. Two categories of homologs are distinguished: *paralogs* and *orthologs* (**2**). Paralogs are homologous genes that have evolved by duplication within a species. Orthologs are genes present in different organisms that have evolved by vertical descent from an ancestral gene between different species. The human α-globin and δ-globin loci are examples of paralogs. The α-globin and β-globin genes of humans and other mammals are examples of orthologs. The adjective *paralogous* refers to nucleotide sequence comparisons.

Further Reading

Brown TA. Genomes. 3rd ed. New York: Garland Science; 2007

Darwin C. The Origin of Species by Means of Natural Selection. London: John Murray; 1859

Eichler EE, Sankoff D. Structural dynamics of eukaryotic chromosome evolution. Science 2003;301: 793–797

Gilbert W. The exon theory of genes. Cold Spring Harb Symp Quant Biol 1987;52:901–905

Miller W, et al. Comparative genomics. Annu Rev Genomics Hum Genet 2004;5:15–56

Ohno S. Evolution by Gene Duplication. Heidelberg: Springer-Verlag; 1970

Rieseberg LH, Livingstone K. Evolution. Chromosomal speciation in primates. Science 2003;300:267–268

Strachan T, Read AP. Human Molecular Genetics. 4th ed. New York: Garland Science; 2011

Yunis JJ, Prakash O. The origin of man: a chromosomal pictorial legacy. Science 1982;215:1525–1530

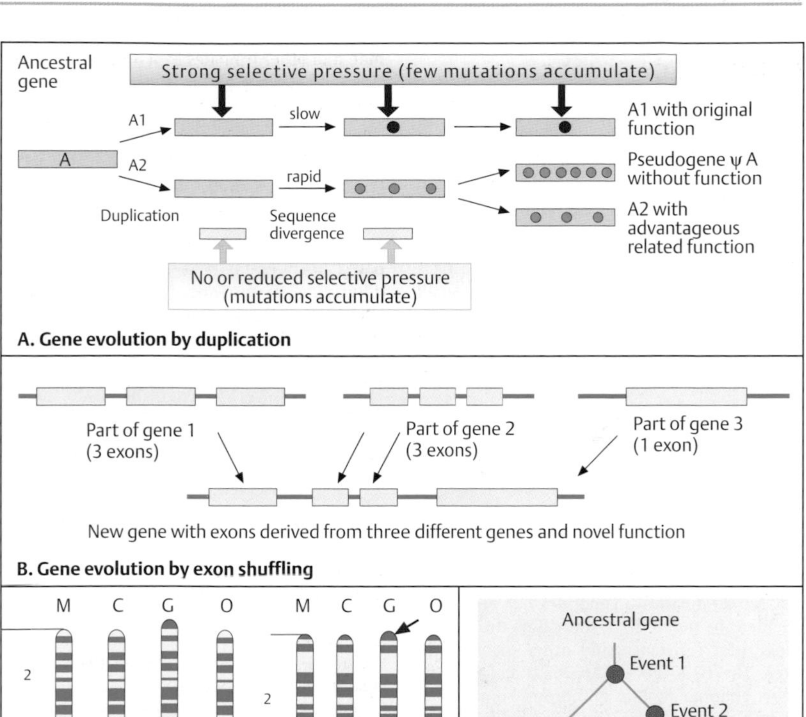

A. Gene evolution by duplication

B. Gene evolution by exon shuffling

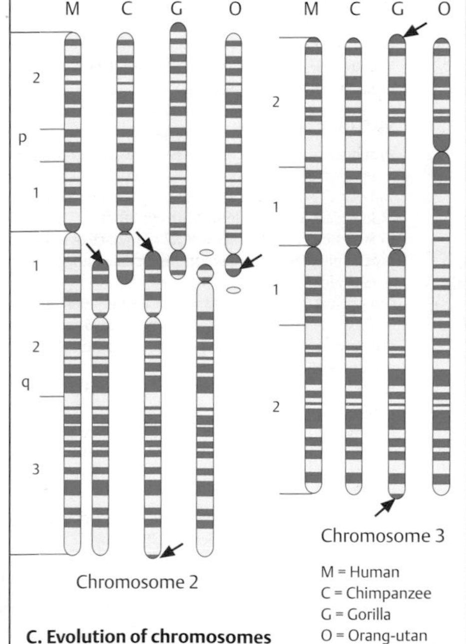

C. Evolution of chromosomes

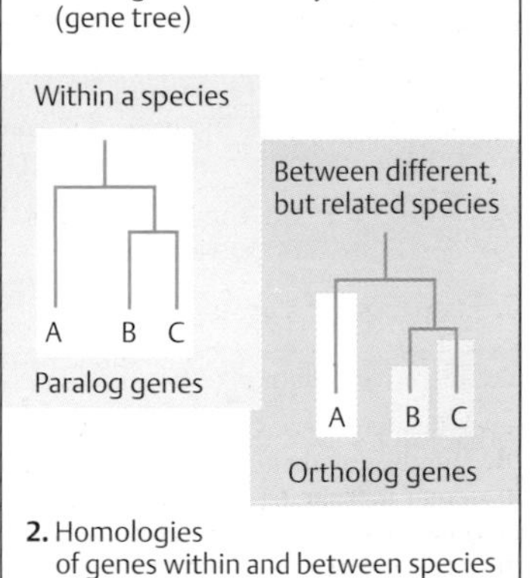

1. Three genes related by evolution (gene tree)

2. Homologies of genes within and between species

D. Molecular phylogenetics and evolutionary tree reconstruction

Comparative Genomics

Comparative genomics refers to a systematic analysis of genomes of different species by comparing their similarities and differences. Genomes of different organisms differ or resemble each other according to their evolutionary relationship. The number and types of differences depend on the time that has passed since their last common ancestor existed and the frequency of genomic changes and mutations that have occurred since they diverged from the ancestor. If two genomes are sufficiently closely related, the order of genes along each chromosome will have been preserved.

Two areas of comparative genomics have practical applications: plant genomes and the search for human disease genes that have homologs in other organisms. For example, wheat, one of the important food plants, has a huge genome of 16 000 Mb (five times larger than the human genome), whereas rice, perhaps more important, has a genome of 430 Mb. Comparative genomics has been used to identify genes that contribute to harvest yield, pest resistance, and other plant attributes. The 6200 genes (approximately) of the yeast genome contain numerous homologs to human disease genes. The same holds true for genes in the genomes of the nematode *Caenorhabditis elegans* and the fruit fly *Drosophila melanogaster.*

A. Homologies of human proteins

The sequence of the human genome has revealed considerable homologies of its proteins with those of other organisms, e.g., 21% with other eukaryotes and prokaryotes. (Figure adapted from IHGSC, 2001.)

B. Chromosome-associated proteins

The chromatin-associated proteins and transcription factors are conserved in evolution. As a consequence, 60% of human, *Drosophila*, and the nematode *C. elegans* chromatin architectures are shared. (Figure adapted from IHGSC, 2001).

C. Chromosomal segments conserved between man and mouse

As a result of evolution from a common ancestor about 80 million years ago, the human genome contains about 340 segments that are shared with the mouse genome. Within these segments of average length of 10 Mb, with a range from 24 kb to 90.5 Mb (IHGSC, 2001), the order of homologous genes is similar. The figure shows segments containing at least two genes in conserved order between man and mouse, indicated in a color corresponding to the mouse chromosome. (Figure adapted from IHGSC, 2001.)

Further Reading

Brown TA. Genomes. 3rd ed. New York: Garland Science; 2007

IHGSC: International Human Genome Sequencing Consortium. Initial sequencing and analysis of the human genome. Nature 2001;409:860–921

Margulies EH, Birney E. Approaches to comparative sequence analysis: towards a functional view of vertebrate genomes. Nat Rev Genet 2008;9:303–313

Rubin GM, et al. Comparative genomics of the eukaryotes. Science 2000;287:2204–2215

Strachan T, Read AP. Human Molecular Genetics. 4th ed. New York: Garland Science; 2011

Waterston RH, et al; Mouse Genome Sequencing Consortium. Initial sequencing and comparative analysis of the mouse genome. Nature 2002;420:520–562

Comparative genomics of four nonvertebrate organisms based on genome sequence data

Organism	Number of genes	Genes arisen by duplication	Gene families (core proteome)	Type of organism
Haemophilus influenzae	1709	284	1425	Bacterium
Saccharomyces cerevisiae	6241	1858	4383	Yeast
Caenorhabditis elegans	18 424	8971	9453	Nematode
Drosophila	13 601	5536	8065	Insect

Data from Rubin et al. (2000).

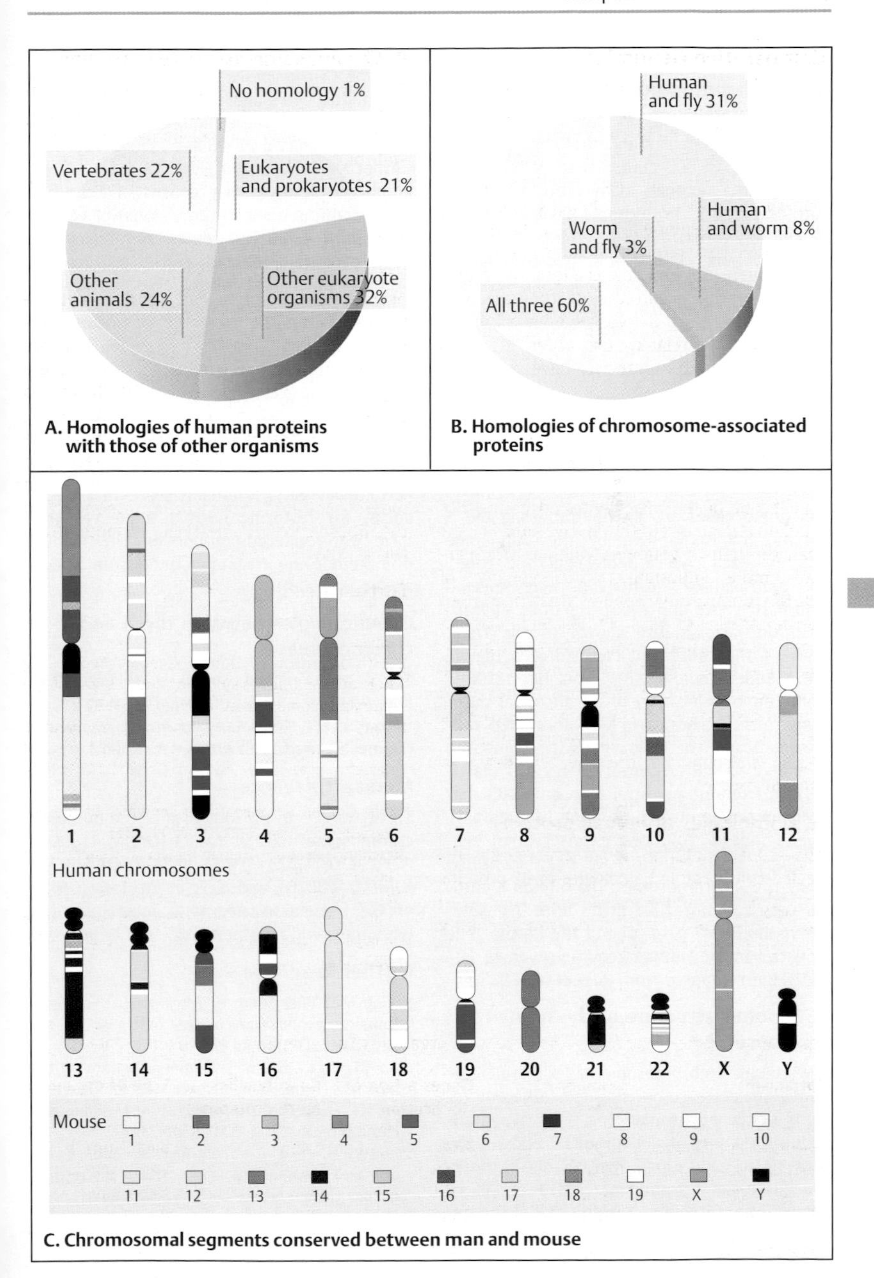

A. Homologies of human proteins with those of other organisms

B. Homologies of chromosome-associated proteins

C. Chromosomal segments conserved between man and mouse

Genomic Structure of the Human X and Y Chromosomes

The human X and Y chromosomes have evolved from a pair of ancestral chromosomes during the past 300 million years. While the X chromosome retained many properties of an autosome, the Y chromosome has lost most of its genes and become greatly reduced in size. Its genetic function is now limited to inducing male development during embryonic development and to maintaining spermatogenesis in adult males. The two chromosomes undergo pairing and recombination at the distal ends of their short arms in the *pseudoautosomal region* (PAR1), whereas all other regions are exempted from recombination.

A. Genomic structure of the human X chromosome

Functional genes are distributed along the X chromosome as shown by the dark blue squares (each square represents one gene). The approximate locations of nine selected landmark genes and their directions of transcription (arrows) are shown to the right. Most of the short arm consists of a region, the X-added region (XAR). This has resulted from the translocation of an ancestral autosome into the short arm (Xp) about 105 million years ago. The long arm (Xq) is composed of a region, XCR (X-conserved region) that is derived from an ancestral autosome. XCR has been conserved during evolution in mammals. Five regions of evolutionary conservation, evolutionary strata S1–S5, have been identified along the X chromosomes. The human X chromosome contains 1098 genes with 7.1 genes per million base pairs, one of the lowest gene densities in the human genome (average 10–13). (Figure adapted from Ross et al., 2005.)

B. Genomic structure of the human Y chromosome

The human Y chromosome has a distinct genomic structure comprising five different regions in the euchromatic part: (i) two pseudoautosomal regions at the distal ends of the short (PAR1) and long arms (PAR2); (ii) the Y-specific male determinant (MSY) region of about 35 kb; (iii) about 8.6 Mb (38% of the euchromatic portion) called *X-degenerate*, derived from an ancestral autosome; (iv) 3.4 Mb derived from former X-linked genes by transposition (*X-transposed*), which occurred about 3–4 million years ago; and (v) 10.2 Mb amplified (*amplionic*) Y-specific sequences. The latter are designated P1–P8 and are derived from former X- and Y-linked genes. They are termed amplionic because they consist of amplified palindromic sequences (amplicons) of various sizes with a marked sequence similarity of 99.9% over long stretches of DNA (tens to hundreds of kilobases). Most genes in the amplionic segments are expressed exclusively in testes.

Since the male-specific sequences on the Y chromosome do not participate in crossing-over, they are deprived of a mechanism for replacing mutations or structural rearrangements with normal sequences. Gene conversion between these palindromic sequences (Y-Y conversion) presumably serves as a mechanism for restoring normal sequences that have been rendered nonfunctional in one arm of a palindrome. (Figure adapted from Skaletsky et al., 2003.)

C. Homologies between the X and Y chromosomes

The X and Y chromosomes share regions of homology as a result of their common evolutionary origin (for details see Ross et al., 2005). (Figure adapted from Ross et al., 2005.)

Medical relevance

Three regions in the long arm of the human Y chromosome: AZFa at Yq1.21 (OMIM 415000), AZFb at Yq11.223 (OMIM 400003), DAZ11.223 (OMIM 400026), and AZFc at Yq11.23 (OMIM 400027) are associated with male infertility when deleted (azoospermia).

Further Reading

Jobling MA, Tyler-Smith C. The human Y chromosome: an evolutionary marker comes of age. Nat Rev Genet 2003;4:598–612

Lahn BT, Page DC. Four evolutionary strata on the human X chromosome. Science 1999;286:964–967

Ross MT, et al. The DNA sequence of the human X chromosome. Nature 2005;434:325–337

Skaletsky H, et al. The male-specific region of the human Y chromosome is a mosaic of discrete sequence classes. Nature 2003;423:825–837

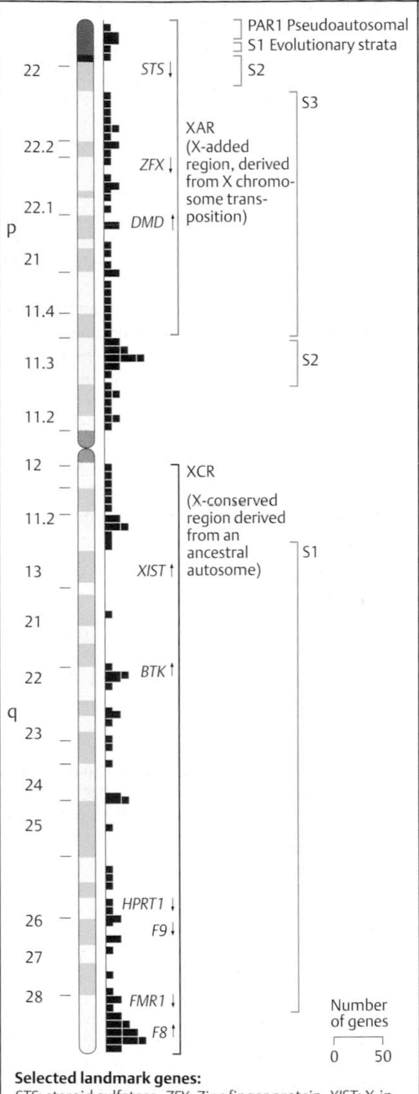

Selected landmark genes:
STS: steroid sulfatase, *ZFX*: Zinc finger protein, *XIST*: X-inactivation-specific transcript, *BTK*: Bruton agammaglobulinemia tyrosine kinase, *HPRT1*: Hypoxanthine-phosphoribosyltransferase 1, *F9* and *F8*: Hemophilia A and B, *FMR1*: Familial Mental Retardation type 1.

A. Genomic structure of the X chromosome

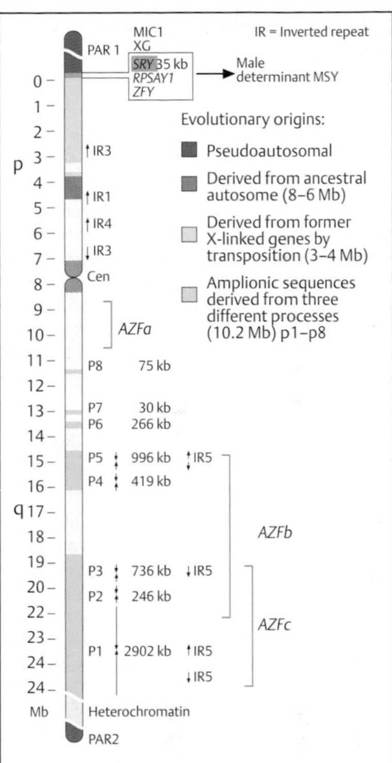

B. Genomic structure of the Y chromosome

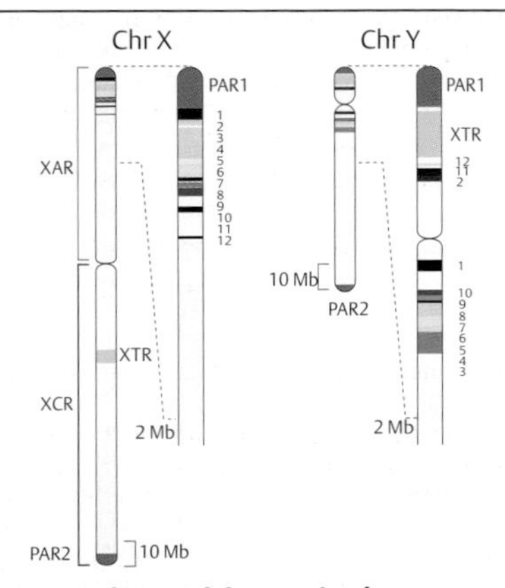

C. Homologies of the X and Y chromosome

Genomic Disorders

Genomic disorders are a group of phenotypically distinct disorders resulting from rearrangements of the underlying structure of the human genome rather than random events. Copy number variations (CNVs), such as segmental duplications, play an important role (p. 224). Their sequence identity predisposes to nonhomologous pairing during meiosis and subsequent nonhomologous nonallelic recombination (NAHR). The resulting duplication/deficiency in the daughter cells causes a local genomic imbalance, which leads to developmental disturbances and disease.

A. Peripheral neuropathy

Hereditary peripheral motor-sensoric neuropathy type 1A (HMSN1A; OMIM 188220; Charcot–Marie–Tooth disease, CMT1A) is the most frequent (1:2500–7500) neurogenic muscle atrophy. It begins in the first two decades of life and progresses with painful sensory perceptions, dissipation of muscle reflexes, and peripheral muscle atrophy of the feet, lower extremities, and hands. A reduced nerve conduction velocity of less than 38 meters per second establishes the diagnosis.

B. Duplication/deletion as cause of different neuropathies

The underlying causes of two clinically distinct neurogenic disorders are two low-copy repeat segmental duplications in the distal short arm of chromosome 17 flanking a gene (**1**). The proximal repeat and the distal repeat are separated by 1.4 Mb genomic DNA at 17p11.2. A tandem duplication involving the *PMP22* gene encoding peripheral myelin protein-22 (OMIM 601097) is the cause of CMT1A in about 50% of patients. This duplication occurs de novo in 75–90% of patients. Point mutations have also been observed. An allelic, reciprocal disorder is hereditary neuropathy with liability to pressure palsies (HNPP; OMIM 162500) due to deletion of *PMP22*.

The PMP22 protein is a component of myelinated peripheral nerve fibers.

C. Nonallelic homologous recombination

Segmental duplications at adjacent regions of the genome (**1**) predispose to nonhomologous pairing (**2**) during meiosis because of their high sequence homology. Nonallelic homologous recombination (NAHR) of two segmental duplications results in two reciprocal states of imbalance: a duplication (**3**) in one daughter cell and a deletion in the other daughter cell (**4**). In the case of CMT1A, the majority of crossovers occur within a limited region known as a recombination hotspot. Numerous other genomic disorders have been recognized.

Further Reading

Boone PM, Wiszniewski W, Lupski JR. Genomic medicine and neurological disease. Hum Genet 2011;130:103–121

Cooper GM, et al. A copy number variation morbidity map of developmental delay. Nat Genet 2011;43: 838–846

Girirajan S, Eichler EE. Phenotypic variability and genetic susceptibility to genomic disorders. Hum Mol Genet 2010;19(R2):R176–R187

Lupski JR, Stankiewicz P. Genomic Disorders. The Genomic Basis of Disease. Totowa, NJ: Humana Press; 2006

Lupski JR, et al. Whole-genome sequencing in a patient with Charcot-Marie-Tooth neuropathy. N Engl J Med 2010;362:1181–1191

Examples of other genomic disorders

Smith–Magenis syndrome (SMS; OMIM 182290) 17p11.2 deletion
Duplication syndrome (17)(p11.2p11.2) 17p11.2 duplication
Deletions causing neurofibromatosis type I (NF1; OMIM 162200) 17q11.2 deletion
Sotos syndrome (MIM 117550) 5q35 deletion
Many microdeletions and microduplications (p. 386)

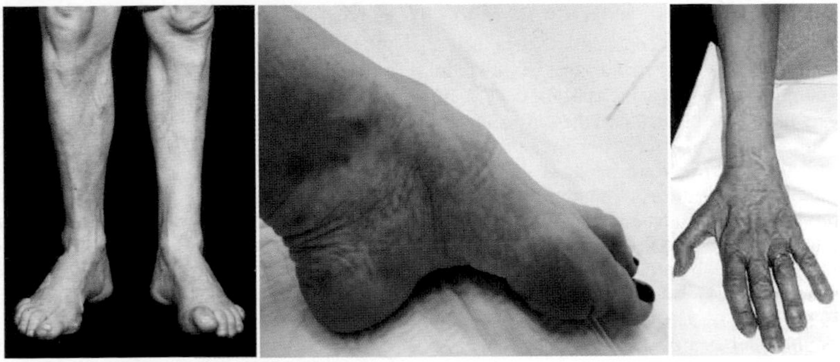

A. Peripheral neuropathy type 1A

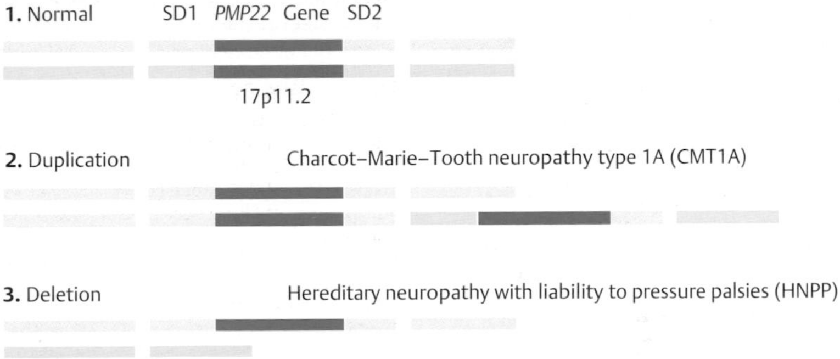

1. Normal SD1 *PMP22* Gene SD2

17p11.2

2. Duplication Charcot–Marie–Tooth neuropathy type 1A (CMT1A)

3. Deletion Hereditary neuropathy with liability to pressure palsies (HNPP)

B. Duplication/deletion as cause of different neuropathies

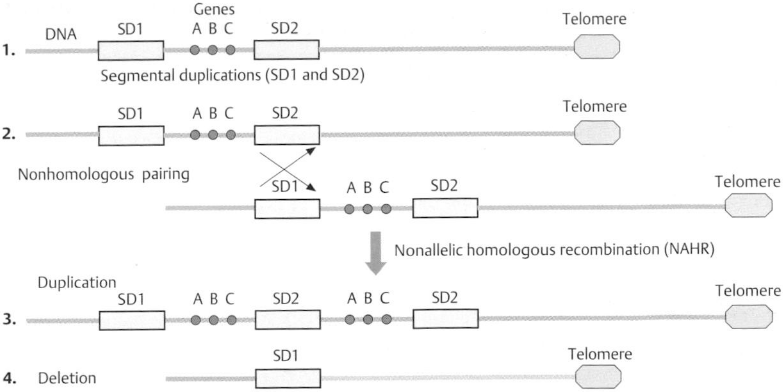

Genes

DNA SD1 A B C SD2 Telomere

1.

Segmental duplications (SD1 and SD2)

SD1 A B C SD2 Telomere

2.

Nonhomologous pairing

SD1 A B C SD2 Telomere

Nonallelic homologous recombination (NAHR)

Duplication

SD1 A B C SD2 A B C SD2 Telomere

3.

SD1 Telomere

4. Deletion

C. Nonallelic homologous recombination

The Mitochondrial Genome of Man

The human genome contains an extranuclear genome, derived from a transfer of a bacterial genome into a eukaryotic cell about 2 × 10^9 years before the present day, the mitochondrial genome. Each human cell contains hundreds of mitochondria. The mitochondrial genome in mammals is small and contains no introns, and in some regions the genes overlap, so that practically every base pair is part of a coding gene. In germ cells, mitochondria are almost exclusively present in oocytes, whereas spermatozoa contain few. Thus, mitochondria are inherited exclusively from the mother (maternal inheritance), without paternal contribution.

A. Mitochondrial genes in man

The size of the human mitochondrial genome, sequenced in 1981 by Andersen et al. is only 16.5 kb (16569 bp). Each mitochondrion contains 2–10 DNA molecules. In the two strands of the mitochondrial DNA (mtDNA) a heavy (H) strand and a light (L) single strand can be differentiated by a density gradient. Human mtDNA contains a total of 37 genes essential for intracellular energy transfer and conversion. Of these genes, 13 are protein-coding genes involved in four metabolic processes: (i) NADH dehydrogenase; (ii) the cytochrome *c* oxidase complex (subunits 1, 2, and 3); (iii) for cytochrome *b*; and (iv) for subunits 6 and 8 of the ATPase complex. Unlike that of yeast, mammalian mitochondrial DNA contains seven subunits for NADH dehydrogenase (ND1, ND2, ND3, ND4L, ND4, ND5, and ND6). Of the mitochondrial coding capacity, 60% is taken up by the seven subunits of NADH reductase (ND).

Most genes are found on the H strand. The L strand codes for a protein (ND subunit 6) and eight transfer RNAs (tRNAs). From the H strand, 14 tRNAs and two RNAs are transcribed: a short one for the 12S ribosomal RNA (rRNA) and a long one for 16S rRNA. A single transcript is made from the L strand. A 7S RNA is transcribed in a counterclockwise manner close to the origin of replication (ORI), located between 11 and 12 o'clock on the circular structure.

B. Cooperation between mitochondrial and nuclear genome

Many mitochondrial proteins are aggregates of gene products of nuclear and mitochondrial genes. About 1100 nuclear genes cooperate with the mitochondrial genome. In the mitochondria, they form functional proteins from subunits of mitochondrial and nuclear gene products. This explains why several mitochondrial genetic disorders show Mendelian inheritance, while purely mitochondrially determined disorders show exclusively maternal inheritance.

C. Evolutionary relationship of mitochondrial genomes

Mitochondria probably evolved from independent prokaryotic organisms that were integrated into eukaryotic cells.

Medical relevance

Numerous maternally transmitted and de novo diseases result from structural changes and point mutations in tDNA. They affect tissues with a high demand for energy, predominantly the brain, muscles, eye, kidney, pancreas, inner ear, metabolic systems, and other organ systems isolated or in combination (see Part III, Mitochondrial Diseases, p. 262 and Appendix p. 414).

Further Reading

Anderson S, et al. Sequence and organization of the human mitochondrial genome. Nature 1981;290: 457–465

Genetics Home Reference. Mitochondrial DNA (http://ghr.nlm.nih.gov/chromosome/MT)

United Mitochondrial Disease Foundation. Available at: http://ghr.nlm.nih.gov/chromosome/MT

Wallace DC. A mitochondrial paradigm of metabolic and degenerative diseases, aging, and cancer: a dawn for evolutionary medicine. Annu Rev Genet 2005;39:359–407

Wallace DC. Bioenergetics and the epigenome: interface between the environment and genes in common diseases. Dev Disabil Res Rev 2010;16:114–119

Wallace DC, Fan W. Energetics, epigenetics, mitochondrial genetics. Mitochondrion 2010;10:12–31

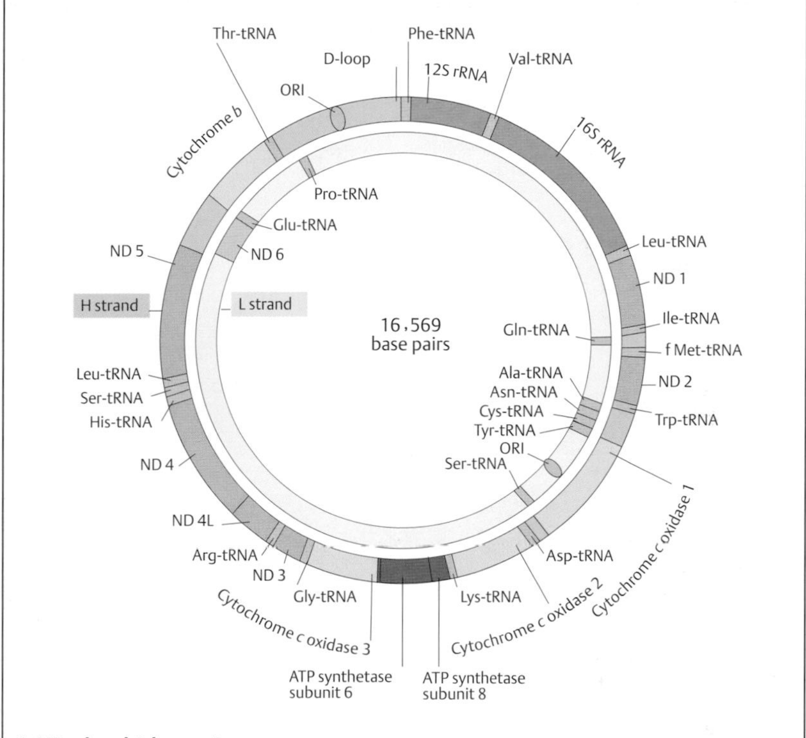

A. Mitochondrial genes in man

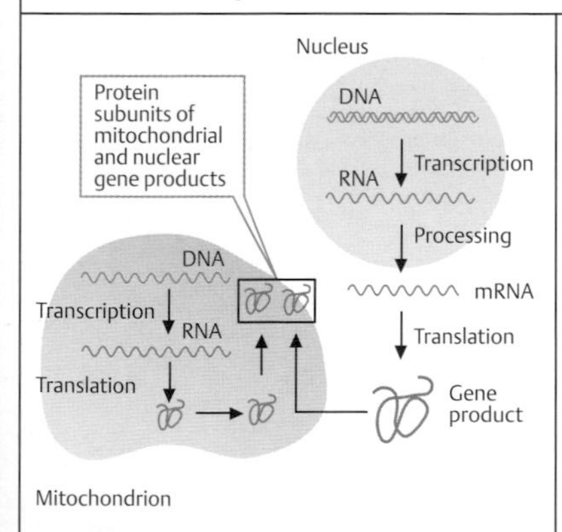

B. Cooperation between mitochondrial and nuclear genome

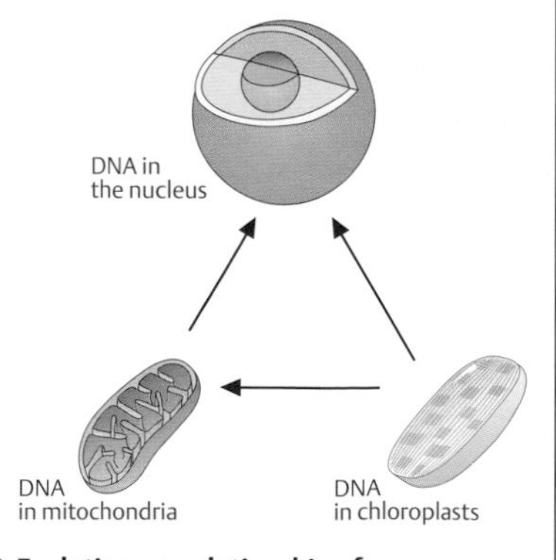

C. Evolutionary relationship of mitochondrial genomes

Genetics and Medicine

Chloride Channel Defects: Cystic Fibrosis

Cystic fibrosis (CF, mucoviscidosis; OMIM 219700) is a highly variable multisystem disorder caused by mutations in the cystic fibrosis transmembrane conduction regulator gene (*CFTR*; OMIM 602421). It is one of the most frequent autosomal recessive hereditary diseases in populations of European origin (about 1 in 2500 newborns). The high frequency of heterozygotes (1:25) is thought to result from a selective advantage in heterozygotes because reduced susceptability to epidemic diarrhea (cholera). Polymorphisms in other genes modify the severity of pulmonary manifestation. Examples of modifying genes are the *TGF1* gene on 19q13.1 (OMIM 190180) and *IFRD1* (OMIM 603502).

A. Cystic fibrosis: clinical aspects

The disease primarily affects the bronchial system and the gastrointestinal tract. Viscous mucus formation leads to frequent, recurrent lung and bronchial infections, resulting in chronic pulmonary insufficiency. The average life expectancy in typical CF is about 30 years. The disease may take a less severe, almost mild, course. Certain mutations occur in males with bilateral absence of the vas deferens (CBAVD; OMIM 277180).

B. Finding the gene for cystic fibrosis

The gene (cystic fibrosis transmembrane conduction regulator, *CFTR*) encodes a chloride channel protein. It was one of the first human genetic diseases identified by positional cloning. Linkage analysis first localized the gene to chromosome 7 at q31.2. A long-range restriction map of about 1500 kb was then narrowed down to 250 kb, and the gene was identified and characterized in a typical manner.

C. The *CFTR* gene and its protein

The large *CFTR* gene spans over 250 kb of genomic DNA, organized into 24 exons (exons 6 and 14 are respectively numbered 6a/6b and 14a/14b), which encode a 6.5-kb transcript with several alternatively spliced forms of mRNA. The protein of 1480 amino acids is a membrane-bound chloride ion channel regulator with five functional domains as shown. Nucleotide-binding domain 1 (NBD1) confers cAMP-regulated chloride channel activity. The most common mutation is a deletion of a phenylalanine codon in position 508 (ΔF508). The R domain contains putative sites for protein kinase A and protein kinase C phosphorylation. *CFTR* is widely expressed in epithelial cells.

The approximately 1900 mutations observed in the *CFTR* gene are classified according to: (i) abolished synthesis of full-length protein, (ii) block in protein processing, (iii) reduced chloride channel regulation, (iv) reduced chloride channel conductance, and (v) reduced amount of normal CFTR protein. The underlying genetic defects include missense mutations, nonsense mutations, RNA splicing mutations, and deletions. The most frequent mutation, ΔF508, accounts for about 66% of patients. In Europe, its contribution ranges from about 88% in the north to 50% in Italy (30% in Turkey). Other relatively frequent mutations are G542X (glycine substituted by a stop codon at position 542; 2.4%), G551D (1.6%), N1303K (1.3%), and W1282X (1.2%). In Ashkenazi Jewish populations, G542X accounts for 12% and G551D for 3% of all patients. To some extent, the type of mutation predicts the severity of the disease. The most severe forms are associated with homozygosity for ΔF508 and compound heterozygotes ΔF508/G551D and ΔF508/G542X. A polymorphic variant (5T) is found in about 40%–50% of patients with CBAVD or disseminated bronchiectasis.

Further Reading

Couzin-Frankel J. Genetics. The promise of a cure: 20 years and counting. Science 2009;324:1504–1507

Cystic Fibrosis Mutation Database. Available at: http://www.genet.sickkids.on.ca/app

Drumm ML, Konstan MW, Schluchter MD, et al; Gene Modifier Study Group. Genetic modifiers of lung disease in cystic fibrosis. N Engl J Med 2005;353:1443–1453

Gu YY, et al. Identification of IFRD1 as a modifier gene for cystic fibrosis lung disease. Nature 2009;458:1039–1042

O'Sullivan BP, Freedman SD. Cystic fibrosis. Lancet 2009;373:1891–1904

Wright FA, et al. Genome-wide association and linkage identify modifier loci of lung disease severity in cystic fibrosis at 11p13 and 20q13.2. Nat Genet 2011;43:539–546

Cystic fibrosis (mucoviscidosis)

Severe progressive disease of the bronchial system and gastro-intestinal tract

Disturbed function of a chloride ion channel by mutations in the *CFTR* gene

Autosomal recessive

Gene locus 7q31.2

Disease incidence approx. 1:2500

Heterozygote frequency approx. 1:25

Mutation ΔF508 in approx. 70%

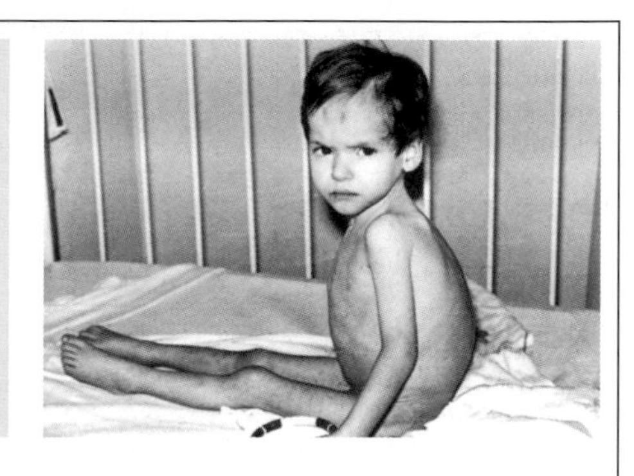

A. Cystic fibrosis: clinical aspects

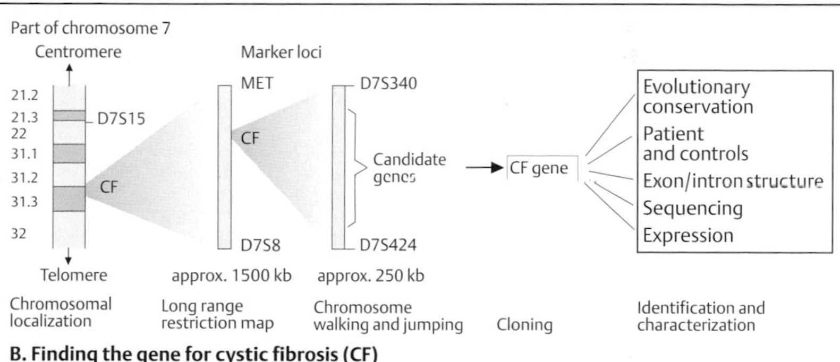

B. Finding the gene for cystic fibrosis (CF)

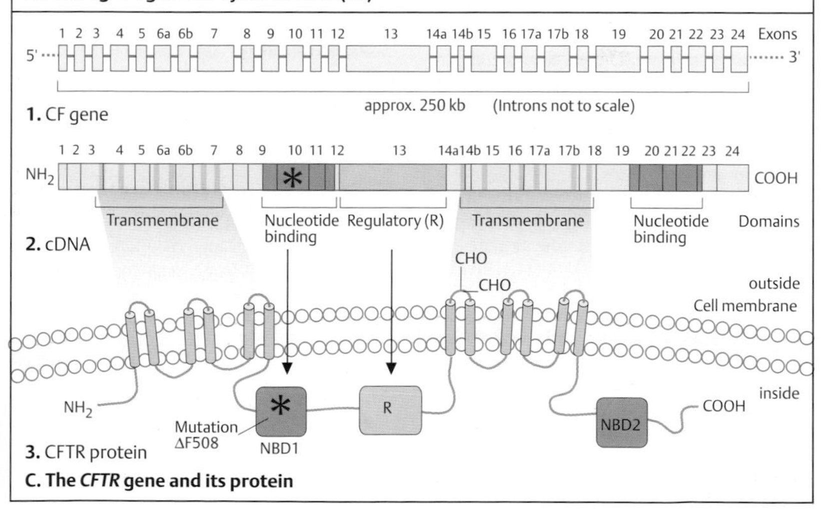

C. The *CFTR* gene and its protein

Genetic Defects in Ion Channels: LQT Syndromes

Mutations in genes encoding ion channels cause more than 30 different monogenic disorders (channelopathies). Ion channels play a prominent physiological role in the functioning of cardiac and striated muscle, neuromuscular junctions, and neurons of the central nervous system, inner ear, and retina. A striking example is the long QT (LQT) syndrome, which comprises 13 genetically different disorders: LQT syndrome types 1–13 (see OMIM 192500), caused by mutations in genes encoding different types of ion channels or their interaction partners. The main manifestation is cardiac arrhythmia leading to cardiac arrest. It is important to distinguish the different types, because the choice of medication differs. About 10% of patients carry a second mutation in the same or another ion channel gene.

A. Long QT syndrome, a genetic cardiac arrhythmia

LQT syndrome is characterized by a prolonged QT interval in the electrocardiogram (>440 milliseconds, corrected for heart rate). Other diagnostic features are T-wave morphology and torsade de pointes ventricular tachycardia (rapid heart beats).

B. Different molecular types of long QT syndrome

Prolongation of the QT interval in the electrocardiogram results from an increased duration of the cardiac action potential (**1**). In phase 0, the myocardiocyte is depolarized by sodium currents ($Na_V1.5$, encoded by the *SCN5A* gene). In plateau phases 1 and 2, the normal potential is supported by calcium currents ($Ca_V1.2$, *CACNA1C*). In phase 3, the resting membrane potential is restored by potassium currents (Kv7.1, *KCNQ1* and Kv11.1, *KCNH2*). LQT1 (**2**, caused by mutations in the *KCNQ1* gene encoding a potassium channel) accounts for about 30% of patients with long QT syndrome. LQT2 (**3**, mutations in the *KCNH2* gene [formerly *HERG*]) account for 10% of patients. *KCNH2* encodes an 1159-amino-acid transmembrane protein of the other major potassium channel that participates in phase 3 repolarization. LQT3 (**4**, *SCN5A*, the gene encoding the main sodium channel protein Na1.5) accounts for 3% of patients. It consists of four subunits (I–IV), each containing six transmembrane domains and several phosphate-binding sites. Other, rare, LQT syndromes are caused by increased sodium currents (LQT3, 9, 10, and 12), augmented calcium currents (LQT4 and 8) or decreased potassium currents (LQT1, 2, 5, 6, 7, 11, and 13).

Syndromic forms of LQT include: the autosomal recessive Jervell and Lange–Nielsen syndrome (OMIM 220400) associated with hearing defects, and caused by homozygous mutations in *KVLQT1* or *KCNE1* (LQT5) genes encoding the α and β subunits of the I_{KS} channel; the autosomal dominant Romano–Ward syndrome (OMIM 192500) caused by mutations in *KVLQT1*, the LQT1 gene; and other syndromic forms (see Appendix, Table 7, p. 413). (Figure adapted from Ackerman & Clapham, 1997.)

Further Reading

Ackerman MJ. Cardiac channelopathies: it's in the genes. Nat Med 2004;10(5):463–464

Ackerman MJ, Clapham DE. Ion channels—basic science and clinical disease. N Engl J Med 1997;336: 1575–1586

Beckmann BM, Pfeufer A, Kääb S. Inherited cardiac arrhythmias: diagnosis, treatment, and prevention. Dtsch Arztebl Int 2011;108:623–634, Quiz 634

Keating MT, Sanguinetti MC. Molecular and cellular mechanisms of cardiac arrhythmias. Cell 2004; 104:569–580

Marks AR. Arrhythmias of the heart: beyond ion channels. Nat Med 2003;9:263–264

Modell SM, Lehmann MH. The long QT syndrome family of cardiac ion channelopathies: a HuGE review. Genet Med 2006;8:143–155

Mohler PJ, et al. Ankyrin-B mutation causes type 4 long-QT cardiac arrhythmia and sudden cardiac death. Nature 2003;421:634–639

Roden DM. Long-QT syndrome. N Engl J Med 2008;358:169–176

Schwartz PJ, et al. Prevalence of the congenital long-QT syndrome. Circulation 2009;120:1761–1767

Sudden Arrhythmia Death Syndromes. Available at: http:(www.sads.org/). Accessed January 27, 2012

Prolonged QT interval in the electrocardiogram

Syncope

Sudden death

Autosomal dominant

13 genes involved (LQT1 - LQT8)

1. Main features

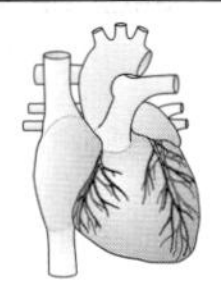

Prolonged QT

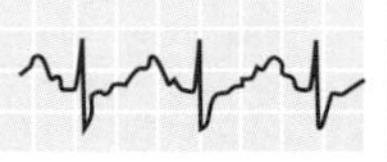

Torsade de pointes

2. Electrocardiogram

Long QT syndromes

Type	Locus	Gene
LQT1	11p15.5	*KCNQ1 (KVLQT1)*
LQT2	7q35–36	*HERG*
LQT3	3p21–24	*SCNA5*
LQT4	4q25–27	*Ankyrin-B*
LQT5	21q22.1	*KCNE1*
LQT6	21q21.1	*KCNE2*
LQT7	17q23	*KCNJ2*
LQT8	12p13.2	*CACNA1c*
LQT9	3p25	*Caveolin-3*
LQT10	11q23	*SCN4B*
LQT11	7p21–q22	*AKAP9*

3. Genetics

A. Long QT syndrome, a genetic cardiac arrhythmia

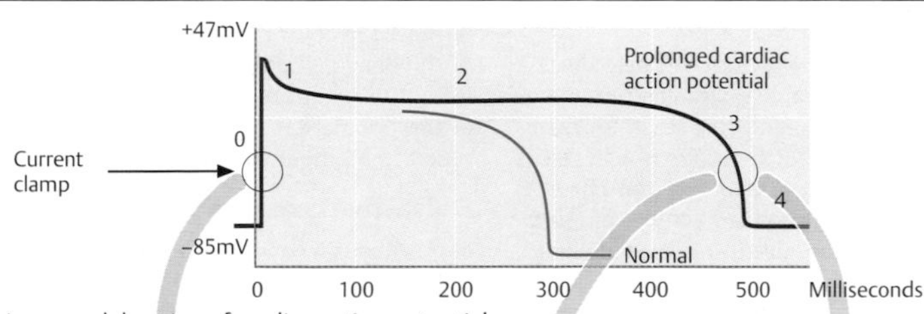

1. Increased duration of cardiac action potential

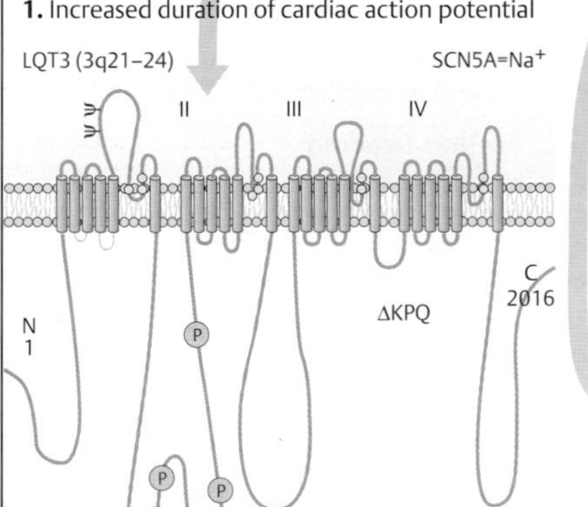

4. Na^+-channel fails to inactivate completely during phase 0

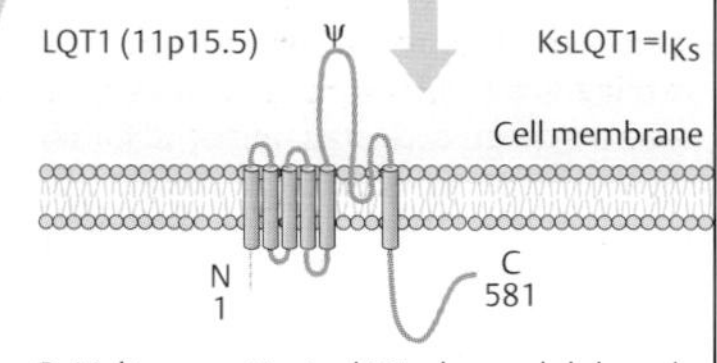

2. Voltage-activated K^+-channel delayed in phase 3

LQT2 (7q35–36) HERG=I_{Kr}

N 1 C 1159

3. Voltage-gated K^+-channel delayed in phase 3

B. Different molecular types of long QT syndrome

α_1-Antitrypsin Deficiency

α_1-Antitrypsin deficiency (AAT; OMIM 613490) is an autosomal recessive disorder of lung function. α_1-Antitrypsin is the main protease inhibitor in blood plasma, first described by Laurell & Ericksson (1963). As a member of the serine protease inhibitor superfamily (SERPIN) it has a major role in activating neutrophil elastase and other proteases to maintain protease/antiprotease balance. It binds to a wide range of proteases, such as elastase, trypsin, chemotrypsin, thrombin, and bacterial proteases. Its most important physiological effect is the inhibition of leukocyte elastase in the bronchial system.

A. α_1-Antitrypsin

Human α_1-antitrypsin is a 52-kDa glycoprotein composed of 394 amino acids, and 12% carbohydrate content.

B. α_1-Antitrypsin deficiency

α_1-Antitrypsin deficiency is a variable disease with chronic obstructive pulmonary emphysema as the main sign, and this becomes manifest by the third or fourth decade of life. This is visible by increased darkness of the chest X-ray. The cause is uninhibited activity of leukocyte elastase on the elastin of the pulmonary alveoli. The concentration of α_1-antitrypsin (α_1-AT) in alveolar fluid is greatly reduced in heterozygotes (red squares) and more so in homozygotes (yellow triangles). This can be corrected by intravenous administration of α_1-antitrypsin. A related, allelic disorder is α_1-antichemotrypsin deficiency (OMIM 107280).
Oxidizing substances have an inhibitory effect and inactivate the molecule. Smokers have a much more rapid course of α_1-antitrypsin deficiency disease (onset of dyspnea at 35 years of age instead of 45–50 years). (Figure kindly provided by N. Konietzko, Essen.)

C. α_1-Antitrypsin mutations

More than 100 allelic variants are known. The α_1-antitrypsin protein is highly polymorphic due to differences in the amino acid sequence and variations in the three carbohydrate side chains at positions 46, 83, and 247. The main clinically important mutations occur at the sites marked by red arrows. It is encoded by the *SERPINA1* gene (OMIM 107400), which spans 12.2 kb DNA with a 1434-bp coding region in five exons on chromosome 14 (14q32.13). The reactive site is located at position 358/359 (methionine/serine). The α_1-antitrypsin alleles are grouped into four classes: (i) normal, (ii) deficiency, (iii) null alleles, and (iv) dysfunctional alleles. The typical normal allele is Pi(M); the most important deficiency alleles are Pi(Z), Pi(P), and Pi(S).
The most frequent deficiency allele, Pi(Z), causes plasma concentrations of α_1-antitrypsin of about 12%–15% of normal in the homozygous genotype Pi(ZZ), and 64% in the heterozygote (Pi*MZ). MS heterozygotes have 86% of the MM homozygote activity.

D. Synthesis of α_1-antitrypsin

The gene encoding α_1-antitrypsin (α_1-AT) is expressed in liver cells (hepatocytes). The gene product is channeled through the Golgi apparatus and released from the cell (secreted). The Z mutation leads to aggregation of the enzyme in the liver cells, with too little of the enzyme being secreted. The S mutation leads to premature degradation. About 2%–4% of the population in central and northern Europe are MZ heterozygotes.

E. Reactive center of protease inhibitors

α_1-Antitrypsin (α_1-AT) is a member of a family of protease inhibitors that show marked homology, especially at their reactive centers. Oxidizing substances have an inhibitory effect and inactivate the molecule.
(Figures adapted from Cox, 2001.)

Further Reading

Cox DW. α_1-Antitrypsin deficiency. In: Scriver CR, et al., eds. The Metabolic and Molecular Bases of Inherited Disease. 8th ed. New York: McGraw-Hill; 2001: 5559–5584. (Online at www.ommbid.com/)

Laurell C-B, Eriksson S. The electrophoretic alpha-1-globulin pattern of serum in alpha-1-antitrypsin deficiency. Scand J Clin Lab Invest 1963;15:132–140

Siekmeier R. Lung deposition of inhaled alpha-1-proteinase inhibitor (alpha 1-PI) – problems and experience of alpha1-PI inhalation therapy in patients with hereditary alpha1-PI deficiency and cystic fibrosis. Eur J Med Res 2010;15(Suppl 2):164–174

Stoller JK, Aboussouan LS. α1-antitrypsin deficiency. Lancet 2005;365:2225–2236

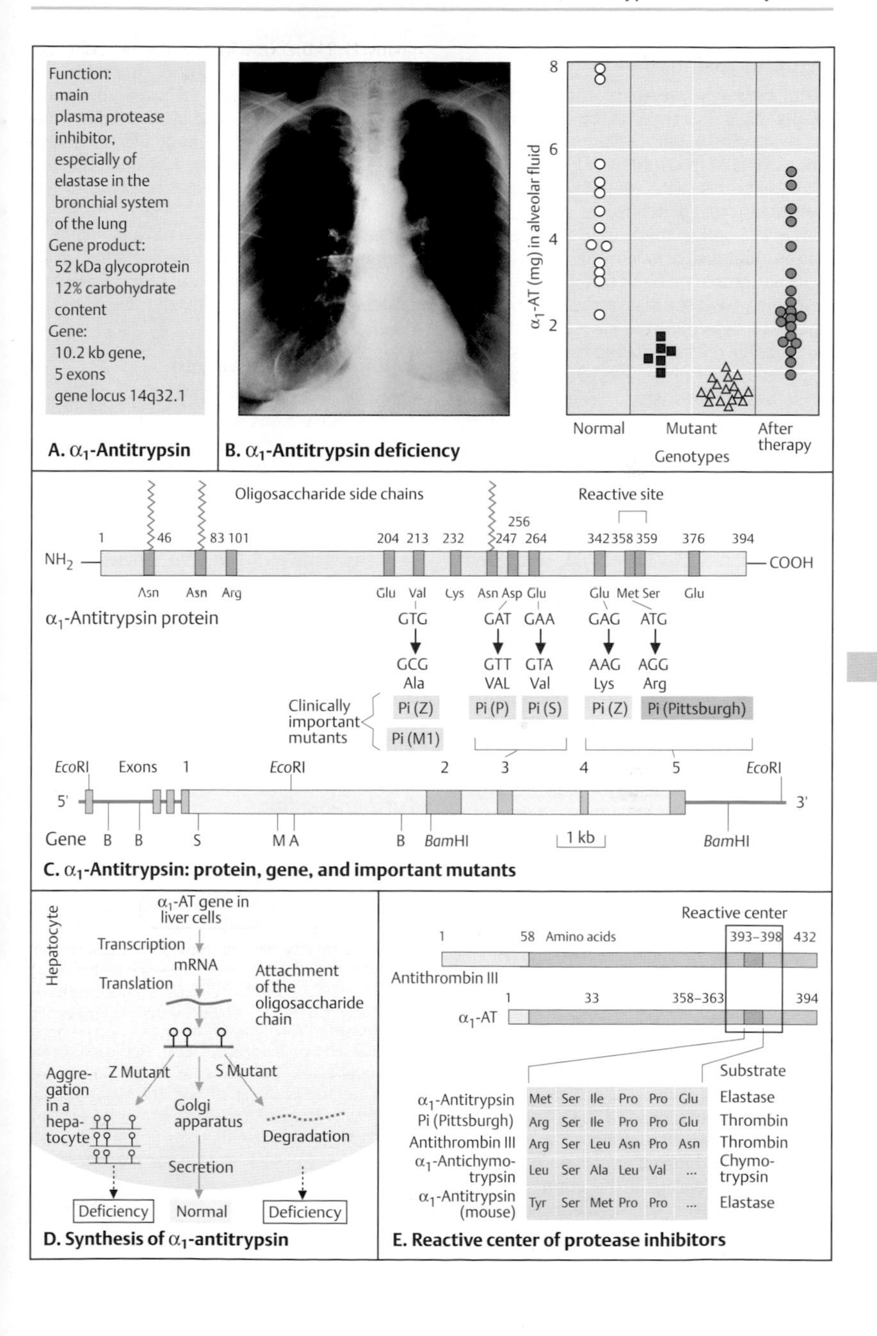

A. α_1-Antitrypsin

B. α_1-Antitrypsin deficiency

C. α_1-Antitrypsin: protein, gene, and important mutants

D. Synthesis of α_1-antitrypsin

E. Reactive center of protease inhibitors

Hemophilia A and B

Hemophilia A is a severe bleeding disorder resulting from inactivity of blood coagulation factor VIII due to different X-chromosomal mutations (OMIM 306700). A related disorder is hemophilia B (OMIM 306900) resulting from X-chromosomal deficiency of factor IX (F9). Factors VIII and IX function in the blood coagulation cascade. Factor VIII functions as a cofactor in the activation of factor X to factor Xa. Hemophilia was the first recognized major disease to be determined to have a genetic cause. The Talmud refers to its increased occurrence in males in certain families. The term hemophilia was introduced by F. Hopff (Würzburg, Germany) in a medical thesis in 1828.

A. Inheritance of hemophilia

The X-chromosomal inheritance is readily apparent in some royal families in Europe. Hemophilia A occurs with a frequency of about 1 in 10 000 males. About 30% of cases result from a new mutation. "Royal hemophilia" is now recognized to be hemophilia B (Rogaev et al., 2009).

B. Blood coagulation factor VIII

Activated factor VIII protein (by thrombin) consists of five subunits (A1, A2, A3, C1, C2) held together by calcium ions (Ca^{2+}; **1**). The inactive factor VIII protein (**2**) contains three domains (A, B, C). Domain A consists of three homologous copies (A1, A2, A3), domain C of two copies (C1, C2), and domain B of one copy. In humans, the *F8* gene (**3**) is located at Xq28, near the gene for factor IX. It consists of 26 exons and spans 186 kb of DNA. Noteworthy are the large exon 14 (3106 bp), which encodes the B domain, and the large intron 22 (32 000 bp) between exons 22 and 23. Most mutations involve the dinucleotide CG in DNA sequences TCGA. This mutates easily to TTGA because the cytosine of this dinucleotide is frequently methylated. Subsequent deamination of methylcytosine leads to a C to T transition. This creates a stop codon (TGA), resulting in a truncated factor VIII protein.

Genetic tests employ a targeted mutation search or sequencing of the entire coding region. RFLP (restriction fragment length polymorphisms) analysis has been utilized (**4**). A polymorphic variant in the recognition sequence for the restriction enzyme *Bcl*I near exons 17 and 18 produces fragments of 879 bp and 286 bp; when the variant is absent, a single fragment of 1165 bp is produced. In the family pedigree (**5**) two affected males (II-1 and III-2) carry the 879-bp fragment. Thus, this fragment indicates presence of the mutation.

Common types of mutation in hemophilia A are nonsense in 14% of patients, small deletions and insertions in 15%, splice mutations in 4%, and inversion flip tip mutations in 42% (see **D**).

C. Clinical manifestations

Frequent acute bleeding episodes (**1**) resulting from minor trauma will result in severe functional consequences such as stiff knee joints (**2**) or elbows, or extensive soft tissue hematoma. (Photographs obtained from: www.pathguy.com/lectures/ (**1**); and Gulnara Huseinova, Baku, Azerbaijan [http://azer.com/aiweb/categories/magazine/73_folder/73_articles/73_hemophilia.html] (**2**); accessed from Google.)

D. Factor VIII inversion

A frequent cause is an inversion between an unrelated gene A in intron 22 and one of two homologous distal genes A, followed by nonhomologous crossing-over. This flip tip inversion results in disruption of the *F8* gene between exon 22 and exon 23.

Further Reading

Dahlbäck B. Blood coagulation. Lancet 2000;355: 1627–1632

Gitschier J, et al. Detection and sequence of mutations in the factor VIII gene of haemophiliacs. Nature 1985;315:427–430

Graw J, et al. Haemophilia A: from mutation analysis to new therapies. Nat Rev Genet 2005;6:488–501

Hopff F. Über die Haemophilie oder die erbliche Anlage zu tötlichen Blutungen. Inaugural Dissertation, Universität Würzburg, 1828

Kazazian HH Jr, Tuddenham EGD, Antonaraksis SE. Hemophilia A: Deficiencies of coagulation factors VIII. In: Scriver CR, et al., eds. The Metabolic and Molecular Bases of Inherited Disease. 8th ed. New York: McGraw-Hill; 2001

National Hemophilia Foundation. (www.hemophilia.org/)

Rogaev EI, et al. Genotype analysis identifies the cause of the "royal disease". Science 2009;326:817

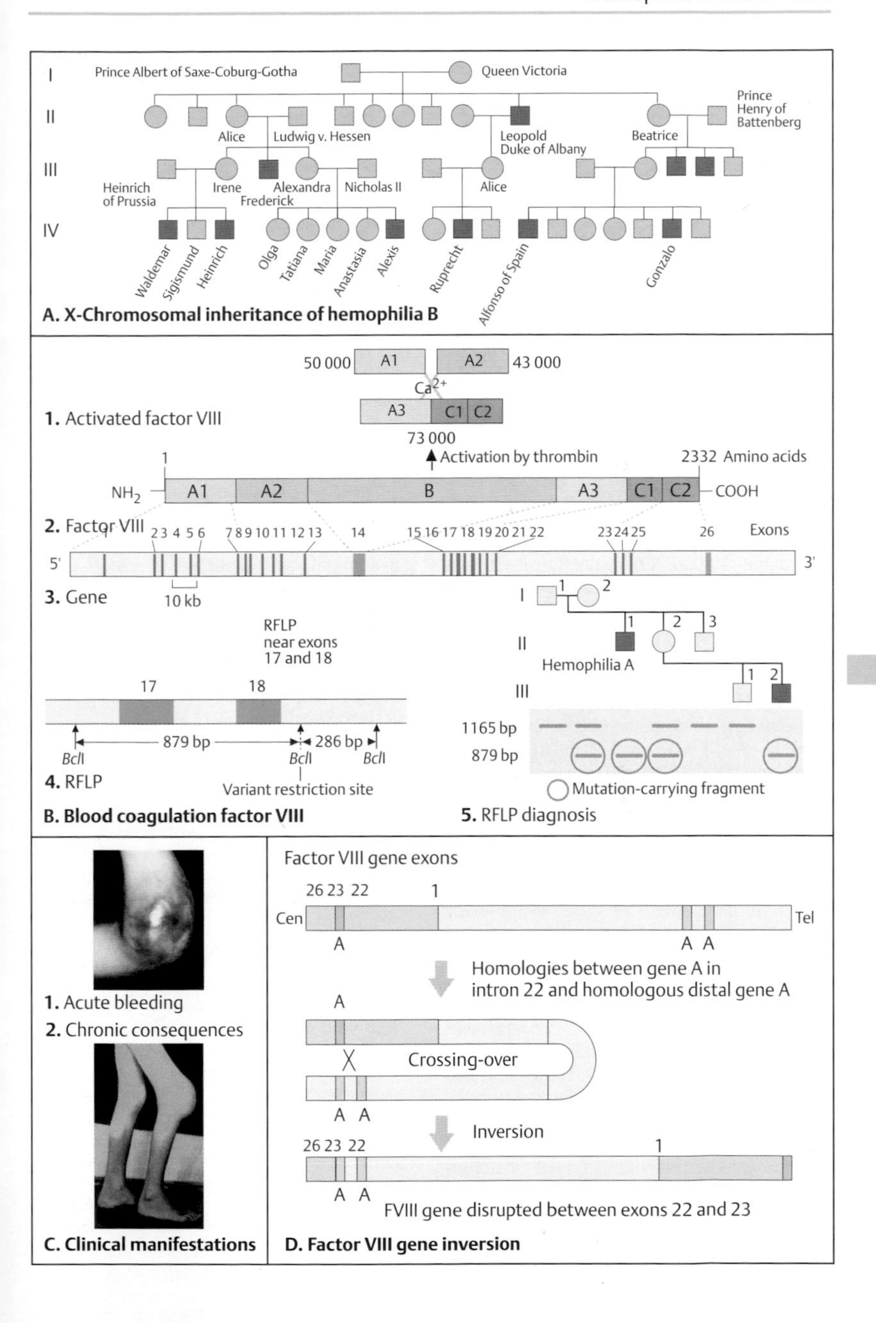
I
II
III
IV
Prince Albert of Saxe-Coburg-Gotha
Queen Victoria
Prince Henry of Battenberg
Alice
Ludwig v. Hessen
Leopold Duke of Albany
Beatrice
Heinrich of Prussia
Irene
Frederick
Alexandra
Nicholas II
Alice
Waldemar
Sigismund
Heinrich
Olga
Tatiana
Maria
Anastasia
Alexis
Ruprecht
Alfonso of Spain
Gonzalo
A. X-Chromosomal inheritance of hemophilia B
50 000
A1
A2
43 000
Ca^{2+}
A3
C1
C2
73 000
1. Activated factor VIII
1
Activation by thrombin
2332 Amino acids
NH_2
B
COOH
2. Factor VIII
1 2 3 4 5 6 7 8 9 10 11 12 13 14 15 16 17 18 19 20 21 22 23 24 25 26
Exons
5'
3'
3. Gene
10 kb
RFLP near exons 17 and 18
17
18
879 bp
286 bp
BclI
BclI
BclI
4. RFLP
Variant restriction site
Hemophilia A
1165 bp
879 bp
Mutation-carrying fragment
5. RFLP diagnosis
B. Blood coagulation factor VIII
1. Acute bleeding
2. Chronic consequences
C. Clinical manifestations
Factor VIII gene exons
26 23 22
Cen
Tel
A
Homologies between gene A in intron 22 and homologous distal gene A
Crossing-over
Inversion
FVIII gene disrupted between exons 22 and 23
D. Factor VIII gene inversion

von Willebrand Bleeding Disease

von Willebrand disease (OMIM 193400) is the most common hereditary bleeding disorder. It results from dysfunction of a complex, multimeric glycoprotein (von Willebrand factor, vWF) in blood plasma, thrombocytes (platelets), and subendothelial mesenchymal tissue of blood vessels. It is a heterogeneous group of disorders, also referred to as von Willebrand–Jürgens syndrome. vWF has two basic biological functions: (i) it binds as a major adhesion molecule to specific receptors on the surface of platelets and subendothelial connective tissue by forming bridges between platelets and damaged regions of a vessel; (ii) it binds to clotting factor VIII, and stabilizes it by prolonging its half-life in circulation. Deficiency of vWF reduces or abolishes platelet adhesion and leads to secondary deficiency of factor VIII. The frequency of hereditary deficiency of vWF is about 1:250 for all forms, and about 1:8000 for severe forms. It was first described in 1926 in a large family on the Åland islands in Finland by Erik von Willebrand.

A. von Willebrand factor glycoprotein

von Willebrand factor (vWF) is formed in endothelial cells, in megakaryocytes, and possibly in some other tissues. It is encoded by a large 178-kb gene with 52 exons of various sizes on chromosome 12p13.3 (OMIM 613160). Along its 8.7-kb cDNA several polymorphic restriction sites (red arrows) exist (**1**). From the mRNA, a primary peptide (prepro-vWF) of 2813 amino acids is translated (**2**). Prepro-vWF is a polypeptide of 2813 amino acids with a signal peptide of 22 amino acids and five repetitive functional domains, A–D and CK, which are distributed from the amino terminal in the sequence D1, D2, D3, A1, A2, A3, D4, B1, B2, B3, C1, C2, and CK. A segment of 741 amino acids in D1 and D2 corresponds to the von Willebrand antigen II. The various domains contain binding sites for factor VIII, heparin, collagen, thrombocytes, and thrombin. A tetrapeptide sequence Arg-Gly-Asp-Ser (RGDS) located near the C-terminus serves as a binding site for vWF. vWF contains 8.3% cysteine (234 of 2813 amino acids), concentrated at the amino and carboxy ends, whereas the three A domains are cysteine-poor (**3**). After posttranslational modification, the mature plasma vWF contains 12 oligosaccharide side chains (**4**), resulting in a carbohydrate weight proportion of 19% of the vWF molecule.

B. Maturation of vWF

A prepropeptide is translated from the vWF mRNA. After the signal peptide is removed in the endoplasmic reticulum, the two pro-vWF units attach to each other at their carboxy ends by means of numerous disulfide bridges to form a dimer. The dimers represent the repetitive units of mature vWF. The pro-vWF dimers are transported to the Golgi apparatus, where the pro-vWF antigen (von Willebrand antigen II [vWAgII]) is removed.

C. Classification

In types I and III, the defect is quantitative; in type II the defect is qualitative. Dominant and recessive (OMIM 277480) phenotypes with vWF deficiency are often difficult to distinguish. Type I with subtypes A and B is the most frequent group (70% of all patients). The bleeding time is prolonged, but coagulation time is normal. Bleeding occurs mainly into mucocutaneous tissues rather than joints.
(Figures adapted from Sadler, 2001.)

Further Reading

Israels SJ, et al. Inherited disorders of platelet function and challenges to diagnosis of mucocutaneous bleeding. Haemophilia 2010;16 Suppl 5:152–159

Konkle B. Disorders of platelets and vessel wall. In: Harrison's Principles of Internal Medicine. 18th ed. DL Longo et al, eds. New York: McGrawHill; 2012:965–973

NIH. The Diagnosis, Evaluation and Management of von Willebrand Disease. (www.guideline.gov/content)

Sadler JE. Von Willebrand disease. In: Scriver CR, et al., eds. The Metabolic and Molecular Bases of Inherited Disease. 8th ed. McGraw-Hill, New York, 2001:4415–4431

von Willebrand EA. Hereditär pseudohemofili. Fin Lakaresallsk Handl 1926;68:87–112

von Willebrand EA, Jürgens R. Über ein neues vererbbares Blutungsübel. Dtsch Arch Klin Med 1933;175:453–483

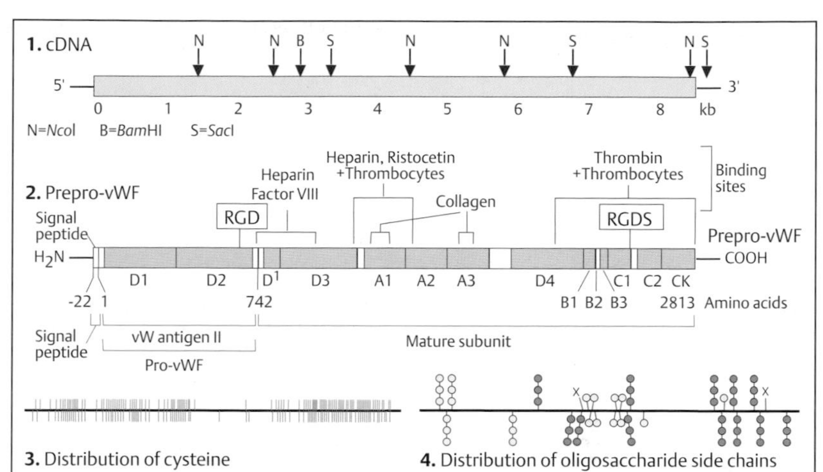

A. von Willebrand cDNA and prepropeptide

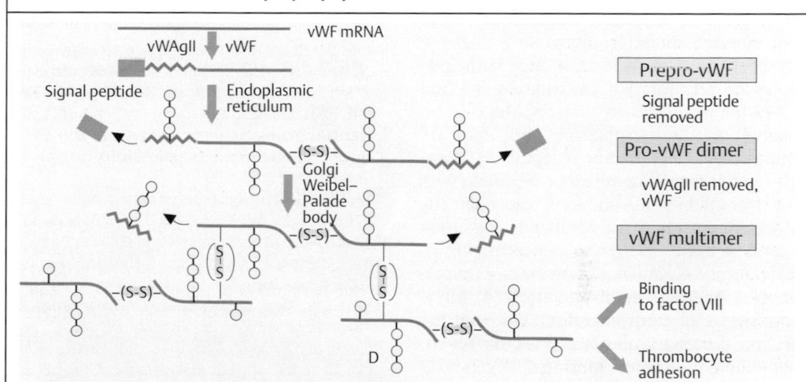

B. Biosynthesis of von Willebrand factors (vWF)

von Willebrand disease	Genetics	vWF antigen	Factor VIII	Multimer structure
Type I	AD	Decreased	Decreased	Normal
IIA	AD	Decreased or normal	Decreased or normal	Large and intermediary absent
IIB	AD	Decreased or normal	Decreased or normal	Large absent in plasma, normal thrombocytes
IIC	AR	Decreased or normal	Decreased or normal	Large absent in plasma and in thrombocytes
IID	AD	Normal	Normal	Large absent
IIE	AD	Decreased	Normal	Large absent
III	AR	Absent	Greatly decreased	Absent

C. Classification of von Willebrand disease

Pharmacogenetics

Multiple individual genetic variants of the genome result in individual differences in pharmacological responses. This may result in either ineffective therapy or adverse reactions, even death. *Pharmacogenetics*, a term introduced by Motulsky (1957) and Vogel (1959), refers to a blend of genetics and pharmacology, now extended to *pharmacogenomics*.

A. Malignant hyperthermia (MH)

Malignant hyperthermia (MH; OMIM 145600) is an autosomal dominant severe, life-threatening complication of anesthesia. It occurs in persons who are hypersensitive to halothane and similar agents used in general anesthesia. Normally a nerve impulse depolarizes the plasma membrane of a nerve ending at the nerve–muscle endplate (**1**). The influx of calcium (Ca^{2+}) into the cell triggers the release of acetylcholine. This temporarily opens the receptor-controlled cation (sodium, Na^{+}) channels. This opens calcium channels located in the sarcoplasmic reticulum of the muscle cell and causes the myofibrils to contract. The calcium channels in the sarcoplasmic reticulum are regulated by the ryanodine receptor (**2**). Mutations in the ryanodine receptor, a protein with four transmembrane domains, cause greatly increased sensitivity to halothane and other anesthetic agents (**3**). They induce muscle rigidity, drastic elevation of temperature (hyperthermia), acidosis, and cardiac arrest (**4**). MH is inherited as an autosomal dominant trait (**5**). MH constitutes a group of at least six genetically different disorders MH1–6. The genes responsible for these disorders are listed in OMIM 145600 under Allelic Series.

B. Butyrylcholinesterase deficiency

Prolonged apnea resulting from excessive muscle relaxation after administration of suxamethonium (succinylcholine) is the main manifestation of this adverse pharmacogenetic reaction. It affects about 1 in 200 individuals, and 1 in 100000 are affected by the severe form. Affected individuals do not degrade suxamethonium because of insufficient activity of serum butyrylcholinesterase (formerly acetylcholinesterase; OMIM 177400). This enzyme hydrolyzes butyrylcholine more readily than acetylcholine (OMIM 100740). Individuals at risk cannot be identified by enzyme activity alone (**1**). However, all three genotypes can be determined after dibucaine, an inhibitory substance, is added (**2**). Individuals at risk (red squares) have only 20% activity of butyrylcholinesterase, heterozygotes have 50%–70%, and homozygous normal persons about 80%. Butyrylcholinesterase is encoded by the *BChE* gene located at 3q26.1. (Figure adapted from Harris, 1975.)

C. Examples of genetically determined adverse reactions to pharmaceuticals

Numerous human pharmacogenetic disorders are known and several examples are shown.

Further Reading

Evans WE, McLeod HL. Pharmacogenomics—drug disposition, drug targets, and side effects. N Engl J Med 2003;348:538–549

Harris H. The Principles of Biochemical Genetics. 2nd ed. Amsterdam: Holland Publishing Company; 1975

Kalow W, Grant DM. Pharmacogenetics. In: Scriver CR, et al., eds. The Metabolic and Molecular Bases of Inherited Disease. 8th ed. New York: McGraw-Hill; 2001:225–255

Meyer UA. Pharmacogenetics - five decades of therapeutic lessons from genetic diversity. Nat Rev Genet 2004;5:669–676

Motulsky AG. Drug reactions enzymes, and biochemical genetics. J Am Med Assoc 1957;165:835–837

Nebert DW. Pharmacogenetics and pharmacogenomics. Nature Encyclopedia Human Genome 2003;4:558–567

Nebert DW, Jorge-Nebert LF. Pharmacogenetics and Pharmacogenomics. In: Rimoin DL, et al., eds. Emery and Rimoin's Principles and Practice of Medial Genetics. 5th ed. Philadelphia: Churchill Livingstone-Elsevier; 2007:456–498

Vogel F. Moderne Probleme der Humangenetik. Ergeb Inn Med Kinderheilkd 1959;12:52–125

Weinshilboum R. Inheritance and drug response. N Engl J Med 2003;348:529–537

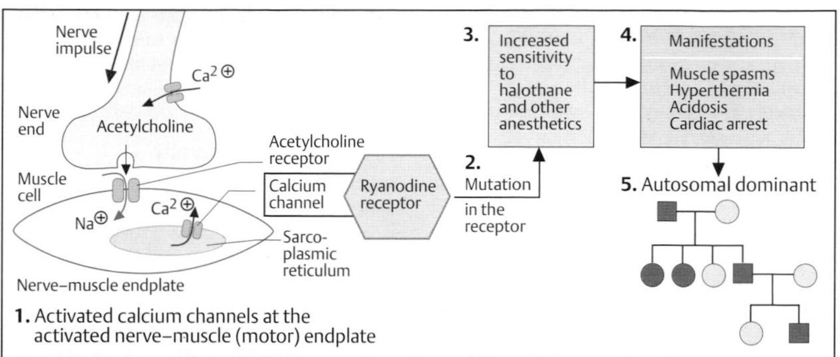

A. Malignant hyperthermia due to a calcium channel disorder in muscle cells

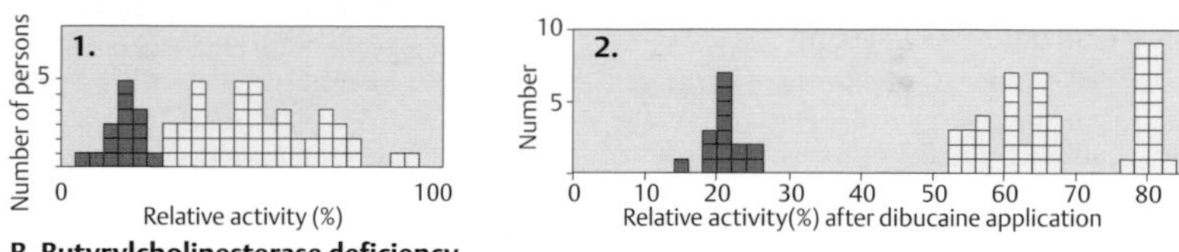

B. Butyrylcholinesterase deficiency

Defect	Relevant chemical substance	Clinical consequence	Frequency	Pathogenesis	Genetics
Coumarin resistance	Coumarin (warfarin)	Ineffective anticoagulation therapy	Rarer than 1:80 000	Increased vitamin K affinity due to enzyme or receptor defect	Autosomal dominant
Increased sensitivity to isoniazid	Isoniazid Sulfamethazine Phenelzine Hydralazine, etc.	Polyneuritis, lupus-like reaction	In about 50%	Decreased activity of liver isoniazid acetylase	Autosomal recessive
Isoniazid ineffective	Isoniazid Sulfamethazine Phenelzine Hydralazine	Reduced antituberculous effect		Increased isoniazid excretion	Autosomal dominant
Glucose-6-phosphate dehydrogenase (G6PD) deficiency	Sulfonamides Antimalarial drugs Nitrofurantoin Vicia faba	Hemolysis	Rare in Europeans, frequent in Africa and parts of SE Asia	G6PD deficiency in erythrocytes	X-chromosomal (many mutant forms)
Hemoglobin Zürich	Sulfonamides	Hemolysis	Rare	Unstable hemoglobin due to point mutation in β-globin (arginine instead of histidine in position 63)	Autosomal dominant
Hemoglobin H	Sulfonamides	Hemolysis	Rare	Unstable hemoglobin of 4 β chains due to deletion of the α loci	Autosomal dominant
Glaucoma in adults (some forms)	Corticoids	Glaucoma	Frequent	Unknown	Possibly autosomal dominant

C. Examples of genetically determined adverse reactions to pharmaceuticals

Cytochrome P450 (CYP) Genes

The cytochrome P450 system refers to a group of numerous genes (CYP genes; OMIM 108330). CYP genes encode enzymes with different functions in detoxification of various chemical substances. They derive their name from their maximal light absorption at 450 nm after binding to carbon monoxide. Cytochrome P450 enzymes form a large, evolutionarily related family of proteins with different enzymatic specificity in mammals. They degrade complex chemical substances, such as drugs or plant toxins, by an oxidation system (monooxygenases) in the microsomes of the liver and mitochondria of the adrenal cortex.

A. Cytochrome P450 system

The cytochrome P450 enzymes carry out phase I of a detoxification pathway (**1**): a substrate (RH) is oxidized to ROH utilizing atmospheric oxygen (O_2), with water (H_2O) formed as a byproduct. A reductase delivers hydrogen ions (H^+) either from NADPH or NADH. In phase II, ROH is further degraded, and eliminated. The P450 enzymes have a wide spectrum of activity (**2**). Characteristically, a single P450 protein can oxidize several structurally different chemical substances, or several P450 enzymes can degrade a single chemical substrate. The enzyme activities of phases I and II have to be well coordinated, since toxic intermediates occasionally arise in the initial stages of phase II.

B. Debrisoquine metabolism

Debrisoquine is an isoquinoline-carboxamidine. It was used to treat high blood pressure until it was found to cause severe side effects in 5%–10% of the population (OMIM 608902). Affected individuals have reduced activity of debrisoquine-4-hydroxylase (CYP2D6). This enzyme degrades several pharmacological substances such as β-adrenergic blockers, antiarrhythmics, and antidepressives. Two groups can be distinguished in the population: those with normal and those with slow degradation (**1**). Individuals with low activity are at increased risk of untoward toxic reactions. Individuals with a slow rate of degradation have an increased ratio of debrisoquine/4-hydrodebrisoquine. This enzyme is encoded by the *CYP2D6* gene (OMIM 124030) located at 22q13.1. Certain mutations of the primary transcript of this gene, with nine exons, produce aberrant splicing (**2**). As a result, variant mRNAs contain an intron and produce proteins with reduced enzymatic activity. (Figure adapted from Gonzales et al., 1988.)

C. CYP gene superfamily

The cytochrome P450 (CYP) genes in mammals consist of a superfamily of genes that resemble each other in exon/intron structure and that encode related enzymes. The CYP gene family arose during the last 1500–2000 million years. The largest P450 family in mammals is CYP2, with 16 genes in humans. It is assumed that the CYP2 family developed in response to toxic substances in plants. At least 30 gene duplications and gene conversions have led to an unusually diverse repertoire of CYP genes. Important enzymes for drug metabolism are CYP2C8, CYP2C9, CYP2C18, and CYP2C19, which together metabolize more than 50 compounds. CYP3A4, CYP2D6, and CYP2C9 are responsible for 50%, 25%, and 5% of drug metabolism, respectively. (Figure adapted from Gonzales et al., 1988, and Gonzales & Nebert, 1990.)

Further Reading

Genetics Home Reference. CYP gene family. (http://ghr.nlm.nih.gov/geneFamily/cyp)

Gonzalez FJ, et al. Characterization of the common genetic defect in humans deficient in debrisoquine metabolism. Nature 1988;331:442–446

Gonzalez FJ, Nebert DW. Evolution of the P450 gene superfamily: animal-plant 'warfare', molecular drive and human genetic differences in drug oxidation. Trends Genet 1990;6:182–186

Lynch T, Price A. The effect of cytochrome P450 metabolism on drug response, interactions, and adverse effects. Am Fam Physician 2007;76:391–396

Nebert DW, Russell DW. Clinical importance of the cytochromes P450. Lancet 2002;360:1155–1162

Nebert DW, Nelson DR. Cytochrome P450 (*CYP*) gene superfamily. Nature Encyc Hum Genome 2003;1:1028–1037

Nelson DR, et al. Comparison of cytochrome P450 (CYP) genes from the mouse and human genomes, including nomenclature recommendations for genes, pseudogenes and alternative-splice variants. Pharmacogenetics 2004;14:1–18

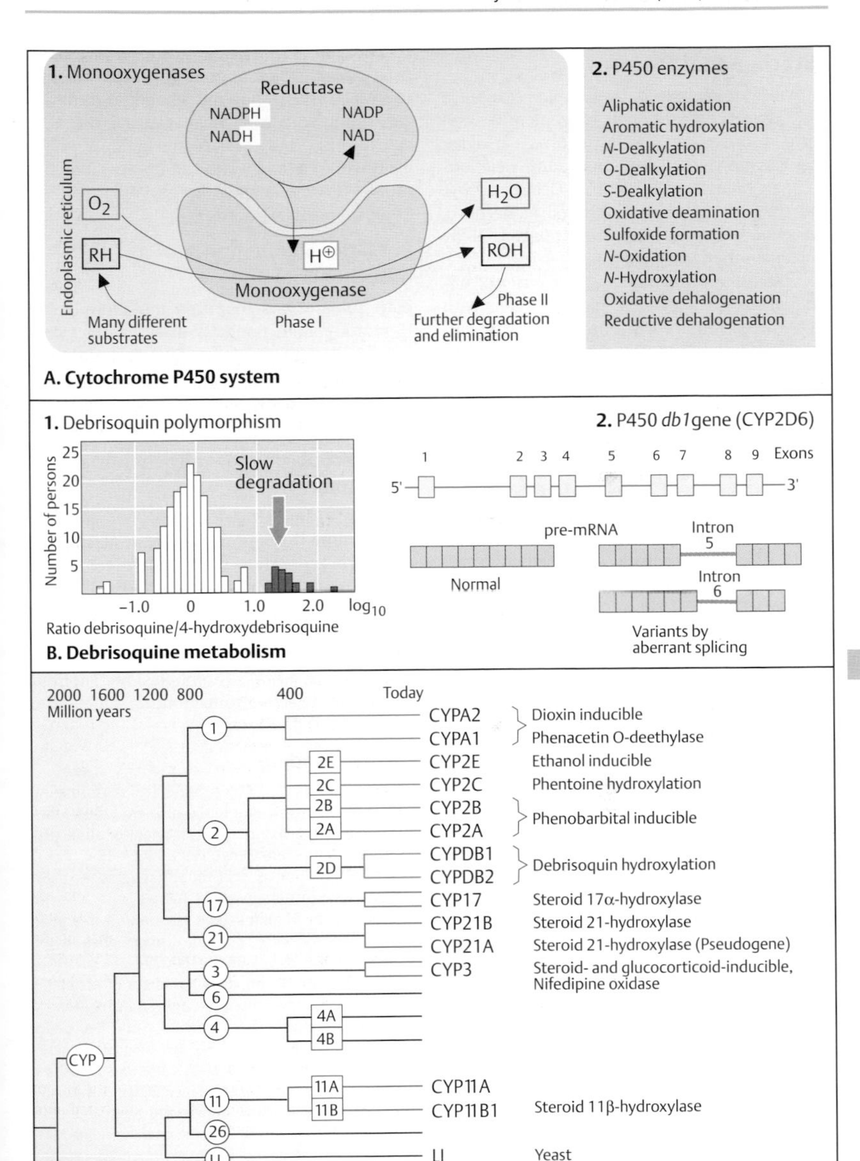

A. Cytochrome P450 system

B. Debrisoquine metabolism

C. CYP gene superfamily

Mitochondrial Diseases

Human mitochondrial dysfunction causes a wide variety of different degenerative and acute disorders. It also is involved in aging and carcinogenesis. Mutations and deletions occur in human mitochondrial DNA (mtDNA), and they can be either inherited or de novo. Mitochondrial mutations are transmitted by maternal inheritance only.

The clinical spectrum and age of onset of mitochondrial diseases vary widely. Organs with high-energy requirements are particularly vulnerable: the brain, heart, skeletal muscle, eye, ear, liver, pancreas, and kidney. These organs are affected by several neurodegenerative and cardiovascular disorders, diabetes mellitus, deafness, and cancer. Acquired mitochondrial mutations accumulate with age.

The mutation rate of mtDNA is 10 times higher than that of nuclear DNA. Mitochondrial mutations are generated during oxidative phosphorylation through pathways involving reactive oxygen molecules. Mutations accumulate because effective DNA repair and protective histones are lacking. At birth most mtDNA molecules are identical (*homoplasmy*); later they differ as a result of mutations accumulated in different mitochondria (*heteroplasmy*).

A. Mutations and deletions in human mitochondrial DNA

Mitochondrial genetic disorders result from deletions, point mutations, and other structural rearrangements. Some are characteristic and recur in different, unrelated patients. The extent of three common deletions of 10.4 kb, 7.4 kb, and 5 kb are shown as partial circles outside the mtDNA map. Inside the site of common mutations are shown: LHON (Leber's optical nerve atrophy; OMIM 535000) at nucleotide positions 11778 and 3460; NARP (neuropathy, ataxia, retinitis pigmentosa; OMIM 551500) at 8993; MERRF (myoclonus epilepsy with ragged red fibers in muscle cells; OMIM 545000) at 8344; MMC (mitochondrial myopathy and cardiomyopathy; OMIM 590050) at 3260; MELAS (mitochondrial myopathy, encephalopathy, lactic acidosis; OMIM 540000) at 3243; and a mutation A to G at 1555. The latter mutation confers high sensitivity to aminoglycosides, causing deafness (see Table 8 on p. 414).

In addition, 175 nuclear-encoded mitochondrial genes are associated with 191 diseases in OMIM. Thus, some mitochondrial disorders can resemble a Mendelian transmission. The age of onset and severity are highly variable, even within a family affected by the same mutation. (Figure adapted from Wallace et al., 2007, and MITOMAP.)

B. Maternal inheritance of a mitochondrial disease

Both somatic and inherited mutations occur. Hereditary mitochondrial diseases are transmitted through the maternal line only, since spermatozoa contain very few, if any, mitochondria. Thus, the disease will not be transmitted from an affected man to his children.

C. Heteroplasmy for mitochondrial mutations

Many mutations or deletions in mitochondria are acquired during the lifetime of an individual. Their proportion may be different in different tissues and increase with age. This difference is referred to as heteroplasmy. It contributes to the considerable variability of mitochondrial diseases. A germline mutation may be present in all cells (homoplasmy). The proportion of defective mitochondria varies after repeated cell divisions.

Further Reading

Estivill X, et al. Familial progressive sensorineural deafness is mainly due to the mtDNA A1555G mutation and is enhanced by treatment of aminoglycosides. Am J Hum Genet 1998;62:27–35

MITOMAP. A human mitochondrial genome database. (www.mitomap.org/MITOMAP)

Scharfe C, et al. Mapping gene associations in human mitochondria using clinical disease phenotypes. PLOS Comput Biol 2009;5:e1000374

Wallace DC. A mitochondrial paradigm of metabolic and degenerative diseases, aging, and cancer: a dawn for evolutionary medicine. Annu Rev Genet 2005;39:359–407

Wallace DC. Mitochondrial DNA mutations in disease and aging. Environ Mol Mutagen 2010;51:440–450

Wallace DC, et al. Mitochondria and neuro-ophthalmologic dieases. In: Scriver CR, et al., eds. The Metabolic and Molecular Bases of Inherited Disease. 8th ed. New York: McGraw-Hill; 2001

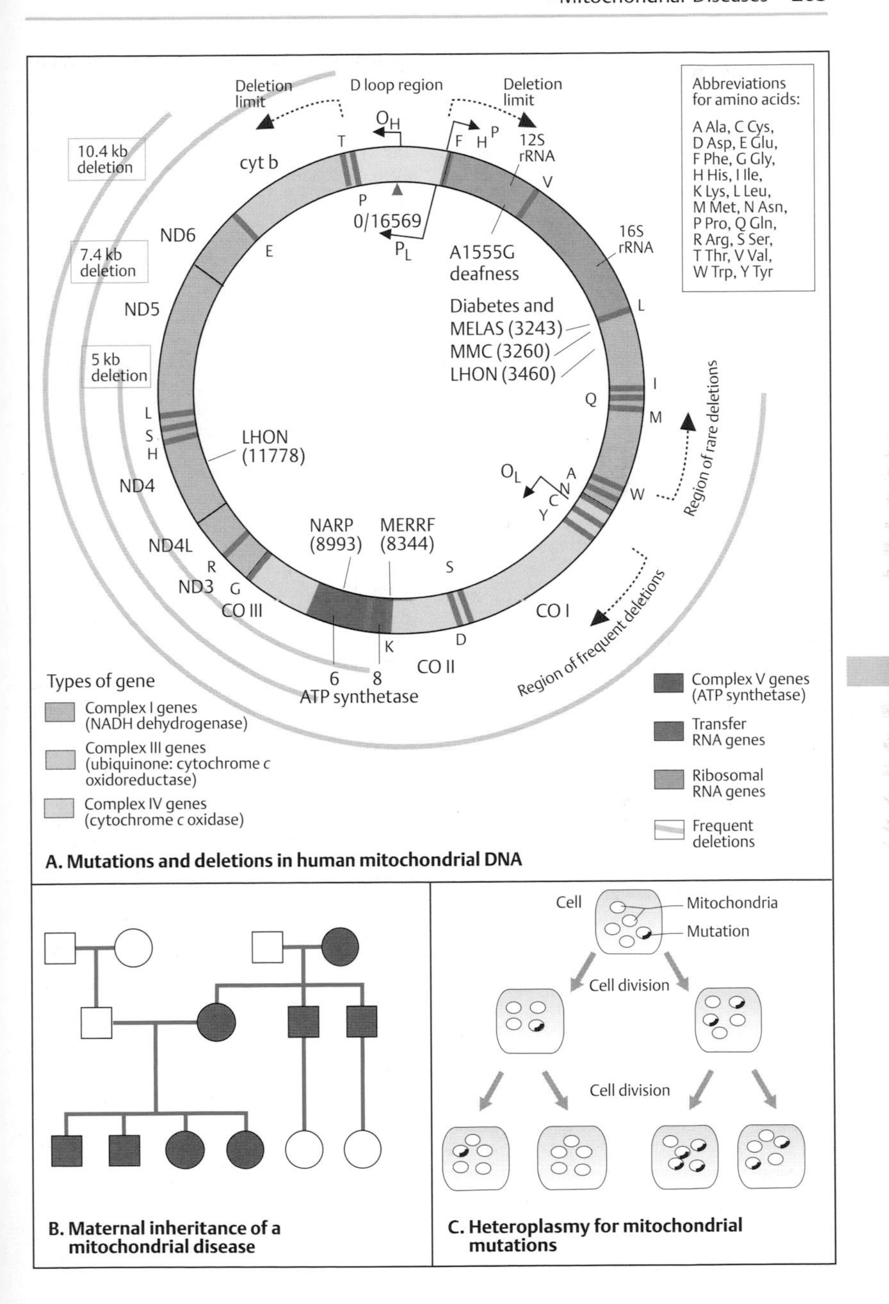

A. Mutations and deletions in human mitochondrial DNA

B. Maternal inheritance of a mitochondrial disease

C. Heteroplasmy for mitochondrial mutations

Diabetes Mellitus

Diabetes mellitus (DM) is an etiologically heterogeneous group of more than 60 individually defined disorders characterized by high fasting levels of glucose in the blood (>125 mg/dl, hyperglycemia). Two principal types of DM, according to the basic pathogenesis, are type 1 (OMIM 222100) and type 2 (OMIM 125853), previously termed insulin-dependent and non-insulin-dependent DM. DM type 1 results from varying degrees of insulin deficiency. DM type 2 is a highly heterogeneous group of disorders with variable insulin resistance, impaired insulin secretion, and increased glucose production. In type 1 an autoimmune or infectious process destroys the beta cells in the pancreas. In type 2, insulin resistance in the peripheral tissues and dysfunction of beta cells with reduced insulin secretion cause hyperglycemia. DM is a common health problem in many parts of the world, affecting up to 1%–2% of the population.

A. Insulin biosynthesis

Human insulin is encoded by a gene (OMIM 176730) with two exons and a signal sequence in the 5′ direction of the gene, located on the short arm of chromosome 11 (11p15.5). It has a beta-cell-specific enhancer and a variable number of tandem repeats (VNTRs) upstream (5′) of the gene. The primary transcript is spliced into mRNA and translated into preproinsulin (1430 amino acids). Then the signal sequence (24 amino acids) and the C-peptide are removed. The A chain and the B chain are joined by two disulfide bonds (see p. 44). The gene is expressed exclusively in the beta cells of the pancreas.

B. Insulin receptor

Insulin exerts its multiple functions through the insulin receptor (INSR; OMIM 147670). INSR is a tetramer of two extracellular α subunits and two transmembrane β subunits, joined by disulfide bridges at specific sites. Distinct functional domains in both extracellular and intracellular parts of the receptor reflect the various functions. Insulin receptor substrate IRS and Shc proteins mediate the principal functions of insulin: activating growth factors, protein synthesis, glycogen synthesis, and glucose transport. The *INSR* gene spans more than 120 kb and has 22 exons, 11 for each of the α and β chains.

C. Diabetes mellitus (simplified model)

DM type 1 is mainly caused by external factors, such as certain viral infections, which directly or by autoimmune reactions cause destruction of beta cells on a background of individual genetic susceptibility. DM type 2 is influenced by lifestyle as a strong environmental component. Monozygotic twins with DM are concordant for type 1 in about 25%, and for type 2 in about 40%–50%. About 2%–7% of first-degree relatives of individuals with DM type 1 are affected, depending on the relationship and age at onset of disease. Several types of autosomal dominant forms of DM type 2 exist in adults (maturity onset diabetes of the young, MODY 1–6; OMIM 125850). DM is a secondary manifestation of several genetically determined diseases involving insulin receptor defects (insulin resistance syndromes).

D. Genetic susceptibility

A major influence on the genetic susceptibility to DM type 1 is exerted by certain alleles of class I MHC genes (see p. 294). Several alleles of the DR3 and DR4 loci are associated with susceptibility to diabetes type I, especially DR3/DR4 heterozygotes. Some alleles of DR2 confer relative resistance to DM. Genetic and genome-wide association studies have identified 36 regions or loci in the human genome conferring susceptibility to DM type 2 (listed in OMIM 125853, November 2011).

Further Reading

Daneman D. Type 1 diabetes. Lancet 2006;367:847–858

Powers AC. Diabetes mellitus. In: Longo DL, et al., eds. Harrisons Principles and Practice of Internal Medicine. 18th ed. New York: McGraw-Hill; 2012: 2968–3009

Sladek R, et al. A genome-wide association study identifies novel risk loci for type 2 diabetes. Nature 2007;445:881–885

Stumvoll M, et al. Type 2 diabetes: principles of pathogenesis and therapy. Lancet 2005;365:1333–1346

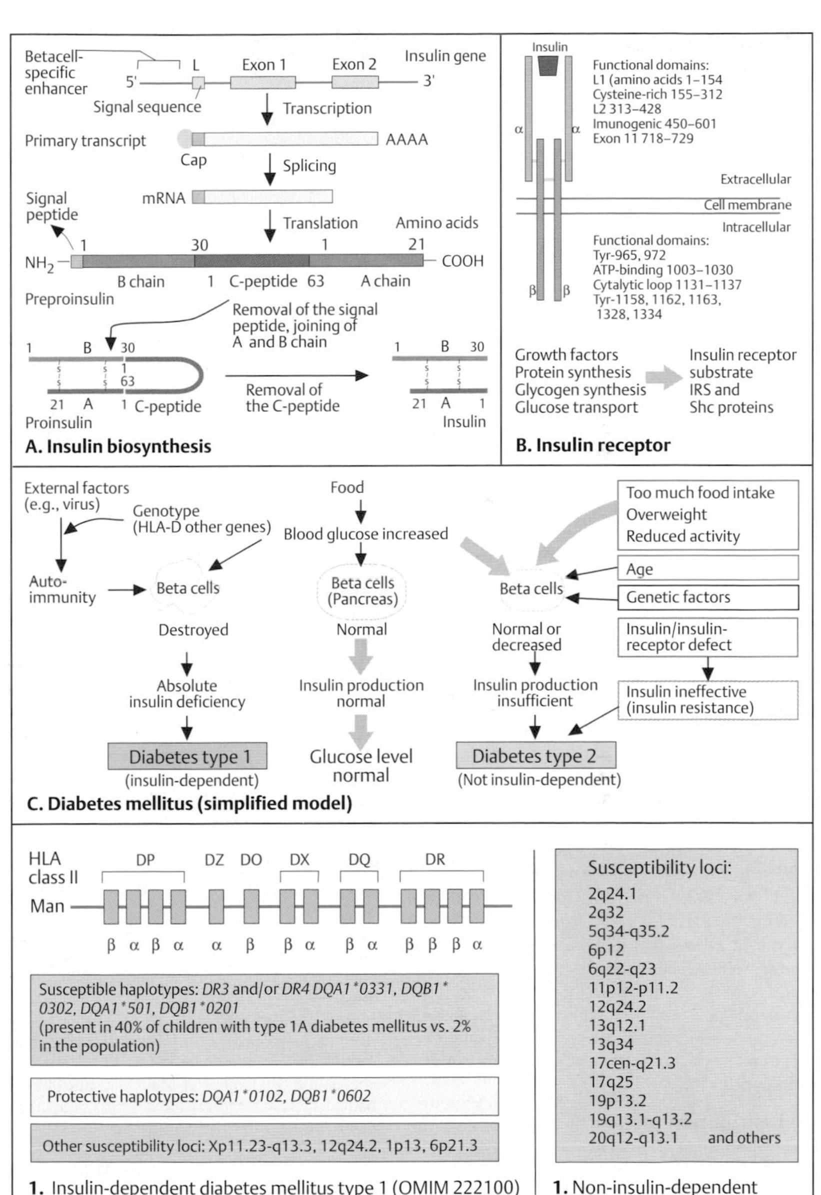
Betacell-specific enhancer
L
Exon 1
Exon 2
Insulin gene
5'
3'
Signal sequence
Transcription
Primary transcript
AAAA
Cap
Splicing
Signal peptide
mRNA
Translation
Amino acids
1
30
1
21
NH2
COOH
B chain
1 C-peptide 63
A chain
Preproinsulin
Removal of the signal peptide, joining of A and B chain
1 B 30
1
63
21 A 1 C-peptide
Proinsulin
Removal of the C-peptide
1 B 30
21 A 1
Insulin
A. Insulin biosynthesis
Insulin
Functional domains:
L1 (amino acids 1–154
Cysteine-rich 155–312
L2 313–428
Imunogenic 450–601
Exon 11 718–729
α
α
Extracellular
Cell membrane
Intracellular
Functional domains:
Tyr-965, 972
ATP-binding 1003–1030
Cytalytic loop 1131–1137
Tyr-1158, 1162, 1163, 1328, 1334
β
β
Growth factors
Protein synthesis
Glycogen synthesis
Glucose transport
Insulin receptor substrate
IRS and
Shc proteins
B. Insulin receptor
External factors (e.g., virus)
Genotype (HLA-D other genes)
Food
Blood glucose increased
Too much food intake
Overweight
Reduced activity
Auto-immunity
Beta cells
Beta cells (Pancreas)
Beta cells
Age
Genetic factors
Destroyed
Normal
Normal or decreased
Insulin/insulin-receptor defect
Absolute insulin deficiency
Insulin production normal
Insulin production insufficient
Insulin ineffective (insulin resistance)
Diabetes type 1
(insulin-dependent)
Glucose level normal
Diabetes type 2
(Not insulin-dependent)
C. Diabetes mellitus (simplified model)
HLA class II
DP
DZ
DO
DX
DQ
DR
Man
β α β α α β β α β α β β β α
Susceptible haplotypes: DR3 and/or DR4 DQA1*0331, DQB1*0302, DQA1*501, DQB1*0201
(present in 40% of children with type 1A diabetes mellitus vs. 2% in the population)
Protective haplotypes: DQA1*0102, DQB1*0602
Other susceptibility loci: Xp11.23-q13.3, 12q24.2, 1p13, 6p21.3
1. Insulin-dependent diabetes mellitus type 1 (OMIM 222100)
Susceptibility loci:
2q24.1
2q32
5q34-q35.2
6p12
6q22-q23
11p12-p11.2
12q24.2
13q12.1
13q34
17cen-q21.3
17q25
19p13.2
19q13.1-q13.2
20q12-q13.1
and others
1. Non-insulin-dependent diabetes mellitus type 2 (OMIM 125853)
D. Genetic susceptibility to diabetes mellitus

Amino Acid Degradation and Urea Cycle Disorders

Mutations in genes encoding the various enzymes involved in amino acid metabolism and urea cycle cause a large group of disorders. Two examples are presented here.

A. Phenylalanine degrading system

Phenylketonuria (PKU; OMIM 261600) is an autosomal recessive deficiency of the enzyme phenylalanine hydroxylase (PAH). This enzyme hydroxylates phenylalanine (Phe) to tyrosine. PKU results from mutations in the *PAH* gene. The complex Phe hydroxylating system includes tetrahydrobiopterin (BH_4) cofactor, which requires several enzymes for recycling, including dihydropteridine reductase (OMIM 261630) and pterin-4α-carbinolamine dehydratase (OMIM 264070). The *PAH* gene (OMIM 612349), located on chromosome 12q23.2, has 13 exons spanning 90 kb of DNA, with complex 5′ untranslated *cis*-acting, *trans*-activated regulatory elements. The gene shows developmental and tissue-specific transcription and translation. The hepatic and renal gene product is a 452-amino-acid polypeptide. Several polymorphic sites (RFLPs and SNPs, and a tetranucleotide short tandem repeat) are in linkage disquilibrium and define haplotypes. The PAH protein has catalytic, regulatory, and tetramerization domains. Modifying genes influence the phenotype.

Hyperphenylalaninemia is defined as a plasma Phe concentration of >120 µM (2 mg/dl). Long-term exposure to Phe concentrations >600 µM results in severe mental retardation, as in classical PKU. This condition, first described in 1934 by Asbjørn Følling in Norway, is the prototype of a treatable metabolic disease. Here, dietary phenylalanine restriction must be instituted in the neonatal period. PKU is detected by neonatal screening. Maternal PKU causes various developmental problems if the maternal Phe level is not well controlled throughout pregnancy.

B. Distribution of PAH mutations

Almost 500 disease-causing mutations have been recorded at the *PAH* locus, many unevenly distributed in different populations. Other mutations are observed in Asian populations (China, Korea). PKU is rare in Finnish, Ashkenazi, American aboriginal, and Japanese populations. (Data from Scriver, 2007, and Zschocke, 2003.)

C. Urea cycle defects

Terrestrial vertebrates synthesize urea by means of the urea cycle, the first metabolic pathway to be described (by Krebs and Henseleit, a medical student, in 1932). Arginine, the immediate precursor of urea, is hydrolyzed by arginase (OMIM 207830) to urea and ornithine. Ornithine transcarbamoylase (OTC; OMIM 310461) transfers carbamoyl phosphate to ornithine, which results in citrulline. Carbamoyl phosphate is synthesized from NH_4^+, CO_2, H_2O, and ATP by carbamoyl phosphate synthetase (OMIM 237300). Argininosuccinate synthetase (OMIM 215700) catalyzes the condensation of citrulline and aspartate to argininosuccinate. This is cleaved by argininosuccinase (OMIM 202900) into arginine and fumarate. Five metabolic disorders result from mutations of genes encoding the enzymes of the urea cycle. They are characterized by high plasma levels of ammonium, which is highly neurotoxic. Affected children may show progressive lethargy and coma leading to death in the neonatal period. The most common form is X-linked OTC resulting from mutations in the *OTC* gene located at Xp21.1. (Figure adapted from Stryer, 1995.)

Further Reading

Blau N, van Spronsen FJ, Levy HL. Phenylketonuria. Lancet 2010;376:1417–1427

Brusilow SW, Horwich AL. Urea cycle enzymes. In: Scriver CR, et al., eds. The Metabolic and Molecular Bases of Inherited Disease, 8th ed. New York: McGraw-Hill; 2001: 1909–1963

Mitchell JJ, Trakadis YJ, Scriver CR. Phenylalanine hydroxylase deficiency. Genet Med 2011;13:697–707

National Society for Phenylketonuria. (www.nspku.org/)

PAH Database. (www.pahdb.mcgill.ca/)

Scriver CR, Levy H, Donlon J. Hyperphenylalaninemia: phenylalanine hydroxylase deficiency. In: The Online Metabolic and Molecular Bases of Inherited Disease. (www.ommbid.com/)

Stryer L. Biochemistry. 4th ed. New York: W. H. Freeman; 1995.

Zschocke J. Phenylketonuria mutations in Europe. Hum Mutat 2003;21:345–356

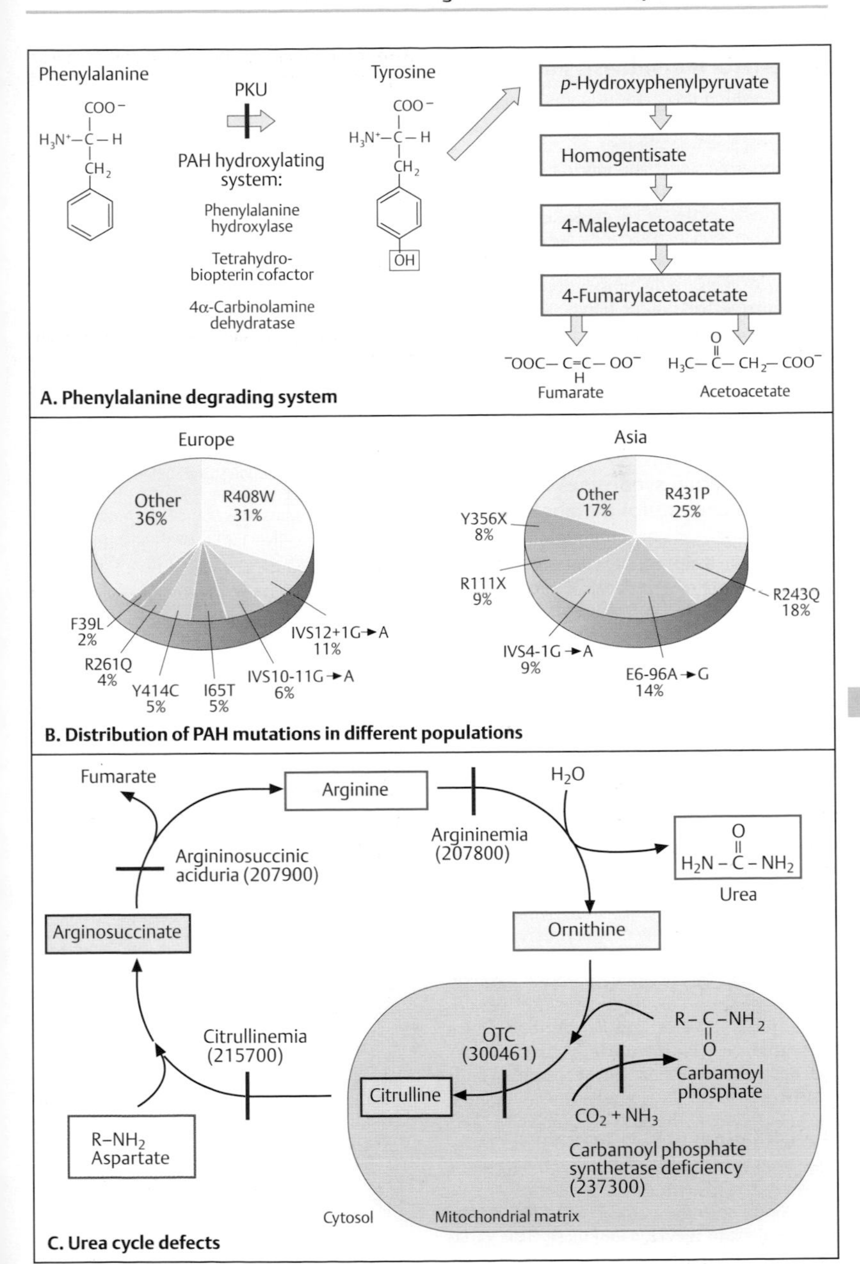
Phenylalanine
PKU
Tyrosine
p-Hydroxyphenylpyruvate
PAH hydroxylating system:
Phenylalanine hydroxylase
Tetrahydro-biopterin cofactor
4α-Carbinolamine dehydratase
Homogentisate
4-Maleylacetoacetate
4-Fumarylacetoacetate
Fumarate
Acetoacetate
A. Phenylalanine degrading system
Europe
Other 36%
R408W 31%
IVS12+1G→A 11%
IVS10-11G→A 6%
I65T 5%
Y414C 5%
R261Q 4%
F39L 2%
Asia
Other 17%
R431P 25%
R243Q 18%
E6-96A→G 14%
IVS4-1G→A 9%
R111X 9%
Y356X 8%
B. Distribution of PAH mutations in different populations
Fumarate
Arginine
H₂O
Argininemia (207800)
Urea
Argininosuccinic aciduria (207900)
Arginosuccinate
Ornithine
Citrullinemia (215700)
OTC (300461)
Citrulline
Carbamoyl phosphate
CO₂ + NH₃
Carbamoyl phosphate synthetase deficiency (237300)
R–NH₂ Aspartate
Cytosol
Mitochondrial matrix
C. Urea cycle defects

Cholesterol Biosynthesis Pathway

Cholesterol is a precursor of many steroid hormones and a major constituent modulating the fluidity of cell membranes in eukaryotes. In 1932, Wieland and Dane elucidated its structure as a monosaturated 27-carbon sterol. The biosynthetic pathway of cholesterol requires about 30 enzymatic reactions regulated by 22 genes in a series including oxidation with molecular oxygen, reductions, demethylations, and alterations in double bonds. Konrad Bloch was awarded the Nobel Prize in 1954 for elucidating this pathway. During the past 10 years, several hereditary diseases resulting from mutations in genes encoding enzymes of the cholesterol biosynthesis pathway have been discovered.

A. Malformation syndromes due to defects in cholesterol metabolism

About six different genetic diseases are known to result from a block of the cholesterol biosynthesis pathway (see p. 270). Three examples are shown: (**1**) the autosomal recessive Smith–Lemli–Opitz syndrome (OMIM 270400); (**2**) X-linked chondrodysplasia punctata type 2 (CDPX2, Conradi–Hünermann syndrome; OMIM 302960); (**3**) and autosomal dominant Greenberg skeletal dysplasia (OMIM 215140). (**1**, Kindly provided by the parents of the child; **2**, courtesy of Dr Richard I. Kelley, Baltimore, Maryland, USA; **3**, courtesy of Dr David L. Rimoin, Los Angeles, California, USA.)

B. Cholesterol biosynthesis overview

Cholesterol biosynthesis begins with acetyl coenzyme A (acetyl CoA), from which all 27 carbon atoms are derived. Acetyl CoA and acetoacetyl CoA condense to 3-hydroxy-3-methylglutaryl-CoA. This is converted by 3-hydroxy-3-methylglutaryl-CoA reductase to mevalonate. This is the precursor of isoprene, which is synthesized in three steps (not shown). Squalene, a 30-carbon linear isoprenoid, is synthesized from six isoprene units. Isopentyl pyrophosphate is the starting point of a reaction C5 → C10 → C15 → C30. The distal (post-squalene) part of the cholesterol biosynthesis pathway begins with squalene.

Mevalonic aciduria (OMIM 251170) results from a block in mevalonate kinase. This variable autosomal recessive disease is characterized by increased urinary excretion of mevalonic acid associated with failure to thrive, psychomotor retardation, vomiting, diarrhea, episodes of fever, and dysmorphic facial features.

C. Squalene to lanosterol

Initially, squalene is circularized through a reactive intermediate, squalene epoxide (not shown), to lanosterol, the first post-squalene sterol intermediate. Squalene epoxide is closed by a cyclase to lanosterol, a 30-carbon sterol. This requires movements of electrons through four double bonds and the migration of two methyl groups. Removal of the 24–25 double bond results in dihydrolanosterol, which is the other precursor of cholesterol.

Further Reading

Farese RV Jr, Herz J. Cholesterol metabolism and embryogenesis. Trends Genet 1998;14:115–120

Fitzky BU, et al. Mutations in the Delta7-sterol reductase gene in patients with the Smith-Lemli-Opitz syndrome. Proc Natl Acad Sci USA 1998;95:8181–8186

Goldstein JL, Brown MS. Regulation of the mevalonate pathway. Nature 1990;343:425–430

Greenberg CR, et al. A new autosomal recessive lethal chondrodystrophy with congenital hydrops. Am J Med Genet 1988;29:623–632

Herman GE. Disorders of cholesterol biosynthesis: prototypic metabolic malformation syndromes. Hum Mol Genet 2003; 12(Spec No 1, R1)R75–R88

Kelley RI, et al. Abnormal sterol metabolism in patients with Conradi–Hünermann–Happle syndrome and sporadic lethal chondrodysplasia punctata. Am J Med Genet 1999;83:213–219

Koo G, et al. Discordant phenotype and sterol biochemistry in Smith-Lemli-Opitz syndrome. Am J Med Genet A 2010;152A:2094–2098

Smith DW, Lemli L, Opitz JM. A newly recognized syndrome of multiple congenital anomalies. J Pediatr 1964;64:210–217

Waterham HR. Inherited disorders of cholesterol biosynthesis. Clin Genet 2002;61:393–403

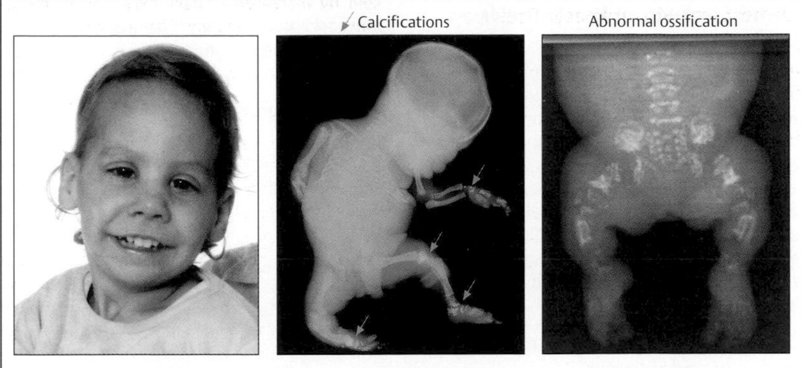

1. Smith–Lemli–Opitz syndrome **2.** Chondrodysplasia punctata **3.** Greenberg dysplasia

A. Malformation syndromes due to defects in cholesterol metabolism (examples)

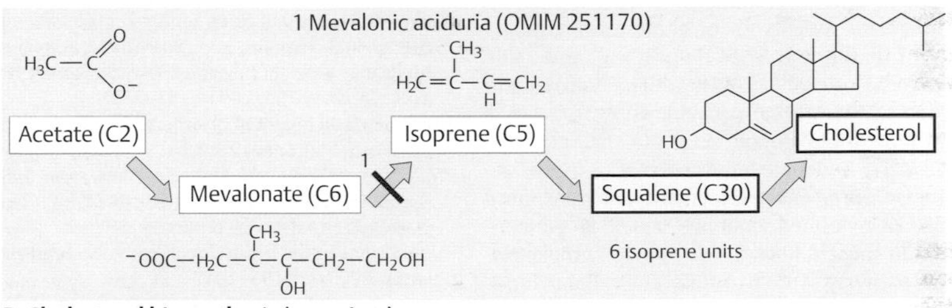

B. Cholesterol biosynthesis (overview)

Squalene

Circularization

Lanosterol

Antley–Bixler syndrome (OMIM 207410) Lanosterol 14-demethylase

Dihydrolanosterol

1a

1b

C. Squalene to lanosterol, the first step in the distal pathway

Distal Cholesterol Biosynthesis Pathway

In the distal part of the cholesterol biosynthesis pathway (post-squalene) lanosterol and dihydrolanosterol (see previous page) are converted to desmosterol and 7-dihydrocholesterol (7-DHC), the immediate precursors of cholesterol. Mutations in genes encoding the enzymes required for cholesterol biosynthesis have been found at all steps. They cause rare genetic diseases characterized by developmental delay and abnormalities of the skeletal and other systems.

A. Distal cholesterol biosynthesis pathway and diseases

Lanosterol and dihydrolanosterol (see p. 268) are converted by four enzymatic reactions through four intermediate metabolites to the immediate precursors of cholesterol, desmosterol (cholesta-5(6),24-dien-3β-ol) and 7-dihydrocholesterol (7-DHC). The enzymatic steps remove three methyl groups at C-4 and C-14, open one double bond at C-24, and shift the C-8–C-9 double bond to C-7–C-8 by an isomerase. Some of the enzymatic reactions must occur in a defined sequence: the Δ^8–Δ^7 isomerization must follow the C-14α demethylation. The pathway is tied to different cellular functions and signaling pathways. Lanosterol and its two subsequent intermediates, 4,4-dimethyl-cholesta-8(9),14,24-trien-3β-ol and 4,4-dimethyl-cholesta-8(9),24-dien-3β-ol, have meiosis-stimulating activity and accumulate in the ovary and testis (see Herman, 2003). 7-DHC is the direct precursor of vitamin D. Hedgehog signaling proteins are modified by cholesterol (see p. 204).

Seven genetic diseases caused by defects in the distal (post-squalene) cholesterol biosynthesis pathway are known. These are (in descending order of the pathway reactions): (1) a proportion of patients with Antley–Bixler syndrome (OMIM 0207410), (2) the prenatally lethal Greenberg skeletal dysplasia (OMIM 215140), (3) X-chromosomal CHILD syndrome (congenital hemidysplasia with ichthyosiform erythroderma or nevus and limb defects; OMIM 308050), (4) X-chromosomal dominant chondrodysplasia punctata type 2 (Conradi–Hünermann syndrome; OMIM 302960), (5) lathosterolosis (OMIM 607330), (6) Smith–Lemli–Opitz syndrome (OMIM 270400), and (7) desmosterolosis (OMIM 602398). The main features of these diseases are summarized in Appendix, Table p. 415.

Further Reading

Farese RV Jr, Herz J. Cholesterol metabolism and embryogenesis. Trends Genet 1998;14:115–120

Greenberg CR, et al. A new autosomal recessive lethal chondrodystrophy with congenital hydrops. Am J Med Genet 1988;29:623–632

Herman GE. Disorders of cholesterol biosynthesis: prototypic metabolic malformation syndromes. Hum Mol Genet 2003; 12(Spec No 1, R1)R75–R88

Herman GE, et al. Characterization of mutations in 22 females with X-linked dominant chondrodysplasia punctata (Happle syndrome). Genet Med 2002;4: 434–438

Kelley RI, Herman GE. Inborn errors of sterol biosynthesis. Annu Rev Genomics Hum Genet 2001; 2:299–341

Kelley RI, Hennekam RCM. Smith–Lemli–Opitz syndrome. In: Scriver CR, et al., eds. The Metabolic and Molecular Bases of Inherited Disease. 8th ed. New York: McGraw-Hill; 2001:6183–6201

Waterham HR. Inherited disorders of cholesterol biosynthesis. Clin Genet 2002;61:393–403

Waterham HR, et al. Autosomal recessive HEM/Greenberg skeletal dysplasia is caused by 3 betahydroxysterol delta 14-reductase deficiency due to mutations in the lamin B receptor gene. Am J Hum Genet 2003;72:1013–1017

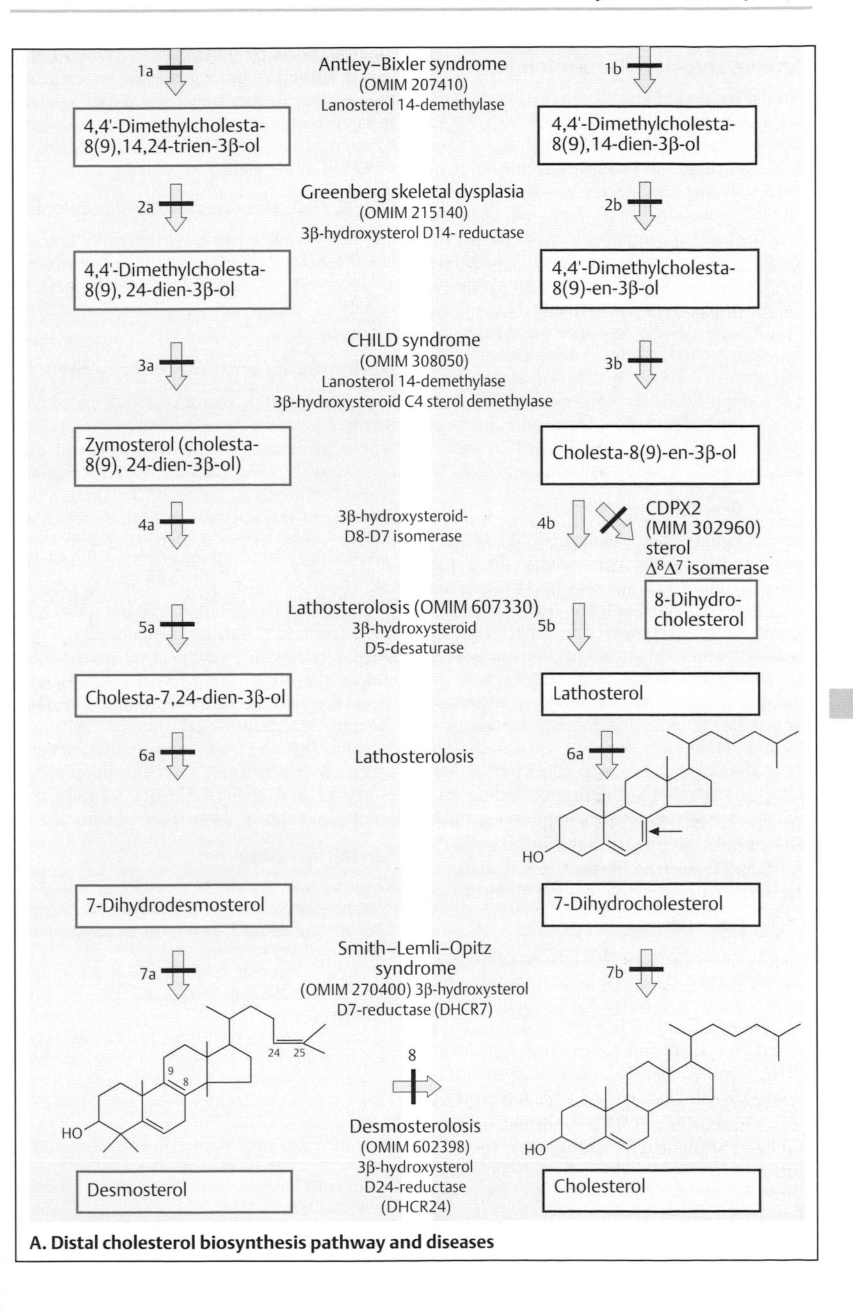

A. Distal cholesterol biosynthesis pathway and diseases

Familial Hypercholesterolemia

Familial hypercholesterolemia (FH) is an autosomal dominant disorder caused by elevated plasma levels of low-density lipoprotein (LDL), the main cholesterol-transporting lipoprotein. Triglyceride levels are normal. Lipoproteins transport cholesterol and triglycerides. The protein components make the hydrophobic lipids accessible to cells. Lipoproteins are classified according to increasing density: chylomicrons, very-low-density lipoproteins (VLDLs), low-density lipoproteins (LDLs), and high-density lipoproteins (HDLs). Different genetic forms of FH are distinguished according to the step of the metabolic pathway involved and the type of mutation. Many other forms of hyperlipidemia exist, often caused by multigenic, monogenic, and environmental factors.

A. The disease phenotype

Familial hypercholesterolemia (OMIM 143890) occurs, in the heterozygous state, in about 1 in 500 individuals (**1**). Plasma cholesterol levels are greatly increased (**2**). The activity of functional LDL receptors per cell is decreased by about 50% and is nearly absent in homozygotes (**3**). Important clinical signs are deposits of cholesterol esters in the tendons, especially the Achilles tendon, and the skin (xanthomas, **4**) and lipid deposits around the iris in the eye (**5**). In the rare homozygous state (1 in 10^6) FH leads to death during the first or second decade of life. Several other related forms at other gene loci are known (OMIM 143890). (Figure parts: **2** and **3**, adapted from Goldstein et al., 2001; **4** and **5**, the author's observations.)

B. The LDL receptor

Different mutations in the LDL receptor gene cause FH (see p. 274). The LDL receptor is a cell surface receptor encoded by a gene located at 19p13.2, consisting of 18 exons spanning 45 kb of genomic DNA and transcribed into 5.3 kb mRNA (OMIM 606945). It is a membrane-bound 160-kDa protein of 839 amino acids, with six distinct functional domains. Seven cysteine-rich units of 40 amino acids each form the ligand-binding region (exons 2–6). The epidermal growth factor (EGF) precursor homology domain (exons 7–14) allows the lipoprotein to dissociate from the receptor in the endosome (see p. 274). The intracellular domain encoded by part of exon 17 and the 5′ end of exon 18 contains signals for localizing the receptor in the coated pits during endocytosis, and targeting within hepatocytes. (Figure adapted from Goldstein et al., 2001, according to Hobbs et al., 1990.)

C. LDL receptor-mediated endocytosis

The LDL receptor mediates the endocytosis of LDL. Receptors loaded with LDL accumulate in a coated pit (**1**), which forms an endocytotic vesicle inside the cell (**2**). (Photographs from Anderson et al., 1977.)

D. Homology with other proteins

The gene encoding mammalian LDL receptor is highly conserved in evolution (90% identical within mammals, 79% identity between man and shark), dating back at least 500 million years. The proximal halves of the extracellular domains of the LDL receptor family are structurally related to the epidermal growth factor (EGF) family. The families of genes encoding LDL receptor and EGF are related to proteins of the blood coagulation system: factors IX and X, protein C, and complement C9.

Other related genes (not shown) are those encoding the VLDL (very-low-density lipoprotein) receptor, the ApoE receptor 2 (ApoER2), the LDL receptor-related protein (LRP), and megalin. LRP and megalin are multifunctional and bind diverse ligands such as lipoproteins, proteases and their inhibitors, peptide hormones, and carrier proteins of vitamins.

Further Reading

Anderson RGW, Brown MS, Goldstein JL. Role of the coated endocytic vesicle in the uptake of receptor-bound low density lipoprotein in human fibroblasts. Cell 1977;10:351–364

Brown MS, Goldstein JL. A receptor-mediated pathway for cholesterol homeostasis. Science 1986; 232:34–47

Goldstein JL, Brown MS, Hobbs HH. Familial hypercholesterolemia. In: Scriver CR, et al., eds. The Metabolic and Molecular Bases of Inherited Disease. 8th ed. New York: McGraw-Hill; 2001:2863–2913. (www.ommbid.com/)

Rader DJ, Hobbs HH. Disorders of lipoprotein metabolism. In: Longo DL, et al., eds. Harrison's Principles of Internal Medicine. 18th ed. New York: McGraw-Hill; 2012:3145–3161

- Low-density lipoprotein (LDL) and cholesterol elevated in blood plasma
- Premature arteriosclerosis
- Xanthoma in skin and tendons
- Decreased life expectancy
- Autosomal dominant
- Mutation in LDL receptor gene

1. General features

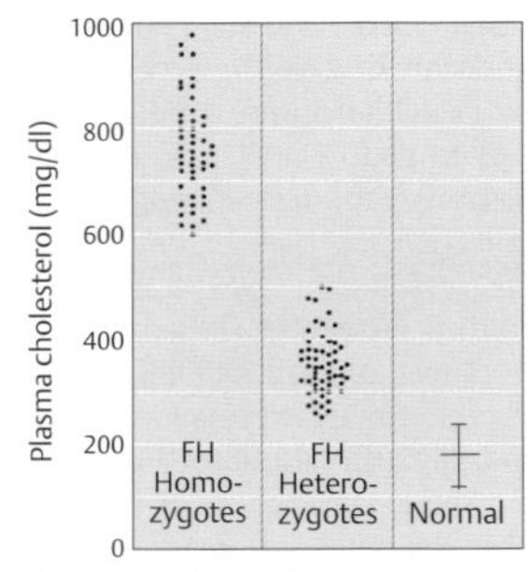

2. Hypercholesterolemia

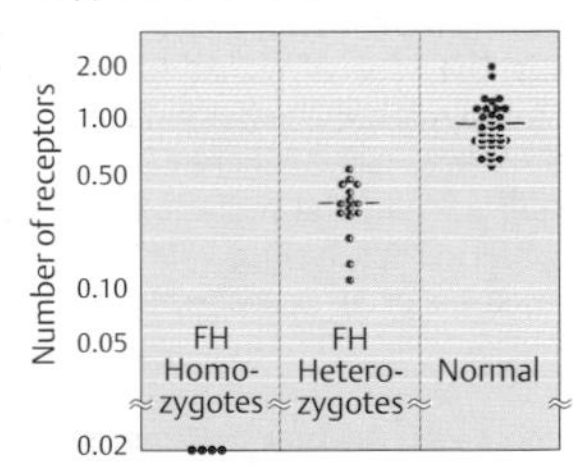

3. LDL receptors decreased

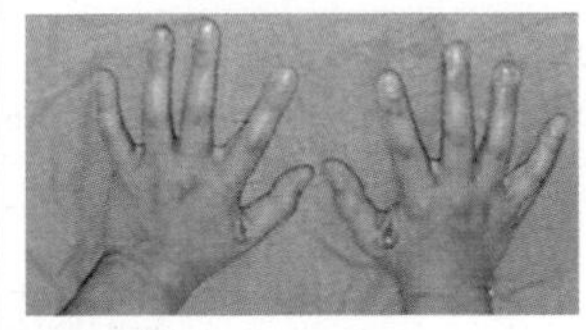

4. Xanthoma formation

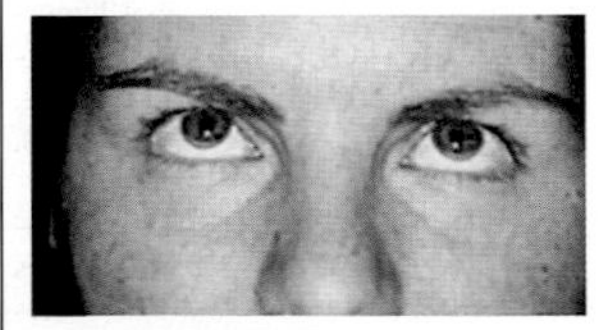

5. Arcus lipoides

A. Familial hypercholesterolemia

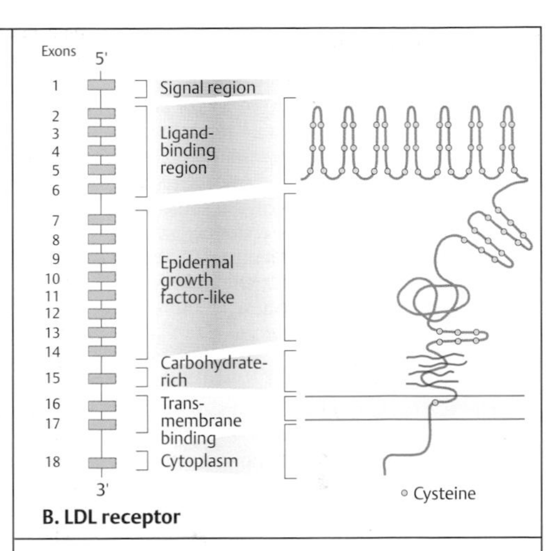

B. LDL receptor

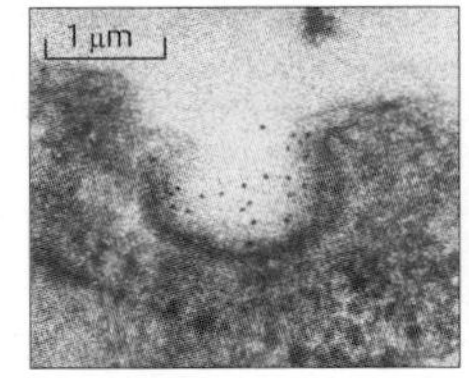

1. Coated pit

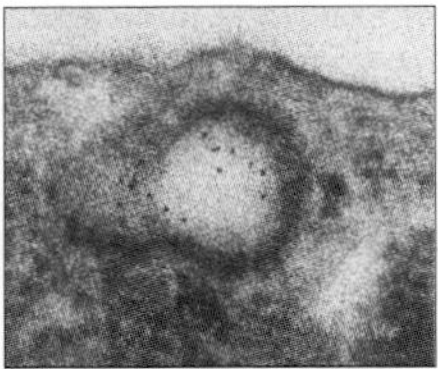

2. Endocytotic vesicle

Electron micrographs of fibroblasts in culture that have taken up LDL molecules (black dots, made visible by binding to ferritin).

C. Receptor-mediated endocytosis of LDL

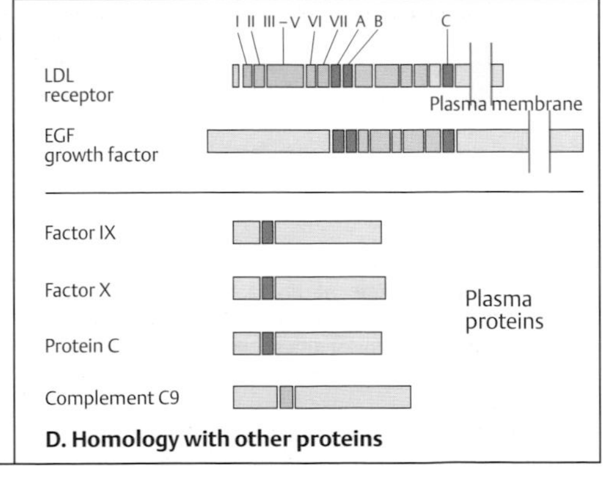

D. Homology with other proteins

LDL Receptor Mutations

Mutations in the low-density lipoprotein (LDL) receptor occur throughout the gene. LDL is the main transport agent of cholesterol in the blood. Its hydrophobic core contains about 1500 esterified cholesterol molecules surrounded by an outer layer of phospholipids and unesterified cholesterols containing a single apoB-100 lipoprotein molecule. LDL delivers cholesterol to peripheral tissues and regulates de novo cholesterol synthesis in these tissues.

A. LDL receptor classes of mutation

The effects of LDL receptor mutations can be understood on the basis of LDL receptor metabolism. The receptor–LDL complex enters the cell by endocytosis (see p. 276) within a coated vesicle. The receptor is separated from the LDL including the apoB-100 in the endosome. The LDL receptor is recycled to the cell surface. In the lysosome the LDL is broken down into amino acids and cholesterol. Free cholesterol activates the enzyme acetyl-CoA cholesterol transferase (ACAT), which catalyzes the esterification. The key enzyme for endogenous cholesterol synthesis is 3-hydroxy-3-methylglutaryl-CoA reductase (HMG-CoA reductase). This enzyme is downregulated by exogenous LDL uptake. LDL receptor mutations interrupt this feedback mechanism and result in increased endogenous cholesterol synthesis.
Five principal classes of LDL receptor mutations can be distinguished: (**1**) receptor null mutations (R^0) resulting from lack of receptor protein synthesis in the endoplasmic reticulum, (**2**) defective intracellular transport to the Golgi apparatus, (**3**) defective extracellular ligand binding, (**4**) defective endocytosis (R^+ mutations), and (**5**) failure to release the LDL molecules inside the endosome (recycling-defective mutations). (Figure adapted from Goldstein et al., 2001.)

B. Mutational spectrum

More than 1000 unique mutations have been recorded in the LDL receptor gene. Of these, 65% are missense mutations caused by substitutions, 24% are small DNA rearrangements, and 11% are large rearrangements. Mutations occur in all parts of the gene, but there is a relative excess of mutations in exons 4 and 9. A high proportion of mutations (74%) located in the ligand-binding domain (exons 2–6) involve amino acids conserved in evolution. Alu repeats may be involved as a cause of intragenic deletions. EGF, epidermal growth factor. (Figure adapted from Goldstein et al., 2001.)

C. A mutation in the LDL receptor gene

In this example, direct sequencing demonstrates a mutation in exon 9. First, exon 9 is amplified by PCR (P1, primer 1;P2 primer 2). The mutation in codon 408, GTG (valine) to GTA (methionine), produces a recognition site (N) for *Nla*III (GATC) that is not normally present. This results in two fragments of 126 bp and 96 bp instead of the usual 222-bp fragment (**1**). Thus, affected individuals (1 and 3 in the pedigree) have two smaller fragments of 126 kb and 96 kb in addition to the 222-kb fragment (**2**). Sequence analysis of the patient (individual 1 in the pedigree) demonstrates the mutation by the presence of an additional adenine (A, yellow circle) next to the normal guanine (**3**). Once the type of mutation has been established in an affected individual, all other members of a family can easily be screened for presence or absence of the mutation after appropriate genetic counseling. (Photographs kindly provided by Dr H. Schuster, Berlin.)

Further Reading

Brown MS, Hobbs HH, Goldstin JL. Familial Hypercholesterolemia. In: Valle D, et al., eds. The Online Metabolic and Molecular Bases of Inherited Diseases 2011 (OMMBID). (www.ommbid.com/OMMBID/)

Familial Hypercholesterolemia (FH) Variant Database. (www.ucl.ac.uk/ugi/fh)

Goldstein JL, Brown MS, Hobbs HH. Familial hypercholesterolemia. In: Scriver CR, et al., eds. The Metabolic and Molecular Bases of Inherited Disease. 8th ed. New York: McGraw-Hill; 2001:2863–2913

Leigh SE, et al. Update and analysis of the University College London low density lipoprotein receptor familial hypercholesterolemia database. Ann Hum Genet 2008;72:485–498. (http://www.ucl.ac.uk/ugi/fh)

Rader DJ, Hobbs HH. Disorders of lipoprotein metabolism. In: Longo DL, et al., eds. Harrison's Principles of Internal Medicine. 18th ed. New York: McGraw-Hill; 2012:3145–3161

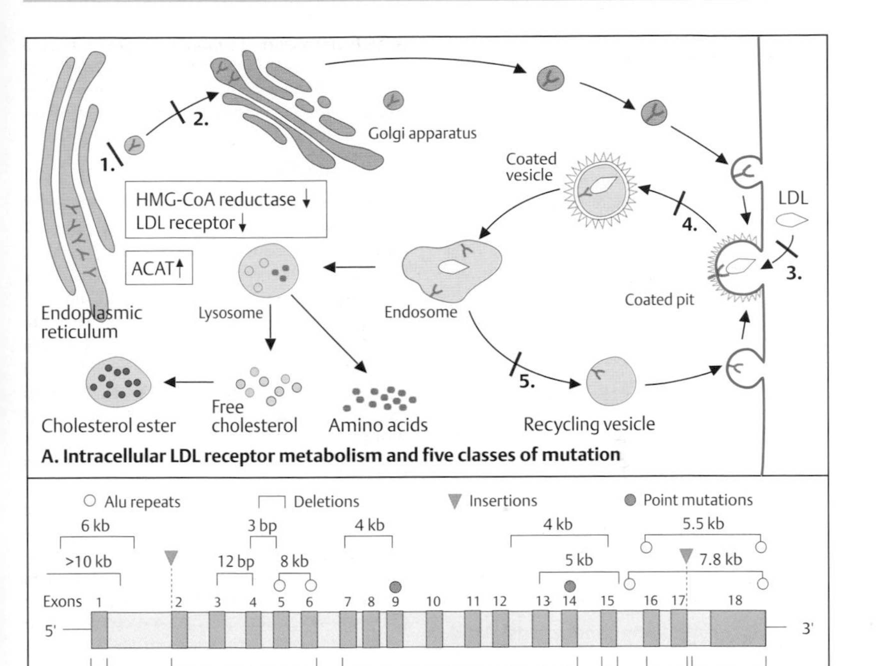

A. Intracellular LDL receptor metabolism and five classes of mutation

B. Mutational spectrum in the LDL receptor gene and effect of mutation function

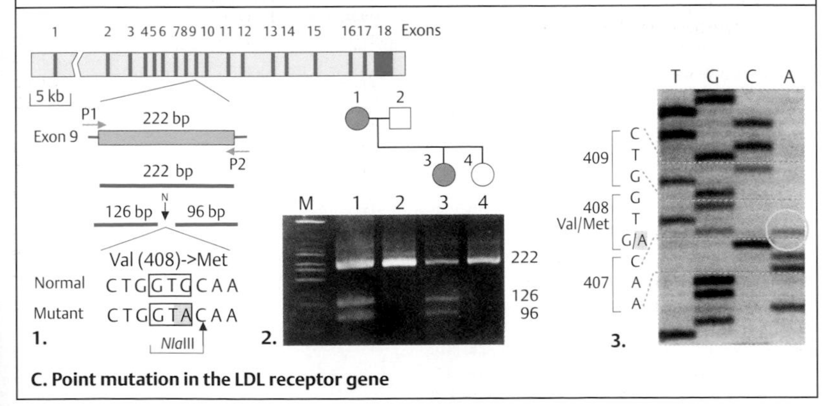

C. Point mutation in the LDL receptor gene

Lysosomal Disorders

Lysosomal disorders are a large group of genetic diseases involving various functions of lysosomes. Lysosomes are membrane-enclosed intracellular vesicles with a diameter of 0.05–0.5 µm, and they are required for the intracellular degradation of large molecules. They contain more than 50 active hydrolytic enzymes (acid hydrolases), such as glycosidases, sulfatases, phosphatases, lipases, phospholipases, proteases, and nucleases (collectively called lysosomal enzymes) in an acid milieu (approximately pH 5). Lysosomal enzymes enter a lysosome by means of a recognition signal (mannose 6-phosphate) and a corresponding receptor.

More than 40 different lysosomal disorders are characterized by abnormal storage of different macromolecules. They are grouped according to the main class of stored material: glycogen, mucopolysaccharides, glycoproteins, glycolipids, sphingomyelin, gangliosides, and others. A therapy using enzyme replacement is being developed for some forms of disorders.

A. Receptor-mediated endocytosis and lysosome formation

Extracellular macromolecules to be degraded are taken into the cell by endocytosis. First, the molecules are bound to specific cell surface receptors (receptor-mediated endocytosis). The loaded receptors are concentrated in an invagination of the plasma membrane (coated pit). This separates from the plasma membrane and forms a membrane-enclosed cytoplasmic compartment (coated vesicle). The cytoplasmic lining of the vesicle consists of a network of a trimeric protein, clathrin. The clathrin coat is removed within the cell, forming an endosome. The receptor and the molecule to be degraded (the ligand) are separated and the receptor is recycled to the cell surface.

A multivesicular body (endolysosome) forms and takes up acid hydrolases arriving in clathrin-enclosed vesicles. Hydrolytic degradation takes place in the lysosome. Parts of the membrane also are also recycled.

A mannose 6-phosphate receptor serves as a recognition signal for uptake into the endolysosome, which will also be recycled back into the Golgi apparatus. The acid milieu in the lysosomes is maintained by a hydrogen pump in the membrane that hydrolyzes ATP and uses the energy produced to move hydrogen ions into the lysosome. Some of the mannose 6-phosphate (mannose 6-P) receptors are transported back to the Golgi apparatus.

B. Mannose 6-phosphate receptors

Two types of mannose 6-phosphate receptor molecules exist. They differ in their binding properties and their cation dependence. They consist of either 2 (cation-independent mannose 6-phosphate receptor, CI-MDR) or 16 (cation-dependent mannose 6-phosphate receptor, CD-MDR) extracellular domains, with different numbers of amino acids. The cDNA of CI-MPR is identical to insulin-like growth factor II (IGF-2). Thus, CI-MPR is a multifunctional binding protein.

C. Biosynthesis

Two enzymes are essential for the biosynthesis of mannose 6-phosphate recognition signals: a phosphate transferase and a phosphoglycosidase. The phosphate is delivered by uridine-diphosphate-*N*-acetylglucosamine (UDP-GlcNAc) to uridine-5′-diphosphate-*N*-acetylglucosamine-glycoprotein-*N*-acetylglucosaminylphosphotransferase (GlcNAc-phosphotransferase). A second enzyme (*N*-acetylglucosamine-1-phosphodiester-*N*-acetyl-glucosaminidase, GlcNAc-phosphoglycosidase) removes the *N*-acetylglucosamine, leaving the phosphate residue at position 6 of the mannose. (Figures adapted from de Duve, 1984, and Sabatini & Adesnik, 2001.)

Further Reading

de Duve C. A Guided Tour of the Living Cell. Vols 1 & 2. New York: Scientific American Books; 1984

Gilbert-Barness E, Barness L. Metabolic Diseases. Foundations of Clinical Management, Genetics, and Pathology. Vols 1 & 2. Natick: Eaton Publishing; 2000

Hopkin RJ, Grabowsi GA. Lysosomal storage diseases. In: Longo DL, et al., eds. Harrison's Principles of Internal Medicine. 18th ed. New York: McGraw-Hill; 2012:3191–3197

Sabatini DD, Adesnik MB. The biogenesis of membranes and organelles. In: Scriver CR, et al., eds. The Metabolic and Molecular Bases of Inherited Disease. 8th ed. New York: McGraw-Hill; 2001:433–517. (www.ommbid.com/)

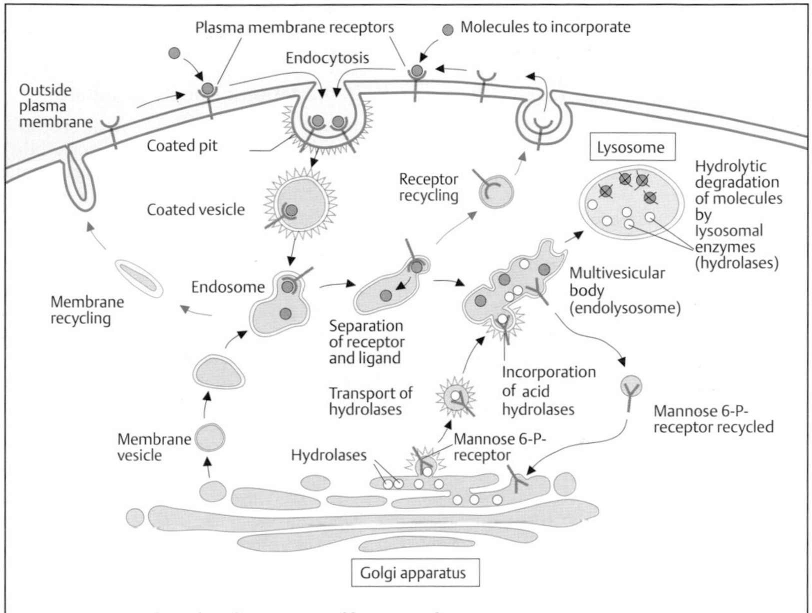

A. Receptor-mediated endocytosis and lysosome biogenesis

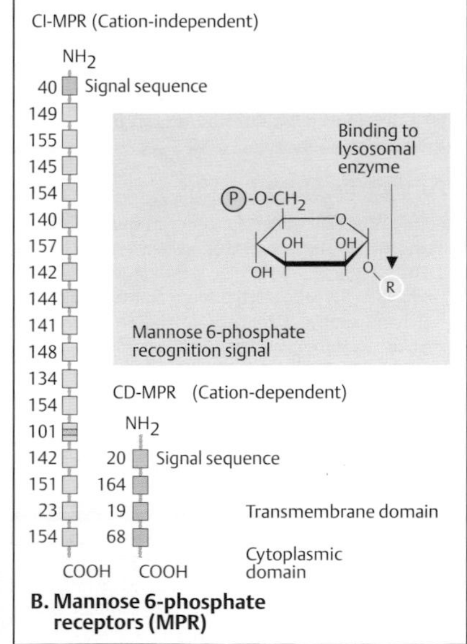

B. Mannose 6-phosphate receptors (MPR)

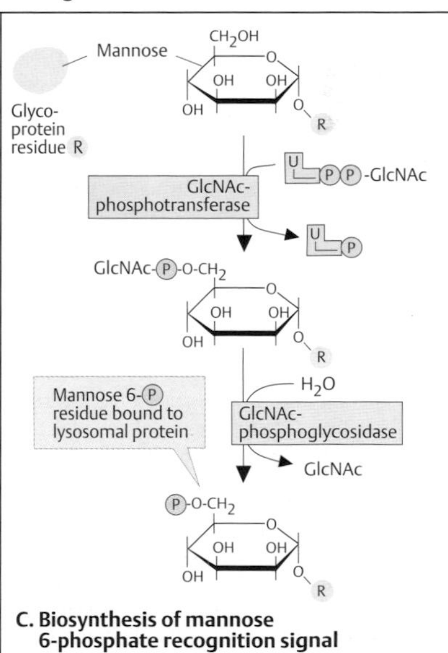

C. Biosynthesis of mannose 6-phosphate recognition signal

Lysosomal Enzyme Defects

Mutations in genes encoding enzymes that degrade complex macromolecules in lysosomes (lysosomal enzymes) cause a large group of different diseases. Their clinical signs and biochemical and cellular manifestations depend on the function normally performed by the enzyme involved. Macromolecules that are usually degraded remain in lysosomes and accumulate in the cell, resulting in *lysosomal storage diseases*. This occurs at different rates, so that each disease has its own characteristic course. Twelve groups of genetically determined disorders of specific lysosomal functions are known, each with about three to ten individually defined diseases.

A. Defective uptake of enzymes into lysosomes: I-cell disease

Mucolipidosis type II alpha/beta, also called *I-cell disease* because of conspicuous cytoplasmic inclusions first described by Leroy & DeMars (1967), is a severe, progressive disorder of abnormal lysosomal transport and protein sorting, manifest in mesenchymal cells (OMIM 252500). The first step of a two-step reaction in the Golgi apparatus is defective because of homozygous or compound heterozygous mutations in the *GNPTAB* gene (OMIM 607840), located at 12q23.2, encoding the lysosomal enzyme *N*-acetylglucosamine-1-phosphotransferase, α and β subunits (GlcNAc-phosphotransferase, see p. 276). This enzyme catalyzes the first step in the synthesis of the mannose 6-phosphate determinant for targeting hydrolases into the lysosome. As a result, the recognition marker that binds mannose 6-phosphate is lacking and mucolipids accumulate in mesenchymal cells (**1**), but not in normal fibroblasts (**2**). The vesicular inclusions consist of hydrolases that cannot enter the lysosomes because the mannose 6-phosphate recognition signal is absent. Lysosomes lack several enzymes, whereas the concentration of these enzymes outside the cells is increased. Mucolipidoses usually become apparent in the first 6 months of life (**3**). The *GNPTA* gene has 21 exons spanning 85 kb DNA. Mutations result in premature translational termination. Two complementation groups have been delineated. Mucolipidosis III alpha/beta is also caused by mutation in the *GNPTAB* gene (OMIM 252600), mucolipidosis III gamma is caused by mutations in the *GNPTG* gene (252605).

B. Degradation of heparan sulfate

Lysosomal enzymes are bond specific, not substrate specific. Thus, they also degrade other glycosaminoglycans, such as dermatan sulfate, keratan sulfate, and chondroitin sulfate (mucopolysaccharides). Ten specific enzyme defects cause the mucopolysaccharide storage diseases (see p. 280). Heparan sulfate is an example of a macromolecule that is degraded stepwise by eight different lysosomal enzymes.
The first step in heparan sulfate degradation is the removal of sulfate from the terminal iduronate group by an iduronate sulfatase. A defect in the gene encoding this enzyme leads to the X-chromosomal mucopolysaccharide storage disease type II (MPS II, type Hunter). All other mucopolysaccharidoses are autosomal recessive. The second enzymatic step removes the terminal iduronate by an α-L-iduronidase. A homozygous mutation in the gene encoding the enzyme responsible for the second step leads to mucopolysaccharidosis type I (MPS I, Hurler/Scheie). Defective function in the next three enzymatic steps causes three of the four subtypes of mucopolysaccharidosis type III (Sanfilippo syndrome: MPS IIIA, MPS IIIC, and MPS IIIB). MPS IIID results from a defect in the last (eighth) step. MPS type VII (Sly) is caused by a defect in β-glucuronidase (step 7), and has a different phenotype to MPS types I, II, and III.

Further Reading

Cathey SS, et al. Molecular order in mucolipidosis II and III nomenclature. (Letter) Am J Med Genet A 2008;146A:512–513

Kornfeld S, Sly WS. I-cell disease and Pseudo-Hurler polydystrophy: Disorders of lysosomal enzyme phosphorylation and localization. In: CR Scriver, et al., eds. The Metabolic and Molecular Bases of Inherited Disease. 8th ed. New York: McGraw-Hill; 2001:3469–3482. (www.ommbid.com/)

Leroy JG, Demars RI. Mutant enzymatic and cytological phenotypes in cultured human fibroblasts. Science 1967;157:804–806

Marschner K, et al. A key enzyme in the biogenesis of lysosomes is a protease that regulates cholesterol metabolism. Science 2011;333:87–90

Tiede S, et al. Mucolipidosis II is caused by mutations in GNPTA encoding the alpha/beta GlcNAc-1-phosphotransferase. Nat Med 2005;11:1109–1112

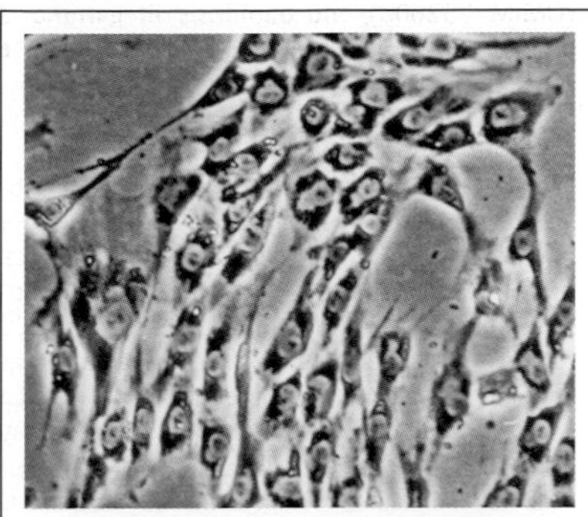

1. Fibroblast culture in I-cell disease

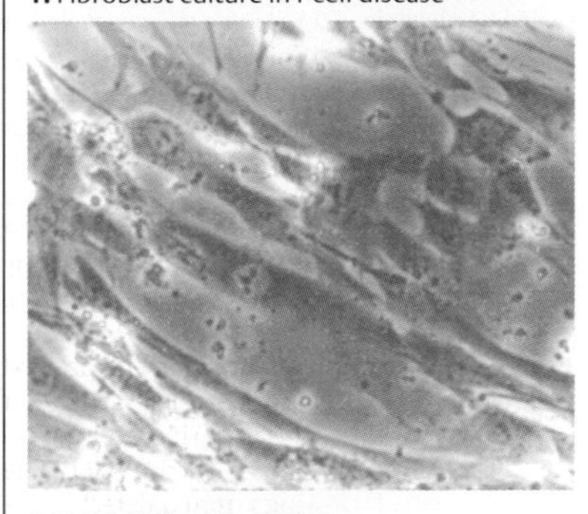

2. Normal fibroblast culture

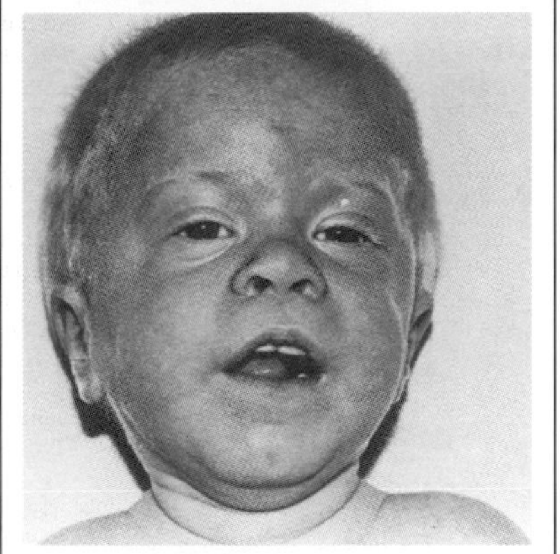

3. Patient with I-cell disease

A. Defective uptake of enzymes in lysosomes: I-cell disease

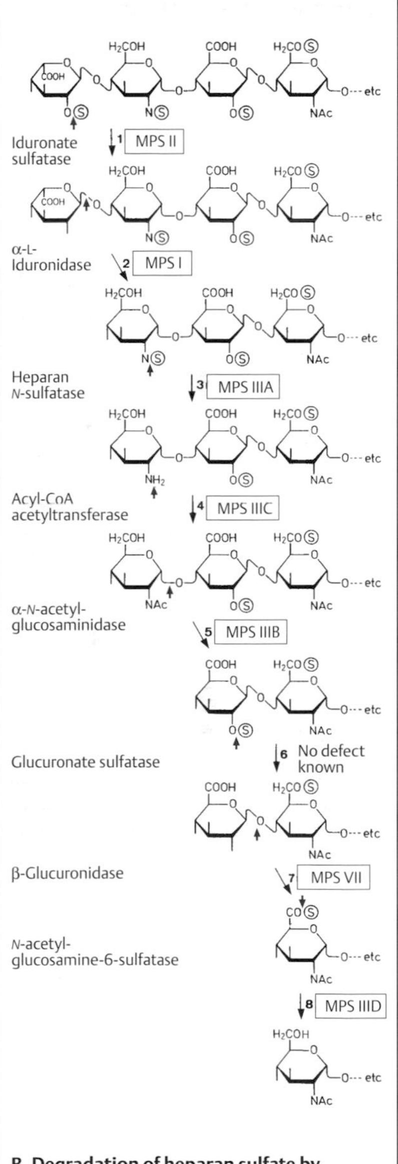

B. Degradation of heparan sulfate by eight lysosomal enzymes

Mucopolysaccharide Storage Diseases

The mucopolysaccharide storage diseases (the mucopolysaccharidoses) are a clinically and genetically heterogeneous group of 19 different lysosomal storage diseases (OMIM 253200) caused by defects in different enzymes for mucopolysaccharide degradation (glycosaminoglycans). All defects are transmitted by autosomal recessive inheritance, except for mucopolysaccharide storage disease type II (Hunter; OMIM 309900), (see Appendix, p. 416).

A. Mucopolysaccharide storage disease type I (Hurler)

Young infants with mucopolysaccharide storage disease type I (Hurler) (MPS IH) (OMIM 252800) seem normal at first. Early signs of the disease occur at about 1–2 years of age, with increasing coarsening of the facial features, retarded mental development, limited joint mobility, enlarged liver, umbilical hernia, and other signs. Radiographs show coarsening of skeletal structures (dysostosis multiplex). The photographs show the same patient at different ages (author's photographs). MPS IS (Scheie) is a clinically different, less severe allelic disease.

B. Mucopolysaccharide storage disease type II (Hunter)

This type of mucopolysaccharidosis is transmitted by X-chromosomal inheritance (OMIM 309900). Four cousins from one pedigree are shown in the diagram. Clinically, the disease is similar to, but less rapidly progressive than, MPS I. (Photographs from Passarge et al., 1974.)

Diagnosis

The diagnosis of MPS is based on the patient's history, clinical and radiological evaluation, and increased urinary concentration of one of several types of glycosaminoglycan, depending on the type of MPS. Molecular genetic diagnosis is possible in most forms.

Further Reading

Hopkin RJ, Grabowsi GA. Lysosomal storage diseases. In: Longo DL, et al., eds. Harrison's Principles of Internal Medicine. 18th ed. New York: McGraw-Hill; 2012:3191–3197

Neufeld EF, Muenzer J. The mucopolysaccharidoses. In: Scriver CR, et al., eds. The Metabolic and Molecular Bases of Inherited Disease. 8th ed. New York: McGraw-Hill; 2001:3421–3452. Available at: http://www.ommbid.com/. Accessed January 27, 2012

OMIM. Online Mendelian Inheritance of Man. (www.ncbi.nlm.nih.gov/omim)

Passarge E, et al. Diseases caused by genetic defects in lysosomal muco-polysaccharide-catabolism. Mucopolysaccharidoses. [Article in German] Dtsch Med Wochenschr 1974;99:144–155

Spranger J. Mucopolysaccharidoses. In: Rimoin DL, et al., eds. Emery and Rimoin's Principles and Practice of Medial Genetics. 5th ed. Philadelphia: Churchill Livingstone-Elsevier; 2007

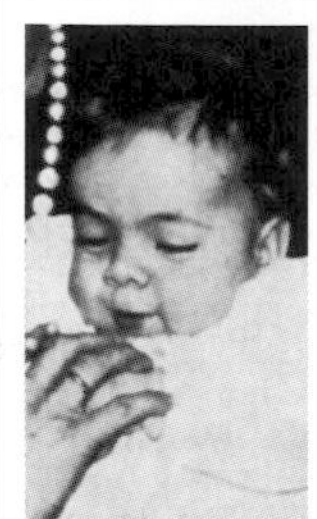

8 weeks

7 months

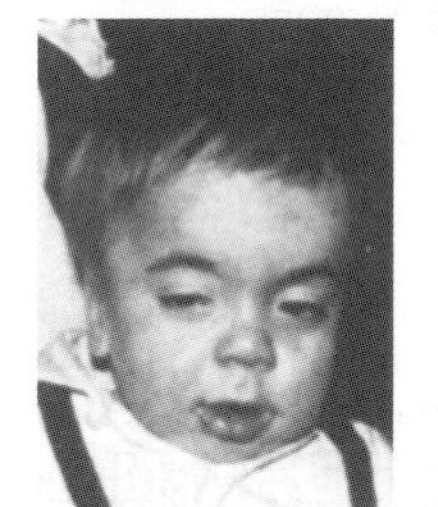

$2\frac{1}{4}$ years

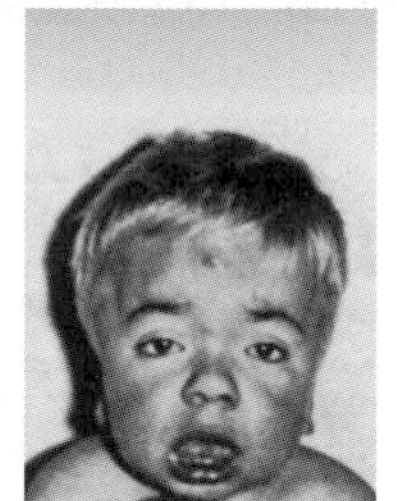

$3\frac{3}{4}$ years

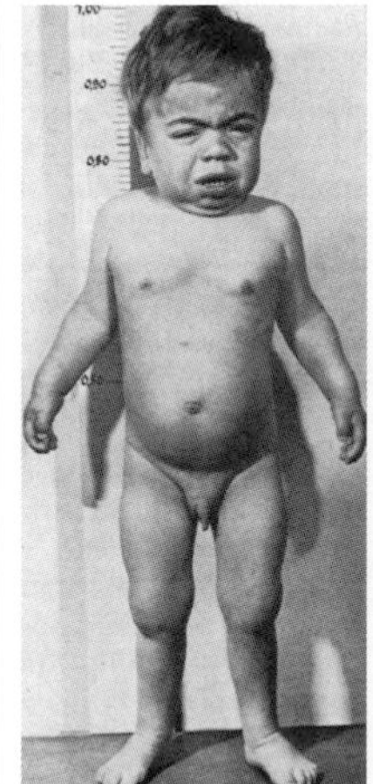

5 years

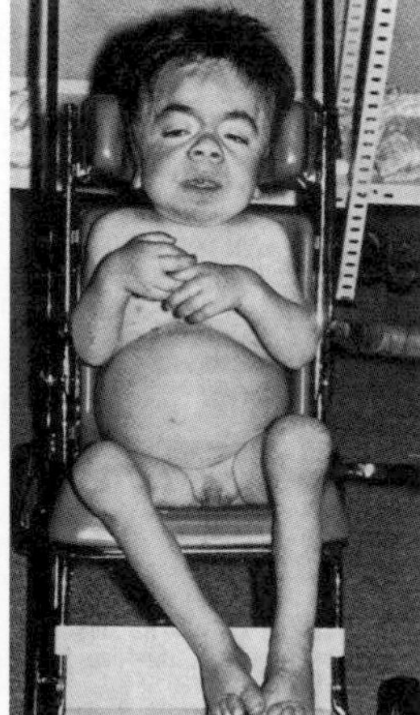

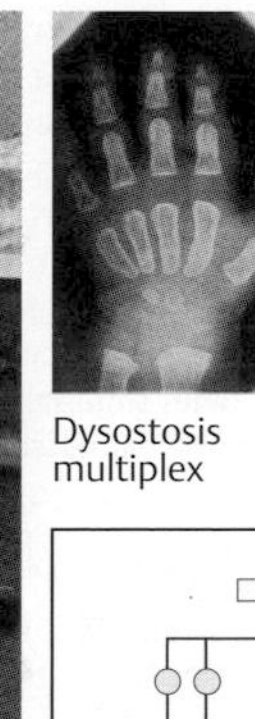

8 years

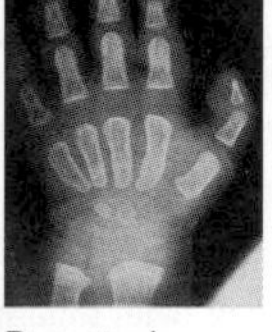

Dysostosis multiplex

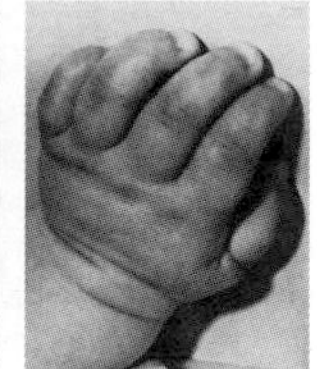

Joint contractures

A. Mucopolysaccharide storage disease type I (Hurler)

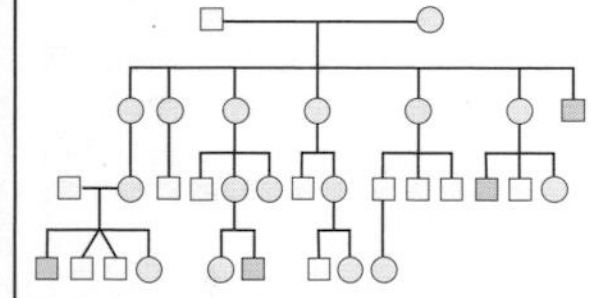

X-Chromosomal inheritance

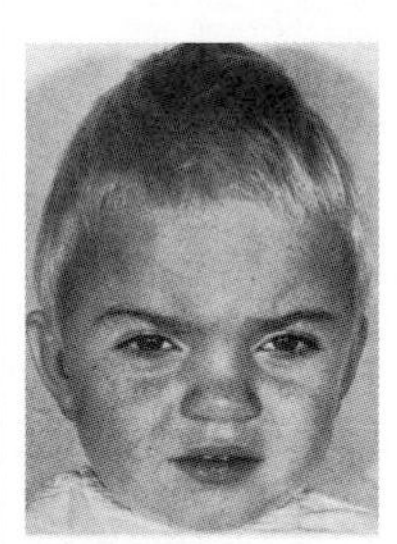

$4\frac{1}{2}$ years

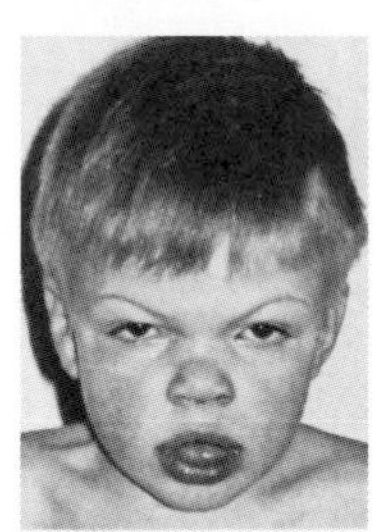

10 years

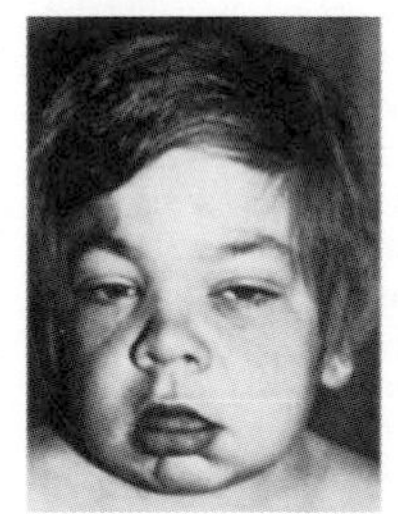

13 years

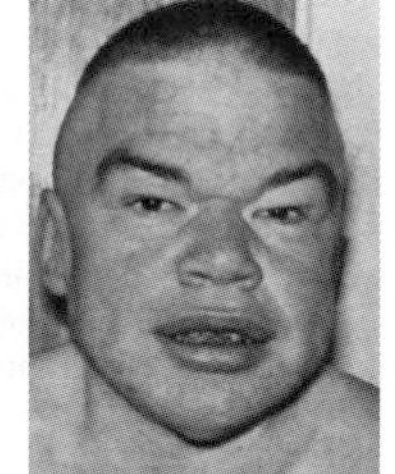

21 years

B. Mucopolysaccharide storage disease type II (Hunter)

Peroxisomal Disorders

Peroxismal disorders are a group of 20 mainly autosomal recessive disorders (OMIM 170993) characterized by neonatal hypotonicity, craniofacial abnormalities, failure to thrive, and other features.

Peroxisomes are small membrane-bound intracellular organelles of about 0.5–1.0 μm diameter, somewhat smaller than mitochondria. Their name is derived from hydrogen peroxide, which is formed as an intermediary product of oxidative metabolism. Most cells, especially in the liver and kidney, contain about 100–1000 peroxisomes. Peroxisomes are surrounded by a single-layer granular matrix, which contains about 50 matrix enzymes involved in anabolic and catabolic metabolic functions, such as β-oxidation of fatty acids, biosynthesis of phospholipids and bile acids, and others. Peroxisome biogenesis involves the synthesis of matrix proteins (peroxines) and their receptor-mediated transfer into the organelle under the control of *PEX* genes and peroxisomal targeting signals (PTS).

A. Biochemical reactions

The electron micrograph (**1**) shows three peroxisomes in a rat liver cell. The dark striated structures within the organelles are urates, a result of an enzyme that oxidizes uric acid. Peroxisomes have both catabolic (degrading) and anabolic (synthesizing) functions (**2**). Two biochemical reactions are especially important: a peroxisomal respiratory chain and the β-oxidation of very-long-chain fatty acids. In the peroxisomal respiratory chain (**3**), certain oxidases and catalases act together. Specific substrates of the oxidases are organic metabolites of intermediary metabolism. Very-long-chain fatty acids are broken down by β-oxidation (**4**) in a cycle with four enzymatic reactions. Energy production in peroxisomes is relatively inefficient compared with that of mitochondria. While free energy in mitochondria is mainly preserved in the form of ATP (adenosine triphosphate), in peroxisomes it is mostly converted into heat. Peroxisomes are probably a very early adaption of living organisms to oxygen. (Photograph from de Duve, 1984.)

B. Peroxisomal diseases

Six important autosomal recessive examples of peroxisomal diseases are listed, together with OMIM numbers. Patients with neonatal adrenoleukodystrophy do not form sufficient amounts of plasmalogens and cannot adequately degrade phytanic acid and pipecolic acid.

C. Zellweger cerebrohepatorenal syndrome

This is a characteristic autosomal recessive disease resulting from mutations in *PEX* genes (OMIM 214100). It is recognized by a characteristic facial appearance (**1–4**), extreme muscle weakness (**5**), and several accompanying manifestations such as calcified stippling of the joints on radiographs (**6**), renal cysts (**7**, **8**), and clouding of the lens and cornea. The severe form of the disease usually leads to death before the age of 1 year. (Photographs **1–5** from Passarge & McAdams, 1967.)

Further Reading

de Duve C. A Guided Tour through the Living Cell. New York: Scientific American Books; 1984

Gould SJ, Raymond BV, Valle D. The peroxisome biogenesis disorders. In: Scriver CR, et al., eds. The Metabolic and Molecular Bases of Inherited Disease. 8th ed. New York: McGraw-Hill; 2001:3181–3217

Muntau AC, et al. Defective peroxisome membrane synthesis due to mutations in human PEX3 causes Zellweger syndrome, complementation group G. Am J Hum Genet 2000;67:967–975

Passarge E, McAdams AJ. Cerebro-hepato-renal syndrome. A newly recognized hereditary disorder of multiple congenital defects, including sudanophilic leukodystrophy, cirrhosis of the liver, and polycystic kidneys. J Pediatr 1967;71:691–702

Peroxisome Database. Available at: http://www.peroxisomedb.org/. Accessed January 27, 2012

Shimozawa N, et al. A human gene responsible for Zellweger syndrome that affects peroxisome assembly. Science 1992;255:1132–1134

Wanders RJ, et al. Peroxisomal disorders. In: Rimoin DL, et al., eds. Emery and Rimoin's Principles and Practice of Medical Genetics. 5th ed. Philadelphia: Churchill Livingstone-Elsevier; 2007

Wanders RJA, Waterham HR. Biochemistry of mammalian peroxisomes revisited. Annu Rev Biochem 2006;75:295–332

0.5 μm

1. Peroxisomes in a rat liver cell

a) Catabolic
- H_2O_2-involving cellular respiration
- β-Oxidation of long-chain fatty acid, Prostaglandins, cholesterol side chains and others
- Purines, urates
- Pipecolic acid, dicarboxy acids
- Ethanol, methanol

b) Anabolic
- Phospholipids (plasmalogen)
- Cholesterol, bile acids
- Gluconeogenesis
- Glyoxalate transamination

2. Function of peroxisomes

O_2 —Oxidases→ H_2O_2 —Catalase→ $2 H_2O$ and heat (→ O_2)

$R\text{-}H_2$ → R; $R'\text{-}H_2$ → R'

R: D- and L-Amino acids, Hydroxy acids, Purines, urates, Oxalate polyamines, Fatty acid derivatives

R': Ethanol, Methanol, Nitrites, Quinones, Formates

3. Peroxisomal respiratory chain

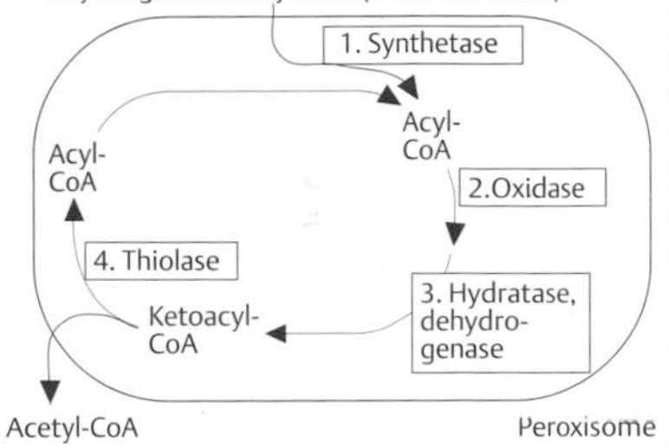

4. β-Oxidation

A. Biochemical reactions in peroxisomes

214100 Zellweger cerebrohepatorenal syndrome
202370 Neonatal adrenoleukodystrophy
266510 Infantile Refsum disease
239400 Hyperpipecolic acidemia
215100 Rhizomelic chondrodysplasia punctata
259900 Primary hyperoxaluria type I
and others

B. Examples of peroxisomal diseases

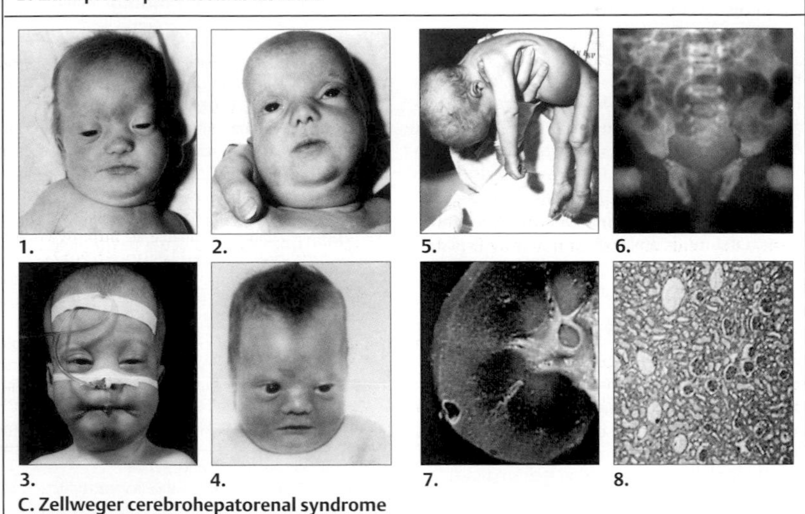

1. 2. 3. 4. 5. 6. 7. 8.

C. Zellweger cerebrohepatorenal syndrome

Components of the Immune System

The immune system consists of an adaptive and an innate system. The adaptive immune system is a relatively young evolutionary immune response mediated by T and by B lymphocytes. Three key properties of the adaptive immune system, which is present in vertrabrates only, are: (i) a highly diverse repertoire of antigen receptors, (ii) a memory function for antigens previously encountered, (iii) immune tolerance to avoid unwarranted damage to the host's own cells. The innate immune system, originated from invertebrates, consists of germline-encoded proteins defending against pathogen-associated molecular patterns. Innate immunity detects and destroys microorganisms immediately, within 4 hours. Infectious organisms that breach these early lines of defense are repulsed by the adaptive immune response. Three principal methods are used by the host to inactivate and eliminate invading foreign molecules: (i) neutralizing extracellular pathogens by antibodies, (ii) destroying a cell that is infected, and (iii) killing bacteria directly by macrophages.

A. Lymphatic organs

The primary lymphoid tissues are the thymus and bone marrow. Secondary lymphoid tissues are the lymph nodes, the spleen, and accessory lymphoid tissue (ALT) including tonsils and appendix.

B. Lymphocytes

There are about 2×10^{12} lymphocytes in the human body, and these are equal in mass to the brain or the liver. Their specific role in adaptive immunity was shown in the late 1950s by irradiation experiments. A mouse irradiated above a certain dose was no longer able to mount an immune response. The adaptive immune response could be restored by lymphocytes from an unirradiated mouse.

C. T cells and B cells

Two functionally different types of lymphocyte exist: T lymphocytes and B lymphocytes. Immature T lymphocytes differentiate in the thymus during embryonic and fetal development (thus designated T cells). B lymphocytes differentiate in the bone marrow in mammals and in the bursa of Fabricius in birds (thus designated B cells). Further maturation and differentiation take place in the lymph nodes (T cells) and in the spleen (B cells). In addition, soluble molecules, a large group of cytokines, interact with specific cellular receptors and mediate inflammatory and immune responses (not shown).

D. Cellular and humoral immune response

The first phase of the immune response to an antigen (e.g., a bacterium, virus, fungus, or foreign protein) is rapid proliferation of B cells (*humoral immune response*). Mature B cells develop into plasma cells, which secrete effector molecules, the antibodies (immunoglobulins). These interact with the antigen by binding to it. The humoral immune response is rapid, but is ineffective against microorganisms that have invaded body cells. These induce a *cellular immune response*, performed by different types of T cells.

E. Immunoglobulin molecules

The basic structural motif of an antibody molecule (immunoglobulin, Ig) is a Y-shaped protein composed of different polypeptide chains. A common type of Ig has two heavy chains (H chains) and two light chains (L chains). Both chains contain regions with variable sequences and constant sequences of amino acids. At defined sites the chains are held together by disulfide bonds.

F. Antigen–antibody binding

An antibody molecule is bivalent by having two identical antigen-binding sites at the end of each arm of the Y shaped end. The structure that an antibody recognizes is an antigenic determinant or epitope. Here a foreign molecule, the antigen, is recognized and firmly bound to six hypervariable regions (three from the light chain and three from the heavy chain). In the hypervariable regions, the amino acid sequences differ from one molecule to the next. As a result the antibodies can bind a wide spectrum of different antigenic molecules.

(Figures adapted from Alberts et al., 2008.)

Further Reading

Abbas AK, Lichtman A, Pillai S. Cellular and Molecular Immunology. 7th ed. Philadelphia, Pa: Elsevier-Saunders; 2012 (with Student Consult Online Acess)

Alberts B, et al. Molecular Biology of the Cell. 5th ed. New York: Garland Science; 2008

Haynes BF, Soderberg KA, Fauci AS. Introduction to the immune system. In: Longo DL, et al., eds. Harrison's Principles of Internal Medicine. 18th ed. New York: McGraw-Hill; 2012

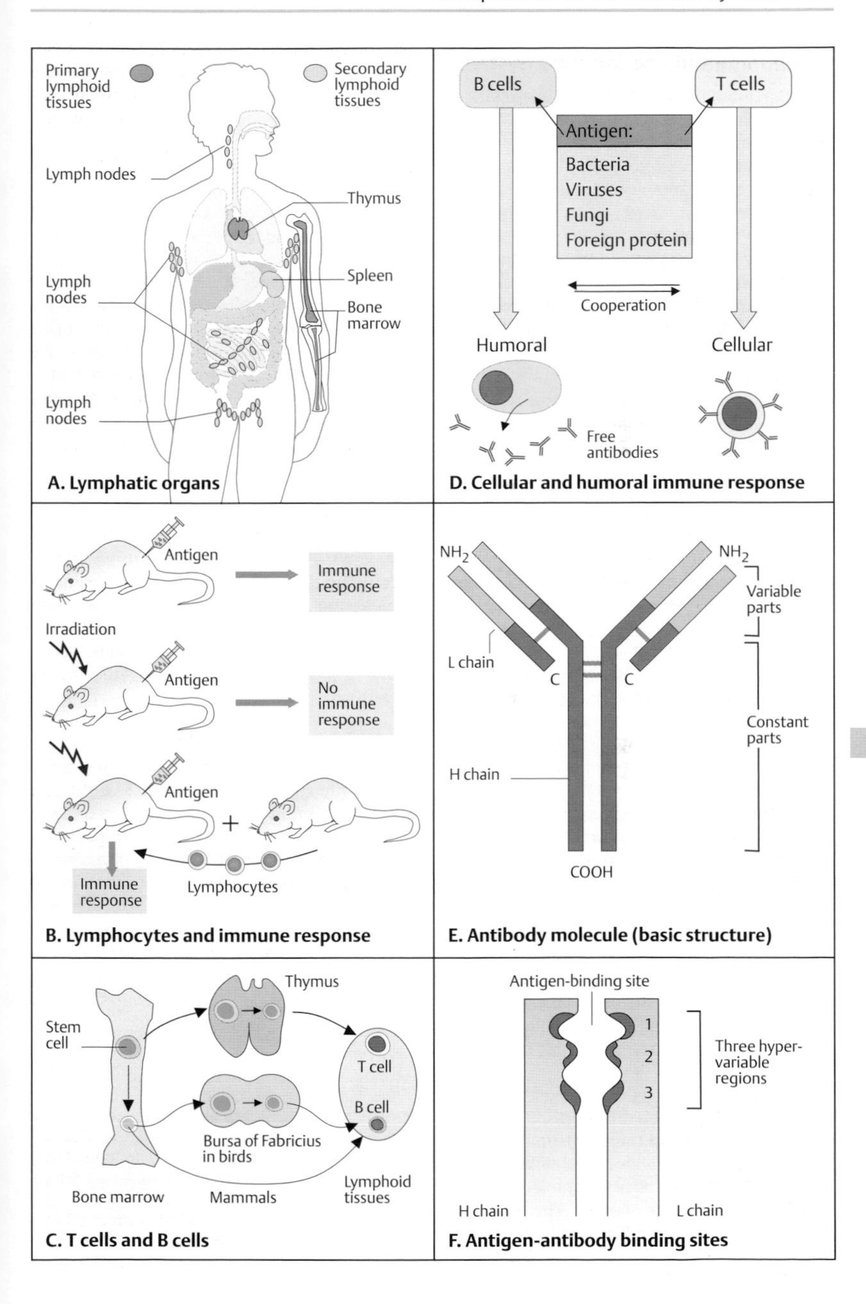
Primary lymphoid tissues
Secondary lymphoid tissues
Lymph nodes
Thymus
Lymph nodes
Spleen
Bone marrow
Lymph nodes
A. Lymphatic organs
B cells
T cells
Antigen:
Bacteria
Viruses
Fungi
Foreign protein
Cooperation
Humoral
Cellular
Free antibodies
D. Cellular and humoral immune response
Antigen
Immune response
Irradiation
Antigen
No immune response
Antigen
+
Immune response
Lymphocytes
B. Lymphocytes and immune response
NH2
NH2
Variable parts
L chain
C
C
Constant parts
H chain
COOH
E. Antibody molecule (basic structure)
Thymus
Stem cell
T cell
B cell
Bursa of Fabricius in birds
Lymphoid tissues
Bone marrow
Mammals
C. T cells and B cells
Antigen-binding site
1
2
3
Three hyper-variable regions
H chain
L chain
F. Antigen-antibody binding sites

Immunoglobulin Molecules

Immunoglobulins are the effector molecules of the immune system. They exist in two basic forms: as membrane-bound cell surface receptor molecules or as free antibodies, each in a vast array of variants. Each molecule has two binding sites for a foreign molecule (the antigen). Each binding site contains regions that vary in their amino acid sequence among individual immunoglobulins. This enables each molecule to bind a foreign protein specifically to a particular epitope. A vast number of different effector molecules provide spectacular diversity. Although they differ in details of their structure and function, they share a relatively simple basic pattern, which is derived from a common ancestral molecule during evolution.

A. Immunoglobulin G (IgG)

Immunoglobulin G is the prototype of secreted antibody molecules in humoral immunity. The molecule is bivalent: it has two H chains and two L chains, held together by disulfide bonds. Each H chain has three constant regions (domains C_H1, C_H2, and C_H3) and one variable region (V_H) with a total of 440 amino acids (110 in the V region). Each L chain (214 amino acids) has one constant (C_L) and one variable (V_L) domain, also with 110 amino acids. The variable regions of each chain contain the antigen-binding sites with three hypervariable regions called complementarity determining regions (CDRs), where the actual physical contact with foreign epitopes takes place. A region called the hinge joins the constant region 1 (C_H1) and constant region 2 (C_H2) of the heavy chains. This allows considerable flexibility of the molecule. The two H chains, and the H and the L chains, are held together by interchain disulfide bonds (–S–S–). Furthermore, there are intrachain disulfide bonds within each of the polypeptide chains. The L chains are of one of two types: kappa (κ) or lambda (λ).

Five classes of immunoglobulin exist: IgA, IgD, IgE, IgG, and IgM. They differ from each other in the constant part of their H chain (C_α, C_δ, C_ε, C_γ). The largest secreted immunoglobulin, IgM, exists as a pentamer of five Ig molecules held together by disulfide bonds. The different types of H chains are referred to as isotypes.

Two proteolytic enzymes cleave Ig molecules into characteristic fragments (**1**). Papain cleaves at the amino terminal disulfide bond (–S–S–) to produce three pieces: two Fab fragments (fragment antigen binding) and one Fc fragment (fragment crystallizable). Pepsin cleaves Ig into one $F(ab')_2$ fragment with an intact disulfide bridge, and one Fc fragment (pFc'), and several small pieces.

B. Structure of an immunoglobulin

Three globular domains of similar size form the Y-shaped antibody molecule (immunoglobulin, Ig). The three regions, called domains, are connected by a flexible tether, called a hinge. The two heavy (H) chains are shown in yellow and grey, the two light (L) chains in brown. The two antigen-binding sites are the two ends of the Y. (Figure adapted from Murphy et al., 2011.)

C. Genes encoding different polypeptides of an immunoglobulin

Each immunoglobulin and receptor molecule is encoded by a different DNA sequence; these sequences belong to a multigene family. The genes for the H chain are located on chromosome 14 at q32 in humans and chromosome 14 in mice. The genes for the κ light chain are located on chromosome 2 at p12 in humans and chromosome 6 in mice. The genes for the λ light chain are located on chromosome 22 at q11 in humans and on chromosome 16 in mice.

Further Reading

Abbas AK, Lichtman A, Pillai S. Cellular and Molecular Immunology. 7th ed. Philadelphia: Elsevier-Saunders; 2012 (with Student Consult Online Acess)

Alberts B, et al. Molecular Biology of the Cell. 5th ed. New York: Garland Science; 2008

Delves PJ, Roitt IM. The immune system. N Engl J Med 2000;343:37–49, 108–117

Haynes BF, Soderberg KA, Fauci AS. Introduction to the immune system. In: Longo DL, et al., eds. Harrison's Principles of Internal Medicine. 18th ed. New York: McGraw-Hill; 2012:2650–2685

Murphy K, Travers P, Walport M. Janeway's Immunobiology. 8th ed. London, New York: Taylor & Francis, Garland Science; 2011

Nossal GJ. The double helix and immunology. Nature 2003;421:440–444

Strominger JL. Developmental biology of T cell receptors. Science 1989;244:943–950

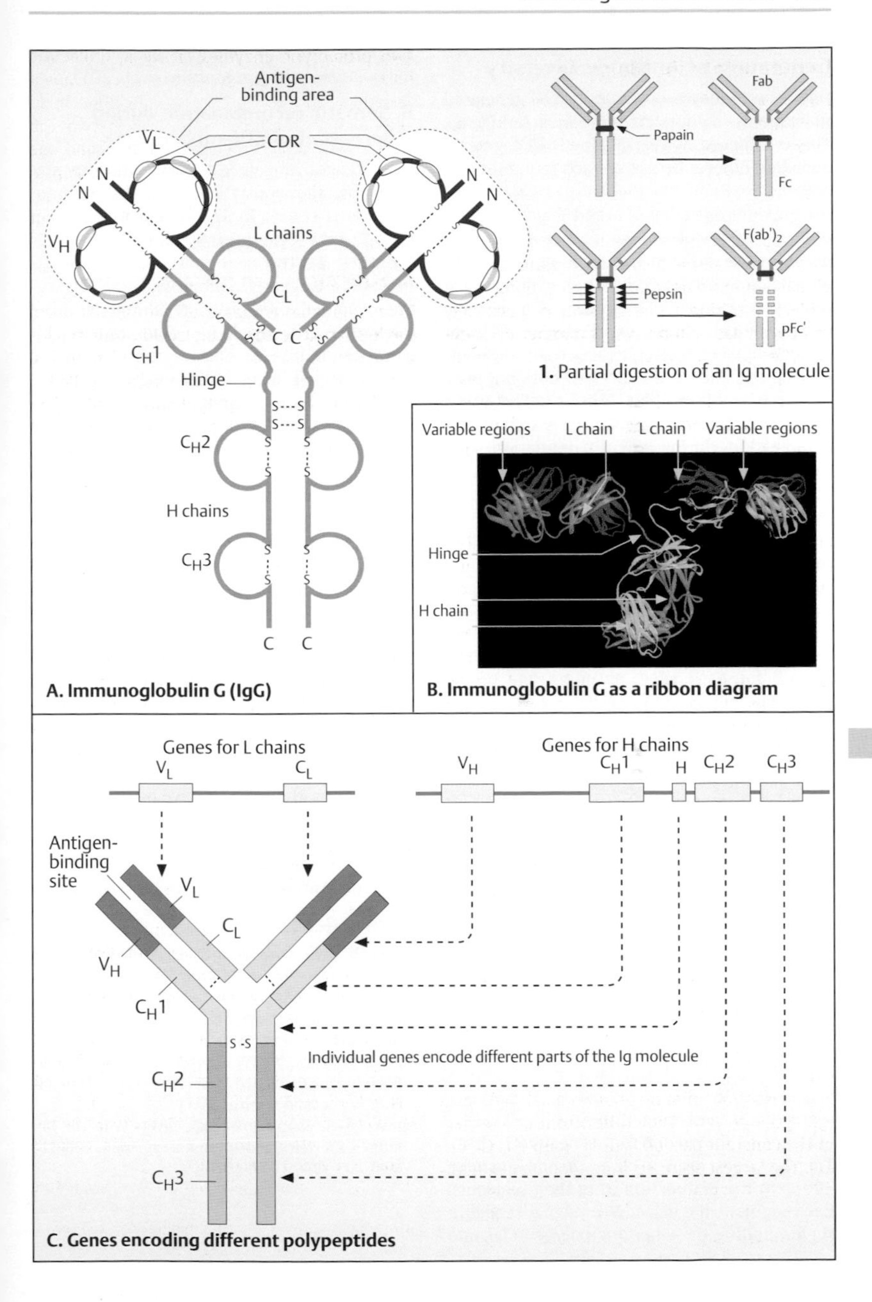

Antigen-binding area
CDR
V_L
V_H
N
L chains
C_L
C C
C_H1
Hinge
S---S
C_H2
H chains
C_H3
C C
A. Immunoglobulin G (IgG)
Fab
Papain
Fc
F(ab')$_2$
Pepsin
pFc'
1. Partial digestion of an Ig molecule
Variable regions
L chain
L chain
Variable regions
Hinge
H chain
B. Immunoglobulin G as a ribbon diagram
Genes for L chains
V_L
C_L
Genes for H chains
V_H
C_H1
H
C_H2
C_H3
Antigen-binding site
V_L
C_L
V_H
C_H1
S -S
C_H2
C_H3
Individual genes encode different parts of the Ig molecule
C. Genes encoding different polypeptides

Generation of Antibody Diversity

Different antibody molecules can be generated in response to foreign molecules (antigens). This is achieved by rearranging DNA segments encoding different parts of each immune molecule.

During differentiation of B and T lymphocytes, different segments of genes present in the germline are randomly selected and arranged in a new combination that is specific for each cell and its progeny. This generates a great diversity of cells, each expressing a different antigen-binding molecule (about 10^{12} different molecules). Each cell expresses only one particular type of receptor. This is called *allelic exclusion*. The main mechanism is somatic recombination during the differentiation of B cells and T cells. For each domain of an Ig molecule several selectable sequences are present in genomic DNA. In lymphocyte DNA, these are combined in a variety of combinations for each molecule. Functional genes are produced by gene rearrangements of different elements (V, D, J) from the germline DNA. For each element, several selectable sequences are present. These sequences are combined in a variety of combinations for each gene, encoding one of the polypeptide chains of the T cell or B cell receptor. In addition, somatic mutations occur in the hypervariable regions, leading to further genetic differences.

A. Organization of immunoglobulin loci in the human genome

Both B cells and T cells pass through a series of defined steps of differentiation to reach their final stage. During this process, somatic recombination occurs within the gene loci encoding the two L chains and the H chain. The corresponding loci consist of coding sequences for each of the following segments: V (variable), J (joining), and C (constant). For the H chain, there are in addition 25 D (diversity) segments (D_H). The number of functional gene segments is about 30 V_λ, 40 V_κ, and 50 V_H. These gene segments are rearranged during B cell differentiation as shown in **B** for the heavy (H) chain. The C_H genes form a large cluster spanning 200 kb in the 3′ direction from the J segments. Each segment has a leader or signal sequence (L) for guiding the emerging polypeptides into the lumen of the endoplasmatic reticulum. (Figure adapted from Murphy et al., 2011.)

B. Somatic recombination during differentiation of lymphocytes

Here the rearrangements by somatic recombination are shown for the locus of the immunoglobulin H chain (**1**). The first rearrangement joins D and J segments in lymphocyte DNA (D–J joining, **2**). The next rearrangement brings one of the V_H genes together with the D–J segments joined previously, resulting in V–D–J joining (**3**). The result is a transcription unit consisting of one V_H, one D_H, one J, and one C gene arranged in this order in the 5′ to 3′ direction (**4**). Each C segment consists of different exons, corresponding to the domains of the complete C region and different isotypes (C_λ, C_δ), etc. The primary transcript is spliced to yield an mRNA consisting of one V, one D, one J, and one C segment (**5**). This is translated into a polypeptide corresponding to the complete H chain (**6**). Following posttranslational modifications, such as glycosylation, the final H chain is produced (**7**).

The result is an array of cells, each with a unique combination of molecules in the antigen-binding site. This provides each molecule with an antigen-binding specificity that differs from that of all other cells. The L chains and the genes encoding the T-cell receptor (see p. 292) are formed in a similar manner. In contrast to the H chains, the L chains have no diversity (D) genes, so that a J gene and a V gene are directly joined by somatic recombination during DNA rearrangement in the lymphocytes. (Figure adapted from Abbas et al., 2012.)

Further Reading

Abbas AK, Lichtman A, Pillai S. Cellular and Molecular Immunology. 7th ed. Philadelphia: Elsevier-Saunders; 2012 (with Student Consult Online Acess)

Alberts B, et al. Molecular Biology of the Cell. 5th ed. New York: Garland Science; 2008

Murphy K, Travers P, Walport M. Janeway's Immunobiology. 8th ed. London, New York: Taylor & Francis, Garland Science; 2011

Schwartz RS. Shattuck lecture: Diversity of the immune repertoire and immunoregulation. N Engl J Med 2003;348:1017–1026

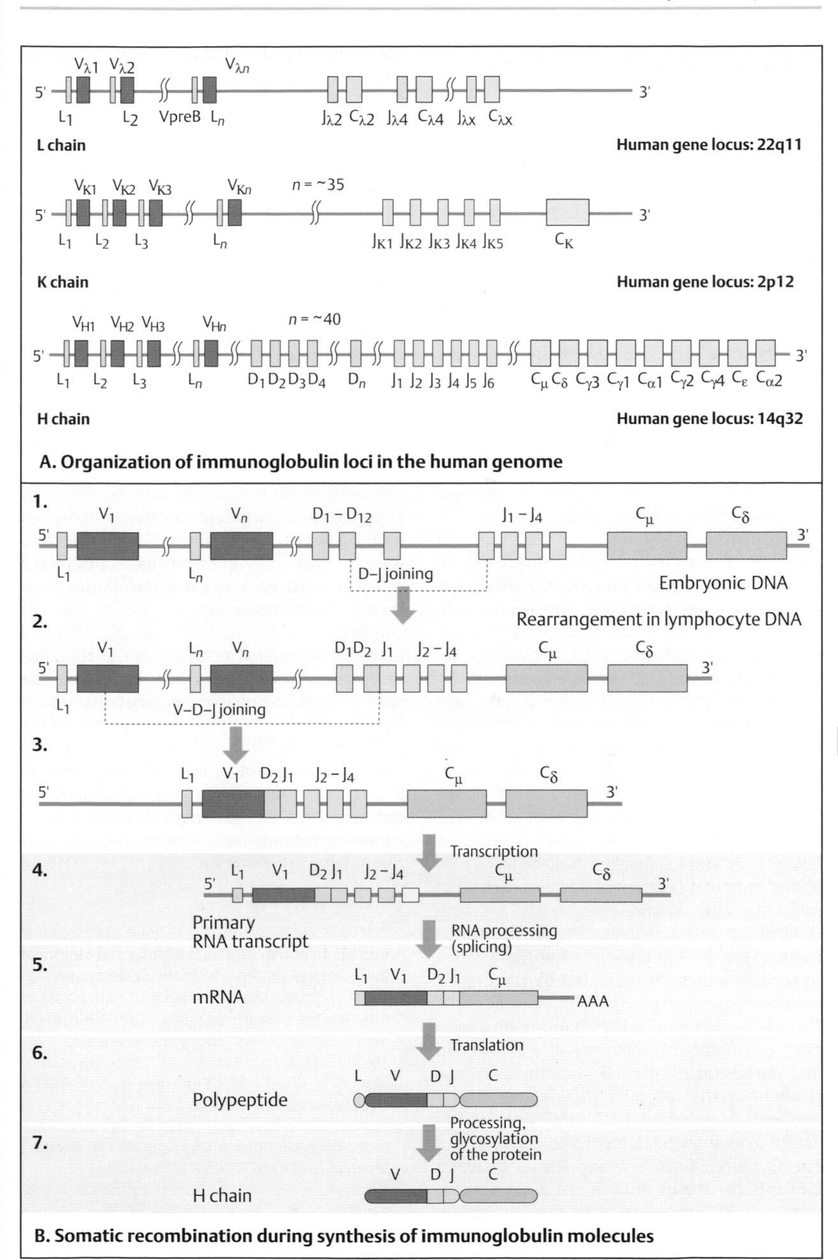

A. Organization of immunoglobulin loci in the human genome

B. Somatic recombination during synthesis of immunoglobulin molecules

Immunoglobulin Gene Rearrangement

The lymphocyte-specific rearrangement of coding DNA segments for immunoglobulins and T-cell receptors follows strict rules under the control of recombination-activating genes (*RAG1*, OMIM 179615; *RAG2*, OMIM 179616). It must be assured that a V gene segment joins a D or J segment, and not another V segment. Conserved noncoding DNA sequences located adjacent to the point of recombination guide this process. The recombinations of V, D, and J segments are performed by lymphocyte-specific DNA-modifying enzymes called V(D)J recombinase. The lymphoid-specific genes *RAG1* and *RAG2* are expressed in pre-B cells and immature T cells. The other enzymes involved are DNA-modifying proteins required for double-stranded DNA repair, DNA bending, and ligation of the ends of broken DNA.

A. DNA recognition sequences

The recognition sequences for the *RAG1* and *RAG2* genes are noncoding, but highly conserved DNA segments of 7 bp (a CACAGTG heptamer) and a 9-bp nonamer (ACAAAAACC). Recombination takes place between segments located on the same chromosome. A gene segment flanked by a recombination signal sequence with a 12-bp spacer can only be joined to one flanked by a 23-bp spacer. This is referred to as the 12/23 rule. As a consequence, a D segment with a 12-bp spacer on both sides must be joined to a heavy chain J segment. Likewise a heavy chain V segment can be joined only to a D segment, but not to a J segment, because V and J segments are both flanked by 23-bp spacers. The RAG proteins induce DNA double breaks at the flanking sequences. Rejoining is mediated by DNA repair enzymes (see p. 86).

During formation of an H chain, nonhomologous pairing of the heptamer of a D segment and the heptamer of a J segment occurs. Lymphocyte-specific recombinase enzymes, RAG-1 and RAG-2, recognize DNA sequences located at the 3′ end of each V exon and the 5′ end of each J exon. The RAG-1 and RAG-2 enzymes cleave both strands of DNA at the recognition sites. The RAG proteins align the recognition site sequences and the endonuclease of the RAG complex cuts both strands of the DNA at the 5′ end. This creates a DNA hairpin in the gene segment-coding region. The D and J segments are then joined (D–J joining) by means of recombination: the spacer of 12 bp or 23 bp and all the intervening DNA form a loop. This loop is excised, and the D and J segments are joined by the nonhomologous end-joining machinery. By pairing and recombination of the recognition sequences at the 5′ end of a D–J segment and the recognition sequence at the 3′ end of a V gene, a V segment is joined to the D–J segment. Diversity of T-cell receptors (see p. 292) is generated in the same manner. (Figure adapted from Abbas et al., 2012.)

B. Antigen genetic diversity

The total diversity, of about 10^{18} possible combinations for all types of immunoglobulins and T-cell receptor genes, is the result of different mechanisms. To begin with, different numbers of variable DNA segments are available for different chains (250–1000 for the H chain, 250 for the L chains, 75 for the α chain of the T-cell receptor TCRα, etc.). The different D and J segments also multiply the number of possible combinations. Finally, DNA sequence changes (somatic mutations) occur regularly in the hypervariable regions, further increasing the total number of possible combinations.

Medical relevance

Mutations in the *RAG1* and *RAG2* genes (OMIM 179615/16) cause severe combined immune deficiency (SCID), see page 298, as well as Omenn syndrome, which is caused by defective V(D)J recombination.

Further Reading

Abbas AK, Lichtman A, Pillai S. Cellular and Molecular Immunology, 7th ed. Philadelphia: Elsevier-Saunders; 2012 (with Student Consult Online Acess)

Agrawal A, Schatz DG. RAG1 and RAG2 form a stable postcleavage synaptic complex with DNA containing signal ends in V(D)J recombination. Cell 1997;89:43–53

Alberts B, et al. Molecular Biology of the Cell. 5th ed. New York: Garland Science; 2008

Matthews AGW, et al. RAG2 PHD finger couples histone H3 lysine 4 trimethylation with V(D)J recombination. Nature 2007;450:1106–1110

Schwarz K, et al. RAG mutations in human B cell-negative SCID. Science 1996;274:97–99

Undergraduate Immunology Class at Davidson College, Davidson, NC 28035 (www.bio.davidson.edu/courses/immunology/ bio307.html)

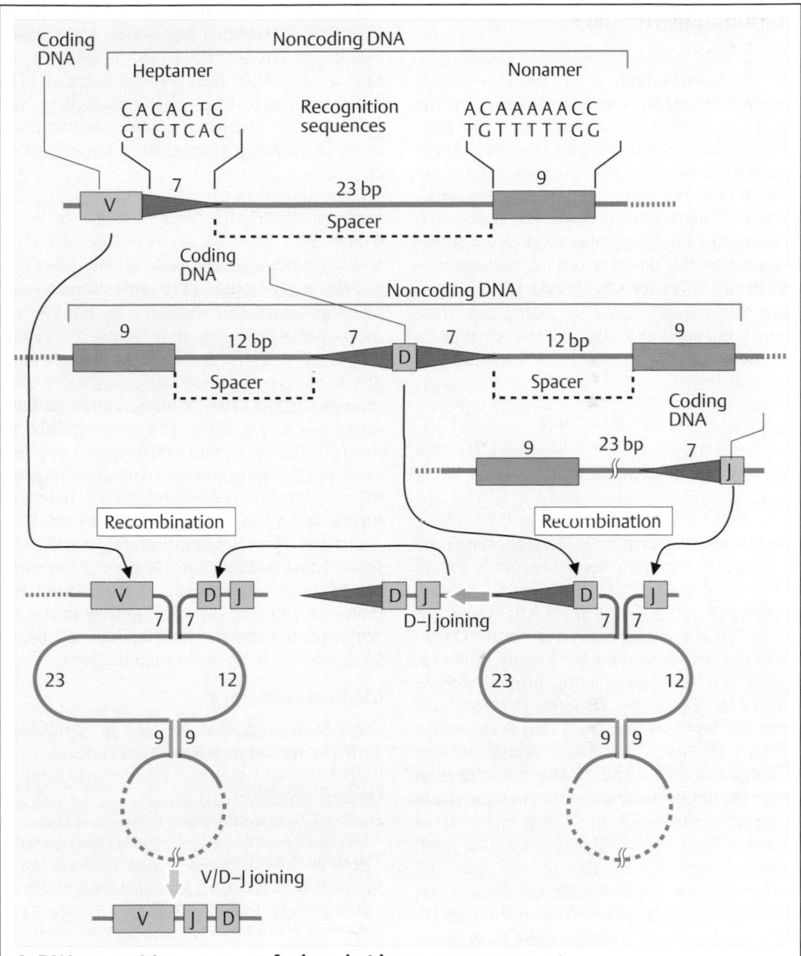

A. DNA recognition sequences for lymphoid gene rearrangement

Mechanism	Immunoglobulin H chain	Immunoglobulin L chain	TCRαβ α	TCRαβ β	TCRγδ γ	TCRγδ δ
Variable domain	250–1000	250	75	25	7	10
Number of D segments	12	0	0	2	0	2
Number of J segments	4	4	50	12	2	2
Variable segment combinations	65 000–250 000		1825		70	
Total diversity	10^{11}		10^{16}		10^{18}	

B. Genetic diversity in immunoglobulin and T-cell receptor genes

T-cell Receptor

T lymphocytes have a membrane-bound protein, the T cell receptor (TCR). The TCR enables T cells to recognize antigens displayed on the surface of a cell from their own body. The antigens are small peptides derived from viruses or intracellular bacteria. They are presented to the TCR by the MHC class I and class II molecules (p. 294). The genes encoding the α/β and δ TCRs are arranged in germline DNA according to the segments they encode, i.e., variable segments (V), diversity (D), joining (J), and constant regions (C). Gene segments are rearranged during maturation in the thymus in the same manner as the genes encoding the immunoglobulins.

A. T-cell receptor structure

The T-cell receptor (TCR) resembles the Fab fragment of an immunoglobulin molecule. It is a heterodimer of one α and one β polypeptide chain covalently linked by a disulfide bridge and produced as an integral membrane protein (**1**). The basic structure is similar to that of the cell surface immunoglobulins. A subtype of TCR consists of a γ chain and a δ chain. The three-dimensional structure (**2**) reveals that the hypervariable regions CDR1, 2, and 3 can be aligned with antigen-binding sites of the antibodies. Genomic DNA contains genes for 50–70 V segments, two D segments, 12–60 J segments, and two C segments. In a given T cell, only one of the two parental genes for a given α chain and β chain is expressed (allelic exclusion). The β chain is the slightly larger chain. The V region of each chain consists of 102–109 amino acids and contains three hypervariable regions, as do the immunoglobulin molecules. Genes for T-cell receptor γ and δ chains exist in addition to those for α and β chains. (**2** adapted from Murphy et al., 2011.)

B. Interaction of T-cell receptor and MHC

The TCR interacts with an antigen-presenting cell carrying a major histocompatibility complex (MHC) molecule, either a human leukocyte antigen (HLA) class I or a class II molecule (**1**). The three-dimensional structure (**2**) shows the tight connection between the interacting molecules. T cells able to destroy an infected cell (cytolytic T lymphocytes, CTLs, or "killer cells") recognize their antigen on MHC class I molecules, while "helper cell" T cells specifically bind to MHC class II molecules. CD8 cells are restricted to MHC class I molecules, and CD4 cells are restricted to MHC class II molecules. (Figure in 2 adapted from Murphy et al., 2011.)

C. Recognition of antigen and T-cell activation

A few of the many molecules involved in T-cell activation are shown. CD4 and CD8 molecules serve as restriction elements. By binding directly to the MHC class II molecule, CD4 stabilizes the TCR interaction with the peptide antigen. CD8 takes this role with cytolytic T lymphocytes (CTLs, "killer cells") binding to MHC class I molecules. Upon antigen recognition by the TCR, the associated CD3 complex is phosphorylated. Complete T-cell activation requires the engagement of costimulatory receptors (CD28, LFA-1) on T cells and their ligands (B-7 and ICAM-1) on antigen-presenting cells. This signal transduction activates the *IL2* (interleukin 2) gene. Interleukin 2 is the main T-cell growth factor of T cells and is responsible for progression in the cell cycle from the G1 to the S phase.

Medical relevance

The gp120 protein of the HIV virus interacts with the second domain of CD4 cells.

Further Reading

Abbas AK, Lichtman A, Pillai S. Cellular and Molecular Immunology. 7th ed. Philadelphia: Elsevier-Saunders; 2012 (with Student Consult Online Acess)

Alberts B, et al. Molecular Biology of the Cell. 5th ed. New York: Garland Science; 2008

Amadou C, et al. Localization of new genes and markers to the distal part of the human major histocompatibility complex (MHC) region and comparison with the mouse: new insights into the evolution of mammalian genomes. Genomics 1995;26:9–20

Fugger L, et al. The role of human major histocompatibility complex (HLA) genes in disease. In: Scriver CR, et al., eds. The Metabolic and Molecular Bases of Inherited Disease. 8th ed. New York: McGraw-Hill; 2001

Jiang H, Chess L. Regulation of immune responses by T cells. N Engl J Med 2006;354:1166–1176

Murphy K, Travers P, Walport M. Janeway's Immunobiology. 8th ed. London, New York: Taylor & Francis, Garland Science: 2011

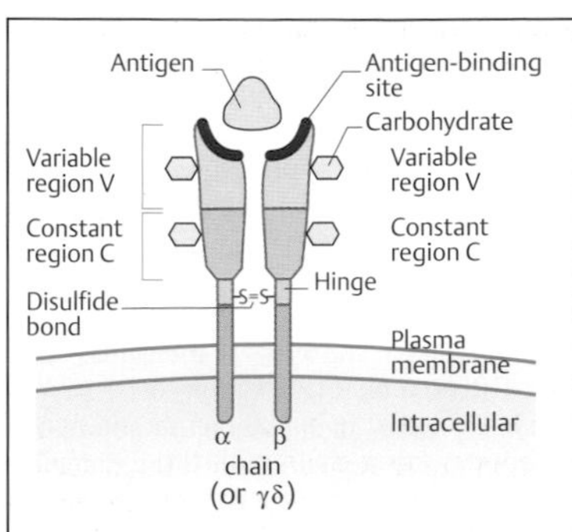

1. Schematic structure

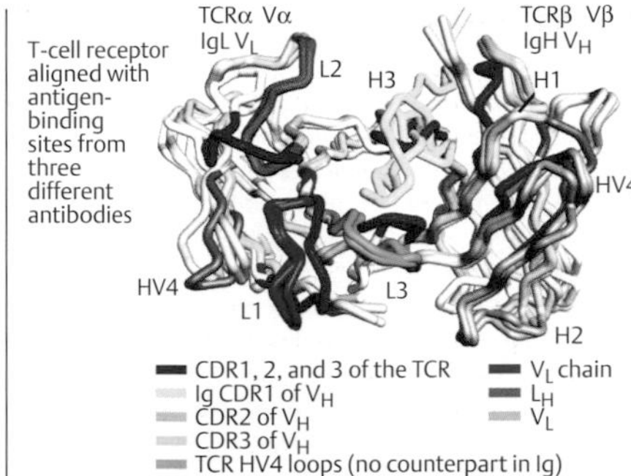

2. Crystal structure at 2.5 Å

A. T-cell receptor

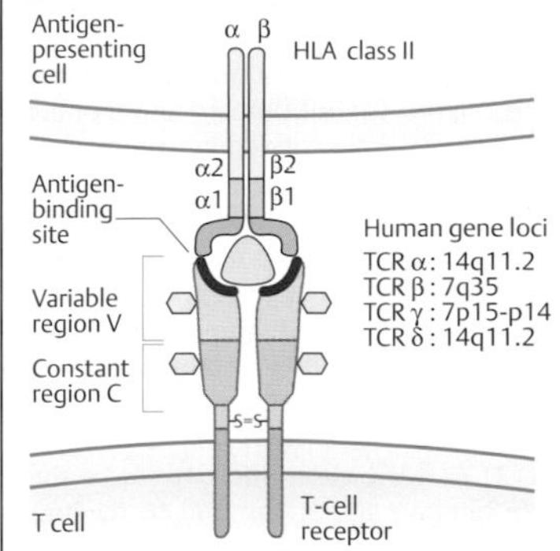

1. Schematic structure

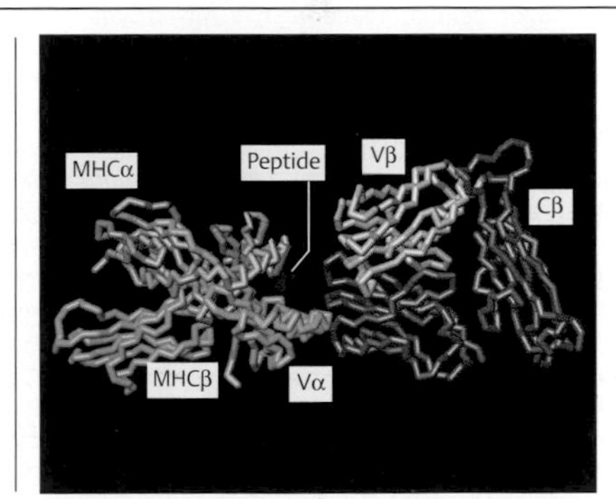

2. Three-dimensional structure

B. Interaction of T-cell receptor and MHC class I or class II molecules

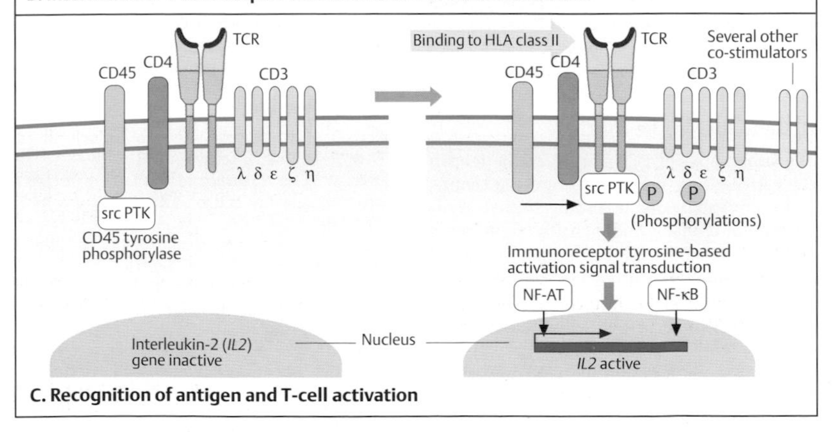

C. Recognition of antigen and T-cell activation

The MHC Region

The major histocompatibility complex (MHC), usually referred to as the human leukocyte antigen (HLA) system (human leukocyte antigen) is a large chromosomal region containing more than 400 genes along nearly 4000 kb (4 Megabases, Mb) of DNA with the highest density of genes in the human genome (six per 100 kb, average 1). It is the most polymorphic region with up to 30–60 alleles at many loci, with more than 400 alleles at HLA-A, more than 700 at HLA-B, and more than 500 at HLA-DRB1. Each MHC molecule is composed of two polypeptide chains and is expressed on the surface of various cells. This can be demonstrated serologically by a binding assay or by a cytotoxic cellular reaction in a mixed lymphocyte test.

A. Genomic organization of the MHC complex

In man, the MHC region is located on the short arm of human chromosome 6 (6p21.3); in the mouse, on chromosome 17. The genes are grouped into three classes: I–III. Class I and class II belong to the HLA system; in mice it is called the H2 system. Class I in humans consists of HLA-A, HLA-B, HLA-C, and several other loci (HLA-E to -J). Mouse class I genes form two groups: D and L at the 3′ end, and K at the 5′ end. Human class II MHC molecules are grouped into DP, DQ, and DR (in mice they are I-A and I-E; letter I, not the roman numeral). Class III includes genes that are not directly involved in the immune system as well as genes involved in the immune response, such as tumor necrosis factor (TFN), lymphotoxin, and others.

B. Gene loci in the MHC

The subgroups DP, DQ, and DR of the human class II HLA loci, oriented toward the centromere, are subdivided according to their composition of α chains and β chains (some genes between DM and DO are not shown). The DR locus contains three β chains, each of which can pair with the single α chain. Thus, three sets of genes can produce four types of DR molecules. The class III loci, located between class I and class II, contain genes encoding complement factors C2 and C4 (*C4A*, *C4B*), steroid 21-hydroxylase (*CYP21B*), and cytokines (tumor necrosis factor, *TNFA*; lymphotoxin, *LTA*, *LTB*). Several other class III genes are located in the class III region (not shown).

The three-dimensional structure shows the tight arrangement of polypeptide chains of the MHC class II protein (**2**). (Figure adapted from Murphy et al., 2011.)

C. Structure of MHC molecules

The MHC class I and class II molecules have distinct structures. Class I molecules, HLA-A, -B, and -C, consist of single membrane-bound polypeptides, an α chain with three domains: α1 at the extracellular N-terminal, α2, and α3 (**1**). The α chain is noncovalently bound to nonpolymorphic β_2-microglobulin, which is encoded by a gene on human chromosome 15. The three extracellular domains of the α chain have about 90 amino acids each. The α1 and α2 regions form the highly polymorphic peptide-binding region. The α3 domain and β_2-microglobulin structurally correspond to an immunoglobulin-like region. Class II MHC molecules (**2**) have two polypeptide chains, α and β, each with two domains, α1 and α2, and β1 and β2; each has 90 amino acids and a transmembrane region of 25 amino acids. The peptide-binding regions α1 and β1 are highly polymorphic. The crystalline structures of the MHC molecules have revealed details of the binding mechanisms (**3**). An MHC class I molecule has a single peptide-binding site at the end of the molecule, and this can accommodate a peptide of 8–10 amino acids. In a class II MHC protein, the antigen-binding groove does not narrow at the end. Therefore, it can bind longer peptides, which are usually 12–20 amino acids long. (Figure from Bjorkman et al., 1987.)

Further Reading

Alberts B, et al. Molecular Biology of the Cell. 5th ed. New York: Garland Science; 2008

Bjorkman PJ, et al. Structure of the human class I histocompatibility antigen, HLA-A2. Nature 1987; 329:506–512

Dausset J. The major histocompatibility complex in man. Science 1981;213:1469–1474

Murphy K, Travers P, Walport M. Janeway's Immunobiology. 8th ed. New York, London: Taylor & Francis, Garland Science; 2011

Nepom GT. The major histocompatibility complex. In: Longo DL, et al., eds. Harrison's Principles of Internal Medicine. 18th ed. New York: McGraw-Hill; 2012

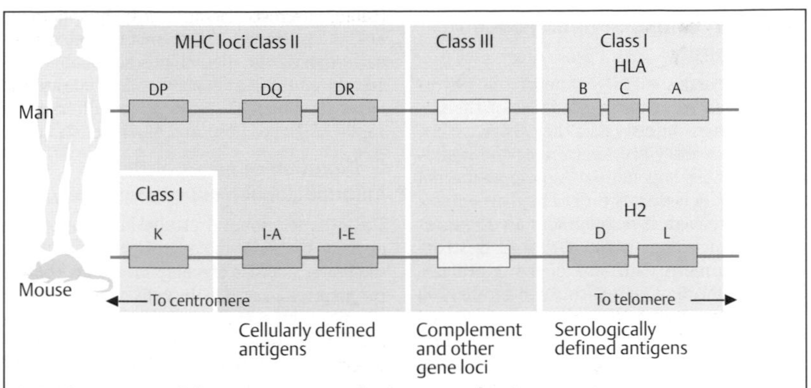

A. Basic structure of the MHC gene complex in man and mouse

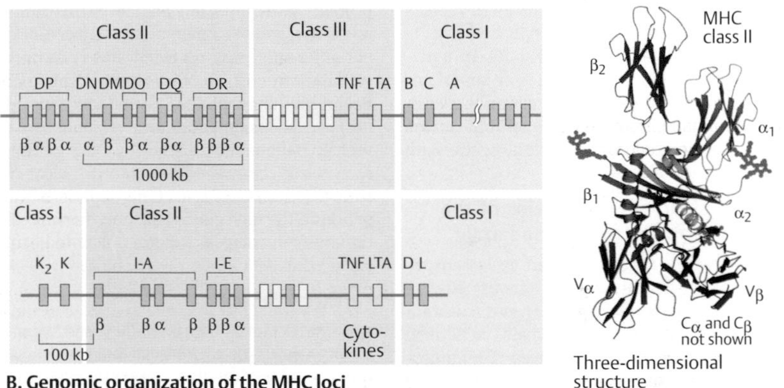

B. Genomic organization of the MHC loci

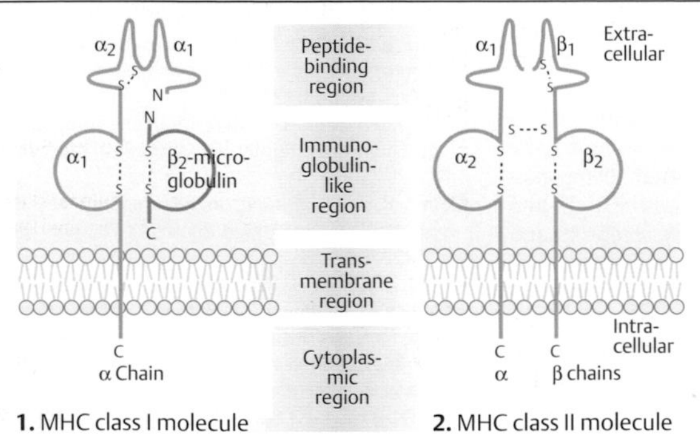

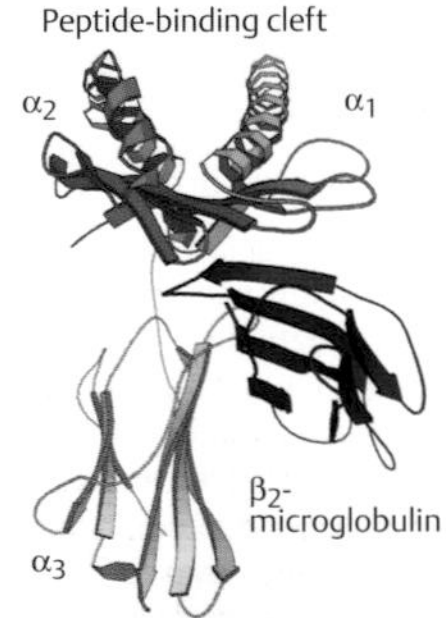

1. MHC class I molecule

2. MHC class II molecule

3. Ribbon diagram of an MHC class I molecule

C. MHC molecules of class I and class II

Evolution of the Immunoglobulin Super Family

The structural similarity of most cell–cell or cell–antigen recognition molecules of the immune system suggests that they share a common evolutionary history, forming the immunoglobulin (Ig) superfamily. A prominent functional feature is their ability to bind other molecules with high specificity. The predominant structural leitmotif is the Y-shaped Ig structure. A prominent attribute of the Ig proteins is their ability to adhere directly to another cell or a foreign protein through specialized cell membrane proteins known as the cell adhesion molecules (CAMs). This large family of molecules is classified into four major families: the Ig superfamily, cadherins, integrins, and selectins. CAMs are usually built up in repeats of domains with different properties. The Ig superfamiliy comprises a large group of molecules with globular domains. A gene superfamily is a collection of genes of common evolutionary origin that arose by gene duplications and subsequently diverged into genes with new, different functions.

A. Basic structure of proteins of the immunoglobulin supergene family

Characteristic structures shared by Igs are repeated domains, usually of about 70–110 amino acids with variable (V) and constant (C) domains (**1**). Each Ig domain is derived from conserved DNA sequences. The prototype, IgG, has three C domains in the heavy chain, one C domain in each of the two light chains, and a variable domain in each chain. The Ig molecules of the T-cell receptors (TCRs) and the class I and class II MHC molecules are basically similar. Although their genes are located on different chromosomes, the gene products form functional complexes with each other. Others, such as the V, D, and J gene segments of all antigen receptors and their genes for the C domain, form gene clusters. Genes of the MHC loci and for the two CD8 chains lie together.

Accessory molecules such as CD2, CD3, CD4, CD8, and thymosine 1 (Thy-1) are members of this family with a relatively simple, but similar structure (**2**). Other members of the Ig superfamily are cell adhesion molecules, such as the Fc receptor II (FcRII); the polyimmunoglobulin receptor (pIgR), which transports antibodies through the membranes of epithelial cells; NCAM (neural cell adhesion molecules); and PDGFR (platelet-derived growth factor receptor) (**3**). (Figure adapted from Hunkapiller & Hood, 1989, and Alberts et al., 2008.)

B. Evolution of genes of the immunoglobulin supergene family

Distinct evolutionary relationships can be recognized by the homology of genes for Ig-like molecules and their gene products. A precursor gene for a variant (V) region and a constant (C) region must have arisen from a primordial cell by duplication, and then subsequently diverged into different cell surface receptor genes. Further duplication events have resulted in structurally related gene segments encoding proteins with repeated domain structures. At an early stage, rearrangements between different gene segments occurred and became the standard for Igs, T-cell receptors, and CD8. Other members of this superfamily evolved into cell adhesion molecules without somatic recombination, such as the thymosine (Thy-1) receptor or the poly-Ig receptor. It is clear that somatic recombination of the genes for antigen-binding molecules has enormous evolutionary advantages. (Figure adapted from Hood et al., 1985.)

Further Reading

Abbas AK, Lichtman A, Pillai S. Cellular and Molecular Immunology. 7th ed. Philadelphia: Elsevier-Saunders; 2012 (with Student Consult Online Access)

Alberts B, et al. Molecular Biology of the Cell. 5th ed. New York: Garland Science; 2008

Hood L, Kronenberg M, Hunkapiller T. T cell antigen receptors and the immunoglobulin supergene family. Cell 1985;40:225–229

Hunkapiller T, Hood L. Diversity of the immunoglobulin gene superfamily. Adv Immunol 1989;44:1–63

Klein J, Sato A. Advances in immunology. The HLA system. New Engl J Med 2000;343:702–709 (part I), 782–786

Klein J, Takahata N. Where do we come from? The Molecular Evidence for Human Descent. Heidelberg: Springer-Verlag; 2002

Shiina T, et al. Molecular dynamics of MHC genesis unraveled by sequence analysis of the 1,796,938-bp HLA class I region. Proc Natl Acad Sci USA 1999;96:13282–13287

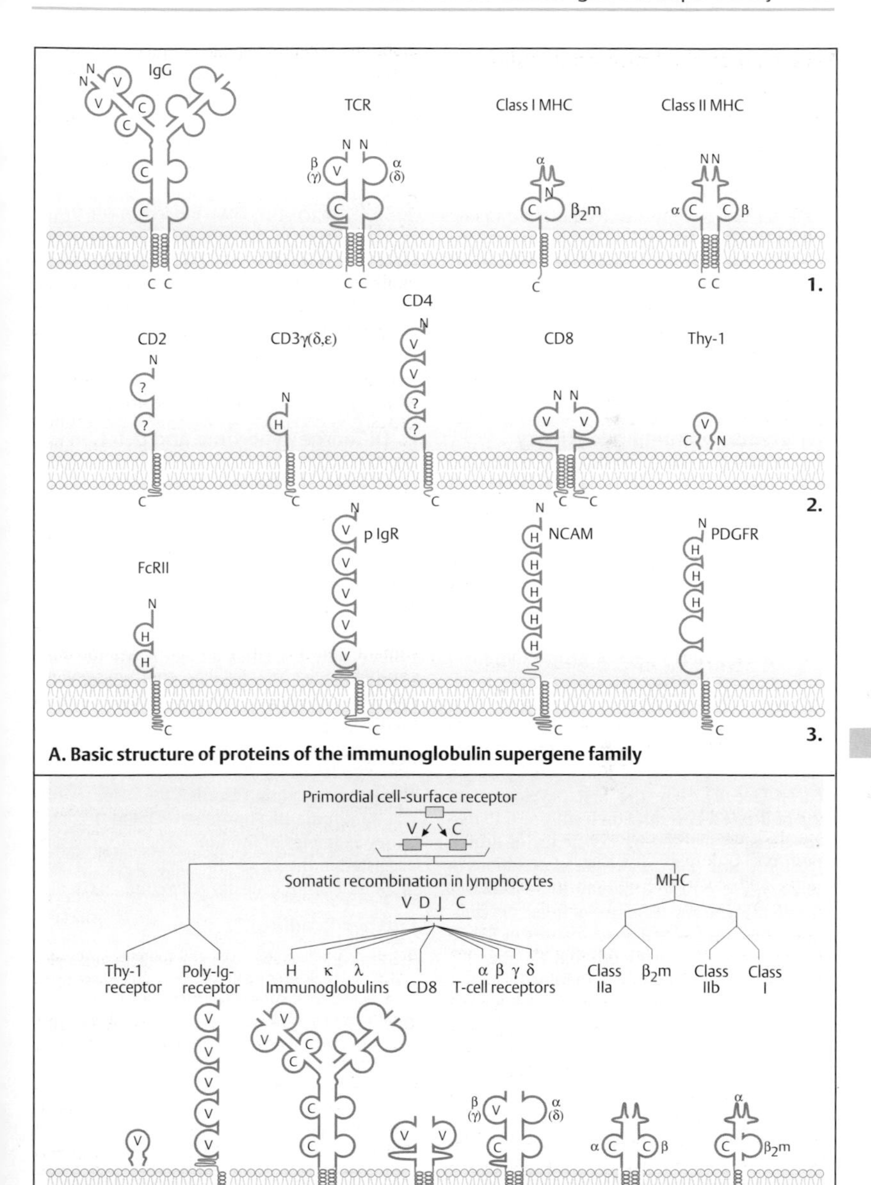

A. Basic structure of proteins of the immunoglobulin supergene family

B. Evolution of genes of the immunoglobulin supergene family

Primary Immunodeficiency Diseases

Primary immunodeficiency diseases comprise a group of about 200 genetic disorders resulting from mutations in about 150 different genes, and they have a prevalence of about 1 in 20000 individuals. The mutational changes in the various proteins involved in the immune system result in severe, often life-threatening diseases. Genetic immunodeficiency diseases occur either isolated or as manifestations of multisystem diseases. Primary immunodeficiency disorders involve innate humoral immunity, innate cell-mediated immunity, or disorders of humoral and cell-mediated adaptive immunity.

A. Hereditary immunodeficiency diseases: overview

This group of diseases can be classified according to which branch of the immune system is primarily or solely involved. One group, severe combined immune deficiency (SCID), results from a genetic block of differentiation of precursor cells before they differentiate into B cells or T cells. Thus, both cell types are involved. It is a heterogeneous group of genetic disorders resulting from various defects in both B-cell and T-cell differentiation.

X-linked agammaglobulinemia type Bruton (OMIM 300300) was the first hereditary immune deficiency described, in 1952 by Ogden Bruton (see **B**). Here the first developmental step of B-cell differentiation from pre-B to mature B cell is blocked by deficiency of Bruton tyrosine kinase (Btk) as a result of mutations in the *BTK* gene on the X chromosome (Xq22.1). This gene is the key regulator of B-cell development. The immunodeficiency results in failure of B lymphocytes to mature and failure of Ig heavy chain rearrangements. Other forms involve later steps of differentiation (variable immune deficiency) or isolated Ig isotype (subclass) deficiencies. Several immune deficiency diseases involve T cells, such as T-cell receptor (TCR) signal transduction, V(D)J recombination (mutations in *RAG1* and *RAG2* genes; OMIM 179615/16), cytokine signal transduction, and apoptosis regulation. Several disorders involve T-cell activation and the function of one or both major subsets of T cells: CD4 or CD8. Predisposition to tumors of the lymphoid system and autoimmune dysfunction is relatively frequent in immunodeficiency diseases. Effective therapy by bone marrow transplantation is possible in some of the diseases.

B. Severe combined immune deficiency

SCID is a heterogeneous group of disorders. The most common is X-linked SCID (OMIM 308380, 300400). SCIDX1 (OMIM 300400) results from mutations in the X-chromosomal gene *IL2RG* on Xq13.1, encoding the γ subunit of the interleukin 2 receptor, gamma-C (IL2Rγ; 308380). This subunit is shared with other interleukin receptors. *IL2RG* is an atypical member of the cytokine receptor family. More than 200 different mutations have been recorded. (Figure adapted from Burmester & Pezzuto, 2003.)

C. DiGeorge syndrome and deletion 22q11

DiGeorge syndrome (OMIM 188400) is a combination of T-cell defects associated with a highly variable spectrum of congenital malformations involving absence of the thymus and other derivatives of the embryonic third and fourth branchial arches. If the parathyroid glands are absent, neonatal hypocalcemia is immediately life threatening. In addition, abnormalities of the face may be present. DiGeorge syndrome is now considered part of an overlapping spectrum of disorders involving various-sized deletions of 22q11 (microdeletion syndromes). This includes a previously recognized distinct disorder, the velocardiofacial syndrome of Shprintzen (OMIM 192430). Other examples of immune deficiency diseases are listed in the Appendix, Table p. 417. (Figure adapted from Burmester & Pezzuto, 2003.)

Further Reading

Belmont JW, Puck JM. T cell and combined immunodeficiency disorders. In: Scriver CR, et al., eds. The Metabolic and Molecular Bases of Inherited Disease. 8th ed. New York: McGraw-Hill; 2001:4751–4783. (www.ommbid.com/OMMBID/)

Burmester G-R. Pezzutto: Color Atlas of Immunology. Stuttgart: Thieme; 2003

Fischer A. Primary immune deficiency diseases. In: Longo DL, et al., eds. Harrison's Principles of Internal Medicine. 18th ed. New York: McGraw-Hill; 2012

Notarangelo LD, et al; International Union of Immunological Societies Expert Committee on Primary Immunodeficiencies. Primary immunodeficiencies: 2009 update. J Allergy Clin Immunol 2009;124: 1161–1178

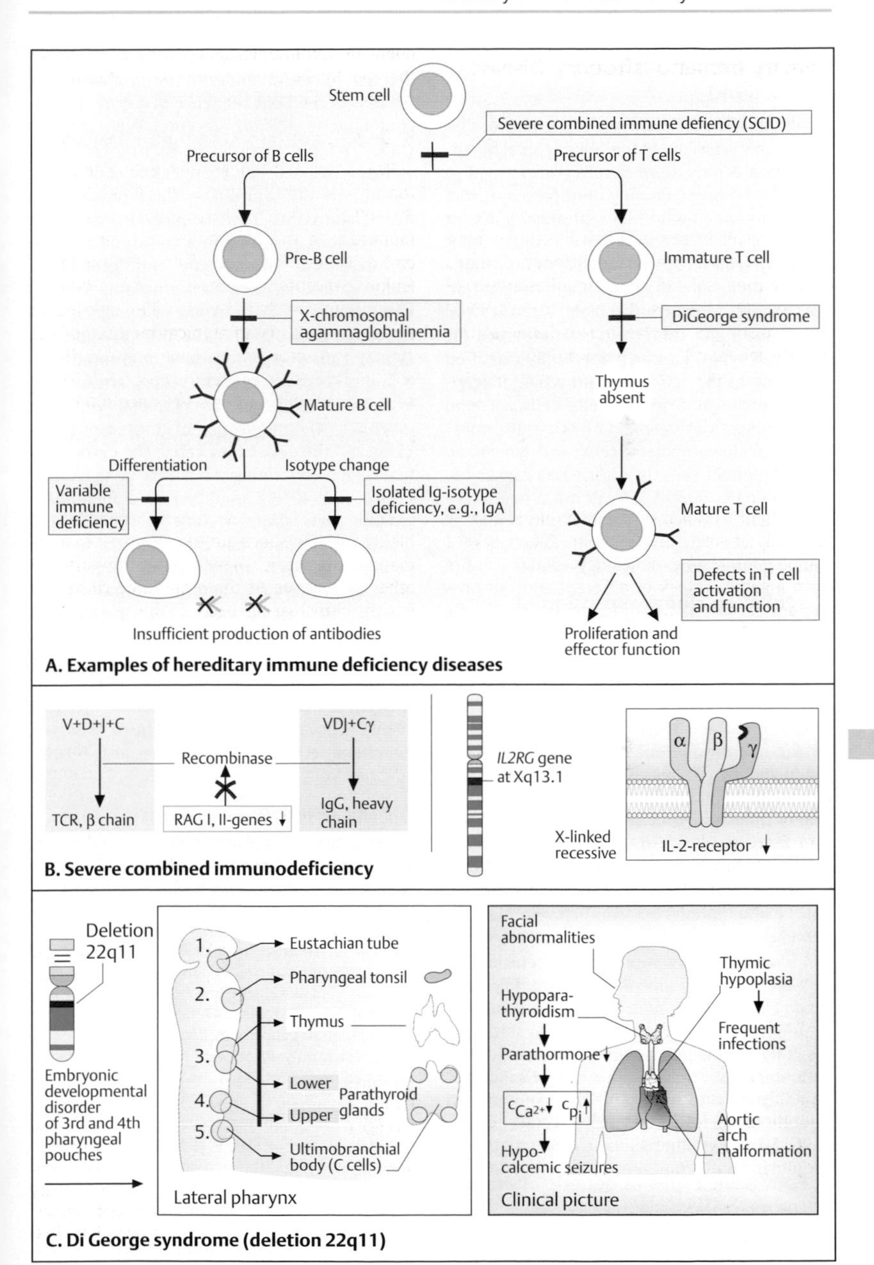

A. Examples of hereditary immune deficiency diseases

B. Severe combined immunodeficiency

C. Di George syndrome (deletion 22q11)

Genetic Causes of Cancer: Background

Cancer is a large, heterogeneous group of disorders characterized at the cellular level by unrestrained proliferation of cells transformed to a malignant state following multiple sequential mutations and genomic rearrangements in specific genes (cancer genes). Cancer cells break two rules imposed on all cells in a multicellular organism: they and their progeny do not adhere to restrained cell division, and they invade and colonize tissues reserved for other cell types. Cancer is medically classified according to the cell type from which it originates: *carcinoma* from epithelial cells, *sarcoma* from connective tissue and muscle cells, *leukemia* from hematopoietic cells, and *lymphoma* from lymphoid cells. In most cases cancer occurs in single individuals without a hereditary component. However, familial clustering is possible when a predisposing mutation in a cancer gene is present in the germline.

A. Multistep clonal expansion of malignancy

Starting usually from a single cell, a population of genetically altered cells acquires properties that lend the cells a selective advantage over their normal neighbors. This resembles a Darwinian microevolutionary process. Undifferentiated dividing cells, arising from stem cells, differentiate into cells that no longer divide, unless they receive a signal for cell proliferation. If such a cell sustains a genetic alteration (**1**), it may divide, which normally would be inhibited. If passing an inhibitory barrier (repair or elimination), a second change may follow (**2**). Again a selective barrier can be passed (**3**). Eventually tumor progression ensues (**4**). Dividing cells are prone to errors in replication. If an error affects a gene involved in control of proliferation and is not recognized and successfully eliminated, it will be transmitted to daughter cells. The estimated rate of spontaneous mutations is about 10^{-6} per gene per cell division (Alberts et al., 2008, p. 1209). (Figure based on Nowell, 1976.)

B. Four basic types of genetic alteration in tumor cells

The genetic alterations affecting growth-controlling genes can be classified into four major categories. (**1**) *Somatic mutation*: here a deletion of two adenines (A) in position 128 is present in the DNA sequence of a growth-controlling gene, *TGFBR2* (receptor type 2 of the transforming growth factor β), in a colorectal cancer (CRC) cell line changes the codon AAG (lysine) to GCC (alanine). This converts the next codons into TGG (tryptophan) and TGA (stop codon), resulting in a truncated protein. (**2**) *Reciprocal chromosome translocation*: a translocation between a chromosome 1 and a chromosome 17 disrupts one or two genes in a neuroblastoma (OMIM 256700) cell line. (**3**) *Loss or gain of a chromosome*: a chromosome 3 and a chromosome 12 (yellow arrows) are lost in a clone of a colorectal cell line (SW837). (**4**) *Amplification* of chromosomal regions and the genes they carry: the chromosomal region containing the N-*myc* gene (*NMYC*; OMIM 164840) is amplified in a metaphase (yellow parts) from a clone of the CRC cell line SW837, which has expanded through 25 generations. Such amplification, occurring in 40% of childhood neuroblastoma (OMIM 256700), indicates a drastically poorer prognosis. Amplifications were first visualized as homogeneously stained regions (HSRs; Biedler & Spengler, 1976). In addition, epigenetic changes or catastrophic chromosome breaking events (chromosthripsis) have been observed (Stephens et al., 2011). (Data and figures adapted from Lengauer et al., 1998.)

Further Reading

Alberts B, et al. Molecular Biology of the Cell. 5th ed. New York: Garland Science; 2008

Biedler JL, Spengler BA. Metaphase chromosome anomaly: association with drug resistance and cell-specific products. Science 1976;191:185–187

Hanahan D, Weinberg RA. Hallmarks of cancer: the next generation. Cell 2011;144:646–674

Lengauer C, et al. Genetic instabilities in human cancers. Nature 1998;396:643–649

Morin PJ, et al. Cancer genetics. In: Longo DL, et al., eds. Harrison's Principles of Internal Medicine. 18th ed. New York: McGraw-Hill; 2012:663-672

Nowell PC. The clonal evolution of tumor cell populations. Science 1976;194:23–28

Stephens PJ, et al. Massive genomic rearrangement acquired in a single catastrophic event during cancer development. Cell 2011;144:27–40

Visvader JE. Cells of origin in cancer. Nature 2011;469:314–322

Weinberg RA. The Biology of Cancer. New York: Garland Science; 2007

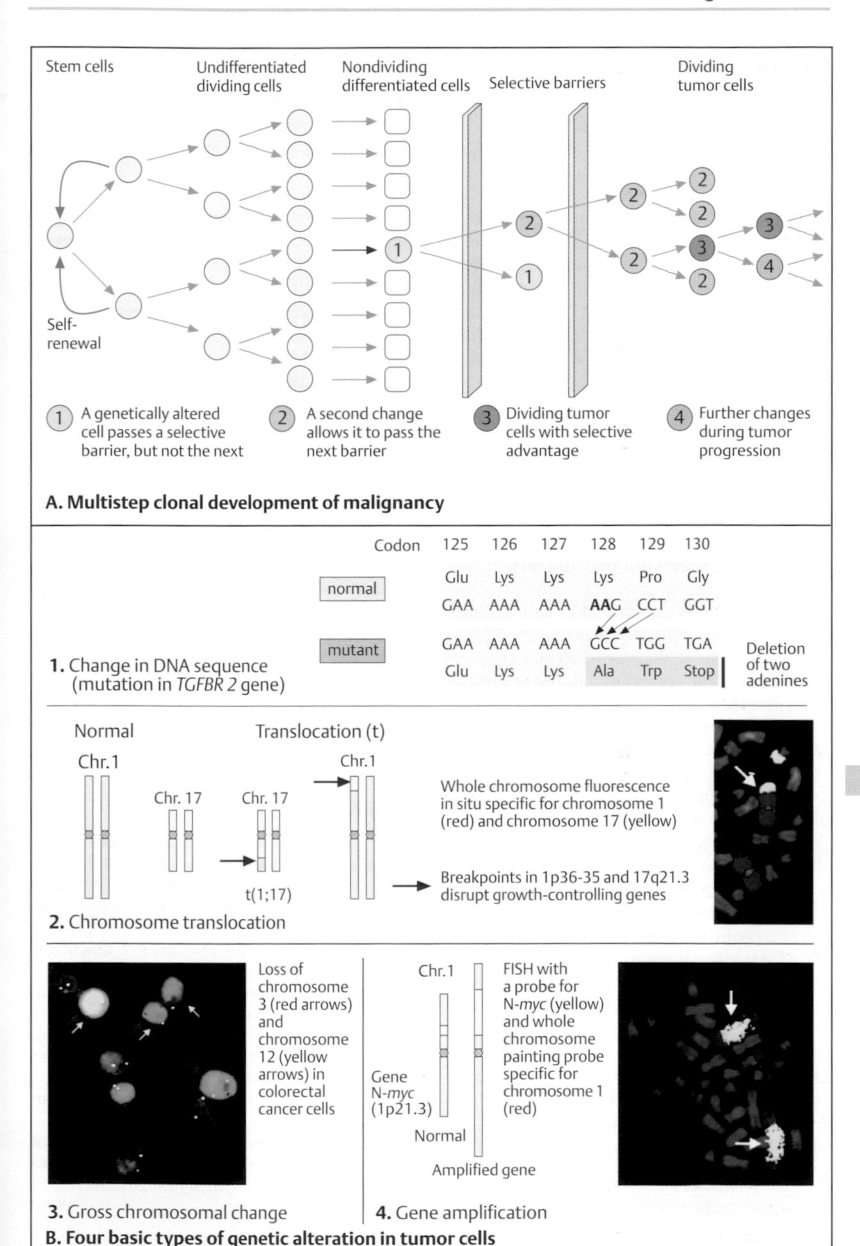

A. Multistep clonal development of malignancy

B. Four basic types of genetic alteration in tumor cells

Categories of Cancer Genes

Cancer affects 1 in every 5 individuals over his or her lifetime. More than 400 genes in the human genome contribute to cancer when altered by mutations (cancer genes). They are classified into three basic categories according to the effects of their mutations: *oncogenes*, *tumor suppressor* genes, and *genome stability* genes. In 85% and more, cancer is limited to somatic cells without a hereditary component (somatic mutations). In up to 15%, depending on the type of cells affected, a predisposing mutation present in the germline represents a hereditary component.

A. Three categories of cancer genes

The first class is *proto-oncogenes* (**1**). Their mutant forms, called *oncogenes*, drive a cell to divide when it normally should not (gain-of-function mutations). A single activating mutation is the first step toward cancer (comparable to a stuck accelerator in a car). The second class is *tumor suppressor genes* (**2**), which require two mutational events to induce tumor development (comparable to a defective brake). The initial mutation predisposes the cell to become a cancer cell. The second mutation then inactivates the other allele (loss-of-function mutation) and results in loss of cell division control. Mutations in the third class of cancer genes (**3**), *stability genes*, affect the stability of the genome by disrupting one of the various repair processes.

B. Oncogene activation

An oncogene positively influences tumor formation following a mutation in one allele. Oncogenes serve in many signal pathways controlling cell division. For example the *Ras* genes (OMIM 190020) encode a family of related cell-growth-controlling proteins. Ras proteins are GTPase-binding proteins functioning as switches, inactive when bound to GDP (guanosyldiphosphate) and active when bound to GTP (guanosyltriphosphate). Ras is activated by a receptor tyrosine kinase, which activates a guanine nucleotide exchange factor (GEF). Mutant forms of Ras are hyperactive and do not respond to GAPs.

C. Tumor suppressor genes

A mutation in a tumor suppressor gene will only initiate tumor formation if both alleles have lost their normal function. Thus, two successive mutational events are required within the same cell (**1**). The first event inactivates one allele without resulting in a cellular tumor phenotype. However, it predisposes the cell to malignant transformation. This occurs, when the other allele is also inactivated by a mutation. Tumor suppressor genes can be assigned to two groups: *gatekeepers* and *caretakers*. Gatekeeper genes directly inhibit tumor growth. Inactivation of caretaker genes leads to genetic instability, indirectly promoting tumor growth.

Loss of one allele in a somatic cell carrying a mutation in the other allele can be visualized by Southern blot analysis (**2**). Whereas somatic cells heterozygous at a marker locus give two signals, tumor cells, which have lost both alleles of the gene, give one signal (loss of heterozygosity, LOH). LOH is a hallmark of a tumor suppressor gene.

Mutations in tumor suppressor genes may be present in the zygote (by transmission or by new mutation) or occur in a somatic cell (**3**). A germline mutation predisposes all cells to develop into tumor cells. A somatic mutation predisposes a single cell. Germline mutations are the basis for hereditary forms of cancer; somatic mutations are the basis for the nonhereditary forms. A germline mutation occurring after the initial division of the fertilized egg may result in a mosaic of mutated and normal cells.

Further Reading

Alberts B, et al. Molecular Biology of the Cell. 5th ed. New York: Garland Science; 2008

Croce CM. Oncogenes and cancer. N Engl J Med 2008;358:502–511

Foulkes WD. Inherited susceptibility to common cancers. N Engl J Med 2008;359:2143–2153

Greenman C, et al. Patterns of somatic mutation in human cancer genomes. Nature 2007;446:153–158

Hanahan D, Weinberg RA. Hallmarks of cancer: the next generation. Cell 2011;144:646–674

Kinzler KW, Vogelstein B. Cancer-susceptibility genes. Gatekeepers and caretakers. Nature 1997; 386:761–763

Morin PJ, et al. Cancer genetics. In: Longo DL, et al., eds. Harrison's Principles of Internal Medicine. 18th ed. New York: McGraw-Hill; 2012:663–672

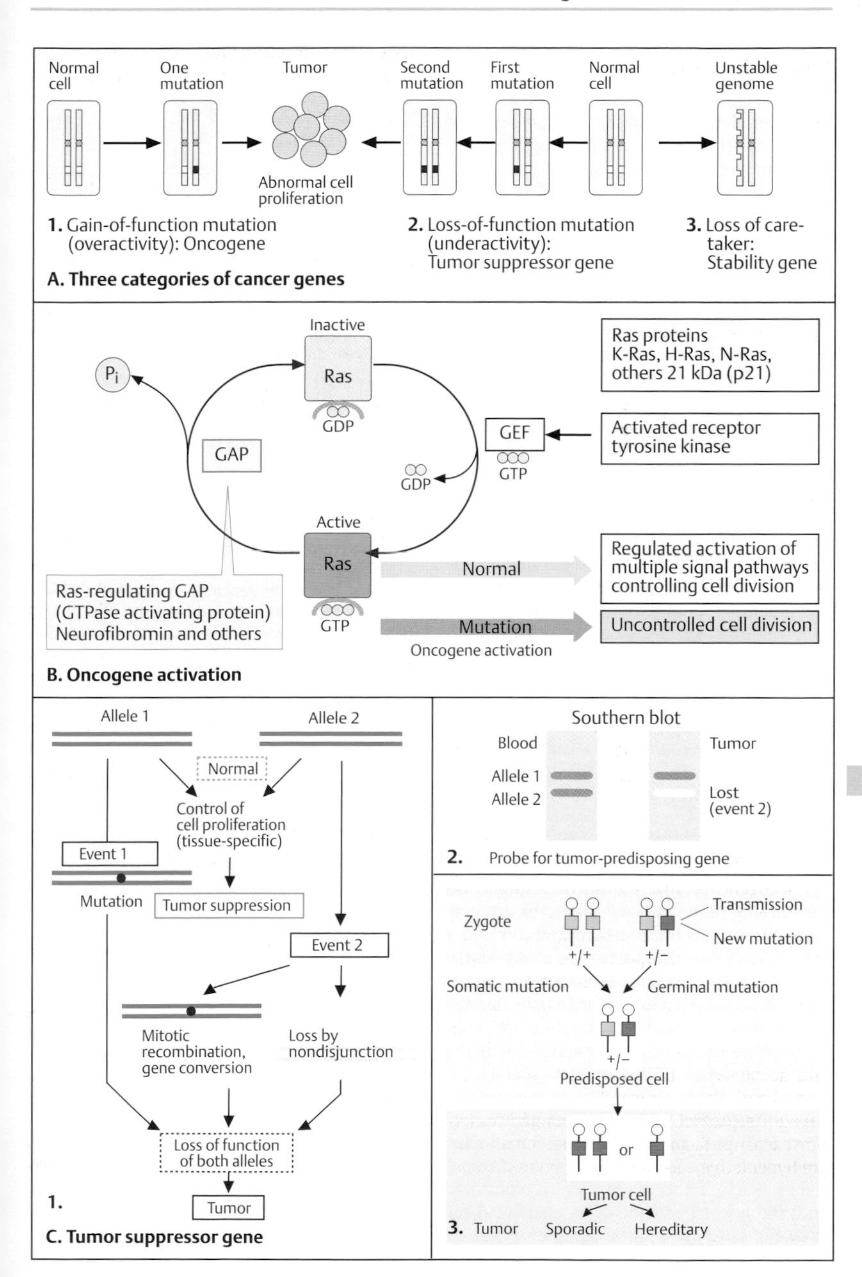
Normal cell
One mutation
Tumor
Abnormal cell proliferation
Second mutation
First mutation
Normal cell
Unstable genome
1. Gain-of-function mutation (overactivity): Oncogene
2. Loss-of-function mutation (underactivity): Tumor suppressor gene
3. Loss of care-taker: Stability gene
A. Three categories of cancer genes
Inactive
Ras
GDP
Pi
GAP
GEF
GTP
GDP
Ras proteins K-Ras, H-Ras, N-Ras, others 21 kDa (p21)
Activated receptor tyrosine kinase
Active
Ras
GTP
Normal
Regulated activation of multiple signal pathways controlling cell division
Mutation
Oncogene activation
Uncontrolled cell division
Ras-regulating GAP (GTPase activating protein) Neurofibromin and others
B. Oncogene activation
Allele 1
Allele 2
Normal
Control of cell proliferation (tissue-specific)
Event 1
Mutation
Tumor suppression
Event 2
Mitotic recombination, gene conversion
Loss by nondisjunction
Loss of function of both alleles
Tumor
1.
C. Tumor suppressor gene
Southern blot
Blood
Tumor
Allele 1
Allele 2
Lost (event 2)
2. Probe for tumor-predisposing gene
Zygote
Transmission
New mutation
+/+
+/−
Somatic mutation
Germinal mutation
+/−
Predisposed cell
or
Tumor cell
3. Tumor
Sporadic
Hereditary

Cancer Genomes

Recent progress in genomic analysis and massively parallel sequencing (p. 68) has provided insight into the genome of cancer cells. At least 400 genes (cancer genes, 1.8% of the 22000 protein-coding human genes) identified to date by the Cancer Genome Project (Cancer Gene Census). A ring-shaped diagram, called a Circos plot, can be display cancer cell genome.

A. Sequence of events from a normal cell to a cancer cell

A normal cell accumulates mutations during lifetime by an intrinsic process, influenced by lifestyle and environment. Early clonal expansion of cells accumulating mutations precedes the early stages of tumor development (mutator phenotype). Driver mutations confer growth advantage. They are positively selected during cancer evolution. The remaining mutations do not confer growth advantage (passenger mutations). Subsequent stages involve the invasion of surrounding tissues and metastasis. Resistance to chemotherapy marks the final stages of a tumor. (Figure adapted from Stratton et al., 2009.)

B. Circos plot

A circus plot depicts the location of all mutations and other changes in a cancer genome. The somatic mutations consist of: substitutions, insertions, or deletions of small or large segments of DNA; rearrangements; copy number variations with increases of up to several hundred copies (gene amplification) from the normal two copies; or loss of DNA sequences at a given site. An outer circle shows the arrangement of chromosomes 1–22, X, and Y. On a second circle on the inside, each mutation is indicated at the site of its location with respect to each chromosome. A third circle, located further inside, indicates changes in DNA copy number, either an increase or a decrease of the normal amount of DNA at each site. In the inner field, chromosomal rearrangements are shown: intrachromosomal rearrangements by a vertical line and interchromosomal rearrangements by a line connecting two chromosomes. Such a cancer genome is compared with the normal genome of an unaffected cell from the same individual. Genome sequences from 25000 tumors are expected to be available in the near future. (Figure adapted from Ledford, 2010, and Stratton et al., 2009.)

C. Genomic profile of an adenocarcinoma

This plot illustrates the key features of the genome of an adenocarcinoma. Red lines indicate interchromosomal structural variations; blue lines represent intrachromosomal structural variations. This figure was created using the Circos program of Krzywinski et al. (2009) (adapted from Lee et al., 2010).
Data presented in a similar way are available for other cancers.

Further Reading

Beroukhim R, et al. The landscape of somatic copy-number alteration across human cancers. Nature 2010;463:899–905

Cancer Gene Census CG. (www.sanger.ac.uk/genetics/CGP/Census)

Ding L, et al. Genome remodelling in a basal-like breast cancer metastasis and xenograft. Nature 2010;464:999–1005

Downing JR. Cancer genomes—continuing progress. N Engl J Med 2009;361:1111–1112

Haber DA, Settleman J. Cancer: drivers and passengers. Nature 2007;446:145–146

Krzywinski M, et al. Circos: an information aesthetic for comparative genomes. Genome Res 2009; 19:1639–1645

Ledford H. Big science: The cancer genome challenge. Nature 2010;464:972–974

Lee W, et al. The mutation spectrum revealed by paired genome sequences from a lung cancer patient. Nature 2010;465:473–477

Stephens PJ, et al. Complex landscapes of somatic rearrangement in human breast cancer genomes. Nature 2009;462:1005–1010

Stratton MR, Campbell PJ, Futreal PA. The cancer genome. Nature 2009;458:719–724

Weinberg RA. The Biology of Cancer. New York: Garland Science; 2007

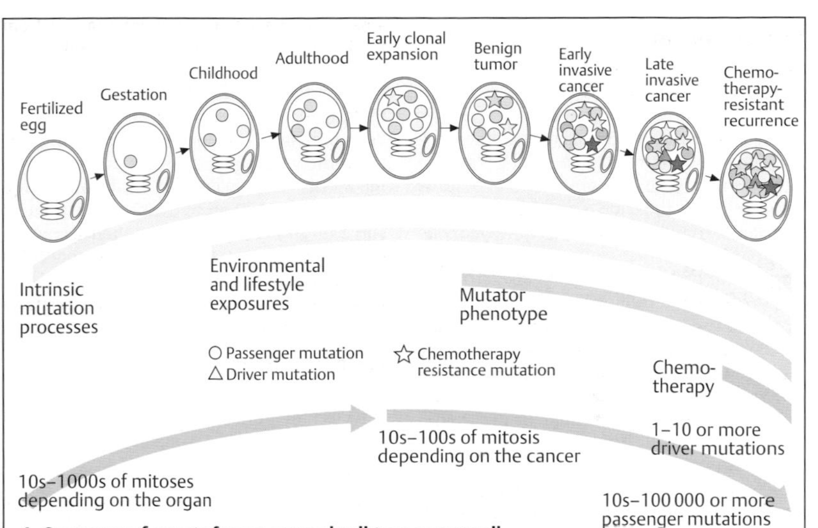

A. Sequence of events from a normal cell to a cancer cell

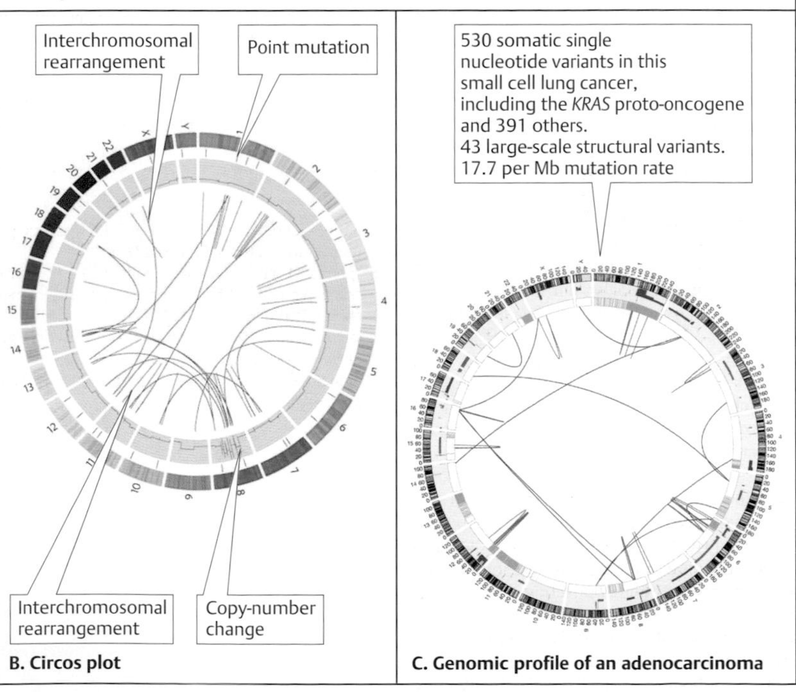

B. Circos plot

C. Genomic profile of an adenocarcinoma

The *p53* Tumor Suppressor Gene

The *p53* tumor suppressor gene (*TP53*; OMIM 191170) plays a central role in cell cycle control, apoptosis, and maintenance of genetic stability. It encodes a 53-kDa nuclear phosphoprotein translated from 2.8 kb mRNA. The gene has 11 exons with two promoters and spans about 20 kb on the short arm of human chromosome 17 (17p13.1). The p53 protein binds to specific DNA sequences and controls the expression of different regulator genes involved in growth. It interacts with other proteins in response to DNA damage and mediates apoptosis (cell death) of the cell when the damage is beyond repair. Its basic function is to control entry of the cell into the S phase (see Cell Cycle Control, p. 108). Somatic mutations in the *TP53* gene occur in about half of all tumors.

A. Model of function of the *TP53* gene

Normally the *TP53* gene is inactive (**1**). This allows the cell to proceed through mitosis. It is activated in response to DNA damage (**2**). If the subsequent DNA repair is successful, the cell may continue through mitosis. If repair fails, it undergoes apoptosis (p. 110). Mutated *TP53* results in damaged cells, and this eventually leads to tumor formation.

Phosporylation at the Mdm2-binding site displaces Mdm2 and activates p53. Phosphorylation is induced by ATM (ataxia-telangiectasia, see p. 318) and probably also the ATR protein. Activated p53 binds to DNA and induces the transcription of p21, a cell cycle controlling protein, which binds to Cdk complexes. As a consequence, the cell cannot enter the S phase. (Figure adapted from Lane, 1992.)

B. The human p53 protein

The active form of the human p53 protein is a tetramer of four identical subunits. It is a transcription factor with several functional domains. Normally p53 is extremely unstable and does not stimulate transcription because it is bound to a protein called MDM2 (OMIM 164785; name derived from Mouse sarcoma cells with Double Minute chromosomes). Phosphorylation at the MDM2-binding site displaces MDM2 and activates p53. It interacts with several other proteins in DNA damage recognition and repair.

Oncogenic mutations in *TP53* have a dominant-negative effect. Each subunit has 393 amino acids with five highly conserved functional domains: I–V. The carboxyl end beyond amino acid 300 has a nonspecific DNA interaction domain and a tetramerization domain. p53 protein function is inhibited by human papilloma virus protein E6, adenovirus protein E1b, SV40, and others. Most mutations are clustered in the conserved domains II–V: codons 129–146 (exon 4), 171–179 (exon 5), 234–260 (exon 7), and 270–287 (exon 8). Six mutations are strikingly frequent: a replacement of conserved amino acids arginine (R) in positions 175, 248, 249, 273, and 282, and glycine (G) in position 245. Mutations are missense, insertions, and deletions.

Knockout mice develop normally, but develop tumors at a high rate. Activated benzopyrene induces mutations at codons 175, 248, and 275 in cultured bronchial epithelial cells. (Figure adapted from Lodish et al., 2007.)

C. Germline mutations of *TP53*

Germline mutations lead to one of many familial cancer syndromes. One form of multiple different cancers is the autosomal dominant Li–Fraumeni syndrome (OMIM 114480) described in 1969 by Li and Fraumeni in families with individuals affected with diverse types of tumors, mainly soft-tissue sarcomas, early-onset breast cancer, brain cancers, cancer of the bone (osteosarcoma), and bone marrow (leukemias), and carcinoma of the lung, pancreas, and adrenal cortex. The pedigree (**1**) shows four individuals (II-2, II-3, III-1, III-2) affected by different types of tumor caused by a mutation in codon 248 of the *TP53* gene (CGG [arginine] to TGG [tryptophan]). The mutation is also present in individuals I-1 and III-5; they have not developed tumors, but are at increased risk. The absence of the mutation in individuals III-3 and III-4 indicates that they do not have an increased risk of cancer (data from Malkin, 2002). A subset of patients with Li–Fraumeni syndrome does not show p53 mutations.

Further Reading

Hanahan D, Weinberg RA. Hallmarks of cancer: the next generation. Cell 2011;144:646–674

Lane DP. Cancer. p53, guardian of the genome. Nature 1992;358:15–16

Malkin D. The Li–Fraumeni syndrome. In: Vogelstein B, Kinzler KW, eds. The Genetic Basis of Human Cancer. 2nd ed. New York: McGraw-Hill; 2002:387–401

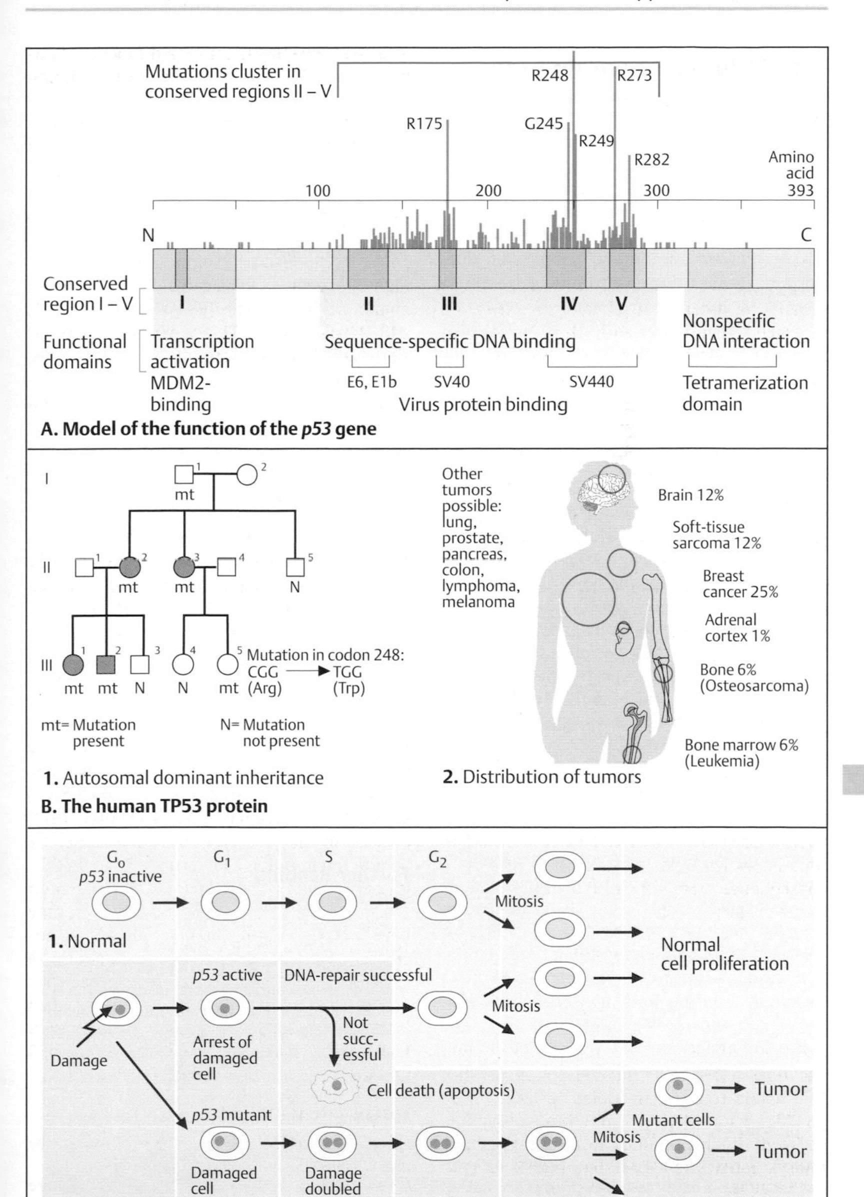

A. Model of the function of the *p53* gene

1. Autosomal dominant inheritance

2. Distribution of tumors

B. The human TP53 protein

C. Mutations of the *p53* gene in familial multiple tumors (Li-Fraumeni syndrome)

The *APC* Gene and Polyposis coli

The *APC* gene (OMIM 611731) encodes a protein with major functions in the WNT/β-catenin signal pathway (p. 202), as well as in cell migration and adhesion, apoptosis, and others. Loss of function of the *APC* gene results in adenomatous polyposis of the colon and, together with other mutations, subsequently leads to cancer progression. Mutant APC protein does not bind to β-catenin. This induces transcription of several growth-controlling genes, including the oncogene *MYC* (OMIM 190080). Germline mutations in the *APC* gene are the main cause of familial adenomatous polyposis of the colon. Cancer of the colon and rectum is the second leading cause of death from cancer.

A. Familial adenomatous polyposis

Familial adenomatous polyposis coli (FAP; OMIM 175100) is an autosomal dominant hereditary disease. In late childhood and early adulthood, up to 1000 and more polyps develop in the mucous membrane of the colon (**1**). Each polyp can develop into a carcinoma (**2**). In about 85% of affected persons, small hypertrophic areas not affecting vision are present in the retina (congenital hypertrophy of the retinal pigment (**3**). (**1** and **2** kindly provided by U. Pfeifer, Bonn, Germany; **3** was kindly provided by W. Friedl, Bonn.)

B. Structure and function of the *APC* gene

The *APC* gene, located on human chromosome 5q22.2, has 8538 bp in 15 exons. It encodes a protein of 2843 amino acids with several alternatively spliced forms. Exon 15 has an exceptionally long open reading frame of 6579 bp. Over 95% of mutations result in a nonfunctional truncated protein with variable loss of the C-terminus resulting from nonsense mutations (40%), deletions (41%), insertions (12%), and splice site mutations (7%). The site of the mutation influences the phenotypic manifestations.

C. Diagnosis in FAP

Indirect DNA analysis can be applied by haplotype analysis using polymorphic DNA marker loci flanking the *APC* locus (**1**). Since two affected individuals (I-1 and II-3) share haplotype 6–8 at loci D5S28 and D5S346, this haplotype must carry the mutation. Individual II-4 also inherited this haplotype and must be considered at risk of developing FAP. The protein truncation test (**2**) detects the abnormal protein that migrates faster than the normal protein because of its smaller size. (Data from W. Friedl, Bonn, Germany).

D. Mutations in colorectal tumorigenesis

Tumor formation progresses through several stages. After the first mutation has occurred in the *APC* gene, a second event may be loss of the normal allele (LOH) or a mutation in the second allele.

Subsequent stages lead to adenoma with less differentiated cells. Early polyps form. Additional mutations in other growth-controlling genes lead to malignant transformation and eventually to tumor development. The order of mutations seems to be important. About half of colorectal cancers have *RAS* mutations (OMIM 603384). Other genes involved include *DCC* (OMIM 120470), *SMAD4* (OMIM 600993), *SMAD2* (OMIM 601366), *p53* (191170), and others. (Figure adapted from Fearon & Vogelstein, 1990.)

Hereditary nonpolyposis colorectal cancer (HNPCC; OMIM 120435) affects about 1 in 200–1000 individuals (3% of all colorectal cancers). It results from germline mutation in one of the DNA-mismatch repair genes *hMSH1*, *hMLH2*, *hPMS1*, or *hPMS2*, or a related gene. Microsatellite instability is an important feature of HNPCC.

Further Reading

Fearon ER, Vogelstein B. A genetic model for colorectal tumorigenesis. Cell 1990;61:759–767

Hanahan D, Weinberg RA. Hallmarks of cancer: the next generation. Cell 2011;144:646–674

Morin PJ, et al. Cancer genetics. In: Longo DL, et al., eds. Harrison's Principles of Internal Medicine. 18th ed. New York: McGraw-Hill; 2012:663–672

Markowitz SD, Bertagnolli MM. Molecular origins of cancer: Molecular basis of colorectal cancer. N Engl J Med 2009;361:2449–2460

Weinberg RA. The Biology of Cancer. New York: Garland Science; 2007

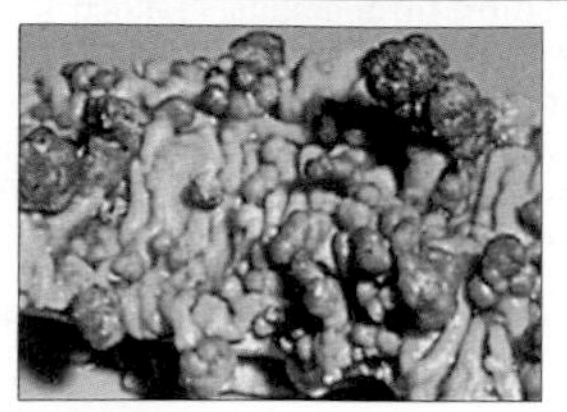

1.

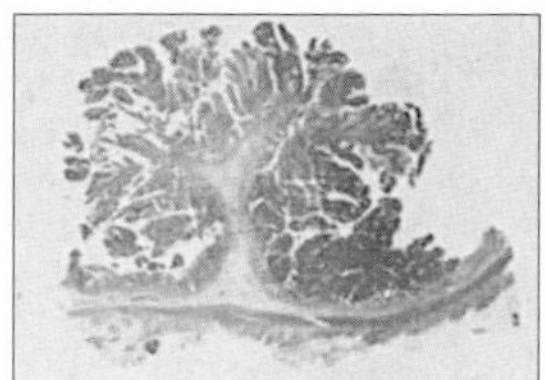

2.

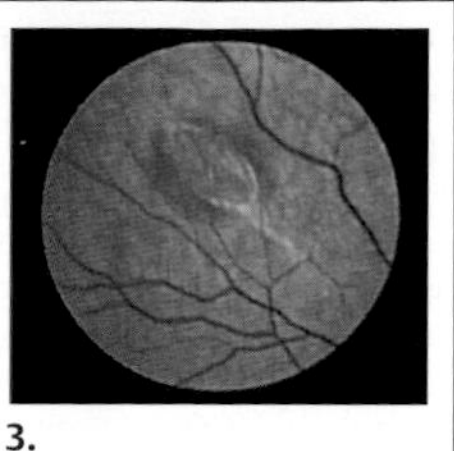

3.

A. Polyposis coli and colon carcinoma

↓ Frequent, recurrent mutations ▼ Splice mutations

Exons (introns not shown)

AP gene 5q21: 1 | 2 | 3 | 4 | 5 | 6 | 7 | 8 | 9 | 9a | 10 | 11 | 12 | 13 | 14 | 15 | 15 | 15

Nucleotides: 1, 135, 422, 645, 933, 1236, 1548, 1743, 1958, 8538

β-catenin binding repeats

APC polypeptide: N … C

Amino acids: 1, 45, 412, 516, 652, 1000, 2843

Interactions with other proteins: γ-catenin, GSK-3β, axin proteins, microtubules, EB-1, hDLG

Distribution of mutations in relation to disease type

Attenuated polyposis

Classic polyposis

CHRPE

Gardner syndrome

B. Structure and function of the *APC* gene

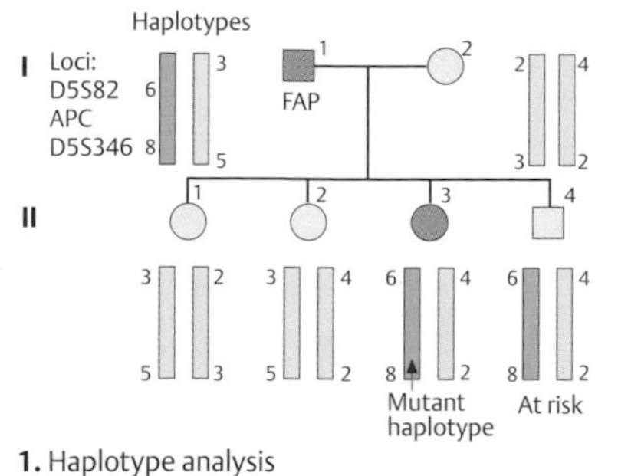

1. Haplotype analysis

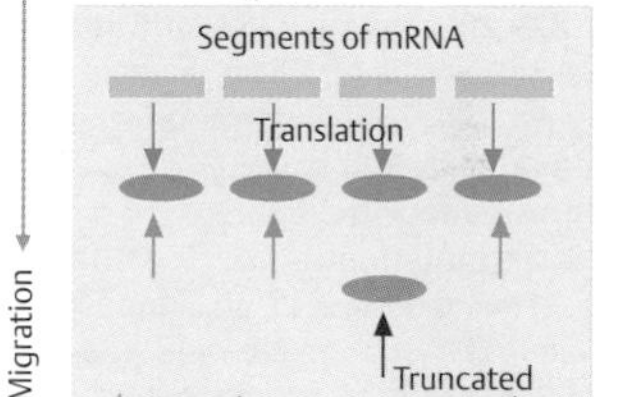

2. Protein truncation test

C. Indirect and direct DNA diagnosis in FAP

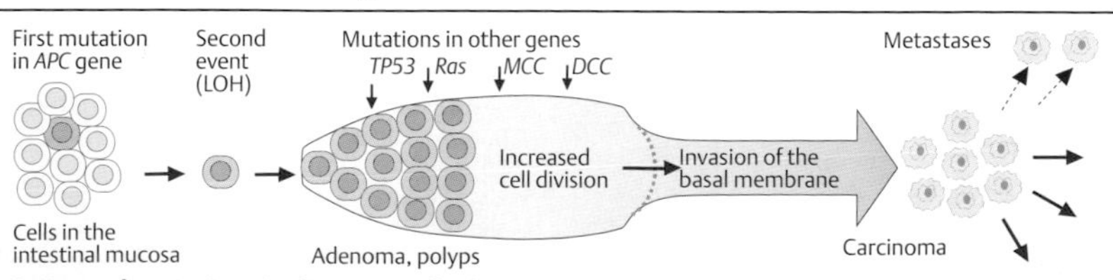

D. Several mutations in the origin of colon carcinoma

Breast and Ovarian Cancer Susceptibility Genes

Breast cancer is one of the most common forms of cancer, accounting for 32% of all cancers in the western world. Two major genes and more than 10 others confer susceptibility to breast and ovarian cancer when mutated. The breast cancer genes *BRCA1* and *BRCA2* encode multifunctional proteins that play important roles in genomic stability, homologous recombination, and double-stranded and transcription-coupled DNA repair (see p. 86). The BRCA1 and BRCA2 proteins interact with other proteins and participate in cell cycle control (see p. 108). Germline mutations are the basis for familial occurrence. The causative role of a particular sequence change may be difficult to assess in a given individual. In addition, polymorphic sequence variants are common. Mutations in other genes are involved: *ATM*, *CHEK2*, *BRIP1*, *PALB2*.

A. The breast cancer susceptibility gene *BRCA1*

The *BRCA1* gene (OMIM 113705, 604370) on chromosome 17 at q21.31 accounts for 20%–30% of inherited, autosomal dominant forms of breast cancer. This gene has 24 exons spanning 8 kb of genomic DNA and it encodes an mRNA transcript of 7.8 kb. Somatic mutations in breast tissue and germline mutations observed in unrelated patients are evenly distributed throughout the gene (red arrows). About 55% of all mutations occur in the large (3.4 kb) exon 11. A deletion of an adenine (A) and a guanine (G) in nucleotide position 185 (185delAG) and an insertion of a cytosine in position 5382 (5382insC) are the most frequent mutations, each accounting for about 10% of mutations. These mutations are particularly frequent in Ashkenazi Jewish population.

In keeping with its multiple functions, the protein (1863 amino acids) has distinct functional domains. Heterodimerization occurs at BARD1 (BRCA1-associated RING domain 1). Three protein-binding domains allow interaction with the p53 protein, the DNA recombination protein RAD51 (a human homolog of the bacterial RecA protein), and an RNA helicase. RAD50 and RAD51 are proteins involved in recombination during mitosis and meiosis, and in recombinational repair of double-stranded DNA breaks. The C-terminus contains a region involved in transcriptional activation and DNA repair. Two nuclear localization signals (NLS) are present at amino acid positions 500–508 and 609–615.

B. The breast cancer susceptibility gene *BRCA2*

Mutations in the *BRCA2* gene (OMIM 600185) at 13q12.3 occur throughout the gene. A deletion of thymine at nucleotide position 6174 (6174delT) is relatively frequent (1%) in the Ashkenazi Jewish population. The BRCA2 protein also has distinct functional domains. A large central domain consists of eight copies of a repeat of 30–80 amino acids; these copies are conserved in all mammalian BRCA2 proteins (BRC repeats). Four of these interact with the RAD51 protein. *BRCA2* is identical to *FANCD1*, a gene in the Fanconi anemia pathway (p. 318). (Figures based on Couch & Weber, 2002, and Welcsh et al., 2000.)

C. BRCA1-mediated effect on TP53

BRCA1 functions as a coactivator of transcription of several genes. Particularly important is the interaction with the p53 protein. Eukaryotic cells have two pathways to deal with double-stranded DNA breaks: nonhomologous end-joining and homologous recombination (HR; the mechanism used for recombination during meiosis, see p. 102). BRCA1 is connected to HR. BRCA1 is involved in apoptosis induced by p53 in response to DNA damage. (Figure redrawn from Hohenstein & Giles, 2003.)

Further Reading

Couch FJ, Weber BL. Breast cancer. In: Vogelstein B, Kinzler KW, eds. The Genetic Basis of Human Cancer. 2nd ed. New York: McGraw-Hill; 2002:549–581

Hohenstein P, Giles RH. BRCA1: a scaffold for p53 response? Trends Genet 2003;19:489–494

Meindl A, et al. Germline mutations in breast and ovarian cancer pedigrees establish RAD51C as a human cancer susceptibility gene. Nat Genet 2010;42:410–414

Welcsh PL, et al. Insights into the functions of BRCA1 and BRCA2. Trends Genet 2000;16:69–74

Zhu Q, Pao GM, Huynh AM, et al. BRCA1 tumour suppression occurs via heterochromatin-mediated silencing. Nature 2011;477:179–184

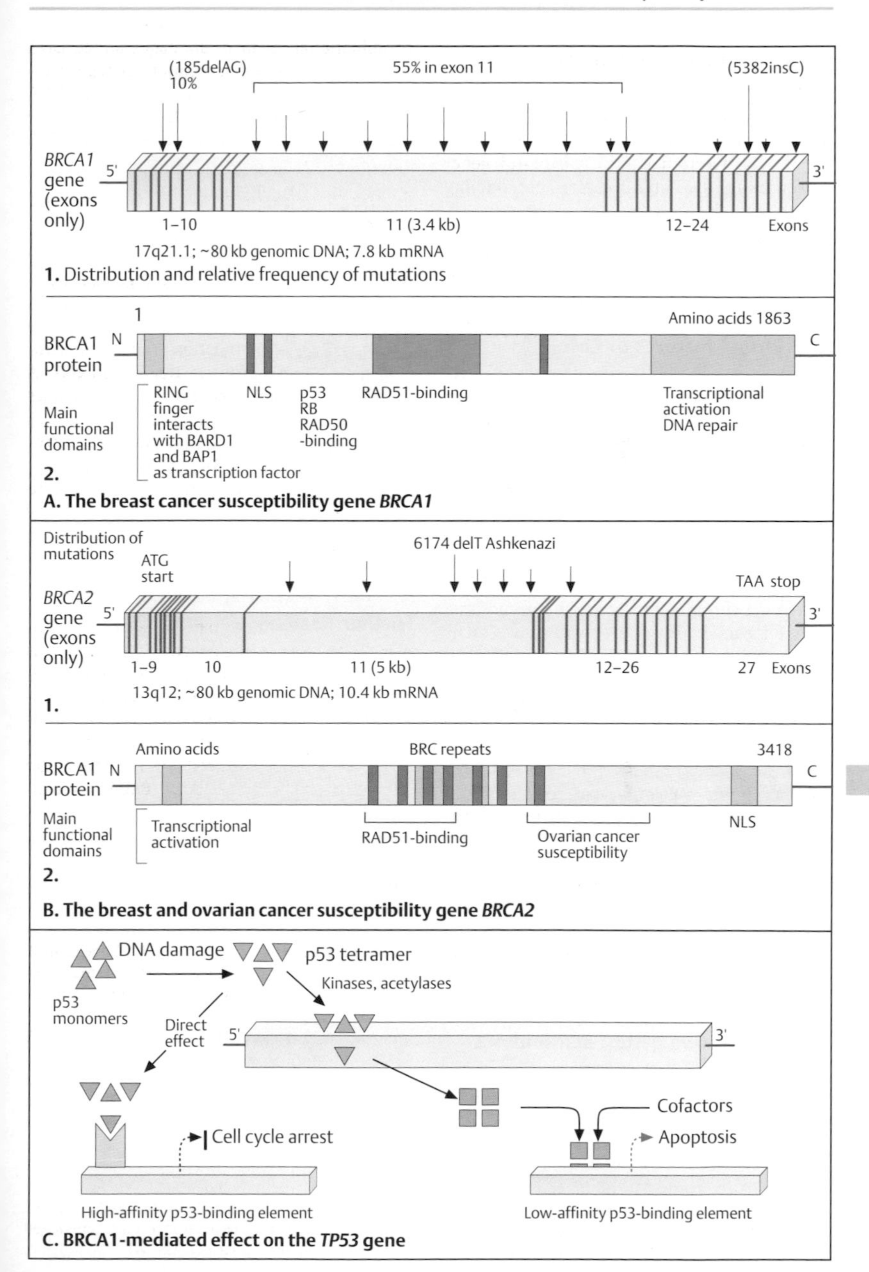

A. The breast cancer susceptibility gene *BRCA1*

B. The breast and ovarian cancer susceptibility gene *BRCA2*

C. BRCA1-mediated effect on the *TP53* gene

Oncogenic Chromosome Translocation

If a reciprocal chromosome translocation in a myelogenous or lymphoid bone marrow stem cell disrupts genes involved in the control of cell proliferation, a leukemia or lymphoma will develop. Chronic myelogenous leukemia (CML; OMIM 608232) is a myeloproliferative tumor in adults (1.5 cases per 100000 per year). About 10 other oncogenic translocations are listed Appendix, Table p. 417.

A. Principal features of CML

A greatly increased number of myelocytes (white blood cells, stained blue) in the peripheral blood (**1**) and a small chromosome 22, which has apparently lost about half of its long arm (small arrow), characterize CML (**2**). The small chromosome is known as the Philadelphia chromosome (Ph^1), named after the city in which it was discovered in 1960 by P. Nowell and D. Hungerford. The Ph^1 chromosome results from a reciprocal translocation between a chromosome 22 and a chromosome 9 (small arrows). The course of chronic CML is interrupted by acute crises intermittently and terminally. (Images by A. Schneider, Essen, Germany, and cmlsupport).

B. Ph^1 translocation

In 1973 J. D. Rowley showed that the Ph^1 chromosome is not a deletion, but results from a reciprocal translocation (**1**). The two breakpoints disrupt two genes on 22q11.21 and 9q34.12 (**2**) as shown by Bartram et al. (1983). About half of the long arm of a chromosome 22 is translocated to the long arm of a chromosome 9, and a very small part of the distal long arm of a chromosome 9 is translocated to a chromosome 22, but this is not visible by light microscopy.

C. Fusion of two genes: *BCR* and *ABL*

The breakpoints are in the *BCR* gene (OMIM 151410) and the *ABL* gene (OMIM 189980). As a result these two genes are fused. In CML the breakpoints cluster in the *BCR* gene to a small region of 5.8 kb (*b*reakpoint *c*luster *r*egion, BCR) in exons 10–12 of the *BCR* gene, and to a small region of 180 kb between exons 1a and 1b in the *ABL* gene. In contrast, in acute Ph^1-positive leukemias (e.g., acute lymphocytic leukemia, ALL) the breakpoints are localized in the 5′ direction in exons 1 or 2.

D. The BCR/ABL fusion protein

The *ABL* gene encodes a tyrosine kinase with functions in cellular growth controls. It is transcribed into two alternative mRNA transcripts of 7 kb (exon 1b, 2–11) and 6 kb (exon 1a, 2–11) length (**1**), and translated into a 145-kDa protein called $p145^{abl}$ (**2**). The fused genes in CML are transcribed into an mRNA transcript of 8.5 kb (**3**). The fusion protein (**4**) is a 210-kDa protein ($p210^{bcr/abl}$) that has an inappropriately active ABL kinase domain. As a result, hematopoietic cells in the bone marrow proliferate excessively. A small molecule, STI571 (**5**), a tyrosine inhibitor, precisely fits into the fusion protein and abolishes its abnormal function (**6**). It is used for therapy in CML under the name Gleevec. Other tyrosine inhibitors exist, such as dasatinib and nilotinib. Resistance to these compounds develops as a result of new mutations in the *BCR/ABL* fusion gene. (**6**, Adapted from Schindler et al., 2000.)

Further Reading

Bartram CR, et al. Translocation of c-ab1 oncogene correlates with the presence of a Philadelphia chromosome in chronic myelocytic leukaemia. Nature 1983;306:277–280

cmlsupport. (www.cmlsupport.org.uk/)

Druker BJ, et al. Efficacy and safety of a specific inhibitor of the BCR-ABL tyrosine kinase in chronic myeloid leukemia. N Engl J Med 2001;344:1031–1037

Melo JV, Barnes DJ. Chronic myeloid leukaemia as a model of disease evolution in human cancer. Nat Rev Cancer 2007;7:441–453

Nowell PC, Hungerford DA. A minute chromosome in human chronic granulocytic leukemia. Science 1960;132:1497

Rowley JD. A new consistent chromosomal abnormality in chronic myelogenous leukaemia identified by quinacrine fluorescence and Giemsa staining. Nature 1973;243:290–293

Savage DG, Antman KH. Imatinib mesylate—a new oral targeted therapy. N Engl J Med 2002;346:683–693

Schindler T, et al. Structural mechanism for STI-571 inhibition of abelson tyrosine kinase. Science 2000;289:1938–1942

Wetzler M, Marucci G, Bloomfield CD. Acute and chronic myeloid leukemia. In: Longo DL, et al., eds. Harrison's Principles of Internal Medicine. 18th ed. New York: McGraw-Hill; 2012:905–918

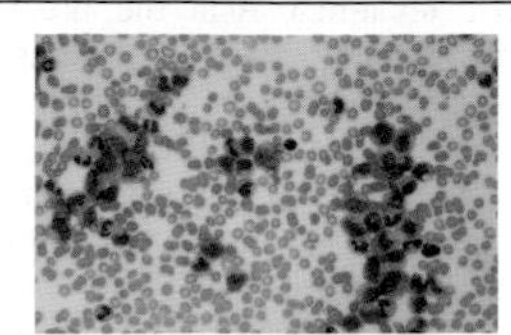

1. Accumulation of white blood cells (blue)

2. The Philadelphia translocation 9q;22q

A. Main features

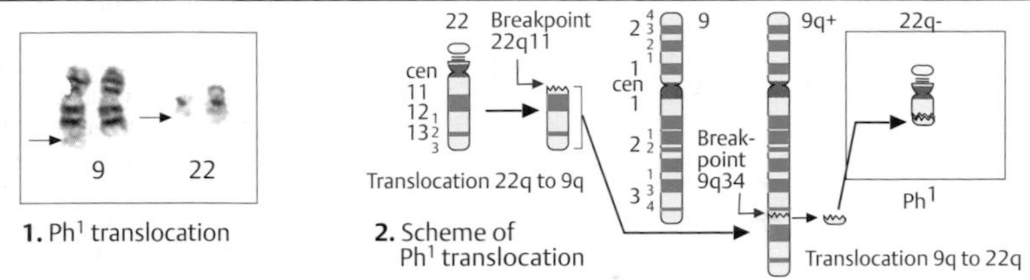

1. Ph^1 translocation

2. Scheme of Ph^1 translocation

B. Ph^1 translocation [t(9;22) (q34;q11)]

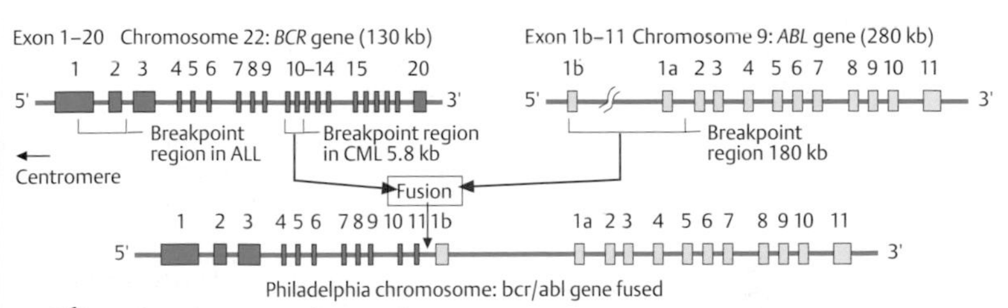

C. Ph^1 translocation causes fusion of two genes

1. Normal ABL mRNA

4. Abnormal fusion protein (210 kDa)

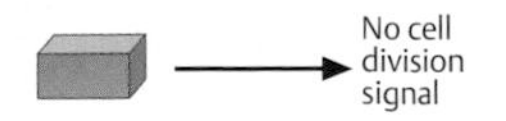

2. Normal ABL protein (145 kDa)

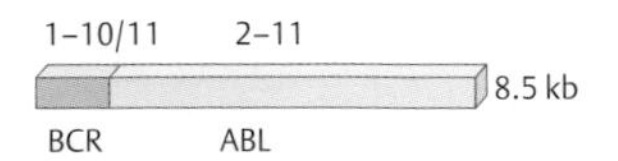

3. Abnormal BCR/ABL fusion

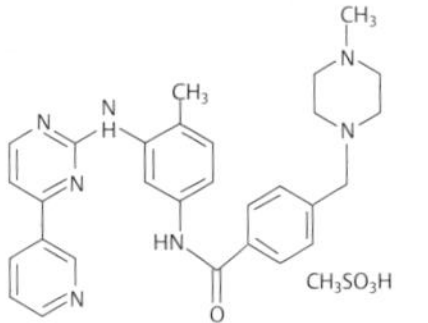

5. Chemical structure of STI571 (Gleevec)

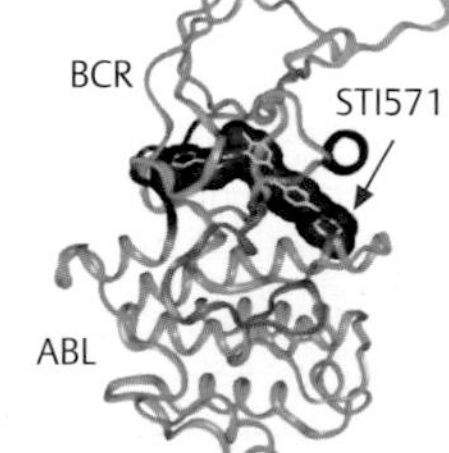

6. Ribbon diagram

D. The BCR/ABL fusion protein

Retinoblastoma

Retinoblastoma (OMIM 180200) is the most frequent malignant tumor of the eye in infancy and early childhood, with an incidence of about 1 in 15 000–25 000 live births. It results from loss of function of both alleles of the retinoblastoma gene *RB1* (OMIM 614041). Two inactivating events are required for tumor initiation, as predicted by A. Knudson in 1971 (two-hit hypothesis). The first mutation predisposes the cell to develop a tumor, and the second initiates tumor formation. The first mutation may be either a somatic mutation in one retinal progenitor cell (retinoblast), or a germline mutation. Patients have an additional risk for osteosarcoma.

A. Phenotype

Retinoblastoma (Rb) may occur in one or both eyes. An early sign is a white shimmer, the "cat's eye" (**1**), and/or rapidly developing strabismus. One or several tumors (unifocal or multifocal, respectively) may be present in the retina of an affected eye (**2**) and progress rapidly (**3**). Early diagnosis and therapy are essential. About 60% of patients have somatic mutations (nonhereditary Rb) and these usually develop unilateral and unifocal Rb. About 40% of patients are heterozygous for an *RB1* mutation that is either transmitted from a parent (10%–15%) or is the result of a new mutation, usually derived from the paternal allele (about 10:1). Heterozygous carriers for an oncogenic *RB1* mutation have a predisposition to Rb, which is transmitted as an autosomal dominant trait. In some families, carriers of an oncogenic mutation do not develop tumors (nonpenetrance). This low-penetrance phenotype is associated with specific *RB1* mutations. Milder phenotypic expression is also observed when the mutation is present in only a proportion of germ cells (mutational mosaicism).

B. Retinoblastoma locus

The *RB1* locus at 13q14.2 was first identified by microscopically visible interstitial deletions.

C. Retinoblastoma gene *RB1* and its protein

The *RB1* gene is organized into 27 exons of different sizes (31–1889 bp) spanning 200 kb genomic DNA (**1**). It is ubiquitously expressed and transcribed into mRNA of 4.7 kb (**2**). The main types of mutation in hereditary retinoblastoma are deletions (about 26%), insertions (about 9%), and point mutations (about 65%), including splice-site mutations, distributed relatively evenly along the gene. Missense mutations have been associated with low-penetrance retinoblastoma.

The gene product (pRB protein), a 100-kDa phosphoprotein with 928 amino acids (**3**), has important functions in the regulation of the cell cycle (p. 108), cell fate specification, and differentiation in the central nervous system. It is activated by phosphorylation at about 12 distinct serine and threonine residues (P) during cell cycle progression from G_0 to G_1. Three functional domains, A, B, and C, bind in a cell-cycle-dependent fashion to transcription factors, including the oncoproteins MDM-2 and c-ABL. A nuclear localization signal (NLS) is located at the C terminus. The *RB1* gene is imprinted (Kanber et al., 2009).

D. Diagnostic principles

In about 3%–5% of patients, an interstitial deletion of 13q14 or a larger 13q area is visible by chromosomal analysis (**1**), usually associated with signs of developmental delay. In familial Rb, formerly indirect DNA diagnosis was used by segregation analysis of DNA markers at the *RB1* locus. An affected girl (II-1) has inherited haplotype a from her unaffected father and haplotype c from her unaffected mother (**2**). Analysis of tumor cells, obtained from the affected eye, shows that only haplotype a is present (loss of heterozygosity, LOH), which must be the allele carrying the mutation. Direct DNA analysis, using methods applied previously, demonstrates a germline mutation in blood cells. Here, the affected individuals I-2 and II-2 (**3**) carry a cytosine to thymine transversion in codon 575 (CAA glutamine to TAA stop codon). (Photographs courtesy of W. Höpping and D. Lohmann, Essen, Germany.)

Further Reading

Kanber D, et al. The human retinoblastoma gene is imprinted. PLoS Genet 2009;5:e1000790

Lohmann DR, Gallie BL. Retinoblastoma: revisiting the model prototype of inherited cancer. Am J Med Genet C Semin Med Genet 2004;129C:23–28

Retinoblastoma Eye Cancer Network. Available at: http://www.eyecancer.com/

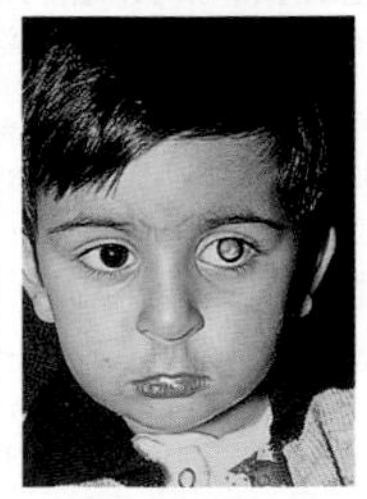

1. So-called cat's eye

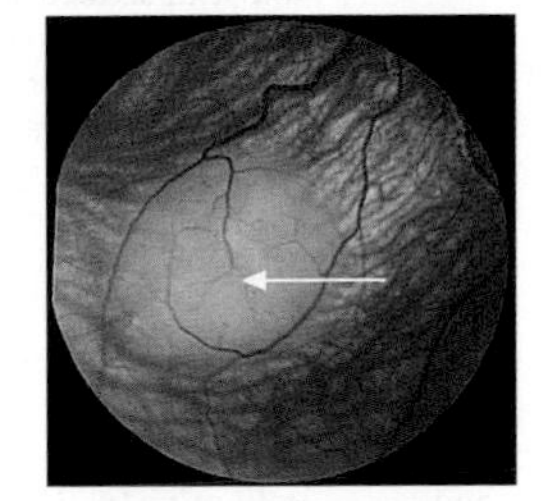

2. Tumor in the retina

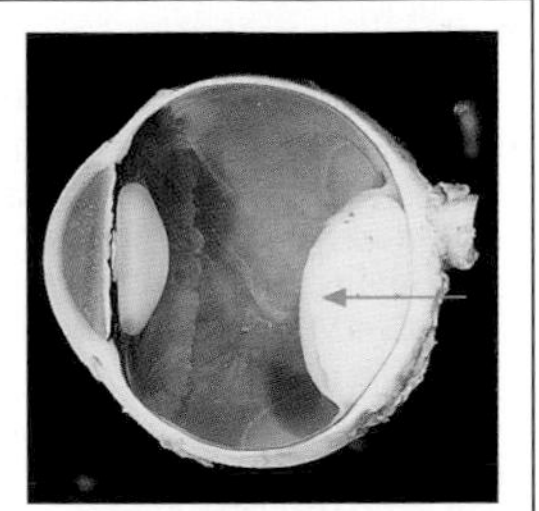

3. Large tumor in the eye

A. Phenotype

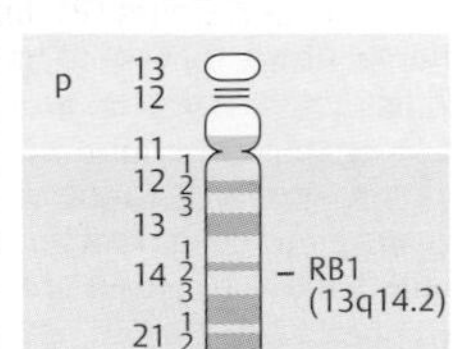

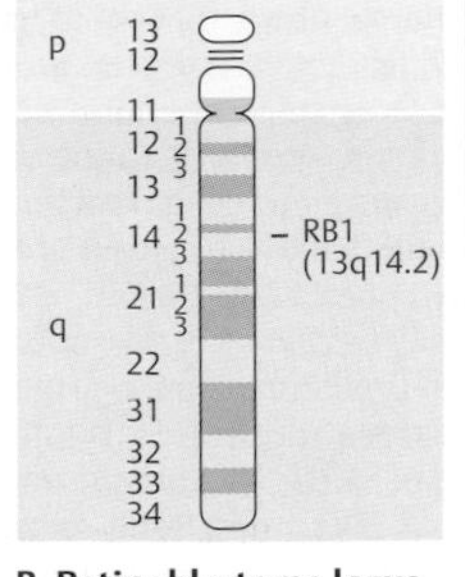

B. Retinoblastoma locus on chromosome 13

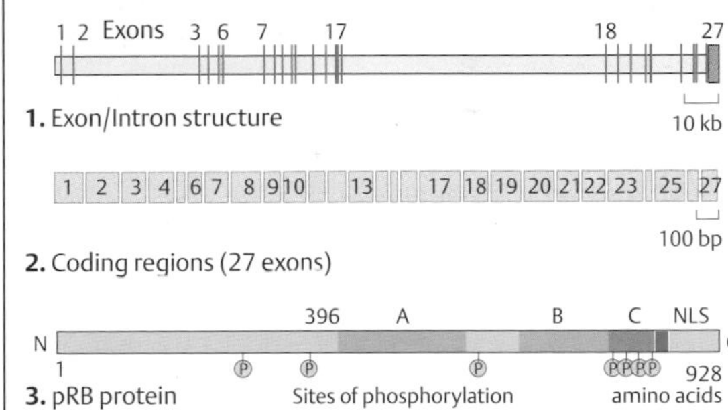

1. Exon/Intron structure

2. Coding regions (27 exons)

3. pRB protein

C. Retinoblastoma protein (pRB1) with functional domains A–C

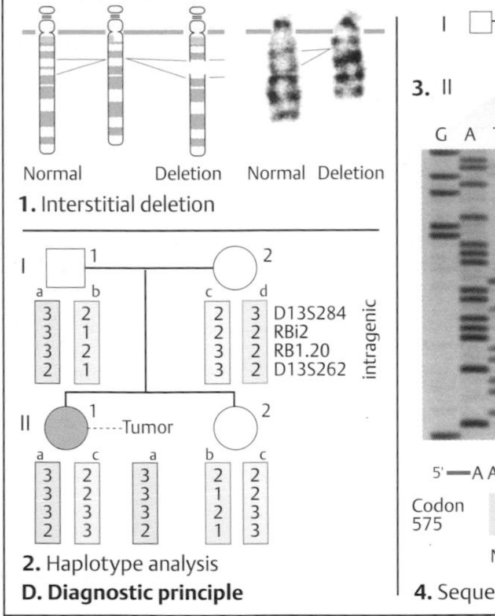

1. Interstitial deletion

2. Haplotype analysis

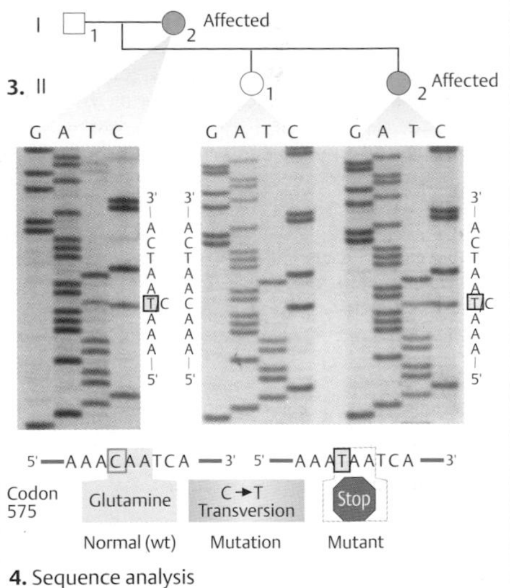

4. Sequence analysis

D. Diagnostic principle

Neurofibromatosis

The neurofibromatoses are a group of clinically and genetically different autosomal dominant diseases that predispose to benign and malignant tumors of the nervous system. Most common among others are neurofibromatosis 1 (NF1; OMIM 162200) and neurofibromatosis 2 (NF2; OMIM 607379).

A. Main manifestations of neurofibromatosis 1

NF1, also called von Recklinghausen disease, affects 1 in 3000 individuals with highly variable manifestations. Characteristic are Lisch nodules of the iris (**1**) in more than 90% of patients, café-au-lait spots (**2**) (more than five spots of more than 2 cm diameter are considered diagnostic) in more than 95% of patients, and multiple neurofibromas (**3**) in more than 90% of patients, usually apparent between the ages of 4 and 15 years. About 2%–3% of patients develop a neurofibrosarcoma or other malignancies. The NF1 phenotype may overlap wth Noonan syndrome (OMIM 601321).

B. The *NF1* gene

The *NF1* gene (OMIM 613113) is large, spanning 350 kb genomic DNA, and organized into at least 59 exons. It is a tumor suppressor gene localized on human chromosome 17 at q11.2. The gene was isolated in 1990 from a 600-kb *Nru*I restriction fragment by positional cloning. Two patients with a translocation involving the long arm of chromosome 17 with breakpoints at 17q11.2 and CpG islands provided anchor points. Three unrelated genes, *OMGP, EVI2B*, and *EVI2 A*, are embedded within the *NF1* gene on the opposite DNA strand. The types of mutations are deletions, insertions, base substitutions, and splice mutations. *NF1* has one of the highest mutation rates, with a new mutation in about 50% of patients. Numerous *NF1* pseudogenes are present in the human genome.

C. *NF1* gene product, neurofibromin-1

The gene product, neurofibromin-1, is a multidomain cytoplasmic protein, which regulates several intracellular signal processes, including the RAS/MAPK pathway (p. 342). The human neurofibromin is a protein of 2818 amino acids. It is transcribed from multiple alternatively spliced transcripts and expressed in many tissues. A region of GAP homology with a gene product in yeast *(Saccharomyces cerevisiae)*, IRA1 (inhibitor of *RAS* mutants), is located between amino acids 840 and 1200. (Figure adapted from Xu et al., 1990.)

D. Neurofibromatosis 2 gene, *NF2*

Neurofibromatosis 2 (OMIM 101000) is characterized by multiple tumors of the eighth cranial nerve, mainly meningioma, often bilateral, with an incidence of 1 in 25 000. The *NF2* gene (OMIM 607174), localized on human chromosome 22 at q12.2, is organized into 17 constitutive exons and one alternatively spliced exon, spanning about 110 kb of genomic DNA. The widely expressed gene product, neurofibromin-2 or schwannomin, regulates cell proliferation, cell–cell adhesion, transmembrane signaling, and other functions. It is a member of the band 4.1 cytoskeleton-associated protein superfamily (see p. 322) called the ERM family (including ezrin, radixin, moesin, and several protein tyrosine phosphatases). The gene was identified within a cosmid contig of YAC (yeast artificial chromosome) clones. Two deletions (Del 1 and Del 2) in unrelated patients were useful in the process of finding the gene. Mutations can be detected in more than 50% of patients (large deletions, including the entire gene or several exons, and small deletions are frequent).

Further Reading

Asthagiri AR, et al. Neurofibromatosis type 2. Lancet 2009;373:1974–1986

Jett K, Friedman JM. Clinical and genetic aspects of neurofibromatosis 1. Genet Med 2010;12:1–11

Riccardi VM, Eichner JE. Neurofibromatosis. Phenotype, Natural History and Pathogenesis. 2nd ed. Baltimore: Johns Hopkins University Press; 1992

Xu GF, et al. The neurofibromatosis type 1 gene encodes a protein related to GAP. Cell 1990;62:599–608

Neurofibromatosis 1 (NF1)
(von Recklinghausen disease)

Autosomal dominant
Frequency 1 in 3000
Gene locus on 17q11.2
Café-au-lait spots
Lisch nodules in the iris
Multiple neurofibromas
Skeletal anomalies
Predisposition to tumors of the nervous system
50% new mutations

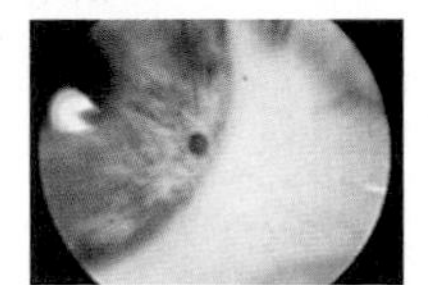

1. Lisch nodule

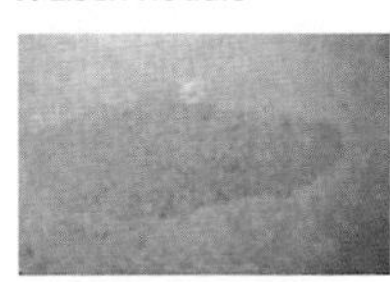

2. Café-au-lait spot

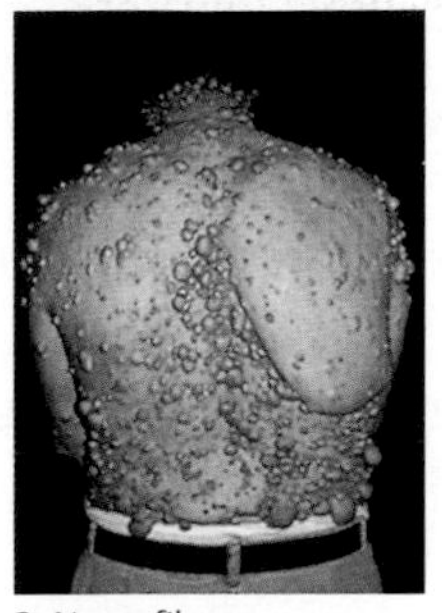

3. Neurofibromas

A. Main manifestations of neurofibromatosis 1

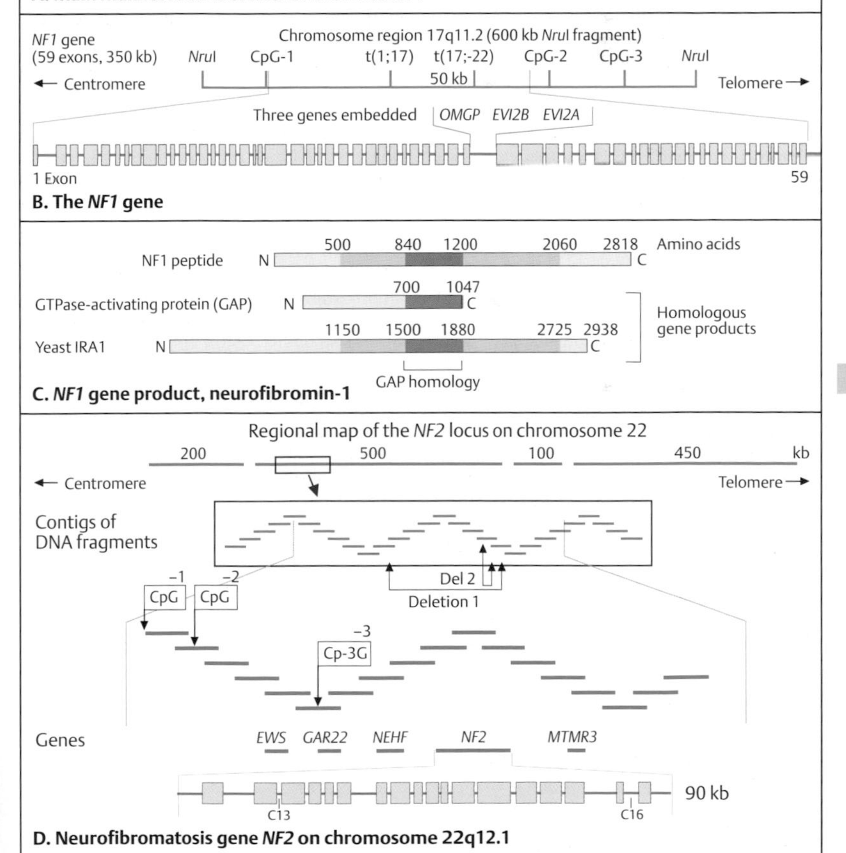

B. The *NF1* gene

C. *NF1* gene product, neurofibromin-1

D. Neurofibromatosis gene *NF2* on chromosome 22q12.1

Genomic Instability Diseases

Genomic instability diseases are a group of distinct, heterogenous disorders of different defects in DNA damage recognition and repair, and signs of instability of the genome. They share a high disposition to different types of cancers and immune deficiency. Different patterns of chromosomal breaks and rearrangements are visible by light microscopy of metaphase cells. Ataxia-telangiectasia, Fanconi anemia, and Bloom syndrome are important examples.

A. Ataxia-telangiectasia (AT)

AT (OMIM 208900) is a variable autosomal recessive disease resulting from mutations in the *ATM* gene at (OMIM 607585) at 11q22.3 with an incidence of 1:40 000. The main manifestations, which develop in early childhood, are immune defects, cerebellar ataxia, and characteristic telangiectasias of the conjunctivae (**1**). Affected individuals are highly sensitive to irradiation and are prone to lymphomas and leukemias. The *ATM* gene has 66 exons spanning 150 kb genomic DNA.

Th ATM protein consists of 3056 amino acids and is a phosphatidylinositol 3-protein kinase. ATM is activated in response to double-strand DNA breaks. It has a central role in a network of proteins that regulate cellular responses to DNA damage and recombination (**2**). Heterozygotes also are predisposed to tumors. Mutations in a functionally related gene result in the Nijmegen breakage syndrome (NBS1; OMIM 251260).

B. Fanconi anemia (FA)

FA (OMIM 227650) is a genetically highly heterogeneous group of 16 autosomal recessive and one X-chromosomal diseases (see Appendix, Table p. 418). It manifests in early childhood as pancytopenia, growth deficiency (**1**), hypoplastic radius often with hypoplastic or absent thumbs (**2**), and other malformations. Eight FA proteins form the nuclear FA complex (FANCA, B, C, E, F, G, L, M) in response to DNA damage and in cooperation with several other proteins (p53, ATM, BRCA1, BRCA2, NBS1, BLM, RAD50/51). The most prevalent mutation is *FA-A* (also called *FANCA*), in about 65% of patients. FA cells are hypersensitive to DNA cross-linking agents, such as diepoxybutane (DEB), which induces chromosomal breaks. *FANCD1* and *BRCA2* are identical. (**3**, Adapted from Schindler & Hoehn, 2007.)

C. Bloom syndrome (BLM)

BLM (OMIM 210900) is a prenatal and postnatal growth deficiency disorder (birth weight 2000 g, birth length 40 cm, adult height approximately 150 cm) with a distinct phenotype (**1**): narrow face, sunlight-induced facial erythema, variable immune deficiency, and a greatly increased risk of different malignancies (about one in five patients). Chemotherapy is very poorly tolerated. The diagnostic hallmark is a 10-fold increase in the spontaneous rate of sister chromatid exchanges (SCE; see Glossary) (**2**, **3**). Breaks in one or both chromatids and exchanges between homologous chromosomes occur in about 1%–2% of metaphase cells.

BLM results from autosomal recessive mutations in the *RECQL3* gene (OMIM 604610) at 15q26.1, encoding a member of the RecQ family of DNA helicases. The mutations are mainly protein-truncating nonsense mutations, and they are distributed fairly evenly along the gene (**4**), but some missense mutations occur. Most distinctive is a founder mutation in populations of Ashkenazi Jewish origin: a 6-bp deletion/7-bp insertion at nucleotide 2281. Homozygosity for mutations in the *BLM* gene results in an increased rate of somatic mutations. The 1417-amino acid BLM protein interacts with the FA complex and is involved in meiotic recombination. It is homologous to yeast Sgs1 (slow growth suppressor) and the human WRN protein (Werner syndrome; OMIM 277700). (**4**, Courtesy of N. A. Ellis & J. German, New York.)

Further Reading

Auerbach AD, et al. Fanconi anemia. In: Vogelstein B, Kinzler KW, eds. The Genetic Basis of Human Cancer. 2nd ed. New York: McGraw-Hill; 2002;289–306

D'Andrea AD. Susceptibility pathways in Fanconi's anemia and breast cancer. N Engl J Med 2010;362:1909–1919

Gatti R. Ataxia-telangiectasia. In: Vogelstein B, Kinzler KW, eds. The Genetic Basis of Human Cancer. 2nd ed. New York: McGraw-Hill; 2002:239–266

German J, Ellis NA. Bloom syndrome. In: Vogelstein B, Kinzler KW, eds. The Genetic Basis of Human Cancer. 2nd ed. New York: McGraw-Hill; 2002:301–315

Schindler D, Hoehn H, eds. Fanconi Anemia. Basel: Karger; 2007

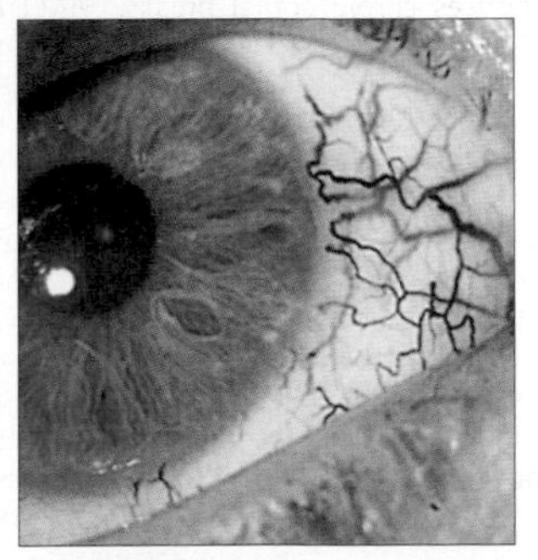

1. Telangiectasias of the conjunctiva

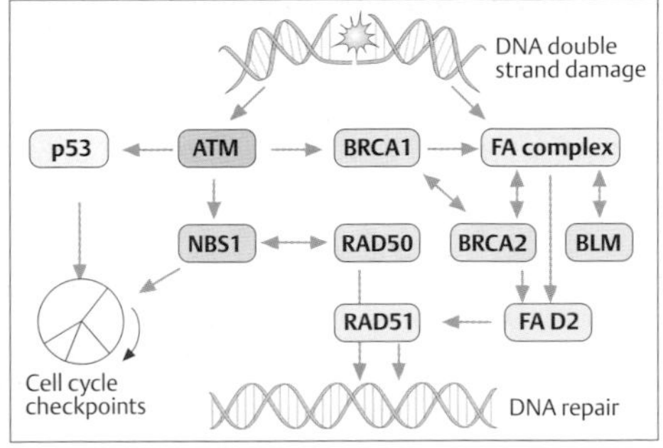

2. Relation of ATM to other proteins maintaining genomic stability

A. Ataxia-telangiectasia (AT)

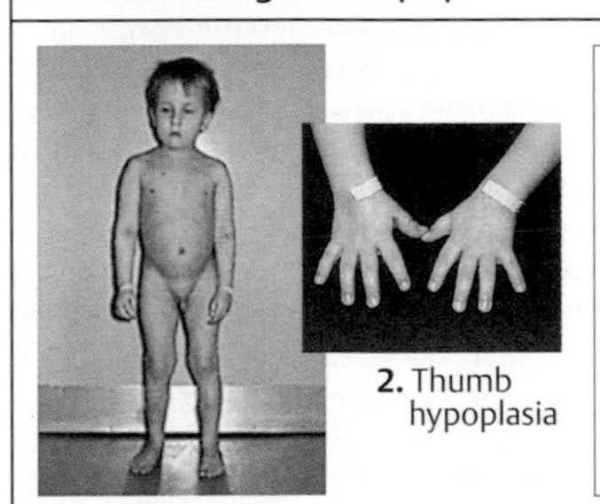

1. Phenotype

2. Thumb hypoplasia

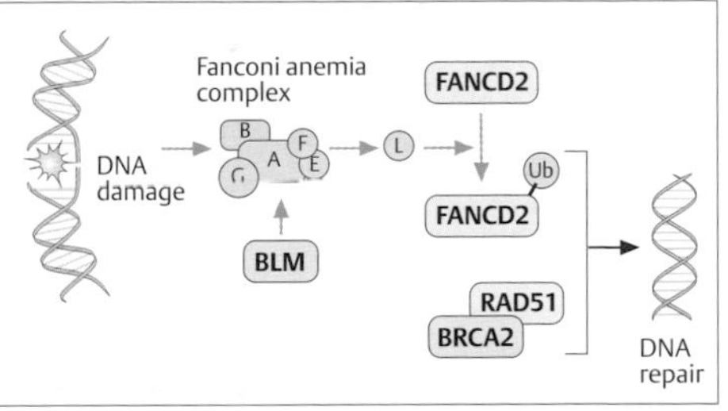

3. Fanconi anemia-associated proteins

B. Fanconi anemia (FA)

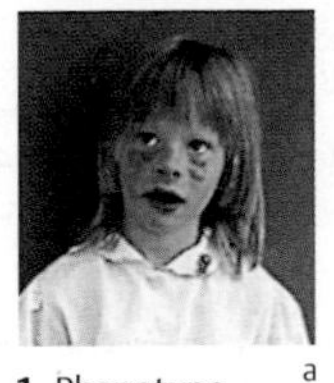

a

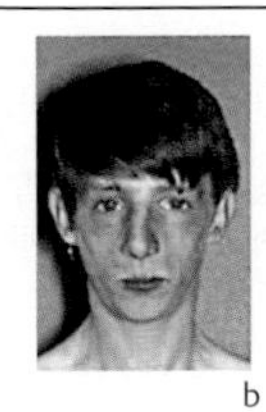

b

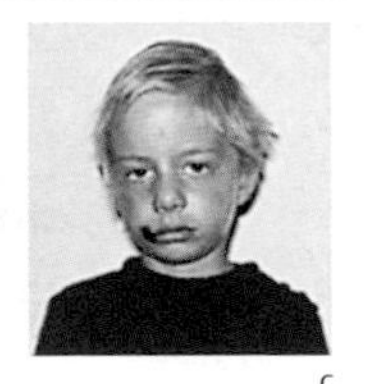

c

1. Phenotype

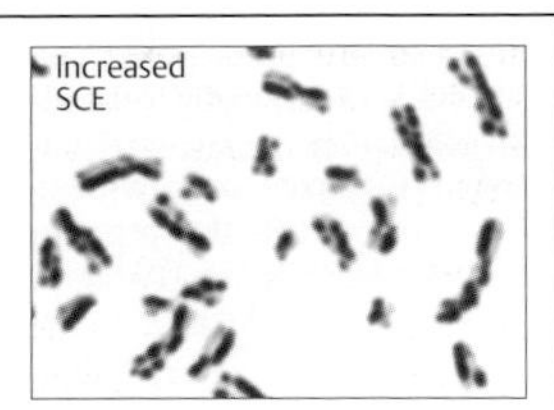

2. Bloom syndrome metaphase

Distribution of nonsense mutations (not all shown)

blm[Ash]

1417 amino acids

1

N

C

Acidic amino acids

RNA Pol II

Helicase domain

NLS

Selected functional domains

4. Diagram of the Bloom protein and distribution of mutations

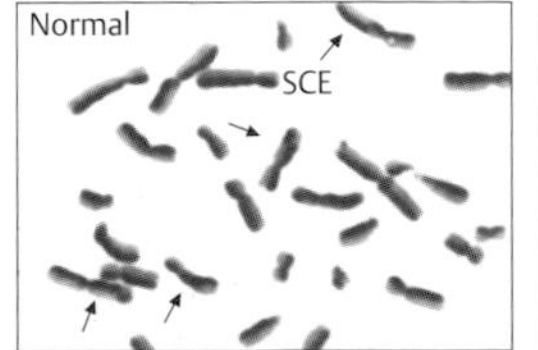

3. Normal metaphase

C. Bloom syndrome

DNA Excision Repair Disorders

Xeroderma pigmentosum (XP, OMIM 278700–80), Cockayne syndrome (OMIM 216400), and Trichothiodystrophy (OMIM 601675) are examples of a genetically heterogenous group of disorders resulting from mutations in different genes involved in nucleotide excision repair.

A. Xeroderma pigmentosum (XP)

The leading signs of XP are dry and pigmented skin in ultraviolet light (UV)-exposed regions of the skin (**1**,**2**), from which the name was derived, when it was first described by F. von Hebra and M. Kaposi in 1874. Unexposed areas are not affected. Thus it is important to protect patients from UV light. An especially important feature is the tendency for multiple skin tumors to develop in the exposed areas (**3**). These may occur as early as childhood or early adolescence. The types of tumor are the same as those occurring in healthy individuals after prolonged UV exposure.

The basic defect is DNA excision repair deficiency, as first shown by Cleaver in 1968. The causes are mainly autosomal recessive mutations in genes at several loci encoding one of the several excision repair proteins: XPA, -B, -C, -D, -E, -F, -G, or XPV. These genes are highly conserved in bacteria, yeast, and mammals.

B. Cellular phenotype of XP

The UV sensitivity is evident in cultured fibroblast cells. When exposed to UV light, XP cells show a distinct dose-dependent decrease in survival rate compared with normal cells (**1**). When cultured in the presence of ^{3}H-thymidine and exposed to UV light, an autoradiograph (a film that, placed on the cells, develops black dots at sites where radiolabeled thymine has been incorporated during DNA synthesis). XP cells show few dots over the nucleus of repair-deficient XP cells. Fused XP cells with different defects can correct each other. Initially, prior to gene identification, different types of XP were asigned to complementation groups based on their ability to complement (correct) each other. The example shows two uncorrected nuclei (XP-D) and a corrected heterokaryon (XP-A/XP-D). (Photograph courtesy of Bootsma & Hoeijmakers, 1999.)

C. Cockayne syndrome (CS)

The leading signs of CS are delayed development with cachectic, progeroid appearance (**1**), cutaneous photosensitivity and atrophy (**2**), and manifestations in other organ systems. Two types of CS can be distinguished, type A and B, caused by different genes at 5q11 (OMIM 609412) and 10q11 (OMIM 133540), respectively. CS overlaps with certain types of XP. (Figures www.cockayne-syndrome.net/images/ and www.dermis.net/dermisroot/de/10 012/image.)

D. Trichothiodystrophy

This heterogeneous group consist of several members (OMIM 234050, 278730, 601675) of photosensitive and a nonphotosensitive form. Ichthyotic skin (**1**) associated with developmental delay, sulfur-deficient brittle hair (**2**) and nails are the phenotypic hallmark of this group. Other abnormal hair structures are apparent by different methods of light microscopy (**3–5**), such as trichoschisis and trichorrhexis nodosa-like fractures. (Figure modified from Ling et al., 2006.)

Further Reading

Boulton SJ. DNA repair: Decision at the break point. Nature 2010;465:301–302

Bootsma DA, Hoeijmakers JHJ. The genetic basis of xeroderma pigmentosum. Ann Génét 1991;34: 143–150

Cleaver JE. Defective repair replication in xeroderma pigmentosum. Nature 1968;218:652–656

Cleaver JE, et al. A summary of mutations in the UV-sensitive disorders: xeroderma pigmentosum, Cockayne syndrome, and trichothiodystrophy. Hum Mutat 1999;14:9–22

Cleaver JE, Lam ET, Revet I. Disorders of nucleotide excision repair: the genetic and molecular basis of heterogeneity. Nature Rev Genet 2009;10:756–768

de Boer J, Hoeijmakers JH. Nucleotide excision repair and human syndromes. Carcinogenesis 2000;21: 453–460

Liang C, et al. Structural and molecular hair abnormalities in trichothiodystrophy. J Invest Dermatol 2006;126:2210–2216

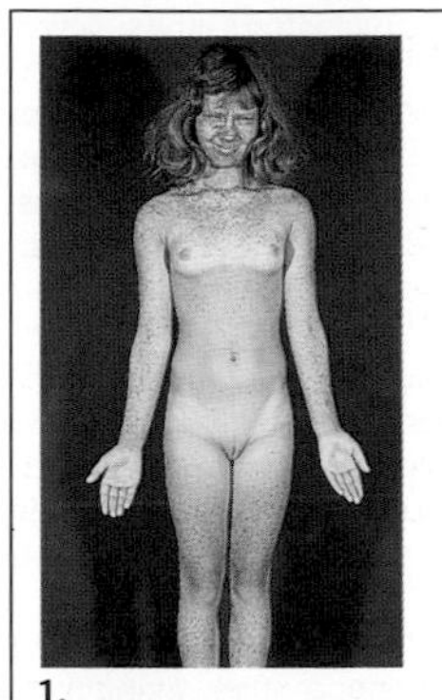
1.

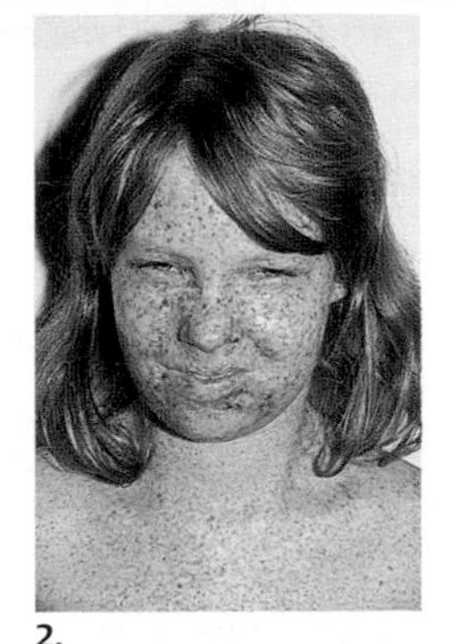
2.

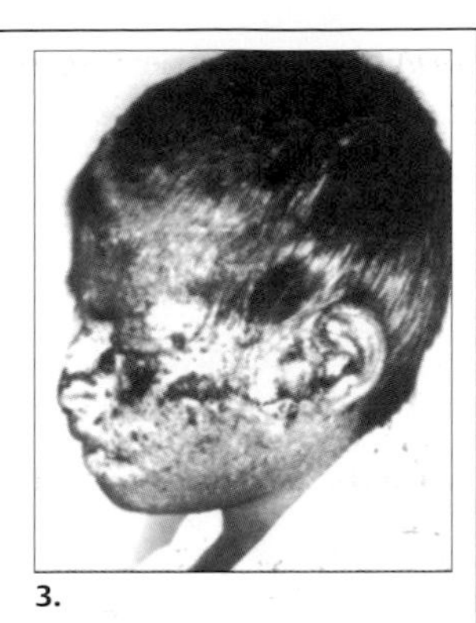
3.

A. Xeroderma pigmentosum: Clinical phenotype

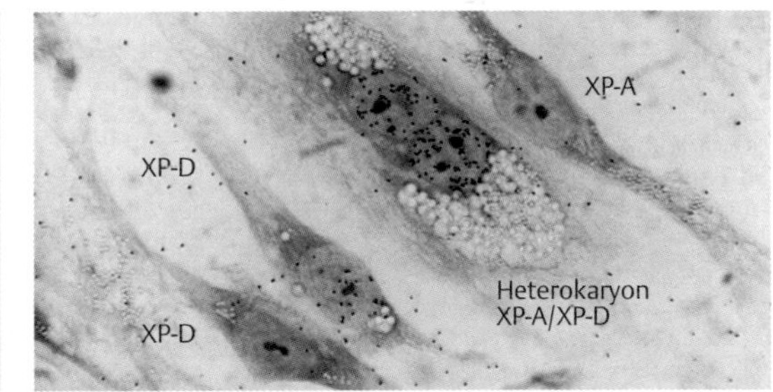

1. UV sensitivity **2.** Complementation following fusion of XP-A and XP-D cells

B. Xeroderma pigmentosum: Cellular phenotype

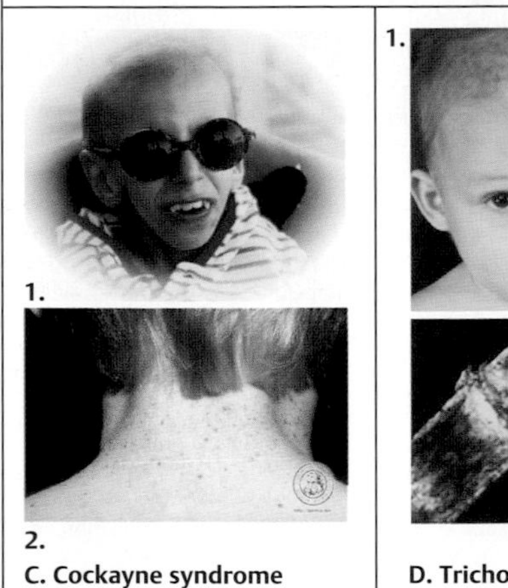
1.
2.

C. Cockayne syndrome

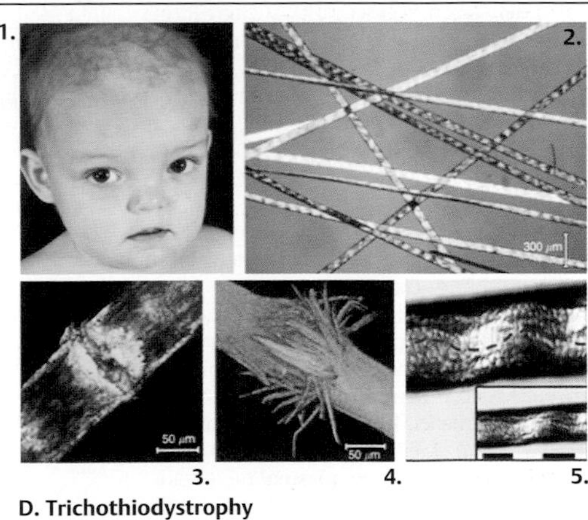

1. 2. 3. 4. 5.

D. Trichothiodystrophy

Cytoskeletal Proteins in Erythrocytes

The cytoskeleton is an intracellular system of proteins with a fibrillar structure. The three main components are microfilaments (7.9 nm diameter), intermediate filaments (10 nm diameter), and microtubuli (24 nm diameter). They consist of polymers of small subunits. Microfilaments and membrane-binding proteins serve as a skeleton of the cell under the plasma membrane. A medium-sized protein of 375 amino acids is *actin*. It is the main cytoskeleton protein (0.5×10^9 molecules) and accounts for 1%–5% of all cellular proteins; in muscle cells it accounts for 10%. The actin cytoskeleton is arranged in bundles and networks that are connected to each other and the cell wall. Actins form an evolutionarily conserved gene family. Erythrocytes have to meet extreme requirements: they traverse small capillaries with diameters less than that of the erythrocytes themselves about a half million times during a 4-month lifespan. Membrane flexibility is also essential for muscle cell function.

A. Erythrocytes

A normal erythrocyte is maintained in a characteristic biconcave discoid form by the cytoskeletal proteins. Genetic defects in different cytoskeletal proteins lead to characteristic erythrocyte deformations: ellipses (elliptocytes), spheres (spherocytes), or as cells with a mouthlike area (stomatocytes) or thornlike projections (acanthocytes). The various forms are the result of defects of different cytoskeleton proteins (see **D**). (The figures are based on scanning electron micrographs from Davies & Lux, 1989.)

B. Proteins in the erythrocyte membrane

Spectrin is a 200-nm long, rod-shaped protein that runs parallel to the erythrocyte plasma membrane. The characteristic shape of an erythrocyte is maintained by attachment of the spectrin–actin cytoskeleton to the erythrocyte plasma membrane. This is accomplished by two specific integral membrane proteins: ankyrin and band 4.1 protein. Ankyrin connects to band 3 protein, which is an anion transport protein in the plasma membrane. Band 4.1 protein binds to glycophorin. Glycophorins (A, B, and C) are transmembrane proteins with several carbohydrate units. Glycophorin A, the major protein marker in the erythrocyte, is a single-pass transmembrane sialoglycoprotein. The anion channels in erythrocytes are important for carbon dioxide transport. (Figure adapted from Luna & Hitt, 1992.)

C. α- and β-Spectrin

Spectrin is the main component of erythrocyte cytoskeletal proteins (**1**). It is a long protein composed of a 260-kDa α chain and a 225-kDa β chain. The chains consist of 20 (α chain) and 18 (β chain) subunits, respectively (**2**), each with 106 amino acids. Each subunit is composed of three α-helical protein strands running counter to one another. Subunit 10 and subunit 20 of the α chain consist of five, instead of three, parallel chains. The individual subunits are assigned to different domains (I–V in the α chain, and I–IV in the β chain). (Figure adapted from Luna & Hitt, 1992.)

D. Erythrocyte skeletal proteins

Sodium dodecyl sulfate (SDS) polyacrylamide gel electrophoresis differentiates numerous membrane-associated erythrocyte proteins. Each band of the gel is numbered, and the individual proteins are assigned to them. The main proteins include α- and β-spectrin, ankyrin, an anion-channel protein (band-3 protein), proteins 4.1 and 4.2, actin, and others.

Medical relevance

Inherited red cell membrane disorders are: five types of spherocytosis (OMIM 182900, 270970) caused by mutations in different genes; elliptocytosis, including the rare pyropoikilocytosis (OMIM 130500); acanthocytosis (OMIM 109270); and stomatocytosis type 1 and type 2 (OMIM 185000). With the exception of an autosomal recessive spherocytosis type 3 (OMIM 270970), all have autosomal dominant inheritance at loci shown in the figure.

Further Reading

Davies KA, Lux SE. Hereditary disorders of the red cell membrane skeleton. Trends Genet 1989;5: 222–227

Delaunay J. The molecular basis of hereditary red cell membrane disorders. Blood Rev 2007;21:1–20

Luna EJ, Hitt AL. Cytoskeleton—plasma membrane interactions. Science 1992;258:955–964

Perrotta S, Gallagher PG, Mohandas N. Hereditary spherocytosis. Lancet 2008;372:1411–1426

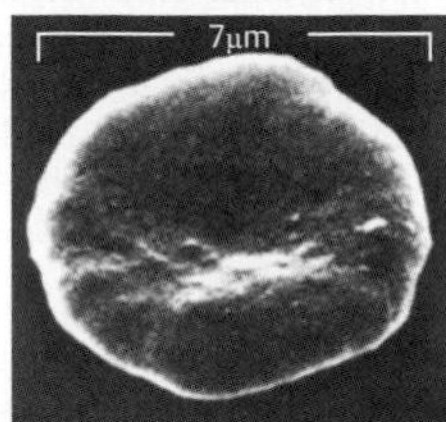

Normal erythrocyte

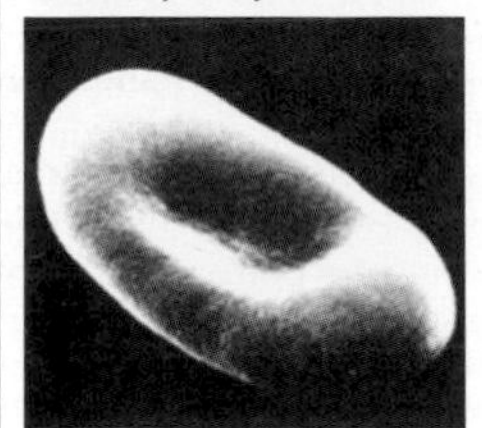

Elliptocyte (spectrin defect)

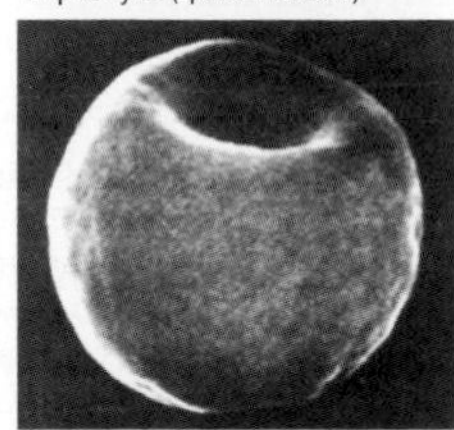

Spherocyte (ankyrin defect)

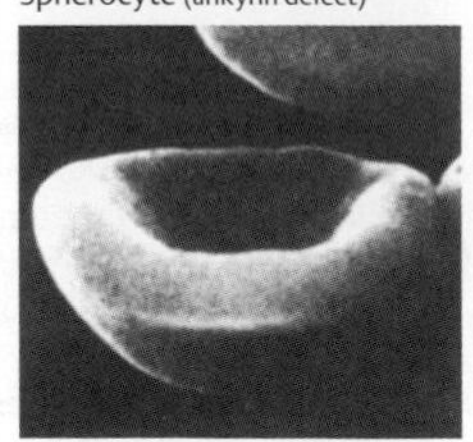

Stomatocyte (tropomyosin defect)

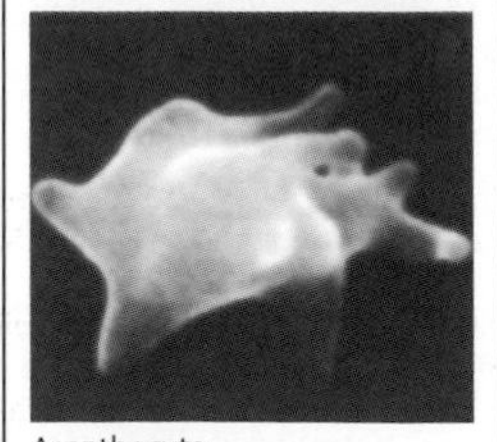

Acanthocyte

A. Erythrocytes

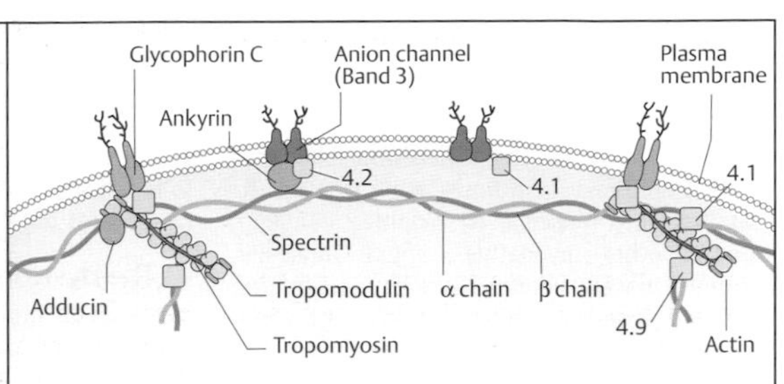

B. Proteins in the erythrocyte membrane

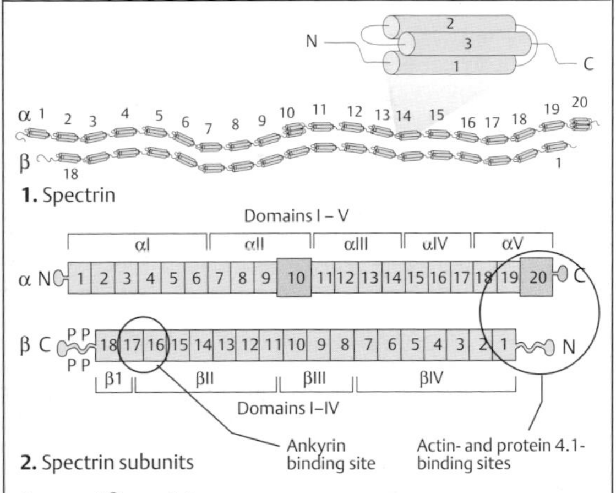

C. α- and β-spectrin

Band	SDS gel	Protein	Chromosomal localization	Disease
1		α–Spectrin	1q22 – 25	Elliptocytosis-1
2		β–Spectrin	14q23 – 24	Spherocytosis-2
		Ankyrin	8p11 – 21	Spherocytosis-1
3		Anion channel	17q21 – q22	Acanthocytosis
4.1		Protein 4.1	1q22 – 25	Elliptocytosis-2
4.2		Protein 4.2		
5		Actin	7pter – q22	
6		Glycerol-aldehyde-3-P dehydro-genase	12p13	
7		Tropomyosin (non-muscle)	1q31 – 41	Stomatocytosis

D. Erythrocyte skeleton proteins

Hereditary Muscle Diseases

Mutations in genes encoding the many proteins in muscle cause a large group of clinically and genetically different neuromusclar diseases. Hereditary neuromuscular diseases are classified into muscular dystrophies, congenital and other myopathies, spinal muscular atrophies, motor neuron diseases, and others. They are genetically heterogeneous and clinically variable, with well over 50 distinct forms listed in OMIM.

A. The dystrophin–glycan complex

The dystrophin–glycan complex is a system of six interconnected proteins bound to the muscle cell plasma membrane (sarcolemma). The proteins belong to the group of dystroglycans and sarcoglycans. Laminins connect with the extracellular matrix. The central protein is dystrophin, a large elongated protein of 175 nm with a specific structure of two subunits, which are connected to the thin myofilament F-actin (filamentous actin) at the N-terminus and to dystrobrevin and syntrophin at the C terminus. Dystrophin provides a bridge between the intracellular cytoskeleton involved in the contractile myofilaments and the extracellular matrix. The largest of the interconnecting proteins, α-dystroglycan, is located outside the cell. It is connected to the extracellular matrix by a heterotrimeric protein, laminin-2. Its partner, β-dystroglycan, is embedded in the sarcolemma and connected to a series of other cytoskeletal proteins, which are divided into the sarcoglycan and syntrophin subcomplexes. Two dystrophin molecules connect neighboring dystrophin–glycan complexes.

Several types of congenital muscular dystrophies are known. The complex group of six types of limb girdle muscular dystrophies (OMIM 607155, 253600) is classified according to the type of sarcoglycan involved.

B. Model of the dystrophin molecule

Dystrophin, the largest member of the spectrin superfamily, is composed of 3685 amino acids, which form four functional domains: (i) the N-terminal actin-binding domain of 336 amino acids; (ii) 24 long repeating units, each consisting of 88- to 126-amino-acid triple-helix segments, as in spectrin; (iii) a 135-amino-acid cysteine-rich domain, which binds to the sarcolemma proteins; and (iv) the C-terminal domain of 320 amino acids with binding sites to syntrophin and dystrobevin. The triple helix segments form the central rod domain, which is 100–125 nm long. (Figure adapted from Koenig et al., 1988.)

C. The dystrophin gene

The human dystrophin gene (*DMD*; OMIM 300377) is located on the short arm of the X chromosome in region 2, band 1.2 (Xp21.2) (**1**). *DMD* is the the largest known gene in man, spanning 2.3 million base pairs (2.3 Mb) in 79 exons (**2**). The large DMD transcript has 14 kb. The dystrophin gene contains at least seven intragenic promoters. The primary transcript is alternatively spliced into a variety of different mRNAs that encode smaller proteins expressed in tissues other than muscle cells, especially tissues in the central nervous system. A related gene encodes utrophin.

D. Distribution of deletions

The most frequent types of disease-causing mutations in the *DMD* gene are deletions, which occur in 60%–65% of patients. They are unevenly distributed. Most frequently involved are exons 43–55 and exons 1–15, roughly corresponding to the F-actin-binding site and the dystroglycan-binding site. Duplications of one or more exons (in 6% of patients) and point mutations also occur. (Data kindly provided by Professor C. R. Müller-Reible, University of Würzburg, Würzburg, Germany.)

Further Reading

Duchenne muscular dystrophy. at: (www.ncbi.nlm.nih.gov/books/NBK22263/)

Koenig M, Monaco AP, Kunkel LM. The complete sequence of dystrophin predicts a rod-shaped cytoskeletal protein. Cell 1988;53:219–228

Muscle Diseases. (www.dmoz.org/)

OMIM. Online Mendelian Inheritance of Man. (www.ncbi.nlm.nih.gov/omim)

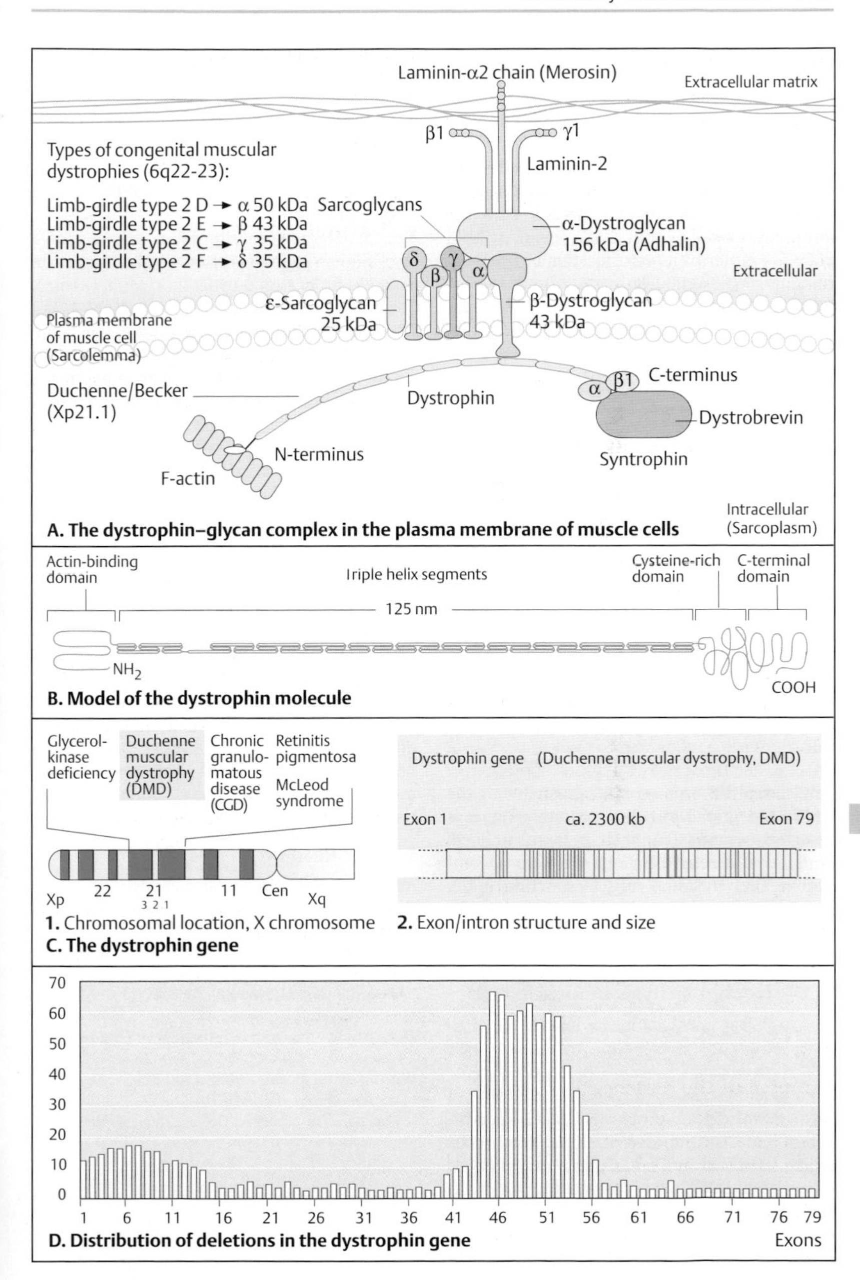
Laminin-α2 chain (Merosin)
Extracellular matrix
β1
γ1
Laminin-2
Types of congenital muscular dystrophies (6q22-23):
Limb-girdle type 2 D → α 50 kDa
Limb-girdle type 2 E → β 43 kDa
Limb-girdle type 2 C → γ 35 kDa
Limb-girdle type 2 F → δ 35 kDa
Sarcoglycans
α-Dystroglycan 156 kDa (Adhalin)
Extracellular
ε-Sarcoglycan 25 kDa
β-Dystroglycan 43 kDa
Plasma membrane of muscle cell (Sarcolemma)
Duchenne/Becker (Xp21.1)
Dystrophin
C-terminus
Dystrobrevin
Syntrophin
N-terminus
F-actin
Intracellular (Sarcoplasm)
A. The dystrophin–glycan complex in the plasma membrane of muscle cells
Actin-binding domain
Triple helix segments
Cysteine-rich domain
C-terminal domain
125 nm
NH2
COOH
B. Model of the dystrophin molecule
Glycerol-kinase deficiency
Duchenne muscular dystrophy (DMD)
Chronic granulo-matous disease (CGD)
Retinitis pigmentosa
McLeod syndrome
Xp
22
21
3 2 1
11
Cen
Xq
1. Chromosomal location, X chromosome
Dystrophin gene (Duchenne muscular dystrophy, DMD)
Exon 1
ca. 2300 kb
Exon 79
2. Exon/intron structure and size
C. The dystrophin gene
70
60
50
40
30
20
10
0
1
6
11
16
21
26
31
36
41
46
51
56
61
66
71
76
79
D. Distribution of deletions in the dystrophin gene
Exons

Duchenne Muscular Dystrophy

Duchenne muscular dystrophy (DMD; OMIM 310200) is the most common of the muscular dystrophies, with a frequency of 1 in 3500 liveborn males. It is named after the French neurologist Guillaume Duchenne (1806–1875), who described this disease in 1861. It is caused by mutations in the *DMD* gene (OMIM 300377), either by a new mutation or by transmission of the mutation from a heterozygous mother. Germline mosaicism has been observed (i.e., a female carrying a *DMD* mutation in a variable proportion of her germ cells). The mutation rate is high.

A. Clinical signs

The age of onset is usually less than 3 years and signs are evident at 4–5 years; the patient requires a wheelchair by 12 years and usually succumbs to the disorder by age 20 years. Progressive muscular weakness of the hips, thighs, and back causes difficulties in walking and in using steps. Lumbar lordosis and enlarged but weak calves (pseudohypertrophy) are visible (**1**). The affected child performs a characteristic series of maneuvers to rise from a kneeling position (Gower's sign, **2**).

A clinically milder variant, Becker muscular dystrophy (BMD; OMIM 300376), is an allelic disorder with a milder course and a later age of onset of the disease. The clinical difference results from the type of rearrangement of the *DMD* gene. In DMD the reading frame is changed, whereas in BMD it is maintained. Thus, even a relatively large deletion is compatible with residual muscle function if the reading frame is not changed. (Drawings by Duchenne, 1861, and Gowers, 1879; from Emery, 2003.)

B. Dystrophin analysis in muscle cells

Dystrophin is normally located along the plasma membrane (sarcolemma) of muscle cells (**1**). In patients it is absent (**2**). Female heterozygotes show a patchy distribution of groups of normal and defective muscle cells (**3**) as a result of X inactivation (see p. 196). (Photographs kindly provided by Dr R. Gold, Department of Neurology, University of Würzburg, Würzburg, Germany.)

C. Investigation of a family with DMD

In about one-third of patients a mutation may not be detectable. Several alternative methods are available. These include single-strand conformational polymorphism, heteroduplex analysis, reverse transcriptase PCR, and the protein truncation test.

The panel shows a simplified example of a two-allele system (marker DXS7) (based on data kindly provided by Dr C. R. Müller-Reible, Würzburg, Germany). Two patients (III-1 and III-2) carry the allele 1. The mothers, II-1 and II-2, can be considered heterozygotes for the mutation, as is the case for their mother, I-2. An unaffected male (II-4) has allele 2. This indicates that allele 2 does not carry the mutation. Two males, III-3 and III-4, are not affected. This can be explained by recombination in their mother, II-5. In current practice one uses a set of several linked markers flanking the disease locus to avoid an erroneous diagnosis due to recombination. Female heterozygotes show mild clinical signs in 8%. About 23% of mothers of isolated patients are not carriers of the mutation.

D. Other forms of muscular dystrophy

Several other forms of genetically determined muscular dystrophy are known in man. Course, diagnosis, and molecular genetic analysis depend on the basic disorder. Selected examples are listed along with their chromosomal location and McKusick (OMIM) no.

Further Reading

Amato AA, Brown RH Jr. Muscular dystrophies and other muscle diseases. In: Harrison's Principles of Internal Medicine. 18th ed. DL Longo, et al., eds. New York: McGraw-Hill; 2012:3487

Emery AEH Duchenne Muscular Dystrophy, 2nd ed. Oxford: Oxford University Press, 2003

Musclar Dystrophy Association. Duchenne Musclar Dystrophy. Available at: http://www.mda.org/disease/dmd.html. Accessed January 28, 2012

Worton RG, et al. The X-linked muscular dystrophies. In: Scriver CR, et al., eds. The Metabolic and Molecular Bases of Inherited Disease. 8th ed. New York: McGraw-Hill; 2001:5493–5523

1. Calf hypertrophy and lordosis

2. Difficulty in rising (Gower's sign)

A. Clinical signs of Duchenne muscular dystrophy

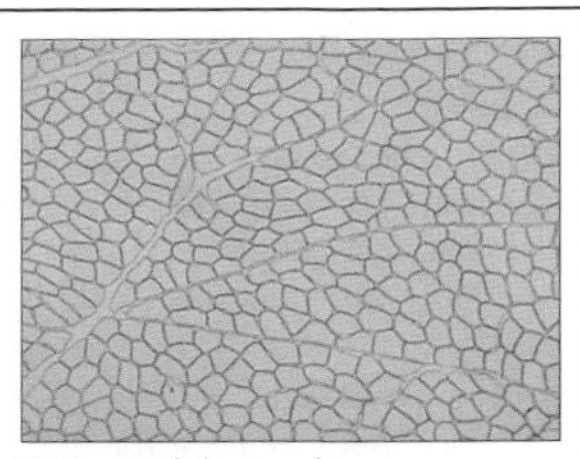

1. Normal dystrophin

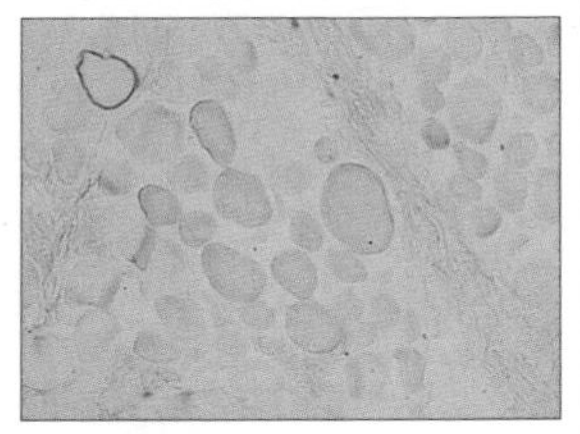

2. Dystrophin absent

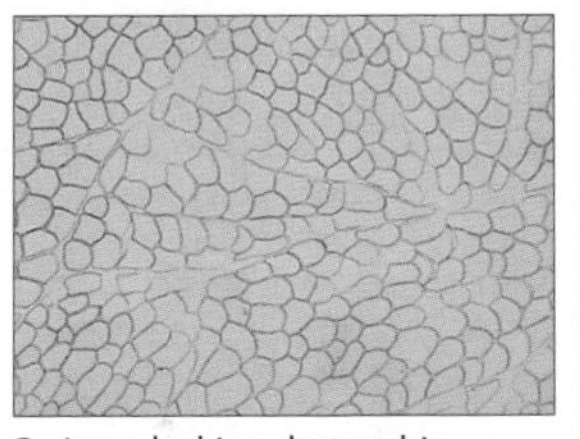

3. Areas lacking dystrophin in heterozygotes

B. Dystrophin analysis in muscle cells

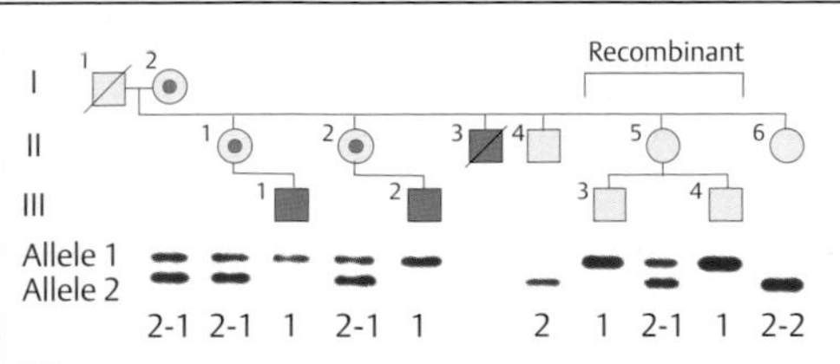

C. Investigation of a family with DMD by DNA markers

Disease	Chromosomal location	McKusick no.
X-chromosomal:		
Duchenne muscular dystrophy	Xp21.2	310200
Becker muscular dystrophy (allelic with DMD)	Xp21.2	310200
Emery–Dreifuss muscular dystrophy	Xq28	310300
Autosomal dominant:		
Myotonic dystrophy (2 types)	19q13; 3q13	160900; 602668
Facioscapulohumeral dystrophy	4q35 – qter	158900
Oculopharyngeal muscular dystrophy	14q11.2 – q13	164300
Autosomal recessive:		
Duchenne-like muscular dystrophy (LGMD2C)	13q12 – 13	253700
Fukuyama congenital muscular dystrophy	9q31 – 33	253800
Limb-girdle muscular dystrophy (several types)	15q15 – q22, other loci	253600

D. Important forms of hereditary muscular dystrophy in man

FGF Receptor Mutations in Skeletal Dysplasias

Skeletal dysplasias are a large group of clinically and genetically different disorders of bone and cartilage development, often associated with short stature. Mutations in one of three fibroblast growth factor receptors (FGFR1, FGFR2, FGFR3; OMIM 136350, 134920, 134934) cause a characteristic phenotypic spectrum of skeletal dysplasias (Table 14, Appendix p. 419). Fibroblast growth factors (FGFs) transmit signals into cells by means of highly specific membrane-bound receptors (FGFRs) that belong to the large family of RTKs (receptor tyrosine kinases, p. 198). Binding of the FGF ligand to their receptors requires both interaction with proteoglycans, such as heparin or heparan sulfate, and dimerization. FGF–FGFR interactions are required to regulate growth in the developing vertebrate limb.

A. Examples of phenotypes

The six examples shown represent distinct types of skeletal dysplasias, which are characterized by their clinical and radiological manifestations. Synostosis of cranial sutures (craniosynostosis) is a typical sign of Crouzon (**1**; OMIM 123500), Pfeiffer (**2**; 101600), and Apert syndromes. Complete syndactyly of fingers and toes are a hallmark feature of Apert syndrome (**3**; OMIM 101200). The combination of short stature (mean adult height 115–145 cm), midface hypoplasia, and other skeletal manifestations are characteristic of achondroplasia (**4**; OMIM 100800). Thanatophoric dysplasia (**5**; OMIM 187600) is a severe, lethal, genetically heterogeneous form. Muenke syndrome (**6**; OMIM 602849) is characterized by coronal craniosynostosis (bilateral or unilateral), tarsal and/or carpal fusion, sensorineural hearing loss, and developmental delay. (Photographs **1**, **2**, and **4–6** courtesy of Dr Maximilian Muenke, National Institutes of Health, USA; photograph **3** from Gilbert, 2010.)

B. Mutations and diseases

The extracellular ligand-binding component of FGFRs, three immunoglobulin (Ig)-like domains (IgI, IgII, IgII), are connected to a single transmembrane domain of 22 amino acids. The cytosolic component has two kinase domains. FGFR-related disorders have a rather specific spectrum of allelic gain-of-function dominant mutations. Mutations in either FGFR1 or FGFR2 cause Pfeiffer syndrome. Two mutations in adjacent exons, S252W (tryptophan, W, replacing serine, S) and P253R (arginine, R, replacing praline, P), account for 66% and 32% of cases of Apert syndrome, respectively. Mutations in FGFR3 cause achondroplasia and thanatophoric dysplasia type 1. Nearly all individuals (97%) with achondroplasia have the same causative mutation, a G380R substitution (glycine replaced by arginine) within the transmembrane domain. Thanatophoric dysplasia type 1 is caused by mutations scattered along different parts of FGFR3 between the IgII-like and IgIII-like domains. The three clinically distinct disorders achondroplasia, thanatophoric dysplasia, and Muenke syndrome (4–6 in **A**) are allelic since they are all caused by mutations in FGFR3. (Figure from Gilbert, 2010.)

C. Effects on the growth plate

Most FGFR mutations increase the affinity of the ligand to the receptor (gain-of-function). The resultant effect on FGFR signaling pathways depends on the type of cell involved. Whereas FGFR activation results in proliferation of fibroblasts, it inhibits growth in chondrocytes. In thanatophoric dysplasia, the normal transition of resting chondrocytes to proliferating chondrocytes, and early osteoblasts in the growth plate (left, normal) of developing bone is totally disorganized and does not contain proliferating chondrocytes (right). (Figure from Gilbert, 2010.)

Further Reading

Gilbert SF. Developmental Biology. 9th ed. Sunderland: Sinauer; 2010

Givol D, Eswarakumar VG. The fibroblast growth factor signaling pathway. In: Epstein CJ, Erickson RB, Wynshaw-Boris A, eds. Inborn Errors of Development. The Molecular Basis of Clinical Disorders of Morphogenesis. 2nd ed. Oxford: Oxford University Press; 2008:449–460

Muenke M, eds. Craniosynostoses. Molecular Genetics, Principles of Diagnosis, and Treatment. Basel: Karger; 2011

Wilkie ACM. FGF receptor mutations: Bone dysplasia, craniosynostosis, and other syndromes. In: Epstein CJ, Erickson RB, Wynshaw-Boris A, eds. Inborn Errors of Development. The Molecular Basis of Clinical Disorders of Morphogenesis. 2nd ed. Oxford: Oxford University Press; 2008:461–473

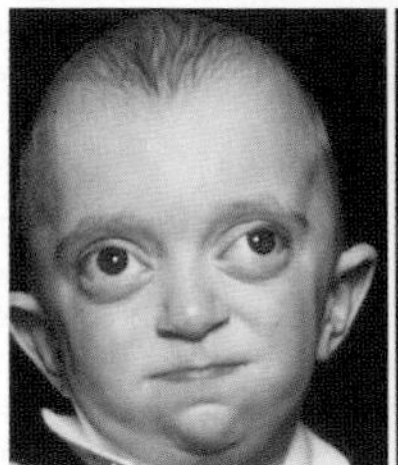

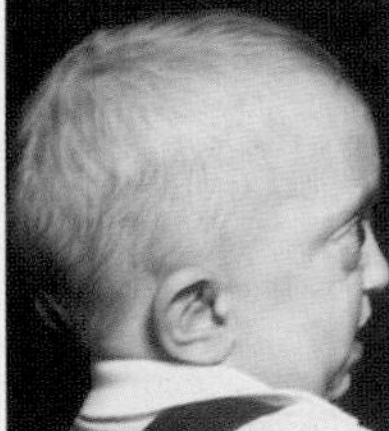

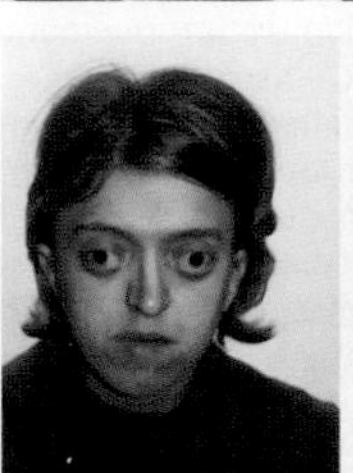

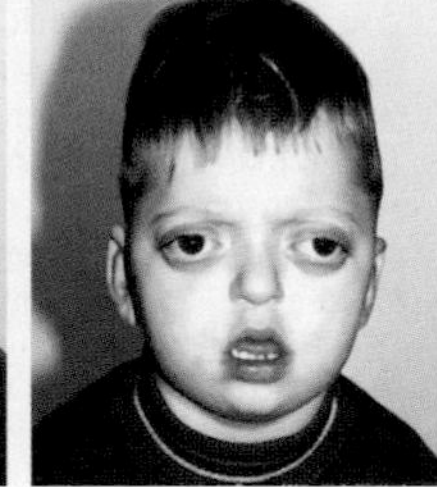

Mother and child
1. Crouzon syndrome

Mother and child affected
2. Pfeiffer syndrome

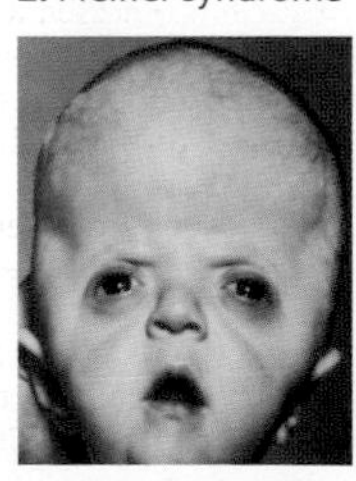

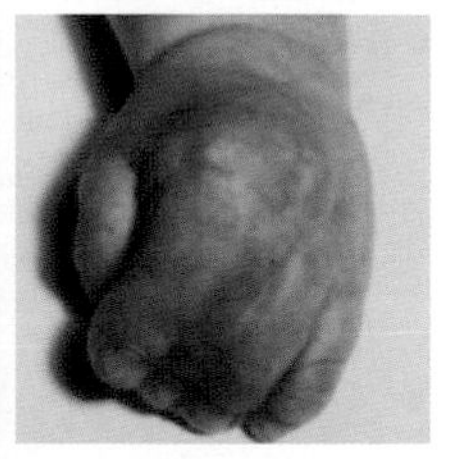

3. Apert syndrome
A. Examples of phenotypes

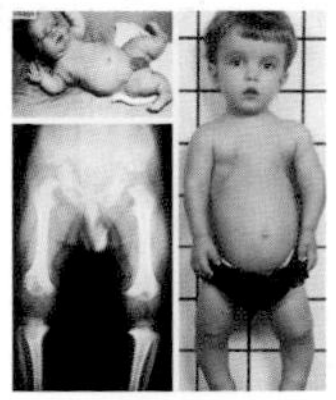

4. Achondroplasia

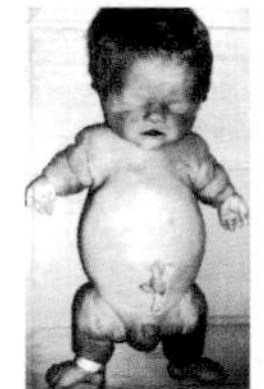

5. Thanatophoric dysplasia

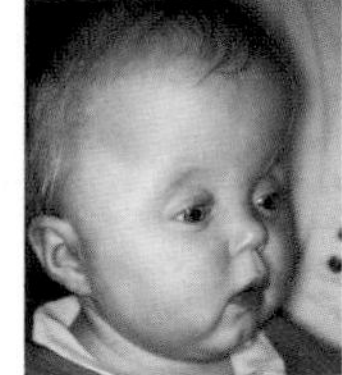

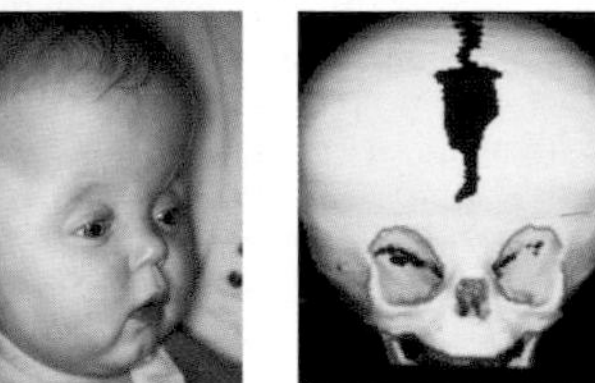

6. Muenke syndrome

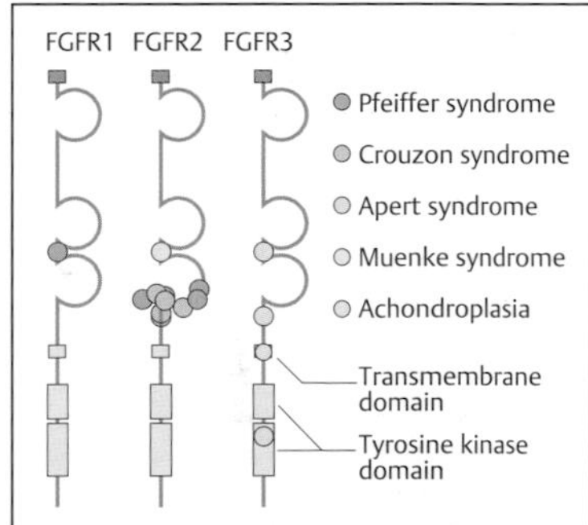

B. Mutations and diseases

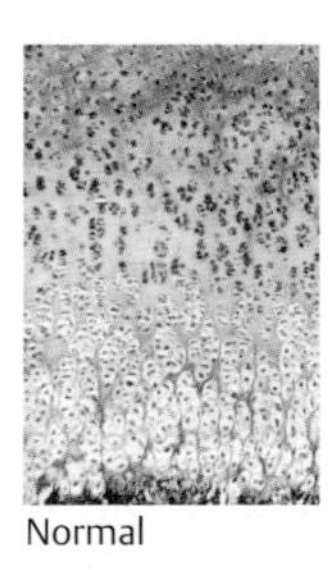

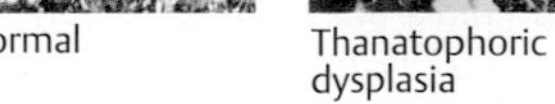

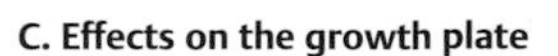

Normal

Thanatophoric dysplasia

C. Effects on the growth plate

Marfan and Loeys–Dietz Syndromes

Marfan syndrome (MFS) and Loeys–Dietz syndrome (LDS) are autosomal dominant syndromes associated with aortic aneurysms and other manifestations. MFS (OMIM 154700), first described in 1896, is caused by mutations in the fibrillin-1 gene *FBN1* (OMIM 134797). Mutations in a similar gene, *FBN2* (OMIM 612570), located at 5q23-q31, cause a related disorder known as congenital contractural arachnodactyly (CCA; OMIM 121050), which may have been the one described by Marfan in 1896.

LDS (OMIM 609192), first described in 2005, is caused by mutations either in the *TGFBR1* gene (transforming growth factor β receptor 1; OMIM 190181) in LDS type 1, or in the *TGFBR2* gene (OMIM 190182) in LDS type 2. The use of an antagonist of angiotensin II in the transforming growth factor β (TGFβ) pathway (see p. 202) is a promising therapeutic approach in these disorders.

A. Marfan syndrome (MFS)

MFS mainly affects the skeleton, the cardiovacular system, and the eye. The visible signs include hyperextensible joints (**1**), long fingers (arachndodytyly, **2**), pectus excavatum (**3**), subluxation of the lens in 50%–80% of cases (**4**), scoliosis, and other variable signs. The main cardiac manifestation is progressive aortic root dilatation and a high risk for aortic aneurysms leading to aortic dissection. β-Adrenergic blockade (propranolol) delays the rate of aortic dilatation. In view of excessive signaling by TGFβ cytokines, blocking the angiotensin II type 1 receptor (AGTR1; OMIM 106165) promises to be an effective therapeutic approach. (Photographs: **1** and **2**, Prime Health Channel [www.primehealthchannel.com/marfan-syndrome]; **3**, Dr Beate Albrecht, Essen, Germany; **4**, Dr M. Siepe, Herz-Kreislauf Zentrum, Freiburg, Germany.)

B. Loeys–Dietz syndrome (LDS)

The main manifestations are facial dysmorphia with hypertelorism, pectus carinatum, scoliosis (**1**), deformed feet (**2**), vascular tortuosity (**3**), cleft palate or bifid uvula in some patients (**4**) associated with mitral valve prolapse, aneuryms in the aorta (5) and other arteries. The phenotype may also overlap with Shprintzen–Goldberg syndrome (OMIM 182212). (Photographs: **1** and **2**, Drera et al., 2009; **3**, Warell et al., 2010; **4** and **5**, Lindsey & Dietz, 2011.)

C. Fibrillin-1 protein (FBN1)

Fibrillins are a group of glycoproteins in the extracellular microfibrils in connective tissue. The FBN1 protein consists of several distinct regions. There are two types of motif: 43 calcium-binding EGF-like (epidermal growth factor-like) motifs, and seven motifs, each with eight cysteine residues; there are two hybrid domains consisting of EGF-like motifs and motifs containing eight cysteine residues. The protein is encoded by the *FBN1* gene located at 15q21.1, consisting of 65 exons extending over 235 kb genomic DNA. The mRNA is 10 kb in length (9749 nucleotides) with three alternatively spliced noncoding 5′ exons.

D. *TGFBR1* and *TGFBR2* genes

These genes encode a serine/threonine kinase transmembrane receptor for TGFβ. Activation by their ligand regulates cell proliferation and differentiation in various developmental processes. Mutations in these two genes lead to increased TGFβ signaling and cause Loeys–Dietz syndrome. (Figure adapted from Loeys et al., 2006.)

E. Increased TGF-β signaling

TGFβ signaling stimulates multiple intracellular pathways. One of these involves extracellular matrix proteins, including fibrillin-1. Inhibition of increased TGFβ signaling ameliorates aortic aneurysms in mice. (Figure based on Habashi et al., 2011.)

Further Reading

Drera B, et al. Loeys-Dietz syndrome type I and type II. Orphanet J Rare Dis 2009;4:24 (www.ojrd.com/)

Habashi JP, et al. Angiotensin II type 2 receptor signaling attenuates aortic aneurysm in mice through ERK antagonism. Science 2011;332:361–365

Judge DP, Dietz HC. Marfan's syndrome. Lancet 2005;366:1965–1976

Lindsay ME, Dietz HC. Lessons on the pathogenesis of aneurysm from heritable conditions. Nature 2011;473:308–316

Loeys BL, et al. Aneurysm syndromes caused by mutations in the TGF-β receptor. N Engl J Med 2006;355:788–798

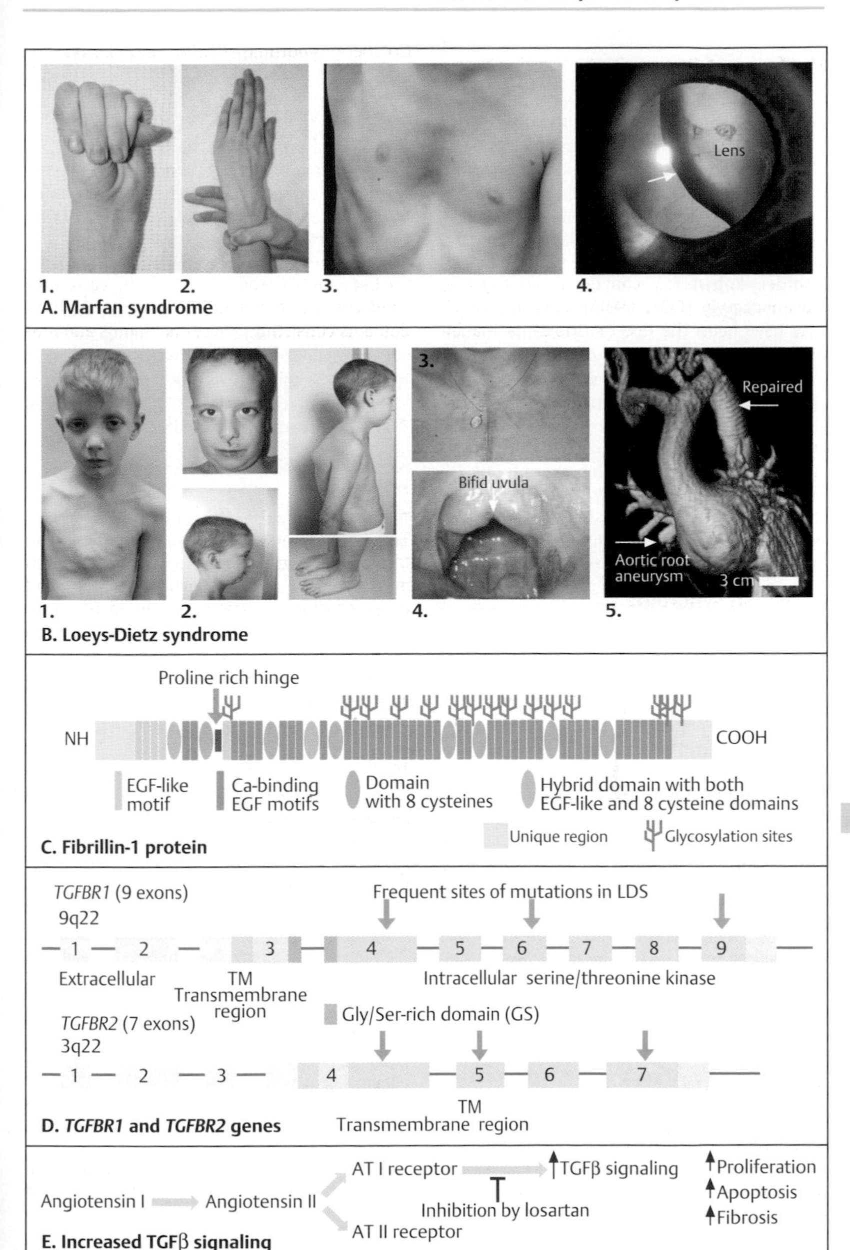
1.
2.
3.
Lens
4.
A. Marfan syndrome
Repaired
Bifid uvula
Aortic root aneurysm
3 cm
1.
2.
3.
4.
5.
B. Loeys-Dietz syndrome
Proline rich hinge
NH
COOH
EGF-like motif
Ca-binding EGF motifs
Domain with 8 cysteines
Hybrid domain with both EGF-like and 8 cysteine domains
Unique region
Glycosylation sites
C. Fibrillin-1 protein
TGFBR1 (9 exons)
9q22
Frequent sites of mutations in LDS
1 2 3 4 5 6 7 8 9
Extracellular
TM
Transmembrane region
Intracellular serine/threonine kinase
Gly/Ser-rich domain (GS)
TGFBR2 (7 exons)
3q22
1 2 3 4 5 6 7
TM
Transmembrane region
D. TGFBR1 and TGFBR2 genes
Angiotensin I
Angiotensin II
AT I receptor
AT II receptor
TGFβ signaling
Inhibition by losartan
Proliferation
Apoptosis
Fibrosis
E. Increased TGFβ signaling

Collagen Molecules

Collagen is a large group of insoluble, extracellular glycoproteins with specialized functions in the maintenance of tissue shape and structure. It is the most abundant protein in mammals, constituting about one-quarter of the total body protein. Collagens occur in skin, bones, tendons, cartilage, ligaments, blood vessels, teeth, basement membranes, and supporting tissues of the internal organs. Most collagens have a remarkable mechanical resilience, with interlinked, insoluble threads (fibrils) of unusual strength. The 27 known types of collagens, numbered I–XXVII, are classified according to their main structural and functional features. The most important collagens in humans are fibrillar collagen types I, II, III, V, and XI, and basal membrane collagen (type IV). Collagen types I, II, and III account for 80%–90% of collagens. Type I is the main component of tendons, ligaments, and bones; type II occurs in cartilage and the notochord of vertebrate embryos; type III occurs in arteries, the intestine, and the uterus; and type IV, in the basal lamina of epithelia, in particular around the glomeruli of the kidney, and in blood capillaries (Appendix Table 15, p. 419).

A. Collagen structure

Collagen is a triple helix of three chains synthesized in seven principal steps. The sequence of the approximately 1000 amino acids of a typical collagen is simple and periodic (**1**). Every third amino acid is glycine (Gly). Other amino acids alternate between the glycines. The general structural motif is $(\text{Gly–X–Y})_n$. X is either proline or hydroxyproline; Y is either lysine or hydroxylysine (**2**). Three chains of procollagen form a triple helix (**3**). In collagen type I, the helix is composed of two identical α1 chains and one α2 chain. First a precursor molecule, *procollagen* is formed (**4**). Procollagen peptidases cleave peptides at the N-terminal and C-terminal ends to form *tropocollagen* (**5**). Tropocollagen molecules, connected by numerous hydroxylated proline and lysine residues, form a *collagen fibril* (**6**). Each fibril consists of staggered, parallel rows of end-to-end tropocollagen molecules, separated by gaps (**7**). Collagen fibrils are visible as transverse stripes under the electron microscope (EM) (**8**). A fiber of 1-mm diameter can hold a weight of almost 10 kg. (Photograph from Stryer, 1995.)

B. Prototype of a gene for procollagen

Procollagen type II consists of a triple helix of three procollagen type II chains, designated as α1[II]. The corresponding gene, *COL2A1* (OMIM 120140), located at 12q13.11, consists of 52 exons of different sizes, each encoding 5, 6, 11, 12, or 18 Gly–X–Y units. The translated part of exon 1 (85 bp) encodes a signal peptide necessary for secretion. The genes for procollagen types I, II, and III differ in that some exons are fused, but otherwise they are similar, especially for the three main fibrillar collagen types (I, II, III). Mutations in *COL2A1* occur in several different disorders (see OMIM 120140).

C. Gene structure of procollagen type α1(I)

Procollagen type I consists of two α1 chains and one α2 chain, designated $\alpha1(\text{I})_2\ \alpha2(\text{II})$. The corresponding genes are *COL1A1* (OMIM 120150) and *COL1A2* (OMIM 120160). The 52 exons of the gene *COL1A1* encoding procollagen type I correspond to the different domains (A–G) of procollagen α1(I). The *COL1A2* gene is about twice as large (about 40 kb) as the *COL1A1* gene because the introns between the exons are twice as long as in *COL1A1*.

Medical relevance

More than 10 distinct human diseases are caused by mutations in one of the genes encoding collagen (see Appendix, Table p. 419).

Further Reading

Bateman JF, et al. Genetic diseases of connective tissues: cellular and extracellular effects of ECM mutations. Nat Rev Genet 2009;10:173–183

Byers PH. Disorders of collagen synthesis and structure. In: Scriver CR, et al., eds. The Metabolic and Molecular Bases of Inherited Disease. 8th ed. New York: McGraw-Hill; 2001:5241–5285. (www.ommbid.com)

Chu M-L, Prockop DJ. Collagen gene structure. In: Royce PM, Steinmann B, eds. Connective Tissue and its Heritable Disorders. 2nd ed. New York: Wiley-Liss; 2002:149–165

Rauch F, Glorieux FH. Osteogenesis imperfecta. Lancet 2004;363:1377–1385

Stryer L. Biochemistry. 4th ed. New York: W. H. Freeman; 1995

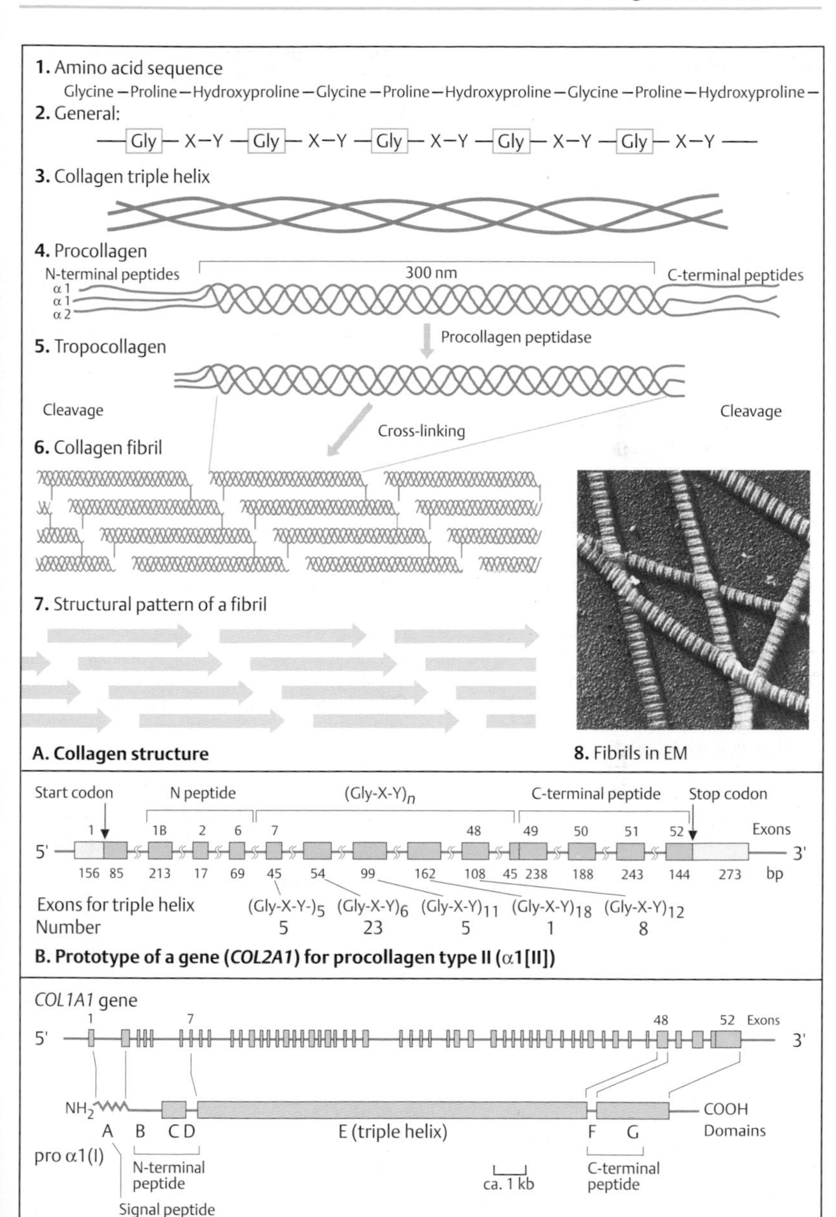

A. Collagen structure

B. Prototype of a gene (*COL2A1*) for procollagen type II (α1[II])

C. Gene structure and procollagen type α1(I)

Osteogenesis Imperfecta

Osteogenesis imperfecta (OI; OMIM 120150), or "brittle bone disease," is a heterogeneous group of 11 clinically and genetically different types of diseases with a total frequency of about 1 in 10000 individuals (OMIM 166200). Common manifestations include spontaneously occurring bone fractures, bone deformity, small stature, defective dentition (dentinogenesis imperfecta), hearing impairment due to faulty formation of the auditory ossicles, and blue sclerae. In most, but not all patients, collagen type I is defective owing to a mutation in one of the two genes encoding the two chains *COL1A1* (OMIM 120150) or *COL1A2* (OMIM 120160). Both autosomal dominant and recessive types of OI exist.

A. Molecular mechanisms

In normal procollagen type I the two chains pro α1(I) and α2(I) are produced in equal amounts (**1**). Some mutations lead to reduced production of pro α1(I) (**2**). The resulting imbalance causes degradation of α2(I) chains. Thus, the amount of procollagen is reduced. Different types of mutations in the genes for pro α1(I) and α2(I) (*COL1A1* and *COL1A2*) cause defective procollagen (**3**) because of a deletion in a *COL1A1* allele; this can be a splicing defect or other type of mutation. Mutations in the *COL1A1* gene are more severe than mutations in the *COL1A2* gene because more defective collagen is formed with the former. (Figure adapted from Wenstrup et al., 1990.)

B. Mutations and phenotype

Generally, mutations in the 3′ region are more serious than mutations in the 5′ region (*position effect*). Mutations of the pro α1(I) chain are more severe than those in the pro α2(I) chain (*chain effect*). The substitution of a larger amino acid for glycine, which is indispensable for the formation of the triple helix, leads to severe disorders (*size effect*). Different types of mutations may occur, such as deletions, mutations in the promoter or enhancer, and splicing mutations. The codons (AAG, AAA) for the amino acid lysine, which occurs frequently in collagen, are readily transformed into a stop codon by substitution of the first adenine by a thymine (TAG or TAA), so that a short, unstable procollagen is formed. Splicing mutations may lead to the loss of exons (exon skipping). (Figure adapted from Byers, 2001.)

C. Different forms of OI

OI is classified into four phenotypes, I–IV, according to the Sillence classification. OI types I and IV are less severe than type II (fatal in infancy) and type III. Three radiographs show: a relatively mild deformity of the tibia and fibula in OI type IV (**1**); severe deformities in the tibia and fibula in OI type III (**2**); and the distinctly thickened and shortened long bones in the lethal OI type II (**3**). Germline mosaicism has been shown to account for rare instances of affected siblings being born to unaffected parents.

Several new types of OI have been described: types V–XXII (see OMIM 166200), some with hyperplastic callus formation.

Further Reading

Basel D, Steiner RD. Osteogenesis imperfecta: recent findings shed new light on this once well-understood condition. Genet Med 2009;11:375–385

Byers PH. Osteogenesis imperfecta. In: Royce PM, Steinmann B, eds. Connective Tissue and its Heritable Disorders. 2nd ed. New York: Wiley-Liss; 2002:317–350

Byers PH. Disorders of collagen synthesis and structure. In: Scriver CR, et al., eds. The Metabolic and Molecular Bases of Inherited Disease. 8th ed. New York: McGraw-Hill; 2001:5241–5285

Forlino A, et al. New perspectives on osteogenesis imperfecta. Nat Rev Endocrinol 2011;7:540–557

Sillence DO, et al. Genetic heterogeneity in osteogenesis imperfecta. J Med Genet 1979;16:101–116

Wenstrup RJ, et al. Distinct biochemical phenotypes predict clinical severity in nonlethal variants of osteogenesis imperfecta. Am J Hum Genet 1990;46:975–982

Willaert A, et al. Recessive osteogenesis imperfecta caused by LEPRE1 mutations: clinical documentation and identification of the splice form responsible for prolyl 3-hydroxylation. J Med Genet 2009;46:233–241

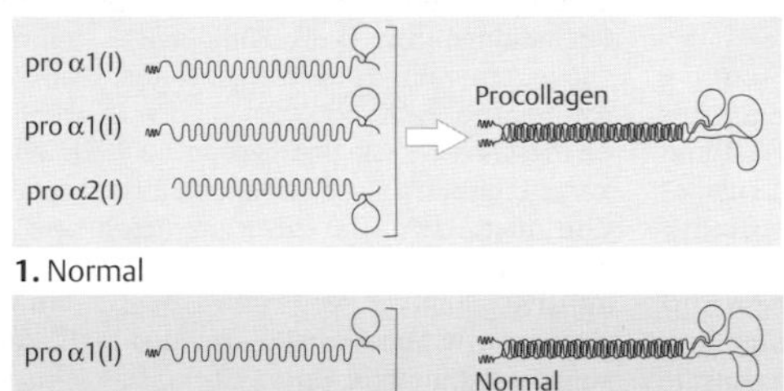

1. Normal

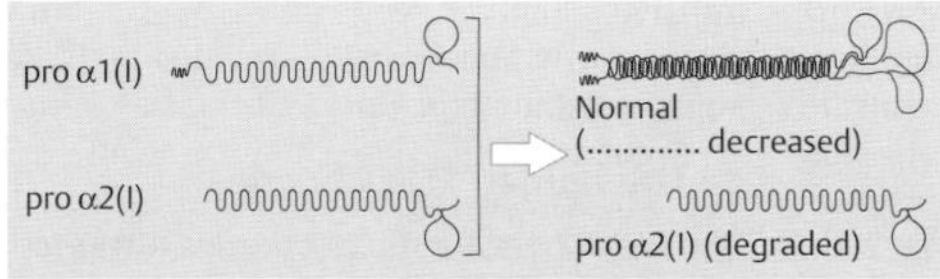

2. Decreased synthesis of procollagen α1(I)

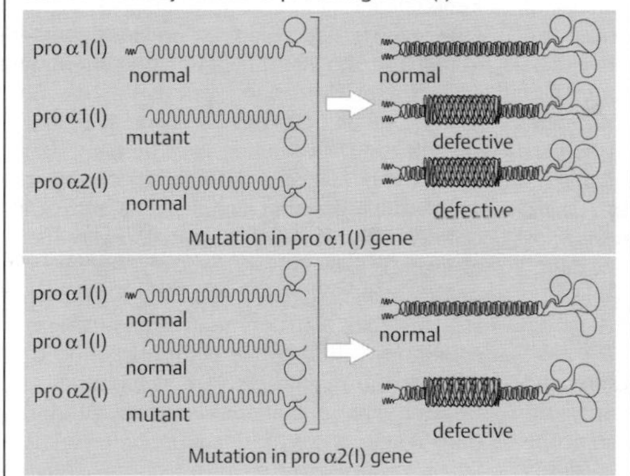

3. Defective procollagen due to a mutation

A. Molecular mechanisms in osteogenesis imperfecta

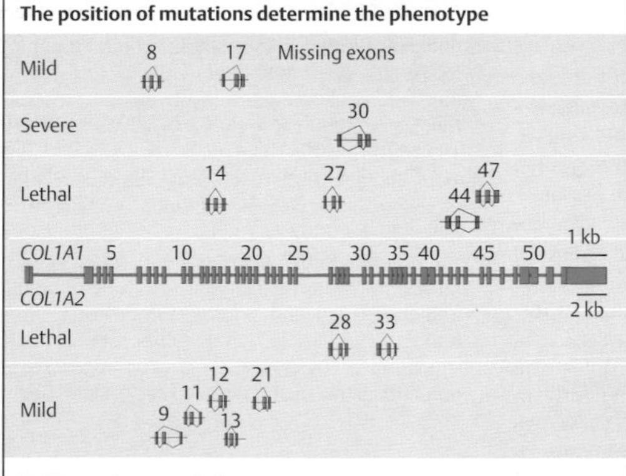

B. Mutations and phenotype

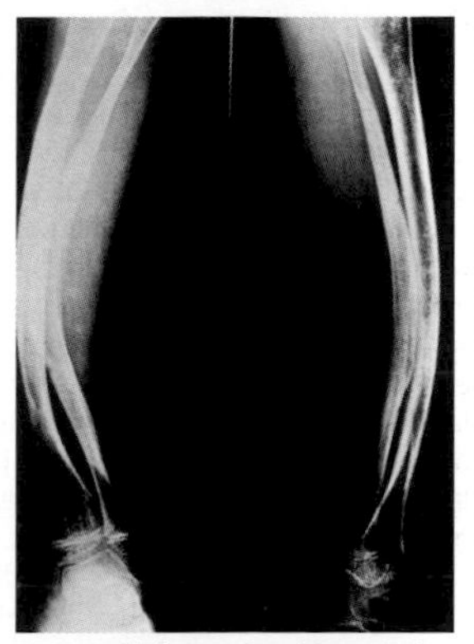

1. Bone deformation (OI type I)

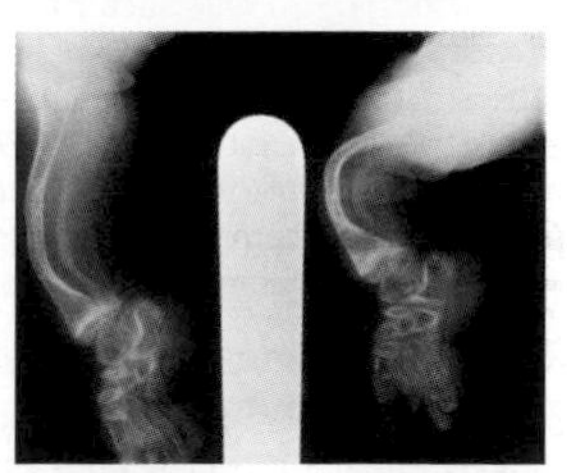

2. Severe deformation (OI type III)

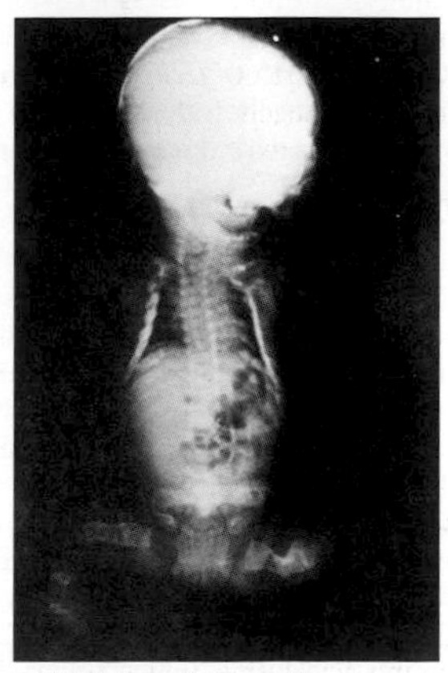

3. Fatal form (OI type II)

C. Different forms of osteogenesis imperfecta

Molecular Basis of Bone Development

Bone develops from three mesodermal cell lineages committed to differentiate into three specialized cell types: *chondrocytes* (cartilage-forming cells), *osteoblasts* (bone-forming cells), and *osteoclasts* (bone-degrading cells). Two major processes form bone (osteogenesis): (i) direct conversion of mesenchymal tissue into bone tissue (*intramembranous* or *dermal ossification*); and (ii) *enchondral ossification*, with cartilage intermediates produced by chondrocytes, which are later replaced by bone cells (osteoblasts). Osteoblasts produce most of the proteins for the extracellular bone matrix and control its mineralization. The osteoblast cell lineage involves osteoblast-specific transcription factors (OSFs). One such transcription factor is a major regulator of osteoblast differentiation in direct intramembranous bone formation: the core-binding factor CBFA1 (runt-related transcription factor, RUNX2; OMIM 600211). Three mammalian *Runx* genes are homologs of the *Drosophila* pair-rule gene *Runt*.

A. Effects of homozygous *Runx2* mutations in mice

Targeted disruption of the *Runx2* gene, located on mouse chromosome 17, in the homozygous state (−/−) results in lack of ossification of the entire skeleton (**1**). Normal calcified skeleton stains red with alizarin (a, c), whereas bones lacking ossification stain blue (b, d, e). These mice are small and die at birth from respiratory failure. The humerus and humeral tuberosity (circle, c) in heterozygous mice (+/−, d) and homozygous mice (−/−) show reduced ossification of the long bones and severe hypoplasia, respectively (d, e). The skull and thorax are also severely affected (**2**). Heterozygous (+/−) mice lack ossification of the skull (b). Normal calcified bone is stained red by alizarin red, here at embryonic day 17.5, three and a half days before birth. Cartilage is stained blue by alcian blue. Heterozygous mice lack clavicles (arrows, d) in contrast to normal mice (c).

B. Cleidocranial dysplasia in humans

Cleidocranial dysplasia (CCD; OMIM 119600) is an autosomal dominant skeletal disease caused by mutation in the human *RUNX2* gene (OMIM 600211), localized at 6p21.1. It is characterized by absence of the clavicles and deficient bone formation of the skull. Radiological findings show generalized underossification. Patients can oppose their shoulders (**1**) due to absence of the clavicles (**2**; photograph by Dr J. Warkany, Children's Hospital Research Foundation, Cincinnati, USA). The calvarium (skull case) is enlarged, with a poorly ossified midfrontal area (**3**).

(Figures in A and B kindly provided by Professor Stefan Mundlos, Berlin, Germany.)

C. The human *RUNX2* gene

The *RUNX2* gene at 6p21 encodes a transcription factor of the core-binding factor (CBF) Runt-related family (OMIM 600211). It has nine exons (not seven as previously determined and shown). It contains two alternative transcription initiation sites with two promoters, P1 and P2. Part of exons 1, 2, and 3 encode the DNA-binding *runt* domain; exons 4, 5, 6, and 7 encode the transcriptional activation and repression domains. The nuclear localization signal (NLS) is located at the 5′ end of exon 3. Exon 6 is alternatively spliced and unique to *RUNX2*. The role of the *RUNX2* gene also includes a major regulatory function in chondrocyte differentiation during endochrondral bone formation. As such, it functions as a "master gene" in bone development. All mutations result in loss of function, i.e., haploinsufficiency causes the CCD phenotype. (Figure kindly provided by J. Warkany, Cincinnati, USA.)

Further Reading

Mundlos S. Cleidocranial dysplasia: clinical and molecular genetics. J Med Genet 1999;36:177–182

Mundlos S, et al. Mutations involving the transcription factor CBFA1 cause cleidocranial dysplasia. Cell 1997;89:773–779

Superti-Furga A, Unger S. Nosology and classification of genetic skeletal disorders: 2006 revision. Am J Med Genet A 2007;143:1–18

Zaidi SK, et al. Runx2 deficiency and defective subnuclear targeting bypass senescence to promote immortalization and tumorigenic potential. Proc Natl Acad Sci USA 2007;104:19861–19866

Zheng Q, et al. Dysregulation of chondrogenesis in human cleidocranial dysplasia. Am J Hum Genet 2005;77:305–312

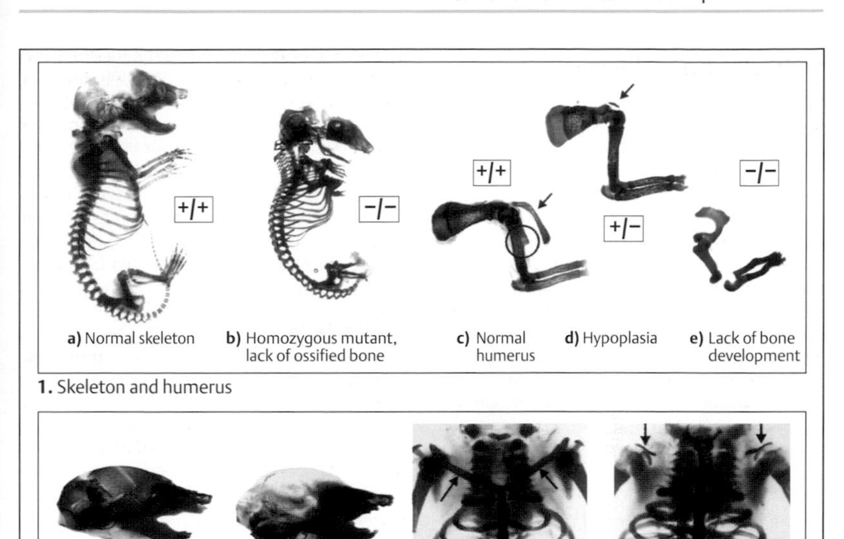

a) Normal skeleton **b)** Homozygous mutant, lack of ossified bone **c)** Normal humerus **d)** Hypoplasia **e)** Lack of bone development

1. Skeleton and humerus

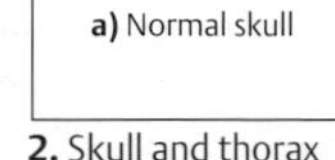

a) Normal skull **b)** Reduced mineralization **c)** Normal thorax **d)** Absent clavicles, hypoplastic humoral tuberosity

2. Skull and thorax

A. Effect of *Runx2* mutations in mice

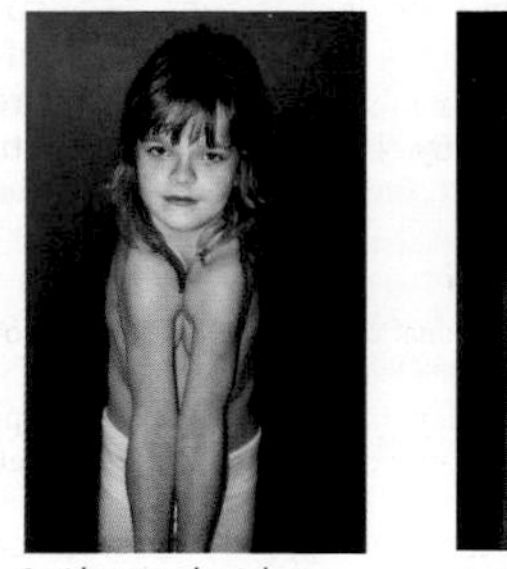

1. Absent clavicles

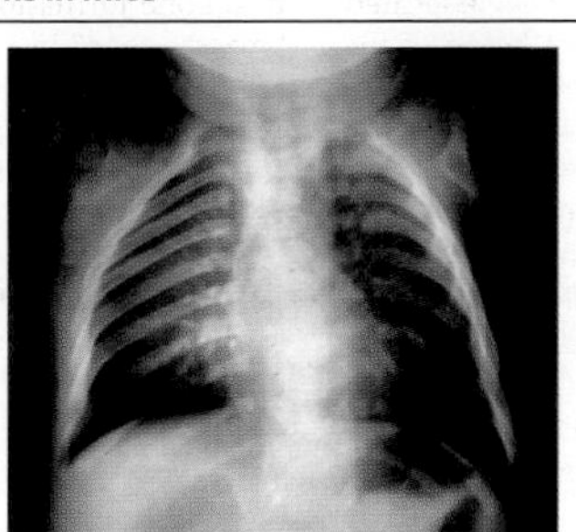

2. X-ray: narrow thorax and absent clavicles

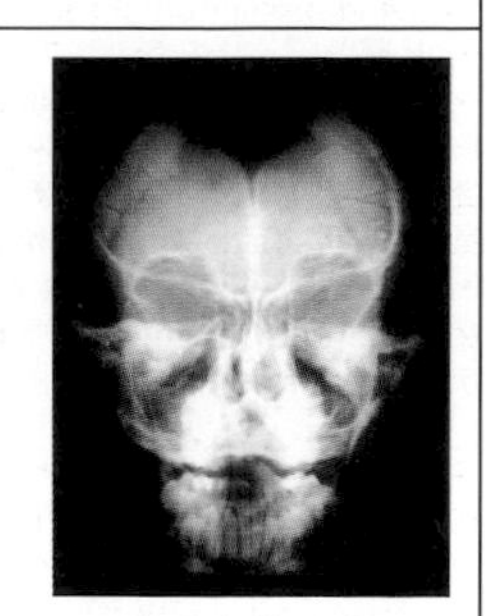

3. Lack of skull ossification

B. Cleidocranial dysplasia in humans

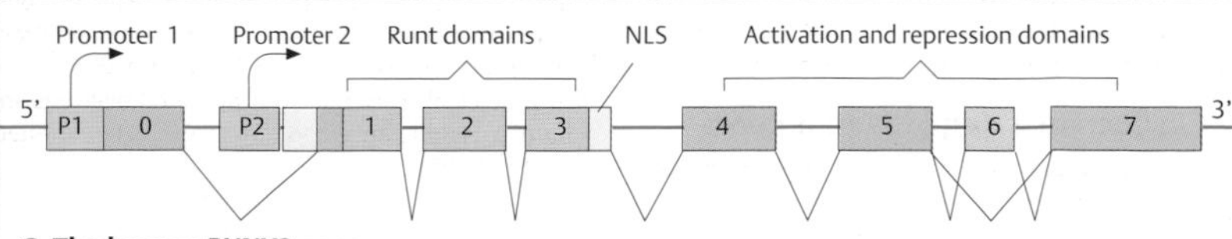

C. The human *RUNX2* gene

Ciliopathies

Ciliopathies have emerged during the past 5–8 years as a group of phenotypically different genetic disorders with a common pathogenetic backgound. They share impaired function of cilia, a specific cell organelle. The wide phenotypic spectrum of ciliopathies results from the wide variety of structures and cells that harbor cilia: the respiratory tract, oviduct and spermatozoa, inner ear, photoreceptors, gallbladder, pancreas, kidney, ependymal lining of the brain, chondrocytes, and others. Cilia are highly conserved in evolution and are almost identical from unicellular green algae to primates. Ciliopathies are an example of a disease classification system based on genotype rather than phenotype.

A. Motile cilia: basic structure

All cilia consist of two parts: a basal body below the cell surface and an axoneme extending from the cell. Perpendicular to a central pair of tubuli that run the length of the cilium, nine radial spokes connect to nine peripheral microtubular doublets. Each doublet is a heterodimer of α and β tubulins assembled into A and B tubules, respectively. Microtubulin-associated proteins form an outer and an inner dynein arm (**1**). Two major classes of dyneins are known: axonemal and cytoplasmic. Axonemal dynein arms are composed of microtubule-associated motor protein complexes consisting of several heavy, light, and intermediate chains. The dynein arms are assembled in the cytoplasm before they are transported to the growing cilia (**2**). The dynein heavy chains derive energy for cilia motility from ATPase activity. About 250 proteins are known to be required to make up a cilium. Intraflagellar transport (IFT) proteins serve to assemble and maintain the structure of cilia, but also are involved in signaling. (Figure from Omran & Olbrich, 2010, kindly provided by the authors.)

B. Primary ciliary dyskinesia (PCD)

PCD, also known as Kartagener syndrome when associated with situs inversus, is a heterogeneous group of hereditary disorders resulting from mostly autosomal recessive mutations in more than 15 known genes (OMIM 244400, see table in Appendix, p. 420). Usually one or both of the dynein arm types are structurally defective. A typical defect occurs in the axonemal dynein arm. In normal cells the outer dynein arm (ODA) can be seen in an electron microscopy cross-section of cilia in normal cells (**1**), whereas it is absent in mutant cells (**2**). Immunofluorescence microscopy of normal respiratory epithelial cells reveals an overlap of acetylated α-tubulin (green) and ODA (red) staining, resulting in a yellow coloration of the entire cilium in the normal cell (**3**), whereas ODA is lacking in the mutant cell (**4**). In addition to PCD, many other syndromic ciliopathies exist. (Figure from Omran & Olbrich, 2010, kindly provided by the authors.)

C. Structure of the *DNAH5* gene

The *DNAH5* gene (OMIM 603335) on 5p14-p15 consists of 80 exons. About 50% of individuals with PCD harbor a mutation in this gene. About 50% of mutations have been observed in exons 34, 50, 63, 76, and 77.

Further Reading

Adams M, et al. Recent advances in the molecular pathology, cell biology and genetics of ciliopathies. J Med Genet 2008;45:257–267

Arnaiz O, et al. Cildb: a knowledgebase for centrosomes and cilia. (http://database.oxfordjournals.org/content/). See also: http://cildb.cgm.cnrs-gif.fr

Badano JL, et al. The ciliopathies: an emerging class of human genetic disorders. Annu Rev Genomics Hum Genet 2006;7:125–148

Ciliary proteome database, v3. (http://v3.ciliaproteome.org/cgi-bin/index.php)

Fliegauf M, Benzing T, Omran H. When cilia go bad: cilia defects and ciliopathies. Nat Rev Mol Cell Biol 2007;8:880–893

Gherman A, Davis EE, Katsanis N. The ciliary proteome database: an integrated community resource for the genetic and functional dissection of cilia. Nat Genet 2006;38:961–962

Novarino G, et al. Modeling human disease in humans: the ciliopathies. Cell 2011;147:70–79

Omran H, Olbrich H. Zilienkrankheiten unter Berücksichtigung der primären ziliären Dyskinesie. Medizinische Genetik 2010;22:315–321

Witt M. *DNAH5* and primary ciliary dyskinesia (Kartagener syndrome). In: Inborn Errors of Development. 2nd ed. Epstein CJ, Erickson RP, Wynshaw-Boris A, eds. Oxford: Oxford University Press; 2008:1338-1349

Zimmermann KW. Beiträge zur Kenntnis einiger Drüsen und Epithelien. Arch Mikrosk Anat 1898; 52:552–706

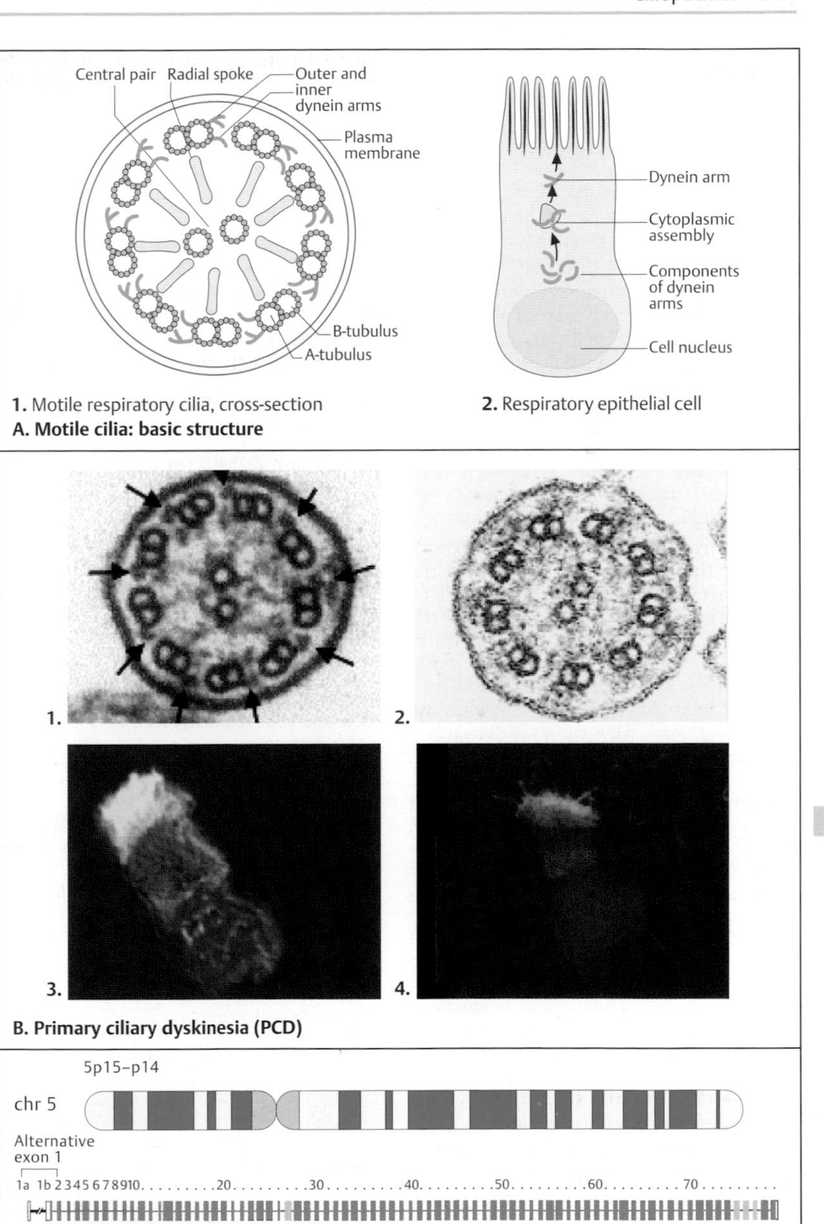

1. Motile respiratory cilia, cross-section

2. Respiratory epithelial cell

A. Motile cilia: basic structure

B. Primary ciliary dyskinesia (PCD)

C. Exon/intron structure of the *DNAH5* gene

Neurocristopathies

The neural crest, first described by Wilhelm His in 1868 as *Zwischenstrang* (cord in between), develops during neurulation in early vertebrate development. From the neural crest, diverse cell lineages of migratory cells develop into melanocytes, enteric neurons, and other cells. Abnormalities in neural crest development resulting from mutations in different genes cause genetically and clinically different human diseases, the *neurocristopathies* (two examples below).

A. Multiple endocrine neoplasias (MEN)

MEN2A (OMIM 171400) and MEN2B (OMIM 162300) are autosomal dominant disorders of endocrine tumors associated with other manifestations. The main tumor types are medullary thyroid carcinoma (MTC; shown in the figure), ganglioneuroma, and pheochromocytoma. Mutations in the *RET* protooncogene (OMIM 164761) are the cause of the tumors, which arise de novo in about 50% of cases. MEN type 2A is an allelic disorder caused by missense mutations in *RET*. Other forms of MEN are type 1 (OMIM 131100) and type 4 (OMIM 610755). (Figures kindly provided by Professor K. W. Schmid, Essen, Germany.)

B. Hirschsprung disease (congenital intestinal aganglionosis)

Hirschsprung disease (HSCR; OMIM 142623), first described by H. Hirschsprung in 1888, is a multigenic neonatal intestinal obstruction syndrome correctable by surgery. It is genetically heterogeneous and clinically variable, ranging from severe in neonates (left photograph) to mild in late infancy and in childhood (middle photograph). The main clinical manifestations are chronic constipation, abdominal distension, and megacolon (right photograph) of variable length resulting from lack of gastrointestinal ganglion cells. The population incidence is 1 in 5000 newborns and the sex ratio is about four males to one female. HSCR is classified according to the length of the aganglionic segment: short-segment Hirschsprung disease (type I or S-HSCR), occurring in 60%–85% of patients; and long-segment Hirschsprung disease (type II or L-HSCR), occurring in 15%–25%. Mutations at nine known, partially interdependent genes are involved (see Appendix, Table p. 420). (Left and middle photographs: own observations; right photograph kindly provided by Professor W. Hirsch, Leipzig.)

C. RET receptor

This is a transmembrane receptor encoded by the *RET* gene (OMIM 164761) located at 10q11.2. It has 1114 amino acids with four functional domains and is transcribed from 21 alternatively spliced exons. Mutations in HSCR are loss-of-function and are distributed along the gene. Mutations in *MEN2A* and *MEN2B* are gain-of-function and tend to cluster in different regions. *RET* mutations account for 15%–35% of sporadic HSCR and 50% of familial cases. Noncoding *RET* changes contribute to HSCR susceptibility.

D. RET and EDNRB receptor signaling

Three signaling pathways in the enteric nervous system are: (i) the RET receptor tyrosine kinase pathway and its ligand, glial cell line derived neurotrophic factor (GDNF; OMIM 600837); (ii) the endothelin type B receptor pathway (EDNRB; OMIM 131244) and its ligand, endothelin-3 (EDN3; OMIM 131242); and (iii) the transcription factor SOX10 (OMIM 602229). The resulting signals exert effects on cell survival, proliferation, migration, and differentiation of neurons in the enteric nervous system. (Figure adapted from Heanue & Pachnis, 2007.)

Further Reading

Amiel J, et al; Hirschsprung Disease Consortium. Hirschsprung disease, associated syndromes and genetics: a review. J Med Genet 2008;45:1–14

Bolande R. The neurocristopathies: A unifying concept of disease arising in neural crest maldevelopment. Hum Pathol 1973;5:409–429

Emison ES, et al. A common sex-dependent mutation in a RET enhancer underlies Hirschsprung disease risk. Nature 2005;434:857–863

Heanue TA, Pachnis V. Enteric nervous system development and Hirschsprung's disease: advances in genetic and stem cell studies. Nat Rev Neurosci 2007;8:466–479

McCallion AS, Chakravarti A. *RET* and Hirschsprung disease. In: Epstein CJ, Erickson RP, Wynshaw-Boris A, eds. Inborn Errors of Development. 2nd ed. Oxford: Oxford University; 2008:512–520

Passarge E. Gastrointestinal tract: Molecular Genetics of Hirschsprung disease. In: Encyclopedia of Life Sciences (ELS). Available at: http://onlinelibrary.wiley.com/book/10.1002/047001590X

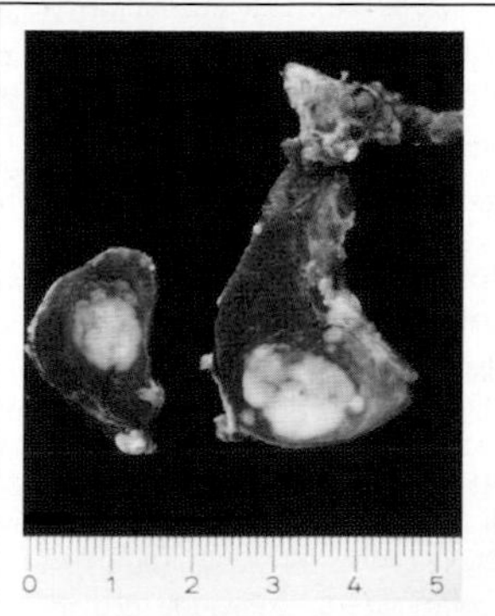

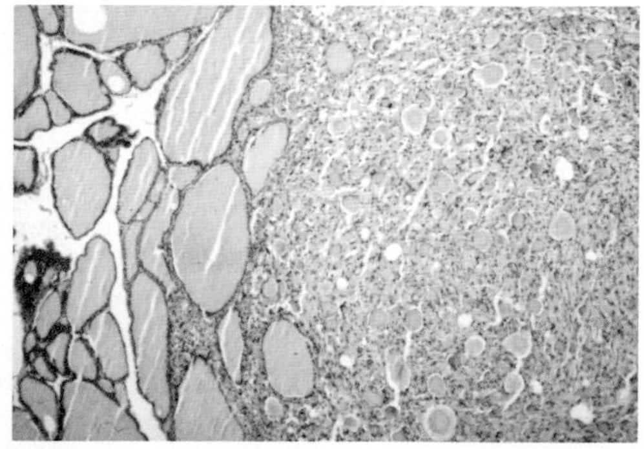

A. Multiple endocrine neoplasias

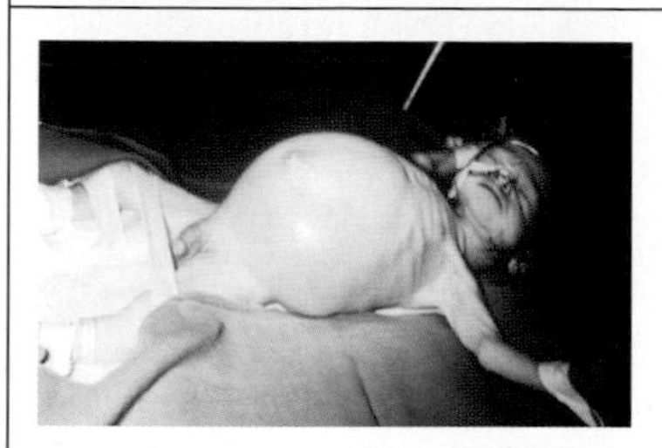

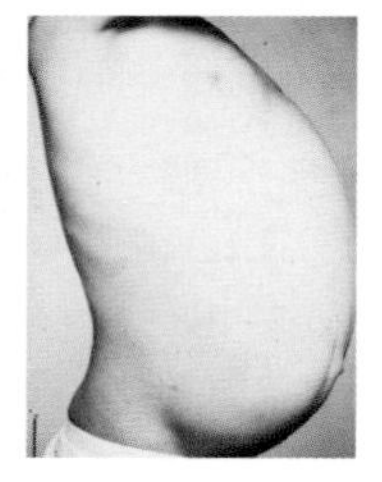

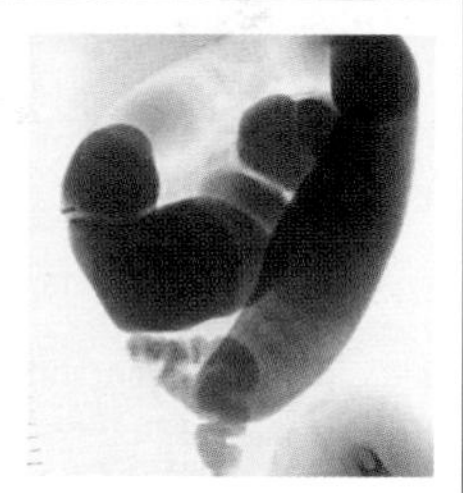

B. Congenital megacolon in Hirschsprung disease

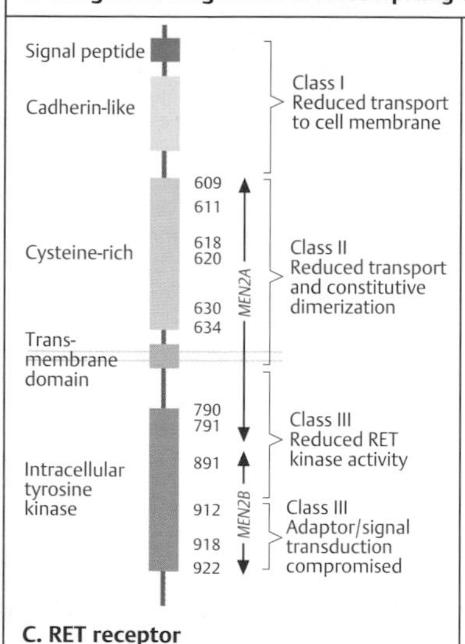

C. RET receptor

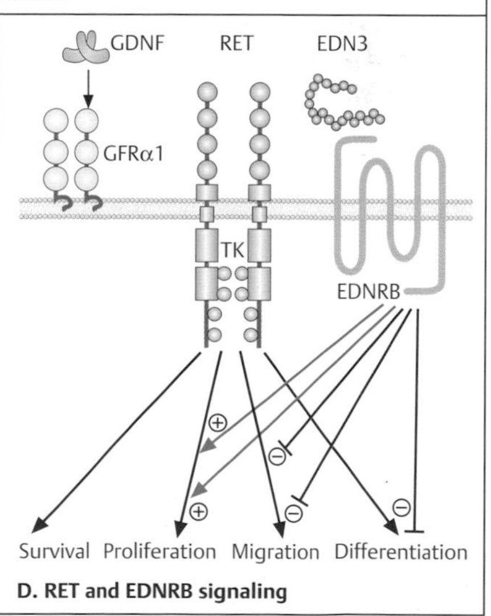

D. RET and EDNRB signaling

Dysregulated RAS-MAPK Signaling Pathway

The RAS protein family (HRAS, KRAS, and NRAS; OMIM 190202) interacts via its effector RAF (OMIM 164760) with mitogen-activated protein kinases (MAPK) in signaling pathways that regulate cell growth and differentiation. A clinically overlapping group of developmental disorders is caused by mutations in genes encoding components or modulators of this pathway: Noonan syndrome (OMIM 163950), Cardio-Facio-Cutaneous syndrome (CFC syndrome; OMIM 115150), Costello syndrome (OMIM 218040), and LEOPARD syndrome (Noonan syndrome with multiple lentigines; OMIM 151100), collectively called *RASopathies*.

A. Clinical phenotypes

Noonan syndrome (left) is the most common (1:2500), while Cardio-Facio-Cutaneous syndrome (middle), and Costello syndrome (right) have characteristic facial features and are associated with congenital heart defects (predominantly pulmonary stenosis, hypertrophic cardiomyopathy, and septal defects), proportionate short stature, and other developmental defects, variable cognitive deficits, and cancer predisposition.

B. The RAS-MAPK signaling cascade

The signaling cascade through RAF, MEK, and ERK (MAP kinases) is an important RAS-dependent pathway. It is initiated by binding of a growth factor to its receptor (receptor tyrosine kinase, RTK, see p. 198), followed by sequentially activated protein kinases. This eventually reaches intranuclear targets and regulates transcription. First, RTK is activated by phosphorylation. CBL (OMIM 165360) is an adaptor protein of RTKs and regulates their degradation by ubiquitination (Ub). SOS1 (son of sevenless; OMIM 182530) acts as a guanine nucleotide exchange factor (GEF) recruited to activated RTKs. It activates RAS by exchanging GDP (guanosyldiphosphate) by GTP (guanosyltriphosphate). SHP2, a tyrosine phosphatase encoded by the gene *PTPN11* (OMIM 176876), modulates signaling via the RAS-MAPK cascade. RAF kinases (BRAF, OMIM 164757; RAF1, OMIM 164760) become activated through binding to active GTP-bound RAS and subsequently phosphorylate MEK kinases (MEK1, OMIM 176872; MEK2, OMIM 601263), which in turn activate ERK (ERK1, ERK2). Phospho-ERK translocates to the nucleus where it phosphorylates nuclear targets. ERK kinases have different effects in different types of cells. Neurofibromin (NF1) and SPRED1 (OMIM 609291) are negative regulators in the RAS-MAPK cascade (p. 316).

C. Mutations in RAS-MAPK pathway components

The RASopathies result from gain-of-function mutations in the genes encoding the signaling proteins, shown **B**. Mutations in the gene *PTPN11* (**1**), protein-tyrosine phosphatase, nonreceptor-type 11, OMIM 176876, encoding SHP2, account for 50% of patients with Noonan syndrome. Most of the unevenly distributed mutations occur in different functional domains of the gene product SHP2. Germline mutations in the *KRAS* gene (**2**) account for about 2% of cases of Noonan syndrome and 5% of cases of CFC syndrome. The common somatic *KRAS* mutations in cancer differ and do not overlap despite clustering in similar regions. (Figures and draft for the text in A–C kindly provided by Professor Martin Zenker, Magdeburg, Germany.)

D. Consequences of RAS mutations

Binding of GTP to RAS induces a conformational change in which two mobile regions (switch 1: yellow, switch 2: cyan) are involved. The inactive state is restituted through the interaction with GTPase-activating proteins (GAPs). Germline KRAS mutations occurring in Noonan and CFC syndromes (affecting residues G60 and P34) and the most common site of oncogenic mutations (G12) are located in this area and impair the RAS–GAP interaction, which results in RAS hyperactivation. (Figure courtesy of Professor M. R. Ahmadian, Düsseldorf, Germany.)

Further Reading

Gremer L, et al. Germline KRAS mutations cause aberrant biochemical and physical properties leading to developmental disorders. Hum Mutat 2011;32:33–43

Swensen J, Viskochil D. The Ras pathway. In: Epstein CJ, Erickson RP, Wynshaw-Boris A, eds. Inborn Errors of Development. The Molecular Basis of Clinical Disorders of Morphogenesis. 2nd ed. Oxford: Oxford University Press; 2008: 605–614

Zenker M, ed. Noonan Syndrome and Related Disorders. A Matter of Dysregulated Ras Signaling. Basel: Karger; 2009

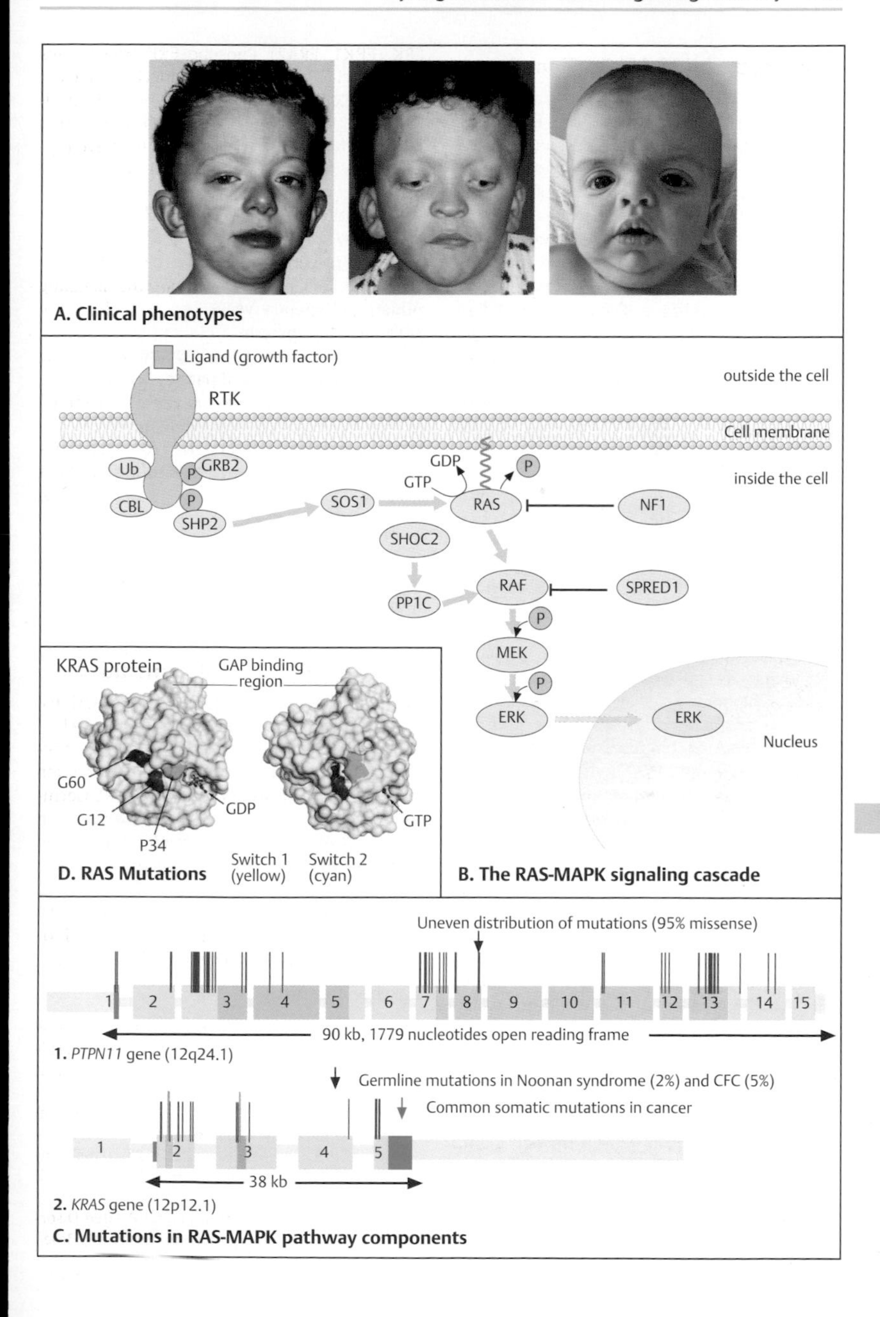
A. Clinical phenotypes
Ligand (growth factor)
RTK
outside the cell
Cell membrane
inside the cell
Ub
GRB2
P
CBL
SHP2
SOS1
GDP
GTP
P
RAS
NF1
SHOC2
PP1C
RAF
SPRED1
MEK
ERK
ERK
Nucleus
B. The RAS-MAPK signaling cascade
KRAS protein
GAP binding region
G60
G12
P34
GDP
GTP
Switch 1 (yellow)
Switch 2 (cyan)
D. RAS Mutations
Uneven distribution of mutations (95% missense)
1 2 3 4 5 6 7 8 9 10 11 12 13 14 15
90 kb, 1779 nucleotides open reading frame
1. PTPN11 gene (12q24.1)
Germline mutations in Noonan syndrome (2%) and CFC (5%)
Common somatic mutations in cancer
1 2 3 4 5
38 kb
2. KRAS gene (12p12.1)
C. Mutations in RAS-MAPK pathway components

Hemoglobin

Hemoglobin is a tetrameric oxgen-transporting molecule, named by Hoppe-Seyler in 1862. It consists of two slightly different pairs of polypeptide chains with a characteristic three-dimensional structure. Each globin chain contains an oxygen-binding site. Hemoglobin evolved from a single-chain oxgen binding molecule, myoglobin. Hemoglobin was the first human protein to be defined by its amino acid sequence (Ingram, 1956) and its three-dimensional structure at 5.5 Å resolution (Perutz et al., 1960). The genes encoding the individual globin chains were the first genes for which the molecular structure was elucidated in 1979. More than 750 genetic variants are known, and many cause severe hemoglobin disorders (see pp. 348–355).

A. Types of hemoglobin

Different types of globin chains are synthesized during embryonic, fetal, and postnatal life (OMIM 141800, 141900). The subunits of adult hemoglobin are designated by the Greek letters α, β, γ, and δ. Specialized globin subunits exist in mammals during embryonic development, designated as ε (epsilon) and ζ (zeta). A hemoglobin molecule is stable only when the two pairs of globin differ. Adult hemoglobin (HbA) is composed of two α and two β chains (α2β2), and fetal hemoglobin (HbF) is composed of two α and two γ chains (α2γ2). A minor adult hemoglobin, HbA_2, has a variant delta chain (α2δ2). (Figure adapted from Lehmann & Huntsman, 1974.)

B. Hemoglobins in thalassemias

The thalassemias are a group of genetically determined disorders of hemoglobin synthesis. They are classified according to which chain is affected, the α chain (α-thalassemias, α-thal; OMIM 141800) or the β chain (β-thalassemia, β-thal; OMIM 613985). β-thalassemias affect HbA; α-thalassemias affect HbA and HbF. Hemoglobins with four identical globin chains are highly unstable and incompatible with life (HbH with four β chains [β4], Hb Bart's with four γ chains [γ4]). (Figure adapted from Lehmann & Huntsman, 1974.)

C. Evolution of hemoglobin

The different types of hemoglobin existing in mammals today arose by duplication of genes encoding the globin chains. A primordial single-chain oxygen-binding protein, present early in evolution, adapted to the rising concentration of oxygen in the earth's atmosphere at an early stage about 1100 million years ago. Several subsequent duplications gave rise to the several different types of globin genes. About 500 million years ago, a hemoglobin gene underwent the first duplication, and from this the genes encoding the α type and the β type hemoglobins arose. During evolution of mammals in the the past 100 million years, hemoglobin evolved from an ancestral β type into different types: δ, two types of γ (γA and γG), and ε, by further duplication events.

D. Globin synthesis during embryonic development

During early embryonic stages transient embryonic hemoglobins are present: Hb Gower 1 (ζ2ε2), Hb Gower 2 (α2ε2), and Hb Portland (ζ2γ2). (Most hemoglobins are named after the town or institution where they were discovered.) The different types of globin chains are synthesized at different developmental stages at different anatomical sites: during the first 6 weeks in humans the embryonic hemoglobins in the yolk sac, from the 6th week until prior to birth (fetal hemoglobin HbF, α2γ2), in the liver and spleen, and during adult life in red blood cell precursors in the bone marrow. They differ in oxygen-binding affinity. Thus, oxygen delivery is optimized for different phases of development. (Figure from Weatherall et al., 2001.)

Further Reading

Ingram VM. A specific chemical difference between the globins of normal human and sickle-cell anaemia haemoglobin. Nature 1956;178:792–794

Lehmann H, Huntsman RG. Man's Hemoglobins. Amsterdam: North-Holland; 1974

Perutz MF, et al. Structure of haemoglobin: a three-dimensional Fourier synthesis at 5.5-A. resolution, obtained by X-ray analysis. Nature 1960;185:416–422

Weatherall DJ, et al. The hemoglobinopathies. In: Scriver CR, et al., eds. The Metabolic and Molecular Bases of Inherited Disease. 8th ed. New York: McGraw-Hill; 2001:179–190. (www.ommbid.com)

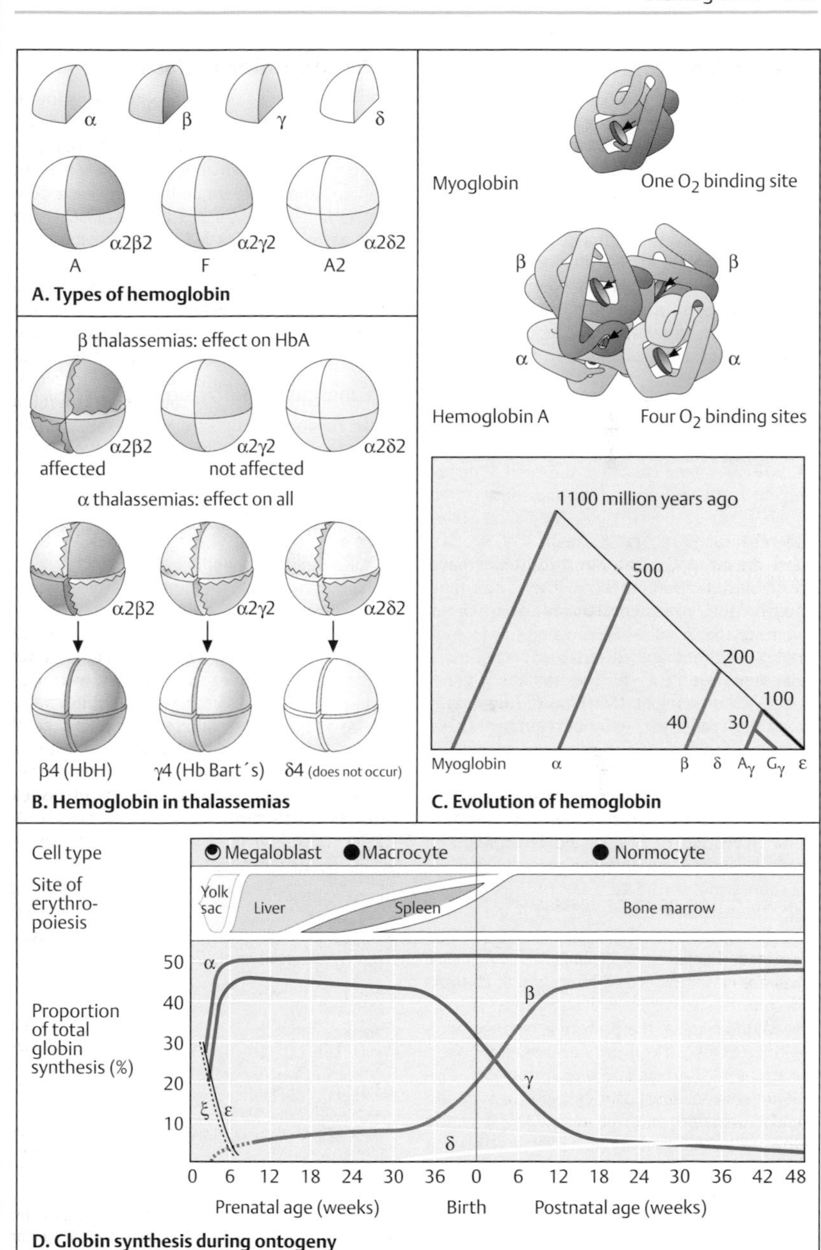

A. Types of hemoglobin

B. Hemoglobin in thalassemias

C. Evolution of hemoglobin

D. Globin synthesis during ontogeny

Hemoglobin Genes

The different hemoglobin chains are encoded by clusters of α-type (two α and one ζ) and β-type globin genes (β, γG, γA, δ, and ε) on human chromosomes 16 and 11, respectively. A specific gene is responsible for each type of different globin polypeptide chain. They are arranged and expressed in a sequence, according to the time of activation during developmental stages (see p. 344). Their structural and DNA sequence similarity predisposes to malalignment during meiosis and unequal crossing-over leading to duplications and deficiencies.

A. The β- and α-globin genes

The β-globin-like genes (β, δ, γA, γG, and ε) of man are located at the Hb β locus HBB (OMIM 141900) as a gene cluster in the 3′ to 5′ direction on the short arm of chromosome 11 in region 1, band 5.5 (11p15.5). They span about 60 000 bp, or 50 kb (kilobases), of DNA (**1**). There are two γ genes, γG and γA, which have arisen by a duplication event. They differ only in codon 136, which encodes an alanine in γA and a glycine in γG. A pseudogene (ψβ1) containing deletions and internal stop codons is located between the γA gene and the δ gene. A locus control region (LCR) located upstream (in the 5′ direction) jointly regulates these genes.

Two α-globin genes are located at the α locus HBA1 (OMIM 141800) and HBA2 (OMIM 141850) on the short arm of human chromosome 16 (16pter-p13.11) on a DNA segment of about 35 kb. A ζ gene, which is active during the embryonic period only, lies in the 5′ direction. Three pseudogenes (ψζ2, ψζ1, ψα2, and ψα1) are located 5′ of the α genes. A further gene locus HBQ1 (theta, Θ; 142240), with unknown function, has been identified in this region.

The β-globin gene, the prototype, spans about 1.6 kb (1600 bp). It consists of three exons separated by a short and a long intron (**2**). The coding sequences of the other β-like genes are also arranged in three exons. The α genes *HBA1* and *HBA2* span about 0.8 kb (800 bp).

B. Tertiary structure of the β-globin chain

The characteristic three-dimensional structures of myoglobin and of the hemoglobin α and β chains, shown schematically, are very similar, although their amino acid sequences correspond in only 24 of 141 positions. The β chain, with 146 amino acids, is somewhat longer than the α chain, with 141 amino acids. Each chain harbors a heme group with an oxygen-binding site inside the molecule, protected from the surroundings. (Figure adapted from Weatherall et al., 2001.)

C. Functional domains of the β chain

Three functional and structural domains can be distinguished in all globin chains. They correspond to the three exons of the gene. Two domains, consisting of amino acids 1–30 and 105–146 (encoded by exons 1 and 3), are located on the outside of the molecule, and consist of mainly hydrophilic amino acids. A third domain, lying inside the molecule (encoded by exon 2), contains the oxygen-binding site and consists of mainly nonpolar hydrophobic amino acids. The hydrophilic amino acids of the two chains forming the outside of the molecule render it flexible and water soluble.
(Figure adapted from Gilbert, 1978.)

Further Reading

Antonarakis SE, Kazazian HH Jr, Orkin SH. DNA polymorphism and molecular pathology of the human globin gene clusters. Hum Genet 1985;69:1–14

Gilbert W. Why genes in pieces? Nature 1978;271: 501–502

Stamatoyannopoulos G, et al. The Molecular Basis of Blood Disease. 4th ed. New York: Saunders; 2001

Weatherall DJ, et al. The hemoglobinopathies. In: Scriver CR, et al., eds. The Metabolic and Molecular Bases of Inherited Disease. 8th ed. New York: McGraw-Hill; 2001:179–190. Available at: http://www.ommbid.com. Accessed January 28, 2012

Weatherall DJ, Clegg JB. The Thalassaemia Syndromes. 4th ed. Oxford: Blackwell Science, 2001

Weatherall DJ. Phenotype-genotype relationships in monogenic disease: lessons from the thalassaemias. Nat Rev Genet 2001;2:245–255

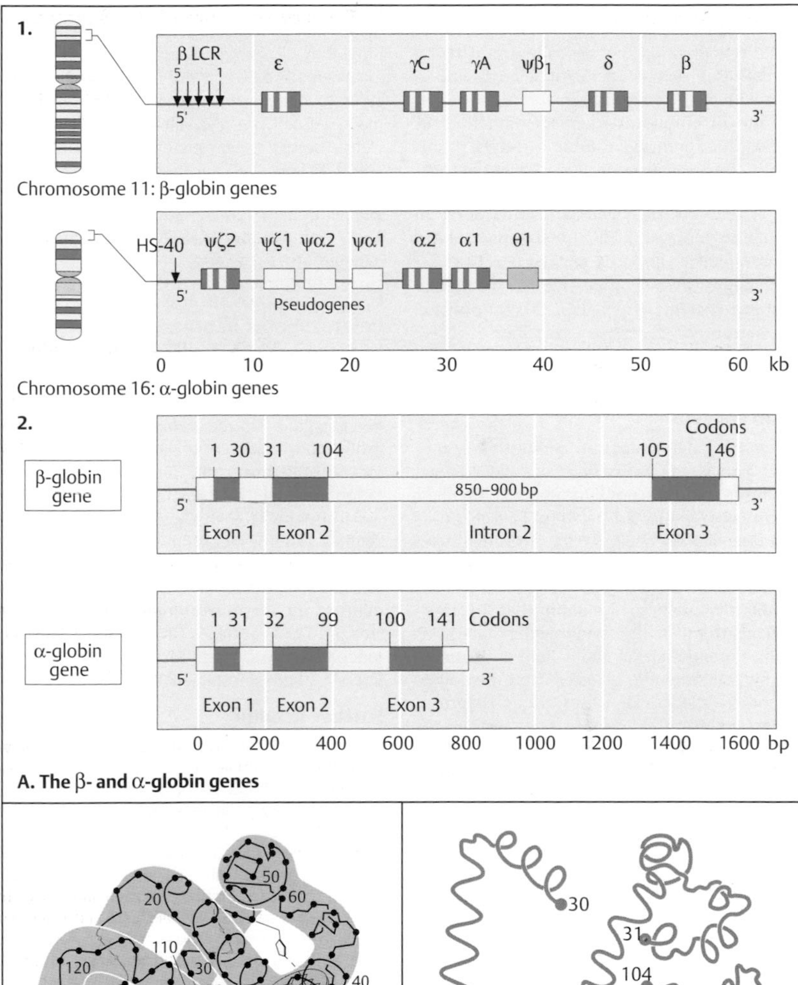

A. The β- and α-globin genes

B. Tertiary structure of the β-globin chain

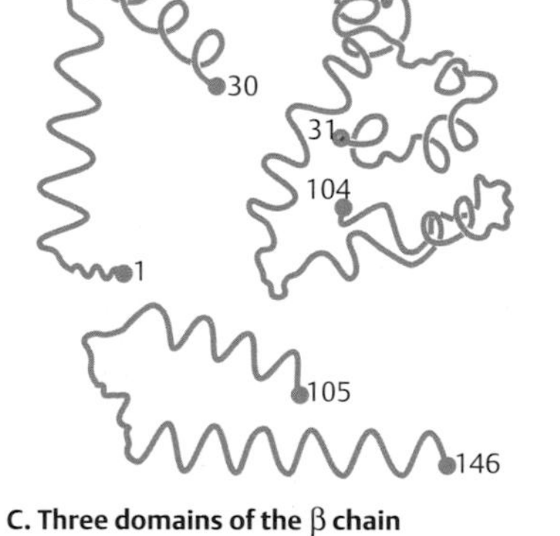

C. Three domains of the β chain

Sickle Cell Disease

Sickle cell disease, first described in 1910 by Herrick, is a severe hemolytic anemia associated with many complications. It is caused by a structurally abnormal hemoglobin (HbS) resulting from homozygosity for a specific mutation in the β-globin gene. It is frequent in tropical regions where malaria is endemic (see p. 138). With a frequency of 1 in 500, it is an important cause of morbidity and mortality in these regions. Sickle cell anemia was the first human disease to be understood at the molecular level (Pauling et al., 1949). It is transmitted by autosomal recessive inheritance (Neel, 1949). Heterozygous carriers can readily be identified.

A. Sickle cells

The leading hematological manifestion is anemia associated with mainly sickle-shaped erythrocytes in a blood smear under the light microscope of affected persons (**1**). Sickle cells lose the pliability needed to pass through small capillary blood vessels and provoke occlusions of small blood vessels (**2**). As a result, the blood supply to many tissues is diminished. In a normal blood smear (**3**), erythrocytes appear as regular round disks of about 7 µm in diameter. In the course of the disease, acute hemolytic episodes, called sickle crises, occur. Heterozygotes (sickle trait) show occasional sickle cells but do not develop sickle crises, and are associated with only very mild signs and symptoms. Other sickle hemoglobinopathies occur when the HbS mutation is present on one allele and a different mutation at the same site is present on the other allele. (**1** and **3**, Dr Daniel Nigro, Santa Ana College, California, USA; **2**, National Heart, Lung and Blood Institute, Bethesda, Maryland, USA).

B. Consequences of the sickle cell mutation

All manifestations of sickle cell anemia can be understood on the basis of the underlying mutation. In 1956 V. M. Ingram determined, by using amino acid sequence analysis, that glutamic acid is replaced by valine in codon 6 of the β-globin gene (mutation E6V). This alteration results from the transversion of an adenine (A) to a thymine (T), which changes codon GAG to GTG. Valine is a hydrophobic amino acid on the outside of the molecule. This makes sickle cell hemoglobin (HbS) less soluble than normal hemoglobin. HbS crystallizes in the deoxy state, forms small rods, and does not allow erythrocytes to deform when passing through small blood vessels. As a result local oxygen deficiency occurs in various organs. Chronic oxygen deficiency of the brain leads to learning disability. Defective erythrocytes are destroyed and cause hemolysis. Chronic anemia results in numerous sequelae such as heart failure, liver damage, and infections.

C. Selective advantage for HbS heterozygotes in areas of malaria

About 1.5–2.5 million children die each year from malaria, mostly in sub-Saharan Africa.
In 1954 A. C. Allison suggested that individuals heterozygous for the sickle cell mutation are protected from severe malaria infection. Erythrocytes of heterozygotes for the sickle cell mutation are a less favorable environment for the malaria parasite than those of normal homozygotes. Heterozygotes have a higher probability of survival and of being able to reproduce. Sickle cell anemia is the best example in humans of a selective advantage in heterozygotes (see p. 138). The sickle cell mutation has arisen independently at least four or five times in different malaria-infested regions.

Further Reading

Allison AC. Polymorphism and natural selection in human populations. Cold Spring Harb Symp Quant Biol 1964;29:137–149

Benz EJ Jr. Disorders of hemoglobin. In: Longo DL, et al., eds. Harrison's Principles of Internal Medicine. 18th ed. New York: McGraw-Hill; 2012:852–861

Herrick JB. Peculiar elongated and sickle-shaped red blood cell corpuscles in a rare case of severe anemia. Arch Intern Med 1910;6:517–521

Ingram VM. A specific chemical difference between the globins of normal human and sickle-cell anaemia haemoglobin. Nature 1956;178:792–794

Neel JV. The inheritance of sickle cell anemia. Science 1949;110:64–66

Pauling L, et al. Sickle cell anemia a molecular disease. Science 1949;110:543–548

Rees DC, Williams TN, Gladwin MT. Sickle-cell disease. Lancet 2010;376:2018–2031

Vernick KD, Waters AP. Genomics and malaria control. N Engl J Med 2004;351:1901–1904

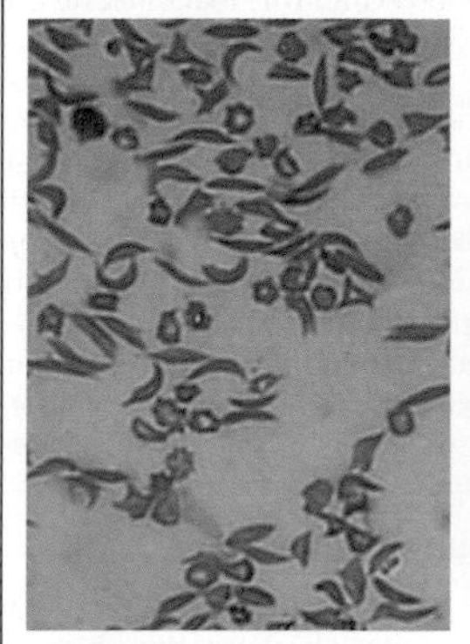

1. Sickle cells in blood smear

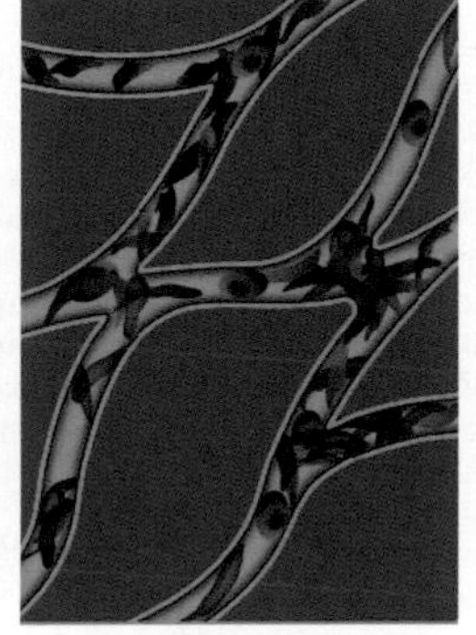

2. Sickle cells in blood vessels

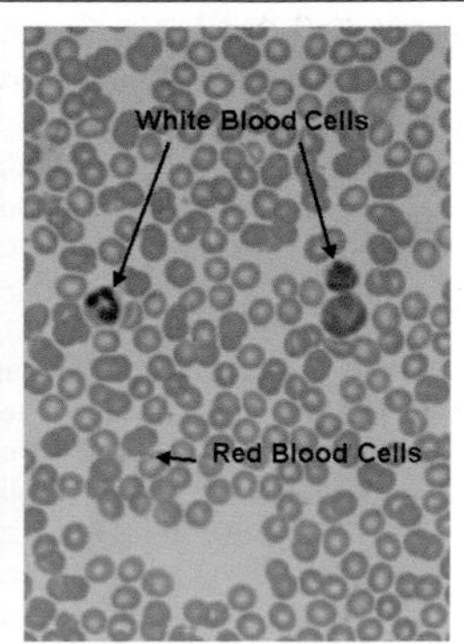

3. Normal blood smear

A. Sickle cells

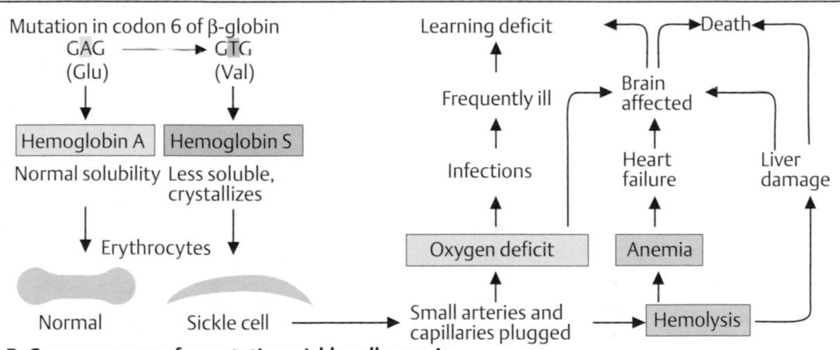

B. Consequences of a mutation: sickle cell anemia

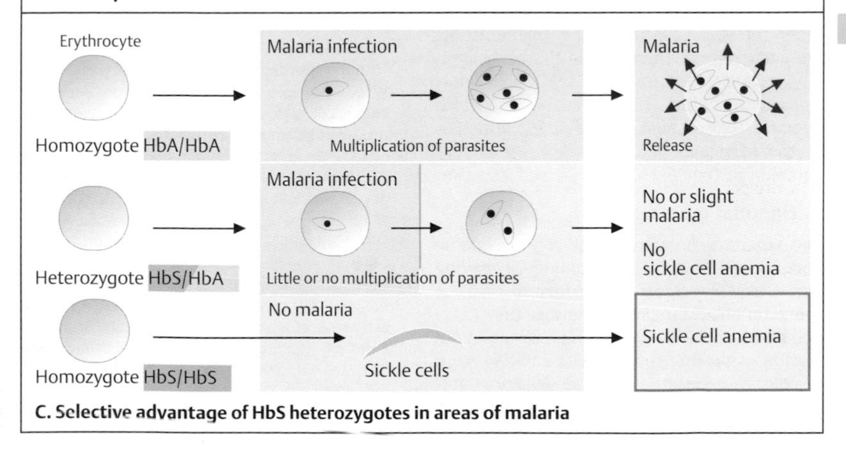

C. Selective advantage of HbS heterozygotes in areas of malaria

Mutations in Globin Genes

Mutations in globin genes cause three different types of hemoglobin disorders: (1) structural hemoglobinopathies, (2) thalassemia syndromes caused by impaired globin synthesis, (3) hereditary persistence of fetal hemoglobin (HPFH). Some variants combine features of structural hemoglobinopathies and thalassemia. More than 750 human hemoglobin variants with single amino acid substitutions in one of the globin chains are known. These are distributed to the types of globin genes as follows: 217 α genes, 362 β genes, 70 γ genes, 32 δ genes, 19 variants invoving two mutations, 27 deletions, 6 insertions, 4 deletion/insertions, and 10 hybrid fusions.

In some cases a globin chain is elongated or shortened or fused with parts of others. The functional consequences of mutations are decreased elasticity of the molecule, altered oxygen affinity, or instability.

A. Structural changes in the β-globin gene

More than 300 point mutations in the β-globin gene and over 100 in one of the α-globin genes have been documented. Two clinically important mutations affect codon 6: the sickle cell mutation, 6 Glu to Val (sickle cell hemoglobin, HbS, see previous page) and 6 Glu to Lys (hemoglobin C, HbC, incorporating lysine instead of glutamic acid in codon 6). Compound heterozygotes with the HbS mutation on one chromosome and the HbC mutation on the other (HbSC) show the sickle phenomenon. The marked methemoglobin formation in $Hb_{Zürich}$ and $Hb_{Saskatoon}$ results from substitutions for histidine (His) in codon 63, which alter the oxygen-binding region of the hemoglobin molecule. HbE ($\alpha 2\beta 2^{26Glu \rightarrow Lys}$) is very common in Thailand, Cambodia, and Vietnam.

B. Unequal crossing-over

The sequence homology between the globin genes may lead to nonhomologous pairing and unequal crossing-over during meiosis. A characteristic example is $Hemoglobin_{Gun\ Hill}$, described in 1968. This variant results from pairing of codon 90 with codon 95, 91 with 96, etc. As a result, codons 91–95 are subsequently deleted in one strand and duplicated in the other (not shown). The result is an unstable hemoglobin, of which more than 90 are known.

C. Fusion hemoglobin

Fused or hybrid hemoglobin variants probably also result from unequal crossing-over involving parts of adjacent genes. The first is Hb_{Lepore}, described in 1962. Here the first 50–80 amino acids of the δ chain are fused with the last 60–90 residues of the normal C-terminal amino acids of the β chain. The complementary situation is Hb anti-Lepore, a δβ fusion gene together with the normal δ and β genes.

D. Hemoglobins with a chain elongation

More than 10 globin chain variants with an elongated chain are known. Single base substitutions in a chain termination codon, frameshift mutations, or mutations affecting the translation initiator methionine are responsible for this type of variant. In $Hb_{Cranston}$ an insertion in codon 145 of the β chain changes UAU (tyrosine) to AGU (serine). This changes the stop codon after position 146 into ACU (tyrosine) and causes a read-through until codon 157, and consequently a chain that is 10 amino acids too long (**1**).

In $Hb_{Constant\ Spring}$ (**2**), the α chain is elongated by a mutation in the stop codon UAA to CAA, which codes for glutamine (Gln).

Further Reading

Baglioni C. The fusion of two peptide chains in hemoglobin Lepore and its interpretation as a genetic deletion. Proc Natl Acad Sci USA 1962;48:1880–1886

Benz EJ Jr. Disorders of hemoglobin. In: Longo DL, et al., eds. Harrison's Principles of Internal Medicine. 18th ed. New York: McGraw-Hill; 2012:852–861

Old J. Hemoglobinopathies and thalassemias. In: Rimoin DL, et al., eds. Emery and Rimoin's Principles and Practice of Medical Genetics. 5th ed. Philadelphia: Churchill Livingstone-Elsevier; 2007:1638–1674

Rieder RF, Bradley TB Jr. Hemoglobin Gun Hill: an unstable protein associated with chronic hemolysis. Blood 1968;32:355–369

Weatherall DJ, et al. The hemoglobinopathies. In: Scriver CR, et al., eds. The Metabolic and Molecular Bases of Inherited Disease. 8th ed. New York: McGraw-Hill; 2001:179–190. (http://www.ommbid.com)

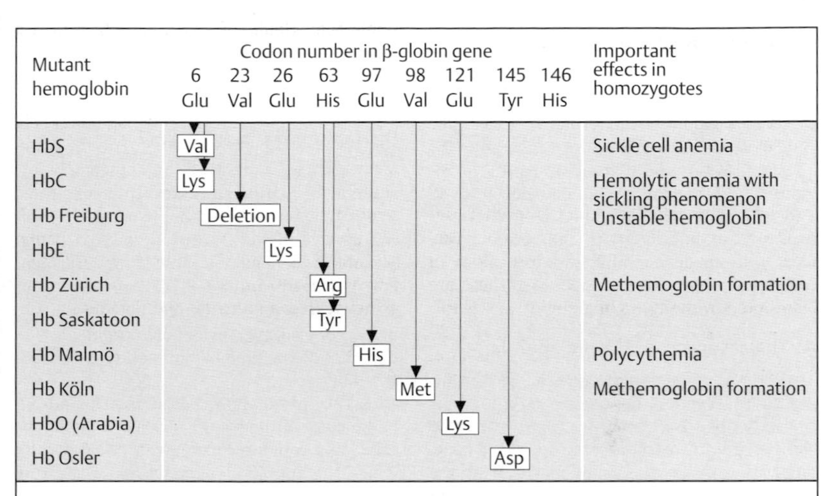

A. Structural changes in the β-globin gene

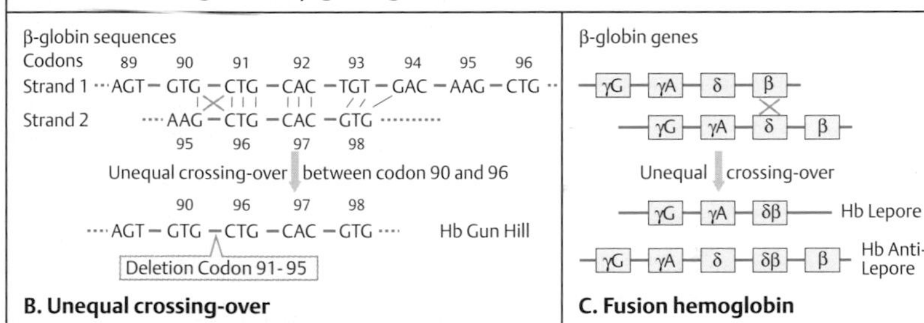

B. Unequal crossing-over

C. Fusion hemoglobin

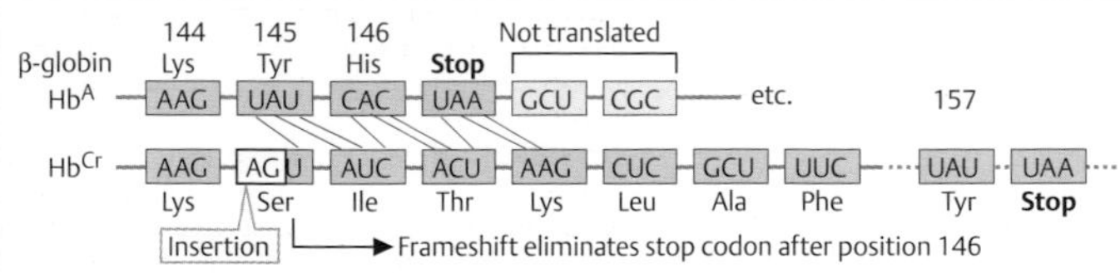

1. Hemoglobin Cranston: chain elongation by frameshift

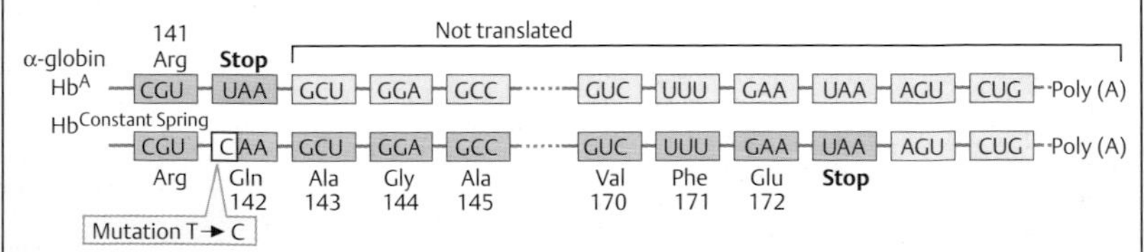

2. Hemoglobin Constant Spring: chain elongation by mutation in the stop codon

D. Hemoglobin with a chain elongation

The Thalassemias

Thalassemias are inherited disorders of α- or β-globin synthesis. Decreased or absent synthesis of a globin chain results in an anemic state of variable, often life-threatening severity. The thalassemias occur predominantly in regions where malaria is endemic (see p. 138). The term is derived from the Greek Θαλασσα (thalassa, sea). More than 300000 children with a severe hemoglobin disorder are born each year, most in low- or middle-income countries (Weatherall, 2010).

A. Thalassemia, a chronic anemia

Thalassemia is a chronic anemia of various grades of severity, depending on the type, first described by Cooley in 1925. It may be associated with massive extramedullary hematopiesis (outside the bone marrow) in the liver and spleen, causing both organs to be enlarged, or bone. Infections, malnourishment, and other signs add to severe illness. Thalassemias are classified into α-thalassemias, β-thalassemias, or unstable δβ fusion globin. (Photographs from Weatherall & Clegg, 2001.)

B. β-Thalassemia and α-thalassemia

The thalassemias have a wide spectrum of different genotypes and phenotypes. In the β-thalassemias (**1**), complete absence of the β chain (β^0) is distinguished from decreased levels of β chains. Since there are four α-globin loci, α-thalassemias (**2**) have a complex pattern of genotypes: αααα/ααα- (silent carrier); a mutation in two α loci on the same chromosome in *cis*: αα/α-α- (thal-1); or a mutation in two α loci on different chromosomes in *trans*: α-/α- (thal-2). Since the α gene loci are located within a 4-kb region of homology interrupted by small, nonhomologous regions, they are prone to mispairing and nonhomologous crossing-over during meiosis, resulting in deletion or duplication. Thal-1 occurs mainly in Southeast Asia, while thal-2 occurs mainly in Africa.

C. Spectrum of mutations in β-thalassemia

Mutations occur throughout the β gene. The entire gene may be deleted. Regulatory regions upstream (5′ direction) may be affected, resulting in decreased transcripion. Nonsense mutations leading to a truncated globin due to a frameshift, missense mutations, and splice mutations occur. A similarly wide spectrum exists for the α-globin genes. (Figure adapted from Antonarakis et al, 1985.)

D. Haplotypes in RFLP analysis

Many common mutations in the β globin gene are in linkage disequlibrium with polymorphic restriction sites in the β globin complex. Thus, certain mutations are linked to a particular haplotype defined by restriction fragment length polymorphisms (RFLPs, see p. 62) or by SNPs (single nucleotide polymorphisms, p. 78). For example, seven polymorphic restriction sites define nine haplotypes, of which five are shown (A: + − − − − + +; B: − + + − + + +; etc.). This information can be used for simplified genetic diagnosis to identify a haplotype carrying a mutation in a given population in which this mutation is prevalent. (Data from Antonarakis et al., 1985.)

α-Thalassemia may be associated with mental retardation. Two different syndromes exist: ATR-16 syndrome (OMIM 141750) associated with a large (1–2 Mb) deletion at 16pter-p13.3 including the α-globin gene cluster; the other is an X-linked disorder (OMIM 301040; gene locus at Xq13).

Further Reading

Antonarakis SE, Kazazian HH Jr, Orkin SH. DNA polymorphism and molecular pathology of the human globin gene clusters. Hum Genet 1985;69:1–14

Benz EJ Jr. Disorders of hemoglobin. In: Longo DL, et al., eds. Harrison's Principles of Internal Medicine. 18th ed. New York: McGraw-Hill; 2012:852–861

Cao A, Galanello R. Beta-thalassemia. Genet Med 2010;12:61–76

Rund D, Rachmilewitz E. β-thalassemia. N Engl J Med 2005;353:1135–1146

Weatherall DJ. The inherited diseases of hemoglobin are an emerging global health burden. Blood 2010;115:4331–4336

Weatherall DJ, et al. The hemoglobinopathies. In: Scriver CR, et al., eds. The Metabolic and Molecular Bases of Inherited Disease. 8th ed. New York: McGraw-Hill; 2001:179–190. (http://www.ommbid.com)

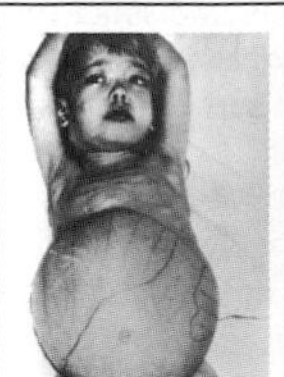
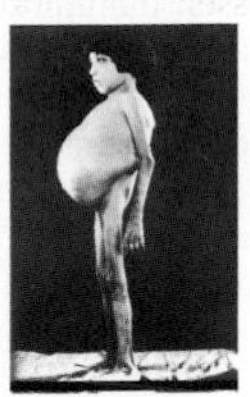
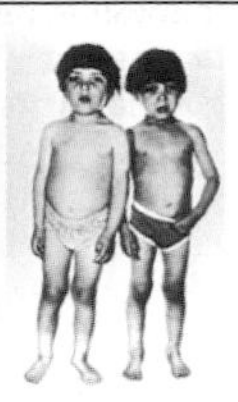

Different forms of thalassemia:

α: Decreased synthesis of α-globin
β: Decreased synthesis of β-globin
δβ: Decreased synthesis of δ- and β-globin

→ Unstable hemoglobin → Chronic anemia

A. Thalassemia, a chronic anemia

1. β-Thalassemias

Genotype		Phenotype
+ / –	β⁰ heterozygote	Thalassemia minor (asymptomatic)
(+) / (+)	β⁺ heterozygote	
(+) / (+)	β⁺ homozygote	Thalassemia intermedia (not transfusion dependent)
+ / –	β⁰ heterozygote	
– / –	β⁰ homozygote (β⁰ Thalassemia)	Thalassemia major (transfusion dependent)
or		
(+) / –	β⁺/β⁰ homozygote (β⁺ Thalassemia)	

2. α-Thalassemias

Genotype	Phenotype
α α / α α	Normal
α – / α α	"Silent carrier" (normal)
– – / α α (thal-1)	Thalassemia
α – / α – (thal-2)	Thalassemia
α – / – –	HbH disease (HbH = β_4)
– – / – –	Hydrops fetalis

B. β-Thalassemia and α-thalassemia

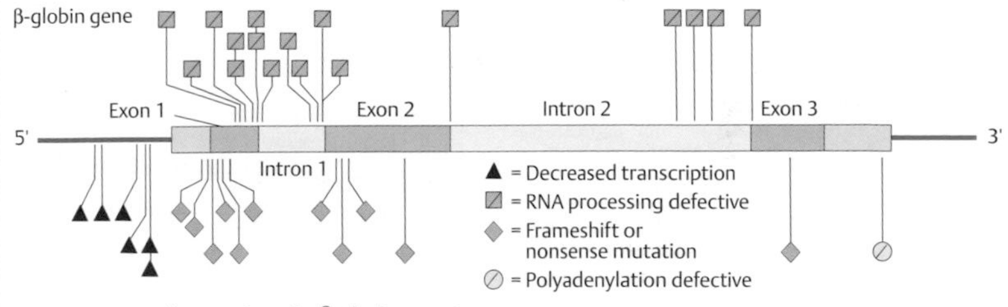

C. Spectrum of mutations in β-thalassemia

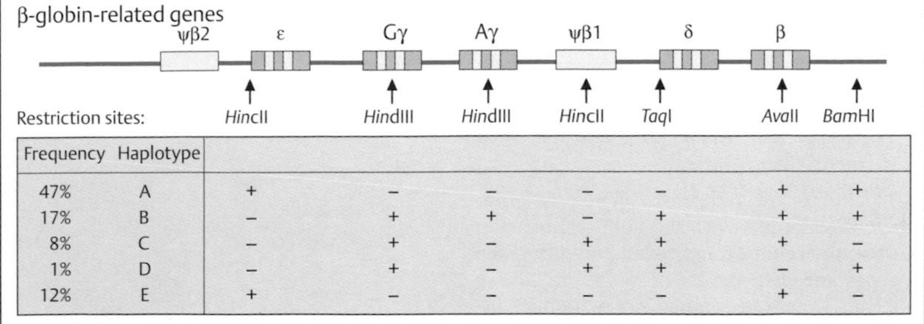

Frequency	Haplotype	HincII	HindIII	HindIII	HincII	TaqI	AvaII	BamHI
47%	A	+	–	–	–	–	+	+
17%	B	–	+	+	–	+	+	+
8%	C	–	+	–	+	+	+	–
1%	D	–	+	–	+	+	–	+
12%	E	+	–	–	–	–	+	–

D. Haplotypes in RFLP analysis

Hereditary Persistence of Fetal Hemoglobin (HPFH)

Hereditary persistence of fetal hemoglobin (HPFH; OMIM 141749) is characterized by high levels of fetal hemoglobin (HbF) in adult life (5%–30% instead of less than 1%). Clinically HPFH is relatively benign, although HbF is not optimally adapted to postnatal conditions. Analysis of HPFH has yielded insight into the control of globin gene transcription and the effects of mutations in noncoding sequences. In some conditions, HbF may be the only β-globin-like gene product formed.

A. Large deletions in the β-globin gene cluster

HPFH can result from deletions within the β-globin gene cluster. Several very large deletions in the β-globin gene cluster region are known, especially at the 3′ end. The deletions are distributed differently in different ethnic populations, reflecting that they originated at different points in time. δβ-Thalassemia and failure of β-globin production have been the result in some cases.

B. Mutations in noncoding sequences of the promoter region

Hereditary persistence of fetal hemoglobin can result from mutations in noncoding sequences of the promoter region upstream (5′ direction) of the β-globin gene cluster (on the 5′ side of the γ-globin genes). Even though the highly conserved sequences CACCC, CCAAT, and ATAAA are not affected, the number of observed mutations substantiates the significance of the remaining noncoding sequences (long-range transcription control). They are probably required for the changes in transcription control in the different gene loci that occur during embryonic and fetal development. (Figure adapted from Gelehrter & Collins, 1990.)

C. Frequent mutations of β-thalassemia in different populations

β-Thalassemia mutations occur in different ethnic populations with different frequencies, as shown here for 10 examples. The difference in types and distribution of mutations allows relatively simple screening tests to identify individuals who might have an increased risk for a severe thalassemia in their offspring. According to estimates of the World Health Organization, about 275 million persons are heterozygous for hemoglobin diseases worldwide. Substantial numbers are due to the β-thalassemias in Asia (over 60 million), α^0-thalassemia in Asia (30 million), HbE/β-thalassemia in Asia (84 million), and sickle cell heterozygosity in Africa (50 million), India, the Caribbean, and the US (about 50 million). At least 200 000 severely affected homozygotes are born annually: about 50% with sickle cell anemia and 50% with thalassemia (Weatherall, 2010).

Further Reading

Antonarakis SE, Kazazian HH Jr, Orkin SH. DNA polymorphism and molecular pathology of the human globin gene clusters. Hum Genet 1985;69:1–14

Benz EJ Jr. Disorders of hemoglobin. In: Longo DL, et al., eds. Harrison's Principles of Internal Medicine. 18th ed. New York: McGraw-Hill; 2012:852–861

Gelehrter TD, Collins F. Principles of Medical Genetics. Baltimore: Williams & Wilkins; 1990

Kan YW, et al. Deletion of the beta-globin structure gene in hereditary persistence of foetal haemoglobin. Nature 1975;258:162–163

Orkin SH, Kazazian HH Jr. The mutation and polymorphism of the human β-globin gene and its surrounding DNA. Annu Rev Genet 1984;18:131–171

Stamatoyannopoulos G, et al., eds. The Molecular Basis of Blood Diseases, 4th ed. Philadelphia: W. B. Saunders; 2001

Weatherall DJ. The inherited diseases of hemoglobin are an emerging global health burden. Blood 2010;115:4331–4336

Weatherall DJ, et al. The hemoglobinopathies. In: Scriver CR, et al., eds. The Metabolic and Molecular Bases of Inherited Disease. 8th ed. New York: McGraw-Hill; 2001:179–190. Available at: http://www.ommbid.com. Accessed January 28, 2012

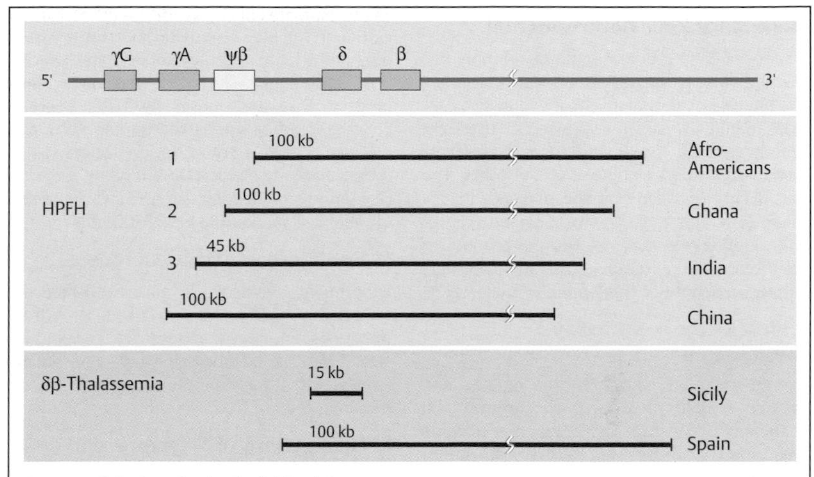

A. Large deletions in the β-globin cluster

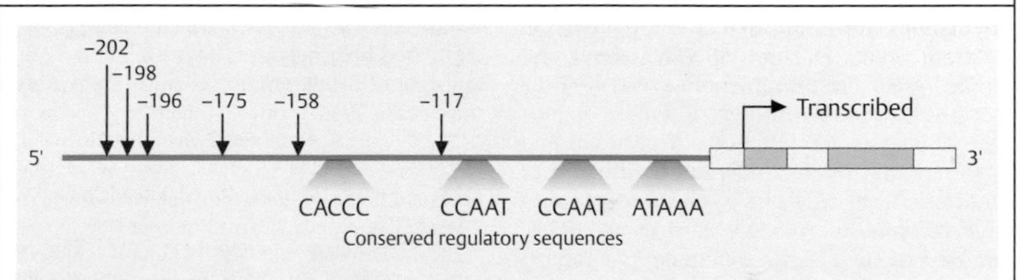

B. Mutations in noncoding sequences in the promoter of γ-globin genes cause hereditary persistence of fetal hemoglobin (HPFH)

β-thal mutation	Frequency	Ethnic group	Type
Intron 1 (110 G→ A)	35%	Mediterranean	β^+
Codon 39 (C→T)	27%	Mediterranean	β^0
TATA-Box (-29 A→G)	39%	Afro-American	β^+
Poly A (T→C)	26%	Afro-American	β^+
Intron 1 (5 G→C)	36%	India	β^+
Partial deletion (619 nt)	36%	India	β^0
Codon 71-72 frameshift	49%	China	β^0
Codon 71-72 frameshift	49%	China	β^0
Intron 2 (654 C→T)	38%	China	β^0
Codon 41-42 -CTTT	frequent	Southeast Asia	β^0

C. Frequent mutations of β-thalassemia in different populations

Mammalian Sex Determination

A series of early prenatal sequential binary developmental processes determine whether anatomical development ensues in male or female mammals. Genes expressed at the right time in relevant tissues initiate each developmental step toward one sex or the other. The first decision is made by the presence or absence of a functional Y chromosome. Thereafter, indifferent stages of the gonads and of the internal and external genital organs differentiate into those of the female or male.

A. Role of the mammalian Y chromosome

The crucial role of the Y chromosome in man became evident in 1959 from chromosomal analysis of two disorders: Klinefelter syndrome and Turner syndrome. Individuals with Klinefelter syndrome have two X chromosomes and a Y chromosome (XXY; Jacobs and Strong, 1959), but clearly have a male phenotype, although this is incompletely developed (p. 384). Even the presence of several X chromosomes does not result in a female phenotype as long as one functional Y chromosome is present. In contrast, individuals with Turner syndrome have only one X chromosome and no Y chromosome (45,XO; Ford et al., 1959), but nevertheless show a female phenotype, although it is incompletely developed and usually accompanied by malformations (p. 384). Thus, the Y chromosome and the fetal testes are required to induce male differentiation.

B. Sex-determining region SRY

In human males only a small region of the distal short arm of the Y chromosome is required to induce male development. This region is named SRY (sex-related Y). SRY (OMIM 480000) is a small region of about 35 kb on the short arm of the human Y chromosome (Yp11.31). It is located just proximal to the pseudoautosomal region 1 (PAR1) within Y-chromosomal interval 1A1.

C. *SRY* gene

The *SRY* gene, located at Yp11.32, consists of a single exon with 841 bp spanning 3.8 kb. It encodes a transcription factor that is a member of the high-mobility group (HMG)-box family of DNA-binding proteins. This gene has a TATAAA motif for binding transcription factor TFIID. It is transcribed into RNA of 1.1 kb. From a coding region of 612 bp, a protein of 204 amino acids is translated (**1**). The conserved HMG motif binds to DNA and causes reversible bending (**2**). The bending opens the double helix and permits access of transcription factors. HMG proteins are nonhistone DNA-binding proteins. (**2** kindly provided by Dr Michael A. Weiss, Cleveland, USA, from Li et al., 2006.)

D. Sry-transgenic male XX mouse

Experimental evidence in mice confirmed the crucial role of Sry. When a 14-kb DNA fragment containing the mouse *Sry* gene is inserted into the blastocyst of a chromosomally female (XX) transgenic mouse, a male mouse develops. (Figure from Koopman et al., 1991.)

E. Time pattern of *Sry* expression

Expression of *Sry* is limited to short period of time during embryonic development. In a mouse embryo with XY chromosomes, *Sry* is expressed only between days 10.5 and 12.5 of embryonic development. (Figure from Koopman et al., 1991.)

Further Reading

Erickson RP. The sex determination pathway. In: Epstein CJ, et al., eds. Inborn Errors of Development. 2nd ed. The Molecular Basis of Clinical Disorders of Morphogenesis. Oxford University Press: Oxford; 2008: 203–211

Knower KC, et al. Failure of SOX_9 regulation in 46XY disorders of sex development with SRY, SOX_9 and SF_1 mutations. PLoS One 2011;11;6:e17751

Koopman P, et al. Male development of chromosomally female mice transgenic for Sry. Nature 1991; 351:117–121

Li B, et al. SRY-directed DNA bending and human sex reversal: reassessment of a clinical mutation uncovers a global coupling between the HMG box and its tail. J Mol Biol 2006;360:310–328

MacLaughlin DT, Donahoe PK. Sex determination and differentiation. N Engl J Med 2004;350:367–378

Sekido R. SRY: A transcriptional activator of mammalian testis determination. Int J Biochem Cell Biol 2010;42:417–420

Sekido R, Lovell-Badge R. Sex determination involves synergistic action of SRY and SF1 on a specific Sox9 enhancer. Nature 2008;453:930–934

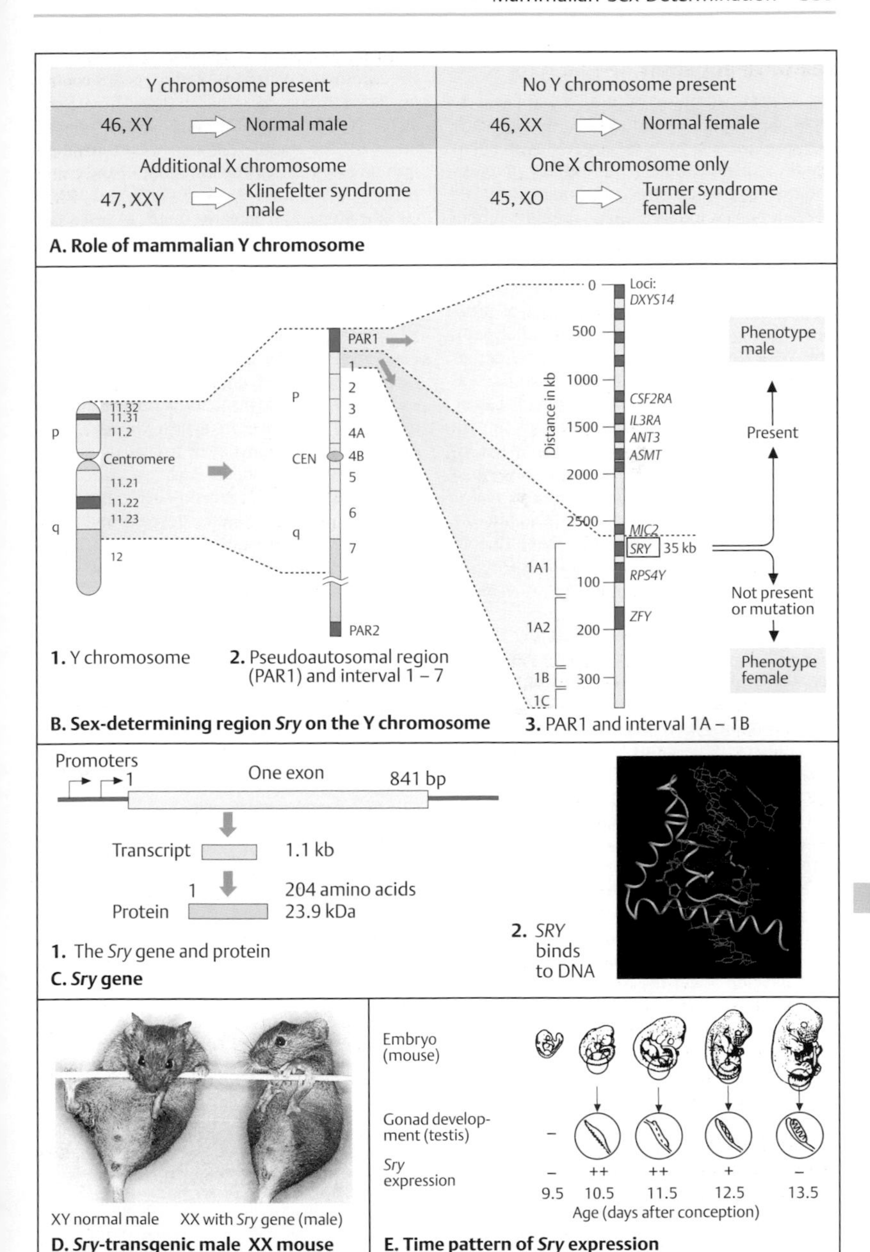

Y chromosome present
No Y chromosome present
46, XY
Normal male
46, XX
Normal female
Additional X chromosome
One X chromosome only
47, XXY
Klinefelter syndrome male
45, XO
Turner syndrome female
A. Role of mammalian Y chromosome
p
q
11.32
11.31
11.2
Centromere
11.21
11.22
11.23
12
PAR1
1
2
3
4A
4B
5
6
7
CEN
PAR2
Loci:
DXYS14
CSF2RA
IL3RA
ANT3
ASMT
MIC2
SRY
35 kb
RPS4Y
ZFY
Distance in kb
0
500
1000
1500
2000
2500
100
200
300
1A1
1A2
1B
1C
Phenotype male
Present
Not present or mutation
Phenotype female
1. Y chromosome
2. Pseudoautosomal region (PAR1) and interval 1 – 7
3. PAR1 and interval 1A – 1B
B. Sex-determining region Sry on the Y chromosome
Promoters
1
One exon
841 bp
Transcript
1.1 kb
Protein
204 amino acids
23.9 kDa
1. The Sry gene and protein
2. SRY binds to DNA
C. Sry gene
XY normal male
XX with Sry gene (male)
D. Sry-transgenic male XX mouse
Embryo (mouse)
Gonad development (testis)
Sry expression
–
++
+
9.5
10.5
11.5
12.5
13.5
Age (days after conception)
E. Time pattern of Sry expression

Sex Differentiation

Sex differentiation occurs in a series of consecutive developmental processes during early mammalian embryogenesis, resulting in either female or male gender. From initially undifferentiated anatomical structures, male or female development is induced under the influence of specific genes.

A. Gonads and external genitalia

At about the end of the sixth week of pregnancy in humans, after the primordial germ cells have migrated to the initially undifferentiated gonads, an inner portion (medulla) and an outer portion (cortex) of the gonads can be distinguished (**1**). The cortex develops into an ovary, the medulla into a testis. Under the influence of estradiol, Müllerian ducts develop in females and Wolffian ducts develop in males. In XY embryos, testes develop at about the 10th week of pregnancy under the influence of a testis-determining factor (TDF), the *SRY* gene (OMIM 480000). If this is not present, ovaries develop.

The excretory ducts differentiate under the influence of the hormones produced by the early gonads (**2**). The early embryonic testis produces two hormones: testosterone, with a male differentiating effect; and the Müllerian inhibition factor MIF (anti-Müllerian hormone, AMH; OMIM 600957). The Müllerian ducts, precursors of the fallopian tubes, the uterus, and the upper vagina, develop when a male differentiating influence is absent. In the male pathway, their development is suppressed by AMH. AMH shares homology with transforming growth factor (TGF)β (OMIM 190180). The Wolffian ducts, precursors of the male efferent ducts (vas deferens, seminal vesicles, and prostate), develop under the influence of testosterone, a male steroid hormone formed in the fetal testis. If testosterone is absent or ineffective, the Wolffian ducts degenerate.

The external genitalia develop after the gonads have differentiated into testes or ovaries. In humans this occurs relatively late, in the 15th to 16th week of pregnancy (**3**). Full development of male external genitalia depends on a derivative of male-inducing testosterone, 5-dihydrotestosterone, a metabolite of testosterone produced by the enzymatic action of 5α-reductase.

The differentiation of the gonadal ridge into the bipotential gonad (**4**) requires genes encoding transcripion factors SF1 (OMIM 601516), WT1 (OMIM 607102), and LHX9 (OMIM 606066). The undifferentiated gonad develops into an ovary under the influence of the genes *DAX1* (OMIM 300473) on Xp21.2 and *WNT4* (OMIM 603490). The early stages of testis development require SRY and SOX9 (OMIM 608160). (**4** adapted from Gilbert, 2010.)

B. Sequence of events in sex differentiation

Four levels of sex differentiation can be defined: (i) genetic, (ii) gonadal, and (iii) anatomical, as prenatal stages, and (iv) psychological, from early childhood on. A fifth, the legal gender, recorded as "female" or "male" in all legal documents, can be added. Each level is reached in a series of time-regulated successive steps. The early embryonic testes develop under the influence of the testis-determining factors (TDF), the *SRY* gene (in humans), or the ovary in its absence. The male differentiating effect of testosterone depends on the function of an intracellular androgen receptor (see p. 360).

Further Reading

Acherman JC, Jameson JL. Disorders of sexual differentiation. In: Longo DL, et al, eds. Harrison's Principles and Practice of Internal Medicine. 18th ed. McGraw-Hill, New York, 2012: 3046–3055

Cox JJ, et al. A SOX9 duplication and familial 46,XX developmental testicular disorder. N Engl J Med 2011;364:91–93

Erickson RP. The sex determination pathway. In: Epstein CJ, et al., eds. Inborn Errors of Development. 2nd ed. The Molecular Basis of Clinical Disorders of Morphogenesis. Oxford: Oxford University Press; 2008: 203–211

Gilbert SF. Developmental Biology. 9th ed. Sunderland: Sinauer; 2010

Goodfellow PN, et al. SRY and primary sex-reversal syndromes. In: Scriver CR, et al, eds. The Metabolic and Molecular Bases of Inherited Disease. 8th ed. New York: McGraw-Hill; 2001: 1213–1221 (Online at www.ommid.com)

Su H, Lau Y-FC. Identification of the transcriptional unit, structural organization, and promoter sequence of the human sex-determining region Y (SRY) gene, using a reverse genetic approach. Am J Hum Genet 1993;52:24–38

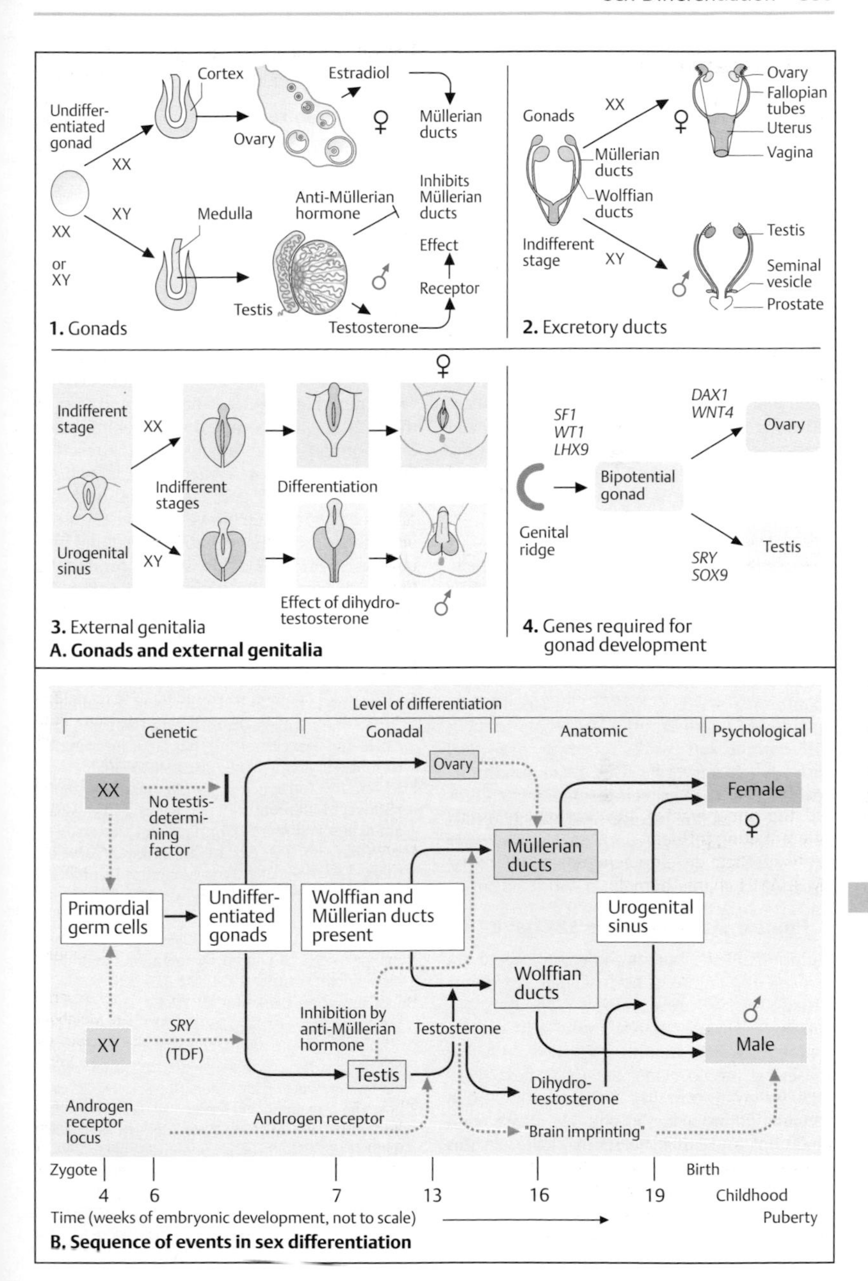

A. Gonads and external genitalia

B. Sequence of events in sex differentiation

Disorders of Sexual Development

Disorders of sexual development may occur at any level of sex determination or differentiation (see previous page). They can be classified according to (1) aberrations of the XY chromosome system, (2) abnormal gonadal development, (3) impaired androgen synthesis or androgen effect, (4) androgen excess, and (5) syndromic forms as part of a generalized disorder involving other tissues or organs (see Appendix, p. 421). In some foms the gonads do not correspond to the genital duct system or external genitalia, or the external genitalia may be ambiguous, i.e., partly female and partly male (pseudohermaphroditism). In true hermaphroditism, gonads contain both testicular and ovarian tissues. Therapy requires that the underlying developmental error be correctly diagnosed.

A. XX males and XY females

These are examples of an impaired XY system. Normally, the male-determining Y-specific DNA sequences (*SRY* gene; OMIM 480000) remain on the Y chromosome during the homologous pairing and crossing-over during meiosis. Since *SRY* is located very close to the pseudoautosomal region 1 (PAR1), crossing-over outside PAR1 transfers the *SRY* region to the X chromosome and results in a male individual with an XX karyotype (XX male syndrome; OMIM 278850). The complementary Y chromosome does not receive *SRY*, which results in a female phenotype with XY chromosomes (XY female gonadal dysgenesis; OMIM 306100).

B. Point mutations in the *SRY* gene

Mutations in the human *SRY* gene lead to sex reversal, i.e., failure of normal male sex differentiation in the presence of a Y chromosome. This gene encodes a 204-amino-acid protein containing a DNA-binding domain of 79 highly conserved amino acids, the HMG box (high-mobility group protein), in the middle section between amino acids 58 and 137. Most point mutations and small deletions cluster in this region. This results in complete or partial gonadal dysgenesis. Sex reversal associated with campomelic dysplasia results from mutations in the *SOX9* gene (SRY-related HMG-box gene, located at 17q24.3-q25.1; OMIM 608160). (Figure adapted from McElreavey and Fellous, 1999.)

C. Androgen receptor

This is an example of a broad spectrum of disorders resulting from insufficient androgen effects owing to genetic defects in the intracellular androgen receptor (OMIM 313700) on Xq11-q12. Testosterone produced by the fetal testes exerts its effect only if it binds to its receptor (**1**). Likewise, dihydrotestosterone (DHT), converted from testosterone in the urogenital sinus by 5α-reductase, requires this receptor. The activated hormone–receptor complex (TR and DR) acts as a transcription factor for genes that regulate the differentiation of the Wolffian ducts and the urogenital sinus. Mutations in the androgen receptor result in androgen insensitivity. Affected XY individuals have a normal *SRY* gene and testes, and they produce testosterone. However, since testosterone cannot exert any effect, the resulting phenotype is female (**2**). This variable condition is the androgen insensitivity syndrome, also called testicular feminization (TFM; OMIM 300068).

Further Reading

Acherman JC, Jameson JL. Disorders of sexual differentiation. In: Longo DL, et al, eds. Harrison's Principles and Practice of Internal Medicine. 18th ed. New York: McGraw-Hill; 2012: 3046–3055

Erickson RP. Introduction to the sex determining pathway: Mutations in many genes lead to sexual ambiguity and reversal. In: Epstein CJ, Erickson RP, Wynshaw-Boris A, eds. Inborn Errors of Development. The Molecular Basis of Clinical Disorders of Morphogenesis. 2nd ed. Oxford: Oxford University Press; 2008: 203–211

Foster JW, et al. Campomelic dysplasia and autosomal sex reversal caused by mutations in an SRY-related gene. Nature 1994;372:525–530

Goodfellow PN, Camerino G. DAX-1, an 'antitestis' gene. Cell Mol Life Sci 1999;55:857–863

Griffin JE, et al. The androgen resistance syndromes: Steroid 5 α-reductase deficiency, testicular feminization, and related disorders. In: Scriver CR, et al, eds. The Metabolic and Molecular Bases of Inherited Disease. 8th ed. New York: McGraw-Hill; 2001: 4117–4146 (Online at www.ommid.com)

MacLaughlin DT, Donahoe PK. Sex determination and differentiation. N Engl J Med 2004;350:367–378

Mendonca BB, et al. 46, XY disorders of sex development (DSD). Clin Endocrinol (Oxf) 2009;70:173–187

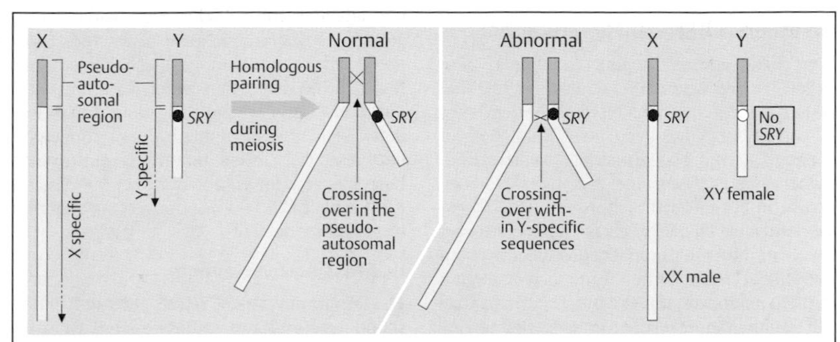

1. *SRY* remains on the Y chromosome **2.** Transfer of *SRY* to the X chromosome

A. XX males and XY females

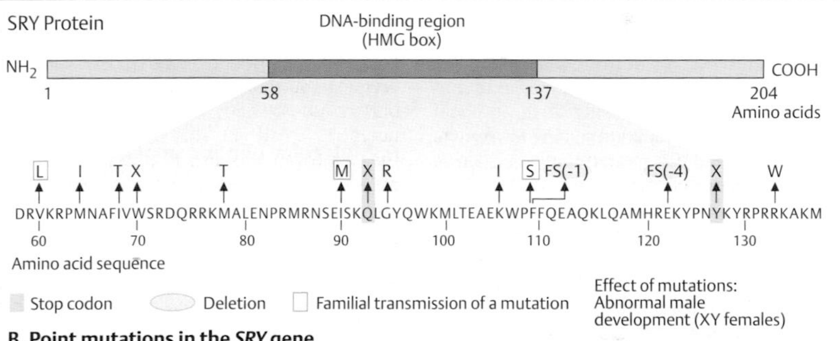

B. Point mutations in the *SRY* gene

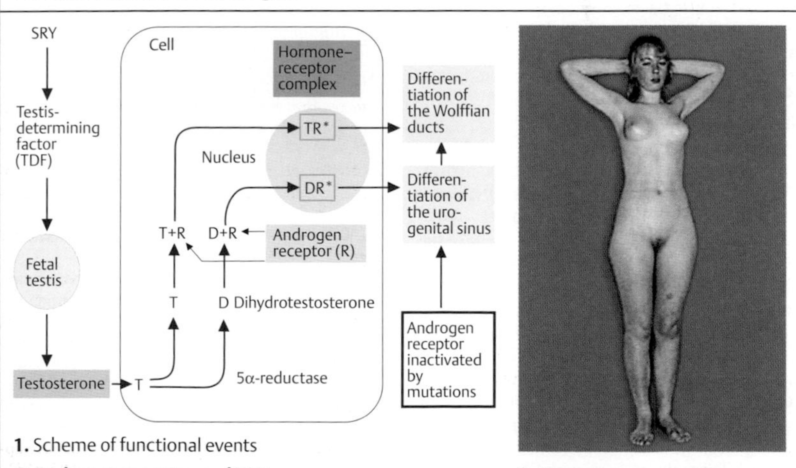

1. Scheme of functional events

2. Phenotype

C. Androgen receptor and TFM

Congenital Adrenal Hyperplasia

Congenital adrenal hyperplasia (CAH), also called adrenogenital syndrome (OMIM 201910), is a genetically determined deficiency of cortisol resulting from insufficient biosynthesis of steroid 21-hydroxylase, a microsomal cytochrome P450 enzyme. A compensatory increase in adrenocortical hormone (ACTH) excretion causes hyperplasia of the adrenal cortex with increased prenatal production of androgenic steroids. With a frequency of about 1 in 5000 newborns, it is the most common form of CAH, the heterozygote frequency being 1:50 in Europe and North America. Its clinical manifestations are highly variable.

A. Overview

The main characteristics of this disease are listed (**1**). CAH occurs in three major manifestations, depending in part on the type of mutation: (i) a severe salt-losing form, in about 60%–75% of patients; (ii) a virilizing form without salt loss; and (iii) an attenuated, late-onset form. The severe salt-losing form is life-threatening in newborns. Ambiguous or virilized genitalia are present in newborn girls (**2**). The biochemical defect results in hyperplasia of the adrenals (**3**). Inheritance is autosomal recessive (**4**). Untreated or poorly treated children show advanced growth and early puberty, but are short as adults because of early closure of the epiphyseal plates. Such girls develop a male physical appearance (**5**).

B. Biochemical defect

The enzymatic defect blocks the conversion of progesterone to deoxycortisol (DOC) by hydroxylation at position 21 (steroid 21-hydroxylase). As a result, the plasma concentration of 17-hydroxy-progesterone is increased.

C. The *CYP21* gene structure

The *CYP21A2* gene (OMIM 613815) encodes the 21-hydroxylase enzyme. It is located at 6p21.3 in tandem with three closely linked genes upstream (in the 5′ direction): the complement 4A (*C4A*) gene (OMIM 120810), a *CYP* pseudogene (*CYP21P*) without function because of a deletion (8 bp) and a nonsense mutation, and the complement 4B (*C4B*) gene (OMIM 120820). These genes lie within the class III HLA genes of the major histocompatibility complex MHC (see p. 294). The *CYP21A2* gene has 10 exons and spans nearly 6 kb. Deletions and nonsense or frameshift mutations result in the severe salt-losing form. Missense mutations occur in the salt-losing and virilizing forms. Most patients are compound heterozygotes. Seven mutations are shown. Deletions account for about 20% of the classical salt-losing form. Duplications occur without clinical consequences.

D. Crossing-over events

The four genes, *C4A*, *CYP21P*, *C4B*, and *CYP21*, share structural and sequence similarity as a result of a duplication event in evolution. This predisposes to mispairing and unequal crossing-over at meiosis. Gene conversion occurs in about 75% of patients. It results from mismatch between the functional *CYP21* gene and the nonfunctional pseudogene *CYP21P*, which converts part of the *CYP21* gene to a pseudogene (*CYP21P*).

E. Molecular genetic analysis

This example shows the detection of an 8-bp deletion by a semiquantitative PCR method. In normal individuals (lanes 2, 4, 6–8) the intensity of the 952-bp fragment (derived from exons 1–3) and the 200-bp fragment is about the same. In heterozygotes (lanes 5 and 9), the 952-bp fragment is less intense than in controls, and it is absent in the patient (lane 3). The fragments at the bottom of each lane represent parts of the β-actin gene for control. (Figure courtesy of Dr Alireza Baradaran, Mashhad, Iran, from Vakili et al., 2005.)

Further Reading

Donohoue PA, et al. Congenital adrenal hyperplasia. In: Scriver CR, et al, eds. The Metabolic Basis of Inherited Disease. 8th ed. New York: McGraw-Hill; 2001: 4077–4115 (Online at www.ommid.com)

Merke DP, Bornstein SR. Congenital adrenal hyperplasia. Lancet 2005;365:2125–2136

Speiser PW, et al. Congenital adrenal hyperplasia due to steroid 21-hydroxylase deficiency. J Clin Endocrinol Metab 2010;95:4133–4160

Vakili R, Baradaran-Heravi A, et al. Molecular analysis of the CYP21 gene and prenatal diagnosis in families with 21-hydroxylase deficiency in northeastern Iran. Horm Res 2005;63:119–124

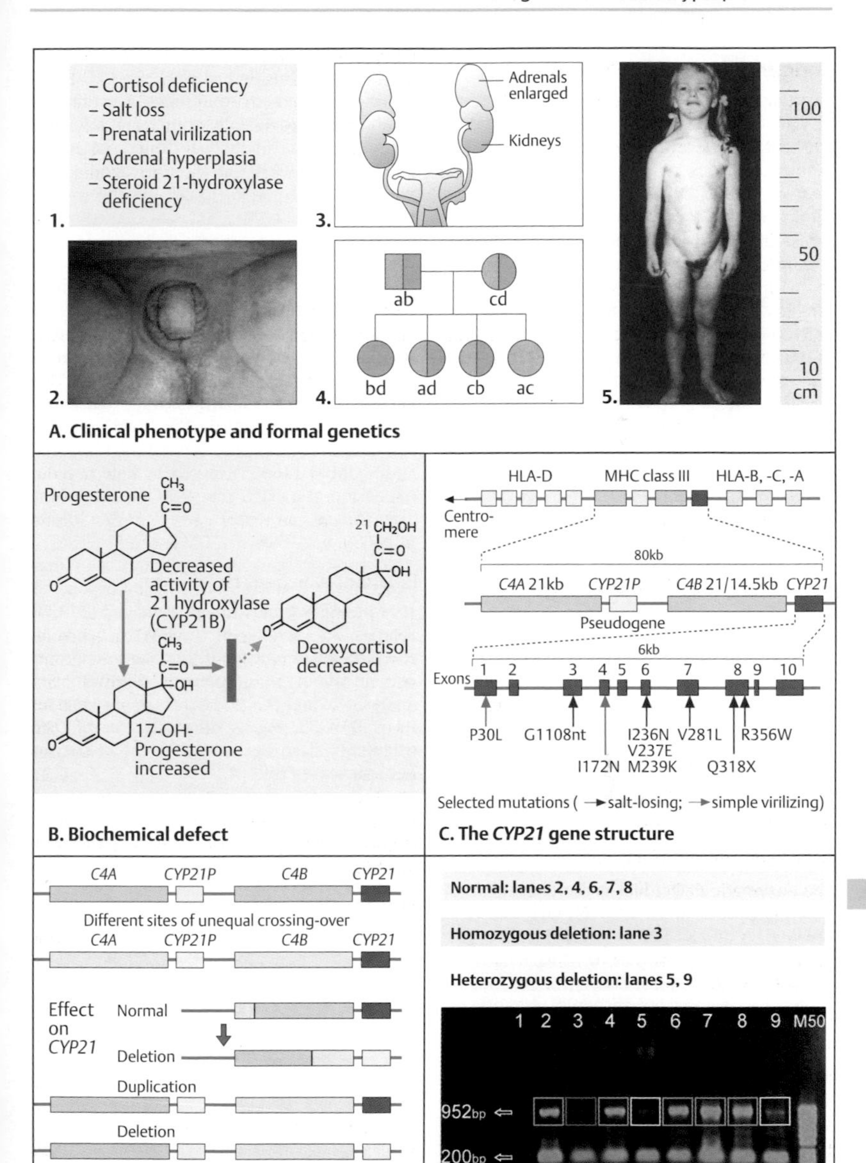

A. Clinical phenotype and formal genetics

B. Biochemical defect

C. The *CYP21* gene structure

D. Crossing-over events in the *CYP21* gene

E. Molecular analysis

Unstable Repeat Expansion

Expansion of unstable nucleotide repeats occur in about 30 neurogenic hereditary disorders (nucleotide expansion disorders). The expanded repeat is localized at a specific region near or within the gene involved (p. 92). In most cases, the repeat consists of three nucleotides. In others, a tetranucleotide or pentanucleotide repeat is expanded.

A. Types of disease

Unstable repeat diseases can be distinguished whether they are located in the untranslated regions of a gene (5′ UTR or 3′ UTR), in an exon, or in an intron. Functional consequences are loss of protein function or gain of an abnormal RNA level or protein, including loss of translational control (FRAXA), signaling (FRAXAE), or mitochondrial function (Friedreich ataxia, FRDA). In about half of these disorders the codon CAG, encoding glutamine, is expanded within a coding region (polyglutamine disorders, or PolyQ), e.g., in Huntington disease.

B. Huntington disease

Huntington disease (HD; OMIM 143100), described 1872 by George Huntington in a family from Long Island, occurs in about 4–7 individuals per 100 000. It is an autosomal dominant late-onset progressive neurodegenerative disease resulting from neuronal cell death, and leading to complete loss of motor control and intellectual abilities within 5–10 years (**1**). It usually begins around the age of 40–50 years with uncoordinated movements (chorea, St Vitus' dance), excitation, hallucinations, and psychological changes. The 180-kb gene (*HTT*; OMIM 613004), located on the distal short arm of human chromosome 4 at 4p16.3 between markers D4S127 and D4S125 (**2**), and has 67 exons. Two transcripts encode a ubiquitously expressed 3144-amino-acid protein, huntingtin, involved in neuron function and survival. The 5′ coding region of the gene contains 4–35 copies of a trinucleotide repeat of cytosine, adenine, and guanine (CAG), which is a codon for glutamine. In patients this is expanded to 40–250 CAG repeats. Incomplete penetrance may occur with 36–40 repeats. The CAG repeat length is inversely correlated with the age of onset. A diagnostic test distinguishes expanded and normal $(CAG)_n$ repeats (**3**). A predictive genetic test is possible many years prior to the onset of the first clinical signs. This requires prior genetic counseling according to established guidelines to ensure that informed consent of the individual tested has been achieved. (**3**, Zühlke et al., 1993, kindly provided by Professor W. Engel, Göttingen, Germany.)

C. Myotonic dystrophy

Myotonic dystrophy is an autosomal dominant neurological disease characterized by (**1**) variable muscular weakness, cataracts, alopecia, abnormal cardiac conduction, testicular atrophy, and a masklike face (**2**). It occurs in two genetic types: DM1 (OMIM 160900) caused by increased CTG repeats in the *DMPK* gene located at 19q13.32 (**3**) (OMIM 60377); and DM2 (OMIM 116955) caused by heterozygous expansion of a CCTG repeat in intron 1 of the zinc finger protein-9 gene (*ZNF9*; OMIM 602668; not shown). DM1 results from a pathogenic increase in the number of CTG repeats in the 3′ untranslated region beyond 50 repeats (**4**). Normal individuals have 5–37 CTG repeats. The severity of the DM1 depends on the number of repeats. Repeat contraction occurs in about 5% of patients. Southern blot analysis using probe pBBO7 at the marker locus D19S95 reveals different sizes of DNA fragments after digestion with the restriction endonuclease *Eco*RI (**4**).

Further Reading

Gatchel JR, Zoghbi HY. Diseases of unstable repeat expansion: mechanisms and common principles. Nat Rev Genet 2005;6:743–755

Harper P, Johnson K. Myotonic dystrophy. In: Scriver CR, et al., eds. The Metabolic and Molecular Bases of Inherited Disease. 8th ed. New York: McGraw-Hill; 2001:5525–5550. (www.ommbid.com)

Hayden MR, Kremer B. Huntington disease. In: Scriver CR, et al., eds. The Metabolic and Molecular Bases of Inherited Disease. 8th ed. New York: McGraw-Hill; 2001:5677–5701. Available at: http://www.ommbid.com)

Orr HT, Zoghbi HY. Trinucleotide repeat disorders. Annu Rev Neurosci 2007;30:575–621

Walker FO. Huntington's disease. Lancet 2007;369: 218–228

Zühlke C, et al. Mitotic stability and meiotic variability of the (CAG)n repeat in the Huntington disease gene. Hum Mol Genet 1993;2:2063–2067

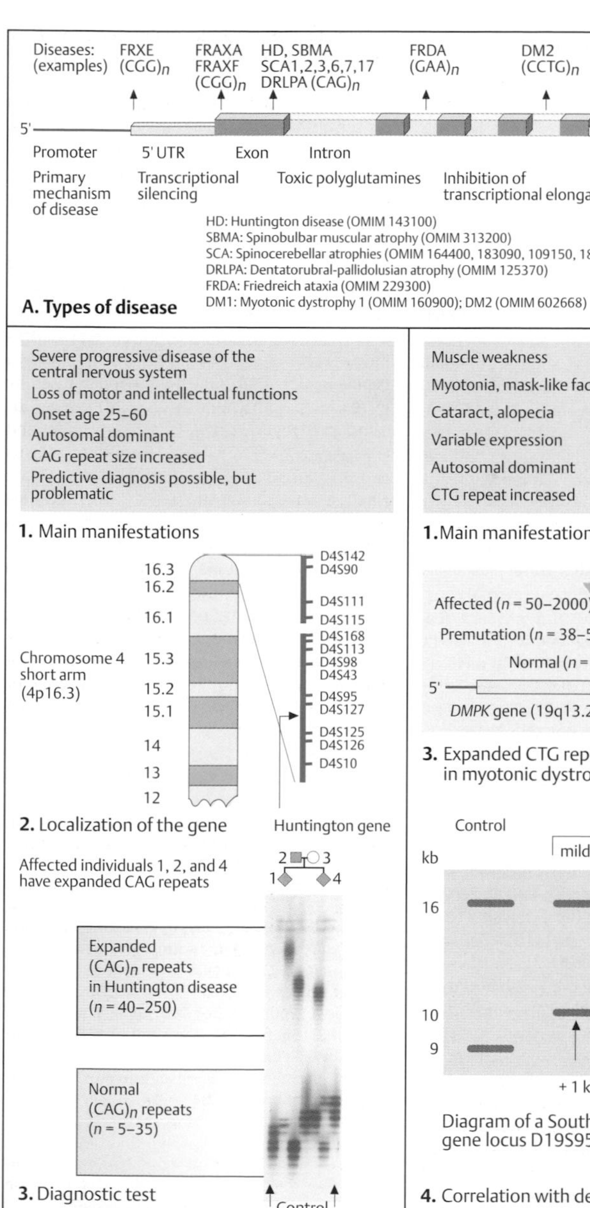
Diseases: (examples)
FRXE (CGG)n
FRAXA FRAXF (CGG)n
HD, SBMA SCA1,2,3,6,7,17 DRLPA (CAG)n
FRDA (GAA)n
DM2 (CCTG)n
DM1 (CTG)n
5'
3'
Promoter
5' UTR
Exon
Intron
3' UTR
Primary mechanism of disease
Transcriptional silencing
Toxic polyglutamines
Inhibition of transcriptional elongation
RNAs with long CUG tracts
HD: Huntington disease (OMIM 143100)
SBMA: Spinobulbar muscular atrophy (OMIM 313200)
SCA: Spinocerebellar atrophies (OMIM 164400, 183090, 109150, 183086, 164500, 607136)
DRLPA: Dentatorubral-pallidolusian atrophy (OMIM 125370)
FRDA: Friedreich ataxia (OMIM 229300)
DM1: Myotonic dystrophy 1 (OMIM 160900); DM2 (OMIM 602668)
A. Types of disease
Severe progressive disease of the central nervous system
Loss of motor and intellectual functions
Onset age 25–60
Autosomal dominant
CAG repeat size increased
Predictive diagnosis possible, but problematic
1. Main manifestations
16.3
16.2
16.1
15.3
15.2
15.1
14
13
12
Chromosome 4 short arm (4p16.3)
D4S142
D4S90
D4S111
D4S115
D4S168
D4S113
D4S98
D4S43
D4S95
D4S127
D4S125
D4S126
D4S10
2. Localization of the gene
Huntington gene
Affected individuals 1, 2, and 4 have expanded CAG repeats
2
3
1
4
Expanded (CAG)n repeats in Huntington disease (n = 40–250)
Normal (CAG)n repeats (n = 5–35)
3. Diagnostic test
Control
B. Huntington disease
Muscle weakness
Myotonia, mask-like face
Cataract, alopecia
Variable expression
Autosomal dominant
CTG repeat increased
1. Main manifestations
2. Phenotype
Affected (n = 50–2000)
Premutation (n = 38–50)
Normal (n = 5–37)
5'
(CTG)n
3'
DMPK gene (19q13.2–q13.3)
3. Expanded CTG repeat in myotonic dystrophy
Control
Affected
mild
severe
congenital
kb
16
10
9
+ 1 kb
+ 2.5 kb
+ 4 kb
Diagram of a Southern blot at gene locus D19S95 (probe pBB0.7)
4. Correlation with degree of severity
C. Myotonic dystrophy

Fragile X Syndrome

Fragile X syndrome (OMIM 300624; synonyms: fragile X mental retardation syndrome, FMR1, or Martin–Bell syndrome) is a form of mental retardation with a prevalence of about 1:3000–6000 males. It is caused by expansions or deletions of a CGG trinucleotide repeat in the 5′ untranslated region of the *FMR1* gene (OMIM 309550) on the distal long arm of the X chromosome at Xq27.3. The concomitant hypermethylation of the CpG island in the promoter of this gene results in transcriptional silencing, change of chromatin to a condensed, closed conformation, absent or dysfunctional FMR1 protein, and defective translational control of neuronal synaptic proteins. It was identified in 1991 as the first human disease that was found to be caused by expansion of an unstable trinucleotide repeat sequence (Oberle et al., 1991; Verkerk et al., 1991).

A. Phenotype

Individuals with fragile X syndrome have varying degrees of intellectual developmental delay associated with behavioral and physical features. Some patients have connective tissue weakness. The testes are usually enlarged.

B. Fragile site FRAXA

This disorder derived it name from a cytogenetically visible fragile site at region Xq27 of the X chromosome in lymphocytes cultured in medium deficient in folic acid. Fragile X tremor/ataxia syndrome (FXTAS; OMIM 300623) results from an abnormal gain-of-function of FMR1 RNA. Other variants with different fragile sites exist, e.g., FRAXE at Xq28 (OMIM 309548).

C. *FMR1* gene and protein

The *FMR1* gene has 17 exons spanning 38 kb. Its transcript is alternatively spliced and translated into at least 20 protein isoforms (FMRPs). The protein (FMRP) selectively binds RNA. At least two functional domains are RNA-binding sites, KH2 and RGG (KH domains consist of 40–60 amino acids with invariant hydrophobic leucine, isoleucine, or methionine; RGG consists of a 20–30 amino acid motif with arginine–glycine–glycine residues). In Drosophila, *Fmr1* is a component of the RNA-induced silencing complex (RISC; see p. 186).

D. Inheritance and genetic testing

Southern blot analysis (**1**) can distinguish individuals with fragile X syndrome (Fra X) carrying a full mutation allele (>200 CGG trinucleotides), a premutation (59–200), or a normal allele (6–50). The visible bands (arrows) are DNA fragments harboring the CGG repeat region, derived from genomic DNA cleaved by restriction endonuclease *Pst*I, and hybridized to the radiolabeled *FMR1* probe Ox0.55. PCR protocols determine the lengths of normal and premutation alleles. Expansion leads to hypermethylation and transcriptional silencing of the *FMR1* gene. (Figure kindly provided by Professor P. Steinbach, Ulm, Germany.)
In the pedigree (**2**) the number of CGG repeats at the *FMR1* locus is shown for each individual. A premutation can be transmitted by a female (I-2, II-3, III-2) or a male (II-2). The premutation allele may expand into a full mutation when passed from a mother to her children. All daughters of a normal male transmitter will be heterozygous. Full mutation males transmit a premutation to all their daughters. Carriers of a premutation allele do not usually have signs of fragile X syndrome, but 50%–60% of girls with a full mutation have significant cognitive deficits.

Further Reading

Jacquemont S, et al. Fragile-X syndrome and fragile X-associated tremor/ataxia syndrome: two faces of FMR1. Lancet Neurol 2007;6:45–55

Oberle I, et al. Instability of a 550-base pair DNA segment and abnormal methylation in fragile X syndrome. Science 1991;252:1097–1102

Penagarikano O, Mulle JG, Warren ST. The pathophysiology of fragile x syndrome. Annu Rev Genomics Hum Genet 2007;8:109–129

Verkerk AJ, et al. Identification of a gene (FMR-1) containing a CGG repeat coincident with a breakpoint cluster region exhibiting length variation in fragile X syndrome. Cell 1991;65:905–914

Warren ST, Sherman SL. The fragile X syndrome. In: Scriver CR, et al., eds. The Metabolic and Molecular Bases of Inherited Disease. 8th ed. New York: McGraw-Hill; 2001:1257–1289

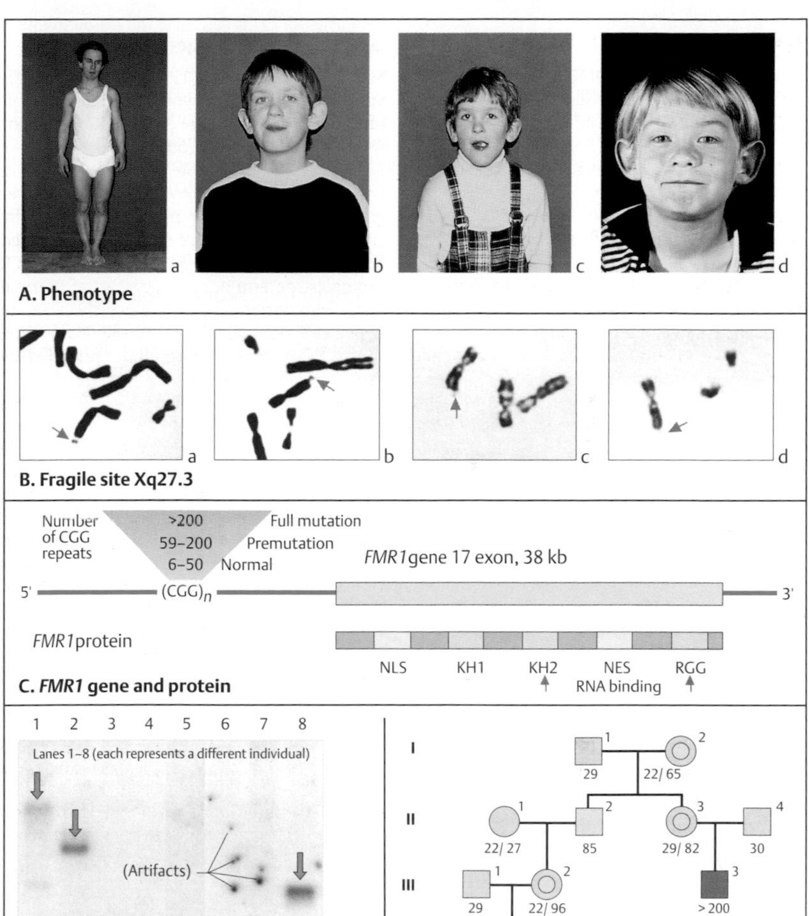

A. Phenotype

B. Fragile site Xq27.3

C. *FMR1* gene and protein

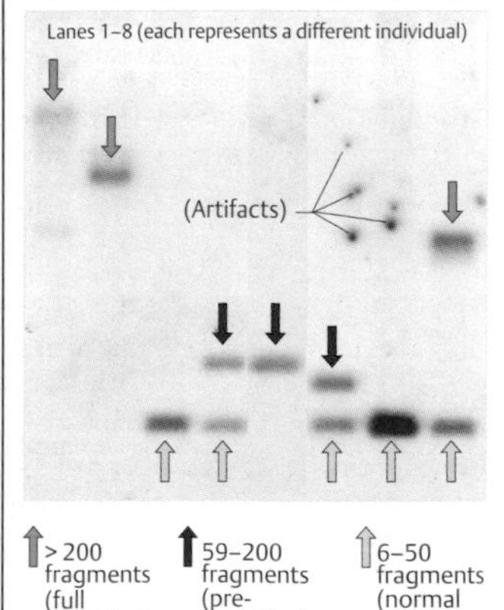

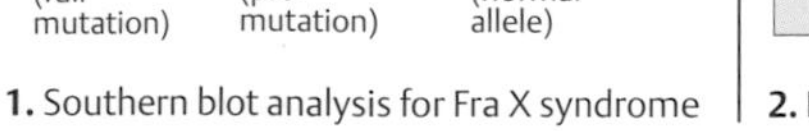

1. Southern blot analysis for Fra X syndrome

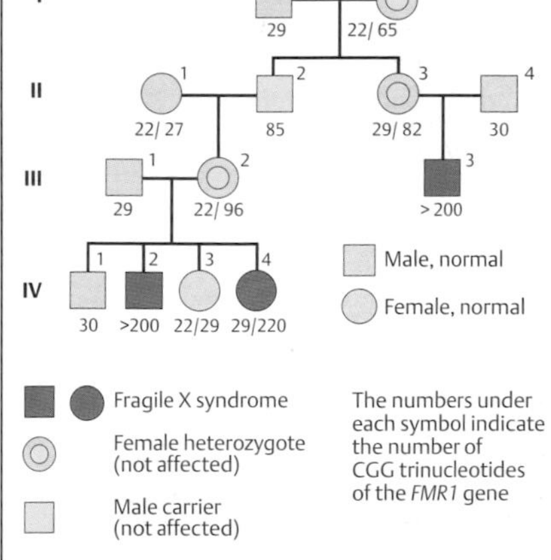

2. Phenotypic effects of expanded CGG repeats

D. Inheritance and genetic testing

Imprinting Disorders

Imprinting disorders result from various defects resulting in faulty imprinting of the particular chromosomal region involved (p. 194). Different mechanisms affect one or more active genes normally expressed in only one parental allele in an imprinted region. Best known are Prader–Willi syndrome (OMIM 176270) and Angelman syndrome (OMIM 105830) at region 15q11-13, and Beckwith–Wiedemann syndrome (OMIM 130650) at 11p15.5.

A. Prader–Willi and Angelman syndromes

Prader–Willi syndrome (PWS) and Angelman syndrome (AS) are neurogenetic developmental disorders resulting from different genetic lesions in an imprinted region of human chromosome 15 (15q11-13). Its effect depends on whether it involves the chromosome 15 of paternal or maternal origin. It results in PWS if the chromosome 15 involved is of paternal origin, but in AS if the lesion involves the chromosome 15 of maternal origin. PWS is characterized by neonatal muscular weakness and feeding difficulties, followed in early childhood by reduced or lack of satiation control, which leads to massive obesity in many patients. In AS, the developmental retardation is usually severe, with nearly complete lack of speech development, an abnormal electroencephalogram, tendency to seizures, and hyperactivity.

B. Types of genetic defects

An interstitial deletion of PWS-critical genes expressed in the alleles of paternal (pat) origin only is by far the most frequent cause of PWS, since only the inactive maternal (mat) alleles are present (**1**). Maternal uniparental disomy (matUPD) is defined by the presence of two maternal alleles and lack of a paternal allele. If the two maternal chromosomes are identical, this is referred to as isodisomy; if they differ, it is referred to as heterodisomy. Different defects involving the imprinting center (IC), which controls the imprint pattern (see p. 194), are rare but important causes. If the breakpoint of a chromosome translocation is located in this imprinted region, it also may be a cause of PWS.

Similarly, AS (**2**) is caused if the lesion involves the AS-critical region with maternal expression of genes in this region. The gene affected is *UBE3A* (OMIM 601623). Defects in the imprinting center and mutations in the *UBE3A* gene may cause familial occurrence of PWS or AS. The functional result of a deletion and of uniparental disomy (UPD) of an imprinted region is the same. (Data for figure kindly provided by K. Buiting and B. Horsthemke, University Hospital, Essen, Germany.)

C. Imprinted chromosomal region

This simplified figure shows the genetic map of the chromosomal region 15q11-13 extending over 2 Mb. Loss of expression of paternally expressed genes (blue) results in PWS. AS results from loss of function of the *UBE3A* gene (ubiquitin-protein ligase E3; OMIM 600012), which is expressed from the maternal copy only (red). The imprinting center, controlling the entire imprinted region, appears to consist of two elements. One is required for the maintenance of the paternal imprint during early embryogenesis, the other for maternal imprinting in the female germline. (Data for figure kindly provided by K. Buiting and B. Horsthemke, University Hospital, Essen, Germany.)

Further Reading

Buiting K. Prader-Willi syndrome and Angelman syndrome. Am J Med Genet C Semin Med Genet 2010;154C:365–376

Horsthemke B. Mechanisms of imprint dysregulation. Am J Med Genet C Semin Med Genet 2010;154C:321–328

Horsthemke B, Dittrich B, Buiting K. Imprinting mutations on human chromosome 15. Hum Mutat 1997;10:329–337

Horsthemke B, Buiting K. Imprinting in Prader–Willi and Angelman syndromes. In: Jorde LB, et al., eds. Encyclopedia of Genetics, Genomics, Proteomics, and Bioinformatics. Vol 1. Chichester: Wiley & Sons; 2005:245–258

Horsthemke B, Buiting K. Imprinting defects on human chromosome 15. Cytogenet Genome Res 2006;113:292–299

Horsthemke B, Wagstaff J. Mechanisms of imprinting of the Prader-Willi/Angelman region. Am J Med Genet A 2008;146A:2041–2052

Nicholls RD, Knepper JL. Genome organization, function, and imprinting in Prader-Willi and Angelman syndromes. Annu Rev Genomics Hum Genet 2001;2:153–175

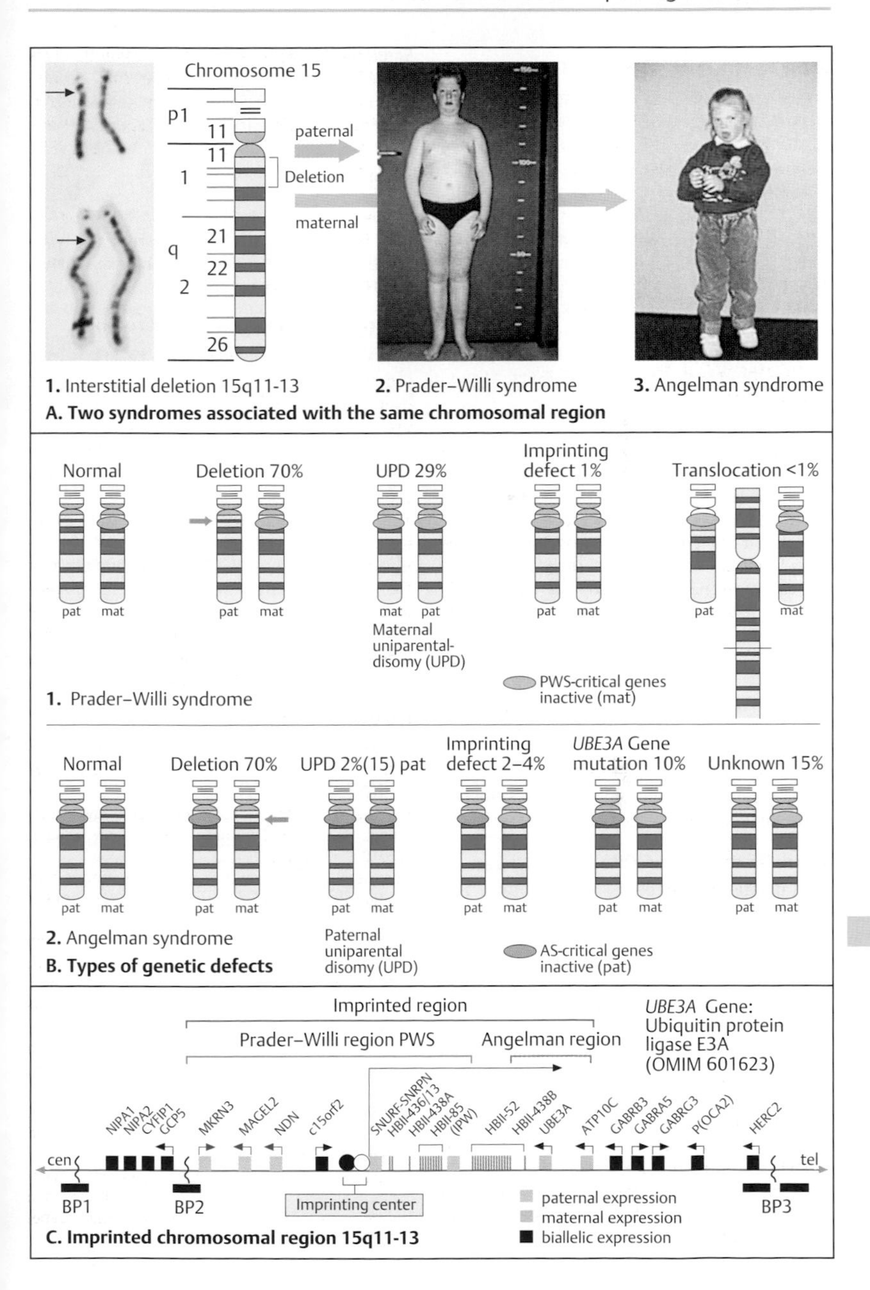

1. Interstitial deletion 15q11-13 **2.** Prader–Willi syndrome **3.** Angelman syndrome

A. Two syndromes associated with the same chromosomal region

1. Prader–Willi syndrome

2. Angelman syndrome

B. Types of genetic defects

C. Imprinted chromosomal region 15q11-13

Rhodopsin, a Photoreceptor

Vertebrates have evolved sensory systems that enable them to perceive light, sound, taste, and smell, and convert this into advantageous behavioral patterns. Two types of photoreceptors in specialized cells of the retina in the eye provide color vision and light perception in the dark. The human retina has about 6 million cone cells (*cones*) for color vision and about 110 million rod cells (*rods*) that operate in weak light. The photoreceptor responsible for weak light is *rhodopsin*. It is a G-protein-coupled receptor activated by light. Only rods contain the trimeric G protein coupled to rhodopsin.

A. Rod cells

A rod cell is a highly specialized cell containing the photoreceptor rhodopsin. An outer segment of a rod contains about 1000 discs with about 4×10^7 molecules of rhodopsin. The discs are folded by the protein peripherin. The trimeric G protein coupled to rhodopsin is called *transducin*. At the other end of a rod is an inner segment with the cell nucleus, endoplasmic reticulum, Golgi apparatus, and mitochondria. Each rod has a synapsis. From here, a signal is transmitted to the optic nerve and from there to the visual cortex of the brain.

B. Photo excitation

In 1958, George Wald and coworkers discovered that light isomerizes 11-*cis*-retinal (**1**) into all-*trans*-retinal (**2**). This structural change in femtoseconds is so great that it triggers a reliable and reproducible nerve impulse. The absorption spectrum of rhodopsin (**3**) corresponds to the spectrum of sunlight, with an optimum at a wavelength of 500 nm. In the dark, all-*trans*-retinal is converted back into 11-*cis*-retinal. All-*trans*-retinal essentially does not exist in the dark. Although vertebrates, arthropods, and mollusks have anatomically different types of eyes, all three phyla use 11-*cis*-retinal for photoactivation.

C. Light cascade

Photoactivated rhodopsin triggers a series of enzymatic reactions, known as the light cascade. First, the photoactivated rhodopsin activates a G-protein-coupled receptor protein, transducin (G_t). Activated transducin is bound to guanosyl-triphosphate (GTP; see p. 200). GTP activates cyclic guanosylmonophosphate phosphodiesterase (cGMP) by binding to its inhibitory γ subunit. This converts cGMP to GMP and closes the cGMP-gated ion channels.

D. Rhodopsin

Rhodopsin is a seven-helix transmembrane protein structure with binding sites for functionally important molecules such as transducin, rhodopsin kinase, and arrestin on the cytosolic side. The binding site for the light-absorbing 11-*cis*-retinal is lysine in position 296 of the seventh transmembrane domain.

E. cGMP as transmitter of light signals

Cyclic guanosine monophosphate (cGMP) is the second messenger in the light signal transduction system. Rod outer segments normally contain a high concentration of cGMP. Light absorption by rhodopsin activates cGMP phosphodiesterase. This hydrolyzes cGMP to 5′-GMP and decreases the cGMP concentration. The high concentration of cGMP in the dark keeps cGMP-gated cation channels open. On exposure to light the channels are closed, the membrane becomes hyperpolarized, and a nerve impulse is triggered. Humans can detect a flash of about five photons (see Lodish et al., 2007, p. 557). (Figures adapted from Stryer, 1995, and Lodish et al., 2007.)

Medical relevance

Mutations in genes encoding the proteins of the light cascade signal pathway are frequent causes of different forms of hereditary blindness resulting from pigmentary degenerative changes to the retina (see p. 372).

Further Reading

Lodish H, et al. Molecular Cell Biology. 6th ed. New York: W. H. Freeman; 2007

Stryer L. Molecular basis of visual excitation. Cold Spring Harb Symp Quant Biol 1988;53(Pt 1):283–294

Stryer L. Biochemistry. 4th ed. New York: W. H. Freeman; 1995

Wald G. The molecular basis of visual excitation. Nature 1958;219:800–807

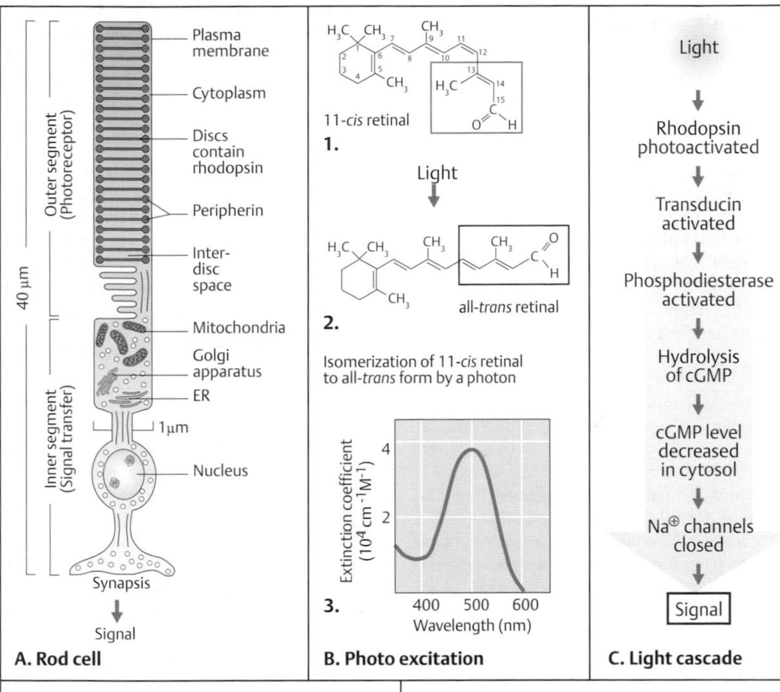

A. Rod cell **B. Photo excitation** **C. Light cascade**

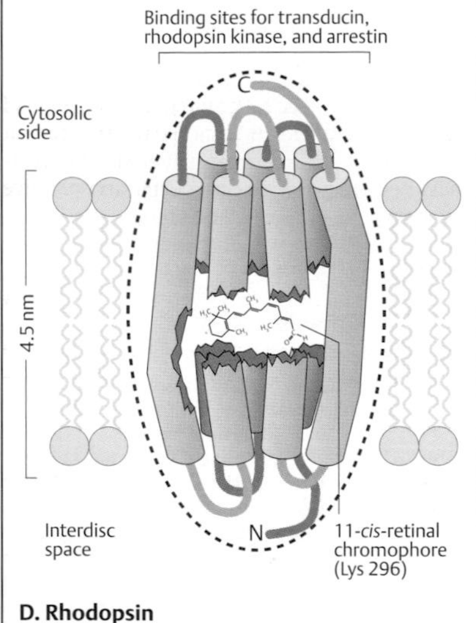

D. Rhodopsin

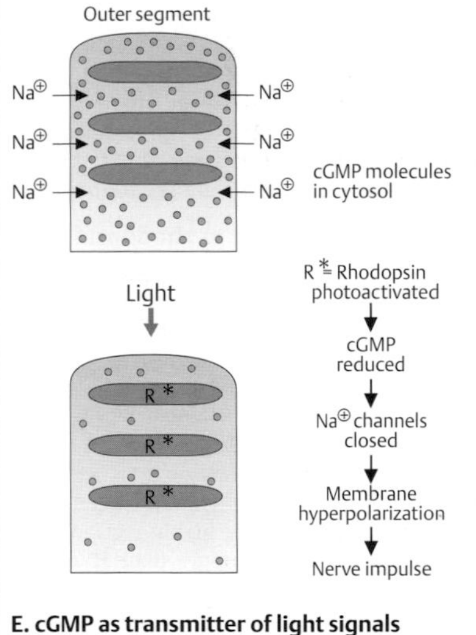

E. cGMP as transmitter of light signals

Pigmentary Retinal Degeneration

Pigmentary degeneration of the retina (retinitis pigmentosa, RP) is a large group of eye diseases (see OMIM 268000 for overview) causing progressive blindness in about 1 in 3–5000 children and adults. Mutations in about 100 gene loci involved in the light cascade signal pathway are causes of retinal degeneration. Aside from rhodopsin these include peripherin, cAMP phosphodiesterase, and others. Retinal degeneration may occur alone (nonsyndromic) or as part of a systemic disorder involving other organ systems (syndromic forms, see OMIM 276900, Usher syndromes).

A. Retinitis pigmentosa

Affected individuals first lose midperipheral vision; then the visual field is progressively reduced to a small island of central vision (tunnel vision). This is associated with the inability to perceive weak light (night blindness). The fundus of the eye shows thin retinal vessels, a pale, waxy yellow optic nerve, and multiple areas of irregular hyperpigmentation and depigmentation. About 30%–40% of cases of RP are the result of autosomal dominant inheritance determined at 40 gene loci (see OMIM 268000). About 50%–60% of cases of RP are inherited as autosomal recessive traits at about 60 genes and additional gene loci (see OMIM 268000 for overview). About 5%–15% of cases of RP result from X-linked inheritance (OMIM 268000). However, in a single case without a family history, the mode of inheritance may remain uncertain. (Photograph kindly provided by Professor E. Zrenner, Tübingen, Germany.)

B. The first mutation in rhodopsin

The first mutation of rhodopsin was described by Dryja et al. (1990). It is a transversion of cytosine (C) to adenine (A) in codon 23 of exon 1. This causes an exchange of a proline for a histidine in codon 23 (CCC for proline to CAC for histidine, designated as mutation P23H). The partial DNA sequence in the banding pattern shows the mutation as an additional band representing the adenine in codon 23. Proline in position 23 is highly conserved in evolution and occurs in more than 10 related G-protein receptors.

C. Mutations in rhodopsin

The gene for human rhodopsin (*RHO*; OMIM 180380) has five exons and spans 5 kb. It is located on the long arm of chromosome 3 in region 2, band 1.4 (3q21.4) and encodes a transmembrane protein of 348 amino acids. Thirty-eight amino acids are invariant in vertebrates. The figure shows the distribution of selected mutations in the rhodopsin molecule. Most are missense mutations leading to an exchange of one amino acid by another. Rearrangements involving a few base pairs cause small deletions.

D. Simplified diagnosis of a mutation

The figure shows an example of a simplified indirect DNA diagnosis based on prior knowledge of the mutation in a given family. Here 13 individuals in three generations are affected (**1**). These are shown by dark shading; unaffected individuals are shown by no shading; males are shown as squares; and females are shown as circles. The mutation is the same (P23H) as shown in **B**. By using allele-specific oligonucleotides and the polymerase chain reaction (**2**), individuals carrying the mutation can be distinguished from those who do not. The oligonucleotide corresponding to the mutant allele has the sequence 3′-CATGAGCTT-CACCGACGCA-5′ (the mutation is the presence of adenine [A] instead of cytosine; compare with the sequence of codons 26–21 in **B**). This mutant allele-specific oligomer hybridizes with the mutant allele only, not with the normal allele. Thus, all affected individuals shown in the pedigree exhibit a hybridization signal, whereas the unaffected individuals do not (II-2, II-12, and III-4 were not examined). (Data from Dryja et al., 1990.)

Further Reading

Dryja TP. Retinitis pigmentosa. In: Scriver CR, et al., eds. The Metabolic and Molecular Bases of Inherited Disease. 8th ed. New York: McGraw-Hill; 2001:5903–5933. (www.ommbid.com)

Dryja TP, et al. A point mutation of the rhodopsin gene in one form of retinitis pigmentosa. Nature 1990;343:364–366

Hartong DT, Berson EL, Dryja TP. Retinitis pigmentosa. Lancet 2006;368:1795–1809

Sheffield VC, Stone EM. Genomics and the eye. N Engl J Med 2011;364:1932–1942

A group of hereditary diseases with degeneration of photoreceptors in the retina

Night blindness

Progressive loss of vision

Frequency about 1:3500

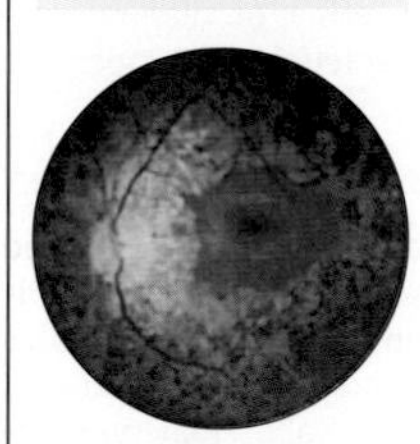

Typical fundus with pigment changes, narrow vessels, and pale, waxy optical nerve

Frequency of the different genetic forms

30–40% autosomal dominant
50–60% autosomal recessive
8% X-chromosomal

Important diagnostic signs

Fundus:
narrow vessels
pale optic nerve
macula changes
widened light reflex
pigment epithelium changes
electroretinogram extinguished

Secondary changes in the anterior chamber:
vitreous body changes

Cataract
Myopia

A. Retinitis pigmentosa

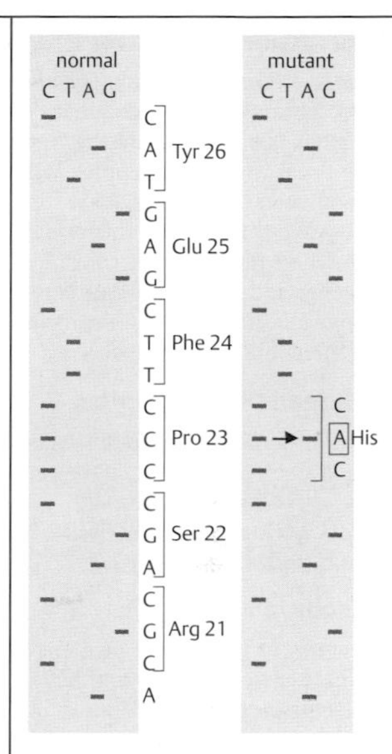

B. Mutation in rhodopsin

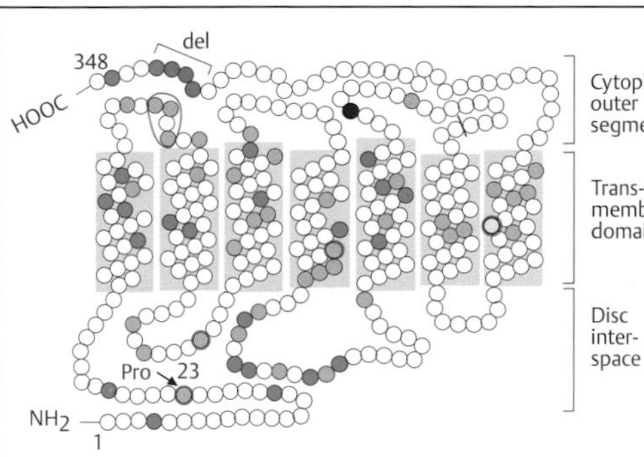

= Invariant amino acid in different vertebrates
= Retinal binding site
= Dominant mutation
= Small deletions
= Autosomal recessive form
= gt → tt Intron 4-donor splice site mutation

C. Mutations in rhodopsin

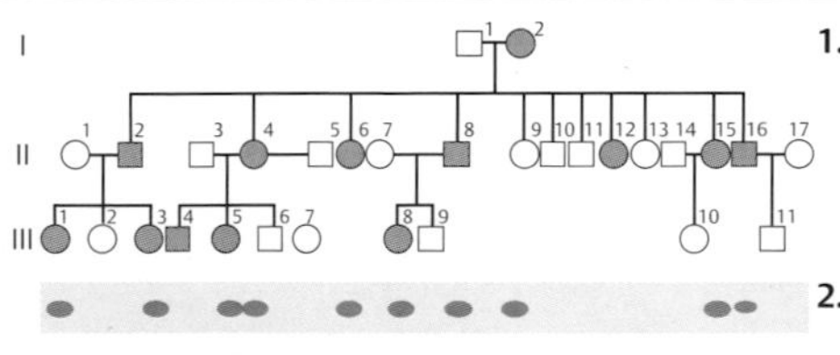

1. 1. Pedigree with autosomal dominant retinitis pigmentosa due to mutation in codon 23 (P23H)

2. 2. Autoradiogram of hybridization of amplified DNA fragments in codon 23 with oligomer 3'-CATGAGCTTCACCGACGCA-5' for the mutant sequence

D. Demonstration of mutation P23H in codon 23 by oligonucleotides after PCR

Color Vision

In 1802, Thomas Young suggested that color vision in humans is trichromatic: red, green, and blue. Cone cells in the retina contain three types of photoreceptors, which respond to long-wave (red), middle-wave (green), or short-wave (blue) light. This was confirmed in 1986 when Nathans and coworkers identified the genes and molecular structure of the color receptors. X-linked red–green color vision defects affect about 8% of males.

A. Human photoreceptors

The overlapping absorption spectra of the three classes of receptors have peaks of sensitivity at 420 nm wavelength (blue; OMIM 613522), 530 nm (green; OMIM 300821) and 560 nm (red; OMIM 300822).

B. Evolution

The genes for the visual pigment photoreceptors have arisen in evolution by duplication of an ancestral gene encoding a light-sensitive protein. About 800 million years ago (mya), an ancestral visual pigment diverged by a duplication event into the rod pigment rhodopsin and another, not yet differentiated, cone pigment. A short-wavelength-responsive (blue) pigment gene and a single mid-wavelength (green–red) gene diverged about 500 million years ago following another duplication event. The split between New World monkeys with a single X-linked receptor and Old World monkeys with two X-linked receptors occurred about 30–40 million years ago when the single green-red pigment gene duplicated. The two copies then evolved into the green and the red pigment genes. Therefore, Old World monkeys and humans have tricolor vision, whereas New World monkeys have dichromatic (blue and green–red) vision.

C. Structural similarity

Four photopigments with identical structures exist as heptahelical (seven helices) transmembrane proteins with similar sequences of amino acids. Nathans et al. (1986) determined the percentage of identical amino acids shown in the figure in pairwise arrangements. Open circles indicate invariant amino acids present in all; dark circles indicate differences in amino acid sequences. (Figure adapted from Nathans et al., 1986.)

D. Polymorphism in the photoreceptor

Subtle differences in red light perception were detected by Motulsky and coworkers (Winderickx et al., 1992). They are due to polymorphic variants at three sites of the receptor (**1**): serine/alanine at amino acid position 180, isoleucine/threonine at position 230, and alanine/serine at position 233 (**2**). The distribution of red light perception showed a difference for the red/red and green mix distribution, depending on whether serine or alanine was present (**3**). (Figure adapted from Winderickx et al., 1992.)

E. Defects in color vision

Defects in color vision can affect any of the three photopigments. The receptor for blue is encoded by an autosomal gene at 7q31.3-q32, the red and green receptors are encoded by X-linked genes on human Xq28. Two types of photoreceptors are involved in defective color vision: blue plus green in protanopia (OMIM 303900) or blue plus red in deuteranopia (OMIM 300822). The tandem arrangement of the genes encoding the red and the green receptors (**1**) predispose to color vision defects. Owing to the sequence similarity, unequal crossing-over occurs frequently in the intergenic region (about 15 kb), which results in deletion/duplication and different degrees of red–green vision defects (**2**) (OMIM 303800). Other vision defects involve the blue pigment (tritanopia; OMIM 190900) or cause complete loss of color vision (achromatopsia; OMIM 216900). Inherited defects in color vision in humans are mentioned in the Talmud.

Further Reading

Motulsky AG, Deeb SS. Color vision and its genetic defects. In: Scriver CR, et al., eds. The Metabolic and Molecular Bases of Inherited Disease. 8th ed. New York: McGraw-Hill; 2001: 5955–5976. (www.ommbid.com)

Nathans J, Thomas D, Hogness DS. Molecular genetics of human color vision: the genes encoding blue, green, and red pigments. Science 1986;232:193–202

Winderickx J, et al. Polymorphism in red photopigment underlies variation in colour matching. Nature 1992;35:431–433

Young T. On the theory of light and colours. Philos Trans R Soc Lond 1802;92:12–48

Autosomal X-chromosomal

Blue Green Red

420 530 560

Absorption

500 600

Wavelength (nm)

A. Photoreceptor proteins in rods

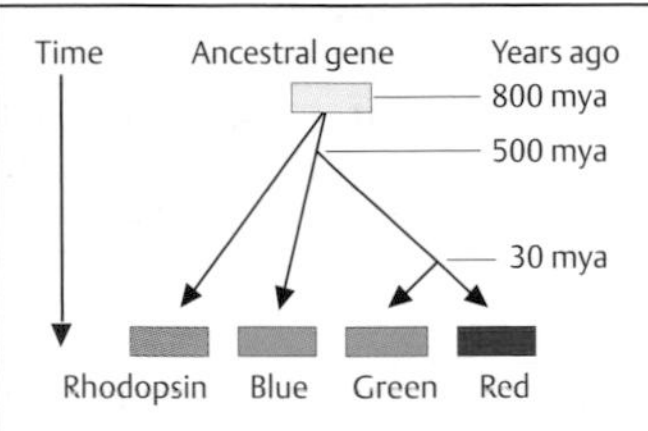

B. Evolution of genes for visual pigment photoreceptors

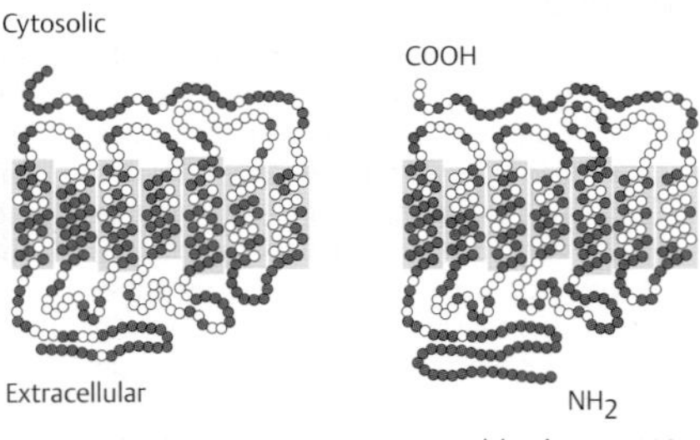

1. Blue/rhodopsin 75% **2.** Green/rhodopsin 41%

3. Green/blue 44% **4.** Green/red 96%

C. Similar structure of visual pigments

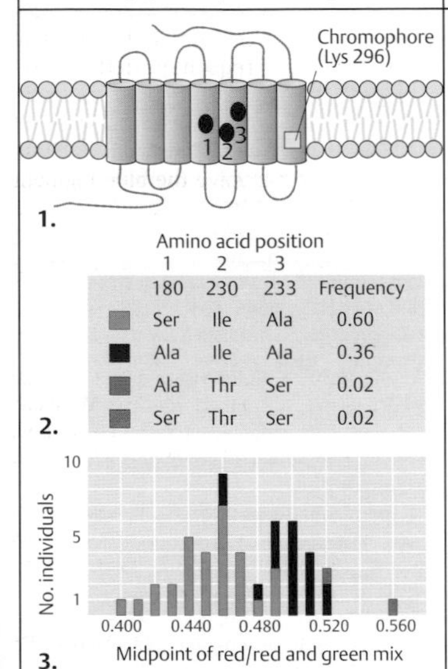

	1 180	2 230	3 233	Frequency
■	Ser	Ile	Ala	0.60
■	Ala	Ile	Ala	0.36
■	Ala	Thr	Ser	0.02
■	Ser	Thr	Ser	0.02

D. Polymorphism in the photoreceptor for red

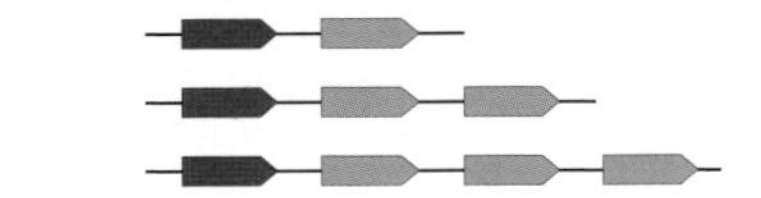

1. Normal arrangement of red and green genes

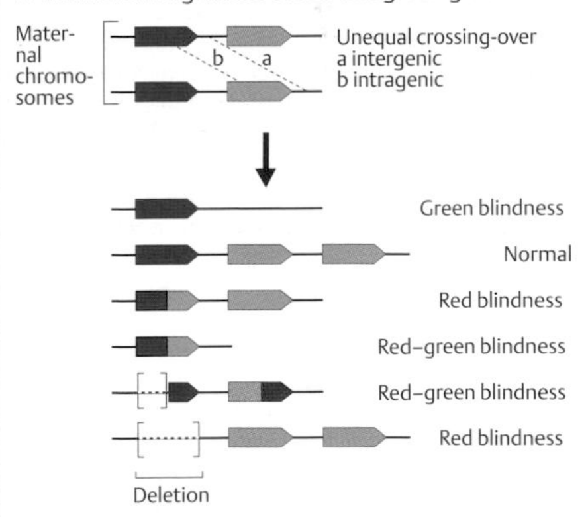

2. Examples of different consequences of unequal crossing-over

E. Normal and abnormal red/green vision

Auditory System

About 1 in 1000 individuals is affected by severe hearing impairment or deafness at birth or in early childhood (prelingual deafness, before development of speech). In 30% of the prelingual forms, other organ systems are involved in addition to the ear (syndromic forms with hundreds of different types, see 11 genes in Usher syndromes; OMIM 276900). About 95% of deaf children are born to parents with normal hearing. Normal hearing is orchestrated by a diverse ensemble of proteins acting in concert. Specialized sensory cells in the inner ear process the incoming sound waves and convert them into nerve impulses to the brain.

A. The main components of the ear

Sound waves induce vibrations in the tympanic membrane, and these are transmitted from the middle ear to the inner ear by a chain of three small movable bones (malleus, incus, and stapes). In the cochlea, the organ of Corti, the acoustic signals are amplified and processed (auditory pathway). In addition, the inner ear harbors the vestibule with the three semicircular canals, utricle, and saccule, where the sense of equilibrium is regulated.

B. The cochlea

The cochlea is a snail-shaped structure that contains three fluid-filled canals: the scala vestibuli, the scala media, and the scala tympani. Between the endolymph of the scala media and the perilymph of the scala vestibuli and scala tympani, there is an electrical potential difference of about +85 mV. Potassium ions (K^+) secreted from the stria vascularis of the scala media into the endolymph are recycled through supporting cells by K^+ channels and gap junctions (e.g., connexin 26 [Cx26]). The cochlea contains two types of sensory cell: one row of inner hair cells and three rows of outer hair cells. Sound-induced vibrations of the tectorial membrane deflect the stereocilia, open mechanosensitive channels, and cause an influx of K^+. The altered membrane potential elicits an impulse at the acoustic nerve, and this is transmitted to the auditory cortex of the brain. (Figure adapted from Kubisch, 2005.)

C. Outer hair cells

The stereocilia of about 50000 outer hair cells (**1**) are arranged in arrays resembling an organ pipe. Their tips are connected by tip links containing myosin and the cell adhesion molecule cadherin-23. The regular arrangement of the outer hair cells (**2**, top) is disrupted by mutations affecting the cytoskeleton (**2**, bottom). (**2**, Adapted from Kubisch, 2005.)

D. Congenital deafness

The more than 100 genes involved in hereditary deafness are localized on nearly every human chromosome (see OMIM 220290, phenotypic series). About 75%–85% of mutations are autosomal recessive (gene loci designated DFNB1–DFNB84 in 2011), 15% autosomal dominant (DFNA1–DFNA51), and 1%–2% X-chromosomal (DFN1–DFN4). The more than 50 identified genes encode a variety of proteins of the acoustic pathway (see Appendix, Table p. 421). The most frequent mutations are in connexin 26 (Cx26/GJB2; OMIM 121011: DFNA3, DFN; B1). Individuals carrying certain mtDNA mutations are highly sensitive to antibiotics of the aminoglycoside class (e.g., streptomycin). Seven genes (*OTC*) are involved in otosclerosis (see OMIM 166800, phenotypic series).

Further Reading

Dodson KM, et al. Vestibular dysfunction in DFNB1 deafness. Am J Med Genet A 2011;155A:993–1000

Hereditary Hearing Loss. (http://hereditaryhearingloss.org/)

Hilgert N, Smith RJ, Van Camp G. Function and expression pattern of nonsyndromic deafness genes. Curr Mol Med 2009;9:546–564

Kubisch C. Genetische Grundlagen nichtsyndromaler Hörstörungen. Dtsch Arztebl 2005;102A:2946–2952

Lalwani AK. Disorders of hearing loss. In: Longo DL, et al., eds. Harrison's Principles of Internal Medicine. 18th ed. New York: McGraw-Hill Medical; 2012:248–255

Petit C, et al. Hereditary hearing loss. In: Scriver CR, et al., eds. The Metabolic and Molecular Bases of Inherited Disease. 8th ed. New York: McGraw-Hill; 2001:6281–6328. (www.ommbid.com)

Smith RJH, Bale JFJr, White KR. Sensorineural hearing loss in children. Lancet 2005;365:879–890

Toriello H, Reardon W, Gorlin RJ, eds. Hereditary Hearing Loss and its Syndromes. 2nd ed. Oxford: Oxford University Press; 2004

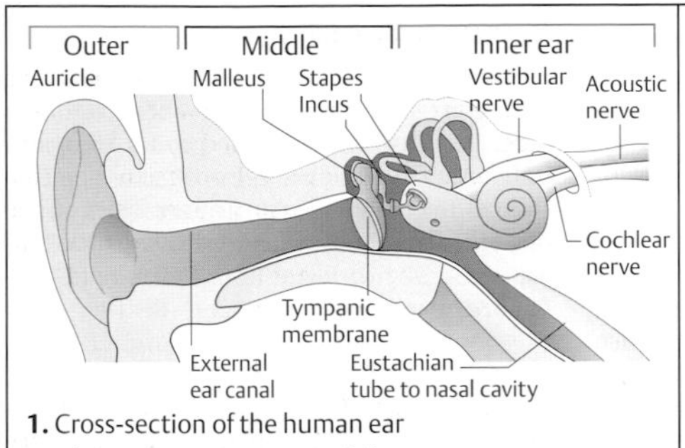

1. Cross-section of the human ear

A. The main components of the ear

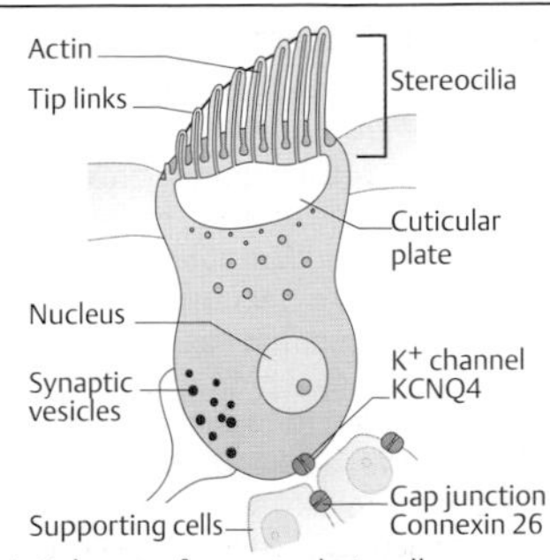

1. Scheme of an outer hair cell

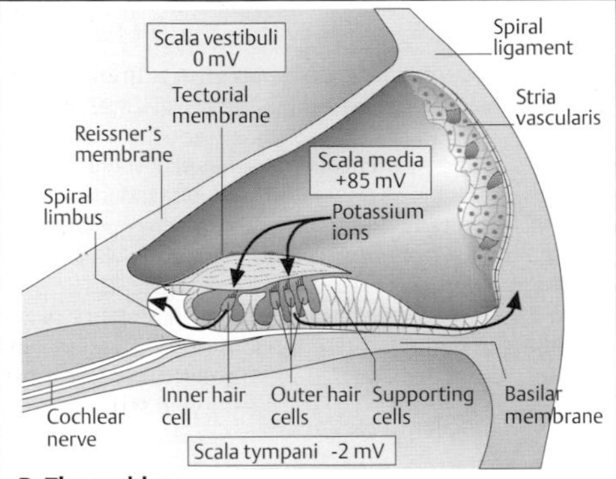

B. The cochlea

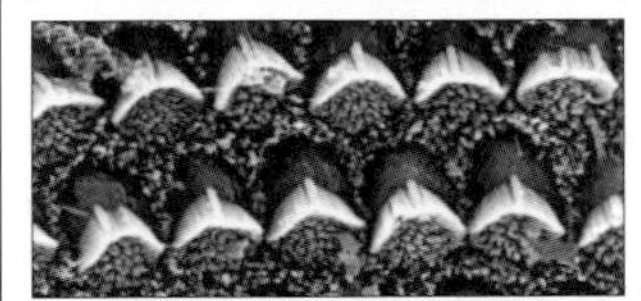

▲ Normal Myosin 7A mutation ▼

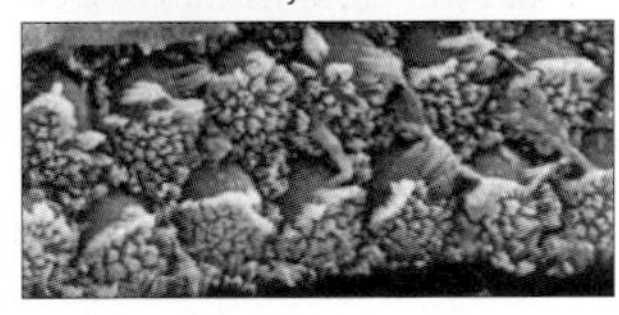

2. Stereocilia of outer hair cells in mice

C. Outer hair cells

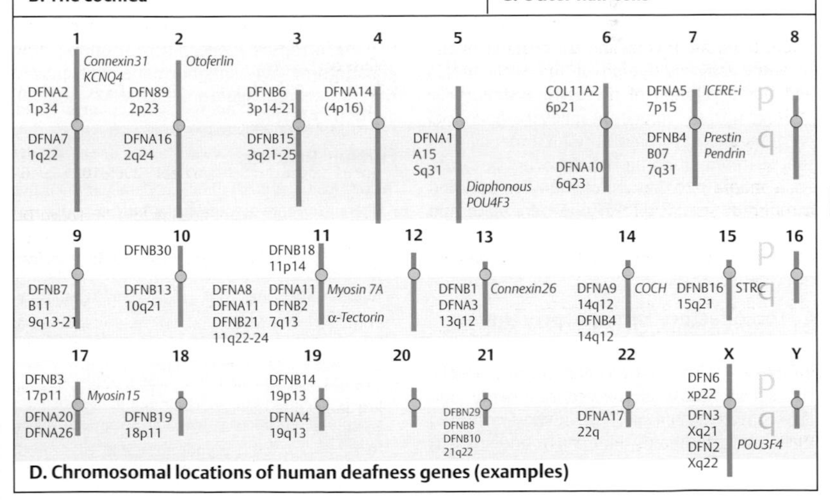

D. Chromosomal locations of human deafness genes (examples)

Odorant Receptors

Vertebrates are able to differentiate thousands of individual odors by means of specific receptors on the cilia of olfactory neurons (odorant receptors, ORs). OR genes have arisen in evolution by many duplication events. Genes of the OR family form the largest family of genes known in mammals, and account for about 3%–4% of all genes. The mammalian genome contains about 1000 OR genes, while the fish genome contains about 100. The rat genome contains 1866 ORs in 113 locations, consisting of 65 multigene clusters and 44 single genes. In humans about 60% of the genes are pseudogenes.

A. Sensory olfactory nerve cells

The peripheral olfactory neuroepithelium of the nasal mucous membrane consists of three cell types: olfactory sensory neurons with axons leading to the olfactory bulb, supporting cells, and basal cells. The latter serve as stem cells that replace olfactory sensory neurons. Each olfactory neuron is bipolar, with olfactory cilia in the lumen of the nasal mucous membrane and a projection to the olfactory bulb. From there, odorant-induced signals are transmitted via the olfactory nerve to the brain.

B. Odorant-specific receptor

The odorant-specific receptor is a GTP-binding protein with a specific stimulatory α-subunit, the G_{olf}. Binding of the odorant ligand to the receptor activates G_{olf}, which in turn activates adenylate cyclase. The increase in cAMP (cyclic 3′,5′-adenosine monophosphate) opens a cAMP-gated ion channel, which depolarizes the cell membrane and elicits a nerve signal. Each receptor in the cilia of the olfactory neurons binds specifically to one odorant ligand only. Thus, signal amplification in olfaction is fundamentally different from phototransduction.

C. The olfactory receptor protein

The odorant receptor is a typical seven transmembrane G-protein-coupled protein. Unlike rhodopsin, the OR proteins contain many variable amino acids, especially in the fourth and fifth transmembrane domains, and this is likely to be related to their function.

D. Exclusive gene expression

Only one allele of an OR gene is expressed and each gene is expressed in a few olfactory neurons only. Receptor-specific probes recognize only very few neurons in the olfactory epithelium of the catfish *(Ictalurus punctatus)*: probe 202 hybridizes to two neurons (two black dots, **1**); probe 32 hybridizes to one neuron (**2**). Olfactory neurons are randomly distributed in the periphery, but their axons project to defined locations in the olfactory bulb. Odors are distinguished in the brain according to which neurons are stimulated. (Figure adapted from Ngai et al., 1993.)

E. Subfamilies within the OR family

Olfactory receptor genes form a large family of related genes. Amino acid sequences derived from partial nucleotide sequences of cDNA clones (F2–F24) are variable, especially in transmembrane domains III and IV (**1**). Within a subfamily, considerable amino acid sequence identity is present. For example, families F12 and F13 differ in only 4 of 44 positions (**2**). This reflects the ability to distinguish a great number of slightly different odorants. (Figures adapted from Buck & Axel, 1991, and Ngai et al., 1993.)

Medical relevance

Olfactory dysfunction affects about 1% of the population below the age of 60 years. Anosmia (lack of ability to smell) is associated with hypogonadotropic hypogonadism resulting from gonadotropin-releasing hormone deficiency in the Kallmann syndrome (OMIM 308700; 147950; 244200).

Further Reading

Buck L, Axel R. A novel multigene family may encode odorant receptors: a molecular basis for odor recognition. Cell 1991;65:175–187

Doty RL, Bromley SM. Disorders of smell and taste. In: Longo DL, et al., eds. Harrison's Principles of Internal Medicine. 18th ed. New York: McGraw-Hill; 2012:241–247

Emes RD, et al. Evolution and comparative genomics of odorant- and pheromone-associated genes in rodents. Genome Res 2004;14:591–602

Keller A, et al. Genetic variation in a human odorant receptor alters odour perception. Nature 2007;449: 468–472

Ngai J, et al. The family of genes encoding odorant receptors in the channel catfish. Cell 1993;72:657–666

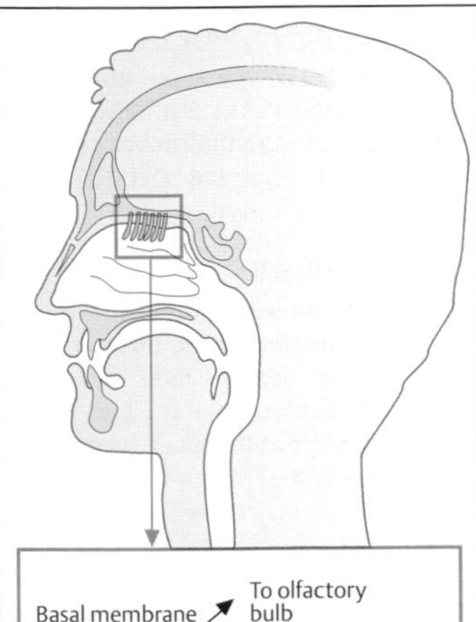

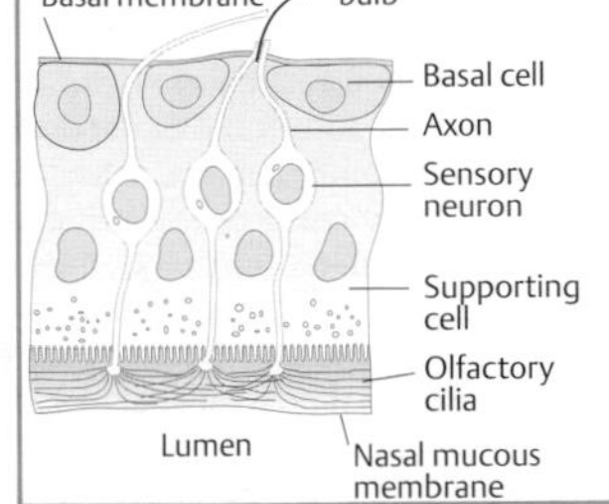

A. Olfactory nerve cells in the nasal mucous membrane

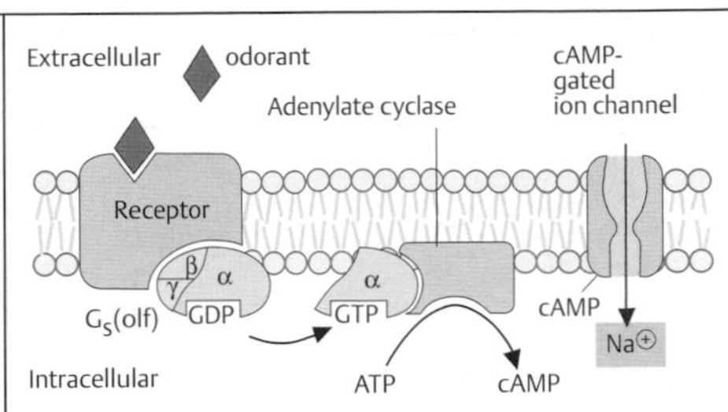

B. Odorant-specific transmembrane receptors

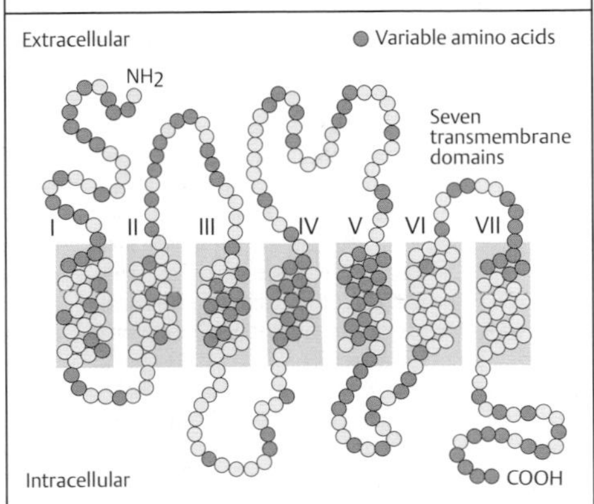

C. The olfactory receptor protein

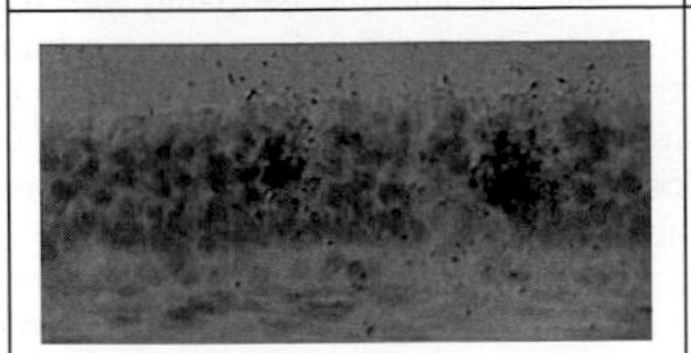

1. Receptor probe 202

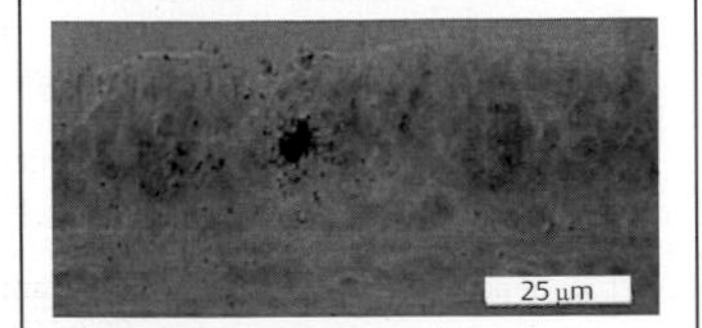

2. Receptor probe 32

D. Exclusive expression of OR genes

cDNA clones		Transmembrane domain	
F2	RVNE	VVIFIVVSLFLVLPFALIIMSYV	RIVSSILKVPSSQGIYK
F3	FLND	LVIYFTLVLLATVPLAGIFYSYF	KIVSSICAISSVHGKYK
F5	HLNE	LMILTEGAVVMVTPFVCILISYI	HITCAVLRVSSPRGGWK
F6	QVVE	LVSFGIAFCVILGSCGITLVSYA	YIITTIIKIPSARGRHR
F7	HVNE	LVIFVMGGIILVIPFVLIIVSYV	RIVSSILKVPSARGIRK
F8	FPSH	LTMHLVPVILAAISLSGILYSYF	KIVSSIRSMSSVQGKYK
F12	FPSH	LIMNLVPVMLAAISFSGILYSYF	KIVSSIHSISTVQGKYK
F13	FPSH	LIMNLVPVMLAAISFSGILYSYF	KIVSSIRSVSSVKGKYK
F23	FLND	VIMYFALVLLAVVPLLGILYSYS	KIVSSIRAISTVQGKYK
F24	HEIE	MIILVLAAFNLISSLLVVLVSYL	FILIAILRMNSAEGRRK

1. Variable amino acid sequences

F12	FPSH	LIMNLVPVMLAAIISFSGILYSYF	KIVSSIHSISTVQGKYK
F13	FPSH	LIMNLVPVMLAAIISFSGILYSYF	KIVSSIRSVSSVKGKYK
F8	FPSH	LTMHLVPVILAAIISLSGILYSYF	KIVSSIRSMSSVQGKYK
I12	FPSH	LIMNLVPVMLGAIISLSGILYSYF	KIVSSVRSISSVQGKHK
F23	FLND	VIMYFALVLLAVVVPLLGILYSYS	KIVSSIRAISTVQGKYK
F3	FLND	LVIYFTLVLLATVVPLAGIFYSYF	KIVSSICAISSVHGKYK

2. Homology within subfamilies

E. Subfamilies within the multigene family

Mammalian Taste Receptors

Humans can distinguish five types of taste: sweet, sour, salt, bitter, and umani (the taste of monosodium glutamate, present in Asian food). Hydrogen ions (H^+) of acids are responsible for sour taste. The salty taste is the result of direct influx of sodium ions (Na^+) from water-soluble salts. In contrast, bitter, sweet, and umani taste perception is mediated by families of G-protein-coupled receptor (GPCR) signaling pathway systems. Bitter taste, especially, is perceived at very low concentrations.

A. Mammalian chemosensory epithelia

The oral and nasal cavities of mammals contain three distinct chemosensory epithelia: (i) the main olfactory epithelium (MOE) containing sensory cells with odorant receptors in the nose (see p. 378); (ii) the taste sensory epithelium of the taste buds of the tongue, soft palate, and epiglottis; and (iii) the vomeronasal organ (VOM, also called Jacobson's organ). The latter is a tubular structure in the nasal septum containing sensory cells with pheromone receptors. The main olfactory bulb (MOB) relays signals from the MOE to the olfactory cortex of the brain. The accessory olfactory bulb (AOB) relays signals from the VOM to areas of the amygdala and hypothalamus.

B. Chemosensory signal transduction

Likewise the mammalian chemosensory receptor cells also belong to three classes of transduction systems: (i) the olfactory system, (ii) the sensory taste system, and (iii) the vomeronasal system, which detects signals arriving as pheromones. A pheromone is a secreted or excreted chemical substance that even in low concentration induces a behavioral response within the same species.

Each neuron of the main olfactory sensory system (**1**) sends an axon to specific glomeruli (mitral cells), and from there to the the main olfactory bulb (MOB) and the olfactory nerve. The odorant receptor (OR) gene family comprises about 1000 members, each encoding a seven-transmembrane cyclic nucleotide-gated channel with distinct odorant specificity (G-olfactory proteins, G_{olf}). The bitter taste sensory system (**2**) connects axonal projections of receptor cells in the taste sensory epithelium of the taste buds to gustatory nuclei in the brain stem. Two families of taste receptors exist, the TIRs (two genes) and T2Rs (50–80 genes of the gustducin class). The putative mammalian pheromone receptors (V1Rs and V2Rs), located in the vomeronasal organ, form two families, encoded by 30–50 genes and over 100 genes, respectively (**3**).

C. Taste receptor gene family

A small family of three G-protein-coupled receptors (T1R1, T1R2, and T1R3) mediate sweet and umani taste perception. Bitter perception is mediated by T2R receptors. Here marked amino acid sequence differences (light fields) between the second (TM2) and third (TM3) transmembrane domains of the predicted amino acid sequences are visible for 23 different T2 receptors (T2Rs) of human (h), rat (r), and mouse (m) origin. In contrast, conserved sequences (dark blue) indicate identity in at least half of the aligned sequences; light blue represents conserved substitutions; and the remaining are divergent regions. The T2R genes cluster on a few chromosomes: human chromosomes 5, 7, and 12, and mouse chromosomes 6 and 15.

D. Expression pattern

Unlike receptor cells of the olfactory system, individual taste receptor cells express multiple receptors (T2Rs). Up to 10 T2R probes hybridize to just a few cells, shown as darkened areas (**1**). Double-label fluorescence in-situ hybridization shows that different receptor genes are expressed in the same taste receptor cell, shown in green for T2R-7 (**2**) and red for T2R-3 (**3**). The T2Rs confer high sensitivity for bitter substances at low concentrations.

(Figures adapted from Adler et al., 2000, and Dulac, 2000.)

Further Reading

Adler E, et al. A novel family of mammalian taste receptors. Cell 2000;100:693–702

Buck LB. The molecular architecture of odor and pheromone sensing in mammals. Cell 2000;100: 611–618

Chandrashekar J, et al. T2Rs function as bitter taste receptors. Cell 2000;100:703–711

Doty RL, Bromley SM. Disorders of smell and taste. In: Longo DL, et al., eds. Harrison's Principles of Internal Medicine. 18th ed. New York: McGraw-Hill; 2012:241–247

Dulac C. The physiology of taste, vintage 2000. Cell 2000;100:607–610

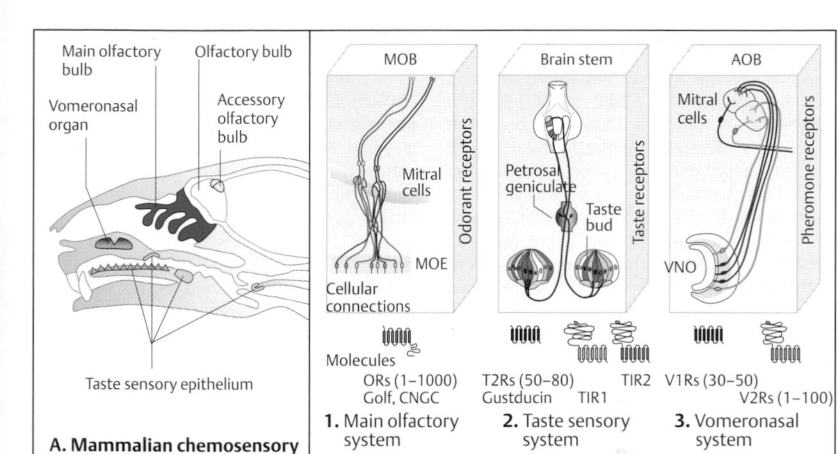

A. Mammalian chemosensory epithelia

B. Chemosensory signal transduction systems

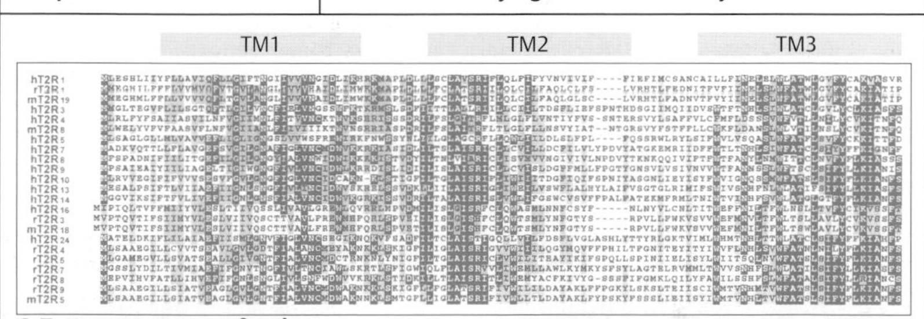

C. Taste receptor gene family

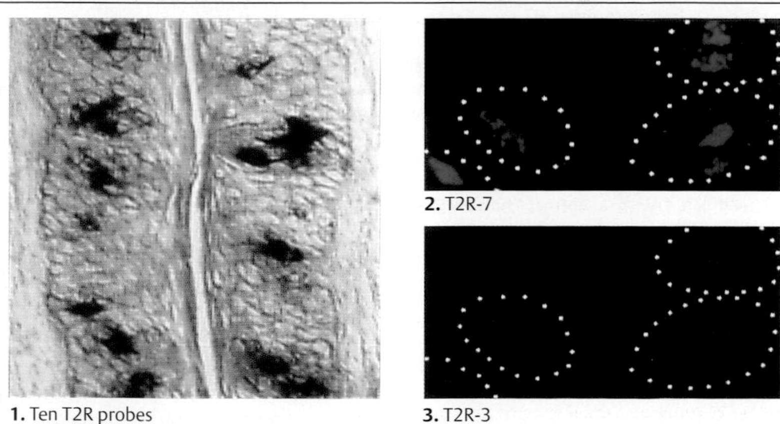

1. Ten T2R probes

2. T2R-7

3. T2R-3

D. Expression pattern of taste receptor genes

Autosomal Trisomies

Three autosomal trisomy syndromes occur in liveborn infants: trisomy 21, trisomy 18, and trisomy 13. Trisomy for all other chomosomes is not compatible with prenatal life until birth and results in spontaneous abortion in the first or second trimester. Trisomy arises during meiosis I or II (see p. 102) by nondisjunction as a *prezygotic* event. It may, less commonly, occur after fertilization as a *postzygotic* event in the early embryo during a somatic cell division (mitosis). In this case the additional chromosome is present only in a certain proportion of cells. This condition is called *chromosomal mosaicism.*

A. Trisomy in *Datura stramonium*

The phenotypic effects of trisomy were first discovered in a plant by A. F. Blakeslee in 1922. He studied jimsonweed plants, also called common thorn apple (*Datura stramonium*), and observed that plants with three copies of one of the 12 chromosomes had a characteristic appearance that was specific for each of the 12 chromosomes of this plant. (Figure adapted from Blakeslee, 1922.)

B. Trisomies in the mouse

Specific phenotypic effects of autosomal trisomies can also be observed in mice. In the 1970s, A. Gropp and coworkers determined a specific developmental profile of morphological changes and malformations for each trisomy in the mouse (**1**). Embryos with a monosomy died very early, within the first 8 days of the 21-day gestation. The intrauterine survival time depended on the chromosome that was trisomic. The examples show a mouse embryo with trisomy 12 (**2**) and the brain of a mouse with trisomy 19 at birth (**3**), each compared with a normal control. Only trisomy 19 is compatible with survival until birth, but the brain is too small. The trisomic mice were generated by breeding mice with translocations among various chromosomes. (**1**, Gropp, 1982; **2** and **3**, courtesy of Dr H. Winking, Lübeck, Germany.)

C. Autosomal trisomies in man

The three autosomal trisomies are associated with a distinctive phenotype and occur with different frequencies. Each has a distinct pattern of congenital malformations associated with variable degrees of mental impairment in trisomy 21 (Down syndrome; OMIM 19068)) and complete lack of mental development in trisomies 18 and 13. Only trisomy 21 is compatible with survival into adulthood, although the overall life expectancy is about half of that of the normal population (Table 20, Appendix p. 422).

D. Nondisjunction as a cause

All three human trisomies occur more frequently with advanced maternal age (**1**). The age of the father has no or very little influence. If nondisjunction occurs in meiosis I, the three chromosomes will be different (1 + 1 + 1), whereas if nondisjunction occurs during meiosis II, two of the three chromosomes will be identical (2 + 1). In humans, about 70% of nondisjunctions occur in meiosis I, and 30% occur in meiosis II. In trisomy 21, the nondisjunction occurs in 90% in maternal meiosis (in meiosis I in 73%, in meiosis II in 25%). The figures for trisomy 18 and 13 are similar.

Further Reading

Blakeslee AF. Variation in Datura due to changes in chromosome number. Am Nat 1922;56:16–31

Boué A, Boué J, Gropp A. Cytogenetics of pregnancy wastage. Adv Hum Genet 1985;14:1–57

Epstein CJ. Down syndrome (trisomy 21). In: Scriver CR, et al., eds. The Metabolic and Molecular Bases of Inherited Disease. 8th ed. New York, NY: McGraw-Hill; 2001:1223–1256. (www.ommbid.com)

Gropp A. Value of an animal model for trisomy. Virchows Arch A Pathol Anat Histol 1982;395:117–131

Miller OJ, Therman E. Human Chromosomes. 4th ed. New York: Springer-Verlag; 2001

Roizen NJ, Patterson D. Down's syndrome. Lancet 2003;361:1281–1289

Schwartz S, Hassold T. Chromosome disorders. In: Harrison's Principle of Internal Medicine. 18th ed. Longo DL, et al., eds. New York: McGraw-Hill; 2012:509–518

Tolmie JJ, MacFaden U. Clinical genetics of common autosomal trisomies. In: Rimoin DL, et al., eds. Emery and Rimoins's Principles and Practice of Medical Genetics. 5th ed. Philadelphia: Churchill Livingstone-Elsevier; 2007:1015–1037

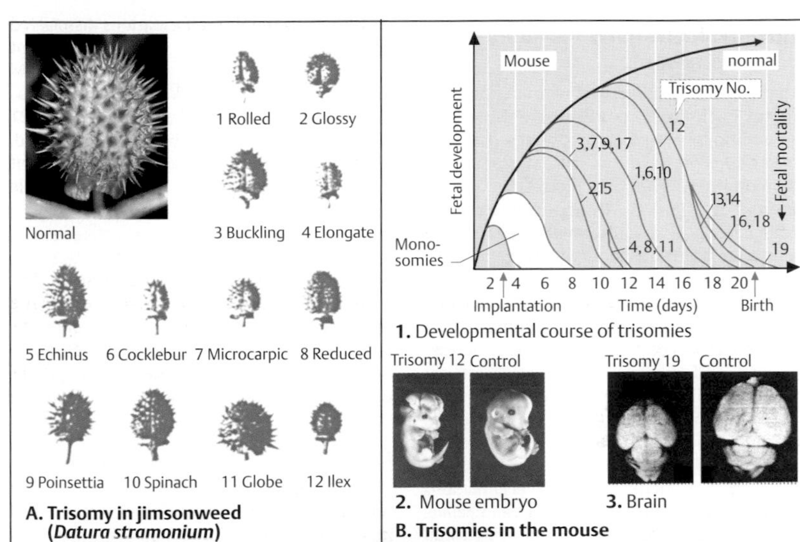

A. Trisomy in jimsonweed (*Datura stramonium*)

1. Developmental course of trisomies

2. Mouse embryo

3. Brain

B. Trisomies in the mouse

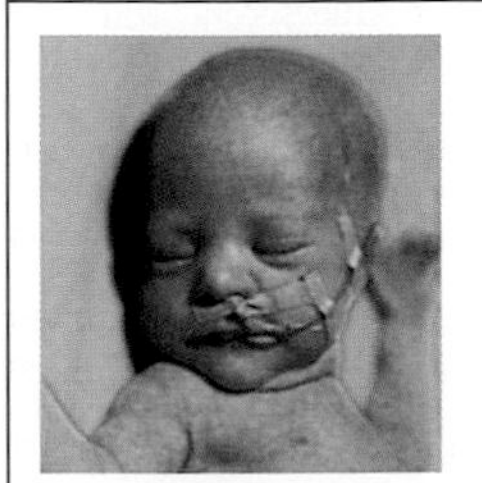

1. Trisomy 21

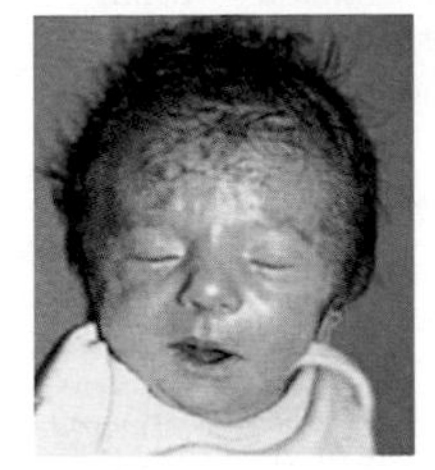

2. Trisomy 18

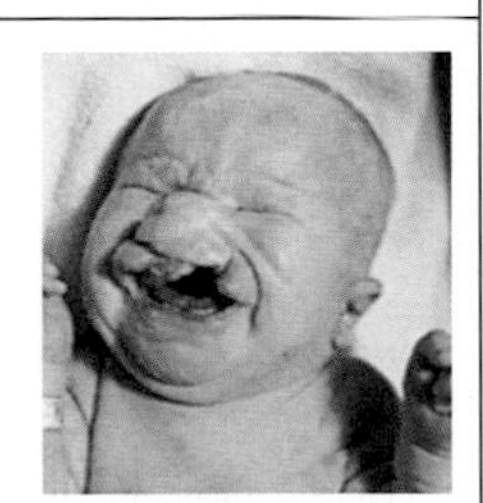

3. Trisomy 13

C. Autosomal trisomies in man

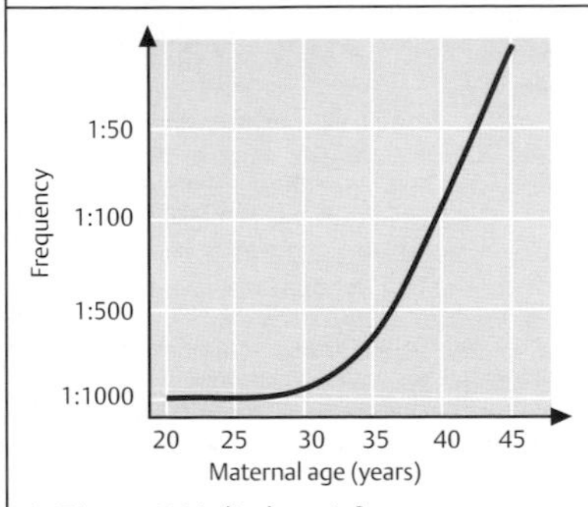

1. Trisomy 21 in liveborn infants

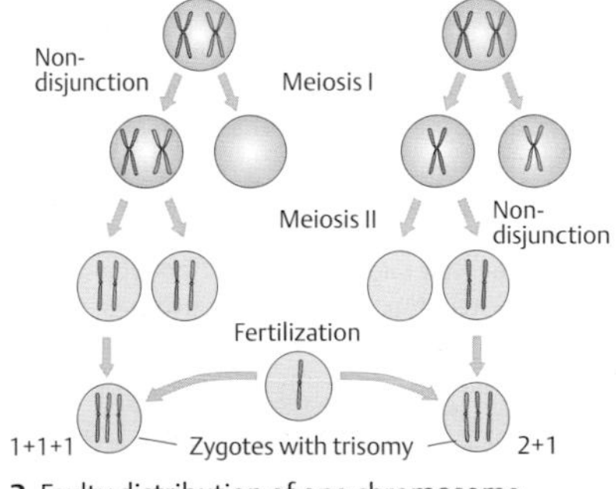

2. Faulty distribution of one chromosome

D. Nondisjunction as cause of trisomy

Triploidy, Monosomy X, Additional X or Y Chromosome

Other conditions associated with an abnormal number of chromosomes involve either an entire additional set of chromosomes (triploidy or tetraploidy) or the X chromosome or Y chromosome. Deviations from the normal number of X or Y chromosome comprise about half of all chromosomal aberrations in man (total frequency about 1:400).

A. Triploidy

In triploidy either one paternal and two maternal sets (karyotype 69,XXY or 69,XXX) or one maternal and two paternal sets are present (69,XXX, 69,XYY or 69,XXY). Triploidy is associated with severe developmental failure and congenital malformations (**1**), which are not compatible with prenatal life until birth. The fetus shows numerous severe malformations (**2**), such as cardiac defects, cleft lip and palate, skeletal defects, and others. Triploidy (**3**) accounts for about 17% of spontaneous abortions. The causes include a diploid spermatocyte, a diploid oocyte, or fertilization of an egg cell by two spermatozoa (dispermy).

B. Monosomy X (Turner syndrome)

Monosomy X (karyotype 45,X0) causes Turner syndrome with a very wide phenotypic spectrum, ranging from severe to very mild. The phenotype during the fetal stage is usually massive lymphedema of the head and neck, and large multilocular thin-walled lymphatic cysts (**1**). Congenital cardiovascular defects, especially involving the aorta, and kidney malformations are frequent. Important is fetal degeneration of the ovaries into connective tissue, as streak gonads. Small stature is always present, with an average adult height of about 150 cm. In many patients the manifestations are mild (**2**), but in others, webbing of the neck (pterygium colli) may be present as a residue of the fetal lymphedema (**3**). Most patients have chromosomal mosaicism 45,X/46, XX, i.e., some of their cells have normal chromosomal complements, a deletion Xp, or an isochromosome for the long arm (i[Xq]). The loss of genes on the short arm of the X chromosome (Xp) is responsible for the phenotype (see *SHOX* genes; OMIM 312865). Monosomy X is frequent, about 5% at conception. However, of 40 zygotes with monosomy X, only one will develop to birth.

C. Additional X or Y chromosome

An additional X chromosome in males (47, XXY) causes Klinefelter syndrome after puberty if untreated (**1**). This includes tall stature, absent or decreased development of male secondary sex characteristics, and infertility due to absent spermatogenesis. Testosterone substitution beginning at puberty is necessary. In contrast, an additional Y chromosome (47, XYY) does not result in a recognizable phenotype (**2**). Girls with three X chromosomes (47, XXX) are physically unremarkable (**3**). However, learning disorders and delayed speech development have been observed in some of these children.

D. Chromosomal aberrations in human fetuses

A wide spectrum of trisomies (and monosomies) occur at conception and lead to spontaneous abortion during the second and third months of pregnancy. The most frequent is trisomy 16, which accounts for about about 5% of all autosomal trisomies. (Data from Lauritsen, 1982.)

Further Reading

Acherman JC, Jameson JL. Disorders of sex development. In: Longo DL, et al., eds. Harrison's Principles of Internal Medicine. 18th ed. New York: McGraw-Hill; 2012:3046–3055

Bondy CA. New issues in the diagnosis and management of Turner syndrome. Rev Endocr Metab Disord 2005;6:269–280

DeGrouchy J, Turleau C. Clinical Atlas of Human Chromosomes. 2nd ed. New York: John Wiley & Sons; 1984

Graham GE, Alanson JE, Gerritsen JA. Sex chromosome abnormalities. In: Rimoin DL, et al., eds. Emery and Rimoins's Principles and Practice of Medical Genetics. 5th ed. Philadelphia: Churchill Livingstone-Elsevier; 2007:1038–1057

Lauritsen JG. The cytogenetics of spontaneous abortion. Res Reprod 1982;14:3–4

Miller OJ, Therman E. Human Chromosomes. 4th ed. Heidelberg: Springer-Verlag; 2001

Otter M, Schrander-Stumpel CTRM, Curfs LMG. Triple X syndrome: a review of the literature. Eur J Hum Genet 2010;18:265–271

Ranke MB, Saenger P. Turner's syndrome. Lancet 2001;358:309–314

1.

Triploidy

- Most frequent chromosomal aberration (15%) in fetuses following spontaneous abortion
- Severe growth retardation, early lethality
- Occasional liveborn infant with severe malformation
- Dispermia a frequent cause

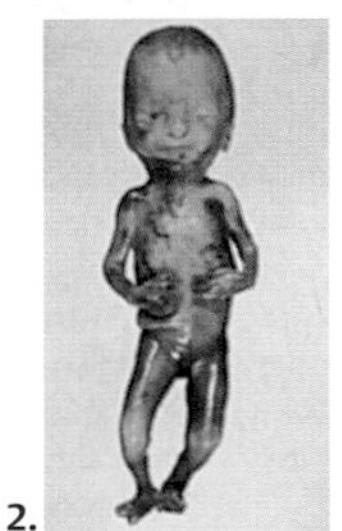

2.

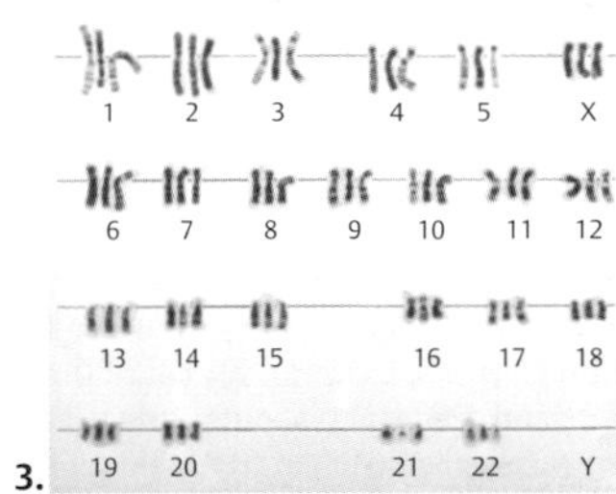

3.

A. Triploidy

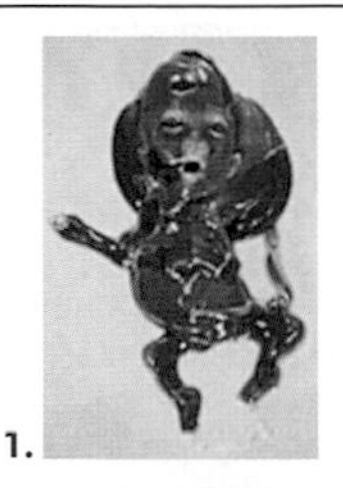

1.

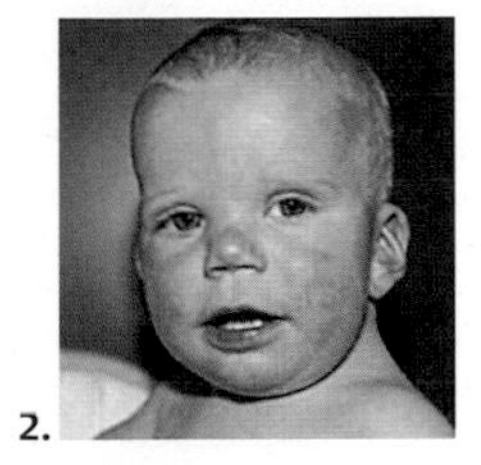

2.

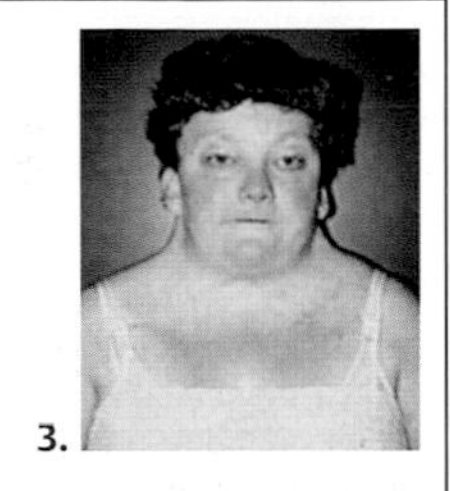

3.

B. Monosomy X (Turner syndrome; 45,XO)

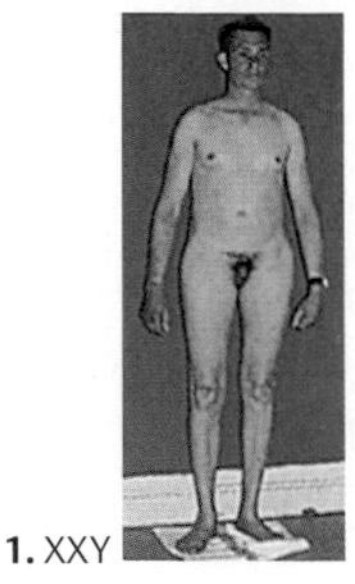

1. XXY

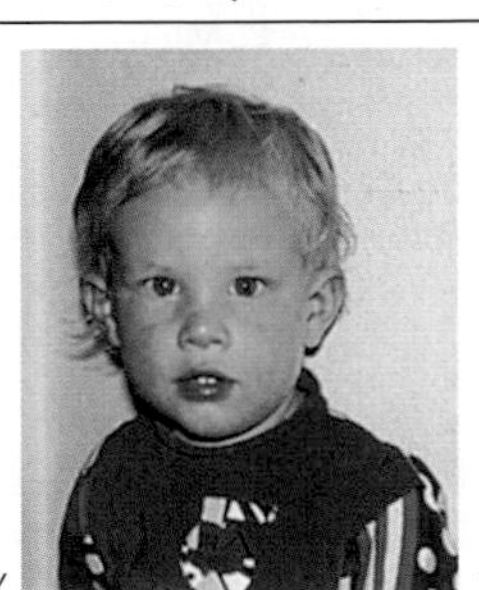

2. XYY

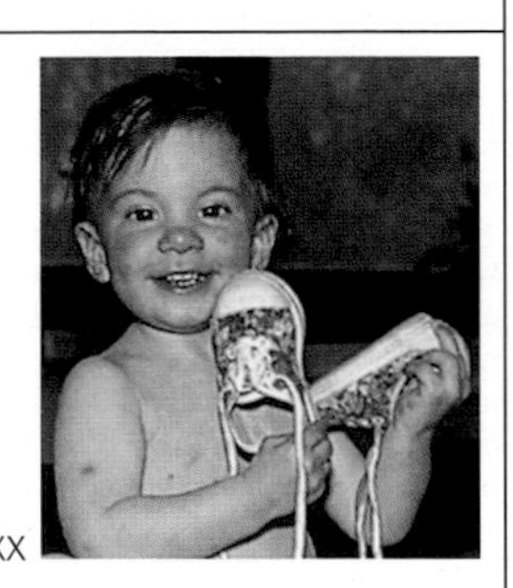

3. XXX

C. Additional X or Y chromosome

Proportion of autosomal trisomies in 669 trisomic fetuses following spontaneous abortion

Proportion (%)

1 2 3 4 5

0.0 4.9 0.6 2.5 0.2 0.5 4.0 3.9 2.7 2.0 0.3 1.0 4.6 4.6 7.7 32.3 0.6 5.1 0.2 2.7 9.4 10.2

1 2 3 4 5 6 7 8 9 10 11 12 13 14 15 16 17 18 19 20 21 22

Trisomic chromosomes

D. Wide spectrum of chromosomal aberrations in human fetuses

Microdeletion Syndromes

Microdeletion syndromes are structural chromosomal disorders that are not detectable by standard metaphase karyotyping because the size of the deletion is below the resolution of the light microscope (below 4 Mb). Microdeletions can be detected by fluorescence in situ hybdridization (FISH, see p. 166) or by array methods (p. 226). Currently about 20 microdeletion and microduplication syndromes have been recognized (Appendix, Table p. 423). Microdeletions are genomic disorders (p. 242), since the deletion or duplication arises from nonhomologous alignment of adjacent DNA segments with high sequence similarity (copy number variations, CNVs). Most deletions occur de novo, but transmission from a parent with a balanced translocation is not rare.

A. Deletion 5p−: cri-du-chat syndrome

This is not a microdeletion syndrome in the strict sense, because the deletion is recognizable by standard metaphase analysis in about 80% of patients. In 1963, Lejeune and coworkers in Paris described children with a partial deletion of the short arm of a chromosome 5 (5p−) and a characteristic pattern of dysmorphic facial features associated with impaired mental development (OMIM 123450). The name is derived from a prolonged high-pitched cry of affected infants, which resembles that of a kitten. In about 12% of children with cri-du-chat syndrome, one of the parents has a translocation involving a chromosome 5. (Photo courtesy of Drs Beate Albrecht and Alma Küchler, University Hospital, Essen, Germany.)

B. Deletion 4 p−: Wolf–Hirschhorn syndrome

This characteristic phenotype (OMIM 194190) results from a partial deletion of variable size of the short arm of a chromosome 4. It was described in 1964 independently by U. Wolf and K. Hirschhorn and their coworkers. Mental and statomotoric retardation is associated with characteristic facial features (**1**, **2**), midline defects (cleft palate, hypospadias), coloboma of the iris, congenital heart defects, and other malformations. In about 80% of cases, the deletion at 4p16.3 is recognizable by light microscopy. In about 10%–15% of cases, one parent carries a balanced translocation, which predisposes to deletion or duplication of this region. (Photo courtesy of Drs Beate Albrecht and Alma Küchler, Essen, Germany.)

C. Deletion 22q11

Several partially overlapping phenotypes result from deletions in region 1, band 1.2 of the long arm of chromosome 22 (22q11.2) (**1**). The most common are the DiGeorge syndrome (OMIM 188400; p. 298) and the velocardial facial syndrome (Shprintzen syndrome; OMIM 192430) with dysmorphic facial features (**2**, **3**). About 80% patients have a large 1.5–3.0 Mb hemizygous deletion of 22q11.2 involving about 45 gene loci in this region, leading to phenotypic differences. (**1** and **2**, courtesy of Dr Diana Mitter, Leipzig, Germany.)

D. Deletion 7q11.23

This deletion is usually associated with a recognizable phenotype, the Williams–Beuren syndrome (WBS; OMIM 194050). Facial and behavioral features associated with supravalvular aortic stenosis are its hallmark. It results from deletions of variable sizes of 1.5–1.8 Mb, involving about 28 gene loci.

E. Microdeletion syndrome 1p36

This is a relatively frequent (1:5000) microdeletion syndrome of the subterlomeric region 1p36 (OMIM 607872) (**1**). Mental retardation associated with a prominent forehead, midfacial hypoplasia (**2**, **3**), microcephaly, muscular hypotonia, and congenital heart defects are among the phenotypic features. The diagnosis can be confirmed by array techniques, here by SNP array analysis (deletion at 1p36, arrow). (**1**–**3**, Courtesy of Prof Dagmar Wieczorek, Essen, **4**, Dr Ludger Klein-Hitpass, Essen, Germany.)

Further Reading

Bassett AS, et al. Practical guidelines for managing patients with 22q11.2 deletion syndrome. J Pediatr 2011;159:332–339, e1

D'Angelo CS, et al. Extending the phenotype of monosomy 1p36 syndrome and mapping of a critical region for obesity and hyperphagia. Am J Med Genet A 2010;152A:102–110

Pober BR. Williams-Beuren syndrome. N Engl J Med 2010;362:239–252

Schwartz S, Hassold T. Chromosome disorders. In: Longo DL, et al., eds. Harrison's Principle of Internal Medicine. 18th ed. New York: McGraw-Hill; 2012:509–518

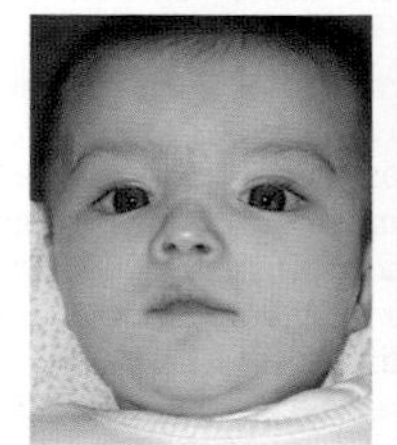

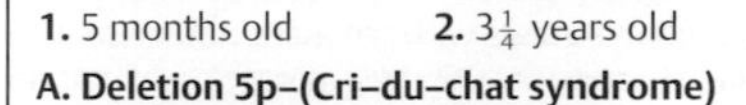

1. 5 months old 2. $3\frac{1}{4}$ years old

A. Deletion 5p–(Cri–du–chat syndrome)

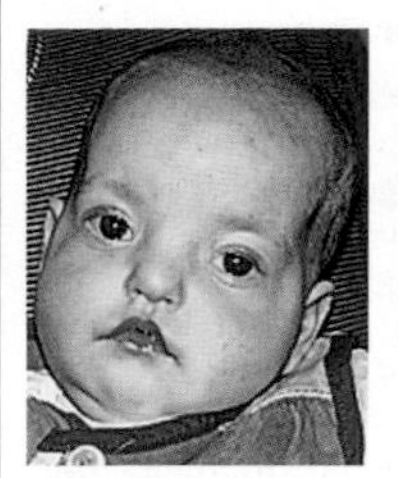

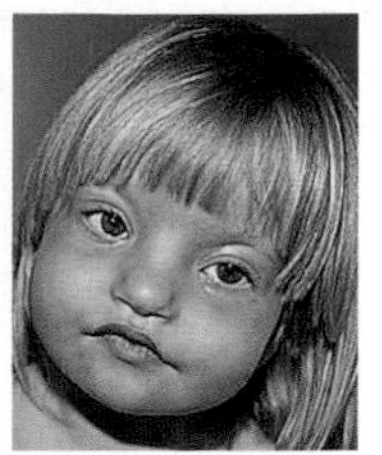

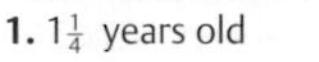

1. $1\frac{1}{4}$ years old 2. 4 years old

B. Deletion 4p–(Wolf–Hirschhorn syndrome)

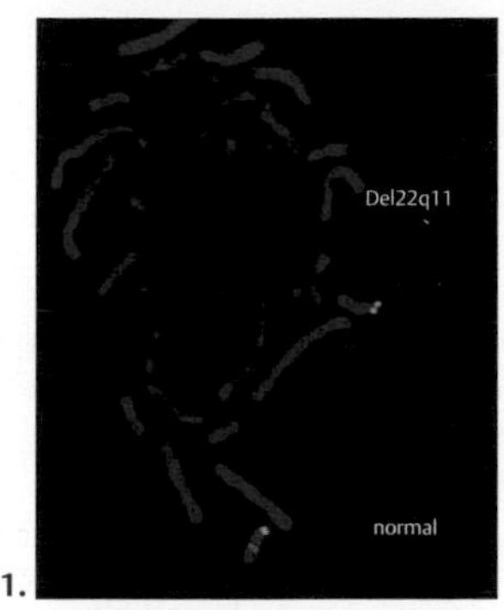

1.

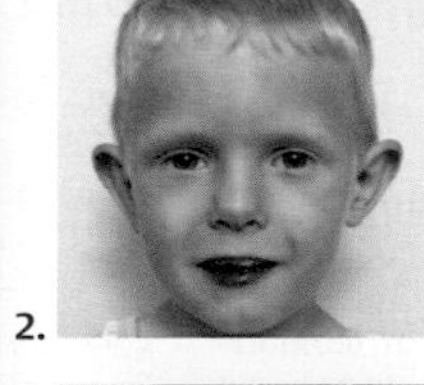

2.

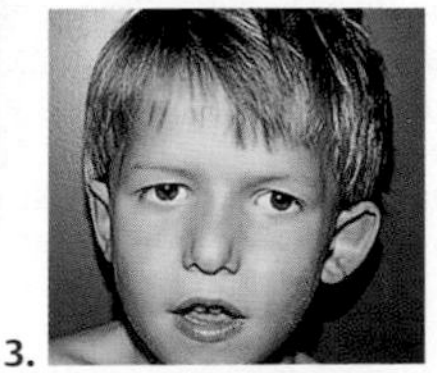

3.

C. Deletion 22q11

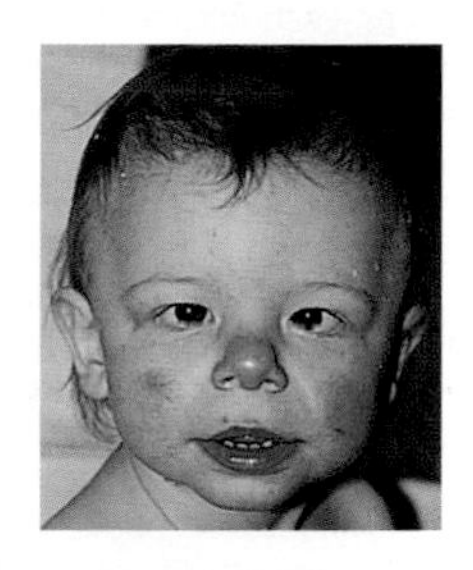

D. Deletion 7q11.23 (Williams–Beuren syndrome)

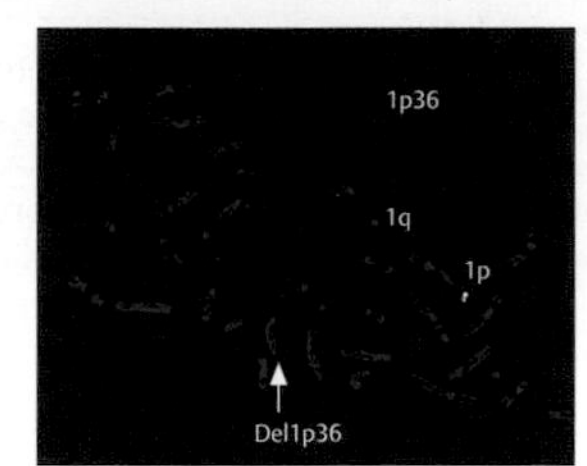

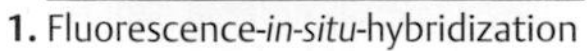

1. Fluorescence-*in-situ*-hybridization

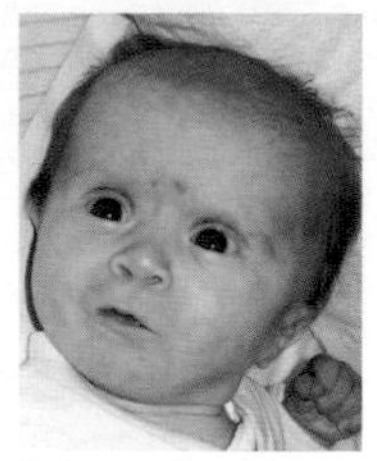

2. 2 months old

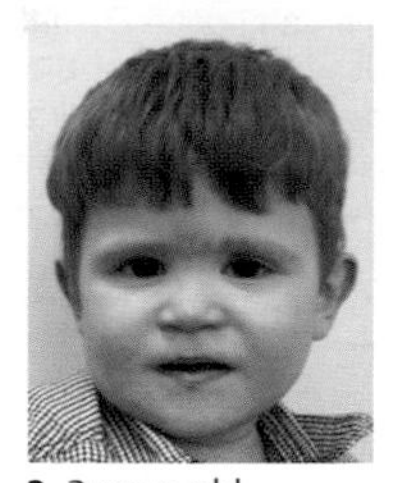

3. 2 years old

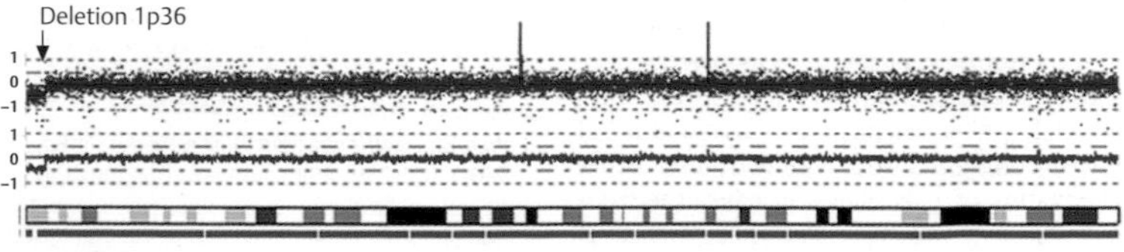

E. Deletion 1p36

4. SNP-array (Affymetrix 259) analysis of chromosome 1

Principles of Genetic Diagnostics

The diagnosis of a genetic disease requires a systematic, interdisciplinary approach that takes many clinical and genetic considerations into account. It begins with an analysis of the phenotype. This is defined at different levels, i.e., recognizable features, age of onset and progression of a disorder, laboratory and X-ray data, etc. Appropriate online systems need to be consulted; for Mendelian disorders, foremost is the OMIM system (McKusick, 1998).

A. Genetic diagnosis, a multistep procedure

A genetic diagnosis is established in a sequence of steps, each with a binary decision. The first decision determines whether a pattern of the manifestations can be recognized. If this is the case, the next decision deals with the presumptive category of disease. Although difficult to establish in practice, this decision is the basis for subsequent steps. Furthermore, as emphasized by Lupski et al. (2011), the underlying causes of genetic diseases represent a continuum, and the assignment to a given category should not be taken too strictly. Since a particular phenotype may be caused by mutations at different loci (*locus heterogeneity*) or by different mutant alleles at the same locus (*allele heterogeneity*), caution has to be exercised not to overlook genetic heterogeneity. All genetic diagnostic procedures should be preceded by genetic counseling, which properly includes obtaining (informed) consent from the persons involved.

B. Genotype analysis by PCR typing

The polymerase chain reaction (PCR) has greatly simplified DNA-based diagnosis in cases where sequencing is not available or suitable. The figure shows how two alleles can be distinguished by the presence of a variant in one allele, but not the other. Here one allele (allele 2) contains a cleavage site and is represented by two fragments of 7 kb and 6 kb. The other allele (allele 1) consists of one 13 kb fragment.

Following amplification, the three possible combinations of two alleles can be visualized as the three genotypes: 1–1, 1–2, and 2–2. Each close relative of the affected individual can thus be genotyped.

C. Protein truncation test (PTT)

This is a test for frameshift, splice, or nonsense mutations that lead to a truncated protein as a result of an early stop codon created downstream of the mutation. The truncated protein is detected by an in vitro translation system. The translation will be interrupted at a premature stop codon resulting from the mutation. The size of the newly translated protein is determined by gel electrophoresis. PTT is useful in studying genes with frequent nonsense mutations, such as the *APC*, *BRCA1*, and *BRCA2* genes.

Further Reading

Aase JM. Diagnostic Dysmorphology. New York: Plenum; 1990

Feero WG, Guttmacher AE, Collins FS. Genomic medicine—an updated primer. N Engl J Med 2010;362: 2001–2011

Harper PS. Practical Genetic Counselling. 7th ed. London: Edward Arnold; 2010

Horaitis R, Scriver CR, Cotton RGH. Mutation databases: Overview and catalogues. In: Scriver CR, et al., eds. The Metabolic and Molecular Bases of Inherited Disease. 8th ed. New York: McGraw-Hill; 2001:113–125. (www.ommbid.com)

Jones KL. Smith's Recognizable Patterns of Human Malformation. 6th ed. Philadelphia: W. B. Saunders; 2006

Jorde LB, Carey JC, Bamshad MJ. Medical Genetics. 4th ed. Philadelphia: Mosby-Elsevier; 2010

Lupski JR, Belmont JW, Boerwinkle E, Gibbs RA. Clan genomics and the complex architecture of human disease. Cell 2011;147:32–43

McKusick VA. Mendelian Inheritance in Man. A Catalog of Human Genes and Genetic Disorders. 12th ed. Baltimore: Johns Hopkins University Press; 1998. (www.ncbi.nlm.nih.gov/omim). Links to diagnostic laboratories at www.ncbi.nlm.nih.gov/sites/GeneTests/

Misfeldt S, Jameson JL. The practice of genetics in clinical medicine. In: Longo DL, et al., eds. Harrison's Principles of Internal Medicine. 18th ed. New York: McGraw-Hill; 2012:519–525

NCBI. GeneTests. Available at: http://www.ncbi.nlm.nih.gov/sites/GeneTests/. Accessed January 28, 2012

Rimoin DL, Pyeritz RE, Korf BR. Emery and Rimoins's Principles and Practice of Medical Genetics. 5th ed. New York: Elsevier Churchill-Livingstone; 2007 (6th ed in press)

Stevenson RE, Hall JG, eds. Human Malformations and Related Anomalies. 2nd ed. Oxford: Oxford University Press; 2006

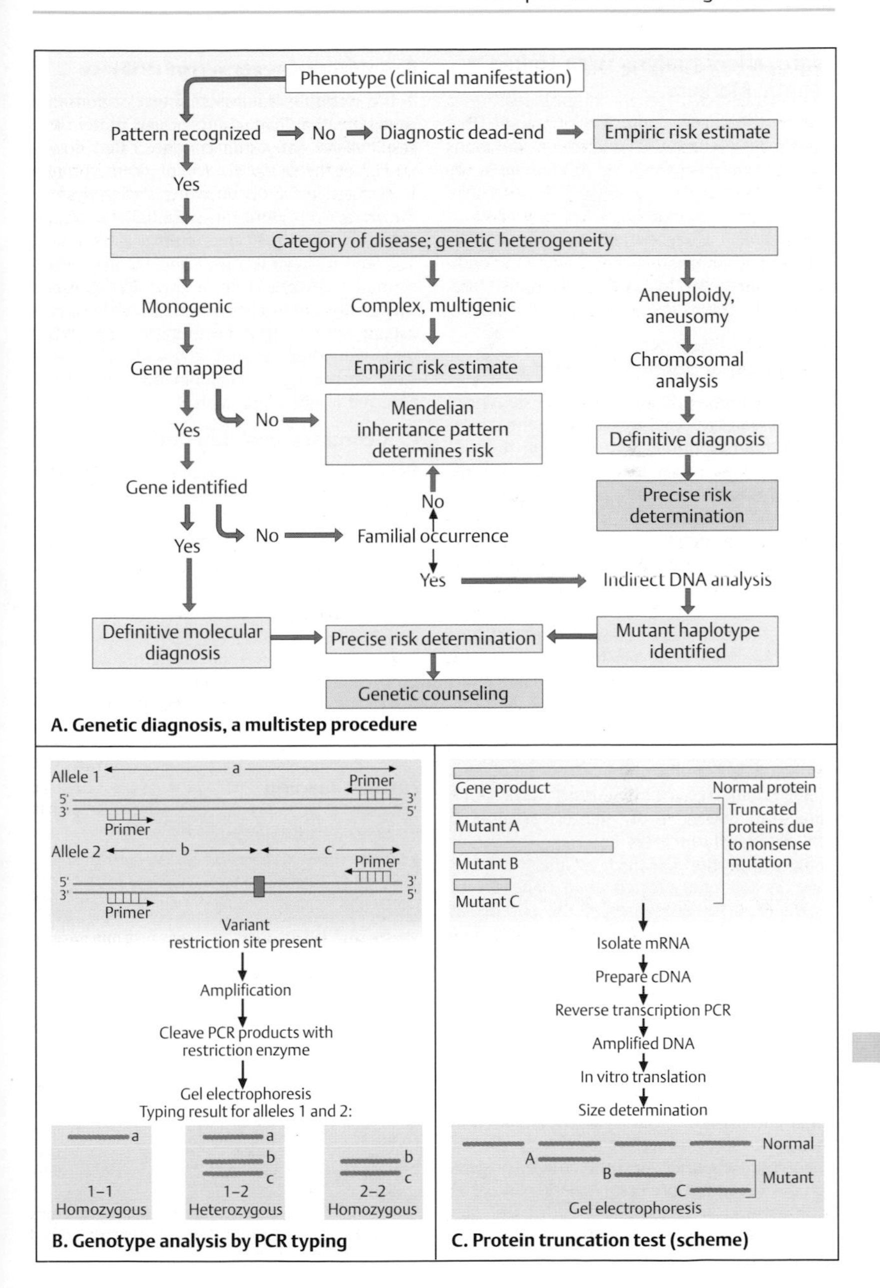

A. Genetic diagnosis, a multistep procedure

B. Genotype analysis by PCR typing

C. Protein truncation test (scheme)

Segregation Analysis with Linked Genetic Markers

Prior to the introduction of widely used DNA sequencing, indirect DNA diagnostics was common. Its principles should still be understood. This approach rests on analysis of known linkage between a disease locus and one or more closely linked DNA marker loci (segregation analysis with linked markers). This can provide information about the genetic risk to members of a family in which a monogenic disease has occurred.

A. Autosomal dominant inheritance

Two pedigrees with an autosomal dominant disorder (red symbols) are shown in the figure: one without recombination between the disease locus and the polymorphic DNA marker locus (**1**), and one with recombination (**2**). Below the pedigree in **1** (left) the lanes of a diagram of a Southern blot or PCR analysis representing each of the eight individuals is shown. The individuals affected are in lanes 2, 4, and 5. The affected mother (lane 2) has two marker alleles, 1 and 2, corresponding to genotype 1–2. Her two affected children have the same marker genotype, 1–2. None of her unaffected children (lanes 3, 6, 7, and 8) has this genotype; they are 2–2. Since the father has the genotype 2–2 and must have transmitted marker allele 2 to all children, the other allele 2 in the unaffected children must have come from the mother. Thus, the diseases locus must be linked to marker 1.

In the pedigree on the right, an affected father (lane 1) and one affected child (lane 4) are shown (**2**). The genotypes of the parents at the marker locus are 1–2 (father) and 1–1 (mother), respectively. Allele 2, inherited from the father, is present in the affected child (1–2). Thus, allele 2 must represent the disease locus. In this family, however, an unaffected sibling (lane 5) also has the genotype 1–2. One can conclude that recombination has occurred between the disease locus and the marker locus. In this case the test result would be misleading. For this reason, very closely linked markers with very low recombination frequencies are used, and preferably marker loci that flank the disease locus.

B. Autosomal recessive inheritance

In this example of autosomal recessive inheritance, the two affected individuals in the pedigree on the left are homozygous for allele 1 (1–1), inherited from each of their parents. Thus one allele 1 of the father and allele 1 of the mother represent the allele that carries the mutation. The unaffected siblings (individuals 3, 5, and 6) have received allele 2 from their mother and allele 1 from their father. Since allele 2 does not occur in the affected individuals, it cannot carry the mutation. Therefore, these individuals are not at risk of. In the pedigree on the right, recombination must have occurred in one child (lane 6).

C. X-chromosomal inheritance

Here an affected son (lane 5) carries the marker allele 1. His sister (lane 6) is the mother of six children (lanes 8–13). She is heterozygous 1–2 at the marker locus. Two of her affected sons also carry the marker allele 1. Her unaffected son (lane 11) carries allele 2. Thus, this allele is probably not linked with the disease locus. However, a third affected son (lane 13) carries allele 2. This can be explained by recombination of the disease locus and the marker locus in this son. In practice one avoids such ambiguity by using several marker loci flanking the disease locus.

Further Reading

Harper PS. Practical Genetic Counselling. 7th ed. London: Edward Arnold; 2010

Jorde LB, Carey JC, Bamshad MJ. Medical Genetics. 4th ed. Philadelphia: Mosby-Elsevier; 2010

Jameson JL, Kopp P. Principles of human genetics. In: Longo DL, et al., eds. Harrison's Principles of Internal Medicine. 18th ed. New York: McGraw-Hill; 2012:486–509

Miesfeldt S, Jameson JL. The practice of genetics in clinical medicine. In: Longo DL, et al., eds. Harrison's Principles of Internal Medicine. 18th ed. New York: McGraw-Hill; 2012:519–525

Nussbaum RL, McInnes RR, Willards HF. Thompson & Thompson Genetics in Medicine. 7th ed. Philadelphia: W. B. Saunders; 2007 (available online)

Turnpenny PD, Ellard S. Emery's Elements of Medical Genetics. 14th ed. Edinburgh: Elsevier-Churchill Livingstone; 2011 (available online)

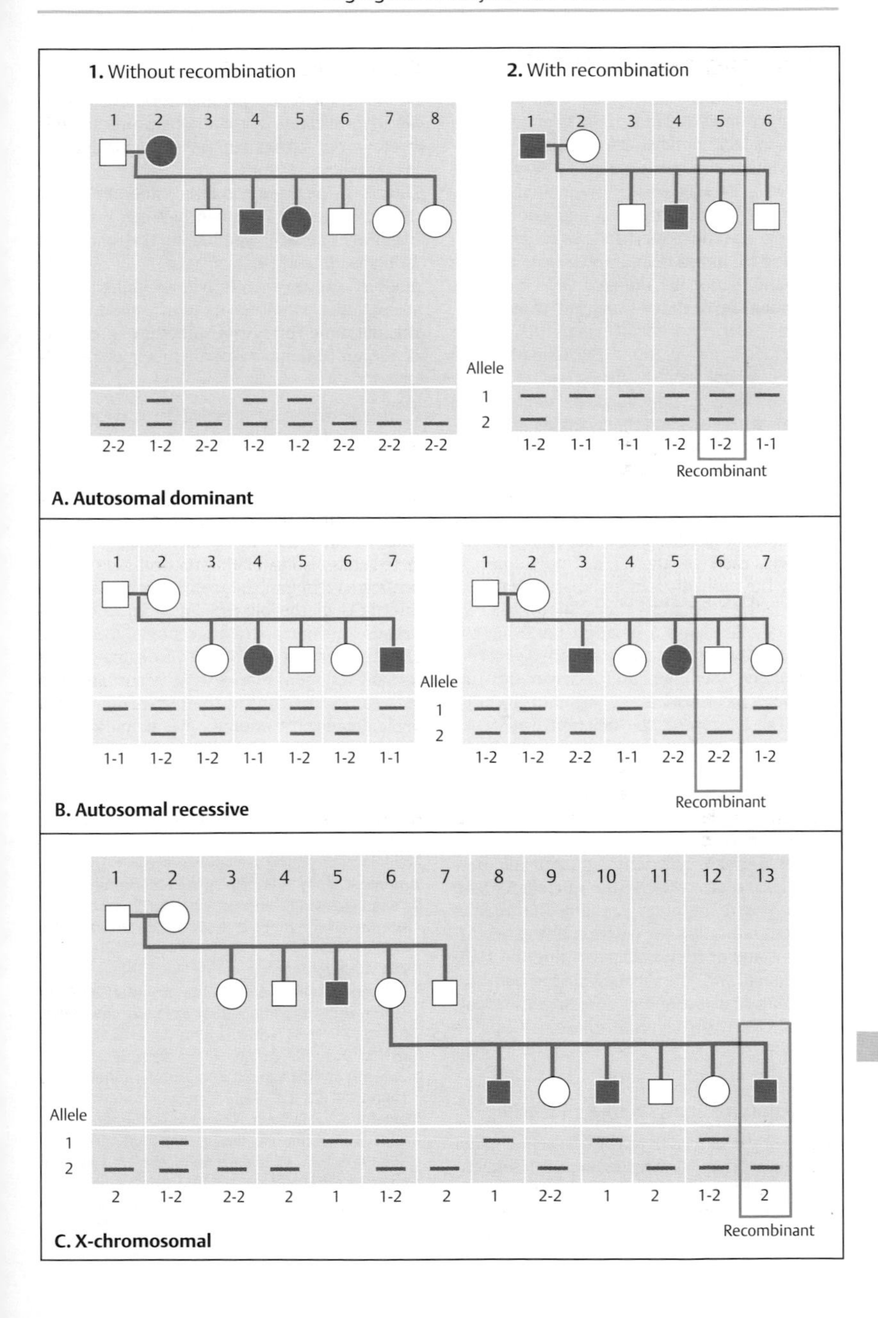
1. Without recombination
2. With recombination
Allele
1
2
2-2 1-2 2-2 1-2 1-2 2-2 2-2 2-2
1-2 1-1 1-1 1-2 1-2 1-1
Recombinant
A. Autosomal dominant
1-1 1-2 1-2 1-1 1-2 1-2 1-1
1-2 1-2 2-2 1-1 2-2 2-2 1-2
Recombinant
B. Autosomal recessive
2 1-2 2-2 2 1 1-2 2 1 2-2 1 2 1-2 2
Recombinant
C. X-chromosomal

Indirect DNA Analysis

Indirect DNA analysis compares the presence or absence of DNA markers in relation to the disease locus as outlined on p. 390.

The techniques of recombinant DNA were first applied to the diagnosis of hemoglobin disorders in 1978, including in prenatal diagnosis and in screening populations with high frequencies of common mutant alleles (Kan & Dozy, 1978). Initially indirect DNA diagnosis was based on haplotype analysis using polymorphic restriction enzyme sites (RFLPs, see p. 62) linked to mutations. Subsequently, the use of the polymerase chain reaction (PCR, see p. 64) facilitated the diagnosis of structural hemoglobin defects and thalassemias (p. 346ff). With so many mutations known, a direct approach is possible in many cases, but the indirect approach continues to be valuable for identifying rare mutations. Here the principle of indirect diagnosis based on RFLP or PCR analysis is illustrated.

A. Indirect detection of a deletion

This example shows a deletion involving the two α-globin genes, α1 and α2. A 14.5-kb DNA restriction fragment (between the two red arrows) that is normally present is reduced to 10.0 kb because of the deletion. The 10-kb fragment represents the deletion, the 14.5-kb fragment represents the normal allele (**1**). A DNA probe recognizing the α2 gene is used to demonstrate the deletion (**2**). Three genotypes can be expected (**3**): both alleles carrying the normal 14.5-kb fragment (normal homozygote); one allele carrying the normal 14.5-kb fragment and the other carrying the 10.5-kb fragment indicating the deletion in the heterozygous state; or both alleles carrying the 10.5-kb fragment, i.e., the deletion in the homozygous abnormal state. The corresponding pattern (**4**) clearly distinguishes these three possibilities and allows a precise diagnosis by genotyping.

B. Haplotype analysis of a mutation

In this situation a difference between a polymorphic marker on the normal allele and the mutant allele is used. The difference results in two fragments of 7 kb and 6 kb in size, respectively, in one allele and one 13-kb fragment in the other allele (**1**). If the analysis of affected and nonaffected family members indicates that the affected individuals all share the 13-kb fragment, this will represent the mutant allele (**2**). Hybridization with a DNA probe for this region will recognize the DNA fragments present.

Therefore, the three possible genotypes can be distinguished (**3**). Although indirect, a specific diagnosis for each member in this particular family is possible (**4**).

The prerequisite for this type of indirect analysis is prior knowledge of which allele carries the mutation. This is established by an analysis of affected and unaffected members of a family (not shown).

C. Recognition of a point mutation by an altered restriction site

A mutation might alter a restriction site by either abolishing an existing one or creating a new one in the mutant allele. For example, the sickle cell mutation in codon 6 of the β-globin gene results in loss of a restriction site (**1**). The mutation changes the recognition sequence CCTNTGG of the enzyme *Mst*II to CCTNAGG because an adenine (A) has replaced a thymine (T) (**2**). Therefore, following *Mst*II digestion, a 1.15-kb fragment represents the normal allele (β^A). A 1.35-kb fragment represents the mutant allele because the mutation has eliminated the restriction site in the middle. Southern blot analysis clearly distinguishes the three genotypes and allows a precise, simple, inexpensive diagnosis (**3**).

Further Reading

Jameson JL, Kopp P. Principles of human genetics. In: Longo DL, et al., eds. Harrison's Principles of Internal Medicine. 18th ed. New York: McGraw-Hill; 2012:486–509

Kan YW, Dozy AM. Polymorphism of DNA sequence adjacent to human beta-globin structural gene: relationship to sickle mutation. Proc Natl Acad Sci U S A 1978;75:5631–5635

Kan YW, Dozy AM. Antenatal diagnosis of sickle-cell anaemia by DNA analysis of amniotic-fluid cells. Lancet 1978;II:910–912

Miesfeldt S, Jameson JL. The practice of genetics in clinical medicine. In: Longo DL, et al., eds. Harrison's Principles of Internal Medicine. 18th ed. New York: McGraw-Hill; 2012:519–525

Old J. Hemoglobinopathies and thalassemias. In: Rimoin DL, et al., eds. Emery and Rimoin's Principles of Medical Genetics. 5th ed. Edinburgh: Churchill-Livingstone; 2007:1638–1674

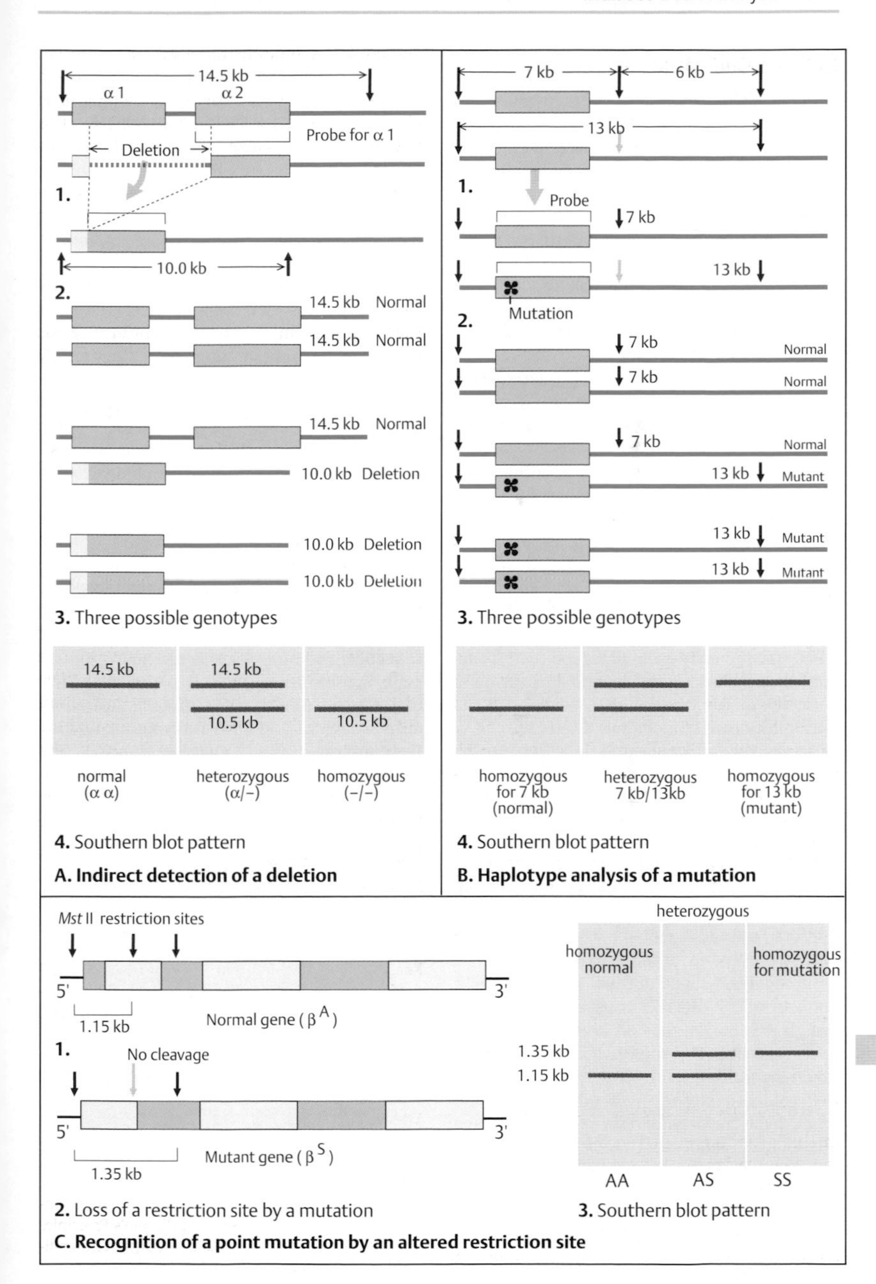

A. Indirect detection of a deletion

B. Haplotype analysis of a mutation

C. Recognition of a point mutation by an altered restriction site

Detection of Mutations without Sequencing

It is still customary to scan a gene for the presence of a mutation before sequencing. Several approaches have been designed to demonstrate the presence of a mutation indirectly. Three examples are illustrated here. Some of these methods are based on differences in the hybridization of mutated and normal segments of DNA resulting from a mismatch.

A. Detection of a point mutation by oligonucleotides

This method is designed to detect a mutation by a difference in hybridization caused by the mutation. If, for example, the normal (**1**) and the mutant (**2**) DNA differ by one nucleotide, here an adenine (A) replacing a guanine (G), a difference in hybridization results when two oligonucleotides are used as probes, one complementary to the normal allele and one complementary to the mutant allele (allele-specific oligonucleotides, ASO). Oligonucleotides, short segments of DNA of about 20 nucleotides, hybridize with the target DNA sequence only when they match each other perfectly. One oligonucleotide (oligonucleotide 1) hybridizes perfectly to the normal DNA because all its nucleotides match, including the G (**3**).
However, this oligonucleotide does not match perfectly to the mutant DNA at A (**4**). In contrast, oligonucleotide 2 hybridizes perfectly with the mutant target DNA (**5**), but not with the normal (**6**) DNA. This difference in hybridization patterns is visualized in a dot-blot analysis (**7**). Hybridization is indicated by a signal, which is a black dot on a roentgen film produced by the radiolabeled probes. If the mutation is present in both strands of DNA (in a homozygote), a single dot will appear after oligonucleotide 2 (ASO 2) is used for hybridization. If it is present in one strand, but not the other (in a heterozygote), two dots will appear when hybridized with both probes. Normal DNA in both strands will produce one dot when hybridized with ASO 1. Thus, all three possibilities (genotypes) can be distinguished.

B. Denaturation gradient gel electrophoresis (DGGE)

This method exploits differences in the stability of DNA fragments with and without mutation. While double-stranded DNA of a control person is completely complementary (homoduplex), a mutation leads to a mismatch at the site of mutation (heteroduplex). This DNA fragment is less stable than completely complementary DNA strands, because it has a lower melting point. If normal DNA (control) and DNA with the mutation are placed in a gel with an increasing concentration gradient of formamide (denaturing gradient gel), the normal fragment is more stable than the mutant fragment. Therefore the normal fragment moves further along in the electrophoretic field than the mutant fragment. Today, automated procedures are used, thus avoiding cumbersome electrophoresis. One such example is denaturing high-performance liquid chromatography (DHPLC). It detects differences between homoduplex (double-stranded normal or mutant) DNA and heteroduplex (mutant/normal) DNA.

C. Ribonuclease A cleavage assay

In the ribonuclease (RNase) A assay, the enzyme RNaseA cleaves a labeled RNA probe at a site of mismatch to a target DNA or RNA sequence. The basis for this method is incomplete hybridization (mismatch) between normal single-stranded DNA (here indicated by an A, adenine) and DNA carrying a mutation (here indicated by a G, guanine). Labeled complementary RNA used as a probe hybridizes completely with the normal DNA, but incompletely with the mutant DNA target. At this site the enzyme cleaves the mutant DNA. Subsequently, the resulting fragments can be separated according to their sizes.

Further Reading

Beaudet AL, et al. Genetics, biochemistry, and molecular bases of variant human phenotypes. In: Scriver CR, et al., eds. The Metabolic and Molecular Bases of Inherited Disease. 8th ed. New York: McGraw-Hill; 2001:3–45 (available online)

O'Donovan MC, et al. Blind analysis of denaturing high-performance liquid chromatography as a tool for mutation detection. Genomics 1998;52:44–49

Strachan T, Read AP. Human Molecular Genetics. 4th ed. New York: Garland Science; 2011

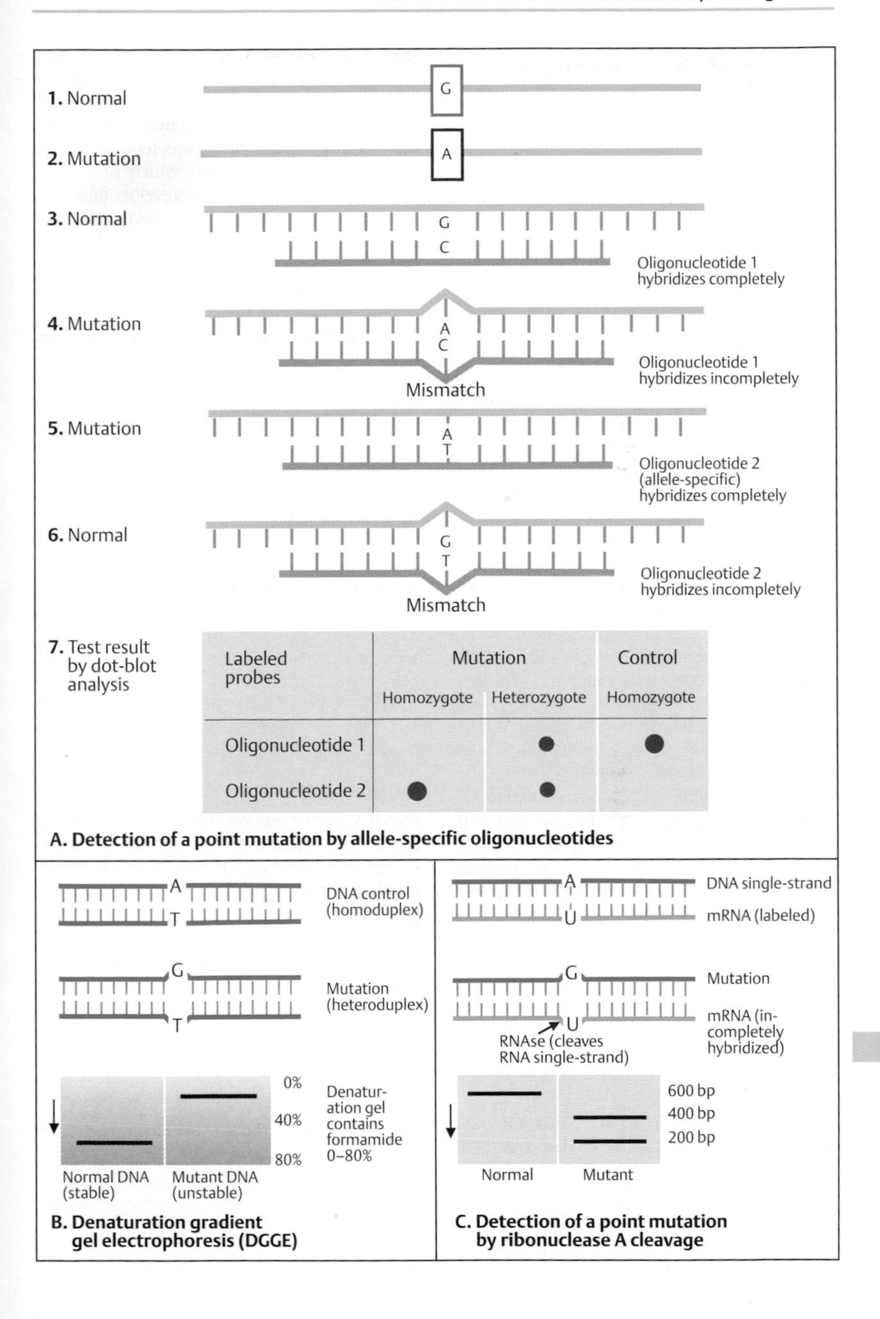

A. Detection of a point mutation by allele-specific oligonucleotides

B. Denaturation gradient gel electrophoresis (DGGE)

C. Detection of a point mutation by ribonuclease A cleavage

Gene and Stem Cell Therapy

Transfer of nucleic acid as the active agent in therapy, rather than small molecules or proteins, is the basis of somatic gene therapy. A different strategy is to replace stem cells for tissue repair. Considerable technical difficulties and side-effects have to be overcome, and ethical considerations have to be taken into account. Therapy of genetic diseases is still limited to selected disorders in about 1700 current clinical trials worldwide.

A. Principles of gene therapy

Somatic gene therapy requires a transport system to deliver a correcting gene to somatic cells to the desired tissues. Two basic forms of somatic gene therapy are *ex vivo* and *in vivo*. In the first, a gene is introduced into cells outside the body. These are subsequently reintroduced into tissues that need to be corrected. In the second, a gene is introduced directly by a viral or a nonviral vector. The advantage of viral vectors is the relative ease with which they can enter the cells of the recipient. However, controlling viral production, carrying capacity, dependence on cell proliferation, and other aspects make application difficult.
An ex vivo approach for introducing a correcting gene into the hematopoietic system is illustrated in the figure. From whole blood (**1**), the red blood cells and white blood cells are separated, and the red blood cells reinfused (**2**). From the white blood cells (**3**), immunologically competent CD34 cells are separated (**4**), placed together with a virus vector carrying the desired normal gene (**5**), and propagated together in a cell culture (**6**). Once the defective cells have incorporated the viral vector and thereby corrected, these cells are reintroduced into the recipient (**7**). G-CSF (granulocyte-colony stimulating factor) is used to stimulate granulocyte formation. (Figure based on Kessler, 2012.)

B. Stem cells

Stem cells are undifferentiated progenitor cells from which differentiated cells derive. Stem cells differ according to their tissue destiny, replicative capacity, and differentiation potential. *Totipotent* stem cells can develop into a complete embryo and form a placenta. This ability is limited to cells derived from the first few divisions of the zygote. *Pluripotent* stem cells can form tissues derived from the endoderm, mesoderm, and ectoderm germinal layers. Stem cells divide symmetrically into two identical stem cells (self-renewal). They form a pool of cells from which progenitor cells of differentiated cells develop following an asymmetric cell division. Such cells lose the ability to undergo cell division and self-renewal.

C. Stem cell therapy

Stem cell therapy would have the advantage of providing the recipient with a permanent supply of cells with the ability to restore damaged tissues that lack the ability of repair. It would be especially important in organs, such as the bone marrow (hematopoietic system) and epithelial cell systems (e.g., in the gastrointestinal tract), where cells are constantly lost and replaced. Different types of stem cells exist. Embryonic stem (ES) cells have the capacity to generate any cell type in the body; however, ES cells have the potential to form teratomas and their use is ethically controversial. Induced pluripotent stem cells (iPS cells) are somatic cells that are reprogrammed into the pluripotent state. The umbilical cord blood contains progenitor cells of the hematopoietic system, but there are limitations for usage. (Figure adapted from Nabel, 2004.)

Further Reading

High KA. Gene therapy in clinical medicine. In: Longo DL, et al., eds. Harrison's Principles of Internal Medicine. 18th ed. New York: McGraw-Hill; 2012:547-551 (with online access)

Hochedlinger K, Jaenisch R. Nuclear reprogramming and pluripotency. Nature 2006;441:1061–1067

Jiang Y, et al. Pluripotency of mesenchymal stem cells derived from adult marrow. Nature 2002;418:41–49

Gene Therapy Clinical Trials Worldwide (updated June 2011). Available at: www.wiley.com/legacy/wileychi/genmed/clinical/

Kessler JA. Applications of stem cell biology in clinical medicine. In: Longo DL, et al., eds. Harrison's Principles of Internal Medicine. 18th ed. New York: McGraw-Hill; 2012:543–551 (with online access)

Nabel GJ. Genetic, cellular and immune approaches to disease therapy: past and future. Nat Med 2004;10:135–141

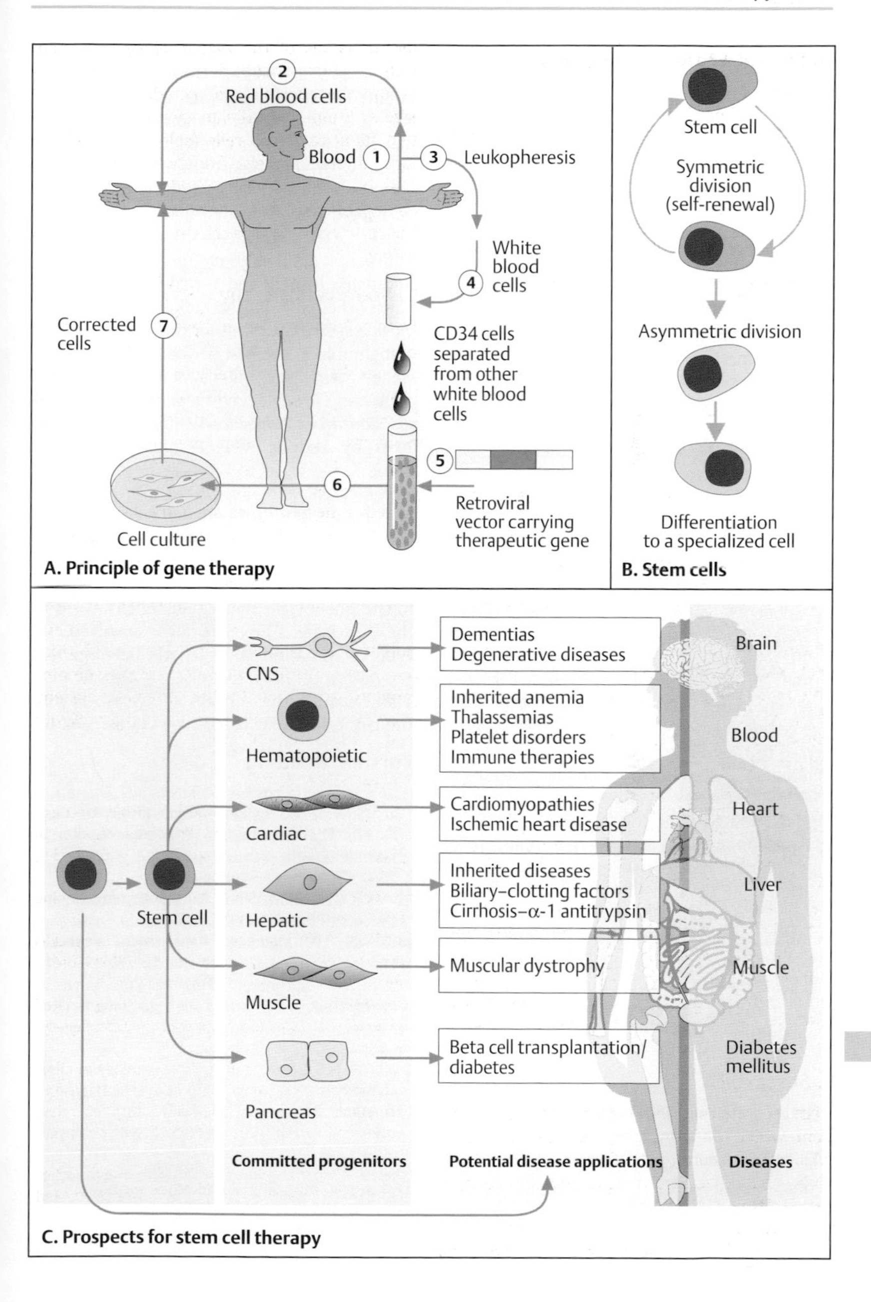

A. Principle of gene therapy

B. Stem cells

C. Prospects for stem cell therapy

Chromosomal Locations of Human Genetic Diseases

The completed DNA sequence of the whole human genome and each of the 24 chromosomes has resulted in precise knowledge of gene loci that are subjected to disease-causing mutation or rearrangements. The following five plates provide a limited overview for selected examples of monogenic disorders. The whole information is no longer accessible in print, but only online at Online Mendelian Inheritance in Man (OMIM, see Further Reading). The systematic collection of information about human genes and phenotypes is documented in 12 published editions of *Mendelian Inheritance in Man. A Catalog of Human Genes and Disorders* (MIM) by Victor A. McKusick, MD (1921–2008) at Johns Hopkins University School of Medicine. Its first edition in 1966 contained a total of 1487 entries. The second edition (1545 entries) in 1968 included the the first autosomal gene to be mapped. The subsequent editions reveal an entry-doubling time of about 15 years (3368 in the 6th edition (1983), 5710 entries in the 10th edition (1992), and 8587 entries in the 12th edition (1998). In 1987, the McKusick catalog became internationally available online as OMIM from the National Library of Medicine. McKusick referred to this disease gene map as *The Morbid Anatomy of the Human Genome* (McKusick, 1998; Amerberger et al., 2001). This map of disease-related gene loci assigned to specific chromosomal sites, *The Morbid Anatomy of the Human Genome*, first appeared in the 3rd edition in 1971 on a single page.

Regularly updated, OMIM is a major source of information on human genes and genetic diseases. Each entry has a unique 6-digit identifying number and is assigned to one of six catalogs according to genetic criteria: (1) autosomal dominant inheritance, (2) autosomal recessive, (3) X-chromosomal, (4) Y-chromosomal, and (5) mitochondrial. Autosomal entries initiated in 1994 begin with the digit 6.

For an understanding of the genetic causes of human diseases, this compilation can be compared with the impact on medicine by the seven volumes of *De humani coporis fabrica libri septa* (die "Fabrica") in 1543 by Andreas Vesalius (1514–1564) and the causal analysis of diseases *De Sedibus et Causis Morborum per Anatomen Indagatis* in 1761 by Giovanni Morgagni (1682–1771). The McKusick catalog provides a systematic basis for the genetics of man comparable to the first periodic table of chemical elements by Dimitri I. Mendeleyev in 1869 or to the *Chronologisch-thematisches Verzeichnis sämtlicher Tonwerke Wolfgang Amadé Mozarts* by Ludwig Alois Ferdinand Köchel in 1862.

The McKusick catalog OMIM also reflects an important difference between customary clinical medicine and medical genetics. Whereas medicine classifies diseases according to their main manifestations and other criteria related to the phenotype, medical genetics focuses on the genotype. The gene locus involved, the type of mutation, and genetic heterogeneity provide the basis for disease classification. This expands the concept of disease beyond the clinical manifestation (see Childs, 1999).

Further Reading

Amberger JS, Hamosh A, McKusick VA. The morbid anatomy of the human genome. In: Scriver CR, et al., eds. The Metabolic and Molecular Bases of Inherited Disease. 8th ed. New York: McGraw-Hill; 2001:47–111

Childs B. Genetic Medicine. Baltimore: Johns Hopkins University Press; 1999

McKusick VA. Mendelian Inheritance in Man. A Catalog of Human Genes and Genetic Disorders. 12th ed. Baltimore: Johns Hopkins University Press; 1998. (www.ncbi.nlm.nih.gov/omim)

OMIM Statistics: total number of entries: 20 994 (16 April, 2012)

13 140 autosomal, 641 X-linked, 48 Y-linked, 35 mitochondrial

3458 phenotypes with known molecular basis: 3164 autosomal, 262 X-linked, 4 Y-linked, 28 mitochondrial

1632 phenotypes without known molecular basis

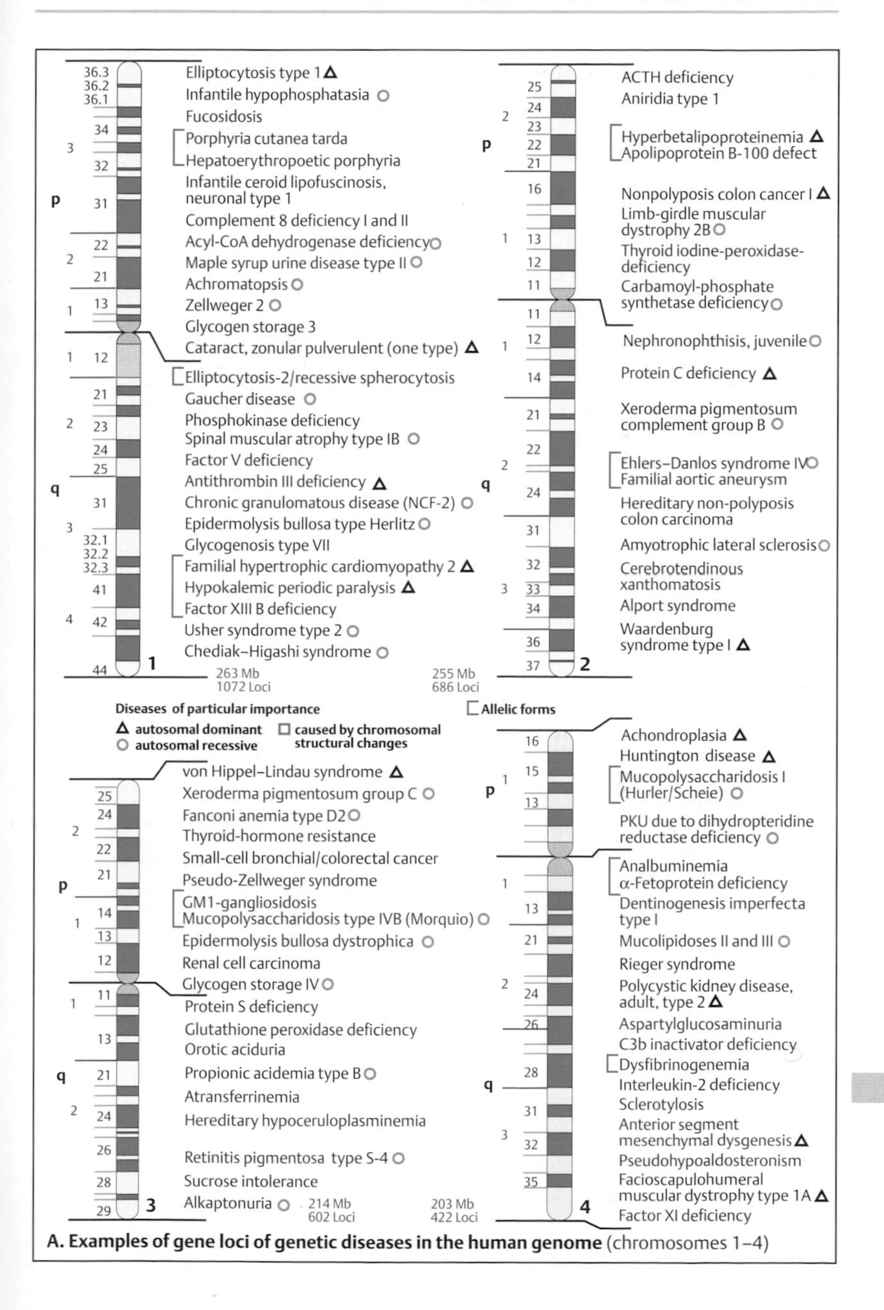

A. Examples of gene loci of genetic diseases in the human genome (chromosomes 1–4)

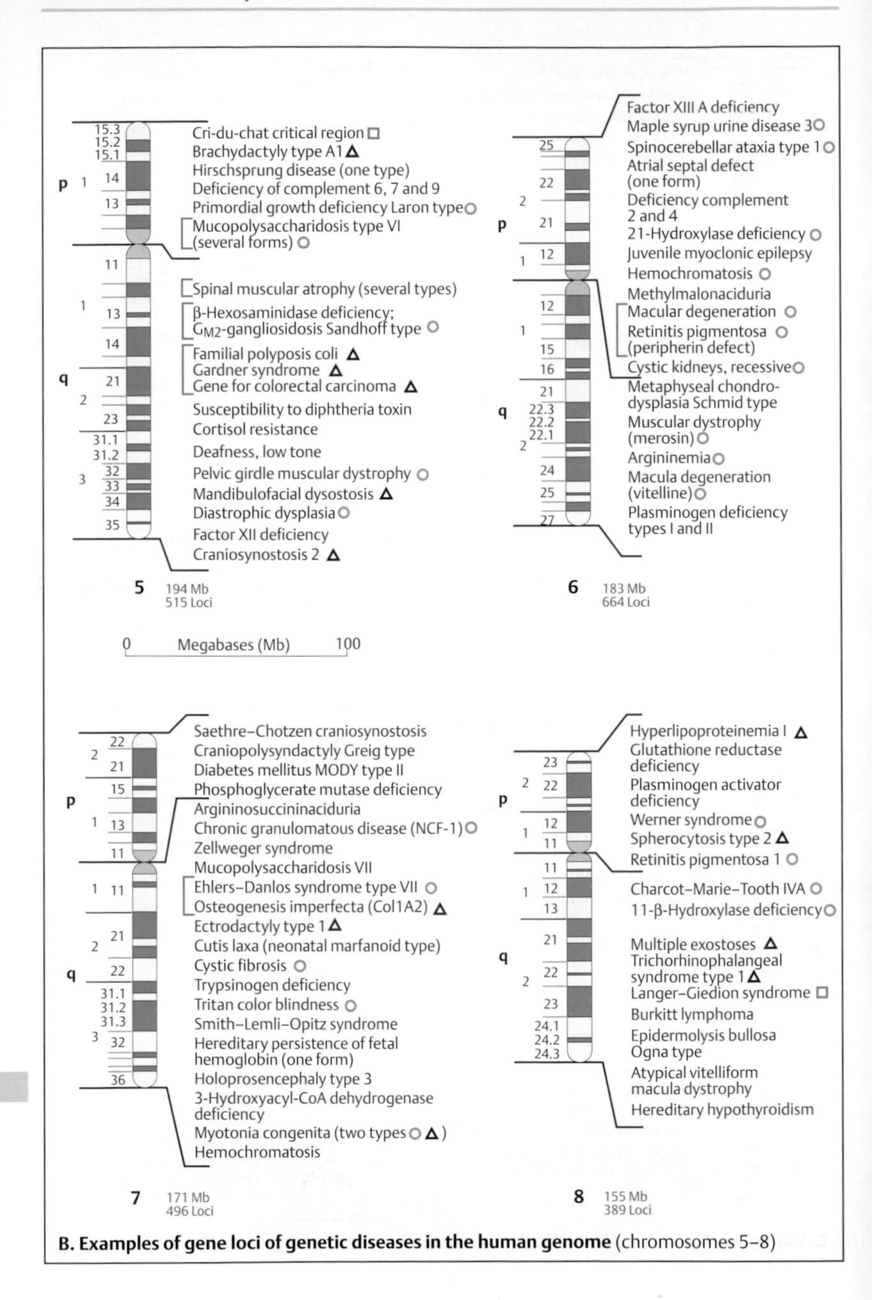

B. Examples of gene loci of genetic diseases in the human genome (chromosomes 5–8)

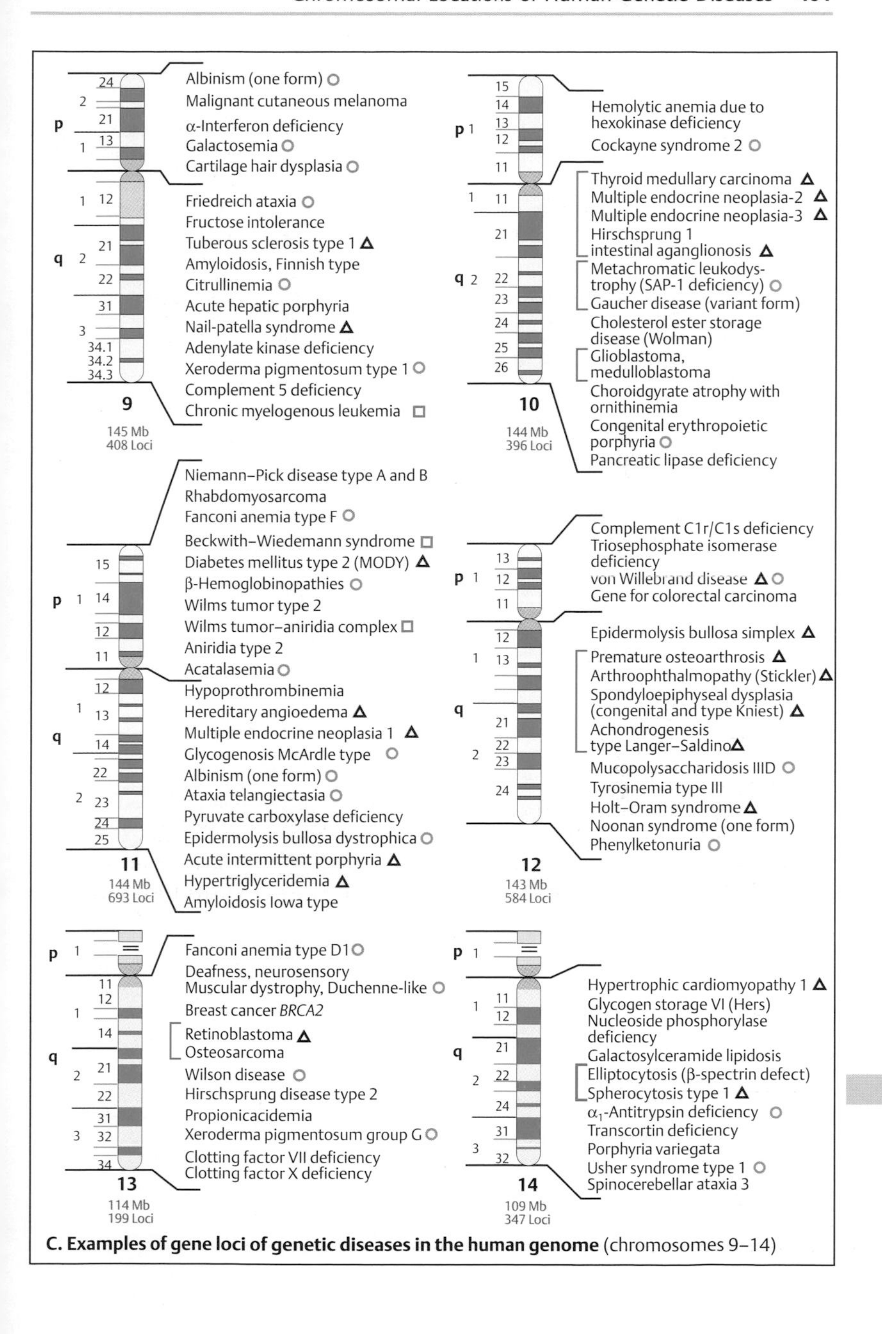

C. Examples of gene loci of genetic diseases in the human genome (chromosomes 9–14)

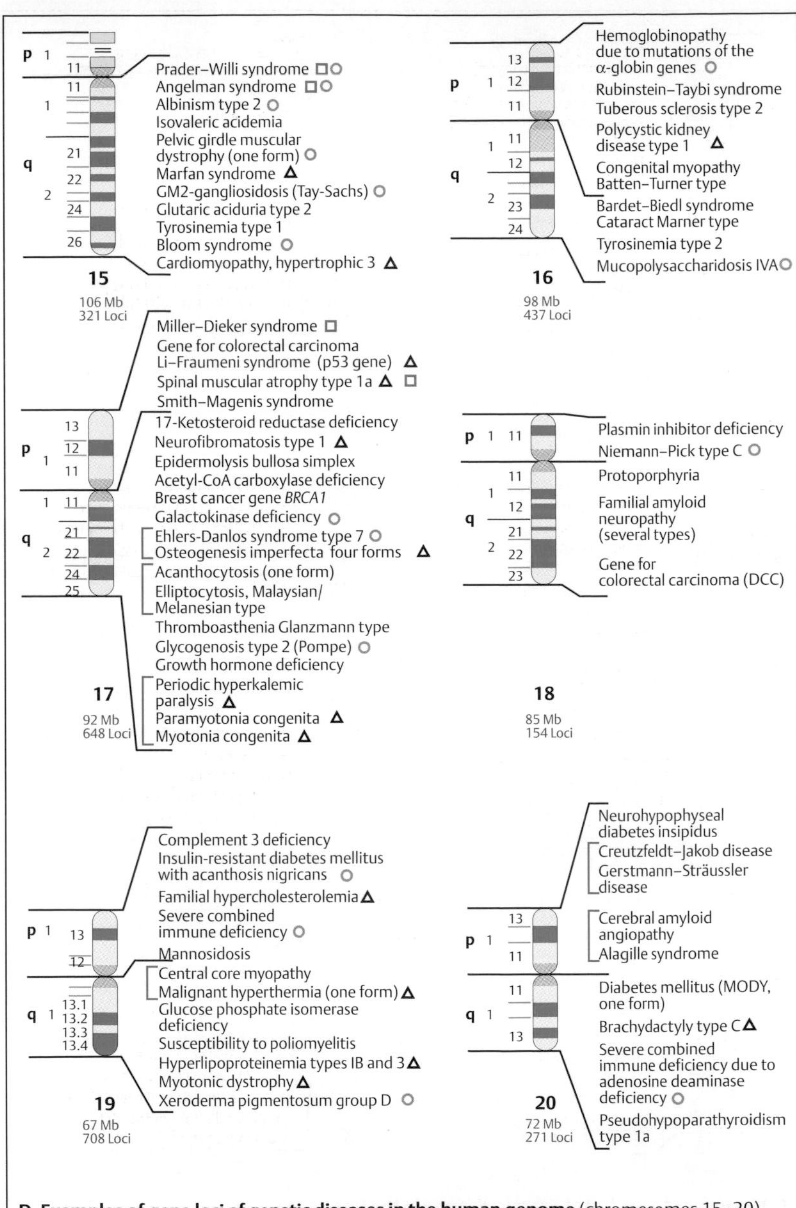

D. Examples of gene loci of genetic diseases in the human genome (chromosomes 15–20)

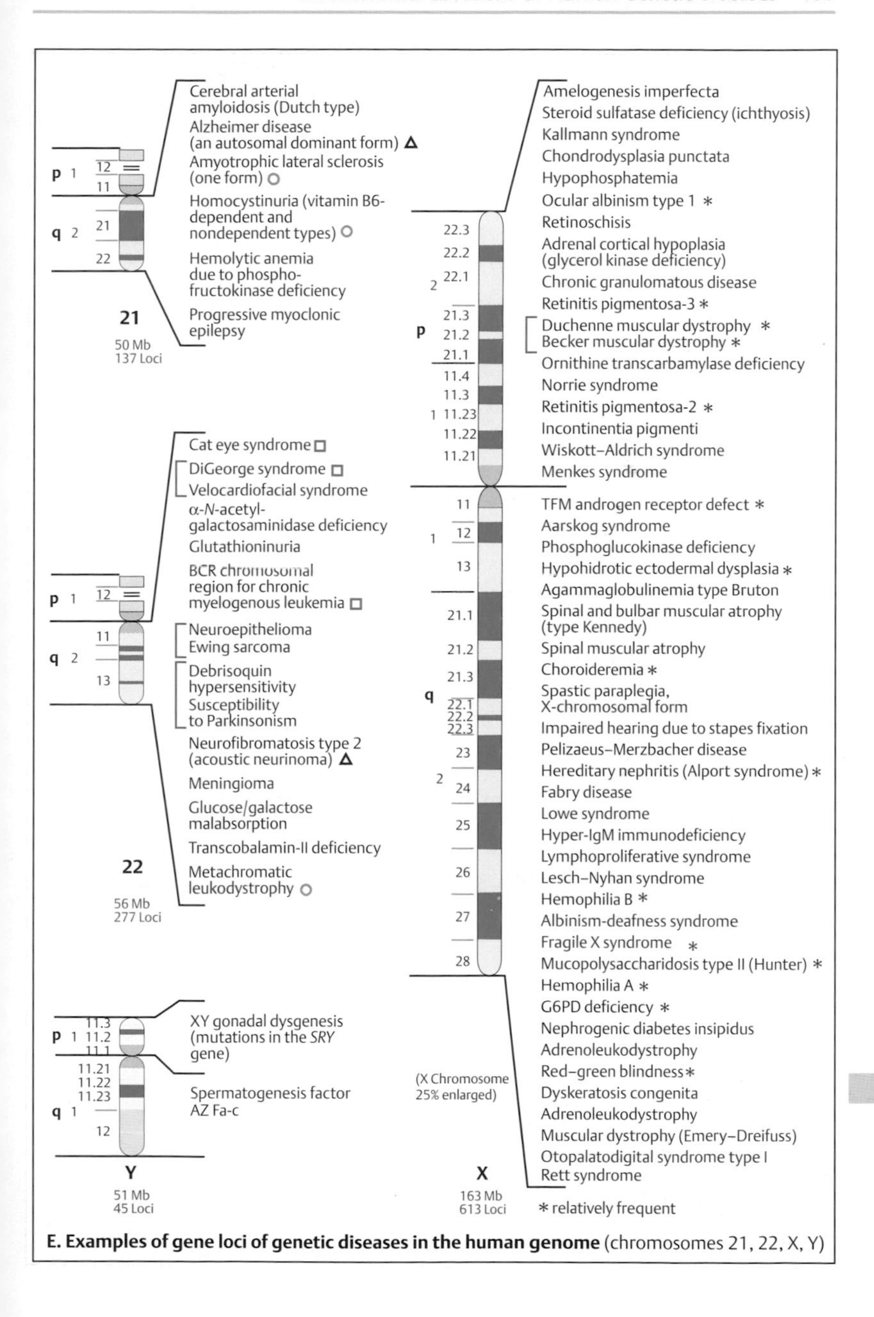

E. Examples of gene loci of genetic diseases in the human genome (chromosomes 21, 22, X, Y)

Chromosomal Locations—Alphabetical List

(Alphabetic list to the maps on pp. 399–403; ch. = chromosome)

Aarskog syndrome (X ch.)
Acanthocytosis (one form) (ch. 17)
Acatalasemia (ch. 11)
Acetyl-CoA carboxylase deficiency (ch. 17)
Achromatopsia (ch. 1)
Achondroplasia (ch. 4)
ACTH deficiency (ch. 2)
Acute hepatic porphyria (ch. 9)
Acute intermittent porphyria (ch. 11)
Acyl-CoA dehydrogenase deficiency (ch. 1)
Adenylate kinase deficiency (ch. 9)
Adrenal cortical hypoplasia with glycerol kinase deficiency (X ch.)
Adrenoleukodystrophy (X ch.)
α-Interferon deficiency (ch. 9)
Agammaglobulinemia (X ch.)
Alagille syndrome (ch. 20)
Albinism (one form) (ch. 9)
Albinism (one form) (ch. 11)
Albinism type 2 (ch. 15)
Albinism—deafness syndrome (X ch.)
Alkaptonuria (ch. 3)
α-*N*-acetylgalactosaminidase deficiency (ch. 22)
α_1-Antitrypsin deficiency (ch. 14)
α-Fetoprotein deficiency (ch. 4)
Alport syndrome (ch. 2)
Alzheimer disease (one form) (ch. 21)
Amelogenesis imperfecta (X ch.)
Amyloidosis, Finnish type (ch. 9)
Amyloidosis, Iowa type (ch. 11)
Amyotrophic lateral sclerosis (one form) (ch. 21)
Amyotrophic lateral sclerosis, juvenile (ch. 2)
Analbuminemia (ch. 4)
Androgen receptor defect (X ch.)
Angelman syndrome (ch. 15)
Aniridia type 1 (ch. 2)
Aniridia type 2 (ch. 11)
Anterior segmental mesenchymal dysgenesis (ch. 4)
Antithrombin III deficiency (ch. 1)
Apert syndrome (ch. 4)
Apolipoprotein B-100 defect (ch. 2)
Argininemia (ch. 6)
Argininosuccinicaciduria (ch. 7)
Arthroophthalmopathy (Stickler syndrome) (ch. 12)
Aspartylglucosaminuria (ch. 4)
Ataxia-telangiectasia (ch. 11)
Atransferrinemia (ch. 3)
Atrial septal defect (one form) (ch. 6)
Atypical vitelliform macular dystrophy (ch. 8)
Bardet—Biedl syndrome (ch. 16)
BCR chromosomal region for chronic myelogenous leukemia (ch. 22)
Becker muscular dystrophy (X ch.)
Beckwith—Wiedemann syndrome (ch. 11)
β-Hemoglobinopathies (ch. 11)
β-Hexosaminidase deficiency; GM2 gangliosidosis type Sandhoff (ch. 5)
11-β-Hydroxylase deficiency (ch. 8)
Bloom syndrome (ch. 15)
Brachydactyly type A1 (ch. 5), type C (ch. 20)
Breast cancer gene *BRCA1* (ch. 17)
Breast cancer gene *BRCA2* (ch. 13)
Burkitt lymphoma (ch. 8)
Carbamoyl phosphate synthetase-I deficiency (ch. 4)
Cardiomyopathy, familial hypertrophic type 3 (ch. 15)
Cartilage—hair dysplasia (ch. 9)
Cat eye syndrome (ch. 22)
C3b-inactivator deficiency (ch. 4)
Central core myopathy (ch. 19)
Cerebral amyloid angiopathy (ch. 20)
Cerebral arterial amyloidosis (Dutch type) (ch. 21)
Cerebrotendinosis xanthomatosis (ch. 2)
Charcot—Marie—Tooth neuropathy type1b and type 2 (ch. 17) (spinal muscular atrophy)
Charcot—Marie—Tooth neuropathy type IVa (ch. 8)
Chediak—Higashi syndrome (ch. 1)
Cholesteryl ester storage disease (Wolman) (ch. 10)
Chondrodysplasia punctata (X ch.)
Choroid gyrate atrophy with ornithinemia (ch. 10)
Choroideremia (X ch.)
Chronic granulomatous disease (NCF-1 deficiency) (ch. 7)
Chronic granulomatous disease (NCF-2 deficiency) (ch. 1)
Chronic granulomatous disease (X ch.)
Chronic myelogenous leukemia (ch. 9)
Citrullinemia (ch. 9)
Clotting factor VII deficiency (ch. 13)

Hepatoerythropoietic porphyria (ch. 1)
Hereditary angioedema (ch. 11)
Hereditary congenital hypothyroidism (ch. 8)
Hereditary hypoceruloplasminemia (ch. 3)
Hereditary nephritis (Alport syndrome) (X ch.)
Hereditary non-polyposis colon carcinoma (ch. 2)
Hereditary persistence of fetal hemoglobin (one form) (ch. 7)
Hirschsprung disease (chs. 10 and 13)
Holoprosencephaly type 3 (ch. 7)
Holt–Oram syndrome (ch. 12)
Homocystinuria (B6-responsive and B6-nonresponsive forms) (ch. 21)
Huntington disease (ch. 4)
3-Hydroxyacyl-CoA dehydrogenase deficiency (ch. 7)
21-Hydroxylase deficiency (ch. 6)
Hyperbetalipoproteinemia (ch. 2)
Hyper-IgM immune deficiency (X ch.)
Hyperlipoproteinemia type 1 (ch. 8)
Hyperlipoproteinemia type 1b (ch. 19)
Hyperlipoproteinemia type 3 (ch. 19)
Hypertriglyceridemia (ch. 11)
Hypertrophic cardiomyopathy (ch. 14)
Hypochondroplasia (ch. 4)
Hypohidrotic ectodermal dysplasia (X ch.)
Hypophosphatemia (X ch.)
Hypoprothrombinemia (ch. 11)
Immune deficiency, severe combined (ch. 19)
Immunodeficiency due to ADA deficiency (ch. 20)
Impaired hearing (lower frequencies) (ch. 5)
Impaired hearing due to stapes fixation (X ch.)
Infantile ceroid lipofuscinosis, neuronal type (ch. 1)
Infantile hypophosphatasia (ch. 1)
Insulin-resistant diabetes mellitus with acanthosis nigricans (ch. 19)
Interleukin 2 deficiency (ch. 4)
Intestinal aganglionosis (Hirschsprung) (ch. 10 and 13)
Isovaleric acidemia (ch. 15)
Juvenile myoclonic epilepsy (ch. 6)
Kallmann syndrome (X ch.)
17-Ketosteroid reductase deficiency (ch. 17)
Lamellar cataract (one type) (ch. 1)
Langer–Giedion syndrome (ch. 8)
Lesch–Nyhan syndrome (X ch.)
Li–Fraumeni syndrome (ch. 17)
Limb–girdle muscular dystrophy 2b (ch. 2)
Lowe syndrome (X ch.)
Lymphoproliferative syndrome (X ch.)
Macular degeneration, vitelline (ch. 6)
Malignant cutaneous melanoma (ch. 9)
Malignant hyperthermia (ch. 19, others)
Mandibulofacial dysostosis (Franceschetti–Klein syndrome) (ch. 5)
Mannosidosis (ch. 19)
Maple syrup urine disease type 2 (ch. 1)
Maple syrup urine disease type 3 (ch. 6)
Marfan syndrome (ch. 15)
Meningioma (ch. 22)
Menkes syndrome (X ch.)
Metachromatic leukodystrophy (ch. 22)
Metachromatic leukodystrophy (SAP-1 deficiency) (ch. 10)
Metaphyseal chondrodysplasia type Schmid (ch. 6)
Methylmalonicaciduria (ch. 6)
Miller–Dieker syndrome (ch. 17)
Morquio syndrome B (ch. 3)
Mucolipidosis types II and III (ch. 4)
Mucopolysaccharidosis type I (Hurler/Scheie) (ch. 4)
Mucopolysaccharidosis type II (X ch.)
Mucopolysaccharidosis type IVa (ch. 16)
Mucopolysaccharidosis type IVb (ch. 3)
Mucopolysaccharidosis type VI (Maroteaux–Lamy) (ch. 5)
Mucopolysaccharidosis type VII (ch. 7)
Multiple endocrine neoplasia type 1 (ch. 11)
Multiple endocrine neoplasia type 2 (ch. 10)
Multiple endocrine neoplasia type 3 (ch. 10)
Multiple exostoses (ch. 8)
Muscular dystrophy, Becker type (X ch.)
Muscular dystrophy, Duchenne type (X ch.)
Muscular dystrophy, Emery–Dreifuss type (X ch.)
Muscular dystrophy, Duchenne-like (ch. 13)
Muscular dystrophy, merosin (ch. 6)
Myotonia congenita (ch. 17)
Myotonia congenita (two types) (ch. 7)
Myotonic dystrophy (ch. 19)
Myotubular myopathy (X ch.)
Nail–patella syndrome (ch. 9)
Nephrogenic diabetes insipidus (X ch.)
Nephronophthisis, juvenile (ch. 2)
Neuroepithelioma (ch. 22)
Neurofibromatosis type 1 (ch. 17)
Neurofibromatosis type 2 (acusticus neurinoma) (ch. 22)
Neurohypophyseal diabetes insipidus (ch. 20)
Niemann–Pick disease type A and type B (ch. 11)
Niemann–Pick type C (ch. 18)

Noonan syndrome (one locus) (ch. 12)
Norrie syndrome (X ch.)
Nucleoside phosphorylase deficiency (ch. 14)
Ocular albinism (X ch.)
Ornithine transcarbamylase deficiency (X ch.)
Oroticacidemia (ch. 3)
Osteogenesis imperfecta (ch. 17)
Osteogenesis imperfecta (COL1 A2) (ch. 7)
Osteosarcoma (ch. 13)
Otopalatodigital syndrome type 1 (X ch.)
Pancreatic lipase deficiency (ch. 10)
Paramyotonia congenita (ch. 17)
Pelizaeus—Merzbacher disease (X ch.)
Pelvic girdle muscular dystrophy (ch. 5 and ch. 15)
Periodic hyperkalemic paralysis (ch. 17)
Phenylketonuria (PKU) (ch. 12)
Phosphoglucokinase deficiency (X ch.)
Phosphoglycerate mutase deficiency (ch. 7)
Phosphokinase deficiency (ch. 1)
PKU due to dihydropteridine reductase deficiency (ch. 4)
Plasmin inhibitor deficiency (ch. 18)
Plasminogen activator deficiency (ch. 8)
Plasminogen deficiency, types I and II (ch. 6)
Polycystic kidney disease (ch. 16 and 4)
Porphyria cutanea tarda (ch. 1)
Porphyria variegata (ch. 14)
Prader—Willi syndrome (ch. 15)
Primordial growth deficiency, Laron type (ch. 5)
Progressive myoclonic epilepsy (ch. 21)
Propionicacidemia type A (ch. 13)
Propionicacidemia type B (ch. 3)
Protein C deficiency (ch. 2)
Protein S deficiency (ch. 3)
Protoporphyria (ch. 18)
Pseudoaldosteronism (ch. 4)
Pseudohypoparathyroidism type 1a (ch. 20)
Pseudo-Zellweger syndrome (ch. 3)
Red—green blindness (X ch.)
Renal cell carcinoma (ch. 3)
Retinitis pigmentosa (peripherin defect) (ch. 6)
Retinitis pigmentosa type 1 (ch. 8)
Retinitis pigmentosa type 2 (X ch.)
Retinitis pigmentosa type 3 (X ch.)
Retinitis pigmentosa type 4 (ch. 3)
Retinoblastoma (ch. 13)
Retinoschisis (X ch.)
Rett syndrome (X ch.)
Rhabdomyosarcoma (ch. 11)
Rieger syndrome (ch. 4)
Rubinstein—Taybi syndrome (ch. 16)
Saethre—Chotzen craniosynostosis (ch. 7)
Sclerotylosis (ch. 4)
Sex reversal (XY females resulting from mutation in the *SRY* gene) (Y ch.)
Small cell bronchial carcinoma/colorectal cancer (ch. 3)
Smith—Lemli—Opitz syndrome (ch. 7)
Smith—Magenis syndrome (ch. 17)
Spastic paraplegia (X chromosomal form) (X ch.)
Spermatogenesis factors AZFa - c (Y ch.)
Spherocytosis type 1 (ch. 14)
Spherocytosis type 2 (ch. 8)
Spinal muscular atrophy (X ch.)
Spinal muscular atrophy Ia (ch. 17)
Spinal muscular atrophy type IB (ch. 1)
Spinal muscular atrophy IVa (ch. 8)
Spinal muscular atrophy Werdnig—Hoffmann and other types (ch. 5)
Spinocerebellar ataxia type 1 (ch. 6)
Spinocerebellar ataxia type 3 (ch. 14)
Spondyloepiphyseal dysplasia (congenital type) (ch. 12)
Spondyloepiphyseal dysplasia (Kniest type) (ch. 12)
Steroid sulfatase deficiency (ichthyosis) (X ch.)
Sucrose intolerance (ch. 3)
Susceptibility to diphtheria toxin (ch. 5)
Susceptibility to Parkinsonism (ch. 22)
Susceptibility to poliomyelitis (ch. 19)
T-cell leukemia/lymphoma (ch. 14)
TFM androgen receptor defect (X ch.)
Thrombasthenia, Glanzmann type (ch. 17)
Thyroid hormone resistance (ch. 3)
Thyroid iodine peroxidase deficiency (ch. 2)
Thyroid medullary carcinoma (ch. 10)
Transcobalamin II deficiency (ch. 22)
Transcortin deficiency (ch. 14)
Trichorhinophalangeal syndrome type 1 (ch. 8)
Triosephosphate isomerase deficiency (ch. 12)
Tritan color blindness (ch. 7)
Trypsinogen deficiency (ch. 7)
Tuberous sclerosis type 1 (ch. 9)
Tuberous sclerosis type 2 (ch. 19)
Tyrosinemia type 1 (ch. 14), type 2 (ch. 16), type 3 (ch. 12)
Usher syndrome type 1 (ch. 14), type 2 (ch. 1)
Velocardiofacial syndrome (ch. 22)
Vitelline macular dystrophy (ch. 6)
von Hippel—Lindau syndrome (ch. 3)
von Willebrand disease (ch. 12)
Waardenburg syndrome type 1 (ch. 2)
Werner syndrome (ch. 8)

Williams–Beuren syndrome (ch. 4)
Wilms tumor–aniridia complex (ch. 11)
Wilms tumor type 2 (ch. 11)
Wilson disease (ch. 13)
Xeroderma pigmentosum complementation group B (ch. 2)
Xeroderma pigmentosum group C (ch. 3)
Xeroderma pigmentosum group D (ch. 19)
Xeroderma pigmentosum group G (ch. 13)
Xeroderma pigmentosum type 1 (ch. 9)
XY gonadal dysgenesis (Y ch.)
Zellweger syndrome (ch. 7)
Zellweger syndrome type 2 (ch. 1)
Zonular cataract (ch. 1)

Caveat: There are numerous similar disorders caused by mutations of genes at other loci, sometimes with other modes of inheritance. This list and the corresponding maps are not complete, but only examples.
For a complete map and list see OMIM (www.ncbi.nlm.nih.gov/Omim).

Appendix—Supplementary Data

The following tables provide supplementary data for selected plates and their accompanying text on the topics listed below.

Table

1. Parallel DNA Sequencing (p. 68)

Examples of massively parallel DNA sequencing methods

Method	Representative	Read length	Output
Sequencing by synthesis of PCR-amplified DNA	Roche GS	> 300 nt	> 0.45 Gb per run, 7 hours
	Illumina	Approx. 100 nt	18 Gb per run, 4 days
Sequencing by ligation of PCR-amplified DNA	ABI	Approx. 100 nt	30 Gb, 7 days
	SOLiD	Variable	50 Gb, 14 days

Modified from Strachan & Read (2011).
Gb, giga base; nt, nucleotide; PCR, polymerase chain reaction.

2. Apoptosis (p. 110)

Examples of the main players in apoptosis

Protein	Synonym	Effect on apoptosis	Gene locus
FAS	APT1, CD95, Apo-1, Fas1	+	10q24
FADD	MORT-1	+	11q13
Caspase 2	ICH1, NEDD2	+	7q35
Caspase 3	CPP32B, NEDD2	+	4q33
Caspase 4	TX, ICH-2, ICE-rel-II	+	11q22
Caspase 6	MCH2	+	4q25
Caspase 7	MCH3, ICE-LAP3, CMH3	+	10q25
Caspase 8	MACH, MCH5, FLICE	+	2q33
Caspase 9	APAF3, MCH6, ICE-LAP6	+	1p36.21
Caspase 10	MCH4	+	2q33
APAF-1	CED4	+	12q23.1
Bcl-2		-	18q21
Bak1	Bcl-2L7	+	6p21
Bax		+	19q13
Bid		+/-	22q11
Bid	NBK	+	22q13.3

Data from: Nagata S. DNA degradation in development and programmed cell death. Annu Rev Immunol 2005;23:821–852.

3. The Banding Pattern of Human Chromosomes (p. 158)

Examples of types of commonly used chromosome bands

Type	Goal of application	Principal method
G	Shows euchromatic light bands and heterochromatic dark bands	Pretreatment with trypsin
Q	Darkfield fluorescence; bright fluorescence of heterochromatin, especially Yq	Quinacrine-induced, Hoechst 33258
R	Reverse of G-bands	Pretreatment with alkali at 80–90°C
C	Preferential stain of centromeres and constitutive heterochromatin	Pretreatment with acid and barium hydroxide
T	Preferential stain of telomeres	Telomere-specific DNA probes
DAPI	AT-specific fluorescence (4′,6-diamidino-2-phenylindole)	Enhanced chromosome fluorescence
SCN	Analysis of sister chromatid exchanges	5-Bromodeoxyuridine for two divisions
NOR	Selective staining of nucleolus organizing regions of acrocentric chromosomes	Silver nitrate staining
DAPI/ distamycin	Selective staining of pericentric regions of chromosomes 1, 9, 15, 16, Yq	Application of DAPI and distamycin

Data from Chromosome Staining and Banding Techniques. Available at: http://homepage.mac.com/wildlifeweb/cyto/text/Banding.html. Accessed January 29, 2012.

4. Nomenclature of Human Chromosome Aberrations

Human chromosome nomenclature (p. 158)

Abbreviation	Explanation in detail
46,XX	46 chromosomes including two X chromosomes (normal female karyotype)
46,XY	46 chromosomes including one X and one Y chromosome (normal male karyotype)
47,XXY	47 chromosomes including two X chromosomes and one Y chromosome
47,XXX	47 chromosomes including three X chromosomes
47, XY, +21	47 chromosomes including an X and a Y chromosome and an additional chromosome 21 (trisomy 21)
13p	The short arm of a chromosome 13
13q	The long arm of a chromosome 13
13p14	Region 1, band 4 of a chromosome 13
13q14.2	Region 1, band 4.2 of a chromosome 13
2q−	Loss of chromosome material from the long arm of a chromosome 2
t(2;5)(q21;q31)	Reciprocal translocation between a chromosome 2 and a chromosome 5 with breakpoints in 2q21 and 5q31
t(13q14q)	Centric fusion between a chromosome 13 and a chromosome 14
der	A derivative chromosome as a result of a rearrangement
dic(Y)	Dicentric Y chromosome (two centromeres)
del(2)(q21-qter)	Deletion in chromosome 2 from q21 until the telomere
fra(X)(q27.3)	Fragile site at q27.3 of an X chromosome
dup(1)	Duplication in a chromosome 1
h	Heterochromatin, constitutive (at centromeres)
i	Isochromosome, e.g., i(Xq) for the long arm of an X chromosome
ins(5;2)(p14; q22q32)	Insertion of the region q22-q32 of a chromosome 2 into region p14 of a chromosome 5
inv (9)(p11q21)	Inversion of a chromosome 9 between p11 and q21 (indicating the breakpoints)
invdup(15)	Inverted duplication in a chromosome 15
mat	Maternal origin
pat	Paternal origin
r(13)	Ring-shaped chromosome 13 (due to a partial deletion)

Data from Shaffer LG, Slovak ML, Campbell LJ, eds. ISCN. An International System for Human Cytogenetic Nomenclature. Basel: Karger; 2009.

5. Diseases and G Proteins (p. 200)

Diseases caused by mutations in G-protein-coupled receptors or G proteins[a]

Disease	OMIM no.	Mutated protein	Type pf inheritance
McCune–Albright syndrome	174800	$G\alpha_s$ subunit of GNAS1 (gain-of-function)	Somatic, 20q13.32
Hypocalciuric hypercalcemia	145980; 601199	Ca^{2+}-sensing receptor; *CASR* gene	AD, 3q21.1
Diabetes insipidus	304800	Arginine vasopressin receptor-2; G-protein subunit $G\alpha_s$ (loss-of-function)	X-chromosomal (Xq28)
Congenital hypothyroidism	275200	$G\alpha_s$ subunit of TSH receptor	AR, 14q31.1
Congenital hypothyroidism	188545	Thyrotropin releasing hormone receptor	AD, 8q23.1

[a] Examples only, genetically heterogeneous. See OMIM. Available at: http://www.ncbi.nlm.nih.gov/omim. Accessed January 29, 2012.
AD, autosomal dominant; AR, autosomal recessive; TSH, thyroid stimulating hormone.

6. Diseases Related to the Hedgehog Signaling Network (p. 204)

Hedgehog and related proteins

Name	Abbreviation	Gene locus	OMIM no.
Sonic hedgehog	SHH	7q36	600725
Indian hedgehog	IHH	2q33-q35	699726
Desert hedgehog	DHH	12q13.1	605423
Hedgehog acetyltransferase	HHAT	1q32	605743
Hedgehog interacting protein	HHIP	4q28-q32	606178
Smoothened	SMOH	7q31-q33	601500
Patched	PTCH	9q22.3	601309

Data from OMIM. Available at: http://www.ncbi.nlm.nih.gov/omim. Accessed January 29, 2012.

Mutations and deletions in more than 10 human genes in the hedgehog gene network result in a group of malformation syndromes: holoprosencephaly (OMIM 236100, 142945, and others, see table below) and accompanying brain malformations. Mutations in *PTCH* (9q22) cause basal cell nevus syndrome, Gorlin–Goltz type (OMIM 109400). Mutations in *SMO* (7q31) are found in some basal cell carcinomas and medulloblastomas. The segment polarity gene *Ci* is an ortholog of the *Gli* gene family in vertebrates. This family consists of three distinct Gli proteins: Gl1, Gl2, and Gl3.

Human *GLI1* is expressed in nearly all isolated basal cell carcinomas (OMIM 139150). *GLI3* mutations cause cephalopolydactyly, Greig type (OMIM 175700, 145400) and Pallister–Hall syndrome (hypothalamic hamartoblastoma; OMIM 146510). Loss-of-function mutations of *Ptch* and *Smo* in *Drosophila* have very similar phenotypes.

Gene loci involved in the causes of holoprosencephaly

Type of holoprosencephaly	Gene locus	OMIM no.
HPE1	21q22.3, 2q37.1-q37.3	236100
HPE2	2p21	157170
HPE3	7q36	142945
HPE4	18p11.3	142946
HPE5	13q32	609637
HPE6	2q37.1-q37.3	605934
HPE7	9q22.3	601309
HPE8	14q13	609408
HPE9	2q14.2	264480

Data from OMIM. Available at: http://www.ncbi.nlm.nih.gov/omim. Accessed January 29, 2012.

7. Inherited Cardiac Arrhythmias (p. 250)

Inherited cardiac arrhythmias (long QT syndromes)[a]

Type	Gene	Locus	Common age of onset	OMIM no.[b]
LQT 1	*KCNQ1, KCNA9*	11p15.5	Childhood (90% by age 20 years)	192500
LQT 2	*KCNH2*	7q35-q36	Young adults (gene formerly known as *HERG*)	152427
LQT 3	*SCN5A*	3p21–24	Young adults	603830
LQT 4	*ANK2*[c]	4q25	Adults	600919
LQT 5	*KCNE1*	21q22	Children	176261
LQT 6	*KCNE2*	21q22	Adults	603796
LQT 7	*KCNJ2*	17q23	Adults (Andersen syndrome)	600681
LQT 8	*CACNA1*	12p13.3	Syndactyly, immune defect;	601005

Continued ►

Table Continued

Type	Gene	Locus	Common age of onset	OMIM no.[b]
LQT 9	*C*	3	children and adults	611818
LQT 10	*CAV3*	3p25.3		611819
LQT 11	*SNC4B*	11q23.3		611820
LQT 12	*AKAP9*	7q21.2		612955
LQT 13	*SNT1; KCNJ5*	20q11.21; 11q24.3		613485

Syndromic forms associated with deafness: Romano–Ward syndrome (OMIM 192500), and Jervell and Lange-Nielsen syndrome (OMIM 220400).
[a] Other forms of inherited arrhythmias exist.
[b] OMIM. Available at: http://www.ncbi.nlm.nih.gov/omim. Accessed January 29, 2012.
[c] Ankyrin-2 (Ank-B in mice).

8. Mitochondrial Diseases (p. 262)

Examples of diseases due to mutations or deletions in human mitochondrial DNA[a]

OMIM no.	Disease name	Abbreviation
530000	Kearns–Sayre syndrome (ophthalmoplegia, pigmentary retinal degeneration, cardiomyopathy)	KSS
535000	Leber hereditary optic atrophy	LHON
540000	Mitochondrial myopathy, encephalopathy, lactic acidosis	MELAS
545000	Myoclonius epilepsy with ragged red fibers in muscle	MERRF
551500	Neuropathy, ataxia, retinitis pigmentosa	NARP
603041	Mitochondrial neurogastrointestinal encephalopathy[b]	MNGIE
557000	Pearson marrow-pancreas syndrome	PEAR
515000	Chloramphenicol-induced toxicity	
580000	Deafness, aminoglycoside-induced (mutation A1555G)	
520000	Diabetes–deafness syndrome, maternally transmitted	

[a] Data from OMIM. Available at: http://www.ncbi.nlm.nih.gov/omim. Accessed January 29, 2012
[b] This may be autosomal recessive.

9. Distal Cholesterol Biosynthesis Pathway (p. 270)

Diseases of the distal (post-squalene) cholesterol biosynthesis pathway

Disease	OMIM no.	Locus	Gene	Main manifestation
Antley–Bixler syndrome (some)	207410, 601637	7q21.2	*CYP51*	Skeletal dysplasias, choanal atresia, radioulnar synostosis
Greenberg dysplasia	215140 600024	1q42.1	*LBR*	Hydrops fetalis, ectopic calcifications, prenatal lethal
CHILD syndrome	308050 300275	Xq28	*NSDHL*	Hemidysplasia, ichthyosiform erythrodermia, limb defects
Chondrodysplasia punctata type 2	302960	Xp11.2	*EBP*	Skeletal dysplasias, calcifications over joints, short extremities
Lathosterolosis	607330	11q13	*SC5DL*	Facial dysmorphism, simlar to SLOS, mental retardation
Smith–Lemli–Opitz syndrome (SLOS)	270400 602858	11q13	*DHCR7*	Skeletal anomalies, facial dysmorphia, malformations
Desmosterolosis	603398	1p31-33	*DHCR24*	Facial dysmorphism, short limbs, perinatal letha

For enzymes involved see page 271.
CYP51, gene encoding cytochrome p51; *DHCR7*, gene encoding 7-dehydrocholesterol; *DHCR24*, gene encoding desmosterol reductase; *EBP*, gene encoding emopamil binding protein; *LBR*, gene encoding laminin receptor B; *NSDHL*, gene encoding sterol-4-alpha-carboxylate 3-dehydrogenase, decarboxylating; *SC5DL*, gene encoding lathosterol oxidase. (For details, see OMIM and GeneReviews. Available at: http://www.ncbi.nlm.nih.gov/omim and http://www.ncbi.nlm.nih.gov/sites/GeneTests. Accessed January 29, 2012.

10. Mucopolysaccharide Storage Diseases (p. 280)

Classification of mucopolysaccharidoses[a]

Type	OMIM no.	Enzyme defect	Gene locus	Main manifestation
IH Hurler	252800	α-L-Iduronidase	4p16.3	Dysostosis multiplex, developmental delay, corneal clouding
IS Scheie	252800	α-L-Iduronidase	4p16.3	Stiff joints, normal development
II Hunter	309900	Iduronate sulfatase	Xq28	Similar to MPS IH, no corneal clouding
MPS III (Sanfilippo)				
IIIA	252900	Heparin-*N*-sulfatase	17q25.3	Progression mental retardation
IIIB	252920	*α-N*-acetyl-glucosaminidase	17q21	Same
IIIC	252930	Acetyl-CoA:α-glucosaminidase	8p11.1	Same
IIID	252940	*N*-acetylglucosamine-6-sulfatase	12q14	Same
IV Morquio				
IVA	253000	*N*-acetylgalactosamine-6-sulfatase	16q24.3	Skeletal abnormalities
IVB	253010	β-Galactosidase	5q13-q14	Short stature
VI (Maroteaux–Lamy)	253200	*N*-acetylgalactosamine-4-sulfate	17q21	Dysostosis multiplex, normal mental development
VII (Sly)	253220	*α-N*-acetyl-glucosaminidase	17q21	Dysostosis multiplex, corneal clouding
IX	601492	Hyaluronidase	3p21.2-p21.3	Soft tissue masses, short stature

Data adapted from: Neufeld EF, Muenzer J. The mucopolysaccharidoses. In: Scriver CR, et al., eds. The Metabolic and Molecular Bases of Inherited Disease. 8th ed. New York: McGraw-Hill; 2001:3421–3452. Available at: http://www.ommbid.com/. Accessed January 27, 2012.

11. Genetic Immune Deficiency Diseases (p. 298)

Hereditary immune deficiency diseases (examples)

Disease	OMIM no.[a]	Gene locus	Gene	Inheritance
Disorders of the complement system	106100 217000 120700	11q11-q13.1 6p21.3 19p13.3-p13.2	*C1NH* *C2* *C3*	AD AR AD
Chronic granulomatous disease	306400 300481	Xp21.1	*CYBB*	X
Adaptive immunity				
Agammaglobulinemia[a], Bruton type	300300	Xq21.3-q22	*TK*	X
Severe combined immunodeficiency disease (T and B cells)	308300 308300	Xq213.3	*L2RG*	X
Adenosine deaminase deficiency	300400	20q13.1	*DA*	AR
Agammaglobulinemia[b] (μ-chain deficiency)	147020	14q32.33	*IGHM*	AR
DiGeorge syndrome	188400	del22q11	Several	Sporadic
Wiskott-Aldrich syndrome	301000	Xp11.23	*AS*	AR
Ataxia-telangiectasia	208900	11q23	*TM*	AR

[a] OMIM. Available at: http://www.ncbi.nlm.nih.gov/omim. Accessed January 29, 2012.
[b] Agammaglobulinemia types 1–6 (see OMIM 601495 phenotypic series).
AD, autosomal dominant; AR, autosomal recessive; X, X-chromosomal.

12. Oncogenic Chromosome Translocations (p. 312)

Examples of oncogenic chromosome translocations[a]

Translocation	Type of tumor	Genes involved
(9;22)(q34;q11)	Chronic myelogenous leukemia	*ABL/BCR*
(14;18)(q32;q21)	Follicular lymphoma	*BCL2, IgH*
(14;19)(q32;q13)	B-cell lymphocytic leukemia	*BCL3, IgH*
(8;14)(q24;q32)	Burkitt lymphoma, B-cell ALL	*MYC, IgH*
(11;14)(q13;q32)	Mantle cell lymphoma	*BCL, IgH*
(1;7)(p34;q35)	T-cell acute lymphocytic leukemia	*LCK, TCRB*
(4;11)(q21;q23)	Acute lymphocytic leukemia	*MLL, ALL1, HRX*
(3;21)(q26;q22)	Acute myeloid leukemia	*AML1, EAP, EV11*
(1;14)(p22;q32)	Mucosa-associated lymphoma (MALT)	*BCL10*
(21;22)(q22;q12)	Ewing sarcoma	*EWS, ERG*
(11;22)(q24;q12)	Ewing sarcoma	*EWS, FL11*

Data from: Morin PJ, Trent JM, Collins FS, Vogelstein B. Cancer genetics. In: Longo DL, et al., eds. Harrison's Principles of Internal Medicine. 18th ed. New York: McGraw-Hill; 2012:663–672.

13. Genes Involved in Fanconi Anemia (p. 318)

Gene	Locus	OMIM no.[a]
FA-A	16q24.3	227650
FA-B	Xp22.2	300514
FA-C	9q22.32	227645
FA-D1[b]	13q13.1	605724
FA-D2	3p25.3	227646
FA-E	6p21.31	600901
FA-F	11p14.3	603467
FA-G	9p13.3	602956
FA-I	15q25-q26	609053
FA-J	17q22	609054
FA-L	2p16.1	608111
FA-M	14q21.3	609054
FA-N[c]	16p12	610832
FA-O[d]	17q22	613390
FA-P	16p13.3	613951
FA-ZF	19q13.1	605859

Data from: Schindler D, Hoehn H, eds. Fanconi Anemia. Basel: Karger; 2007.
[a] OMIM. Available at: http://www.ncbi.nlm.nih.gov/omim. Accessed January 29, 2012.
[b] Same as *BRCA2*.
[c] Same as *PALB2*.
[d] Same as *RADC51*.

14. Genetic Diseases and FGF Receptors (p. 328)

Examples of genetic disorders due to mutations in FGF receptors[a]

Gene	Location	Disease	Main manifestation	OMIM no.[b]
FGFR1	8p11.2	Pfeiffer syndrome	Craniosynostosis, broad thumbs	101600
FGFR2	10p25.3	Apert syndrome	Craniosynostosis, fused digits	101200
		Crouzon syndrome	Craniosynostosis, ocular proptosis	123500
		Pfeiffer syndrome	Craniosynostosis, broad thumbs	101600
FGFR3	4p16.3	Achondroplasia	Short stature, bone dysplasias	100800
		Hypochondroplasia	Mild form of achondroplasia	146000
		Muenke syndrome	Asymmetric coronal stenosis	602849

[a] Many more exist; some with unique mutations.
[b] OMIM. Available at: http://www.ncbi.nlm.nih.gov/omim. Accessed January 29, 2012.

15. Collagen Molecules and Diseases (p. 330)

Important types of human collagen and diseases caused by mutations in their genes

Type	Molecular structure	Gene	Gene locus	Disease	OMIM no.[a]
I	$[\alpha1(I)_2\alpha2(II)]$	*COL1A1*	17q21-22	Osteogenesis imperfecta	120150
		COL1A2	7q22	Ehlers-Danlos syndrome	130000
II	$[\alpha1(II)_3]$	*COL2A1*	12q13.1	Stickler syndrome	108300
				Spondyloepiphyseal dysplasia	183900
				Achondrogenesis	200600
				Others	
III	$[\alpha1(III)_3]$	*COL3A1*	2q31	Ehlers-Danlos syndrome IV	225350
IV	$[\alpha1(IV)\alpha2(IV)]$ and others	*COL4A1*, *A2*	13q34 2q36	Alport syndrome autsomal	203780
		A3,A4 *A5,A6*	Xq22	Alport syndrome X-chromosomal	301050
V	$[\alpha1(V)_2\alpha2(V)]$	*COL5A1* *COL5A2*	9q34.2 2q31	Ehlers-Danlos syndrome I and II	130000

Data from: Byers PH. Disorders of collagen synthesis and structure. In: Scriver CR, et al., eds. The Metabolic and Molecular Bases of Inherited Disease. 8th ed. New York: McGraw-Hill; 2001:5241–5285. Available at: http://www.ommbid.com. Accessed January 28, 2012.
[a] OMIM. Available at: http://www.ncbi.nlm.nih.gov/omim. Accessed January 29, 2012.

16. Ciliopathies (p. 338)

Examples of genetic disorders with impaired ciliary function

Name	OMIM no.[a]	Gene	Locus
Alström syndrome	606844	*ALMS1*	2p13
Asphyxiating thoracic dysplasia	611177	*IFT80*	3q24-q26
Bardet–Biedl syndromes	209901	*BBS1–18*	18 Autosomal loci
Polycystic kidney diseases	173910	*PKD1*	16p13.3-p13.12
		PKD2	4q21-q23
AR polycystic kidney disease	606702	*PKHD1*	6p12
Joubert syndrome group	213300	*JBTS1–10*	10 Autosomal loci
Meckel syndrome group	249000	*MKS1–6*	6 Autosomal loci
Nephronophthisis group	256100	*NPHP1–11*	10 Autosomal loci

[a] OMIM. Available at: http://www.ncbi.nlm.nih.gov/omim. Accessed January 29, 2012.
AR, autosomal recessive.

17. Neurocristopathies (p. 340)

Genes involved in the causes of Hirschsprung disease

Gene	Location	Main effect	Penetrance (%)	OMIM no.
RET	10q11.2	Dominant, loss-of-function	50–72	164761
GDNF	5p13.1	Dominant/recessive	Unknown	600837
EDNRB	13q22	Recessive	8–85	131244
EDN3	20q13	Recessive	Unknown	131242
SOX10	22q13	Dominant/recessive	>80	602229
ECE1	1p36	Dominant/recessive	Unknown	600423
NTN, NRTN	19p13	Unknown	Unknown	601880
ZEB2 (formerly ZFHX1B SIP1)	2q22	Sporadic	Unknown	605802
PHOX2B	4p12	Congenital hypoventilation	Unknown	603851

In addition, susceptibility loci have been identified at 3p12, 4q31.3; 9q31, and 19q12.

18. Disorders of Sexual Development (p. 360)

Overview of disorders of sex differentiation

1	Defects of sex determination caused by mutation or structural aberration of *SRY* (XX males, XY gonadal dysgenesis, and others)
2	Disorders of testis development (*SF1*, *DAX1*, *WNT4*, *SOX9*, and other genes)
3	Defects of androgen biosynthesis (e.g., 21-hydroxylase deficiency)
4	Defect in steroid 5α-reductase (dihydrotestosterone deficiency)
5	Defects of androgen receptor (testicular feminization)
6	Defects of the Müllerian inhibition factor (hernia uteri syndrome)
7	XO/XY gonadal dysgenesis
8	Turner syndrome (45,X), Klinefelter syndrome (47,XXY)
9	True hermaphroditism XX/XY

19. Genes involved in Inherited Deafness (p. 376)

Examples of genes and proteins involved in hereditary deafness

Type of protein	Main function	Gene	DFN type	OMIM no.[a]	Mouse mutant
Cytoskeletal proteins					
Myosin 6	Motor protein	*MYO6*	DFNB37/A22	600970	Snell's waltzer
Myosin 7A	Motor protein	*MYO7A*	DFNB2/A11	276903	Shaker-1
Myosin 15	Motor protein	*MYO15*	DFNB3	600316	Shaker-2
Ion transporters					
Connexin 26	Gap junction	*GJB2/ CX26*	DFNB1/A3	22029	
Connexin 30	Gap junction	*BJB6/CX30*	DFNB1/A3	604418	
KCNQ4	K^+ channel	*KCNQ4*	DFNA2	600100	
Pendrin	Iodide–chloride	*SLC26A4*	DFNB4	605646	
Structural proteins					
α-Tectorin	Tectorial membrane	*TECTA*	DFNB21/A8/ A12	602574	
Collagen XI	Extracellular matrix	*COL11 A2*	DFNB53	609706	
Cochlin	Extracellular matrix	*COCH*	DFNA9	603196	
POU3F4		*POU3F4*	DFN3	300039	

Continued ►

Table Continued

Type of protein	Main function	Gene	DFN type	OMIM no.[a]	Mouse mutant
Mitochondrial					
12S RNA			DFNA5	600994	
Unknown					
Diaphanous	Actin polymerization in hair cells = DIAPH1		DFNA1	602121	

Data from: Petit C, et al. Hereditary hearing loss. In: Scriver CR, et al., eds. The Metabolic and Molecular Bases of Inherited Disease. 8th ed. New York: McGraw-Hill; 2001:6281–6328. Available at: http://www.ommbid.com. Accessed January 28, 2012; and Petersen MB, Willems PJ. Nonsyndromic, autosomal-recessive deafness. Clin Genet 2006;69:371–392.

[a] OMIM. Available at: http://www.ncbi.nlm.nih.gov/omim. Accessed January 29, 2012.

20. Common Numerical Human Chromosomal Disorders[a] (p. 382)

Examples of common numerical human chromosomal disorders

Aberration	Frequency (liveborn)	Main manifestations
Trisomy 21	1:650	Muscular hypotonia, flat face, oblique eye lids, congenital heart defect (60%), other congenital malformations, developmental delay
Trisomy 18	1:5000	Characteristic face, muscular hypertonia, congenital malformations (heart 90%), severe developmental impairment, lethality during first year 90%
Trisomy 13	1:8000	Arrhinencephaly, anophthalmia, microphthalmia, cleft lip/palate, heart defect, severe developmental impairment, lethality during first year 90%
Monosomy X females (mosaic 45X/46,XX common) Turner syndrome	1:2500 females	Short stature, congenital lymphedema and/or webbed neck (variable), coarctation aortae (45%), streak ovaries, other malformations; other manifestations, highly variable
47,XXY Klinefelter syndrome	1:600 males	Postpubertal delay of sexual maturation, small testes, hypogonadism, infertility (highly variable)
47,XXX	1:1000 females	No recognizable phenotype; variable, usually mild neuromotor deficits
47,XYY	1:1000 males	No recognizable phenotype; variable, mild developmental delay

[a] Several other, rare numerical aberrations occur with variable manifestations and developmental impairment: XXXX, XXXXX, XXYY, XXXY, XXXXY (see references on p. 384).

21. Microdeletion/Microduplication Syndromes (p. 386)

Examples of microdeletion/microduplication syndromes[a]

Chromosomal region	Name	Comments	OMIM no.[b]
1p36	1p36 Deletion syndrome	See p. 386	607872
4p16.3	4p- or Wolf–Hirschhorn syndrome	See p. 386	194190
5p15.2-15.3	5p- or cri-du-chat syndrome	Familial occurrence by translocation in 12%–15%	123450
5q35	Sotos syndrome	Overgrowth, retardation, seizures; *NSD1* gene deletions; frequent in Japan	117550
7q11.23	Williams–Beuren syndrome	Elastin gene and another gene involved; deletion shown in 70%	194050
8q24.12	Langer–Giedion syndrome (I and II)	Sparse hair, bulbous nose, mental retardation	190350
11p13	Wilms tumor–aniridia–genitourinary anomalies (WAGR)	*WT1* and *PAX6* genes involved	194072
11p15.5 dup	Beckwith–Wiedemann syndrome	Macrosomia, macroglossia	130650
15q11-13	Prader–Willi syndrome	Paternal chromosome 15 involved (see p. 368)	176270
15q11-13	Angelman syndrome	Maternal chromosome 15 involved, UBE3A gene mutations (see p. 368)	105830
16p13.3	Rubinstein–Taybi syndrome	Gene encoding CREB binding protein involved	180849
16pter-p13.3	ATR-16 syndrome	Mental retardation, deletion hemoglobin-α genes	141750
17p13.3	Miller–Dieker syndrome	Lissencephaly, L1S1 gene deletion in about 90%	247200

Continued ►

Table Continued

Chromosomal region	Name	Comments	OMIM no.[b]
17p11.2	Smith–Magenis syndrome	Complex malformation syndrome	182290
20p12.1	Alagille syndrome	Arteriohepatic dysplasia and other systemic manifestations, *JAG1* gene mutations	118450
22q11	DiGeorge/Shprintzen syndrome	Immune defects, neonatal hypocalcemia, congenital heart defects, wide clinical spectrum, *TBX1* gene deletion in 70%–90%,	192430
Xq21.1	ATRX syndrome	Mental retardation, α-thalassemia, dysmorphias	301040

[a] Other microdeletions have been observed at 1q21.1, 1q41-42, 2p15-16.1, 2q23.1, 5q14.3, 12q14, 15q13.3, 15q24.1, 15q26, 16p11.2, 16p13.11, 17q21.3, 19p13.11, 16q11.2, 17q12, 17q21.31, 18q21. (See: Girirajan S, Campbell CD, Eichler EE. Human copy number variation and complex genetic disease. Ann Rev Genet 2011;45:203–226. Also see: DECIPHER v5.1: Syndromes. Available at: https://decipher.sanger.ac.uk/syndromes. Accessed January 29, 2012.)
[b] OMIM. Available at: http://www.ncbi.nlm.nih.gov/omim. Accessed January 29, 2012.

Definition of Terms in Genetics and Genomics—Glossary

Acentric Refers to a chromosome or chromatid without a centromere.

Acrocentric (White 1945) Refers to a chromosome with a centromere that lies very close to one of the ends, dividing the chromosome into a long arm and a very short arm.

Actin A structural protein interacting with many other proteins. In muscle cells, as F actin, it interacts with myosin during contraction.

Allele (Johannsen 1909) **or allelomorph** (Bateson and Saunders 1902) One of several alternative forms of a gene at a given gene locus.

Allelic exclusion Expression from one allele only.

Allozygous Refers to a gene locus with alleles of independent origin (see autozygous).

Alternative splicing Production of different mRNAs from one transcript.

Alu sequences A family of related DNA sequences that are each ~300 base pairs long and containing the recognition site for the Alu restriction enzyme; ~1.2 million copies of Alu sequences are dispersed throughout the human genome.
A full-length Alu repeat is a dimer of about 280 bp: 120 bp for each monomer, followed by a short sequence rich in A residues. They are asymmetric: the repeat to the right contains an internal 32-bp sequence, while the other does not.

Amber codon The stop codon UAG (a word play on the discoverer Bernstein [amber in German]).

Ames test A mutagenicity test performed with a mixture of rat liver and mutant bacteria.

Amino acid An organic compound with an amino ($-NH_2$) and a carboxyl ($-COOH$) group.

Aminoacyl tRNA A transfer RNA carrying an amino acid.

Amplification Production of additional copies of DNA sequences.

Amplicon Fragment of DNA amplified by a specific PCR reaction.

Anaphase (Strasburger 1884) A stage of mitosis and of meiosis I and II. Characterized by the movement of homologous chromosomes (or sister chromatids) toward opposite poles of the cell division spindle.

Ancestral origin Refers to genetic material of common origin from previous generations.

Aneuploidy (Täckholm 1922) Deviation from the normal number of chromosomes by gain or loss (see trisomy and monosomy).

Aneusomy Deviation from the normal presence of homologous chromosomal segments. Aneusomy by recombination refers to the duplication/deficiency resulting from crossing-over within an inversion (inverted region).

Anneal Hybridizing complementary single strands of nucleic acid to form double-stranded molecules (DNA with DNA, RNA with RNA, or DNA with RNA).

Antibody A protein (immunoglobulin) that recognizes and binds to an antigen as part of the immune response.

Anticipation A tendency of a disease to increase in severity in successive generations.

Anticodon A trinucleotide sequence in tRNA that is complementary to a codon for a specific amino acid in mRNA.

Antigen A molecule that induces an immune response.

Antiparallel Double-stranded DNA or RNA running in opposite directions.

Antisense RNA An RNA strand that is complementary to normal mRNA. Natural antisense RNAs made from the nontemplate strand of a gene may regulate gene expression by preventing it from being used as a template for normal translation.

Apoptosis Programmed cell death (see p. 110).

Archea Archaebacteria, one of the three evolutionary lineages of organisms living today, a domain distinct from prokaryotes and eukaryotes.

ARMS Amplication refractory mutation system. A genetic test using allele-specific PCR.

Array CGH See CGH. A genetic test based on a comparative analysis by hybridization of a large number of DNA or RNA samples of different origin (test sample and reference sample) on a small surface (array) in the search for deletions or duplications in the genome (see copy number variation).

Ascertainment The mode of identification of individuals for a genetic study.

Ashkenazi The Jewish population descended from Eastern Europe and medieval Jewish communities that lived along the river Rhine and in the Alsace region (Ashkenaz is the medieval Hebrew name for this region). (Compare with Sephardim.)

ASO Allele-specific oligonucleotides (see p. 394).

Association Different alleles occurring together more often than expected from their individual frequencies (as opposed to linkage).

Assortative mating Nonrandom selection of mates based on shared phenotypic features, as opposed to panmixis.

Attenuator A DNA sequence regulating the termination of transcription, involved in controlling the expression of some operons in bacteria.

Aurignacian Tools and artwork associated with upper Paleolithic fossils of ancestral modern humans in Europe. Named after Aurignac in the Pyrenees, where it was first discovered. (Compare with Mousterian.)

Australopithecus (Dart 1924) Ancestral forms of early hominids in Africa (see p. 30). Derived from australis (southern) and pithecus (ape).

Autonomously replicating sequence (ARS) A DNA sequence that is required to induce replication.

Autoradiography (Lacassagne and Lattes 1924) Photographic detection of a radioactive substance incorporated into cells or tissue. The distribution of the radioactively labeled substance can be demonstrated, e.g., in tissue, cells, or metaphase chromosomes, by placing a photographic film or photographic emulsion in close contact with the preparation.

Autosome (Montgomery 1906) Any chromosome except a sex chromosome (the latter is usually designated X or Y). Autosomal refers to genes and chromosomal segments that are located on autosomes.

Auxotrophy (Ryan and Lederberg 1946) Refers to cells or cell lines that cannot grow on minimal medium unless a certain nutritive substance is added (compare with Prototrophy).

Backcross Cross of a heterozygous animal with one of its homozygous parents. In a double backcross, two heterozygous gene loci are involved.

BAC Bacterial artificial chromosome; a synthetic DNA molecule that contains bacterial DNA sequences for replication and segregation (see YAC).

Bcl-2 family A family of proteins localized to mitochondria involved in regulation of apoptosis.

Bacteriophage A virus that infects bacteria. Usually abbreviated to phage.

Banding pattern (Painter 1939) A specific staining pattern of a chromosome consisting of alternating light and dark transverse bands. Each chromosomal segment of homologous chromosomes shows the same banding pattern, characterized by the distribution and size of the bands, which can be used to identify that segment. The term was introduced in 1939 for the linear pattern of bands in polytene chromosomes of certain diptera (mosquitoes, flies). Each band is defined relative to its neighboring bands. The sections between bands are interbands.

Barr body See X chromatin.

Base pair (bp) In double-stranded DNA, bp refers to two nucleotide bases—one a purine, the other a pyrimidine—lying opposite each other and joined by hydrogen bonds. Normal base pairs are AT and CG. Other pairs can form in ribosomal RNA.

B cells B lymphocytes.

Bimodal distribution A frequency distribution curve with two peaks. It may indicate two different phenotypes distinguished on a quantitative basis.

Bivalent (Haecker 1892) Pairing configuration of two homologous chromosomes during meiosis in the pachytene stage. As a rule the number of bivalents corresponds to half the normal number of chromosomes in diploid somatic cells. Bivalents are the prerequisite for recombination by crossing-over of nonsister chromatids. A trisomic cell forms a trivalent of the three chromosomes.

Breakage–fusion–bridge cycle Refers to a broken chromatid that fuses to its sister, forming a bridge.

Breakpoint Site of a break in a chromosomal alteration, e.g., translocation, inversion, or deletion.

Cadherins Dimeric cell adhesion molecules.

Carcinogen A chemical substance that can induce cancer.

Caspase A member of the family of specialized cysteine-containing aspartate proteases involved in apoptosis (programmed cell death).

CT (or CAAT) box A regulatory DNA sequence in the 5′ region of eukaryotic genes; transcription factors bind to this sequence.

Catenate A link between molecules.

C-bands Specific staining of the centromeres of metaphase chromosomes.

cDNA Complementary DNA synthesized by the enzyme reverse transcriptase from RNA as the template.

CD region Common docking, a region involved in binding to a target protein.

Cell cycle (Howard and Pelc 1953) Life cycle of an individual cell. In dividing cells, the following four phases can be distinguished: G_1 (interphase), S (DNA synthesis), G_2, and mitosis (M). Cells that do not divide are in G_0 phase.

Cell hybrid A somatic cell generated by fusion of two cells in a cell culture. It contains the complete or incomplete chromosome complements of the parental cells.

Cell organelle Defined structural and functional unit within a cell, e.g., mitochondrion, ribosome, endoplasmic reticulum, Golgi apparatus, lysosome.

CEPH A set of three-generation families with known DNA marker genotypes (Centre d'Étude du Polymorphisme Humain in Paris), introduced by J. Dausset 1986 (see Chakravarti A. Information content of the Centre d'Etude du Polymorphisme Humain (CEPH) family structures for linkage studies. Hum Genet 1991;87:721–724).

Centimorgan A unit of length on a linkage map (100 centimorgans, cM = 100 Morgan). The distance between two gene loci in centimorgans corresponds to their recombination frequency expressed as percentage, i.e., one centimorgan corresponds to 1% recombination frequency. Named after Thomas H. Morgan (1866–1945), who initiated the classic genetic experiments on Drosophila in 1910.

Centriole Small cylinder of microtubules.

Centromere (Waldeyer 1903) Chromosomal region to which the spindle fibers attach during mitosis or meiosis. It appears as a constriction at metaphase. It contains chromosome-specific repetitive DNA sequences.

CGH Comparative genomic hybridization (see there).

Chaperone A protein needed to assemble or fold another protein correctly.

Chiasma (Janssens 1909) Cytologically recognizable region of crossing-over in a bivalent. In some organisms the chiasmata move toward the end of the chromosomes (terminalization of the chiasmata) during late diplotene and diakinesis (see meiosis). The average number of chiasmata in autosomal bivalents is ~52 in human males, and ~25–30 in females. The number of chiasmata in man was first determined in 1956 in a paper that confirmed the normal number of chromosomes in man (Ford CE, Hamerton JL. The chromosomes of man. Nature 1956;178:1020).

Chimera (Winkler 1907) An individual or tissue that consists of cells of different genotypes of prezygotic origin.

ChIP Chromatin immunoprecipitation. A method to detect DNA sequences that bind to a specific protein in chromatin.

Chromatid (McClung 1900) Longitudinal subunit of a chromosome resulting from chromosome replication; two chromatids are held together by the centromere and are visible during early prophase and metaphase of mitosis, and between diplotene and the second metaphase of meiosis. Sister chromatids arise from the same chromosome; nonsister chromatids are the chromatids of homologous chromosomes. After division of the centromere in anaphase, the sister chromatids are referred to as daughter chromosomes. A chromatid break or a chromosomal aberration of the chromatid type affects only one of the two sister chromatids. It arises after the DNA replication cycle in the S phase (see Cell cycle). A break that occurs before the S phase affects both chromatids and is called an isolocus aberration (isochromatid break).

Chromatin (Flemming 1882) The stained material observed in interphase nuclei. It is a general term for packaged DNA, composed of DNA, basic chromosomal proteins (histones), nonhistone chromosomal proteins, and small amounts of RNA.

Chromatin remodeling The energy-dependent displacement or reorganization of nucleosomes for transcription or replication.

Chromosome (Waldeyer 1888) The gene-carrying structures that are composed of chromatin and are visible during nuclear division as threadlike or rodlike bodies. Polytene chromosomes (Koltzhoff 1934, Bauer 1935) are a special form of chromosomes in the salivary glands of some diptera larvae (mosquitoes, flies).

Chromosome painting A staining method to distinguish individual chromosomes.

Chromosome walking Sequential isolation of overlapping DNA sequences to find a gene on the chromosome studied.

Cis-acting Refers to a regulatory DNA sequence located on the same chromosome (cis), as opposed to trans-acting over a distance from other locations.

Cis/trans (Haldane 1941) In analogy to chemical isomerism, refers to the position of genes of double heterozygotes (heterozygotes at two neighboring gene loci) on homologous chromosomes. When two certain alleles, e.g., mutants, lie next to each other on the same chromosome, they are in cis position. If they lie opposite each other on different homologous chromosomes, they are in a trans position. The cis/trans test (Lewis 1951, Benzer 1955) uses genetic methods (genetic complementation) to determine whether two mutant genes are in the cis or in the trans position. With reference to genetic linkage, the expressions cis and trans are analogous to the terms coupling and repulsion (q.v.).

Cistron (Benzer 1955) A functional unit of gene effect as represented by the cis/trans test. If the phenotype is mutant with alleles in the cis position and the alleles do not complement each other (genetic complementation), they are considered alleles of the same cistron. If they complement each other they are considered to be nonallelic. This definition by Benzer was later expanded by Fincham in 1959: accordingly, a cistron now refers to a segment of DNA that encodes a unit of gene product. Within a cistron, mutations in the trans position do not complement each other. Functionally, the term cistron can be equated with the term gene.

Clade A group of organisms evolved from a common ancestor.

Clathrin (Pearse 1975) A protein interacting with adaptor proteins to form the coated vesicles that bud from the cytoplasm during endocytosis.

Clone (Webber 1903) A population of molecules, cells, or organisms that have originated from a single cell or a single ancestor and are identical to it and to each other.

Clonal selection and expansion Specific increase in number of a single cell, e.g., lymphocytes with a specific antigen receptor or cancer cells.

Cloning efficiency A measure of the efficiency of cloning individual mammalian cells in culture.

Cloning vector A plasmid, phage, or bacterial or yeast artificial chromosome (BAC, YAC) used to carry a foreign DNA fragment for the purpose of cloning (producing multiple copies of the fragment).

Coalescence time The time going back to the most recent common ancestor (see Divergence time).

Coding strand of DNA The strand of DNA bearing the same sequence as the RNA strand (mRNA) that is used as a template for translation (sense RNA). The other strand of DNA, which directs synthesis of the mRNA, is the template strand (see antisense RNA).

Coding RNA RNA involved in transcription (see noncoding RNA).

Codominant Expression of two dominant traits together, e.g., the AB blood group phenotype, see Dominant.

Codon (Brenner and Crick 1963) A sequence of three nucleotides (a triplet) in DNA or RNA that codes for a certain amino acid or for the terminalization signal of an amino acid sequence.

Coefficient of inbreeding An expression for the proportion of loci that are homozygous by descent from an ancestor, observed by parental consanguinity.

Coefficient of relationship The proportion of loci that is homozygous for the same allele by descent from a common ancestor.

Co-factor A molecule or metal required for a biological activity of a protein.

Cohesin A protein complex of four subunits (Scc1, Scc3, Smc1, Smc3) regulating the separation of sister chromatids during cell division.

Colinear The 1:1 representation of triplet nucleotides in DNA and the corresponding sequence of amino acids.

Common disease—common variant hypothesis Genetic influences on common diseases may be attributed to common variants present in 1–5% of the population.

Comparative genomic hybridization (CGH) A test comparing the genome of an individual with a control to assess differences in the amount of DNA that result from a duplication or a deletion in one or several chromosomes (see Array CGH).

Complementary Two nucleic acids with identical sequences in opposite directions. This enables them to anneal and form a double-stranded molecule.

Complementation, genetic (Fincham 1966) Complementary effect of (restoration of normal function by) double mutants at different gene loci (see complementation groups in xeroderma pigmentosum, or Fanconi anemia, p. 320).

Compound heterozygote Two different mutant alleles at the same gene locus.

Concatemer An association of DNA molecules with complementary ends linked head to tail and repeated in tandem. Formed during replication of some viral and phage genomes.

Concordance The occurrence of a trait or a disease in both members of a pair of twins (monozygotic or dizygotic) (see Discordance).

Condensation Refers to the density of chromatin in a chromosome.

Condensin Protein complexes (condensin I and condensin II), involved in the chromosome assembly for cell division (chromosome condensation).

Conformation The three-dimensional structure of a complex molecule required for biological function.

Conjugation (Hayes; Cavalli, Lederberg, Lederberg 1953) The transfer of DNA from one bacterium to another.

Consanguinity Blood relationship. Two or more individuals are referred to as consanguineous (related by blood) if they have one or more ancestors in common. A quantitative expression of consanguinity is the coefficient of inbreeding (q.v.).

Consensus sequence A corresponding or identical DNA sequence in different genes or organisms.

Conserved in evolution Refers to genes or parts of chromosomes that have been maintained in the course of evolution because of their functional importance.

Contig A series of overlapping DNA fragments (contiguous sequences).

Contiguous gene syndrome A disease resulting from pathological changes in adjacent genes with different functions.

Copy number variation (CNV) Differences in the number of a defined DNA sequence repeats in the genome of different individuals.

Cosmid A plasmid carrying the cos site (q.v.) of a phage in addition to sequences required for division. Serves as a cloning vector for DNA fragments up to 40 kb.

Cos site A restriction site required of a small strand of DNA to be cleaved and packaged into the λ phage head.

Coupling (Bateson, Saunders, Punnett, 1905) A cis configuration (q.v.) of double heterozygotes.

Covalent bond A stable chemical bond holding molecules together by sharing one or more pairs of electrons (as opposed to a noncovalent hydrogen bond).

CpG island A stretch of 1–2 kb in mammalian genomes that is rich in unmethylated CpG doublets; usually at the 5′ end of a gene.

Crossing-over (Morgan and Cattell 1912) The exchange of genetic information between two homologous chromosomes by chiasma formation (q.v.) in the diplotene stage of meiosis I; leads to genetic recombination of neighboring (linked) gene loci. Unequal crossing-over (Sturtevant 1925) results from mispairing of the homologous DNA segments at the recombination site. It results in structurally altered DNA segments or chromosomes, resulting in a duplication in one and a deletion in the other. Crossing-over may also occur in somatic cells (Stern 1936).

Cyclic AMP (cAMP) Cyclic adenine monophosphate, a second messenger produced in response to stimulation of G-protein-coupled receptors.

Cyclin A protein involved in cell cycle regulation.

Cytokine A small secreted molecule that can bind to cell surface receptors on certain cells to trigger their proliferation or differentiation.

Cytoplasmic inheritance Transmission of genetic information located in mitochondria. Since sperm cells do not contain mitochondria, the information transmitted is of maternal origin.

Cytoskeleton Network of stabilizing protein in the cytoplasm and cell membrane.

Cytosol The content of the cytoplasm of a cell, excluding cell organelles.

Dalton A unit of atomic mass, approximately equal to the mass of a hydrogen atom (1.66×10^{24} g).

Deficiency (Bridges 1917) Loss of a chromosomal segment resulting from faulty crossing-over, e.g., by unequal crossing-over or by crossing-over within an inversion (q.v.) or within a ring chromosome (q.v.). It arises at the same time as a complementary duplication (q.v.). This is referred to as duplication/deficiency.

Deletion (Painter and Muller 1929) Loss of part of or a whole chromosome or loss of DNA nucleotide bases.

Denaturation Reversible separation of a double-stranded nucleic acid molecule into single strands (see Renaturation).

Diakinesis (Haecker 1897) A stage during late prophase I of meiosis.

Dicentric (Darlington 1937) Refers to a structurally altered chromosome with two centromeres.

Dictyotene A stage of fetal oocyte development during which meiotic prophase is interrupted. In human females, oocytes attain the stage of dictyotene ~4 weeks before birth; further development of the oocytes is arrested until ovulation, at which time meiosis is continued.

Differentiation The process in which an unspecialized cell develops into a distinct, specialized cell.

Diploid (Strasburger 1905) Cells or organisms that have two homologous sets of chromosomes, one from the father (paternal) and one from the mother (maternal).

Diplotene A stage of prophase I of meiosis.

Direct repeat Repeated DNA sequences oriented in the same direction (see Inverted repeat).

Discordance The occurrence of a given trait or disease in only one member of a pair of twins (see Concordance).

Disomy, uniparental (UPD) Presence of two chromosomes of a pair from only one of the parents. One distinguishes UPD due to isodisomy, in which the chromosomes are identical, and heterodisomy, in which they differ.

Dispermy The penetration of a single ovum by two spermatozoa.

Divergence time The time elapsed since two populations have split from a common ancestor (see Coalescence time).

Dizygotic Twins derived from two different zygotes (fraternal twins), as opposed to monozygotic (identical) twins, derived from the same zygote.

D-loop Displacement loop; a DNA sequence that is formed in a part of opened DNA double helix, e.g., in mitochondrial DNA (p. 244) or in the telomere (p. 156).

DNA (deoxyribonucleic acid) The molecule containing the primary genetic information in the form of a linear sequence of nucleotides in groups of threes (triplets; see codon).

Dynamic mutation A change in DNA characterized by expansion of nucleotide repeats (see p. 92).

Satellite DNA (sDNA) (Sueoka 1961, Britten and Kohne 1968) Contains tandem repeats of nucleotide sequences of different lengths. sDNA can be separated from the main DNA by density gradient centrifugation in cesium chloride, after which it appears as one or several bands (satellites) separated from that of the main body of DNA. In eukaryotes, light (AT-rich) and heavy (GC-rich) satellite DNA can be distinguished. Microsatellites are small (2–10) tandem repeats of DNA nucleotides. Minisatellites are tandem repeats of ~20–100 bp; classical satellite DNA consists of large repeats of 100–6500 bp (see p. 224).

DNAase An enzyme that cleares bonds in DNA.

DNA library A collection of cloned DNA molecules comprising the entire genome (genomic library) or of cDNA fragments obtained from mRNA produced by a particular cell type (cDNA library).

DNA chip or microarray A set of many thousands of different nucleotide sequences on a surface. Used to determine the gene expression pattern of thousands of genes simultaneously.

DNA methylation The presence of methyl groups at specific sites by changing cytosine to 5-methylcytosine (see p. 190).

DNA polymerase A DNA-synthesizing enzyme. To begin synthesis, it requires a primer of RNA or a complementary strand of DNA.

DNase (deoxyribonuclease) An enzyme that cleaves DNA.

Domain A distinctive functional region of the tertiary structure of a protein.

Dominant (Mendel 1865) Refers to a genetic trait that can be observed in the heterozygous state. This depends in part on the accuracy of observation (see Codominant and Recessive).

Dominant negative A mutant allele that produces an undesirable effect resembling loss of function (see gain-of-function mutation).

Dorsal Refers to the back of an animal (for opposite, see ventral).

Dosage compensation (Muller 1948) Refers to mechanisms that balance a difference in activity of alleles.

Downstream The 3′ end of a gene.

Drift, genetic (Wright 1921) Random changes in gene frequency over generations. in small populations. Under some conditions an allele may disappear completely from a population or be fixed, i.e., present in all individuals.

Duplication (Bridges 1919) Addition of a chromosomal segment resulting from faulty crossing-over (see Deficiency). It may also refer to additional DNA nucleotide base pairs. Duplication of genes (gene duplication) played an important role in the evolution of eukaryotes.

Ectoderm One of the three primary cell layers of an embryo giving rise to epidermal tissues, the nervous system, and external sense organs (see Endoderm and Mesoderm).

Effector A protein exerting a specific effect.

Electrophoresis (Tiselius 1937) Separation of molecules by utilizing their different speeds of migration in an electrical field. As a support medium, substances in gel form such as starch, agarose, acrylamide, etc. are used. Further molecular differences can be detected by modifications such as two-dimensional electrophoresis (electric field rotated 90 degrees for the second migration) or cessation of migration at the isoelectric point (isoelectric focusing).

Elongation Addition of amino acids to a polypeptide chain.

Elongation factor (EF) One of the proteins that associates with ribosomes while amino acids are added. EF in prokaryotes and eEF in eukaryotes.

Embryonic stem (ES) cells Undifferentiated pluripotent cells derived from an embryo.

Endocytosis Specific uptake of extracellular material at the cell surface. The material is surrounded by an invagination of the cell membrane, and this fuses to form a membrane-bound vesicle containing the material.

Endoderm The inner of the three primary cell layers of an embryo; gives rise to the gastrointestinal system and most of the respiratory tract (see Ectoderm and Mesoderm).

Endonuclease A heterogeneous group of enzymes that cleaves bonds between nucleotides of single-stranded or double-stranded DNA or RNA.

Endoplasmic reticulum A complex system of membranes within the cytoplasm.

Endoreduplication (Levan and Hauschka 1953) Chromosome replication during interphase without actual mitosis. Endoreduplicated chromosomes in metaphase consist of four chromatids lying next to each other, held together by two neighboring centromeres.

Enhancer (Baneriji 1981) A cis-acting regulatory DNA segment that contains binding sites for transcription factors. An enhancer is located at various distances from the promoter. It causes an increase in the rate of transcription (~10-fold).

Enzyme (Büchner 1897) A protein that catalyzes a biochemical reaction. Enzymes consist of a protein part (apoenzyme), responsible for the specificity, and a nonprotein part (coenzyme), needed for activity. Enzymes bind to their substrates, which become metabolically altered or they combine with other substances during the train of the reaction. Most of the enzymatically catalyzed chemical reactions can be classified into one of six groups: (1) hydrolysis (cleavage with the addition of H_2O), by hydrolases; (2) transfer of a molecular group from a donor to a receptor molecule, by transferases; (3) oxidation and reduction, by oxidases and reductases (transfer of one or more electrons or hydrogen atoms from a molecule to be oxidized to another molecule that is to be reduced); (4) isomerization, by isomerases (rearranging the position of an atom or functional group within a molecule); (5) joining of two substrate molecules to form a new molecule, by ligases (synthetases); (6) nonhydrolytic cleavage with formation of a double bond on one or both of the two molecules formed, by lyases.

Epigenetics (Waddington 1942) The study of heritable changes in gene expression caused by mechanisms that do not alter the DNA sequence, often by DNA methylation. A portmanteau (blended word) of epigenesis and genetics.

Epigenome Refers to the overall epigenetic state of a genome.

Episome (Jacob and Wollman 1958) A DNA sequence, e.g., in plasmid (q.v.), that can exist either independently in the cytoplasm or as an integrated part of the genome of its bacterial host.

Epistasis (Bateson 1907) Nonreciprocal interaction of genes at the same gene locus (allelic) or at different gene loci (nonallelic) that alter the phenotypic expression of a gene.

Epitope The part of an antigen molecule that binds to an antibody.

EST (expressed sequence tag) A sequenced site from an expressed gene that "tags" a stretch of unsequenced cDNA next to it; used to map genes (see STS).

Eubacteria A major class of prokaryotes (see p. 26).

Euchromatin (Heitz 1928) Chromosome or chromosomal segment that stains less intensely than heterochromatin (q.v.). Euchromatin corresponds to the genetically active part of chromatin that is not fully condensed in the interphase nucleus.

Eukaryote (Chatton 1937) Cells in animals and plants that contain a nucleus and organelles in the cytoplasm (see Eubacteria and Prokaryote).

Euploid (Täckholm 1922) Refers to cells, tissues, or individuals with the complete normal chromosomal complement characteristic of that species (compare with Aneuploidy, Heteroploid, Polyploid).

Excision repair Repair of bulk lesions in DNA in which a stretch of nucleotides (~14 in prokaryotes and ~30 in eukaryotes) is excised from the affected strand and replaced by the normal sequence (resynthesis).

Exocytosis Specific process by which nondiffusable particles are transported through the cell membrane to be discharged into the cellular environment.

Exome All exons in a genome.

Exon (Gilbert 1978) A segment of DNA that is represented in the mature mRNA of eukaryotes (compare with Intron).

Exonuclease An enzyme that cleaves nucleotide chains at their terminal bonds only, at either the 5′ or 3′ end (compare with Endonuclease).

Expression The observable effects of an active gene.

Expression vector A cloning vector containing DNA sequences that can be transcribed and translated.

Expressivity (Vogt 1926) Refers to the degree of phenotypic expression of a gene or genotype. Absence of expressivity is also called nonpenetrance.

Fibroblast Type of connective tissue cell. Can be propagated in culture flasks containing suitable medium (fibroblast cultures).

Fingerprint, genetic A characteristic pattern of small polymorphic fragments of DNA or proteins.

FISH (fluorescence *in-situ* hybridization) Identification of a DNA stretch on a surface (in situ) by fluorescent markers.

Fitness, biological Refers to the probability (between 0.0 and 1.0) that a gene will be transmitted to the next generation. For a given genotype and a given environment, the biological (or reproductive) fitness is determined by survival rate and fertility.

Fixation An allele that becomes permanently present in all members of a population.

Fluorophore A macromolecule bound to a fluorescent molecule with a defined light wavelength that allows recognition at the site of hybdridization (see FISH).

Founder effect Presence of a particular allele in a population as a result of a mutation in a single ancestor.

Fragile site A specific site on a metaphase chromosome that appears to be broken because of a local difference in chromosome condensation. Inducible by culture medium conditions.

Frameshift mutation A mutation that alters the reading frame of a triplet in mRNA. As a consequence an incorrect amino acid is inserted or a stop codon is created.

Gain-of-function A mutation that causes a new type of function, usually undesirable (see Dominant negative mutation).

Gamete (Strasburger 1877) A haploid germ cell, either a spermatozoon (male) or an ovum (female). In mammals, males are heterogametic (XY) and females homogametic (XX). In birds, females are heterogametic (ZW) and males homogametic (ZZ).

G-bands A type of banding pattern of metaphase chromosomes used for their identification.

Gene (Johannsen 1909) A hereditary factor that constitutes a single unit of genetic information. It corresponds to a segment of DNA that encodes the synthesis of a single polypeptide chain (compare with Cistron).

Gene amplification (Brown and David 1968) Selective production of multiple copies of a given gene without proportional increases of other genes.

Gene bank A collection of cloned DNA fragments that together represent the genome from which they are derived (see Gene library).

Gene cluster (Demerec and Hartman 1959) A group of two or more neighboring genes of similar function, e.g., the HLA system or the immunoglobulin genes.

Gene conversion (Winkler 1930, Lindgren 1953) Nonreciprocal transfer of genetic information. One gene serves as a sequence donor, remaining unaffected, while the other gene receives sequences and undergoes variation.

Gene desert A region along a chromosome that contains fewer genes than other regions.

Gene dosage Refers to the quantitative degree of expression of a gene.

Gene family A set of evolutionarily related genes by virtue of identity or great similarity of some of their coding sequences.

Gene flow (Berdsell 1950) Transfer of an allele from one population to another.

Gene frequency The frequency of a given allele at a given locus in a population (allele frequency).

Gene locus (Morgan, Sturtevant, Muller, Bridges 1915) The position of a gene on a chromosome.

Gene map The position of gene loci on chromosomes. A physical map refers to the absolute position of gene loci, with their distance from each other expressed as the number of base pairs between them. A genetic map expresses the distance of genetically linked loci by their frequency of recombination.

Gene pool All genes present in a particular population.

Gene product The polypeptide or ribosomal RNA encoded by a gene (see Protein).

Genetic code The information contained in the triplets of DNA nucleotide bases used to incorporate a particular amino acid into a gene product.

Genetic distance The distance on a genetic map as defined by the frequency of recombination (recombination fraction), measured in centimorgans (cM).

Genetic marker A polymorphic genetic DNA sequence that can be used to distinguish the parental origin of alleles.

Genetics (Bateson 1906) The science of heredity and variation. The hereditary basis of organisms; derived from the Greek word genesis (origin).

Genome (Winkler 1920) All of the genetic material of a cell or of an individual.

Genome browser An electronic data bank providing a graphical representation of genes along a stretch of DNA (see UCSC Genome Bioinformatics. Available at http://genome.ucsc.edu/. Accessed January 30, 2012).

Genomic disorders A group of disorders resulting from certain structural features of the human genome (see p. 242).

Genome scan A search with marker loci on all chromosomes for linkage with an unmapped locus.

Genome-wide association study (GWAS) A search for the sites of an association of disease susceptibility with polymorphic DNA markers (usually SNPs) in the whole genome (see p. 232).

Genomics The scientific field dealing with the structure and function of entire genomes (see Part II, Genomics).

Genotype (Johannsen 1909) All or a particular part of the genetic constitution of an individual or a cell (compare with Phenotype).

Germ cell A cell able to differentiate into gametes by meiosis (as opposed to somatic cells).

Germinal Refers to germ cells, as opposed to somatic cells.

Germline The cell lineage giving rise to germ cells.

Germline mosaic The presence of a proportion of germ cells in a gonad carrying a mutation.

G protein Guanine-nucleotide-binding proteins involved in signal transduction.

G6PD Glucose-6-phosphate dehydrogenase.

Growth factor A protein, usually a small peptide acting as a ligand that activates a receptor

Gyrase A topoisomerase that unwinds DNA.

Haldane's rule Hybrid sterility or inviability preferentially affecting the heterogametic sex.

Haploid (Strasburger 1905) Refers to cells or individuals with a single chromosome complement; gametes are haploid.

Haploinsufficiency Refers to a diploid gene that does not exert normal function in the haploid state (e.g., following loss of the homologous gene locus).

Haplotype (Ceppellini et al. 1967) A series of alleles at two or more closely linked gene loci on the same chromosome, e.g., in the HLA system (q.v.).

Hardy–Weinberg equilibrium The distribution of frequencies of genotypes in a population (see p. 136).

Heat map A graphical display of gene expression in array data. A row represents a gene, a column represents a sample of genes. The color of each gene represents the gene expression (high activity usually in red, low in blue).

Heavy strand Refers to differences in density of DNA that result from differences in A and G, and C and T bases, in contrast to the light strand.

Helicase An enzyme that unwinds and separates the two strands of the DNA double helix by breaking the hydrogen bonds during transcription or repair.

Helix–loop–helix A structural motif in DNA-binding proteins, such as some transcription factors (p. 180).

Hemizygous Refers to genes and gene loci that are present in only one copy in an individual, e.g., on the single X chromosome in male cells (XY) or because a homologous locus has been lost.

Heritability (Lush 1950, Falconer 1960) The ratio of additive genetic variance to the total phenotypic variance. Phenotypic variance is the result of the interaction of genetic and nongenetic factors in a population.

Heterochromatin (Heitz 1928) A chromosome or chromosomal segment that remains darkly stained in interphase, early prophase, and late telophase because it remains condensed, as all chromosomal material is in metaphase. This contrasts with euchromatin, which becomes invisible during interphase. Heterochromatin corresponds to chromosomes or chromosome segments showing little or no genetic activity. Constitutive and facultative heterochromatin can be distinguished. An example of facultative heterochromatin is the heterochromatic X chromosome resulting from inactivation of one X chromosome in somatic cells of female mammals. An example of constitutive heterochromatin is the centric heterochromatin at centromeres that can be demonstrated as C bands.

Heterodisomy Presence of two homologous chromosomes from one parent only (compare with Isodisomy and UPD).

Heteroduplex Refers to a region of a double-stranded DNA or DNA/RNA molecule with noncomplementary strands that originated from different duplex DNA molecules.

Heterogametic (Wilson 1910) Refers to the two different types of gametes (q.v.), e.g., X and Y in (male) mammals or Z and W in female birds.

Heterogeneity, genetic (Harris 1953, Fraser 1956) An apparently uniform phenotype caused by two or more different genotypes.

Heterokaryon (Ephrussi and Weiss 1965, Harris and Watkins 1965, Okada and Murayama 1965) A cell having two or more nuclei with different genomes.

Heterereoplasmy Differences in the presence or absence of mutations in mitochondria within a single cell.

Heteroploid (Winkler 1916) Refers to cells or individuals with an abnormal number of chromosomes.

Heterosis (Shull 1911) Increased reproductive fitness of heterozygous genotypes compared with the parental homozygous genotypes, in plants and animals.

Heterozygous (Bateson and Saunders 1902) Having two different alleles at a given gene locus (compare with Homozygous).

Hfr cell A bacterium that possesses DNA sequences that lead to a high frequency of DNA transfer at conjugation.

HGPRT Hypoxanthine-guanine-phosphoribosyl transferase. An enzyme in purine metabolism that is inactive in Lesch–Nyhan syndrome.

Histocompatibility Tissue compatibility. Determined by the major histocompatibility complex MHC (see HLA).

Histone (Kossel 1884) Chromosome-associated protein of the nucleosome. Histones H2A, H2B, H3, and H4 form a nucleosome (q.v.).

Histone code A pattern of chemical modifications in histones (e.g., methylation, acetylation, etc.) that are related to the regulation of gene activities (see p. 192).

HLA (Dausset and Terasaki 1954) Human leukocyte antigen system A. HLA is also said by some to refer to Los Angeles, where Terasaki made essential discoveries.

HMG proteins High-mobility group proteins that exhibit high mobility during electrophoresis. Nuclear proteins involved in different DNA interactions.

Hogness box A nucleotide sequence (5′-TATAAA-3′) located ~25 bp upstream (in 5′ direction) of the promoter region in eukaryotic genes (usually called TATA box).

Holoprosencephaly A wide range of congenital malformations involving the midline embryonic forebrain, ranging from a single eye (cyclopia) without nose formation at the most severe end of the spectrum to a single incisor tooth and a flat face with closely set eyes (hypotelorism) as result of mutations in several genes (p. 413).

Homeobox A highly conserved DNA segment in homeotic genes.

Homeosis Transformation of one body part into another.

Homeotic gene One of the developmental genes in Drosophila that can lead to the replacement of one body part by another by mutation.

Homologous Refers to a chromosome or gene locus of similar maternal or paternal origin.

Homozygosity mapping Mapping genes by identifying chromosomal regions that are homozygous as a result of identity by descent from a common ancestor in consanguineous matings (see Identity by descent, IBD).

Homozygous (Bateson and Saunders 1902) Having identical alleles at a given gene locus.

Hormone (Bayliss and Starling 1904) An organic compound able to induce a specific response in target cells. Derived from the Greek "to spur on").

Hotspot A site on a chromosome where recombination frequently and preferentially occurs.

Housekeeping gene A gene carrying information about basic functions in a cell.

Hox genes Clusters of mammalian genes containing homeobox sequences. They are important in embryonic development.

Hybridization In genetics, the fusion of two single complementary DNA strands (DNA–DNA hybridization) DNA and RNA strands (DNA–RNA hybridization), or the in vitro fusion of cultured cells of different species (cell hybridization).

Hybridoma A clone of hybrid cells.

Hydrogen bond A noncovalent weak chemical bond between an electronegative atom (usually oxygen or nitrogen) and a hydrogen atom; important in stabilizing the three-dimensional structure of proteins or base pairing in nucleic acids.

Hypomorph An allele that produces a reduced amount or activity of a gene product.

Identity by descent (IBD) Refers to homozygous alleles at one gene locus that are identical because they are inherited from a common ancestor (see Consanguinity).

Immunoglobulin An antigen-binding molecule.

Imprinting, genomic Different expression of a gene depending on the parental origin.

Inbreeding coefficient (Wright 1929) Measure of the probability that two alleles at a gene locus of an individual are identical by descent, i.e., that they are copies of a single allele of an ancestor common to both parents (IBD, identity-by-descent). Also, the proportion of loci at which the individual is homozygous.

Incidence A measure of the rate of occurrence of a disease in a population in a specified period of time (compare with Prevalence).

Indel Insertion/deletion variant in DNA.

Index patient The patient who first calls attention to a family and leads to ascertainment.

Inducer A molecule that induces the expression of a gene.

Initiation factor A protein that associates with the small subunit of a ribosome when protein synthesis begins (IF in prokaryotes, eIF in eukaryotes).

Insertion Insertion of chromosomal material of nonhomologous origin into a chromosome without reciprocal translocation (q.v.).

Insertion sequence (IS) A small bacterial transposon carrying genes for its own transposition (q.v.).

In silico A process taking place within a computer; e.g., analysis of biological data.

Intercalating agent A chemical compound that can occupy a space between two adjacent base pairs in DNA.

***In-situ* hybridization** Hybridization of complementary single-stranded DNA or RNA in the original or natural position (see FISH).

Interphase The period of the cell cycle between two cell divisions (see Mitosis).

Intron (Gilbert 1978) A segment of noncoding DNA within a gene (compare with Exon). It is transcribed, but removed from the primary RNA transcript before translation.

Inversion (Sturtevant 1926) Structural alteration of a chromosome through a break at two sites with reversal of direction of the intermediate segment and reattachment. A pericentric inversion includes the centromere in the inverted segment. A paracentric inversion does not involve the centromere.

Inverted repeat Two identical, oppositely oriented copies of the same DNA sequence. They are a characteristic feature of retroviruses.

In vitro A biological process taking place outside a living organism or in an artificial environment in the laboratory.

In vivo Within a living organism.

iPS cells (iPSCs) (Yamanaka 2006) Induced pluripotent stem cells. Adult cells that have been reprogrammed to express genes (Oct3, Oct4, SOX2, c-Myc, KLF4) that are active in pluripotent stem cells only, such as embryonic stem cells.

Isochromatid break A break in both chromatids at the same site.

Isochromosome (Darlington 1940) A chromosome composed of two identical arms connected by the centromere, e.g., two long or two short arms of an X chromosome. Implies duplication of the doubled arm and deficiency of the absent arm. An isochromosome may have one or two centromeres.

Isodisomy Presence of two identical chromosomes from one of the parents (compare with Heterodisomy).

Isolate, genetic (Waklund 1928) A physically or socially isolated population that has not interbred with individuals outside of that population (no panmixis).

Isotype Closely related chains of immunoglobulins.

Isozyme or isoenzyme (Markert and Möller 1959, Vesell 1959) One of multiple distinguishable polymorphic variants of enzymes of similar function in the same organism. Isoenzymes are a biochemical expression of genetic polymorphism.

Jewish genetic diseases As a result of relative isolation, several autosomal recessive diseases occur more frequently in Jewish populations (see Jewish Genetic Disease Consortium. Available at: http://www.jewishgeneticdiseases.org/. Accessed January 30, 2012).

Karyogram The visual display of all chromosomes of a cell arranged in pairs of homologous chromosomes.

Karyotype (Levitsky 1924) The complete chromosome set of a cell, an individual, or a species.

Kilobase (kb) 1000 Base pairs.

Kinetochore Chromosomal structure at the centromere, where the spindle fibers attach during cell division.

Knockout (of a gene) Intentional inactivation of a gene in an experimental organism to obtain information about its function (same as targeted gene disruption).

Lagging strand of DNA The new strand of DNA replicated from the 3′ to 5′ strand. It is synthesized in short fragments in the 5′ to 3′ direction (Okazaki fragments), and these are subsequently joined together.

Lamins Intermediate filament proteins forming a fibrous network, the nuclear lamina on the inner surface of the nuclear envelope.

Lampbrush chromosome (Rückert 1892) A special type of chromosome found in the primary oocytes of many vertebrates and invertebrates during the diplotene stage of meiotic division and in Drosophila spermatocytes. The chromosomes show numerous lateral loops of DNA that are accompanied by RNA and protein synthesis.

Lariat An intermediate form of RNA during splicing when a circular structure with a tail is formed by a 5′ to 3′ bond.

Leader sequence A short N-terminal sequence of a protein that is required for directing the protein to its target.

Leaky mutant A mutation causing only partial loss of function.

Leptotene A stage of meiosis (q.v.).

Lethal equivalent (Morton, Crow, and Muller 1956) A gene or combination of genes that in the homozygous state is lethal to 100% of individuals. This may refer to a gene that is lethal in the homozygous state, to two different genes that each have 50% lethality, or to three different genes each with 33% lethality, etc. It is assumed that each individual carries about five or six lethal equivalents.

Lethal factor (Bauer 1908, Hadorn 1959) An abnormality of the genome that leads to death in utero, e.g., numerous chromosomal anomalies.

Leucine zipper A specific DNA-binding protein that resembles a zipper and serves as a transcription factor.

Library See DNA library.

Ligand A molecule that can bind to a receptor and induce a signal in the cell.

Ligase An enzyme that closes a gap in a DNA strand.

LINE (long interspersed nuclear element) Long interspersed repetitive DNA sequences (see p. 224).

Linkage disequilibrium (Kimura 1956) Nonrandom association of alleles at closely linked gene loci that deviates from their individual frequencies as predicted by the Hardy–Weinberg equilibrium (see p. 136).

Linkage, genetic (Morgan 1910) Localization of gene loci on the same chromosome close enough to cause deviation from independent segregation.

Linkage group Gene loci on the same chromosome that are so close together that they usually are inherited together without recombination.

Linker DNA A synthetic DNA double-stranded fragment that carries the recognition site for a restriction enzyme and that can bind two DNA fragments. Also, the stretch of DNA between two nucleosomes.

Locus Same as gene locus.

Loss-of-function mutation A mutation that causes total or partial loss of function of a gene product (see gain-of-function mutation).

Loss of heterozygosity (LOH) Hemizygosity or homozygosity in a tumor resulting from loss of a heterozygous gene locus or chromosome (see p. 302).

Long terminal repeat (LTR) A repeat DNA sequence of up to 600 bp flanking the coding regions of retroviral DNA and viral transposons.

Lymphocyte Cell of the immune system, of one of two general types: B lymphocytes from the bone marrow and thymus-derived T lymphocytes.

Lysogeny The ability of a phage to integrate into the bacterial chromosome.

Lysosome (de Duve 1955) Small cytoplasmic organelle containing hydrolytic enzymes.

Lytic infection A phage infecting a bacterium and causing lysis.

Map distance Distance between gene loci, expressed in either physical terms (number of base pairs, e.g., kb [1000 bp] or Mb [million bp]) or genetic terms (recombination frequency, expressed as cM, centimorgan. One cM corresponds to 1%).

Mapping Various methods to determine the position of a gene on a chromosome (physical map) or its relative distance to other gene loci and their order (genetic map).

Marker, genetic An allele used to recognize a particular genotype.

Megabase (Mb) 1 million base pairs.

Meiosis (Strasburger 1884) The special division of a germ cell nucleus that leads to reduction of the chromosome complement from the diploid to the haploid state. Prophase of the first meiotic division is especially important and consists of the following stages: leptotene, zygotene, pachytene, diplotene, diakinesis.

Mendelian inheritance (Castle 1906) Inheritance according to the laws of Mendel as opposed to extrachromosomal inheritance, under the control of cytoplasmic hereditary factors (mitochondrial DNA).

Metabolic cooperation (Subak-Sharpe et al. 1969) Correction of a phenotype in cells in culture by contact with normal cells or cell products.

Metacentric Refers to chromosomes that are divided by the centromere into two arms of approximately the same length.

Metaphase (Strasburger 1884) Stage of mitosis in which the contracted chromosomes are readily visible.

MHC (major histocompatibility complex) (Thorsby 1974) The principal histocompatibility system, consisting of class I and class III antigen genes of the HLA system, including the class II genes.

Microfilament Cytoskeletal fiber of ~7 nm diameter that results from polymerization of monoglobar (G) actin (actin filament).

microRNA (miRNA) Small RNA molecules of 21–23 nucelotides that associate with multiple proteins in an RNA-induced silencing complex (RISC; see RNAi).

Microsatellite DNA Small tandem repeats of ~1–6 bp (e.g., CACACACA... etc), usually less than 0.1 kb in total length; also referred to as short tandem repeats (STRs).

Microsatellite instability Instability of microsatellites in some forms of cancer, especially in the colon.

Mismatch repair A DNA repair mechanism to repair improperly paired DNA bases.

Missense mutation A mutation that alters a codon to one for a different amino acid (see Nonsense mutation).

Mitochondrion Plural: mitochondria. Cell organelles containing DNA.

Mitosis (Flemming 1882) Nuclear division during the division of somatic cells, consisting of prophase, metaphase, anaphase, and telophase.

Mitosis index (Minot 1908) The proportion of cells present that are undergoing mitosis.

Mixoploidy (Nemec 1910, Hamerton 1971) A tissue or individual containing cells with different numbers of chromosomes (chromosomal mosaic).

MLC (Bach and Hirschhorn, Bach and Lowenstein 1964) Mixed lymphocyte culture. MLC is a test for differences in HLA-D phenotypes.

Mobile genetic element A DNA sequence that can change its position (see Transposon).

Modal number (White 1945) The number of chromosomes of an individual or a cell.

Monoclonal antibody An antibody representing a single antigen specificity, produced from a single progenitor cell.

Monolayer (Abercrombie and Heaysman 1957) The single-layered sheet of cultured diploid cells on the bottom of a culture flask.

Monosomy (Blakeslee 1921) Absence of a chromosome in an otherwise diploid chromosomal complement (see Trisomy).

Monozygotic Twins with identical sets of nuclear genes (compare with Dizygotic).

Morphogen A protein present in embryonic tissues in a concentration gradient that induces a developmental process.

Mosaic Tissue or individuals made up of genetically different cells, as a rule of the same zygotic origin (compare with Chimera).

Mousterian Middle Stone Age tools mainly associated with Neandertals (compare with Aurignacian).

mRNA (Brenner, Jacob, and Meselson 1961, Jacob and Monod 1961) Messenger RNA.

mtDNA Mitochondrial DNA.

Multigene family A group of genes related by their common evolution.

Mutagen A chemical or physical agent that can induce a mutation.

Mutation (de Vries, 1901) A heritable structural and functional alteration of the DNA sequence. Different types include point mutations from exchange, loss, or insertion of base pairs within a gene (see Missense mutation and Nonsense mutation).

Mutation nomenclature Mutations are designated according to an international nomenclature system (JT den Dunnen & S Antonarakis, Human Mutat 2000;15:7–12). It distinguishes whether a mutation is described at the DNA level or at the protein level. Substitutions at the DNA level are designated by the sign “>,” deletions by “del” after the deleted interval, insertions by “ins” followed by the inserted nucleotides. Thus, 1994G>T denotes a change of G (guanine) to T (thymine) at nucleotide (nt) position 1994. 1994–1996delTTC denotes a deletion of two T and one C (cytosine). 1994–1995insAG denotes an insertion of an A (adenine) and G between nt position 1994 and 1995. Intron changes are designated by “IVS” followed by the intron number or the cDNA position. g.10G>C describes a change of G to C at nt position 10 in relation to genomic DNA; c.10G>C describes this change in cDNA and m.10G>C in mitochondrial DNA. Mutations designated at the protein level indicate the deduced consequence preceded by “p,” e.g., p. R25S designates a change of arginine (R) to serine (S) at amino acid position 25. Deletions and insertions at the protein level are designated by “del” or “ins” after the positions of the amino acids involved: R98–S101del means a deletion between R at position 98 to S at 101. R98–S99insR means an insertion of R between R at 98 and S at 99. Note that at the protein level the deduced consequence is described, but not the nature of the mutation. For more complex changes see den Dunnen & Antonarakis (2000).

Mutation rate The frequency of a mutation per locus per individual per generation.
The average mutation rate in man is estimated to be ~2.5×10^{-8} (MW Nachman & SL Crowell, Genetics 2000;156:297–304).

Myosin A class of motor proteins.

Necrosis Cell death resulting from tissue damage

Neural crest A structure parallel to the dorsal neural tube containing a network of migratory cell populations that give rise to different cell types and anatomical structures.

Neurotransmitter Extracellular signaling molecules at the nerve–muscle junction.

N-linked oligosaccharides Branched oligosaccharide chains attached to the amino group side chain of asparagine in glycoproteins.

Non-allelic homologous recombination (NAHR) Recombination between misaligned DNA repeats on the same chromosome or on sister chromatids. The consequences are deletions or duplications (see Genomic disorders).

Noncoding RNA (ncRNA) RNA not involved in transcription and translation, but with different functions (see Coding RNA).

Noncovalent bond A (weak) chemical bond between an electon-negative atom (usually oxygen or nitrogen) and a hydrogen atom (see hydrogen bond), but not involving sharing of electrons.

Nondisjunction (Bridges 1912) Faulty distribution of homologous chromosomes at meiosis. In mitotic nondisjunction, the distribution error occurs during mitosis.

Nonpolar Molecules lacking a net electric charge or having an asymmetric distribution of positive and negative charges; generally insoluble in water.

Nonsense codon A codon that does not have a normal tRNA molecule. Any of the three triplets (stop codons) that terminate translation (UAG, UAA, UGA).

Nonsense-mediated mRNA decay Degraded mRNA molecules containing a premature stop codon (usually <50 nucleotides upstream of the last splice junction).

Nonsense mutation A mutation that results in lack of any genetic information, e.g., a stop codon (see Missense mutation).

Northern blot A word play on Southern blot. Transfer of RNA molecules to a membrane by a procedure similar to that of a Southern blot (q.v.) or Western blot.

Nucleic acid A molecule such as DNA and RNA that can store genetic information.

Nucleolus The site in the nucleus where ribosomal RNA is transcribed.

Nucleolus organizer (NOR) The satellite stalks of human acrocentric chromosomes 13, 14, 15, 21, and 22 containing ribosomal RNA that can be specifically stained with silver.

Nucleoside Compound of a purine or a pyrimidine base with a sugar (ribose or deoxyribose) (compare with Nucleotide).

Nucleosome (Navashin 1912, Kornberg 1974) A subunit of chromatin consisting of DNA wound around histone proteins in a defined spatial configuration.

Nucleotide Single monomeric building block of a polynucleotide chain that makes up nucleic acid. A nucleotide is a phosphate ester consisting of a purine or a pyrimidine base, a sugar (ribose or deoxyribose as a pentose), and a phosphate group.

Ochre codon The stop codon UAA (see Amber codon, and Stop codon UAG).

Okazaki fragments Short (<1000 nucleotides) single-stranded DNA fragments that are synthesized on the lagging strand of DNA during replication (q.v.). They are rapidly joined by DNA ligase into a continuous new DNA strand.

O-linked oligosaccharides Attachment to the hydroxyl groups of the side chains of serine and threonine in glycoproteins (see N-linked).

Oncogene (Huebner and Todaro 1969) Originally a DNA sequence of viral origin that can lead to malignant transformation of a eukaryotic cell after being integrated into the cellular DNA. Now used to designate a gene involved in control of cell proliferation (formerly proto-oncogene).

Open reading frame (ORF) A DNA sequence of variable length that does not contain a stop codon within any of the triplet reading frames. Any ORF beginning with a start codon and extending for 100 or more codons probably encodes a protein, and is therefore meaningful because it can be translated.

Operator (Jacob and Monod 1959) The recognition site of an operon at which the negative control of genetic transcription takes place by binding to a repressor.

Operon (Jacob et al. 1960) In prokaryotes, a group of functionally and structurally related genes that are regulated together.

Organelle A membrane-enclosed subcellular structure in a eukaryotic cell (see Cell organelle).

Origin of replication (ORI) Site of start of DNA replication.

Ortholog A homologous DNA sequence or a gene that has evolved by speciation from a common ancestor between related species, e.g., the α- and β-globin genes (see paralog).

Pachytene (de Winiwarter 1900) Stage of prophase meiosis I.

Palindrome (Wilson and Thomas 1974) A sequence of DNA nucleotides that reads the same in the 5′ to 3′ direction as in the 3′ to 5′ direction. This occurs in recognition sequences of restriction enzymes, e.g., 5′-AATG-3′ in one strand and 3′-GTAA-5′ in the other.

Panmixis (Weismann 1895) Pairing system with random partner selection, as opposed to assortative mating.

Paracrine Refers to secreted signaling molecules acting over a short distance on neighboring cells (see p. 94).

Paralog A DNA sequence or a gene that has evolved by duplication from a common ancestor within a species, e.g., the two α-globin gene loci in humans (see ortholog).

Parasexual (Pontocorvo, 1954) Refers to genetic recombination by nonsexual means, e.g., by hybridization of cultured cells (see Hybridization).

PCR (polymerase chain reaction) (Mullis 1985) Technique for in vitro propagation (amplification) of a given DNA sequence. It is a repetitive thermal cyclic process consisting of denaturation of genomic DNA of the sequence of interest, annealing the DNA to appropriate oligonucleotide primers, and replication of the DNA segment complementary to the primer (see p. 64).

Penetrance (Vogt 1926) The frequency or probability of expression of an allele (compare with Expressivity) and become visible in the phenotype.

Peptide A compound of two or more amino acids joined by peptide bonds.

Phage Abbreviation for bacteriophage.

Phagocytosis Refers to cells that incorporate foreign cells, such as bacteria.

Pharmacogenetics Study of the individual influence of certain genes on the metabolism and function of pharmaceutical substances (see p. 258).

Pharmacogenomics Study of genetic variation in response to pharmaceutical substances on a whole genome basis.

Phenocopy (Goldschmidt 1935) A nonhereditary phenotype that resembles a genetically determined phenotype.

Phenotype (Johannsen 1909) The observable effect of one or more genes on an individual or a cell.

Pheromone A signaling molecule that can alter the behavior or gene expression of other individuals of the same species.

Phosphodiester bond The chemical bond linking adjacent nucleotides of DNA or RNA.

Phytohemagglutinin (PHA) A protein substance obtained from kidney beans (Phaseolus vulgaris). Nowell (1960) discovered its ability to induce blastic transformation (see Transformation) and cell division in lymphocytes. This is the basis of phytohemagglutinin-stimulated lymphocyte cultures for chromosomal analysis (see p. 162).

Plasmid (Lederberg 1952) Autonomously replicating circular DNA structures found in bacteria. Although they are usually separate from the actual genome, they may become integrated into the host chromosome.

Plastids Any of several types of organelles found in plant cells, e.g., chloroplasts.

Pleiotropy (Plate 1910) Expression of a gene with multiple, seemingly unrelated phenotypic features.

Point mutation Alteration of the genetic code within a single codon. The possible types are the exchange (substitution) of a base: a pyrimidine for another pyrimidine (or a purine for another purine), i.e., a transition (Frese 1959), e.g., thymine for cytosine (or adenine for guanine); or the exchange of a pyrimidine for a purine or visa versa: transversion, i.e., thymine by adenine or visa versa (Frese 1959). In addition to the two types of exchange, a point mutation may be due to the insertion of a nucleotide base or the deletion of one or several base pairs.

Polar Refers to molecules with a net electric charge or a symmetric distribution of positive and negative charges.

Polar body (Robin 1862) An involutional cell arising during oogenesis that does not develop further as an oocyte.

Polyadenylation The addition of multiple adenine residues at the 3′ end of eukaryotic mRNA after transcription.

Polycistronic messenger mRNA including coding regions from more than one gene (in prokaryotes).

Polygenic (Plate 1913, Mather 1941) Refers to traits that are based on several or numerous genes whose effects cannot be individually determined. The term multigenic is sometimes used instead.

Polymerases Enzymes that catalyze the polymerization of new DNA or RNA nucleotides from a template as in transcription and DNA replication.

Polymorphism, genetic (Ford, 1940) Existence of more than one variant of a normal allele at a gene locus. By definition, frequency must exceed 1%. Polymorphisms exist at several levels, i.e., sequence of DNA, amino acids, certain structures of chromosomes, or phenotypic traits.

Polypeptide See peptide.

Polyploid (Strasburger 1910) Refers to cells, tissues, or individuals having more than two copies of the haploid genome, e.g., three (triploid) or four (tetraploid). Polytene (Koltzoff 1934, Bauer 1935) refers to a special type of chromosome resulting from repeated endoreduplication of a single chromosome. Giant chromosomes arise in this manner (compare with Chromosome).

Population (Johannsen 1903) Individuals of a species that interbreed and constitute a common gene pool (compare with Race).

Position effect (Sturtevant 1925) An effect on the expression of a gene depending on its location on a chromosome (Sturtevant AH. The effects of unequal crossing over at the bar locus in Drosophila. Genetics 1925;10:117–147).

Positional cloning Identification of a gene based on prior information of the position of its locus.

Premutation A state leading to a mutation (see dynamic mutations repeat expansion disorders (p. 92).

Premature chromosome condensation (Johnson and Rao 1970) Induction of chromosomal condensation in an interphase nucleus after fusion with a cell in mitosis. Condensed S-phase chromosomes appear pulverized (so-called chromosome pulverization).

Prevalence The total number of individuals with a given disease in a population at a given time (compare with Incidence).

Pribnow box Part of a promoter (TATAAT sequence 10 bp upstream of the gene) in prokaryotes.

Primary transcript The RNA transcribed from a eukaryotic gene before processing (splicing, addition of the cap, and polyadenylation).

Primer A short DNA or RNA oligonucleotide of ~15–25 bases that specifically binds to a target sequence and initiates synthesis of a complementary strand.

Primordial germ cells (PGCs) Cells in the gonads that give rise to germ cells in the ovary or testis.

Prion Proteinaceous infectious particles that cause degenerative disorders of the central nervous system.

Proband The individual who first calls attention to a family with a given genetic disorder (same as Propositus or Index patient).

Probe A defined single-stranded DNA or RNA fragment, labeled for detection with a fluorophore, to identify complementary sequences by specific hybridization.

Prokaryote Single-celled microorganisms without a cell nucleus or intracellular organelles, such as bacteria and archea (compare eukaryote).

Promoter A defined DNA region at the 5′ end of a gene that binds to transcription factors and RNA polymerase during the initiation of transcription. The 10 sequence is the consensus sequence TATAATG ~10 bp upstream of a prokaryotic gene (Pribnow box).

Prophage A viral (phage) genome integrated into the bacterial (host) genome.

Prophase An early stage of mitosis or meiosis.

Propositus, proband The individual in a pedigree that has brought a family to attention for genetic studies.

Protein A molecule with a distinct biological function, consisting of one or more polypeptide subunits each with a defined amino acid sequence folded into a specific three-dimensional structure.

Proteome The complete set of all protein-encoding genes or all proteins produced.

Proteomics The study of all proteins in a cell or an organism.

Proto-oncogene See oncogene.

Prototrophy Refers to cells or cell lines that do not require a special nutrient added to the culture medium (compare with Auxotrophy).

Provirus Duplex DNA derived from an RNA retrovirus and incorporated into a eukaryotic genome.

Pseudoautosomal region (PAR) Regions at the end of the X- and the Y-chromosome that participate in recombination as in autosomes (see p. 240).

Pseudogene DNA sequences that closely resemble a gene but are without function because of an integral stop codon, deletion, or other structural change. A processed pseudogene consists of DNA sequences that resemble the mRNA copy of the parent gene, i.e., it does not contain introns.

Pseudohermaphroditism A condition in which an individual has the gonads of one sex and phenotypic features of the opposite sex.

qPCR Quantitative PCR

Quadriradial figure The configuration assumed when homologous segments of chromosomes involved in a reciprocal translocation pair at meiosis. Rarely, such a figure may occur during mitosis.

Race A population (q.v.) that differs from another population within the same species in the frequency of some of its gene alleles (Dunn LC. Heredity and Evolution in Human Populations. Cambridge: Harvard University Press; 1967). Accordingly, the concept of race is flexible and relative, defined in relation to the evolutionary process. The term race erroneously divides humans into distinct categories defined by external phenotypic features such as skin color that are the result of adaptation to climate conditions during evolution. It is often associated with discriminatory reactions for which no genetic basis exists. The classification of individuals according to race is uncertain and of dubious biological value.

Reading frame Sequence of DNA nucleotides that can be read in triplets in the genetic code (compare with Open reading frame).

Receptor A transmembrane or intracellular protein involved in transmission of a cell signal.

Recessive (Mendel 1865) Refers to the genetic effect of an allele (q.v.) at a gene locus that is manifest as phenotype in the homozygous state only (see Dominant).

Reciprocal translocation Mutual exchange of chromosome parts.

Recombinant DNA An artificial DNA molecule constructed of fragments of different origins. Widely used in molecular genetic analysis.

Recombination (Bridges and Morgan 1923) The formation of new combinations of chromosomes and genes as a result of crossing-over between homologous chromosomes during meiosis. Recombination is the main source of genetic variation.

Recombination frequency Frequency of recombination between two or more gene loci. Expressed as the recombination fraction theta (Θ) value. A Θ of 0.01 (1% recombination frequency) corresponds to 1 centimorgan (cM).

Regulatory gene A gene encoding a protein that regulates other genes.

Renaturation of DNA Combining of complementary single strands of DNA to form double-stranded DNA (compare with Denaturation).

Repair (Muller 1954) Correction of structural and functional DNA damage.

Replication Identical duplication of the DNA double helix prior to cell division.

Replication fork The unwound region of the DNA double helix in which replication takes place.

Replicon (Huberman and Riggs 1968) An individual unit of DNA in eukaryotic DNA that is capable of replication.

Reporter gene A gene used to analyze another gene, especially in the regulatory region of the latter.

Repressor A protein that suppresses gene function.

Repulsion (Bateson, Saunders, and Punnett 1905) Term to indicate that the mutant alleles of neighboring heterozygous gene loci lie on opposite chromosomes, i.e., in trans configuration (see cis/trans).

Resistance factor A gene in plasmids causing antibiotic resistance.

Restriction enzyme, or restriction endonuclease (Meselson and Yuan 1968) An endonuclease that cleaves DNA at a specific base sequence (restriction site or recognition sequence).

Restriction fragment length polymorphism (RFLP) A DNA polymorphism that creates or abolishes a restriction site. This can be used to identify the parental origin of a given DNA fragment and used in diagnostics.

Restriction map A segment of DNA characterized by a particular pattern of absence or presence of a restriction site.

Restriction site A particular sequence of nucleotide bases in DNA at which a particular restriction enzyme cleaves a DNA molecule.

Retrotransposon A mobile DNA sequence that can insert itself at a different position by using reverse transcriptase (see Transposon).

Retrovirus A virus with a genome consisting of RNA that replicates by copying its RNA genome into cDNA prior to integration into a chromosome of the host cell.

Reverse transcriptase An enzyme complex present in RNA viruses that can synthesize DNA from an RNA template.

RFLP Restriction fragment length polymorphism.

Rho factor A protein involved in termination of transcription in Escherichia coli.

Ribonuclease (RNAase) Enzymes (nucleases) that can cleave RNA.

Ribosome (Dintzis et al. 1958, Roberts 1958) Complex molecular structure in prokaryotic and eukaryotic cells consisting of specific proteins and ribosomal RNA in different subunits. The translation of genetic information occurs in ribosomes.

Ring chromosome A circular chromosome. In prokaryotes the normal chromosome usually is ring shaped. In mammals a ring chromosome represents a structural anomaly and implies that chromosomal material has been lost.

RNA (ribonucleic acid) A polynucleotide with a structure similar to that of DNA except that the sugar is ribose instead of deoxyribose.

RNA editing A change of RNA sequence following transcription; a mechanism of gene regulation (see p. 174).

RNAi (RNA interference) Inhibition of transcription by antisense RNA (see p. 186).

RNA polymerase An enzyme that synthesizes RNA from a DNA template.

RNA silencing Downregulation or suppression of a gene by antisense RNA. The ability of double-stranded RNA to suppress a gene.

RNA splicing The processing of the primary transcript in mRNA (in eukaryotes).

rRNA Ribosomal RNA. Any of many large RNA molecules in ribosomes.

RTK Receptor tyrosine kinase; membrane bound proteins (G proteins) involved in signal transduction.

Satellite (Navashin 1912) Small mass of chromosomal material attached to the short arm of an acrocentric chromosome (q.v.) by a constricted appendage or stalk. It is involved in the organization of the nucleolus (see Nucleolus organizer).

Satellite DNA (sDNA) (Kit 1961, Sueoka 1961, Britten and Kohne 1968) DNA that forms additional minor band in density gradient centrifugation because it contains either heavier (GC-rich) or lighter (AT-rich) than the main DNA (see DNA). Not to be confused with the satellite regions of acrocentric chromosomes.

SCE (sister chromatid exchange) (Taylor 1958) An exchange between the two chromatids of a metaphase chromosome. After two replication cycles in a cell culture in the presence of a halogenated base analogue (e.g., 5-bromodeoxyuridine), both DNA strands of one chromatid will be substituted with the halogenated base analogue, whereas only one DNA strand of the other chromatid will be substituted. As a result, the two chromatids differ in staining intensity and it is possible to determine where somatic crossing-over of the two chromatids has occurred. An increased rate of SCEs is a definitive diagnostic test in Bloom syndrome (see p. 318).

Segmental duplication Duplicated DNA block with the same or highly related sequence (see p. 224).

Segregation (Bateson and Saunders 1902) The separation of alleles at a gene locus at meiosis and their distribution to different gametes. Segregation accounts for the 1:1 distribution of allelic genes to different chromosomes.

Segregation analysis A set of different methods to determine the mode of inheritance.

Selection (Darwin 1858) Preferential reproduction or survival of different genotypes under different environmental conditions. Positive selection is in favor of a given genotype as opposed to negative selection.

Selection coefficient Quantitative expression (from 0 to 1) of the disadvantage that a genotype has (compared with a standard genotype) in transmitting genes to the next generation. The selection coefficient (s) is the quantitative expression by which biological fitness (1–s) is decreased; i.e., a selection coefficient of 1 indicates a complete lack of biological fitness.

Selective medium A medium that supports growth of cells in culture containing a particular gene.

Senescence Aging of cells in culture or referring to aging in general.

Sense strand The DNA strand used for transcribing into RNA. It is complementary to the antisense strand. Gene DNA sequences are always written in the 5′ to 3′ direction of the sense strand.

Semiconservative (Meselson and Stahl 1958) The normal type of DNA replication. One DNA strand is completely retained; the other is synthesized completely anew.

Sephardic The Jewish population primarily from Spain, North Africa, parts of France and Spain, and Turkey. They were expelled from Spain in 1492. The name is derived from the Hebrew word Sepharad for Spain (see Ashkenazi).

Sequencing The determination of the sequence of nucleotides in genes or amino acids in protein (see p. 66).

Serpentine A seven-helix transmembrane protein.

Signal sequence The N-terminal amino acid sequence of a secreted protein; required for transport of the protein to the right destination in the cell.

Silencer sequence A eukaryotic DNA sequence that blocks the access of gene activity proteins required for transcription, by forming condensed heterochromatin in that particular area.

SINE (short interspersed nuclear element) Short repetitive DNA sequences (compare with LINE).

SMADS Intracellular proteins that transmit extracellular signals from the transforming growth factor β ligands to the cell nucleus, where they induce downstream TGFβ gene transcription. The name is derived from a protein Sma in Caenorhabditis elegans and a Drosophila protein Mad (mothers against decapentaplegic).

SNP Single nucleotide polymorphism. A variant in the DNA sequence resulting from a difference in a single nucleotide (see p. 78).

snRNA Small nuclear RNA. One of several small RNAs located in the nucleus; five are components of the spliceosome.

snRNPs (small nuclear ribonucleoprotein particles) Complexes of small nuclear RNA molecules and proteins.

Somatic Refers to cells and tissues of the body, as opposed to germinal (referring to germ cells).

Somatic cell Any cell of an organism that does not undergo meiosis and does not form gametes (as opposed to germ cells).

Somatic cell hybridization Formation of cell hybrids in culture.

Southern blot (Southern 1975) Method of transferring DNA fragments from an agarose gel to a membrane after the fragments have been separated according to size by electrophoresis. (Southern E. Detection of specific sequences among DNA fragments separated by gel electrophoresis. J Mol Biol 1975;98:503–517).

Speciation (Simpson 1944) Formation of species during evolution. One of the first steps toward speciation is the establishment of a reproductive barrier against genetic exchange. A frequent mechanism is chromosomal inversion.

Species A natural population with interbreeding of the individuals, which share a common gene pool.

S phase (Howard and Pelc 1953) Phase of DNA synthesis (DNA replication) between the G1 and the G2 phase of the eukaryotic cell cycle.

Spliceosome An aggregation of different molecules that can splice RNA.

Splice junction The sequences at the exon/intron boundaries.

Splicing A step in processing a primary RNA transcript in which introns are excised and exons are joined.

Stem cell An undifferentiated cell able to renew itself by division, retaining the potential for differentiation within a particular developmental pathway. Omnipotent (totipotent) stem cells can differentiate into any cell type; pluripotent stem cells are descendants of totipotent stem cells, which can differentiate into a wide variety of cell types, but not all.

Steroid receptor A transcription factor that responds to a steroid hormone.

Stop codon A codon that terminates translation (UAG, UAA, UGA), originally called nonsense codons.

STS (sequence tagged site) A short segment of DNA of known sequence.

Submetacentric Refers to a chromosome consisting of a short and a long arm because of the position of its centromere (see Metacentric and Acrocentric, p. 144).

SUMO Small ubiquitin-related modifier. Attaches to other proteins (SUMOylation) to modify their function (Meulmeester E, Melchior F. Cell biology: SUMO. Nature 2008;452:709–711).

Superfamily A set of genes or proteins related to each other by evolution.

Synapse A region at the junction of a nerve and a muscle cell or between two nerve cells.

Synapsis (Moore 1895) The pairing of homologous chromosomes during meiotic prophase.

Synaptonemal complex (Moses 1958) Parallel structures associated with chiasmata formation during meiosis, visible under the electron microscope (see p. 104).

Syndrome Within human genetics, a group of clinical and pathological characteristics that are etiologically related, regardless of whether the details of their relationship have yet to be identified.

Synteny (Renwick 1971) Refers to gene loci that are located on the same chromosome, whether or not they are linked.

Tandem duplication Short identical DNA segments adjacent to each other.

TATA box A conserved, noncoding DNA sequence ~25 bp in the 5′ region of most eukaryotic genes. It consists mainly of sequences of the TATAAAA motif. Also known as Hogness box (compare with Pribnow box, in prokaryotes).

T cells T lymphocytes.

Telocentric (Darlington 1939) Refers to chromosomes or chromatids with a terminally located centromere, without a short arm or satellite. They do not occur in man.

Telomerase (Greider and Blackburn 1987) A ribonucleoprotein enzyme that adds nucleotide bases at the telomere.

Telomere (Muller 1938) The terminal areas of both ends of a chromosome containing specific consensus sequences.

Template The molecule that determines the nucleotide sequence for the formation of another, similar (complementary) molecule (see DNA and RNA).

Teratogen (Ballantyne 1894) Chemical or physical agent that leads to disturbances of embryological development and malformations.

Termination codon One of the three triplets signaling the end of translation (UAG, UAA, UGA).

Tetraploid (Nemec 1910) Having a double diploid chromosome complement, i.e., four of each kind of chromosome are present (4n instead of 2n).

Topoisomerase A class of enzymes that unwinds DNA and can control the three-dimensional structure of DNA during replication by cleaving one DNA strand, rotating it about the other, and resealing it (class I) or cleaving and resealing both ends (class II). Used to unwind the DNA double helix at transcription.

Trait An observable phenotype.

Trans-acting A regulatory element acting from a distance.

Transcript An RNA copy of a segment of the DNA of an active gene.

Transcription The synthesis of messenger RNA (mRNA), the first step in relaying the information contained in DNA (see Translation).

Transcription factor DNA-binding proteins that regulate transcription activity.

Transcription unit All of the DNA sequences required to code for a given gene product (operationally corresponding to a gene). Includes promoter and coding and noncoding sequences.

Transduction (Zinder and Lederberg 1952) Transfer of genes from one cell to another (usually bacteria) by special viruses, the bacteriophages.

Transfection Introduction of pure DNA into a living cell (compare with Transformation).

tRNA Transfer RNA, the intermediate between mRNA and protein synthesis carrying a specific amino acid to the growing polypeptide chain in the ribosome.

Transformation A term with several different meanings in biology. In genetics, three main types of transformation are distinguished: (1) malignant transformation, the transition of a normal cell to a malignant state with loss of control of proliferation; (2) genetic transformation (Griffith 1928, Avery et al. 1944), a change of genetic attributes of a cell by transfer of genetic information; and (3) blastic transformation, the reaction of lymphocytes to mitogenic substances (e.g., phytohemagglutinin or specific antigens) leading to cell division.

Transgene A cloned gene introduced into a plant or animal and transmitted to subsequent generations.

Transgenic Refers to an animal or a plant into which a cloned gene has been introduced and stably incorporated. It reveals information about the biological function of the (trans)gene.

Transition A mutation by replacement of a purine with another purine or a pyrimidine with another pyrimidine (see Transversion).

Translation The second step in the relay of genetic information. Here the sequence of triplets in mRNA is translated into a corresponding sequence of amino acids to form a polypeptide as the gene product (see Transcription).

Translocation Transfer of all or part of a chromosome to another chromosome. A translocation is usually reciprocal, leading to an exchange of nonhomologous chromosomal segments. A translocation between two acrocentric chromosomes that lose their short arms and fuse at their centromeres is called a fusion type translocation (Robertsonian translocation).

Transmembrane protein A protein located in the cell membrane with domains inside and outside the cell, usually involved in signal transduction.

Transposition Movement of a genetic sequence, a transposon, from one location in the genome to another (see p. 90).

Transposon A DNA sequence with the ability to move and be inserted at a new location of the genome.

Transversion A mutation with replacement of a purine with a pyrimidine or vice versa (see Transition).

Triplet A sequence of three nucleotides comprising a codon of a nucleic acid and representing the code for an amino acid (triplet code, see Codon).

Trisomy (Blakeslee 1922) An extra chromosome in addition to a homologous pair of chromosomes (see Monosomy).

Tumor suppressor gene A gene involved in control of cell division. It suppresses tumor development when one allele retains its normal function (see Oncogene).

Ubiquitin A small protein that inactivates other proteins.

Ubiquitination Inactivation of a protein by attachment to ubiquitin

Unclassified variant (UV) A variant in the DNA sequence without known functional consequence.

Untranslated region (UTR) Regions at the 5′ or 3′ end of a gene that are not translated into a gene product (5′ UTR, 3′ UTR).

UPD Uniparental disomy. Refers to the presence of two chromosomes derived from the same parent, either as isodisomy when both chromosomes are identical, or heterodisomy when different (see p. 368).

Upstream 5′ Direction of a gene (see Downstream).

Variation The differences among related individuals, e.g., parents and offspring, or among individuals in a population.

Variegation Different phenotypes within a tissue.

Vector A molecule that can incorporate and transfer DNA.

Virion A complete extracellular viral unit or particle.

Virus DNA or RNA of defined size and sequence, enclosed in a protein coat encoded by its genes and able to replicate in a susceptible host cell only.

VNTR Variable number of tandem repeats; a type of DNA polymorphism.

Voltage-gated channel An ion channel that is opened or closed by a gradient of electric current.

Western blot Technique to identify antibodies to proteins, in principle similar to the Southern blot method (q.v.).

Wild-type Refers to the genotype or phenotype of an organism found in nature or under standard laboratory conditions, meaning "normal." Not used when referring to humans.

X chromatin (formerly called Barr body or sex chromatin) (Barr and Bartram 1949) Darkly staining condensation in the interphase cell nucleus representing an inactivated X chromosome

Xenogenic Refers to transplantation between individuals of different species.

X-inactivation (Lyon 1961) Inactivation of one of the two X chromosomes in somatic cells of female mammals during the early embryonic period by formation of X chromatin (see p. 196).

X-linked Refers to genes on the X chromosome.

YAC (yeast artificial chromosome) A yeast chromosome into which foreign DNA has been inserted for replication in dividing yeast cells. YACs can incorporate relatively large DNA fragments, up to ~1000 kb (see BAC).

Y chromatin (F body) (Pearson, Bobrow, Vosa 1970) The brightly fluorescent long arm of the Y chromosome visible in the interphase nucleus.

Yeast two-hybrid system A technique to identify genes or proteins that interact in function.

Z DNA Alternate conformation of DNA. Unlike normal B DNA, (Watson–Crick model), the helix is left-handed and angled (zigzag, thus Z DNA, see p. 52).

Zinc finger A finger-shaped region found in many DNA-binding regulatory proteins, the "finger" being held together by a centrally placed zinc atom (see p. 178).

Zoo blot A Southern blot containing conserved DNA sequences from related genes of different species. It is taken as evidence that the sequences are coding sequences from a gene.

Zygote (Bateson 1902) The new diploid cell formed by the fusion of the two haploid gametes, an ovum and a spermatozoon, at fertilization. The cell from which the embryo develops.

Zygotene (de Winiwarter 1900) A stage of prophase of meiosis I.

References used in Glossary

King RC, Stansfield WD. A Dictionary of Genetics. 7th ed. Oxford: Oxford University Press; 2006

Lodish H, et al. Molecular Cell Biology. 6th ed. New York: W. H. Freeman; 2007

Passarge E. Definition genetischer Begriffe (Glossar). In: Elemente der Klinischen Genetik. Stuttgart: G. Fischer; 1979:311–323

Rieger R, Michaelis A, Green MM. Glossary of Genetics and Cytogenetics. 5th ed. Heidelberg: Springer-Verlag; 1979

Strachan T, Read AP. Human Molecular Genetics. 4th ed. New York: Garland Science; 2011

Web sites

Genetics Home Reference. Available at: http://ghr.nlm.nih.gov/; Accessed January 30, 2012

Glossary of Genetic Terms. Available at: http://www.kumc.edu/gec/glossnew.html. Accessed January 30, 2012

Human Genome Project Information. Genome Glossary. Available at: http://www.ornl.gov/sci/techresources/Human_Genome/glossary/. Accessed January 30, 2012

National Cancer Institute. NCI Dictionary of Genetics Terms. Available at: http://www.cancer.gov/cancertopics/genetics/genetics-terms-alphalist/a-e. Accessed January 30, 2012

Picture Credits

Page 31, C (right):
Conard NJ, Malina M, Münzel SC. New flutes document the earliest musical tradition in southwestern German. Nature 2009;460: 737–740. Adapted by permission from Macmillan Publishers Ltd.

Page 233, A
Plenge RM, et al. TRAAF1-CS as a risk locus for rheumatoid arthritis – a genomewide study. N Engl J Med 2007;357:1199–1209. Reprinted with permission.

Page 233, B
Wang K, et al. Common genetic variants on Sp14.1 associated with autism spectrum disorders. Nature 2009;459:528–533. Adapted by permission from Macmillan Publishers Ltd.

Page 305, A
Stratton MR, Campbell PJ, Futreal PA. The cancer genome. Nature 2009;458:719–724. Adapted by permission from Macmillan Publishers Ltd.

Page 305, B
Ledford H. Big science: The cancer genome challenge. Nature 2010;464:972–974. Adapted by permission from Macmillan Publishers Ltd.

Page 321, C 1
Reprinted with kind permission of the Share & Care Cockayne Syndrome Network.

Page 329, A4
Reprinted from Horton WA, Hall JG, Hecht JT. Achondroplasia. Lancet 2007;370:162–172. With permission from Elsevier.

Page 329, A5 and C
Courtesy of Enid Gilbert-Barness, MD

Page 329, B
Reprinted from Yamaguchi TP, Rossant J. Fibroblast growth factors in mammalian development. Curr Opin Genet Devel 1995;5:485–491. With permission from Elsevier.

Page 331, A
From: http://www.primehealthchannel.com/marfan-syndrome-pictures-symptoms-treatment-and-life-expectancy.html

Page 331, B1
Courtesy of Dr. Beate Albrecht, Essen, Germany.

Page 331, B2 and B3
From Drera et al. Loeys-Dietz syndrome type I and type II: clinical findings and novel mutations in two Italian patients. Orphanet J Rare Dis 2009;4:24. With permission Biomed Central Ltd.

Page 331, B4
Reprinted with kind permission from Dr. Matthias Siepe, Herz-Kreislauf-Zentrum Universitätsklinikum Freiburg, Germany.

Page 331, B5
From Lindsay ME, Dietz HC. Lessons on the pathogenesis of aneurysm from heritable conditions. Nature 2011;473:308–316. Reprinted by kind permission of Macmillan Publishers Ltd.

Page 339, A–C
Reprinted with kind permission from Springer Science+Business Media. From: Omran H, Olbrich H: Zilienkrankheiten unter Berücksichtigung der primären ziliären Dyskinesie. Med Genet 22: 315–321, 2010.

Index

Note: Glossary entries are not included in the index